ENGINEERING DRAWING

N.S. Parthasarathy

Former Professor of Mechanical Engineering
Head – Engineering Design Division
and
Director of AU-FRG Inst. for CAD/CAM
College of Engg., Guindy
Anna University
Chennai

Vela Murali

Professor of Mechanical Engineering
and
Former Head – Engineering Design Division
College of Engg., Guindy
Anna University
Chennai

OXFORD
UNIVERSITY PRESS

OXFORD
UNIVERSITY PRESS

Oxford University Press is a department of the University of Oxford.
It furthers the University's objective of excellence in research, scholarship,
and education by publishing worldwide. Oxford is a registered trade mark of
Oxford University Press in the UK and in certain other countries.

Published in India by
Oxford University Press
22 Workspace, 2nd Floor, 1/22 Asaf Ali Road, New Delhi 110002, India

First published in 2015
Digitally Printed in 2024

ISBN-13: 978-0-19-945539-3
ISBN-10: 0-19-945539-2

Typeset in Times
by Archetype, New Delhi 110063
Printed in India by Manipal Technologies Limited, Manipal

Dedicated to the memory of
My dear parents
who inspired the spirit of teaching in me

N.S. Parthasarathy

Dedicated to the memory of
My spiritual Guru
Shri Parthasarathi Rajagopalachari (Chariji)
Shri Ram Chandra Mission, Babuji Memorial Ashram, Manapakkam, Chennai
for his guidance, love, and affection

and

My mother-in-law
Shrimathi M. Kasthuri
for her great affection in caring for me and my family

Vela Murali

Preface

Imagination stimulates, knowledge nurtures and skills perfect, but still the imagination holds the key to success.

– Anonymous

The communication of ideas through pictorial sketches is an age-old practice of mankind, even before we learnt how to write formal scripts. These sketches were initially used to depict the shapes of animals and humans, which later got improvised to convey emotions too. When human beings realized the need to communicate their ideas and explain the shape of the objects needed for everyday use, the informal sketches became 'drawings'. When the need further arose to design and manufacture the products, such drawings got classified as *engineering drawings* and *technical drawings*. Engineering drawings speak about the geometrical features of an object and its positional references, whereas technical drawings add further details like the tolerances on the dimensions that describe the geometry and prescribe the surface finish that are necessary for the manufacture of the product. Thus engineering drawing became the basic language of communication among the engineers and continues to remain so, irrespective of the inner disciplines or specializations.

This art of communication transcended from the ancient ages in which the drawing sketches were made on caves, walls, wood, pottery surfaces, and animal skins. During the Renaissance period, drawing became the foundation work in all arts and the invention of paper as a medium for drawing not only supported its existence, but further helped it grow more methodically. Leonardo da Vinci (1452–1519) is considered to be one of the first graphic artists. By combining scientific interest with artistic ability, he was able to merge visual art with science and invention. Many scientists and mathematicians started to emphasize the need for drawings with greater precision to incorporate Euclidean geometry and other developments in mathematics, which transformed drawing from being an artistic communication to a need-based product. The early books written in 1435 paved way for this and proposals for multiple views started emerging. With the development of descriptive geometry and allied researches, the formal engineering drawing began to evolve in the 18th century. The industrial revolution of the 19th century gave a technical face to engineering drawing and the book on *A History of Engineering Drawing* by Peter Booker distinguished the technical practice from the earlier craft practices.

Drawing skills ranging from manual sketching to instrumented drafting became essential for technical drawings and many drafting tools including the Universal Drafting Machine were invented to support the preparation of the drawings. As drawings became the mode of communication among engineers across the world, the need for standardization of the drawings was required and International standards for representation of drawings came into existence.

When computers came into existence, they had an effect on the engineering practice too. Computer-aided drafting (CAD) procedures started emerging to produce more accurate and precise drawings in less time and they also overcame the transfer and storage problems of the manual drawings. The rapid growth in mathematical procedures and computing algorithms in the later part of the 20th century not only enabled the preparation of three-dimensional (3D) solid models of an object in the computer, but also helped them to be edited, altered, and viewed in any orientation in full or cut modes, with or without animation effects. When such developments are linked with the production line practice, the engineering prototypes emerge from paper to product form in a remarkably less time and this once again rationalized the continued existence of drawing as a basic need in all disciplines of science and engineering.

In order to prepare the young and budding engineers to obtain a holistic understanding of this indispensable subject, the authors have attempted to bring out this book with greater emphasis on kindling the imagination capabilities and imparting sound content. The authors firmly believe that only when these two qualities are properly nurtured in the young minds, will the development of drafting skills through any software make any meaningful impact on them.

ABOUT THIS BOOK

This book has been carefully prepared and pedagogically designed to meet the requirements of first year engineering students who come from varied backgrounds. Every chapter has been structured with a clear objective, a basic introduction necessitating the need for the chapter, its link with the previous chapter(s), detailed discussions of the underlying principles, through varieties of illustrative worked examples with step- by-step procedures, commonly observed engineering applications, and associated problems, and ends with a recapitulation section revisiting the salient discussions in that chapter. It is hoped that this will give a comprehensive understanding of the chapter learnt and hence the subject as a whole.

A special chapter on *Visualization Concepts and Freehand Sketching* has been included to motivate the students. In this chapter, a 3D object has been considered to be made up of a base unit with many features in the form of cut-outs, slots, holes, blocks with sloping and curved surfaces, etc., added to it, to explain its views. The projections are explained by the sequential additions or deletions of the basic view. This novel idea of visualizing a 3D object as being made up of a base unit with many features built over it, enables the observation of the views of the object in a simpler and faster way. This chapter has been purposely introduced before discussing any projection principles, to help condition the mind to freely observe the features in a natural way. With this in mind, even the practice of these views has been advised in freehand sketching mode, as this is a quicker means of transformation with no gadgets or tools. Emphasis is given to preserve and improve this natural talent present in oneself in order to kindle curiosity and pave way for innovation.

KEY FEATURES

Systematic knowledge building   Every chapter has been constructed on a systematic knowledge-building pattern, clearly defining the need, providing detailed explanations of the principles and illustrations with a number of worked examples, built with increasing order of involvement. The pictorial sketches of an object in the reference plane system have been illustrated in box mode to explain its projections and alignment of views clearly.

Consolidation of knowledge   A comprehensive idea of what is to be learnt in each chapter has been spelt out clearly in the learning objectives. The concepts discussed in each chapter have been consolidated at the end in the form of the recapitulation section. This section is not only intended to summarize the learnings of the chapter, but it will also serve as a quick review for assessments and examinations.

Enhancement of student output   The practice problems are arranged in multiple levels to enable the student to obtain a better performance and output. Practice of problems in Work Practice Level I helps the students to get familiar with the methodology of working. After gaining sufficient output in this, the students are advised to go to the next stages, which tunes them for complete management of time as expected in any university examination. The objective-type questions serve as the first level knowledge output, mostly consolidating the principles with one or two words.

User friendliness   The entire book is prepared based on a student-friendly approach, bringing the basic observations or inferences, hints for solving the problems, and with detailed step-by-step procedures. For instance, chapters on Points, Straight Lines, Planes and Solids contain inferences or hints added under each worked example, before the procedural working commences. These will serve as a good foundation for other chapters too.

Kindling the imagination and visualization abilities   A dedicated chapter has been developed on the topic *Visualization Concepts and Freehand Sketching* to kindle interest among the students while visualizing objects to observe their views.

Logical reasoning and observation   The projection of objects in a reference plane system, though a methodical constructional formality, is explained in this book with a logical thinking and reasoning approach. This has been used in Chapter 6 on Projection of Points, in which the views in the multiple quadrant system are exposed. The student, after becoming familiar with the projection of points with respective to one quadrant, has been shown the easier way to extend it other quadrants logically to avoid re-visualizing the plane system to obtain its views.

Illustrative diagrams   All illustrative diagrams and drawings have been prepared by adopting the line thickness and dimensioning procedures as per Bureau of Indian

Standards (BIS) conventions. Several high-quality drawings are presented with lucid text material.

Easier visualization with common objects as examples   The textbook has been prepared with utmost care to make this subject easier for the students. Objects that are commonly encountered in our everyday life are cited as examples along with their projections. Live day-to-day observations, logical thinking, and extension of ideas to similar problems are the key techniques followed in this book to make the subject feel lively.

Online Resources

The book is supplemented with the following online resources:

For instructors:

- Solutions for select exercise problems in the book
- Additional MCQs

For students:

- Multiple-choice questions
- 19 select example problems from the book with step-by-step drawing visualization and supporting audio voice-over in PowerPoint presentation format

CONTENTS

This book is divided into 16 chapters. A brief summary of each chapter is presented here for ready reference.

Chapter 1 presents the historical evolution of engineering drawing as a refined communication tool among the engineers. The manual drafting tools that are available to produce drawings are explained along with their limitations. The specifications of the drawing sheet, size and its layout, representation of various lines in the drawing as per different applications, dimensioning and lettering practice, etc., are dealt with in detail as recommended by BIS.

Chapter 2 deals with the basic geometrical constructions of curves used in engineering practice such as conic sections, cycloids, and involutes. It also focusses on the path traced by point(s) in the mechanisms when their members occupy different orientations in the working environment.

In *Chapter 3*, the need for scales in engineering is presented along with the various types of customized scales.

Chapter 4 stresses the importance of the visual abilities of the observer or viewer as the prime asset to extract details from any pictorial representation of a 3D object. A novel idea of considering a 3D object to be made up of primitive shapes or features and hence observe its views from the individual shapes has been demonstrated with a number of examples. This concept enables any complex-shaped object to be mentally split into simple shapes and then be assembled.

Chapter 5 outlines the basic principles of orthographic projection and demonstrates the layout of the views of an object as per accepted methods of projection.

Chapter 6 discusses the projection of points situated in different quadrants in a more logical way than in a procedural way. The facts that are known with the projection of points in the first quadrant are logically extended to other quadrants. The preparation of the side view is also explained in a novel way by marking the distance of the object and the vertical projection (VP), directly on the extension lines from the front view, rather than resorting to obtaining this from the top view and transferring it by a 45° line or using quarter circular arcs.

Chapter 7 discusses the projection of straight lines located in the first quadrant space in different orientations. Simplified procedures are discussed to obtain the projections of a straight line in any orientation and also to obtain the orientations from its projections. Two popular methods—rotating line method and trapezoidal method—are highlighted for this purpose. A comprehensive coverage of problems involving traces has been made with many engineering application problems.

Chapter 8 extends the concepts of orthographic projections to two-dimensional surfaces, that is, planes, distinguishing the basic requirement of tilting about a corner or an edge. The inference of the shape of the object from its projections is also discussed.

Chapter 9 addresses 3D solids as being built from the planar base and discusses the orthographic projections of various types of solids in simple and inclined positions by change of position and change of reference line (auxiliary line) methods. Typical examples of the inclination of the axis to one and both the reference planes are considered for study.

Chapter 10 discusses the engineering need for sectioning of a solid by a cutting plane and explains the various types of section planes and the changes that take place in the views of the solids when they are cut. The cases where the solids are in simple positions and the cutting planes are in inclined orientations and when both are inclined are elaborately discussed.

Chapter 11 extends the concepts of sectioning a solid and a plane to when the solids are cutting each other. The intersection cases where prismatic and cylindrical solids are vertical and the intersecting solids are in horizontal and inclined orientations are discussed, when their axes are intersecting and not intersecting. The corresponding cases when the vertical solids are pyramidal or conical are also addressed with due changes. Finally a typical case of a cone and a cone intersection is also presented.

Chapter 12 presents the development of lateral surfaces of solids when they are cut by the section planes in the form of a straight line, arc, circular hole, etc. The importance of this topic in sheet metal working and packaging industries is also highlighted.

Chapter 13 addresses the methodology of obtaining a single pictorial projection, where all the sides and faces of a solid are reduced and related by the same proportion. This is the next preferred projection to orthographic projection as it facilitates all the three dimensions in a single picture without any ambiguity. The constructional schemes of this projection, commonly known as isometric projection, are discussed with a variety of problems.

Chapter 14 uses principles similar to those adopted in Chapter 4, but help to transform the views as per the principle of orthographic projection.

Chapter 15 exposes to the reader to perspective projection, which uses the concept of linking the observer and the object in one platform, while drawing the projections. The influence of the relative distances between the object and the observer, and their dimensions or size creates a single pictorial representation, often resembling the one seen by a human eye. Though the picture representation works on the principle of optical illusion and is unreal to observe any dimension details, it has a very high appeal among the common public and is often used in architectural engineering and advertising media.

While Chapters 1–15 deal with the principles for manual drafting, *Chapter 16* is completely dedicated to the preparation of drawings using one of the most popular and commonly used 2D software known as AutoCAD. The salient features of the software, file generating commands, commands for drawing various entities, their editing and display means, are all explained elaborately with snapshots of the computer windows. The complete step-by-step commands of drawing 2D views of 3D objects and obtaining 3D pictures from their orthographic projections are explained in detail.

Readers are welcome to share feedback and suggestions for further improvement of the book with the authors at parthan.nsp@gmail.com and velamurali@gmail.com.

ACKNOWLEDGEMENTS

We wish to express our sincere and heartfelt gratitude to Prof. Dr M. Rajaram, the respected Vice-Chancellor, Anna University, Chennai, who in his broader vision to help the academic community, has always encouraged and supported us to bring out this book, apart from encouraging us to disseminate this knowledge through various web and EDUSAT lectures of Anna University.

We thank Prof. Dr S. Ganesan, the Registrar, Anna University, Chennai, for his sustained interest in the publication of this book to benefit undergraduate students of Engineering programmes.

We express our sincere thanks to the beloved Prof. Dr P. Narayanasamy, the Dean, College of Engineering, Guindy campus, Anna University for his constant encouragement.

We are indebted to all the past and present Directors of the Centre for Academic Courses, Centre for Faculty Development, Chairmen of Faculty of Mechanical Engineering, Anna University, and former Heads of the Department of Mechanical Engineering, College of Engineering, Guindy, for providing continuous opportunities and support to deliver training lectures on Engineering Drawing to the faculty of Anna University colleges.

We thank Prof. L. Karunamoorthy, Chairman, Faculty of Mechanical Engineering, Anna University and Prof. Dr B. Mohan, Head of Department of Mechanical Engineering, College of Engineering, Guindy, for constantly instilling interest in us to develop this book.

We sincerely thank all our colleagues in the Department of Mechanical Engineering, who have shown their interest and support at every stage of the development of this book.

We are deeply grateful to former professors—Dr G. Ramaiyan, Dr K. Chandrasekaran, Thiru P. Bagavathiperumal, Dr K. Srinivasan, and Dr G. Thanigaiyarasu—for their continued interest in this book.

Dr Latha Nagendran, Director AUFRG Institute for CAD/CAM deserves a special mention for her help in reading through the manuscript during different stages and offering useful suggestions.

We wholeheartedly thank our students, Mr V. Ranjith Kumar, Mr M. Sutherson, Mr V. Sasindhar, who have helped in the development of drawings for this book and special thanks are due to Mr A. Ramachandran, who has helped in the development of the online resources containing PowerPoint presentations with audio support.

We sincerely thank the complete editorial team of Oxford University Press, India for their patient cooperation in bringing out this book.

This book would not have taken its life but for the critical reviews, comments, and suggestions by the esteemed reviewers, whose role though invisible, was vital. We owe our sincere thanks to them.

Prof. N.S. Parthasarathy wishes to place on record his respect and gratitude to his professor and mentor Dr G. Ramaiyan for his timely advice and inspiration to prepare this book. He also recalls with a note of thanks, all his students and colleagues who have reposed faith in him to bring out this book on the subject he has taught for more than 24 years. He fondly acknowledges his PhD students Dr B.K. Gnanavel and Dr G. Shibu for all their help in the preparation of drawings for the exercises in this book. And, last but not the least, he expresses his heartfelt thanks to his beloved wife, Mrs K. Chandra and his dear son, daughter-in-law, daughter, son-in-law, and his grandchildren, without whose patience, encouragement, love, and support, this book wouldn't have become a reality.

Prof. Vela Murali likes to thank all his colleagues both teaching and non-teaching in Engineering Design Division, Department of Mechanical Engineering, College of Engineering, Guindy, Anna University, Chennai for all their support during the preparation of this book. He likes to acknowledge his PhD student, Mr A. Arockia Julias for his help in the preparation of some of the drawings related to straight lines. He also likes to thank his wife, Mrs Lakshmi, son, Tharun Kumar, and daughter, Tharanya, for their patience during the time he spent in writing this book. Special thanks are due to Tharun Kumar for his help in taking certain photographs related to the examples of sphere and various parts of the computer.

N.S. Parthasarathy
Vela Murali

Features of the Book

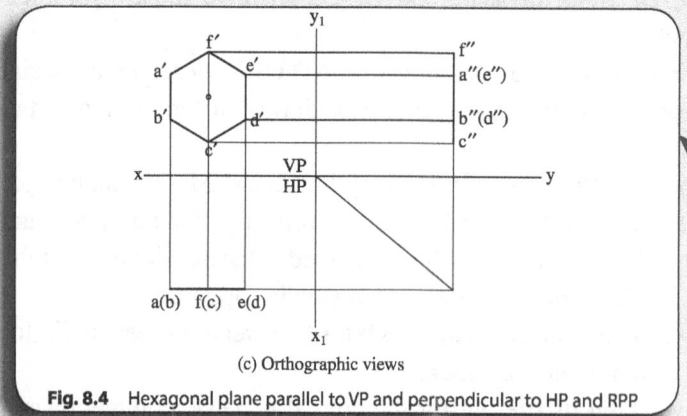

(c) Orthographic views

Fig. 8.4 Hexagonal plane parallel to VP and perpendicular to HP and RPP

Illustrates concepts with over 450 drawings spread over 16 chapters

Presents more than 450 drawings illustrating various types of objects and varied projections

Detailed chapter on freehand sketching and visualization

Presents a detailed chapter on freehand sketching, which aids students in enhancing their visualization capabilities

Visualization Concepts and Freehand Sketching

4

WORK PRACTICE LEVEL – 1

Problems on Geometrical Constructions

2.1 Divide a straight line of length 110 mm into nine equal parts.

2.2 Divide a straight line of length 130 mm into proportionate parts of one-half, one-third, one-

2.9 Two points C and D are 110 mm apart. The third point E is 80 mm from C and 65 mm from D. Draw an ellipse passing through the points C, D, and E and measure the focal distance.

2.10 The major axis of an ellipse is 130 mm long and minor axis is 90 mm long. Draw one half of the

WORK PRACTICE LEVEL – 2

Problems on Involutes

2.1 A string is wound around the circumference of the regular hexagon of side 40 mm completely. Holding one end free, the string is unwound completely such that the string is always tightly

line. Draw the curve traced by the point P which is initially lying at a point of contact of the horizontal line and the circle.

2.6 A circle of 50 mm diameter rolls outside on another circle of 110 mm diameter and completes one revolution. Draw to full scale the locus of a

WORK PRACTICE LEVEL – 3

Problems on Engineering Applications (Conics, Involutes, and Cycloids)

2.1 The distance between two straight parallel fencings of a plot is 120 m. An elliptical

swimming pool is to be constructed inside the plot with major axis length of 90 m. Locate the foci and the two ends of the major axis by considering the two straight parallel

'Work Practice' modelled according to difficulty

Provides chapter-end exercises graded according to the difficulty and complexity of the problems

Learning objectives and recapitulation

All chapters begin with a set of learning objectives and a complementary chapter-end recapitulation for quick summary of discussions

Over 130 multiple-choice questions with answers

Every chapter ends with a set of multiple-choice questions with final answers provided

Exclusive chapter on AutoCAD

An exclusive chapter on AutoCAD 2015 discusses methods of drawing by step-by-step procedure through the software

Computer-Aided Design 16

16.1 INTRODUCTION

After studying all topics of engineering graphics, it is understood that the first course on 'Engineering Drawing' plays a dominant role in all the phases and steps in the process of engineering design. When computers came into existence, they changed the engineering practice by adopting the design methodologies too. In order to produce repeatable drawings with accuracy in a simple and faster manner, the need for computer-based graphic system was realized, since the manual drawings require large workspace and hardware, consume more time and labour, make editing more strenuous, and pose many

Section on Engineering Applications

Every chapter ends with a brief section describing the engineering applications of the concepts learnt in that chapter

Brief Contents

Detailed Contents

Basics of Engineering Drawing

1

OBJECTIVES

This chapter will help the reader to understand the following:

- Evolution of various modes of communication and the emergence of engineering drawing or graphics as a refined communication tool to manufacture and develop an engineering product
- International and national standards of practice for uniform presentation of drawings
- Manual drafting tools that are available for making the drawings and the ways of handling them
- Need to incorporate lines of different intensities in a drawing and hence to distinguish the different features of an object as per the standards
- Develop the text writing in a drawing by following a methodical lettering practice as per the standards
- Detailed dimensional procedures to make the drawing meaningfully communicate and enable the manufacture
- Preparation of a holistic technical drawing and to read and convey unambiguously to the fellow engineers across the world

1.1 INTRODUCTION

The pictorial representation of an object is one of the simplest and common communication modes adopted by the mankind. When the need arose to develop and manufacture a product for human use, the formal pictorial representations of the products got refined as engineering drawing and became a common communication tool among the engineers. The guidelines for engineering drawings were standardized universally, and hence, the International and National Standards were formulated. The drawings can be made traditionally either by manual means or by the digital means due to the recent advancement of computers. In order to practice and understand the drawing effectively, this chapter discusses the preparation of engineering drawings with manual drafting tools with incorporation of the dimensioning and lettering formalities as recommended in Bureau of Indian Standards (BIS).

1.2 MODES OF COMMUNICATION

Humans tried to develop different modes of communication with each other since they started to exist. Early humans tried to live in groups, in order to protect them from attack. To communicate among their groups, the early people needed a mode of communication. Initially, sound was used as the communication media. The sound produced by one group was understood by the other and was replied again in similar manner. However, during nights, the light source in the form of fire became effective to convey the presence of one group to the other. When they moved closer to each other and when their civilization improved, communication through signs started evolving. The practice and development of sign conventions with sound components added later, turned out to be the language and became vernacular depending on the groups that practised it. Apart from these modes, the mankind had started communicating with the sketches or pictures of the animals, the nature, landscape, and the environment in which they lived and used the available tools, such as charcoal. The walls of public places or tombs or places of worship were used for this purpose. Understanding the sizes and shapes of the objects and incorporating them into the drawings with due proportions has led to the development of paintings and sculptures. The artists further improvised them to convey the feel or the emotion. This had received wide popularity among the mankind. Though, vernacular languages were developed later for effective communication, they got restricted to that group of people only. Hence the pictorial representations were the only common mode of communication because of their uniqueness, simplicity, and reach to the entire mankind, universally, irrespective of the differences in the topography, language, and knowledge level of the people.

The power of pictorial communication had been understood by the engineers and was further improvised to reveal the shape and sizes of the engineering objects to be created. The three-dimensional objects are presented in two-dimensional drawings, which are known as *engineering drawings*. Since the main purpose of engineering drawing is to create or manufacture the object for human use, these drawings are to be realistic with the true dimensions, unlike the other pictorial drawings created by the artist, whose focus is only towards the appreciation and not the realism. The engineering drawings can be further tuned to make the shapes aesthetically pleasing and ergonomically viable. Since at the universal level, the facts conveyed by means of such drawings are to be the same, the drawings are drawn using the universal standardization procedures and specifications.

1.3 ENGINEERING DRAWING AS A COMMUNICATION TOOL

1.3.1 History and Evolution

The history of drawing is as old as the history of mankind. People started sketching or drawing much before they learnt how to write, as that helped them to communicate. The oldest known cave art paintings are nearly 35,000 years old. In the prehistoric age, charcoal and ochre were used to create the images of animals. Ancient Egyptians (about 3000 BC) decorated the walls of their temples and tombs with the drawings of Gods. Greeks during 800 BC recorded their drawings on pottery surfaces and later, after 500 BC, made them more realistic with natural proportions and details and such figures were known as silhouettes. Artists made drawings on prepared animal skins such as vellum or parchment and on tablets of wood, wax, or slate, before paper was invented. Sometimes, the drawings were also made directly on the walls to be painted. Roman artists painted scenes from the stories and the 'great scrolls' on the walls of their homes.

For centuries, engineering was focusing on war, either building defensive fortifications or the machines to attack these fortifications. The first non-military engineering discipline came to exist was known as the 'civil engineering'. The sketches of war machines that existed during that period were made on parchment or scratched on clay tablets. However, there is no record of sketches that existed for the magnificent construction of the Colosseum in Rome or the Greek Pantheon.

In early 15th century, the concept of graphic projections was well understood by Italian architects and the drawing took a rebirth, and at the same period, the invention of paper began to replace the parchment as a drawing medium.

During the Italian Renaissance (1400–1525), Flemish Renaissance (1440–1540), German Renaissance (1400–1540), and High Renaissance (1500–1520) periods, drawing became the foundation work in all arts. Art students were trained first in drawing, before taking up painting, sculpture, and architectural works. Portrayal of human figures became more realistic due to this. Adoption of the prominent Golden Rectangle or Golden Mean Ratio (3:5) by Leonardo da Vinci, the mathematician gave more realistic drawings.

Most of the drawings drawn in these periods are housed in European museums and libraries and are restricted to academic researchers. These drawings are mostly in the form of sketches, are not to scale, and do not have dimensions but contain exhaustive textual descriptions. Some of the best known engineering drawings are that of Leonardo da Vinci. Though he is well known for the Monalisa painting, he was a designer of military machines, a mathematician, an artist, and a scientist. His design works were artistic in nature and did not have multiple views, but still craftsmen were able to construct models from his single-view sketches.

Simultaneously, people who practised drawing realized the need to make it stand on its own merits and wanted to incorporate greater precision. In 1435–1436, Leon Battista Alberti wrote two books and explored the need to incorporate mathematics as a common ground into contemporary drawings. Proposals for drawings with multiple views emerged. With the works of Rene Descartes (1596–1650) and Gaspard Monge (1746–1818) on the development of descriptive geometry, engineering drawing began to evolve in the 18th century and picked up speed with industrial revolution of the 19th century. Peter Booker in his book *A History of Engineering Drawing*, distinguished the industrial/technical practices and the earlier craft practices.

The growth of the patent process really became the catalyst to develop technical drawing. When the need arose for submission of multiple copies of drawings for obtaining patent, the invention of the blue printing device (cyamotype process) took place in 1842 by Sir John Herschel. Industrial drawings were prepared using pencil, T-square, set squares, and scales, permanent drawings, and tracings were made by ink. Then, the attention of lettering was realized and lettering templates came into existence. After World War I, the first American Drafting Standard was prepared in 1935, with this focus.

A major advance took place on the drafting device that led to the Universal Drafting Machine. This device was basically developed to combine the T-square, set squares, scales, and protractors, which had allowed different engineering disciplines to develop their own approaches to design and drafting. While architects followed an appreciable style of drafting for their works, aeronautical engineers used a different style with the purpose of accuracy. Not only the drafting quality, but the speed with which the drawings were produced, also became important. After the Second World War, together with a new generation of reproduction machines, the time taken to prepare drawings was substantially reduced.

1.3.2 Emergence of Engineering Graphics

When computers came into existence, they changed the engineering practice too. As the manual drawings require large workspace and hardware, consume more time and labour, need strenuous effort to edit the drawings, and pose a lot of storage or transfer problems, the computer-based graphic system was realized to produce fast, simple, accurate, and repeatable engineering drawings.

The Computer-Aided Drafting (CAD) procedures facilitate all these benefits. Due to the rapid growth in mathematical procedures and computing algorithms, three-dimensional models are drawn with ease. Using primitive shape of basic solids and Boolean combinations, three-dimensional solid models are created. The software 'solid modelling' is used to achieve these drafting. From the solid modelling, the required two-dimensional drawings of objects for any specified position can be obtained. In addition, the application of graphical techniques help the solid models to get rotated or turned to any position, amenable for the addition and deletion of various parts and obtaining the corresponding two-dimensional drawings. The method of preparing and altering drawings through computers using graphical techniques, including animation effects is known as 'Engineering Graphics'.

1.4 MANUAL DRAWING TOOLS AND ACCESSORIES (HARDWARE)

This section deals with the tools and accessories that are necessary to make engineering drawings, manually.

1.4.1 Drawing Sheets

Drawing sheets are uniformly thick, strong, tough, and smooth white-coloured papers that serve as the medium on which the pencil and the ink drawings are made. The drawing sheets should retain the impressions of the drawings made on them and their fibres should not disintegrate when erasers are used. When the drawings are made in ink, the ink should not spread. They should be made from good quality papers that have fibres to enable them to stack or to roll.

1.4.2 Drawing Board

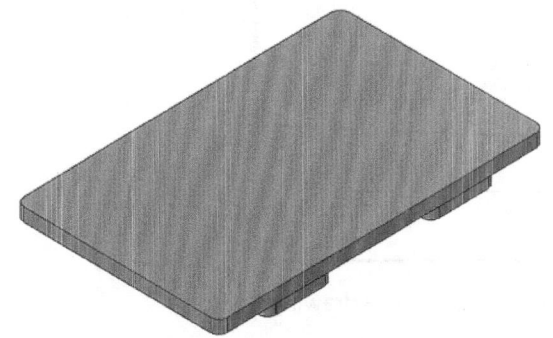

Fig. 1.1   Drawing board

To provide support to the drawing sheets or papers, the drawing boards are used, and the top (working) surface of the board should be smooth in order to prepare quality drawings. The drawing boards are made of fine-quality lightwood in platens or strips and are connected at the bottom by means of battens, as shown in Fig. 1.1. The left edge of the board is known as the working edge, and it should be perfectly straight to enable the T-square (a tool used to draw long continuous horizontal lines) to slide to different positions. In some boards, this working edge is provided with a groove so that the T-square is perfectly guided while sliding. For working convenience in the class rooms, the drawing boards are held on the table top with the required slope suitable to the user.

1.4.3 T-square

The T-square is one of the oldest devices made of wood or plastic used to draw long and continuous horizontal lines in a drawing sheet. It consists of two portions, namely a stock and a blade, joined together at right angles by means of screws and pins and resembles the alphabet 'T'. The stock is held at the working edge of the drawing board, while the blade lies on the surface of the drawing sheet. A bevelled working edge is provided in the blade to have perfect seating on the drawing sheet while drawing. The T-square is used in drawing offices and by architectural engineers, while its use has become limited in the colleges due to the popular usage of the mini drafter or drafting machine. Figure 1.2 shows the arrangement of T-square, drawing sheet, and drawing board.

1.4.4 Set Squares

The set squares is one of the most common drawing instruments, often made of transparent acrylic plastic and is often used in conjunction with T-square, to draw parallel, inclined, and perpendicular lines. The two types of set squares, namely 45° set square and 30°–60° set square are right-angled triangular strips, bevelled outside to have perfect seating. Figure 1.3 shows the two types of set squares. The perpendicular edges are usually graduated to enable measurements while drawing. The size of the 45° set square is in the range of 150–200 mm, while the other set square is of 200–250 mm

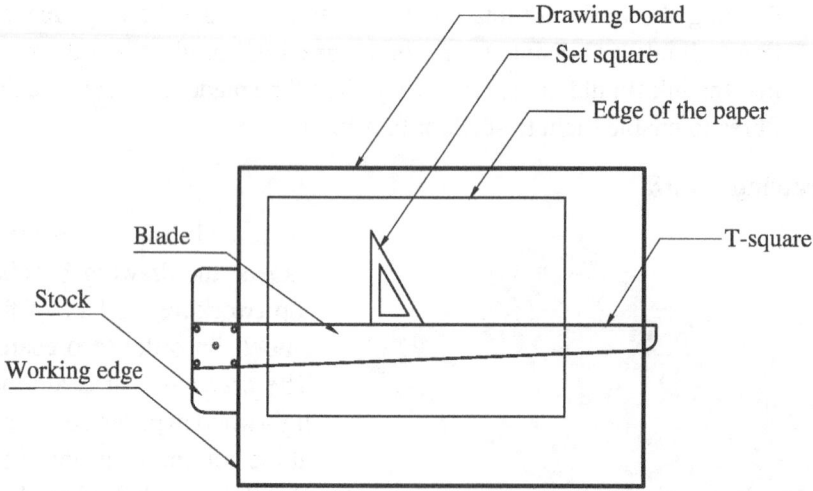

Fig. 1.2 Arrangement of T-square in a drawing board

size. The set squares can also be held with the working edge of the T-square to draw inclined and perpendicular lines as shown in Fig. 1.2.

With a proper and judicial assembly of both the set squares, straight lines in multiples of 15° can be generated.

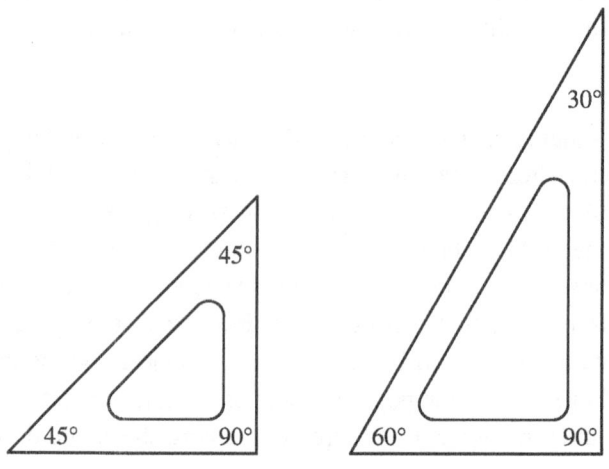

Fig. 1.3 Set squares

1.4.5 Protractor

Similar to set squares, the protractor is also a very popular drawing instrument used from the school level itself. They are made of transparent acrylic plastic in semicircular or circular shapes, bevelled at the circumferential edges. They are graduated in degrees, measurable with a least count of up to 1° and are used to set or measure any angle. Figure 1.4 shows the sketch of a 180° protractor. Sometimes a 180° protractor is also provided in some of the 45° set squares.

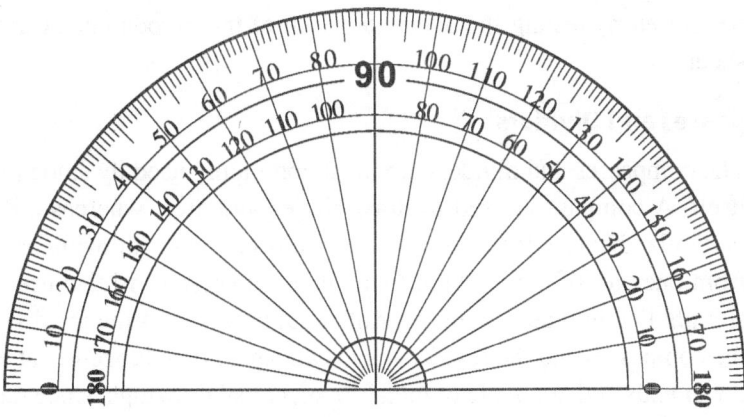

Fig. 1.4 Protractor

1.4.6 Mini Drafter

The mini drafter, a miniaturized version of a drafting machine, is a modern equipment that combines the advantages of T-square, set squares, protractors, and scales which becomes a convenient drafting equipment for college students. In industries, it is used by professional draftsmen. It consists of a fixed clamp (used to fix the equipment with the drawing board) and two sets of parallelogram of bars connected by pivot plate. The free end of the parallelogram of bars has a set of graduated acrylic or metallic scales held at 90° and attached to a protractor head with an index plate and a locking screw knob. The two perpendicular arms along with the graduated scales can swivel to any angle and can be moved to any location in the drawing sheet with the help of parallelogram of bars. Thus, straight lines can be drawn at any inclinations, including the horizontal and the vertical modes.

Figure 1.5 shows a typical arrangement of the mini drafter on the drawing board. While fixing the drafter to the drawing board, a suitable location of the fixing clamp can

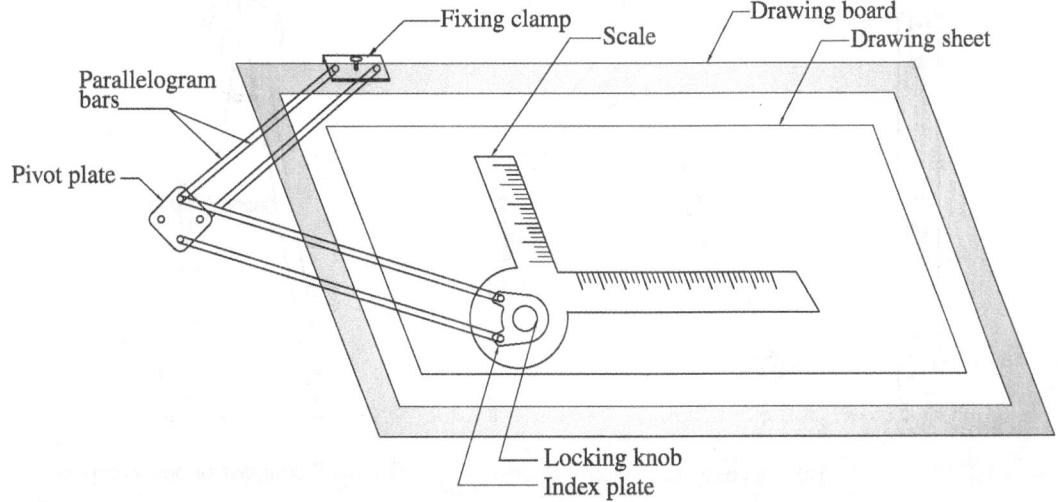

Fig. 1.5 Mini drafter with its various parts

be chosen by testing the adequate reach of the perpendicular arms within the drawing sheet.

1.4.7 Compasses and Dividers

The compasses and dividers are most commonly used by students from the school level itself. A compass is used to draw circles and arcs, whereas a divider is used to set a particular distance from the scale and transfer it to the drawing or transfer the dimensions from one part of the drawing to another. The divider is also used to divide a circle or a straight line into desired number of equal parts by setting the distances repeatedly. Both the compasses and the dividers have two legs, hinged together at the top and free at their lower ends and are made of metal. In technical drawings, since radii or arcs are required with a wide range for the use, the compasses and the dividers are available in two sizes, namely large and small.

The large-sized compasses are provided with adjustable legs for drawing radii or arcs up to 120 mm, while the lengthening bars are additionally added to draw radii beyond 120 mm. One of the free ends of the compasses has a pointed needle fitted with a knee joint to hold the compass on the drawing sheet. The other leg is adjustable with a knee joint and an interchangeable holder for wooden pencil leads or inking pens for making the respective drawings. For drawing larger radii, a lengthening bar is inserted between this holder and the remaining portion of that leg which is connected with a knee joint. The knee joints enable the legs to be bent through the required angle, while the needle portion and the pencil or ink holders are held perpendicular to the surface of the drawing

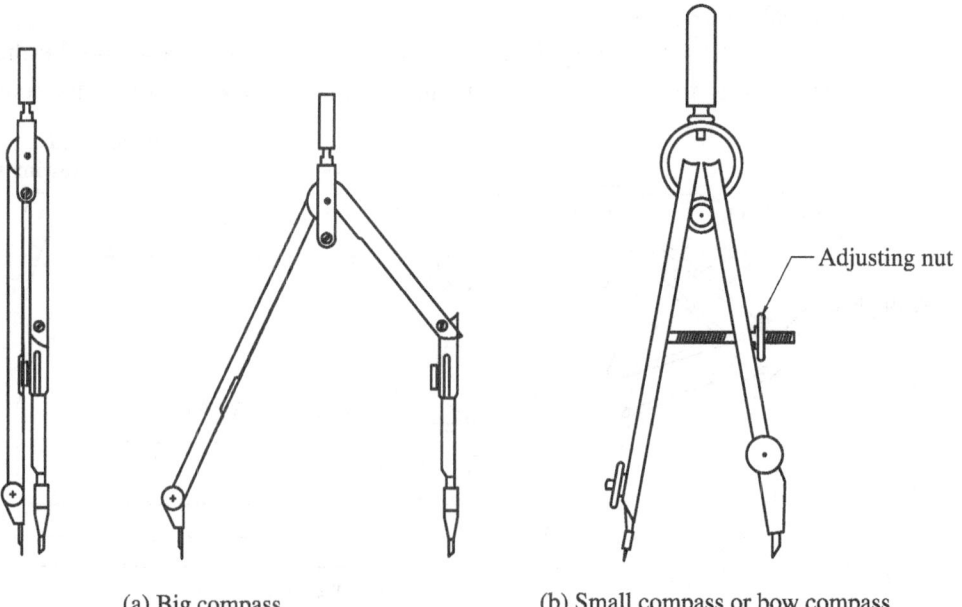

(a) Big compass   (b) Small compass or bow compass

Fig. 1.6   Compass fixed with sharpened lead

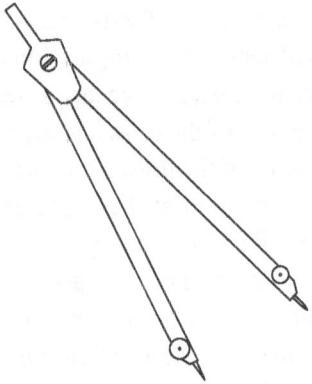

Fig. 1.7   Divider

sheet, while negotiating a curve or an arc. The needle leg is slightly lower than the other leg by 1 mm to insert that into the drawing sheet. The dividers are fitted with needles at both the free ends of the legs.

The construction of the small compass is also similar to that of the larger ones, except the legs are made straight with no knee joints, as the radii of the curves or arcs required to be made with them are small, usually up to 25 mm. Further, the legs are interconnected with an adjustable screw and nut arrangement to ensure that the radius set is not getting altered while drawing. The pencil- or ink-holding facility is provided in one of the legs, whereas a needle arrangement is provided in other leg. The small compasses are also known as bow compasses.

The large and small or bow dividers have their legs fitted with needles or spikes at their ends. All the above-mentioned compasses and their attachments, dividers, etc. are nicely arranged inside the velvet-lined case known as the instrument box and made with proper slots to protect the instruments. The instrument box is regarded as a pride jewel box for an engineer. Figures 1.6 and 1.7 show the sketches of the big compass, small compass, and divider.

1.4.8 French Curves

Fig. 1.8   French curve

The French curves are a set of curves of different curvatures made on a thin wood or plastic or celluloid sheet. Some set squares also contain these curves cut in their middle. French curves are used for drawing curves that have different radii and curvatures which cannot be drawn with a compass. A continuous smooth curve required through a set of points that do not lie on a straight line or on a circle can be drawn by judiciously joining them with the French curves. Figure 1.8 shows one of the French curve templates made on a plastic sheet.

1.4.9 Roll and Draw

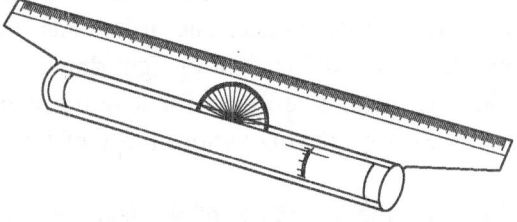

Fig. 1.9   Roll-N-Draw

It is a simple handy ruling machine, consisting of a roller, a graduated scale, and a protractor, used to draw rapidly a set of lines in horizontal, vertical, or inclined directions. The instrument is about 16 cm in length and is also known as Roll-N-Draw (refer Fig. 1.9).

1.4.10 Sheet-holding Devices

These are simple devices that help to secure the drawing sheet on the drawing board. After setting the drawing sheet on the drawing board, the sheet is fixed on their corners with

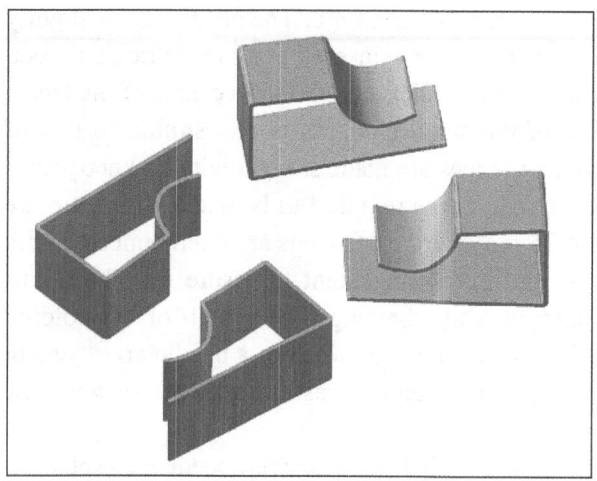

Fig. 1.10   Drawing clips

the help of metallic pins or adhesive tapes. Since the pins will leave their impressions on the drawing board, the adhesive tapes are preferable. If the edges of the drawing sheet are closer to the edges of the drawing board, then the sheets can be held at these places by means of metallic clips that embrace the sheet and the board. The flat portion of the clips should be used on the drawing sheet, while the curved portion of the clip will be used beneath the board, to enable the movement of the drafter on the sheet without interruption. Figure 1.10 shows the drawing clips that are used to secure the drawing sheet.

1.4.11 Pencils and Accessories

Pencils are the primary tools in engineering drawing, used by engineers to communicate their ideas through text or drawings. The pencils are not only handy and cheap but also facilitate noting down the changes, incorporation of new ideas, etc. as they are amenable for erasing or making alterations, without causing permanent impressions as caused by other devices such as pens. Lead pencils are available in holders that appear like pens and can decorate the pockets of the engineers with the same esteem like any other possessions. The quality and the clarity of the drawings are decided by the grade of the pencils used and the graphite content present in them. The pencils are classified into two major categories as soft and hard, according to European system and are represented by the English alphabets B (Blackness) and H (Hardness), respectively. The degrees of the blackness or hardness are indicated by a numeral preceding these letters. Thus, the increasing order of soft pencils (more graphite deposits or blackness) is represented by 1B, 2B, 3B, etc., and the increasing order of hardness (more impressive but less graphite deposit) is represented by 1H, 2H, 3H. etc. with the HB representing the standard grade writing pencil. The grade of a pencil lead is usually marked near one of its ends. In engineering drawings, H and 2H pencils are used for making faint or light drawings and are made thick using 3H pencils. H and HB pencils are used for lettering and dimensioning. The different grades of pencils used for drawing various types of lines are explained in the following section.

The quality of the drawing is also decided by the sharpness of the leads while mending the pencil. Two distinct forms, namely (a) the conical point and (b) the chisel edge, are used in pencil leads while mending them. The conical point is used for sketch work and lettering, and the chisel edge is used for drawing long thin lines. The chisel edge is obtained by rubbing the lead on a sand paper block and making it flat gradually.

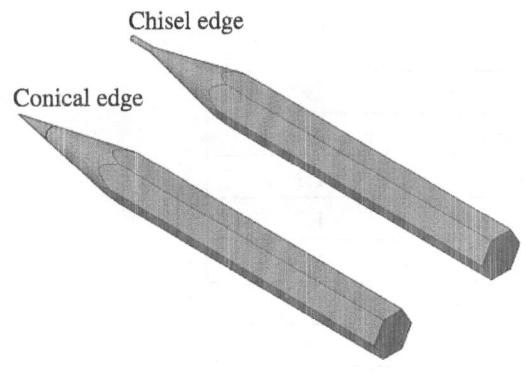

Chisel edge

Conical edge

Fig. 1.11   Pencils

Figure 1.11 shows the sharpened pencil edges of the conical and the chisel types. The pencil edges can be sharpened with hand-held or table-mounted sharpeners. Mechanical clutch pencils with different lead sizes are available, and they do not need sharpening, like the wooden pencils. The size of the available diameters of the clutch pencils conform to ISO: 9175(Part I)-1988. The drawings prepared with pencils, when often to be modified or erased, good quality soft erasers are to be used. If drawings are made with soft pencils or with more graphite deposits, the graphite particles will smear and then create a shabby appearance, which is to be avoided.

1.4.12 Scales and Scale Sets

Scales are used to measure the required lengths and set them in the drawing. They are made of plastic or cardboard for making technical drawings and are available in 15 cm or 30 cm lengths usually. The width will range from 2 cm to 3 cm. They are usually 1 mm thick, and the edges will be bevelled for greater thicknesses. Both the longer edges of the scale are marked in centimetres and are subdivided into millimetres or one longer edge in centimetre or millimetre units, whereas the other longer edge contains inches and its subunits.

A scale box consisting of a set of metric scales is usually made of cardboard for selecting the lengths of large-sized to small-sized components. When large-sized objects such as buildings and bridges in civil engineering, and machines and automobiles in mechanical engineering have their dimensions in metres, small-sized miniaturized components such as IC chips, electronic components, and tiny mechanical parts have their dimensions in millimetres or even in fractions thereof. When the drawing sheets and the normal scales do not cater to these dimensions, a set of scales are to be prepared using representative fraction (RF) known as a numerical scale.

A *scale* is defined as the ratio of a linear dimension of an object represented in the drawing sheet to the real linear dimensions of the original object itself. A scale is known as full scale, where the ratio is 1:1, and a scale with the ratio of 1:2, 1:3, etc. is known as reduced scale, where the original object is twice or thrice the length represented in the drawing. Similarly, a scale of 2:1, 3:1, etc. represent an enlarged scale where the drawing represents a length which is twice or thrice the original length of the object. A set of reduction scales classified as M1, M2, …, M8 is shown in Table 1.1 as recommended by Bureau of Indian Standard specifications. On each side of a scale, two different scale settings are marked. For example, in M1, the full scale is marked on one side, whereas a reduction scale of 1:2 is marked in the rear side. By noting the actual dimensions of the

Table 1.1   Recommended standard scales set

S. no.	Designation	Scale	
		Description	Value
1.	M1	Actual size	1:1
		50 cm to a metre	1:2
2.	M2	40 cm to a metre	1:2.5
		20 cm to a metre	1:5
3.	M3	10 cm to a metre	1:10
		5 cm to a metre	1:20
4.	M4	2 cm to a metre	1:50
		1 cm to a metre	1:100
5.	M5	0.5 cm to a metre	1:200
		0.2 cm to a metre	1:500
6.	M6	0.33 cm to a metre	1:300
		0.66 cm to a metre	1:600
7.	M7	0.25 cm to a metre	1:400
		0.125 cm to a metre	1:800
8.	M8	1 mm to a metre	1:1000
		0.5 mm to a metre	1:2000

object, the same can be set in the drawing sheet by choosing the concerned scale. This avoids the manual conversion of a given object dimension as per the scale chosen but helps to mark it in the drawing sheet quickly. Various other types of scales are discussed in Chapter 3.

1.4.13 Drafting Templates

Templates are thin plastic sheets with the shapes of different polygons, circles, ellipses of different sizes cut on them. This will enable the reproduction of these shapes in the drawings, particularly when they are repeated. The use of the templates will save valuable time in the drawing of standard figure and symbols. Electrical, hydraulic and pneumatic equipment, springs, fasteners, gears, and architectural symbols are also available in certain templates.

1.5 SOFTWARE TOOLS

Due to the advances in mathematics and development of computing techniques, the preparation of engineering drawings have undergone a thorough change from its manual mode to that of a completely computer-operated mode. The importance of surface geometry and geometric modelling became significant. Creation of digital models using Pierre Bezier curves and NURBS technology has led to advanced geometric modelling. Many solid modelling software and two-dimensional drafting packages have been developed since 1980. Some of the common three-dimensional modelling packages are ProE, Unigraphics, IDEAS, CATIA, Solidworks, etc. The popular two-dimensional drafting packages such as AutoCAD and Solid Edge are widely used for drafting.

1.6 DRAWING STANDARDS

1.6.1 International Standards

International Organization for Standardization (ISO) is the leading world's largest voluntary developer for International Standards and is responsible to specify the standards for products, services, and good practice, to ensure quality, safety, and efficiency. These standards are referred and followed throughout the world for international trading of any product and its service. This organization was initially started in the year 1947 at Geneva, Switzerland and has published more than 19,500 International Standards covering almost every industry, from technology to food safety, agriculture, and healthcare.

The International Standard (referred by ISO and the standard no.) is a document that provides requirements, specifications, guidelines, or characteristics that can be used consistently to ensure that materials, products, processes, and services are fit for their purpose.

The benefits that can be derived by following the standards are that they become the strategic tools that reduce costs by minimizing waste and errors, and increase productivity, and help companies to access new markets and interact with the developing countries for their global trade.

1.6.2 Bureau of Indian Standards

The Bureau of Indian Standards (BIS) is the national standards body of India working under the support of Ministry of Consumer Affairs, Food, and Public Distribution, Government of India. It is established by the Bureau of Indian Standards Act, 1986 which came into effect on 23 December 1986. It was formerly known as the Indian Standards Institution (ISI), set up under the Resolution of the then Department of Industries and Supplies No. 1 Std.(4)/45, dated 3 September 1946. The ISI was registered under the Societies Registration Act, 1860. As a corporate body, it has 25 members drawn from central or state governments, industry, scientific and research institutions, and consumer organizations. Its headquarters is located in New Delhi, with regional offices in Kolkata, Chennai, Mumbai, Chandigarh, and Delhi and 20 branch offices. The Indian Standards are referred by the letter 'IS', followed by the standard no. Some of the standards that are followed for the practice of engineering drawing under BIS are listed in Table 1.2 along with their respective International Standard numbers.

Table 1.2 Various standards for the practice of engineering drawing under BIS and their respective codes

S. No.	IS no.	Part no.	Year of publication	Title	Relevant international standard
1.	IS 1444	—	1989 (Reaffirmed 1999)	Engineer's pattern drawing boards-specification (second revision)	—
2.	IS 3221	—	1966 (Reaffirmed 2006)	Sets for drawing instruments	—
3.	IS 8930	1	1995	Technical product documentation—vocabulary—Part 1: Terms relating to technical drawings: General and types of drawings	ISO 10209-1: 1992
4.	IS 8976	—	1978	Guide for preparation and arrangement of sets of drawings and parts lists	—
5.	IS 9609	0	2001	Technical product documentation—lettering—Part 0: General requirements	ISO 3098-0: 1997
6.	IS 9609	1	2006	Technical product documentation—lettering—Part 1: Latin alphabet, numerals and marks	ISO 3098-2: 2000
7.	IS 10711	—	2001	Technical product documentation—sizes and layout of drawing sheets	ISO 5457: 1999
8.	IS 10712	—	1983	Presentation of item references on technical drawings	ISO 6433: 1981
9.	IS 10713	—	1983	Scales for use in technical drawings	ISO 5455: 1979
10.	IS 10714	—	1983	General principles of presentation on technical drawings	ISO 128: 1982
11.	IS 10714	20	2001	Technical drawings—general principles of presentation—Part 20: Basic conventions for lines	—
12.	IS 10714	21	2001	Technical drawings—general principles of presentation—Part 21: Preparation of lines by CAD systems	ISO 128-21: 1997
13.	IS 11065	1	1984	Drawing practice for axonometric projections—Part 1: Isometric projection	—
14.	IS 11663	—	1986	Conventional representation of common features and materials on technical drawings	—
15.	IS 11664	—	1986	Folding of drawing prints	—
16.	IS 11665	—	1985	Technical drawings—Title block	ISO 7200: 1984
17.	IS 11666	—	1985	Technical drawings—Item lists	ISO 7573: 1983

(Contd)

Table 1.2   *(Contd)*

S. No.	IS no.	Part no.	Year of publication	Title	Relevant international standard
18.	IS 11669	—	1986	General principles of dimensioning on technical drawings	ISO 129: 1985
19.	IS 11670	—	1993	Technical drawings—Abbreviations and symbols for use in technical drawings	—
20.	IS 15021	3	2001	Technical drawings—Projection methods—Part 3: Axonometric representations	—
21.	IS 15021	1	2001	Technical drawings—Projection methods—Part 1: Synopsis	—
22.	IS 15021	2	2001	Technical drawings—Projection methods—Part 2: Orthographic representations	—
23.	IS 15021	4	2001	Technical drawings—Projection methods—Part 4: General projection	—
24.	IS 15057	—	2001	Computer-Aided Design (CAD) technique—Use of computers for the preparation of construction for drawings	—
25.	IS 15093	—	2002	Construction drawings—Spaces for drawing and for text, and title books on drawing sheets	ISO 9431: 1990

1.7 BIS CONVENTIONS AND SPECIFICATIONS ON DRAWING

1.7.1 Size, Layout, and Folding of Drawing Sheets

Size of Drawing Sheets

Table 1.3   Different types of drawing sheets (Main ISO-A series – SP.46.2003)

Sheet designation	Trimmed size (mm)	Untrimmed size (mm)
A0	841 × 1189	880 × 1230
A1	594 × 841	625 × 880
A2	420 × 594	450 × 625
A3	297 × 420	330 × 450
A4	210 × 297	240 × 330
A5	148 × 210	165 × 240

The drawing sheets for technical drawings should confirm to the trimmed sizes as recommended by BIS and its special publications SP46-2003 (IS No. 10711-2001) and are shown in Table 1.3. The sheets are classified into six sizes A0–A5, the largest size being A0. The surface area of the basic A0 size drawing sheet is one square metre, and its length-to-width ratio is maintained as $\sqrt{2}$:1. This will yield the length and the width dimensions of the A0 size drawing sheets as 1189:841 mm, respectively. The successive sheet sizes are also obtained in the same ratio and by halving the length dimension such that its area is half the previous

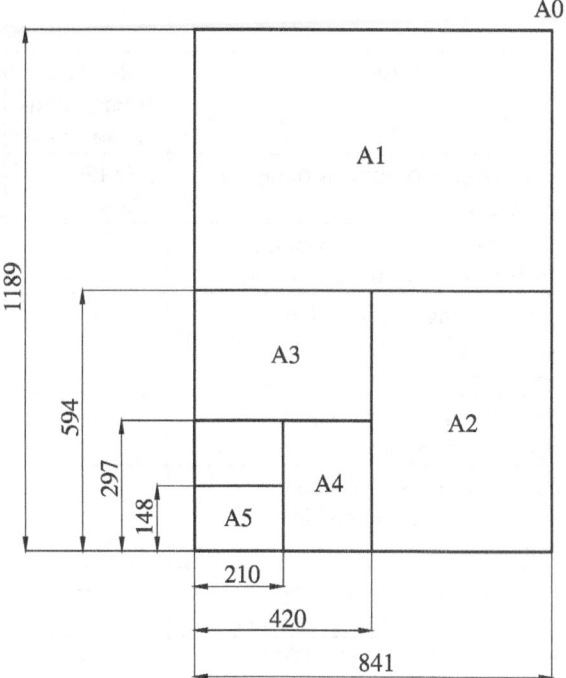

Fig. 1.12 Drawing sheet sizes

A0 sheet area. For example, one dimension of A1 size drawing sheet is obtained by halving the length of A0 size, whereas the other dimension confirms to the width of A0 size, that is, 841 mm. These dimensions will have to satisfy the length to width ratio of $\sqrt{2}$:1 and surface area of A1 size has to be half of that of A0 size, thus obtaining the length and the width of A1 size drawing sheet as 841 × 594 mm. Table 1.3 explains the dimensions of the drawing sheets of various sizes and Fig. 1.12 shows the methodology of obtaining the sizes of each drawing sheet. A2-sized drawing sheets are used by college students for drawing practice. The A3- and A4-sized sheets are usually known as drawing papers and are used for home work. A0- and A1-sized sheets are used for industrial drawings and civil and architectural engineering drawings. The standard sizes of the drawing boards as recommended by BIS are shown in Table 1.4 for different paper sizes. The thickness of the drawing boards is generally around 20–25 mm. In drawing offices and industries, D0- and D1-sized boards are used, whereas in colleges, D2-sized boards are used, and D3-sized boards are adopted in the schools.

Table 1.4 Standard sizes of drawing boards

S. no.	Designation	Use of recommended drawing sheet	Length × width (dimensions are in mm)
1.	D0	A0	1500 × 1000
2.	D1	A1	1000 × 700
3.	D2	A2	700 × 500
4.	D3	A3	500 × 350

Fastening the Drawing Sheet

To begin a drawing practice, the drawing sheet should be firmly fastened or fixed to the drawing board properly. This is ensured by spreading the drawing sheet on the drawing board and checking its edges aligned with the vertical and horizontal scales of the mini drafter, for the college practice. When T-square is used, the top edge is to be aligned with the working edge of the T-square. Except A4-sized drawing sheets, all the other sheets are positioned horizontally. The general procedure of fastening the drawing sheet is explained as follows:

1. Loosen the clamping screw of the mini drafter, place it at any convenient location on the drawing board (preferably near the left corner), and tighten it to the board.
2. Move the drafter scale to the centre of the sheet and check its reach along the drawing sheet.
3. Set the scale screw against 0° mark and tighten it.
4. Move the drafter scale and align it with the top edge of the drawing sheet, also check and align the left edge of the drawing sheet with the vertical scale. After checking these alignments, fix a tape on the top left corner and the bottom right corner of the drawing sheet. After spreading the sheet gently, fix the tape at the other corners.
5. While doing this alignment, if the bottom and the right edges of the drawing sheet are fixed closely to the respective edges of the drawing board, then the drawing clips can be inserted in the corners of these edges and the cellotape can be affixed to the top left corner alone. This location reduces the hand and trunk movements of the user to the minimum and is preferable.

Layout of the Drawing Sheet

1. After fastening the drawing sheet with the drawing board, the formal work on the drawing sheet commences. It begins with a proper layout of the drawing sheet. The layout involves the drawing of border lines and the provision of a title block that reveals in a nut shell, the salient information of the drawing, such as its title, author, identification number, and documentation references if any.
2. The border lines are drawn at a distance of 10 mm from all the trimmed edges and at 20 mm from the left edge to have a proper filing margin.
3. A title block of length 170 mm and height 60 mm is included at the right bottom corner, from the border lines. For class room practice, the title block is divided into five parts horizontally. The details of the author, that is, the student name, register number, class, and section are furnished in the first two compartments. The title of the drawing is printed in the third compartment. The sheet number and date of preparation are marked in the fourth compartment. The last compartment reveals the scale and the units adopted for dimensions. The method of the angle of projection is also represented with its due symbol. Figure 1.13 shows the layout of the drawing sheet with the border lines, title block details, etc.
4. The space between the border lines and the title block becomes the working or drawing space.

Folding of the Drawing Sheet

After the drawings are made, the drawing sheet is to be folded as per the methods shown in the IS specification. All sheets are to be folded such that they reduce to A4 size, with a height of 297 mm and a width of 210 mm, including the filing margin. The reader should familiarize with the folding procedure which explains the number of folds and the distances between different folds as mentioned in IS.

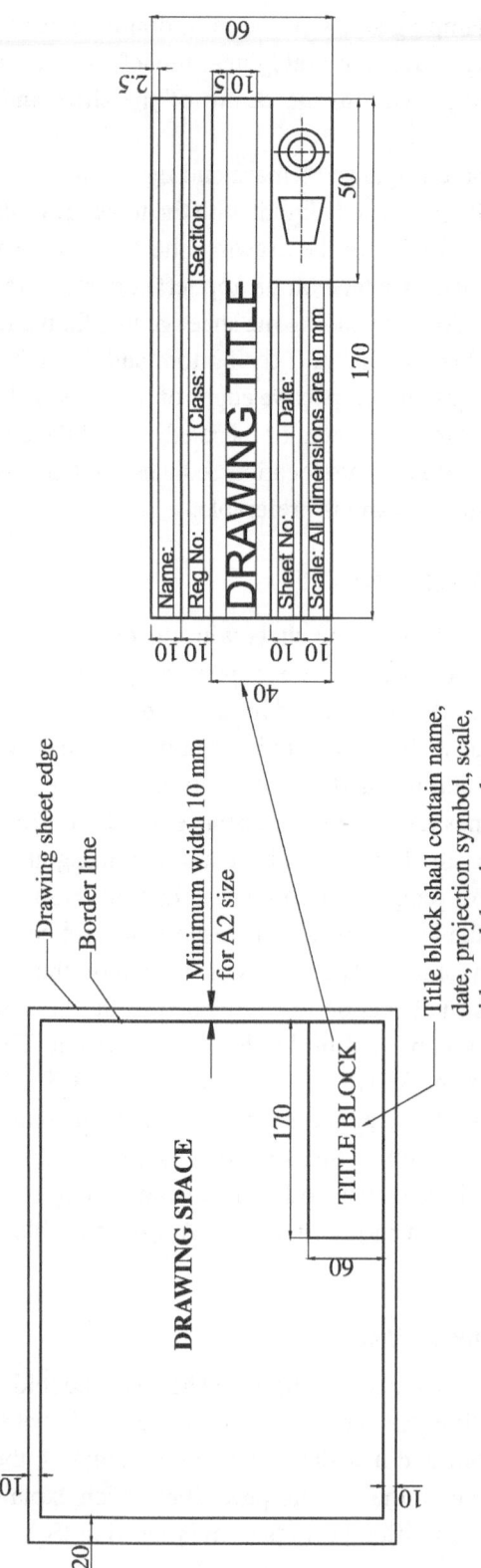

Fig. 1.13 Layout of a drawing sheet

1.7.2 Lines and their Representation in Drawings

Need

Lines give life to the drawings and create a feel of the object. Since they are used to represent the external and internal appearances and the different features of an object, the lines are to be represented with different intensities that distinguish their usage.

Hence, standardized representation on lines becomes essential to convey the same thought to all the readers and users. This section deals with the types of lines, applications, and their representation in engineering drawings as recommended by the latest BIS codes IS10714 (part 20 and 21)–2001 and Special Publications SP46:2003 confirming to ISO128, parts 20–24.

Types of Lines

A line representing an external surface or internal surface may be straight, curved, continuous, or segmented and may take any orientation to suit the contour of the surface of the object. The objects whose dimensions are symmetric about an imaginary line known as the axis needs a different representation in the drawings. To explain the details of certain inside features, the object has to be imaginarily cut and such representations call for the inclusion of section or cutting plane lines in the drawings. Sometimes certain parts or locations of an object are to be addressed with a pointing mark for certain specific identities, and hence, the pointing lines or leader lines are to be marked differently. The lines that are used for bordering, title block, and dimensioning should appear distinctly different. The formal lines that are used for the basic constructions of certain shapes, though drawn, are to be less visible in the overall appearance. Therefore, all these lines are to be represented in different ways such that the resulting contrast makes the drawing easier to convey its meaning quickly and completely. There are 15 basic types of lines used in engineering practice which are identified by their designating numbers as per BIS specifications. Table 1.5 shows some of the relevant types of lines required for usage at this level, along with their designating numbers and their applications.

Table 1.5 Types of lines (as per IS.SP.46.2003, Clause: 6.3.1)

Designation no.	Basic type of line category	Commonly used sub types of lines and their designation no.	General representation	Application
01	Continuous line	Continuous narrow line (1.1) Thickness = 0.13 mm	———	1.1.1 Construction lines 1.1.2 Projection lines 1.1.3 Short center lines 1.1.4 Extension lines 1.1.5 Dimension lines 1.1.6 Dimension line terminations 1.1.7 Leader lines 1.1.8 Hatching/section lines 1.1.9 Imaginary lines of intersection 1.1.10 Out lines of revolved sections

(Contd)

Table 1.5 (*Contd*)

Designation no.	Basic type of line category	Commonly used sub types of lines and their designation no.	General representation	Application
		Continuous wide line (1.2) Maximum width or thickness = 0.25 mm	——————	1.2.1 Visible outlines of parts in cut and section, when hatching is used. 1.2.2 Visible outlines of parts in view 1.2.3 Reference lines in projection 1.2.4 Ground lines 1.2.5 Arrow lines for making of views, cuts and sections. 1.2.6 Border lines, title block etc.
		Continuous extra wide line (1.3) Maximum width or thickness = 0.5 mm	━━━━━	1.3.1 Visible outlines 1.3.2 Visible outlines of parts in cut and section, when hatching is not used. 1.3.3 Lines of special importance.
		Continuous line, free hand narrow (1.4)	———⌐—	1.4.1 Short break lines 1.4.2 Limits of partial or interrupted views
		Continuous lines, Zig zag hand narrow (1.5)	—\/\——	1.5.1 Long break lines 1.5.2 Limits of partial or interrupted views
02	Dashed lines	Dashed narrow line (2.1)	....	2.1.1 Hidden lines or edges 2.1.2 Hidden outlines
		Dashed wide line (2.2)	-------------	2.2.1 Hidden lines or edges
03	Dashed spaced line		— — — — — — —	
04	Long dashed dotted line	Long dashed dotted narrow line (Chain narrow line) (4.1)	— · —	4.1.1 Center lines 4.1.2 Lines of symmetry 4.1.3 Pitch circle of holes 4.1.4 Pitch circle of gears 4.1.5 Indication of cutting planes
		Long dashed dotted wide line (Chain wide line) (4.2)	— · —	4.2.1 Cutting plane representations at the ends and at locations of change of direction 4.2.2 Surface treatment indication locations
05	Long dashed double dotted line	Long dashed double dotted narrow line (5.1)	— ·· —	5.1.1 Centroidal lines 5.1.2 Locus lines 5.1.3 Alternate positions of movable parts 5.1.4 Extreme positions of movable parts 5.1.5 Out lines of adjacent parts

Width of Lines

The width 'd' of all types of lines is decided by the mathematical series which bears the common ratio $1:\sqrt{2}$ (1:1.414). The width of the line, here, means the thickness of the line, and it remains constant throughout the whole line. The general width (thickness) adopted for the lines are as follows: 0.13 mm, 0.18 mm, 0.25 mm, 0.35 mm, 0.5 mm, 0.7 mm, 1.0 mm, 1.4 mm, and 2.0 mm. Based on the width, any type of line can be classified into three subcategories such as narrow (thin), wide (medium), and extra wide (thick), and their width ratios are 1:2:4. The line width shall be chosen according to the type, size, and the scale of the drawing; the requirements at microcopying and other methods of reproduction. When the line is designated by a particular designation number, it will be further subclassified as per the thickness used. For example, when 01 denotes a continuous line, 01.1, 01.2, and 01.3 represent the narrow, wide, and extra-wide type lines, respectively, in that category. The lines are grouped based on the line widths as shown in Table 1.6. In pencil drawings, the narrow and wide lines are used, and the line group 0.25 is the most preferred one. The thickness (or the width) of the lines can be varied by changing the pressure exerted on the pencil by the user.

Table 1.6   Line widths

Line group (mm)	Narrow line (mm)	Wide line (mm)	Extra-wide line (mm)
0.25	0.13	0.25	0.5
0.35	0.18	0.35	0.7
0.5	0.25	0.5	1
0.7	0.35	0.7	1.4
1	0.5	1	2

Line Elements

An element of a line is a single part of a non-continuous line. For example, the line types 02 to 05 in Table 1.5 are non-continuous lines, and they consist of dots, dashes, and gaps (which vary in length), and these are known as the *elements*. The lengths of the elements are decided by Table 1.7 when its configurations are shown.

Table 1.7   Configuration of lines

Line element	Line type no.	Length
Dots	4.1, 4.2, 5.1	≤ 0.5 d
Gaps	2.1, 2.2, 4.1, 4.2, 5.1	3 d
Dashes	2.1, 2.2, 3.0	12 d
Long dashes	4.1, 4.2, 5.1	24 d
Spaces	3.0	18 d

Application of Various Line Types

The usage of the various types of lines shown in Table 1.5 has been explained in the following paragraphs and through drawings in Fig. 1.14.

Line Type 1.1: Representation with Continuous Narrow (Thin) Lines

The general width (thickness) of the lines is 0.13 mm and is used for the following:

Line 1.1.1: Construction lines These lines are made for the construction of geometrical features or location of some points of the object that should be drawn very thin and faint and hardly visible in the finished drawing. These are only the supporting lines to facilitate the shape and do not represent the main edges of the object.

Line 1.1.2: Projection lines These lines are emanating from the various points of an object when it is projected to the planes of reference and help to connect the views such as the front view, the top view, and the side views (refer Fig. 1.14a). These are explained in the Chapter 5 on Orthotropic Projection and are adopted in the other chapters. However, when the reader develops sufficient familiarity, these are used very thin and are not present in industrial drawings.

Line 1.1.3: Short centre lines These are used to mark the centre of small circles represented in the drawings and are marked with a + (plus) sign (refer Fig. 1.14a).

Line 1.1.4: Extension lines These lines are used to extend outside the drawing from the feature to be dimensioned and is perpendicular to it (refer Fig. 1.14a).

Line 1.1.5: Dimension lines The *dimension line* represents the dimensions of the feature in the drawing and is generally a continuous line drawn parallel to the outer edge. The numerical value of the dimension is written above this line (refer Fig. 1.14a).

Line 1.1.6: Dimension line terminations The ends of the dimension lines are usually indicated by arrows. Sometimes instead of arrows, oblique lines that cut the extension lines are marked at the ends. Such lines are known as *termination lines* (refer Fig. 1.14a).

Line 1.1.7: Leader lines The leader lines are used to pin point a location or a feature in a drawing. They are drawn usually at an angle to the representation and touch the feature with an indication of an arrow (refer Fig. 1.14a).

Line 1.1.8: Hatching lines (section lines) These are used to indicate a surface, which is cut. A series of parallel lines is marked in the cut surface in an inclined fashion. They have to maintain a uniform spacing among them (refer Fig. 1.14a).

Line 1.1.9: Imaginary lines of intersection These are lines drawn as intersecting diagonals inside a polygon surface to represent it as an open surface. When an object intersects with another object, these intersecting lines drawn from both the objects, meet in common (refer Chapter 11 on Intersection of Surfaces).

Line 1.1.10: Outlines of revolved sections When the cross section of a long object is represented within the view representing its length, it is called a revolved section. The outlines of these sections are drawn as per Line 1.1.10. The spokes of the wheels and the cross section of a connecting rod of an internal combustion engine are the examples of this representation (refer Fig. 1.14e).

Line Type 1.2: Representation with Continuous Wide or Medium Thick Lines

These lines are drawn with a uniform thickness of 0.25 mm and are explained in the following applications:

Line 1.2.1: Visible outlines of parts in cut and section, when hatching is used   The outlines of cut section boundaries which become visible during sectioning are shown by these lines (refer Fig. 1.14a).

Line 1.2.2: Visible outlines of parts in view   The extreme outline boundaries of the object are shown in continuous wide medium thick lines. When certain objects have protruding portions or parts such as steps, pillars, or risers, the edges corresponding to them are also marked in continuous thick lines, inside the outer boundaries, and such lines fall under this category (refer Fig. 1.14a).

Line 1.2.3: Reference lines in projection   Reference lines are the lines used to distinguish the views or the lines separating the front view, the top view, and the side views of an object. These lines are shown in all the drawings from Chapter 5 onwards (refer Fig. 1.14a).

Line 1.2.4: Ground lines   These are also like reference lines, but are used to denote the representation of the ground. Refer Chapter 5 representing the Third Angle Projection and Chapter 15 on Perspective Projection.

Line 1.2.5: Arrow lines for marking the views of cut sections   When the section plane is represented as a straight line, its ends are provided with arrow lines, which will indicate the direction of viewing after the cutting or sectioning is done. Such lines are shown Fig. 1.14(a).

Line 1.2.6: Border lines and title block lines   These are continuous lines denoting the border and the title block border lines in the drawing sheet before the commencement of any drawing work.

Line Type 1.3: Representations with Continuous Extra-wide or Extra-thick Lines

These lines are drawn with a uniform thickness of 0.5 mm and are explained in the following applications:

Line 1.3.1: Visible outlines   All the visible edges and the outline boundaries of an object are shown in continuous lines with the maximum thickness of the line group (refer Fig. 1.14a).

Line 1.3.2: Visible outlines of parts in cut and section, when hatching is not used   When the surfaces of an object are cut or sectioned and the cut section is represented only by its boundaries (not with hatching lines inside), such outer lines are represented as per this clause and as shown in Fig. 1.14(a). Sectioning of thin walls or ribs or thin but wide pieces (known as standard parts) is not represented by hatching lines inside, though their surfaces are cut. The boundaries of such pieces are represented by the arrangement shown in Fig. 1.14(a).

Line Type 1.4: Representation with Continuous Free Hand and Narrow Lines

Line 1.4.1: Short break lines   These lines are used to indicate a short break in moderately lengthy objects such as bars and channels that are of uniform cross section (regular or irregular) and are marked in wavy pattern with freehand as indicated in Fig. 1.14(e).

Line 1.4.2: Lines to indicate partial or interrupted views   These are similar to the short break lines that are used to indicate the limits of partial or interrupted views of a drawing.

Line Type 1.5: Representation with Continuous Zigzag and Narrow Lines

Line 1.5.1: Long break lines   These lines are used to indicate a break in lengthy objects such as bars and channels that are of uniform cross sections (regular or irregular) and are marked in zigzag wavy pattern with freehand as indicated in Fig. 1.14(d).

Line 1.5.2: Lines to indicate partial or interrupted views   These are similar to the long break lines and are used to indicate the limits of partial or interrupted views of a drawing.

Line Group 2.0: Representation with Dashed Lines

Line Type 2.1: Representation with Narrow Dashed Lines

Line 2.1.1: Hidden lines or edges   These are used to mark the invisible or hidden edges or features of an object. They are made up of short dashes of approximately equal lengths of 2 mm with a gap of 1 mm. These are also called as dotted lines (refer Fig. 1.14a).

Line 2.1.2: Hidden out lines   When the free ends of an object are hidden by another surface, the edges corresponding to that object are shown as hidden out lines as shown in Fig. 1.14(g).

Line Type 2.2: Representation with Dashed Wide Lines

Line 2.2.1: Hidden lines or edges   Same as 2.1.1 or 2.1.2 with thickness as demanded by the size and the type of the object (refer Fig. 1.14a).

Line Group 3.0: Representation with Dashed Spaced Lines

These lines are similar to the lines mentioned in type 2.2, with the space between the dashes almost equal to the length of the dashed part itself.

Line Group 4.0: Representation with Long Dashed Dotted Lines

Line Type 4.1: Long Dashed Dotted Narrow Lines

The long dashed narrow lines are thin long chain lines consisting of long dashes of 9–12 mm with dots spaced in between approximately 1 mm apart.

Line 4.1.1: Centre lines   These are drawn to represent the axis of the cylindrical, conical, and spherical features, including the holes in these bodies and are marked protruding slightly beyond the outlines of the object. These lines are also used to show the centres of the circles and arcs (refer Fig. 1.14a). However, it is also a long term practice to mark centre lines as long and short dashed narrow lines.

Line 4.1.2: Lines of symmetry   These are marked similar to the centre lines and are used to represent the axis of symmetrical objects such as regular prisms, pyramids, and similar bodies, solid and hollow. The axes of the spokes or arms of the wheels are also marked by these lines (refer Fig. 1.14a).

Line 4.1.3: Lines to represent pitch circle of holes   When holes or any features are laid along a circular path, equidistant from the axis of a cylindrical disc, the imaginary circular path of the centre of the holes is known as the pitch circle and is represented by the line arrangement as shown in Fig. 1.14(b).

Line 4.1.4: Lines to represent the pitch circle of gears   The various teeth in gear are equally spaced along the circumference of a circular blank. The shape of the gear tooth is identified as the addendum and dedendum portions above and below an imaginary circle, known as the pitch circle. The pitch circle is represented similar to the centre line representation, except that it is arranged along a circle as shown in Fig. 1.14(c).

Line 4.1.5: Lines to indicate the cutting planes   The above line arrangement is also used to represent the location of the imaginary cutting or section planes as shown in Fig. 1.14(a). However, the ends of the lines are drawn thick and also at the corner locations where these lines change their path, owing to sectioning at different parallel planes as shown in Fig. 1.14(a). However, it is also a long term practice to mark the cutting or section planes as long and short dashed narrow lines with the ends marked thick.

Line Type 4.2: Representation of long dashed wide lines

Line 4.2.1: Lines to indicate the ends of the cutting planes and the corner locations   These line representations are similar to 4.1.5 (refer Fig. 1.14a).

Line 4.2.2: Lines to indicate the surface treatment   These line representations are used to indicate the location of the areas of an object requiring special surface treatment or surface finish for its functional improvement or improvement in the appearance of its surface.

Line Group 5.0: Representation of Long Dashed Double Dotted Lines

Line Type 5.1: Representation of Long Dashed Double Dotted Narrow Lines

These line representations are made with alternate dashes of 9–12 mm long, with two dots in between and spaced at 1 mm apart among them from the ends of the dashes.

Line 5.1.1: Centroidal lines   These are similar to the imaginary axis lines but passing through centroid or the centre of gravity (CG) of the object (refer Fig. 1.14d).

Line 5.1.2: Locus lines   These lines are used to indicate the path followed by a moving point or part of a mechanism. A sample representation of this line is shown in Fig. 1.14(f).

Line 5.1.3: Lines to represent alternate positions of moving parts   When a circular wheel revolves about an axis, its spokes or arms move to different positions periodically. The instantaneous or alternate positions of a crank, connecting rod, and the piston of an internal combustion engine are shown in Fig. 1.14(f).

Line 5.1.4: Lines to indicate extreme positions of movable parts   The extreme positions in a movement—rotation, oscillation, or reciprocation—of an object are indicated by this line representation. When the piston of an internal combustion engine moves back and forth, its extreme positions are indicated by this line representation as shown in Fig. 1.14(f).

Line 5.1.5: Lines to indicate outlines of adjacent parts   The boundary lines of other parts adjoining or surrounding the part under study are indicated by this line representation. Figure 1.14(h) shows one such representation.

1.7.3 Lettering Practice

Introduction

A technical drawing is made not only to convey the shape of an object but also should be capable of revealing all the details connected with the object so that the object can be manufactured with no ambiguity. Such details generally include the dimensions of the object, the material particulars, notes, and manufacturing instructions on the quality of the surfaces that exhibit their functional and assembly requirements. These instructions can be incorporated in the drawing by using alphabets, texts, numerals, and symbols and are usually known as *lettering*. Since the lettering is not ornamental, but an essential embodiment of a drawing, it should be made clear, unambiguous, legible, uniform in style, and simple enough to be practised rapidly by freehand or sometimes with minimum stencils or templates. The BIS in conjunction with International Standards has given its detailed recommendations in IS9609:2001 and the Special Publications SP-46:2003 and the relevant extracts of these information are presented in this section.

Definition

The lettering in the drawing refers to the incorporation of a set of graphic characters that includes letters of an alphabet (upper case and lower case), numerals, diacritical marks, punctuation marks, and additional graphical symbols. The lettering practice is the art of arranging them in appropriate proportions to maintain clarity and uniformity as recommended in the standards.

Types of Lettering

The letters are classified into two categories—single stroke and double stroke. Indian Standards recommends single-stroke letters in the drawing practice, which are obtained with uniform thickness, equal to that of the lines produced by the tip of a pencil or a pen. In ink drawings, stencils are used for writing the letters or numerals or character sets, the size of the tip of the pen selected ensures this uniform thickness. Indian Standards further classify the lettering into four categories as 'Lettering A', 'Lettering B', 'Lettering CA', and 'Lettering CB'. The first two classifications use pencil and ink drawings, while the other two use the applications of numerically controlled drafting through CAD. The style of lettering can be vertical or inclined. Inclined lettering is done at an angle of 15° to the vertical or 75° to the horizontal.

Dimensions of Letters

Nominal size or height   As per the BIS, the nominal size of the lettering is defined by height (h) of the outline contour of the upper case or capital letters. Even when lower case or small letters are used, the sum of the heights of their stem and tail portions is

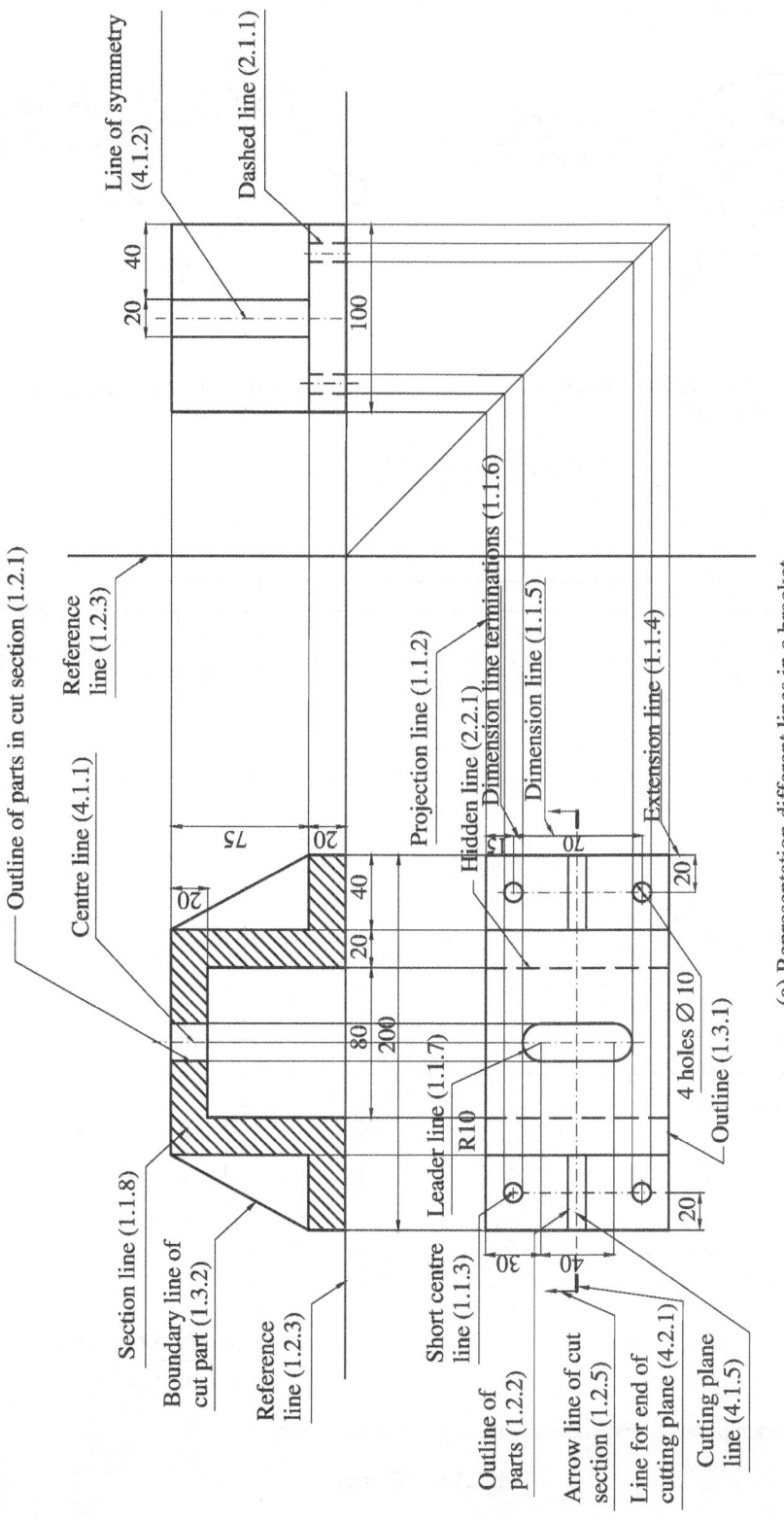

(a) Representation different lines in a bracket

Fig. 1.14   Various types of lines in engineering drawing (*Contd*)

8 holes ∅10

Line type for Pitch
circle of holes (4.1.3)

Line type for Pitch
circle of a gear (4.1.4)

(b) Representation of pitch circle holes
in an annular part

(c) Representation of pitch circle of a gear

Long break lines (1.5.1)

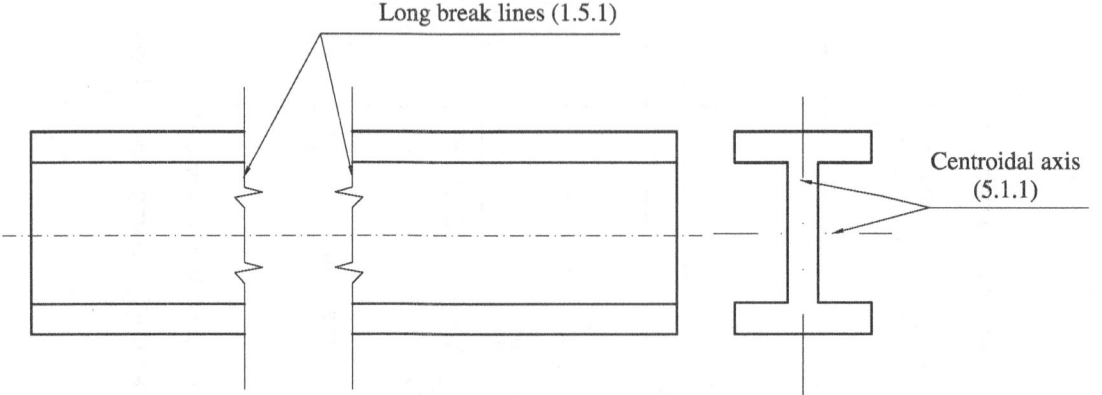

Centroidal axis
(5.1.1)

(d) Representation of centroidal axes and long break Lines

Short break lines (1.4.1)

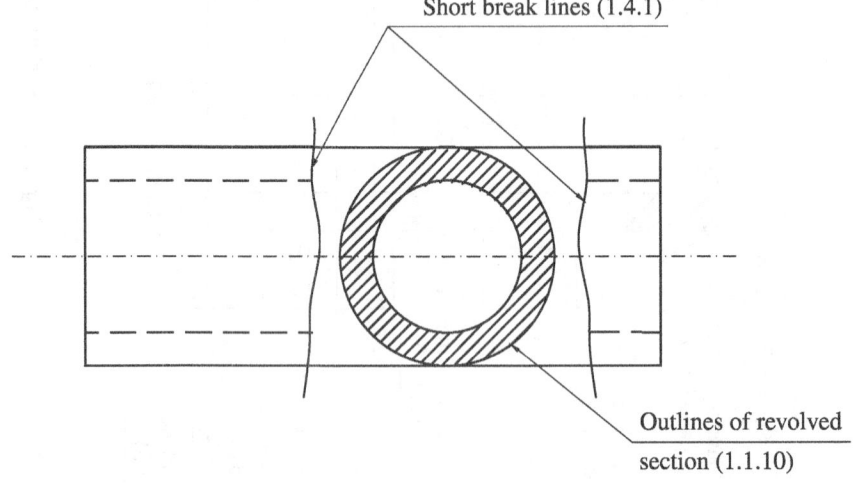

Outlines of revolved
section (1.1.10)

(e) Representation of revolved section and short break Lines

Fig. 1.14 *(Contd)*

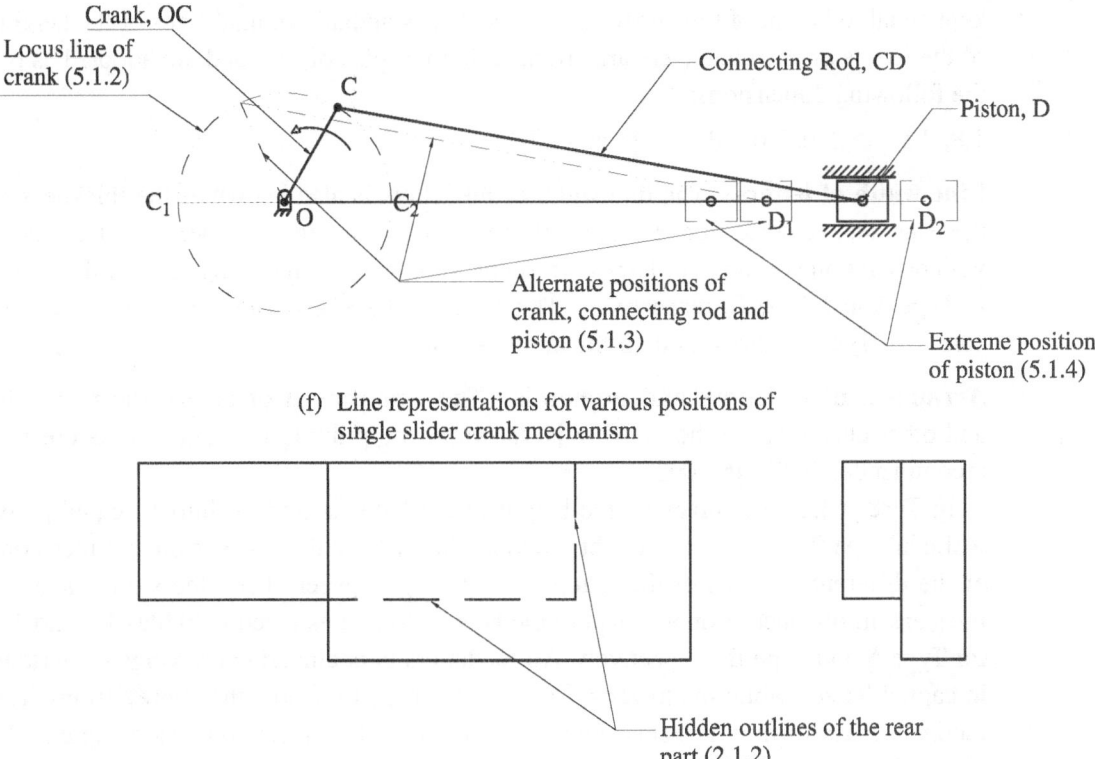

Crank, OC

Locus line of
crank (5.1.2)

Connecting Rod, CD

Piston, D

C

C_1

O

C_2

D_1

D_2

Alternate positions of
crank, connecting rod and
piston (5.1.3)

Extreme positions
of piston (5.1.4)

(f) Line representations for various positions of
single slider crank mechanism

Hidden outlines of the rear
part (2.1.2)

(g) Line representations hidden edges

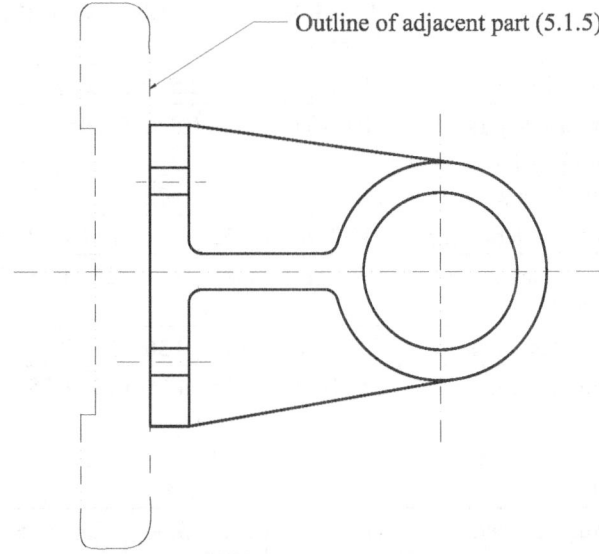

Outline of adjacent part (5.1.5)

(h) Representation of adjoining member of a
cantilever bracket

Fig. 1.14   (*Contd*)

kept equal to height of the upper case letters. The standard nominal sizes or the heights of the letters or any characters are arranged in multiples of $\sqrt{2}$ and are adopted as per the following dimensions:

1.8, 2.5, 3.5, 5.0, 7.0, 10.0, 14.0, and 20.0 mm

Line width of letters   The line width of the letters is also known as the thickness of the letters, and it corresponds to the uniform size of the letters or numerals that are made with pencil point or ink pen. They correspond 1/14 or 1/10 times the height of the letters in Type A and Type B, respectively. The Type B letters are generally thick, stout, and dark than Type A letters, as their thickness is more.

Arrangement of letters and numerals   The arrangement of letters, the numerals, and other characters in the drawing is known as the printing of letters and is done as recommended in SP-46:2003.

In Type A lettering practice, the height of the letter is divided into 14 equal parts, while in Type B lettering, the height is divided into 10 equal parts. The other dimensions of the different portions of the letters, the spacing between them, the words, and the lines are all obtained proportionate to the height (h) and as listed in Tables 1.8 and 1.9 for Type A and Type B, respectively. All the letters in engineering drawings are written in capital letters, while the lower case letters are used to denote the abbreviations. The inclined letters are arranged at an angle of 15° to the vertical on the right side. Figure 1.15 shows the proportion of the letters, words, and line spacing.

Table 1.8   Dimensioning of lettering type A (as per IS.SP.46.2003, Clause: 7.4.1, 7.4.2, 7.4.4, and 7.4.5)

S. no.	Characteristic	Multiple of h	Dimensions (in 'mm')
1.	Lettering height (h)	(14/14) h	1.8 2.5 3.5 5 7 10 14 20
2.	Height of lower-case letters Lettering height (x-height) (c1)	(10/14) h	1.3 1.8 2.5 3.5 5 7 10 14
3.	Tail of lower-case letters (c2)	(4/14) h	0.52 0.72 1 1.4 2 2.8 4 5.6
4.	Stem of lower-case letters (c3)	(4/14) h	0.52 0.72 1 1.4 2 2.8 4 5.6
5.	Area of diacritical marks (Upper-case letters) (f)	(5/14) h	0.65 0.9 1.25 1.75 2.5 3.5 5 7
6.	Spacing between characters (a)	(2/14) h	0.26 0.36 0.5 0.7 1 1.4 2 2.8
7.	Minimum spacing between baselines (1) (b1)	(25/14) h	3.25 4.5 6.25 8.75 12.5 17.5 25 35
8.	Minimum spacing between baselines (2) (b2)	(21/14) h	2.73 3.78 5.25 7.35 10.5 14.7 21 29.4
9.	Minimum spacing between baselines (3) (b3)	(17/14) h	2.21 3.06 4.25 5.95 8.5 11.9 17 23.8
10.	Spacing between words (e)	(6/14) h	0.78 1.08 1.5 2.1 3 4.2 6 8.4
11.	Line width (d)	(1/14) h	0.13 0.18 0.25 0.35 0.5 0.7 1 1.4

Figure 1.16 shows the constructional arrangement or writing of the upper case, lower case letters, the numerals, and the Roman letters using a grid format. Two horizontal guide lines are drawn equal to the height of the letters chosen as recommended in BIS. For writing letters as per Type A, this vertical distance is divided into 14 equal parts and vertical lines are drawn at one part spacing. The height–to–width ratio of the letters is approximately maintained as 7:5 or 7:6, though it varies from letter to letter. The letter I and 1 are narrowest while the letter W is the widest.

Table 1.9   Dimensioning of lettering type B (as per IS.SP.46.2003, Clause: 7.4.1, 7.4.2, 7.4.4, and 7.4.5)

S. no.	Characteristic	Multiple of h	Dimensions (in 'mm')
1.	Lettering height (h)	(10/10) h	1.8 2.5 3.5 5 7 10 14 20
2.	Height of lower-case letters Lettering height (x-height) (c1)	(7/10) h	1.26 1.75 2.5 3.5 5 7 10 14
3.	Tail of lower-case letters (c2)	(3/10) h	0.54 0.75 1.05 1.5 2.1 3 4.2 6
4.	Stem of lower-case letters (c3)	(3/10) h	0.54 0.75 1.05 1.5 2.1 3 4.2 6
5.	Area of diacritical marks (Upper-case letters) (f)	(4/10) h	0.72 1 1.4 2 2.8 4 5.6 8
6.	Spacing between characters (a)	(2/10) h	0.36 0.5 0.7 1 1.4 2 2.8 4
7.	Minimum spacing between baselines (1) (b1)	(19/10) h	3.42 4.75 6.65 9.5 13.3 19 26.6 38
8.	Minimum spacing between baselines (2) (b2)	(15/10) h	2.7 3.75 5.25 7.5 10.5 15 21 30
9.	Minimum spacing between baselines (3) (b3)	(13/10) h	2.34 3.25 4.55 6.5 9.1 13 18.2 26
10.	Spacing between words (e)	(6/10) h	1.08 1.5 2.1 3 4.2 6 8.4 12
11.	Line width (d)	(1/10) h	0.18 0.25 0.35 0.5 0.7 1 1.4 2

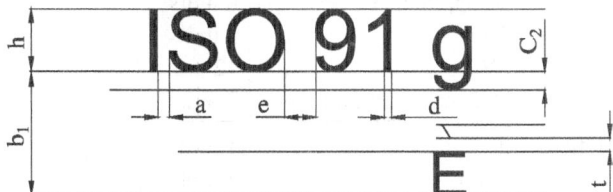

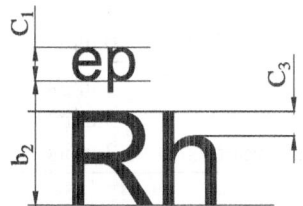

Fig. 1.15   Proportions of the letters and their spacing [as per IS 9609 (Part 0): 2001/ISO 3098-0:1997/IS.SP.46.2003]

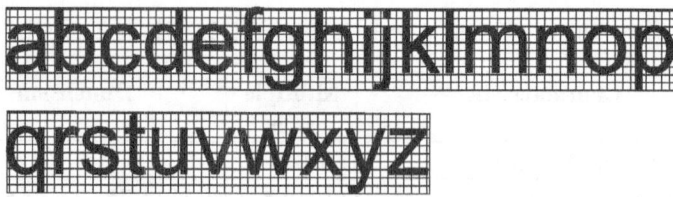

Fig. 1.16 Type A lettering—upper case, lower case, numbers, and Roman letters

Table 1.10 shows the width of various letters as per Type A lettering scheme.

Table 1.10 Width of various letters as per Type A lettering

Type of letter	Letter	Width
Capital or upper case letters	I	1 unit
	J, L	5 units
	C, E, F	6 units
	A, Q, V, X, Y	8 units
	M	9 units
	W	12 units
	All other capital letters	7 units
Lower case or small letters	i	1 unit
	j, l, r	3 units
	f, t	4 units
	m	9 units
	w	10 units
	All other small letters	6 units
Numerals	1	4 units
	3, 5	5 units
	All other numerals	7 units

The proportions of Type A inclined letters are the same as that of Type A vertical letters and numerals.

Table 1.11 Width of various letters as per Type B lettering

Type of letter	Letter	Width
Capital or upper case letters	I	1 unit
	J	4 units
	C, E, F, L	5 units
	A, M, Q, V, X, Y	7 units
	W	9 units
	All other capital letters	6 units
Lower case or small letters	i	1 unit
	l	2 units
	c, f, j, r, t	4 units
	All other small letters	5 units
Numerals	1	3 units
	4	6 units
	All other numerals	5 units

In Type B lettering, the height is divided into 10 equal parts. The letters and the numerals are wider than Type A and the line width or thickness of the letters is also more. Hence, they will appear stout and bold or dark compared to Type A letters.

Table 1.11 shows the width of various letters as per Type B lettering scheme.

Since the letters and the numerals are wider in Type B, the lettering done with smaller heights is advisable. The body height of smaller or lower case letters is approximately 0.7 times the height of the capital letters. The stem and the tail portions are approximately 0.3 times the height of the capital letters, and the total height of a lower case letter is approximately the same as that of a capital letter.

In a drawing sheet, the title block, the name of the institution or the college, and the student name can be written with 10 mm size letters. The main titles can be written with 6–8 mm size letters. The notes, the dimensioning details, the material list, the scale, etc., can be written with 3–5 mm size letters. The drawing number can be printed with numerals of 10 mm size.

A sample printing of few statements in capital and small letters are shown for a height of 10 mm and 5 mm, respectively, in Fig. 1.17.

DRAWING IS THE LANGUAGE OF ENGINEERS

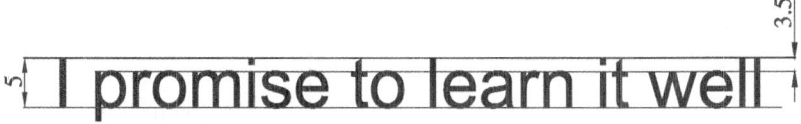

Fig. 1.17 Printing of statements on the drawing sheet

1.7.4 Dimensioning Practice

Introduction

A technical drawing without dimensions incorporated in it is similar to a person without an identity and address, and it is not only incomplete but also meaningless. The dimensions of an object illustrate the physical description of its length, width, height, etc. and are expressed in numerals in appropriate length units of measurement. In addition, the details that enable the production or inspection and functioning of the object are also to be furnished in the drawing. Dimensioning involves the above activities. As per the definitions BIS (SP: 46-2003), the dimension is a numerical value expressed in appropriate units of measurement and is indicated graphically on technical drawings with lines, symbols, and notes.

Therefore, the dimensioning procedure involves the incorporation of numerical values of different sizes of the objects and its various features by referring them with appropriate lines and indicating and writing down the information that are essential for the manufacturing of the object.

BIS classifies the dimensions into three categories, namely functional, non-functional, and auxiliary dimensions. The functional dimension of an object is essential for its functioning, while the non-functional dimension is not mandatory for its functioning. The auxiliary dimension can be derived from the other values shown in the drawing. The functional and the non-functional dimensions are to be indicated in the drawing, while the auxiliary dimension can be calculated from them and need not be expressed in the drawing but can be shown in brackets. This will be discussed in detail while drawing the machine components.

Elements of Dimensioning

The various constituents or elements of the dimensioning are explained in Fig. 1.18(a) and (b) and are listed as follows:

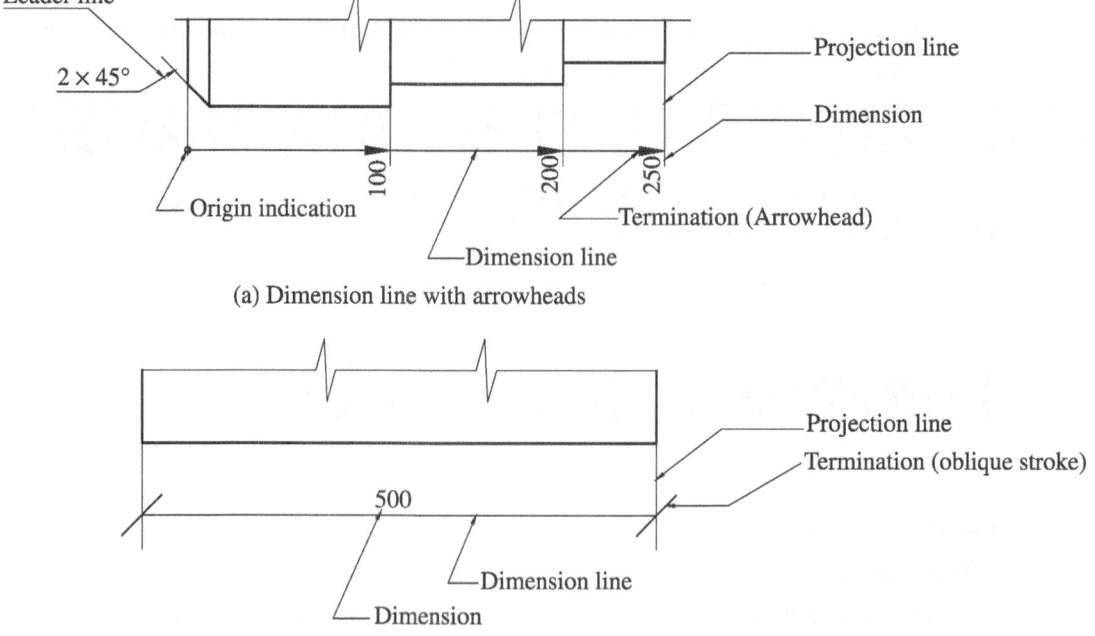

(a) Dimension line with arrowheads

(b) Dimension line with oblique strokes

Fig. 1.18 Elements of dimensioning

Projection line or extension line   This is a thin short line drawn perpendicular to the feature to be dimensioned and should project slightly beyond the respective dimension line. These lines may be drawn parallel (Fig. 1.18a) or oblique (Fig. 1.19a). The projection lines are generally outside the drawing, but to dimension some features that are inside, they can also emanate from them. In Fig. 1.19(b), the diameter of the circular feature that is inside is represented by the projection lines emanating from inside.

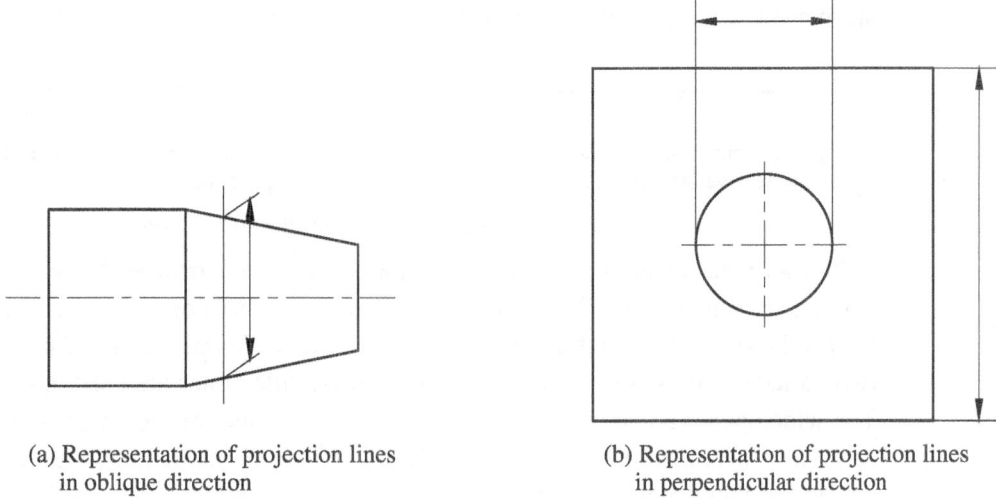

(a) Representation of projection lines
in oblique direction

(b) Representation of projection lines
in perpendicular direction

Fig. 1.19   Representation of dimension projection lines

Dimension line   This is a continuous narrow line (Table 1.5) which is drawn parallel to the edge or the surface whose length is to be dimensioned and is placed outside the view wherever possible. It is drawn in between the concerned extension lines with proper end terminations. Both the end terminations of a dimension line can contain arrows (Fig. 1.19b) or one end can be an origin indication and the other end can be with an arrowhead termination (Fig. 1.18a) or both the ends can be terminated with oblique representations as in Fig. 1.18(b).

Leader line   A leader line is a line used to point a feature (dimension, object, and outline), that is, on the outline of an object or any point inside it. The leader lines should terminate with an arrowhead, if they end on the outline of an object; or with a dot, if they end within the outlines of an object; or without dot or arrow head, if they end on a dimension line. The leader line can be inclined at any convenient angle (usually 30°, 45°, or 60°) to the horizontal. The tail end of the leader line will terminate on a horizontal line to carry the details of a note or dimension with relevant symbols. In Fig. 1.18(a), a leader line is shown to point an inclined surface and carries the details connected with that. The leader lines are usually used to indicate the diameter or the radius of a circular feature.

Dimension line terminations   The dimension lines will have terminations in the form of arrowheads (Fig. 1.18a) or oblique strokes (Fig. 1.18b). The oblique strokes can be used where the space is too small to draw an arrowhead.

Arrowheads   The arrowheads are used to terminate the dimension line ends and are also drawn at one end of the leader line. The arrowheads can be placed at the ends of successive dimension lines as shown in Fig. 1.20(a) or where the space is limited on a dimension line, the arrowheads may be shown outside the limits or the adjoining line arrowheads will account for this small dimension line also as in Fig. 1.20(b).

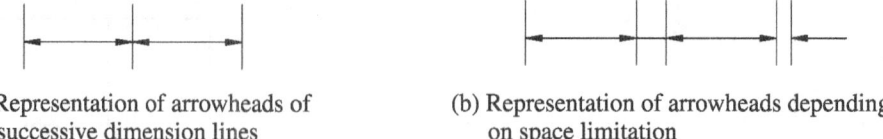

(a) Representation of arrowheads of
successive dimension lines

(b) Representation of arrowheads depending
on space limitation

Fig. 1.20   Representation of dimension lines

The arrowheads are drawn such that the barbs of the arrowheads form an included angle between 15° to 90°. The arrowheads may be open (Fig. 1.21a-i), closed (Fig. 1.21a-ii), closed and filled (Fig. 1.21a-iii), or wide open (Fig. 1.21a-iv). The first two varieties are used in pencil drawings, and the filled ones are used in ink drawings. The wide open type arrowhead is drawn where the length of the dimension is not long enough to accommodate the other varieties. When the short lines are used with the arrowhead inclined at 45°, if it becomes an oblique stroke as shown in Fig. 1.21(b) and is used, where the space is too small for drawing arrowheads. The proportion of the arrowhead (except wide open type) is such that its length is at least three times its width and one such proportionate arrangement is shown in Fig. 1.21(c).

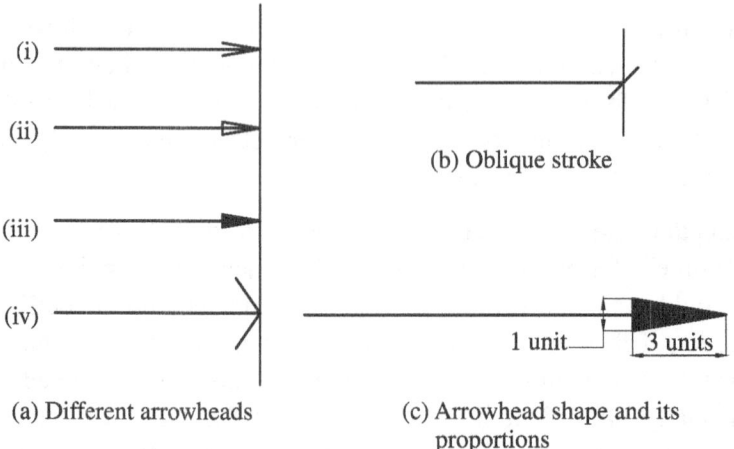

(a) Different arrowheads

(b) Oblique stroke

(c) Arrowhead shape and its
proportions

Fig. 1.21   Representation of arrowheads

Dimension   A dimension is a numerical value of the length or the angle, written above the dimension line or written by breaking the dimension line. All the dimensions of the drawing should be expressed in the same units. Usually, the millimetre unit of measurement is adopted. In Imperial System of Measurement, inches are used to express the dimensions.

When all the drawings bear the dimensions in millimetre unit, there is no need to write along with each numerical value of dimension, but a foot note containing the text 'All dimensions in mm' can be written. If other units are used, similar foot note is prepared. The angles are expressed in degrees.

Methods of Dimensioning

The dimensions are indicated on a drawing as per one of the following methods as recommended in BIS (SP:46-2003).

Method I (aligned method)   As per this method, the dimensions are placed parallel to the dimension line and above it and preferably in the middle. The numerical values are placed on the horizontal dimension lines such that they can be read from the right-hand side of the drawing (which is left to the viewer), while on the vertical dimension lines, they should read from the bottom of the drawing. Figure 1.22 (a) shows the dimension arrangements as per the orientations of the dimension lines mentioned in clause 12.3.4.1, SP 46:2003. When the dimensions are horizontal, dimensions are placed above the dimension line and when they are vertical, the dimensions are placed left to the dimension line (as referred with respect to the viewer). When the dimension lines are in inclined orientation (usually referred as oblique lines), the dimensions are placed by rotating the vertical dimension line to the required inclination along with its associated dimension line. Figure 1.22(b) explains the various orientations of the dimension lines and the placement of the respective dimensions.

Angular dimension values are placed either as shown in Fig. 1.22(c) as per the line scheme indicated in Fig. 1.22(a) or can be indicated as shown in Fig. 1.22(d), where the dimension values are placed in horizontal direction so that they can be read from the right of the drawing.

Figure 1.22(e) shows the horizontal dimension lines and their associated representations. These dimensions which involve cylindrical parts are indicated by their diameter values, preceded by the symbol Ø.

Method II (unidirectional method)   In this method, the dimensions are indicated so that they can be placed in only horizontal direction. Non-horizontal dimension lines (vertical and inclined) are interrupted, and the dimension values are inserted perpendicular to such lines as shown in Fig. 1.22(b)-(i), (ii), and (iii) for linear, angular, and circular features, respectively. This system of dimensioning is generally adopted in large sized drawings such as buildings, air-crafts, and automobiles.

NOTE   *Any one of the methods indicated previously can be used in the drawing sheet, but not to be mixed.*

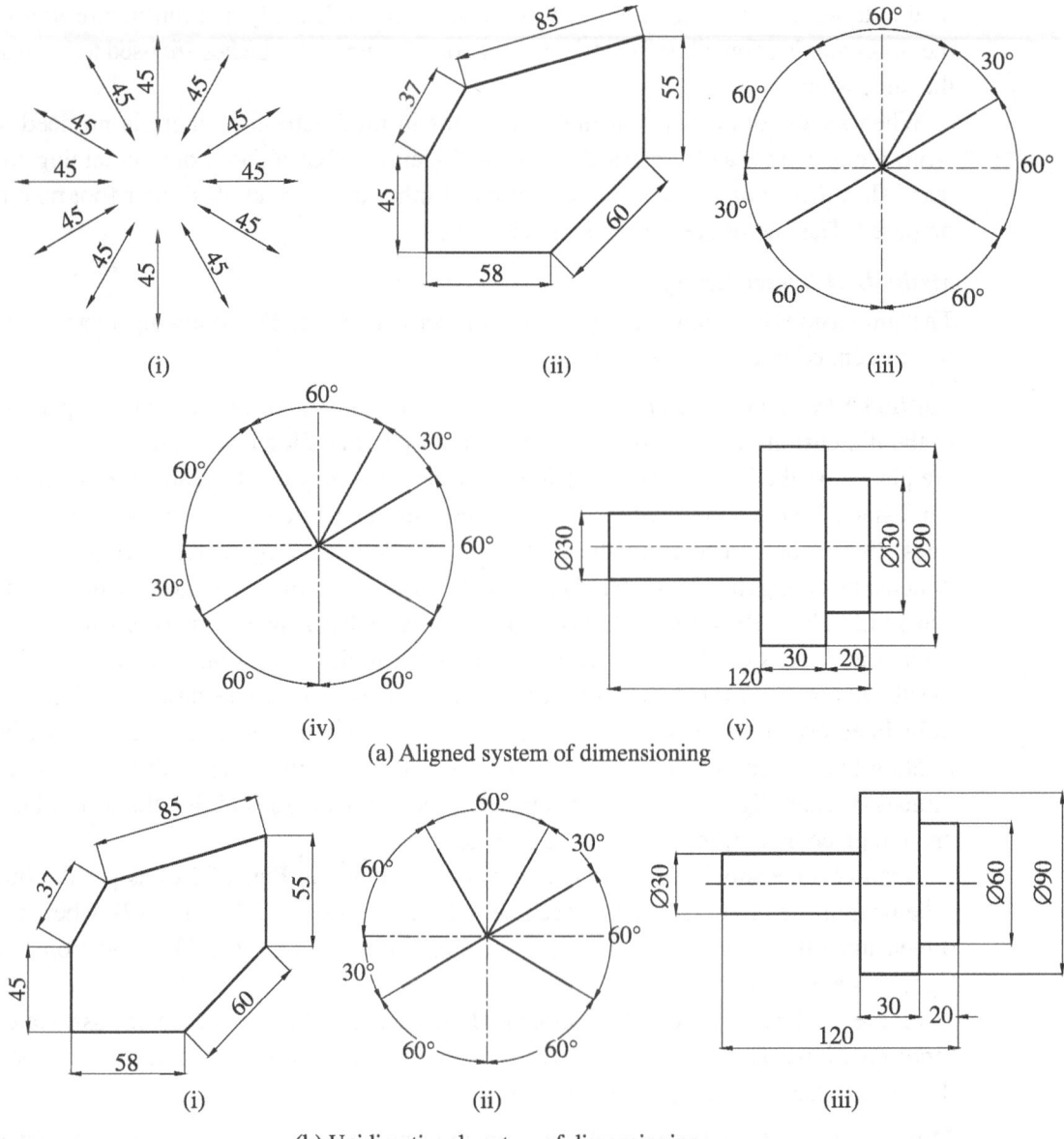

(i) (ii) (iii)

(iv) (v)

(a) Aligned system of dimensioning

(i) (ii) (iii)

(b) Unidirectional system of dimensioning

Fig. 1.22 Methods of dimensioning

Types of Dimensioning

Dimensioning has to be done on many features. The following possibilities or types are recommended.

Chain dimensioning All the single-dimension lines of the different features are aligned continuously along a chain, such that the successive arrowheads touch each other as shown in Fig. 1.23(a). This is also known as continuous dimensioning. It is

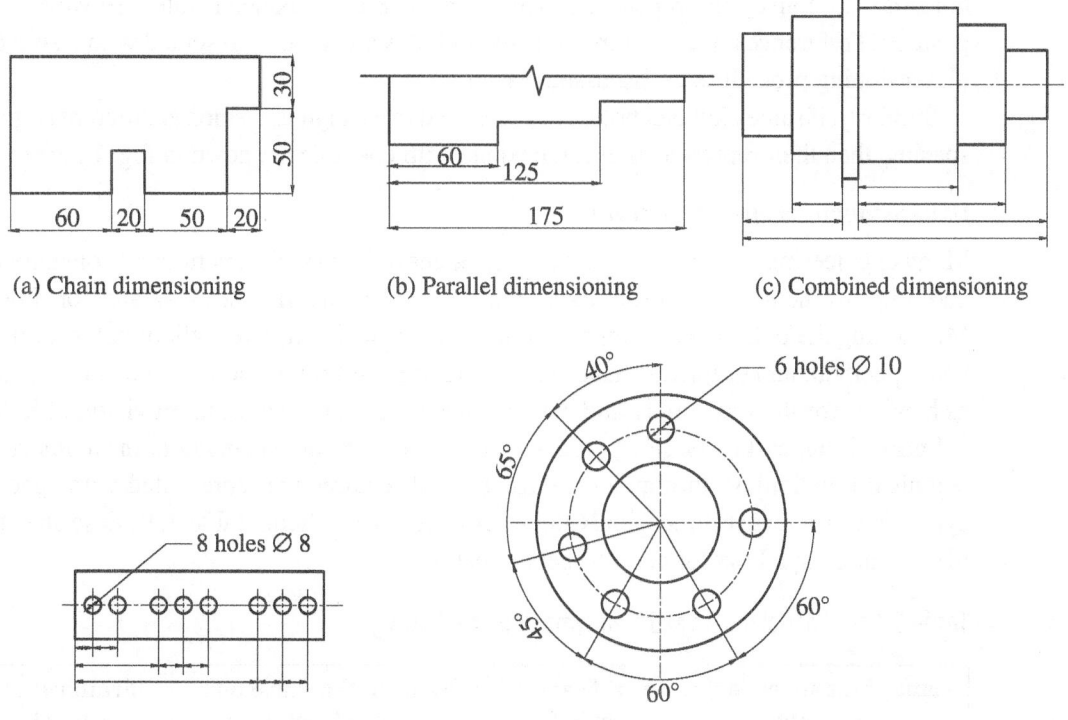

(a) Chain dimensioning   (b) Parallel dimensioning   (c) Combined dimensioning

(d) Representation of repeated features   (e) Representation of angular dimensioning

Fig. 1.23   Types of dimensioning

customary to indicate the overall length apart from the subdimensions and in that case one of the subdimensions mentioned in the chain can be omitted.

Parallel dimensioning   All the single-dimension lines of different features are placed parallel to one another and successively one below the other or one adjacent to the other, from a common extension line as a reference. This helps to refer all the dimensions from a common datum in a progressive manner as shown in Fig. 1.23(b). This is also known as progressive dimensioning.

Combined dimensioning   Certain drawings may require the combination of single dimensions, chain dimensioning, and parallel dimensioning procedures to stress the importance of certain feature and referring other features from it. In Fig. 1.23(c), the central collar in the stepped shaft has been identified as a special feature and other features to its right and left are marked as per the parallel dimensioning procedure.

Dimensioning of Repeated Features

When certain features of the same shape and size are repeated in the drawing at different locations, the dimensions can be referred in one feature with a symbol or note to describe the other repeated ones. Figure 1.23(d) shows eight circular holes of same diameter of 8 mm. One of the holes can be represented with a leader line and a note '8 holes φ8' to convey the repeated arrangement of 8-mm-diameter holes. However, the positional references of the centre of different holes are to be represented with a suitable dimensioning procedure as discussed earlier.

Similarly, if same diameter holes are positioned in an angular fashion at different angular spacing, then their representation can be done with one hole as shown in Fig. 1.23(e).

Dimensioning of Special Features

Many engineering objects do not have sharp edges or corners for operational convenience and improvement in aesthetic look, but are provided with smooth arcs or fillets. Mentioning the radii of such arcs is important in drawing. Square, cylindrical, spherical, hemispherical, and elliptical shapes are provided in certain portions of the objects, such as handles, for their assembly and convenience of human operation. Provision of holes, cylindrical shapes, tapers, slots, grooves, threads, etc. become mandatory in an object for technical functioning. In drawings, these special features are represented with specific symbols and relevant dimension values associated with them. Table 1.12 describes the identification symbols for certain special features.

Table 1.12   Special shapes and their symbols in the drawing

Name of the shape (or) special feature	Identification symbol	Name of the shape (or) special feature	Identification symbol
Radius	R	Pitch circle diameter	PCD
Spherical radius	SR	Equi-spaced	EQSP
Diameter	Ø	Counter sunk	CSK
Spherical diameter	SØ	Counter bore	C'BORE
Square	□ or Sq	Metric thread	M

Figure 1.24 shows some special shapes and their representation in the drawings.

General Guidelines on Dimensioning

The general guidelines adopted for dimensioning procedure are as follows:

1. All dimensions are to be made as per aligned method or unidirectional method only and not to be mixed up.
2. The same unit of length should be used for all dimensions in the drawing. The units need not be written after each dimension value but should be mentioned with a common note near the title block.
3. Dimension lines are usually drawn parallel to the length or boundary of the feature to be dimensioned and should be marked outside the view at a minimum distance of 6 mm to 8 mm (refer Fig. 1.25).
4. Dimension lines can be drawn parallel to each other, representing the smaller length feature first and closer to the object outline and others drawn parallel at a distance of 6 to 8 mm among them. In Fig. 1.25, the vertical distance of the centre of hole

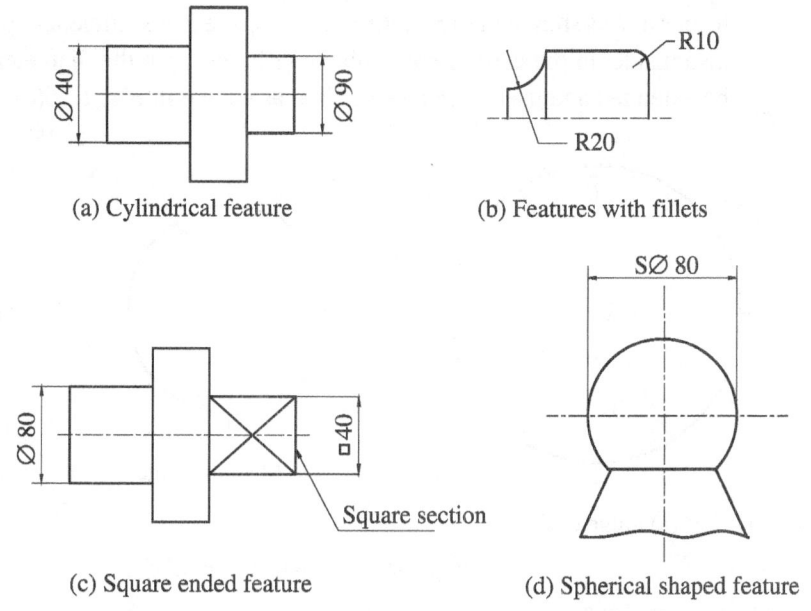

(a) Cylindrical feature   (b) Features with fillets

Square section

(c) Square ended feature   (d) Spherical shaped feature

Fig. 1.24   Shape identification symbols

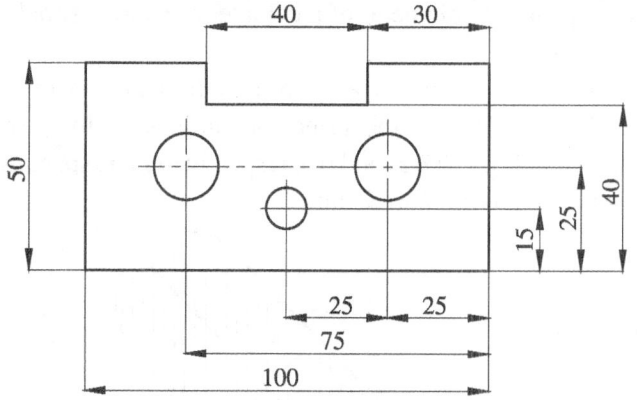

Fig. 1.25   Representation of dimensions of an object

nearer to bottom boundary is marked first and that of the other holes and the top boundary are marked parallel to each other at an approximate distance of 6 to 8 mm among them.

5. Dimension lines and extension lines should not cross other lines or intersect among themselves.

6. The centre lines or the outlines of a part should not be used as dimension lines but may be used in the place of a projection or extension line. In Fig. 1.25, the centre lines of the holes are used as extension lines to mark the dimensions.

7. Each dimension should be referred only once and not to be repeated.

8. All dimensions should not be staggered in one view, but to be distributed properly in all the views.

9. Dimension should not be marked on hidden lines or dotted lines.

10. A circle may be dimensioned by marking an inclined diameter line (approximately 30 to 45° to the horizontal) to touch the boundaries. Arrowheads may be placed at its ends and the diameter value may be indicated with the letter ϕ as shown in Fig. 1.26(a), if the diameter is large enough to print that value inside. If the space available inside the circle is insufficient to accommodate the dimension, then it can be indicated by a

leader line as shown in Fig. 1.26(b). If the space is insufficient to place the arrowheads also inside, in the case of small diameter holes, then the diameter representation can be extended and used with a leader line as shown in Fig. 1.26(c).

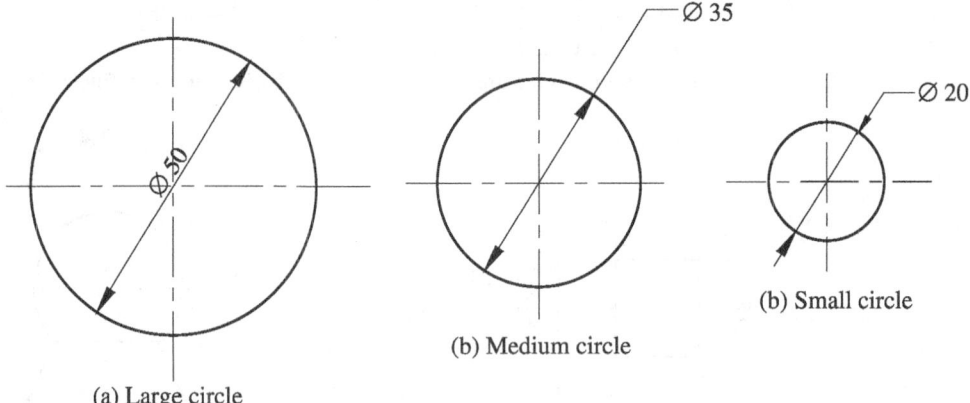

(a) Large circle

(b) Medium circle

(b) Small circle

Fig. 1.26 Dimensioning of circles

11. Cylindrical features should be dimensioned as far as possible in the views where they appear as rectangles. However, the diameter of circular holes should be represented in the views where the circles appear. The diameter of a cylinder or a circle should not be written in terms of its radius.

12. The sectional views of the shapes are marked by a set of parallel hatching lines (continuous narrow lines) inclined at 45° to the principal outlines or lines of symmetry of the sections as shown in Fig. 1.27(a) and (b), respectively. The spacing between the hatching lines should be around 2 to 3 mm.

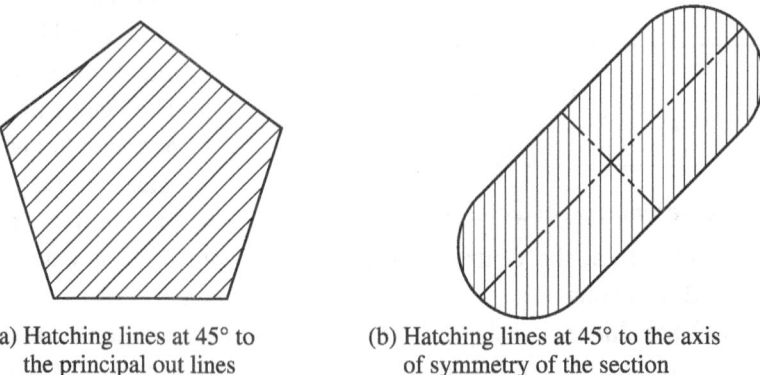

(a) Hatching lines at 45° to the principal out lines

(b) Hatching lines at 45° to the axis of symmetry of the section

Fig. 1.27 Hatching lines

13. Figure 1.28 shows the different types of lines used in the projections of the object along with their dimensioning details.

NOTE *Since the indication of dimensions is mandatory in a drawing, it should be practised more judiciously to enhance clarity and quick readability with the purpose of manufacturing the product.*

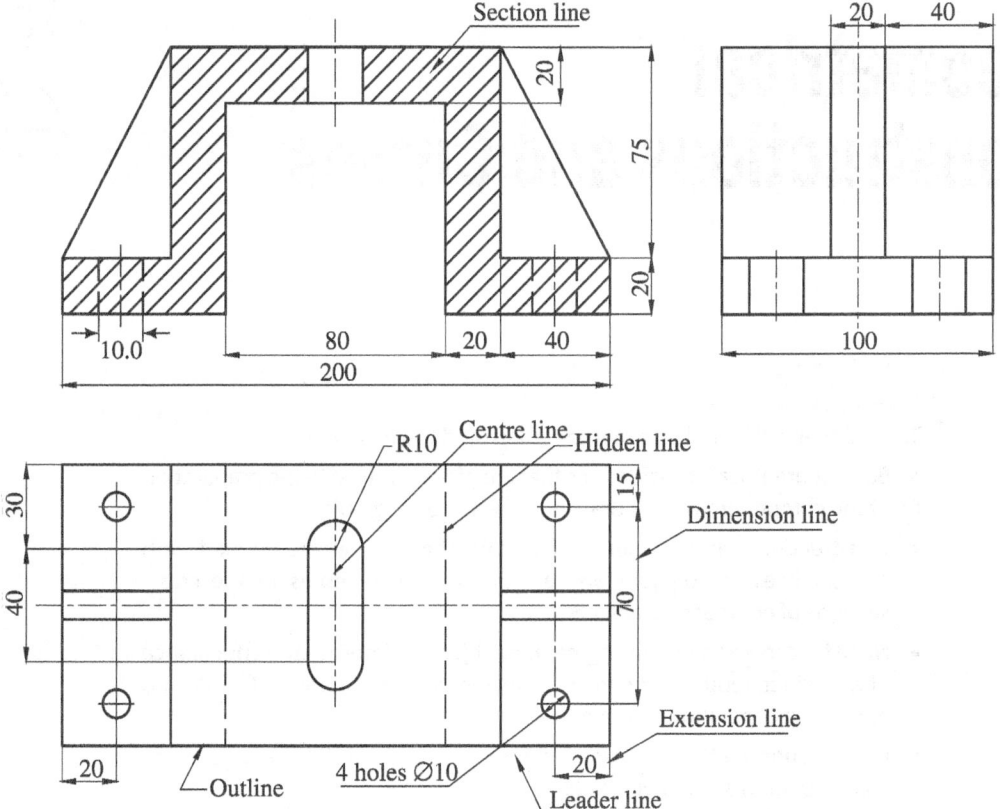

Fig. 1.28   Sketch showing various types of lines used in engineering drawing

RECAPITULATION

- The pictorial drawings have stayed with the mankind as an age old universal mode of communication, as they are not restricted to any language, place, knowledge, and boundary.

- The engineering or technical drawing is a pictorial drawing, depicting the various dimensions of the object in multiple two-dimensional drawings so that the object can be manufactured. They are drawn as per the guidelines stipulated in International and Indian Standards.

- These drawings can be prepared with manual tools on drawing sheets or with software tools in the computer.

- The major manual drafting tools are as follows: mini drafter, set squares, protractor, scales, and associated accessories such as compass, divider, and pencils.

- The representation of drawings with lines of different intensities is adopted to represent different applications.

- The lettering with different English alphabets, numerals, and special characters are done as per Type A or Type B, single-stroke scheme.

- Dimensioning a drawing is the incorporation of physical dimensions of the objects shown in length units (usually 'mm') and is done by drawing extension lines, dimension line, and leader line (as applicable) with suitable arrow heads.

- The dimensions are indicated as per the aligned system where they are placed above the dimension lines that are not broken. In horizontal lines, they are read from the right-hand side of the drawing sheet, while on vertical dimension lines, they are read from the bottom of the drawing sheet.

Geometrical Constructions and Curves

2

OBJECTIVES

This chapter will help the reader to understand the following:

- Basic geometrical constructions that are often involved in the preparation of the component or part drawings of engineering objects
- Constructional arrangements of certain shapes that are formed with straight lines or polygons and are often used to represent the cross-sections of engineering objects
- Need for curved shapes in engineering objects to improve the appearance of a product, reduce the stresses involved during its period of work, and enhance its operational convenience
- Construction methodologies of commonly used curved shapes such as ellipse, parabola, and hyperbola
- Path traced by a point on certain shapes, when it is unwound from the periphery
- Path traced by a point on circles when they roll on a fixed straight line or on a fixed curve externally or internally
- Locus of a point when it moves under a specified constraint on a plane
- Need to understand the movement of different members or parts of any machine or mechanism in general, and the movements of various points in the respective members in particular
- Path traced by a point on the links or members of a slider crank and a four-bar mechanism
- Engineering applications of certain specific curved shapes

2.1 INTRODUCTION

In engineering practice, during the design and drawing stages, the configurations of the objects are often made with certain basic regular shapes and are modified further based on the engineering needs and aesthetics. Sharp corners or the edges of the objects are often smoothened by rounding them in the form of an arc or a fillet to avoid damage or injury to the user and also for reducing the stresses in the members in these regions. Certain basic constructions explaining the division of a line, angle, or curve into a set of desired parts are often required to be effected at any stage during the construction of a component drawing. The construction features such as a closed polygon, a circle, or a sector becomes fundamental during the drawing stage of any component. These procedures not only save time but also help improve the accuracy. This section deals with the geometrical constructions of dividing a straight line, a circle, or an angle into desired parts and then discusses the step-by-step construction stages of regular polygons, circles, and so on.

2.2 BASIC GEOMETRICAL CONSTRUCTIONS

2.2.1 Dividing a Straight Line into the Desired Number of Equal Parts

PRINCIPLE

A straight line can be divided into a set of equal parts through a scale if the length of the line is a whole number and is a direct multiple of the length of the number of parts involved. Otherwise, a new straight line with a known whole number length that is a direct multiple of the length of the number of parts involved can be assumed and divided. These division points can be transferred proportionately to a straight line that is to be divided, resulting in dividing the given line into a number of equal parts.

Example 2.1

Divide the given straight line PQ into 12 equal parts.

PROCEDURE (Refer Fig. 2.1)

Step 1: Draw the line PQ for the given length.

Step 2: Set a line PR at any convenient angle and mark 12 known equal lengths, as shown by P–1 = 1–2 = ... = 11–R.

Step 3: Join the end point R of the selected line with the end Q of the given line and draw parallel lines to it from 1, 2, etc., to meet PQ, respectively, at 1′, 2′, etc.

Step 4: P1′, 1′2′, ..., 11′ Q are the desired 12 equal parts of the line PQ.   ▲

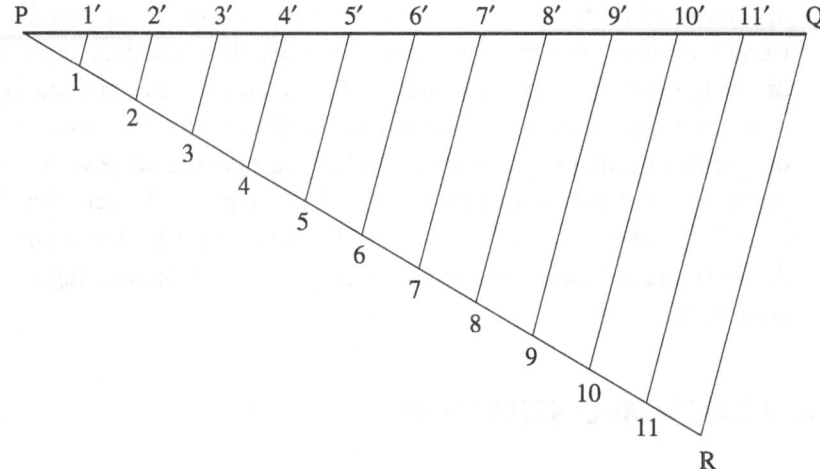

Fig. 2.1 Division of a line into equal number of parts

2.2.2 Dividing a Straight Line into Unequal/Proportionate Parts

> **PRINCIPLE**
>
> The diagonals of a rectangle constructed at the given line, divide the line by one-half. By selecting points of one-third, one-fourth, and other required fractions of lengths on the diagonal, the line can be divided accordingly.

Example 2.2

Divide the given straight line PQ into one-half, one-third, one-fourth, and so on, by a common procedure.

PROCEDURE (Refer Fig. 2.2)

Step 1: Draw the line PQ for the given length and construct a rectangle PQRS below it.

Step 2: Join the diagonals PR and QS and drop a perpendicular from their intersecting point 1 to meet the given line PQ at 1′. This divides the line by one-half.

Step 3: Join the point S with 1′ to meet PR at 2. The perpendicular drawn from 2 meets the line PQ at 2′ and this divides the line by one-third.

Step 4: Repeat the same procedure to get 3′, 4′, etc., that divide the line by one-fourth, one-fifth, and so on, as shown in Fig. 2.2.

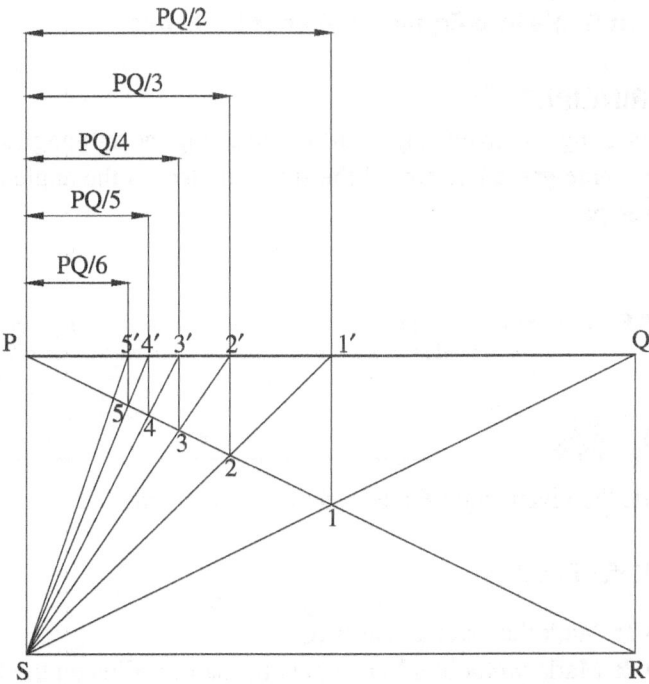

Fig. 2.2   Division of a line into proportionate parts

2.2.3 Dividing an Angle into Two Equal Parts

> **PRINCIPLE**
>
> Marking two arbitrary points of the same radius on the two legs of the angle given, and marking two arcs of the same length with these points as centres, result in an intersecting line that bisects the given angle.

Example 2.3

Divide an angle into two equal parts (or) bisect the given angle.

PROCEDURE   (Refer Fig. 2.3)

Step 1: Mark the given angle POQ.

Step 2: With O as centre, mark two points M and N with the same radius on the legs OP and OQ.

Step 3: With M and N as centres, draw arcs of the same or any other radius to intersect at O_1.

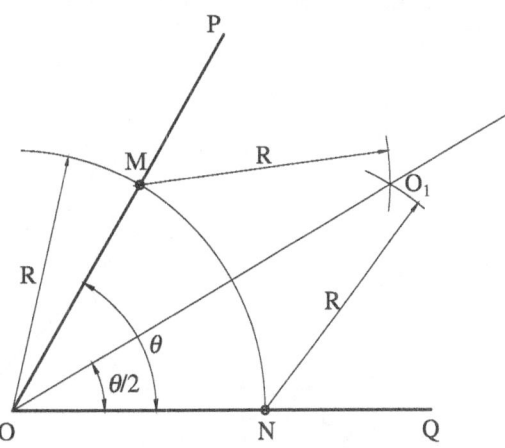

Fig. 2.3   Bisecting a given angle

Step 4: Join the line $O-O_1$ that bisects the given angle θ.

2.2.4 Dividing an Angle into Equal Number of Divisions

> **PRINCIPLE**
> Bisecting the given angle, further bisecting the sub-angles obtained, and repeating the same procedure for all the sub-parts leaves the angle divided into many equal even parts.

NOTE *This procedure is applicable only for dividing an angle into an even number of divisions. Trial and error procedure can be used to divide an angle into an odd number of divisions.*

Example 2.4

Divide the given angle θ into eight equal divisions.

PROCEDURE (Refer Fig. 2.4)

Step 1: Mark the given angle POQ.
Step 2: Mark two points M and N at the same radius on the legs OP and OQ.
Step 3: With M and N as centres, draw arcs of equal radius to intersect at A.
Step 4: The line OA bisects the angle θ.
Step 5: Consider one-half and repeat the same procedure by marking M and 4 on OP and OA and obtain the line OB as indicated. This results in bisecting the angle POA. Repeat this procedure until all the eight divisions are obtained as shown in Fig. 2.4. ▲

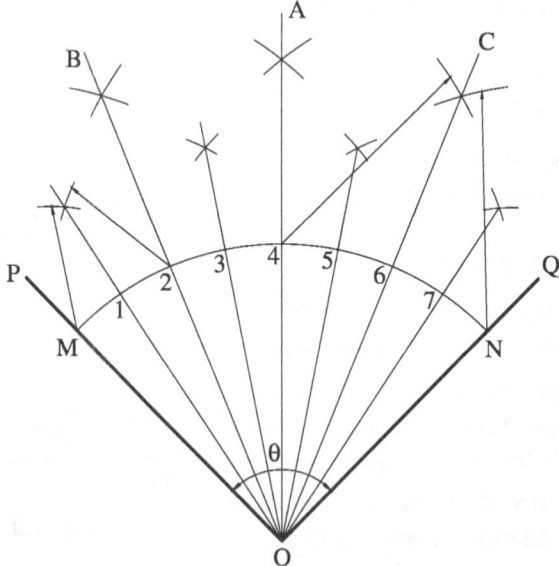

Fig. 2.4 Dividing an angle into even number of divisions

2.2.5 Construction of a Regular Polygon of any Number of Sides, Given the Length of One Side

Method of Using Exterior or Interior Angles

> **PRINCIPLE**
>
> A regular polygon of n sides subtends n exterior angles whose sum is equal to 360°. We have to mark one side and set an exterior angle at one of its ends. Following this, marking the length of the side and repeating the same will result in the desired polygon. As the interior angle is 180° minus one exterior angle, similar construction can be done by setting the interior angles also. The polygon can be drawn with one side horizontal, vertical, or at any inclination by setting the first side to that orientation and by marking the other sides as per the interior or the exterior angles. In this method, as the angle is set at different orientations and the length of the side is marked every time depending on the number of the sides, there could be measurement errors. Hence, other methods that use this principle and obtain the centre of the circle circumscribing the polygon are used very often.

Method of Using Interior Angles and Centre of the Circumscribing Circle

> **PRINCIPLE**
>
> A regular polygon of n sides subtends a total angle of 360° at the centre, with n isosceles triangles meeting at it. The identical sides of each isosceles triangle are equal to the radii of the circle circumscribing the polygon and they subtend semi-internal angles at each end of the sides of the polygon. This fact is used to draw the circumscribing circle and mark the edges of the desired polygon.

Example 2.5

Construct a regular pentagon of side of length 30 mm.

PROCEDURE (Refer Fig. 2.5)

Step 1: Mark the line PQ equal to the length of the given side of 30 mm.

Step 2: Subtend at each end, semi-internal angles of 54°, obtain the centre O, and draw a circle of radius OP or OQ.

Step 3: Mark arcs on the circle of radius equal to the side of the polygon and obtain the points R, S, and T.

Step 4: Join the points P to T that give the desired pentagon.

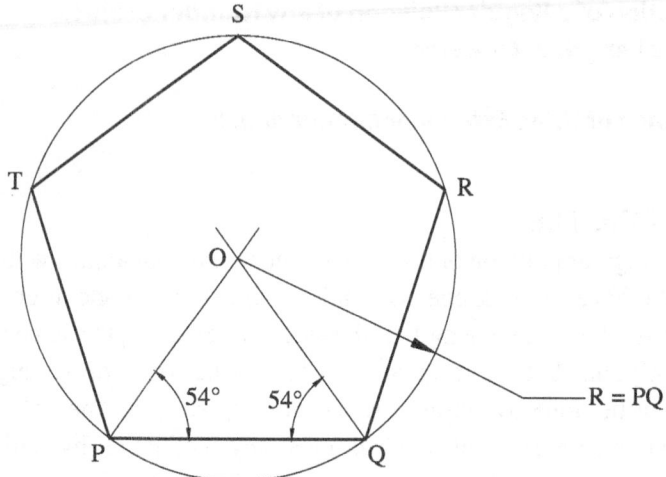

Fig. 2.5   Pentagon with internal angles method

Method of Using Circumscribing Circle

PRINCIPLE

The basic principle lies with the fact that the radius of the circle circumscribing a square (a four-sided polygon) is equal to the semi-length of the diagonal, and the centre lies on the perpendicular bisector of its side. For a regular hexagon (a six-sided polygon), the radius of the circumscribing circle is equal to its side and the centre lies on the perpendicular bisector of its side. Logically the centre of the pentagon (a five-sided polygon) should also lie on the perpendicular bisector of its side and is located in the middle of the centres of the square and the hexagon, when all these sides are marked on a common reference line. The centres of the other polygons can also be marked on the perpendicular bisector at successive intervals. The polygons can be obtained by drawing the respective circumscribing circles and marking their sides on them.

Example 2.6

Construct a set of regular polygons with four, five, six, seven, and eight sides of equal lengths.

PROCEDURE   (Refer Fig. 2.6)

Step 1: Mark the line PQ equal to the length of the given side and draw its perpendicular bisector.

Step 2: Erect a perpendicular QA at the end Q and join PA. This becomes the diagonal for the square with side equal to PQ. The centre 4 of the square lies on this diagonal where the perpendicular bisector meets it. Draw a circle with 4 as centre and radius equal to 4P or 4Q. Cut successive arcs equal to the side PQ. This gives a four-sided polygon or a square as shown in Fig. 2.6(a).

Step 3: As the radius of a hexagon is equal to the length of its side, draw an arc with P or Q as centre and PQ as radius. When this cuts the perpendicular bisector at 6, the centre of the circle circumscribing the hexagon is obtained. Draw a circle with 6 as centre and radius equal to 6P or 6Q and cut successive arcs equal to the side PQ and obtain the hexagon (refer Fig. 2.6a).

Step 4: The centre of the pentagon is located at 5, being the mid-point of the line joining the points 4 and 6. Draw a circle with 5 as centre and radius equal to 5P or 5Q and cut successive arcs equal to the side PQ and obtain the pentagon (refer Fig. 2.6b).

Step 5: The centres of other polygons with 7, 8, 9, and 10 sides are also successively located on the perpendicular bisector by setting distances equal to 4–5 or 5–6. With the respective centre and radius equal to the line joining that centre and the end point P or Q, circles can be drawn, and by cutting arcs equal to the side PQ, the respective polygons may be obtained. Seven- and eight-sided polygons (heptagon and octagon) are shown in Fig. 2.6(a). Similar procedure can be adopted to draw polygons of any sides. ▲

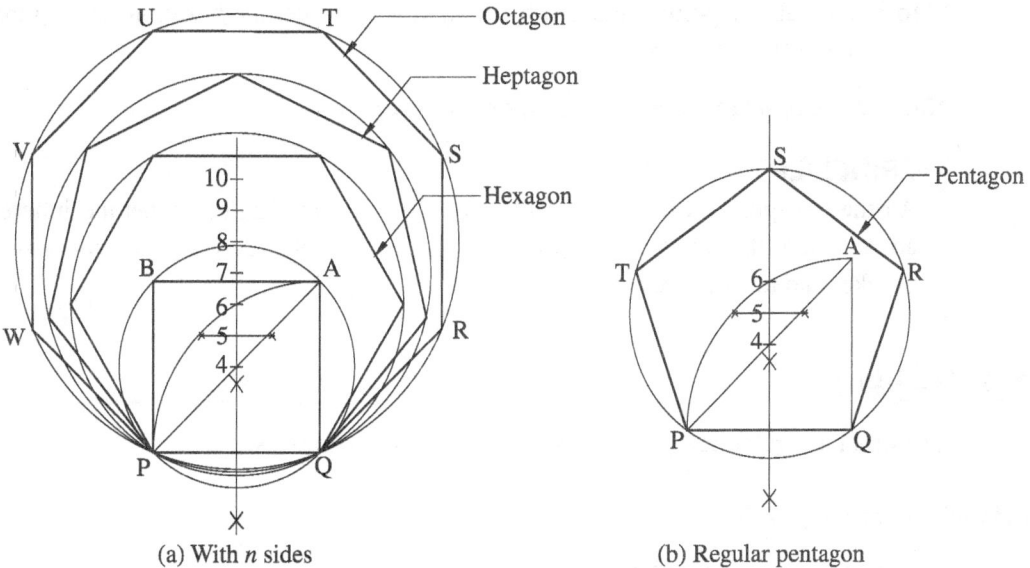

(a) With *n* sides   (b) Regular pentagon

Fig. 2.6   Construction of regular polygon

Method of Using Circumscribing Circle for Different Orientations of a Hexagon

PRINCIPLE

Very often in drawing, polygons have to be oriented such that one side is horizontal, vertical, or inclined. This can be achieved by drawing that side to the required orientation and setting the interior or exterior angles as outlined in the method of using exterior or interior angles. However, for the hexagon, as the diameter of the circumscribing circle becomes parallel to one side, it can be drawn with the diameter as the reference, as noted in Example 2.7.

Example 2.7(a) ▲

Construct a regular hexagon of side 30 mm with one side in (i) vertical, (ii) horizontal, and (iii) inclined orientations.

PROCEDURE (Refer Fig. 2.7a)

Step 1: Draw a circle of radius equal to the side of the hexagon and mark its diameter as vertical, horizontal, and inclined at $\theta°$ to the horizontal, as required in (i), (ii), or (iii), as shown in Fig. 2.7(a).

Step 2: With the ends of the respective diameters as centres, draw arcs of lengths equal to the side of the hexagon on either side of the diameter.

Step 3: Join all the points obtained by straight lines to get the hexagons as required in the different cases. ▲

Method of Using Set Squares for Drawing a Hexagon

PRINCIPLE

As the hexagon has its sides inclined at 60° (exterior angle), it has the inherent advantage of the usage of 30°–60° set square for its construction. This will provide quicker and accurate results.

Example 2.7(b) ▲

Construct a regular hexagon of side 30 mm using set squares.

PROCEDURE (Refer Fig. 2.7b)

Step 1: Mark one side PQ for a length of 30 mm.

Step 2: Hold the 60° set square at one end (say Q) and draw the line QR for a length of 30 mm.

Step 3: Reverse the set square and repeat the same operation through P and obtain PU.

Step 4: Repeat similar operations on the right and left through the points R and U and obtain the other points and complete the hexagon. ▲

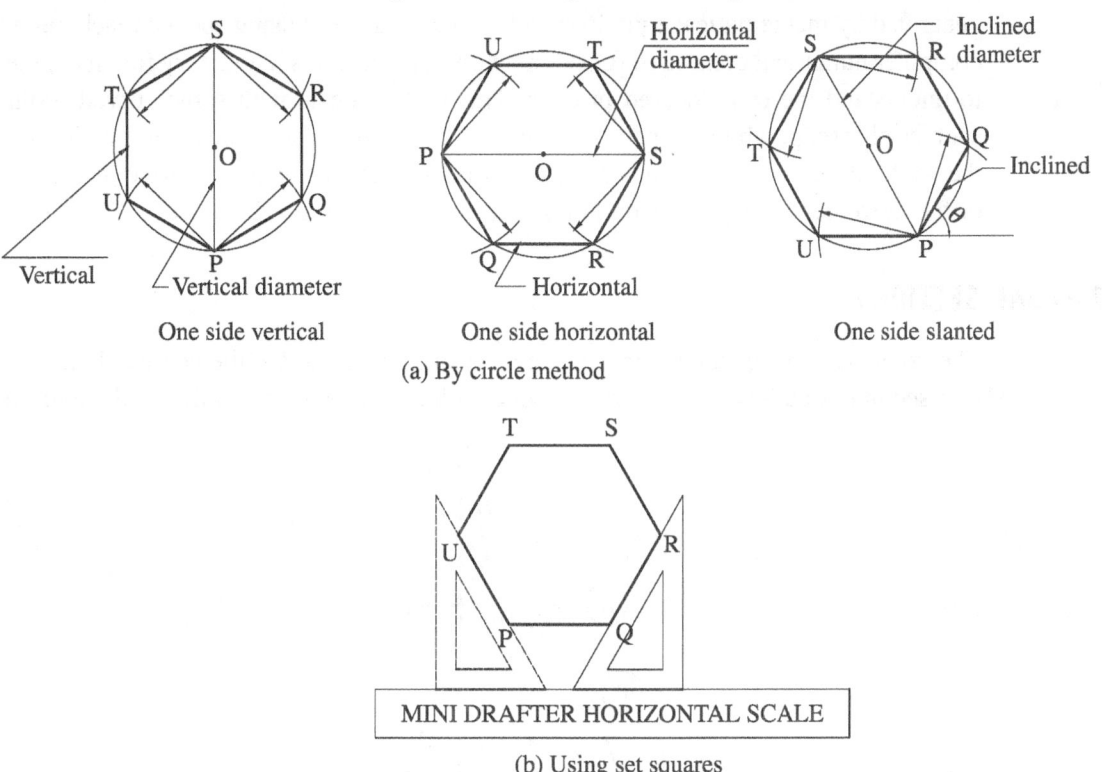

One side vertical   One side horizontal   One side slanted

(a) By circle method

(b) Using set squares

Fig. 2.7   Construction of hexagon

2.3 CURVES USED IN ENGINEERING PRACTICE

In engineering practice, the shapes of some objects have to be changed to curves or specific profiles to suit the functional, aesthetic, or ergonomic needs. While the profiles of satellite dishes, reactors, containers, household utensils, internal combustion engines, interiors of machine parts, etc., have to be designed to meet functional needs, the external profiles adopted for automobiles, ships, airplanes, and others have to be designed from an aerodynamic point of view to maximize their performance. The external appearance of consumer products (e.g., television sets, refrigerators, mobile phones, furniture, drawing room items, and fancy and gift items), and civil engineering structures (e.g., buildings and other associated products) will be enhanced only when their profiles suit and attract users. Hence, they must be presented aesthetically. Apart from the functional and aesthetic aspects, the products and gadgets used by humans, such as steering wheels, knobs, handles of machines, and domestic appliances, also have to be user-friendly when they are operated or used. Naturally, the shapes of such objects have to conform to the dimensions of the human parts that use them, and hence, have to be ergonomically suitable. To meet all these practical needs, the shapes or profiles of the objects cannot be straight, but have to be curved. These curved profiles are generated by mathematical equations and are manufactured using special machines and computer numerical controlled (CNC) machines. Hence, it is mandatory for an engineer to understand the basic curved profiles. This section deals with some curved profiles such as ellipse, parabola, hyperbola, cycloids, epi- and hypocycloids, and involutes. As the study of curved profiles deals with the tracing methods using the principle of locus of the movement of a point, a unit on generation of loci is also included.

2.4 CONIC SECTIONS

The conic sections or the conics, as they are simply called, are the curves obtained by intersecting or cutting a right circular cone with a section, or by cutting a plane located

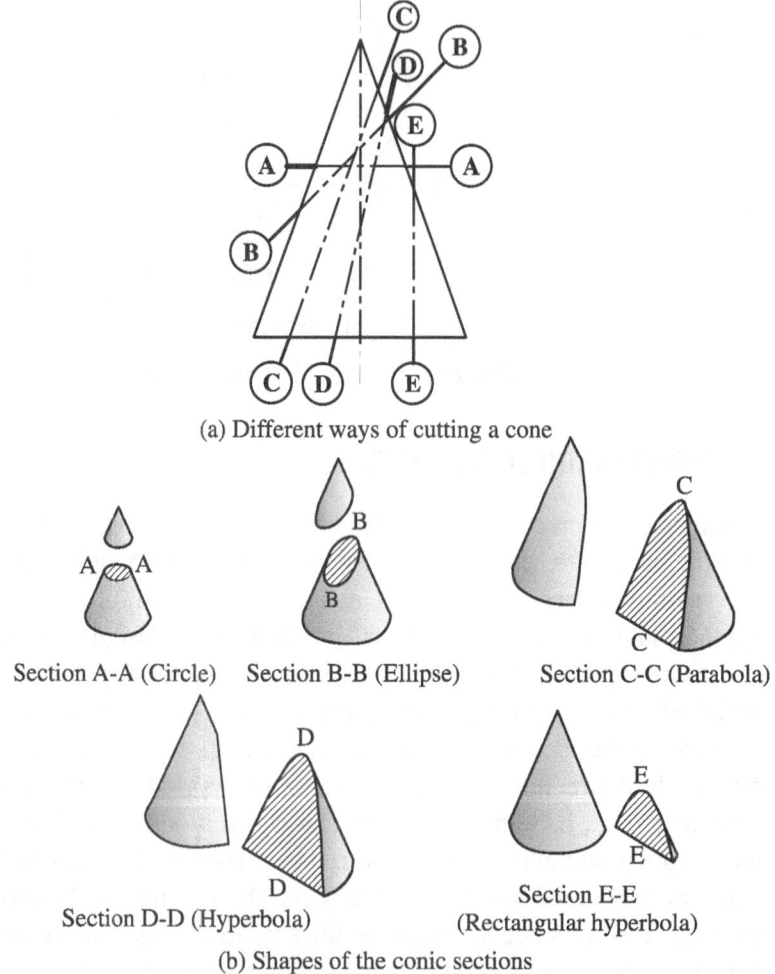

(a) Different ways of cutting a cone

Section A-A (Circle)   Section B-B (Ellipse)   Section C-C (Parabola)

Section D-D (Hyperbola)

Section E-E
(Rectangular hyperbola)

(b) Shapes of the conic sections

Fig. 2.8   Conic sections

in different orientations. Figure 2.8(a) shows the different ways of cutting the cone. Figure 2.8(b) shows the shape of the respective cut portions obtained by splitting the cone into two portions as required. For clarity, the shape of the cut portion is indicated by shading with the narrowly spaced section lines.

When the cutting plane is parallel to the base and perpendicular to the axis of the cone as indicated by A–A, the conic section obtained is a circle.

When the cutting plane is inclined to the base and the axis, and cuts all generators of the cone as indicated by B–B, the conic section obtained is an ellipse.

When the cutting plane is inclined to the base and the axis and becomes parallel to one of the end generators, thereby cutting a few generators of the cone as indicated by C–C, the conic section obtained is known as a parabola.

When the cutting plane makes an angle with the axis smaller than what the end generators make with the axis, as indicated by D–D, the conic section obtained is known as a hyperbola. When this angle becomes zero or if the section plane becomes parallel to the axis itself, as shown by E–E, the hyperbola then obtained is a rectangular hyperbola.

Mathematically, a conic is defined as the locus of a point moving in a plane such that the ratio of its distances from a fixed point and a fixed straight line is always constant. The fixed point is known as the focus and the fixed line is known as the directrix. The ratio of these distances is known as eccentricity. The eccentricity e is less than 1 for ellipse, is equal to 1 for parabola, and is greater than 1 for hyperbola.

2.4.1 Terminology of Conic Sections

The technical terminology of conics can be explained by referring to their sketches as shown in Fig. 2.9. As the ellipse is obtained by cutting all the generators of the cone, it is a closed curve, and will have two focal points. The other conics—parabola and hyperbola—are obtained by cutting only a few generators. They are not closed curves and will have only one focus.

Focus This is the fixed point responsible for generating the conics. It is denoted by F. As described earlier, two fixed points—the foci F_1 and F_2—are used to define a closed curve like the ellipse.

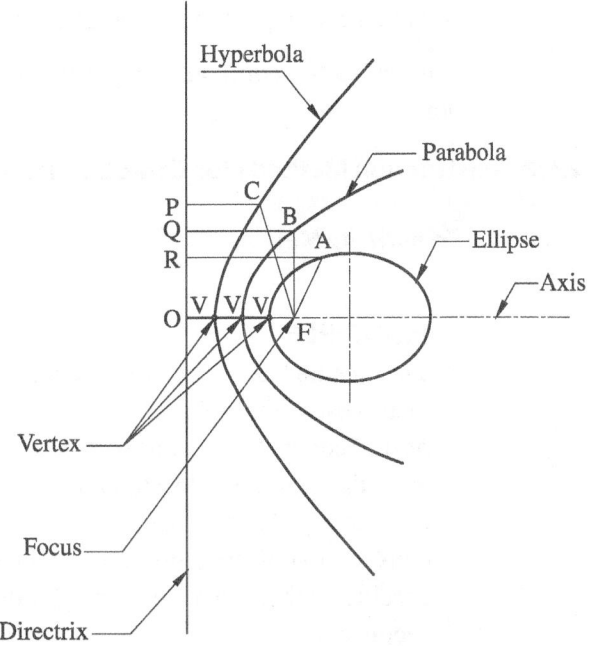

Fig. 2.9 Conic sections terminology

Directrix This is (usually) the fixed vertical straight line responsible for generating the conics.

Transverse axis or axis This is (usually) the horizontal line passing through the focus and is perpendicular to the directrix. This cuts the directrix at O.

Vertex This is the point at which any conic curve intersects the axis. As it is a point on the curve and takes part in tracing the curve like any other point, its distance from the focus and the directrix is often used as a vital data to draw the curve. This is denoted by V.

Eccentricity This is the ratio of the distance of any point on the curve (including the vertex) from the focus to that of the directrix. It is usually denoted by e.

As per Fig. 2.9,

for ellipse, $e = AF/AR < 1$

for parabola, $e = BF/BQ = 1$

for hyperbola, $e = CF/CP > 1$

Latus rectum This is the chord of a conic perpendicular to the axis and passing through the focus.

Tangent to the conic at any point This is the tangent at any point on the curve and forms a right-angled triangle at the focus; one of the two perpendicular sides being the line joining the given point and the focus, and the other being the line joining the focus and the meeting point of the tangent line on the directrix.

Normal to the conic at any point This is perpendicular to the tangent made at any point.

2.4.2 Construction Methods for Conic Sections

Eccentricity Method

PRINCIPLE

This method uses the basic definition of eccentricity of a conic, and uses the distances of the point from the focus and the directrix to locate various points on the conic. As the vertex is also a point on the conic curve, it is located first and other points are located on a convenient scale. This gives the required ratio of eccentricity. As this method is common for all types of conics, the examples on construction of the ellipse, parabola, and hyperbola are explained here. However, specific methods that are unique for different conics will be explained in successive sections.

Example 2.8

Construct an ellipse with the distance of focus from the directrix as 50 mm and eccentricity as 2/3. In addition, draw a normal and tangent to the curve at a point 40 mm from the directrix.

PROCEDURE (Refer Fig. 2.10)

Step 1: Draw the horizontal axis line AB and the vertical directrix line CD and mark the intersection A.

Step 2: Mark the given distance of the focus F_1, 50 mm from A, and divide it into five equal parts (corresponding to the sum of the numerator and the denominator of the given eccentricity).

Step 3: Mark the vertex V being the third division point (corresponding to the denominator, which represents that part of the distance of the vertex from the directrix).

Step 4: Construct a scale that will give the required eccentricity ratio directly by drawing a perpendicular VE equal to VF_1 at V. Join AE and extend it. As the ratio VE/VA (or VF_1/VA) satisfies the eccentricity ratio, any perpendicular erected at a convenient point on the axis to meet the line AE which is extended, will give the same eccentricity ratio.

Step 5: Mark convenient points 1, 2, 3, etc., on the axis line, and erect perpendiculars from them to meet AE extended at points 1′, 2′, 3′, etc.

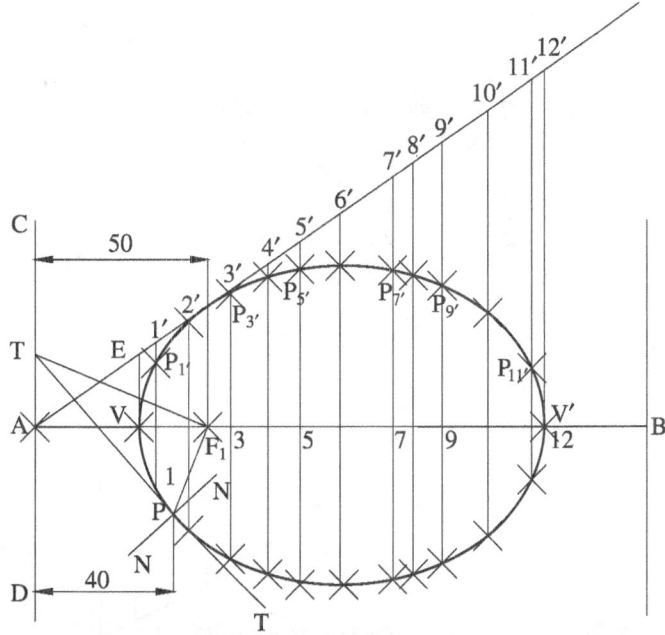

Fig. 2.10 Ellipse by eccentricity method

Step 6: With F_1 as centre and radius as $11'$, cut an arc to intersect the perpendicular through 1 at points p_1' simultaneously above and below the axis line.

Step 7: Repeat this procedure for all the other points 2, 3, etc., selected.

Step 8: Join the points P_1', P_2', P_3', etc., located above and below the axis by a smooth curve that results in the ellipse. It must be noted that when many points are selected on the axis line and the procedure is repeated, the closed curve will evolve with the second focus F_2. In addition, the right-hand side directrix is seen symmetrical to the left-hand side directrix.

Step 9: Locate the given point P on the curve at a distance of 40 mm from the directrix. Join P with focus F_1 and set 90° at F_1 to get the point T on the directrix. Join TP and extend it to a short distance and mark the end as T. This gives the tangent (TPT) at P. Through P, draw a line N–N perpendicular to TP. This gives the normal to the curve at point P. ▲

Example 2.9 ◢

Construct a parabola with the distance of focus from the directrix as 50 mm. Draw a normal and a tangent to the curve at a point 40 mm from the directrix.

PROCEDURE (Refer Fig. 2.11)

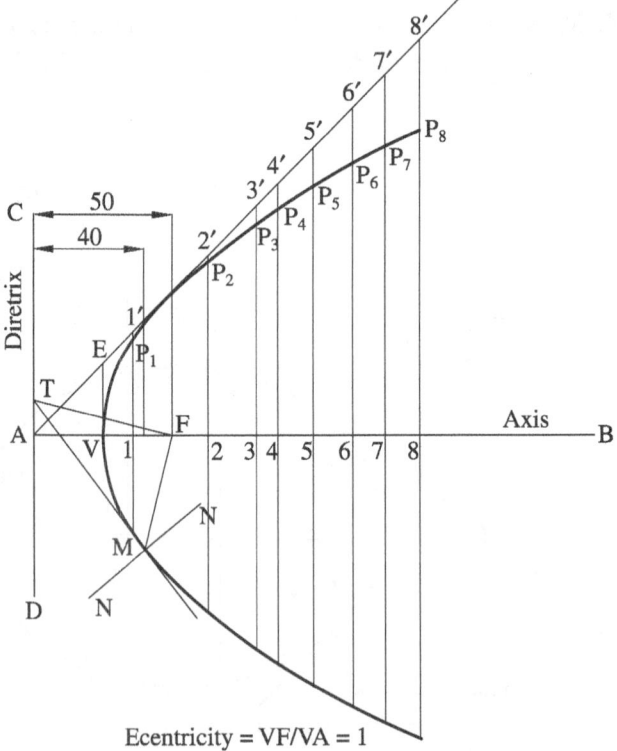

Ecentricity = VF/VA = 1

Fig. 2.11   Parabola by eccentricity method

Step 1: The focus F can be marked at 50 mm and the line AF can be divided into two equal parts (as eccentricity is 1) and the vertex V can be located.

Step 2: All other procedures are similar to Example 2.8, including the drawing of the tangent and the normal. It can be noted that the selected points 1, 2, etc., do not give a closed curve, and hence, the other focus, F_2, and the right-hand side directrix will not emerge. ▲

Example 2.10 ▲

Construct a hyperbola with the distance between the focus and the directrix as 50 mm and eccentricity as 3/2. In addition, draw normal and tangent to the curve at a point 30 mm from the directrix.

PROCEDURE (Refer Fig. 2.12)

Step 1: The focus F can be marked at 50 mm and the line AF can be divided into five equal parts. The vertex V can be located at the second division point (corresponding to the denominator, which represents that part of the distance of the vertex from the directrix).

Step 2: All other procedures are similar to Example 2.8, including the drawing of the tangent and the normal. It can be noted that the selected points 1, 2, etc., do not give a closed curve, and hence, the other focus, F_2, and the right-hand side directrix will not emerge. ▲

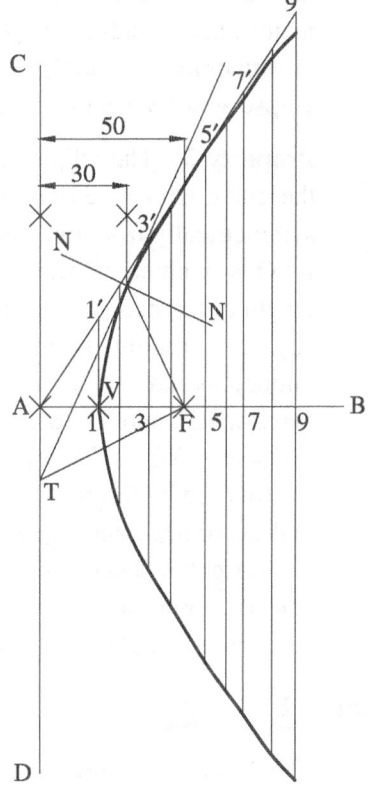

Eccentricity = (VF/VA) >1

Fig. 2.12 Hyperbola by eccentricity method

Specific Construction Methods for Ellipse

The general properties of ellipse are mentioned here. The construction methods suggested in this section make use of one or many of these properties.

Property 1   As the ellipse is a closed curve with two foci, F_1 and F_2, it can be described as the locus of a point moving in such a way that the sum of its distances from the foci is always a constant. In Fig. 2.13, if A is a point on the ellipse, then $F_1A + F_2A = $ constant. This property is used when two foci and one point on the ellipse are given or known.

Property 2 As the ellipse is a closed curve, it cuts the axis line passing through the two foci at two extreme points. The distance between these two extreme points is known as the major axis. A straight line bisecting the major axis and perpendicular to it forms the other two extreme points. Here, the curve takes a turn and the length of this line is known as the minor axis. The ends of the major axis, P and Q, and that of the minor axis, R and S, lie on the ellipse as shown in Fig. 2.13.

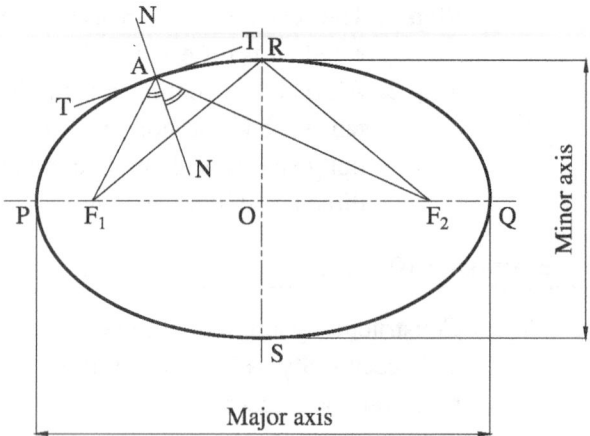

Fig. 2.13 Parameters of the ellipse

Property 3 The elliptical curve is symmetrical about the major and minor axes. As the curve is symmetrical, the two extreme points on the major axis are also located symmetrically from the two foci. Here, we use the first property, $PF_1 + PF_2$ = constant. As Q is symmetrical from F_2, $PF_2 = QF_1$. Therefore, $PF_1 + PF_2 = PF_1 + F_1Q = PQ$ = the length of the major axis. Therefore, it can be concluded that the sum of the distance of any point from the two foci is equal to the length of the major axis. As RS lies on the minor axis and also on the ellipse, $RF_1 + RF_2$ or $SF_1 + SF_2 = PQ$ = the length of the major axis. As RF_1 and RF_2 are symmetrical and equal, RF_1 or RF_2 = the length of the semi-major axis. With this property, if the major and minor axes of the ellipse are given, then the foci can be located and the usual procedure mentioned in property 1 can be adopted to draw or trace the ellipse.

Hence, the data of the length of the major and the minor axes form the basis for the construction of an ellipse as explained in the subsequent examples. Depending on the procedures adopted, these methods are also named uniquely.

Example 2.11

The distance between the two foci of an ellipse is 90 mm and the major axis is 120 mm long. Draw the ellipse using pin and thread method or loop method and measure the length of the minor axis.

PRINCIPLE

As the major axis and the foci distances are given, the minor axis can be located. The two foci and the extreme points of the minor axis can be used to draw the ellipse as shown by property 1 given in previous section.

PROCEDURE (Refer Fig. 2.14)

Step 1: Draw the major axis PQ for a length of 120 mm and locate the foci, F_1 and F_2, symmetrically at a distance of 45 mm from the centre O.

Step 2: Draw a perpendicular bisector at O and mark two points R and S on it, such that F_1R or F_1S is equal to the semi-major axis. The points R and S lie on the ellipse, and RS is the minor axis.

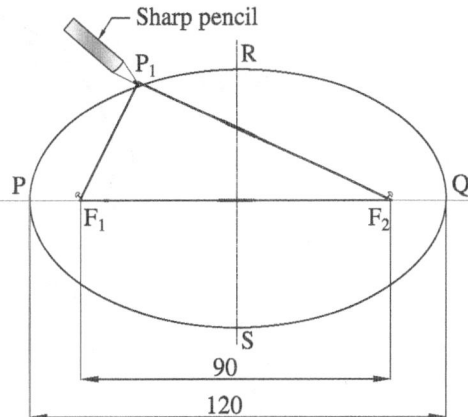

Step 3: As the point R is on the ellipse, the sum of the distances F_1R and F_2R is constant. Every point on the curve obeys this property.

Fig. 2.14   Ellipse by pin and thread method

Step 4: In order to achieve other points on the ellipse, insert a pin at each focus point F_1 and F_2, and keep a pencil point at R. Tie a thread taut passing through these points.

Step 5: Move the pencil around the two foci by keeping the thread taut and maintain an even tension. Obtain another point P_1 on the ellipse as shown.

Step 6: Similarly, move the pencil point repeatedly to other positions by keeping the thread taut and join these points by a smooth curve, which is the required ellipse.   ▲

Example 2.12

Using intersecting arcs method draw an ellipse having a major axis of length 120 mm and its two foci being 90 mm apart. Find the minor axis length.

> **PRINCIPLE**
> This method uses the principle of locating the points on the ellipse such that the sum of their distances from the two foci is always constant and is equal to the length of the major axis, which is given.

PROCEDURE (Refer Fig. 2.15)

Step 1: Draw the major axis PQ for a length of 120 mm and locate the foci, F_1 and F_2 symmetrically at a distance of 45 mm from the centre O.

Step 2: Mark points 1, 2, etc., on the major axis PQ, arbitrarily.

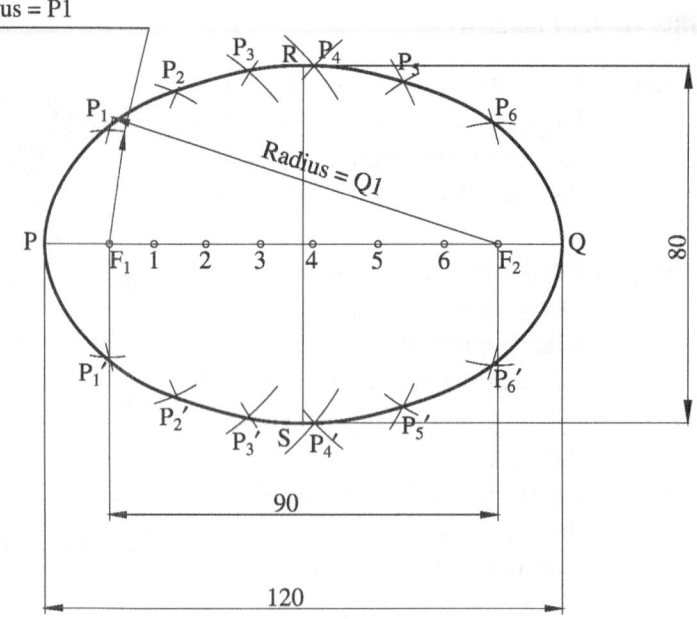

Fig. 2.15 Ellipse by intersecting arcs method

Step 3: With the two foci F_1 and F_2 as centres and radius equal to P1, draw arcs on either side of PQ.

Step 4: With the same centres and radius equal to Q1, draw arcs on either side of PQ to cut the previously mentioned arcs at P_1, P_1', P_6, and P_6'.

Step 5: Repeat the same procedures with the other points 2, 3, etc., and obtain the other points P_2, P_2', P_5, P_5', etc. Join all these points by a smooth curve, which is the required ellipse. ▲

Example 2.13 ◣

The major and minor axes of an ellipse are 120 mm and 80 mm, respectively. Draw an ellipse using concentric circles method.

PROCEDURE (Refer Fig. 2.16)

Step 1: Draw the given major axis AB and minor axis CD. Locate the centre O and obtain two concentric circles with them as diameters.

Step 2: Draw radial lines O–1′–1, O–2′–2, ..., O–12′–12 by dividing the circles into equal parts.

Step 3: Draw vertical lines from 1, 2, ..., 12, which lie on the major axis circle, and horizontal lines from 1′, 2′, ..., 12′, which lie on the minor axis circle, and obtain their meeting points as P_1, P_2, ..., P_{12}. Join these points to get the required ellipse. ▲

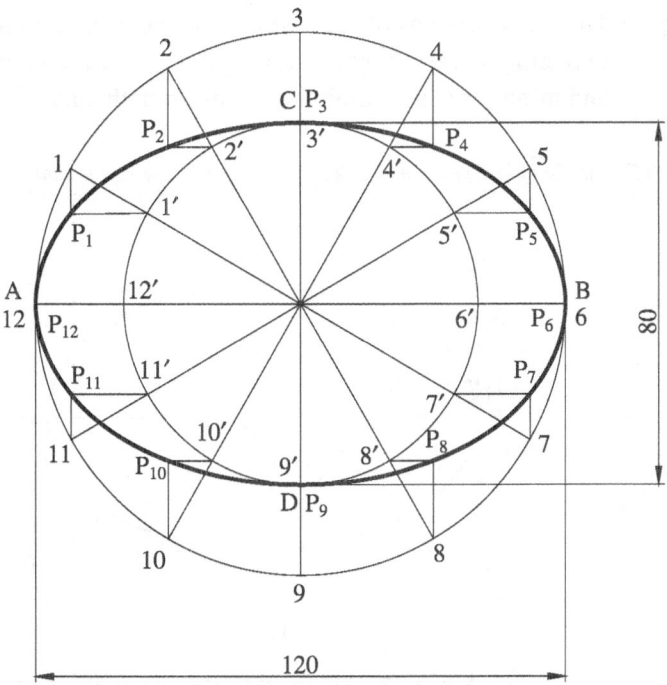

Fig. 2.16 Ellipse by concentric circles method

Example 2.14

Draw an ellipse of major and minor axes of 140 mm and 85 mm, respectively, using rectangle method or oblong method.

PROCEDURE (Refer Fig. 2.17)

Step 1: Mark the major and minor axes and obtain their meeting point O.

Step 2: Construct a rectangle or oblong ABCD that makes four quadrants.

Step 3: The ellipse can be constructed in any one quadrant P–O–R–D as follows:

(a) Divide the semi-major axis PO into a number of equal parts as shown by the points 1, 2, and 3, starting from P.

(b) Divide the semi-minor axis OR also into the same number of equal parts. Mark them on the parallel line PD in the boundary as shown by $1_1, 2_1, 3_1$, etc., also starting from P.

(c) Draw lines joining $1_1, 2_1$, and 3_1 with R (the end of the minor axis). Similarly join 1, 2, and 3 with S, the other end of the minor axis, and extend them to meet the lines from R at P_1, P_2, P_3, etc.

(d) The smooth curve joining P, P_1, …, R gives one-quarter of the ellipse.

Step 4: The other portions of the ellipse can be constructed in the remaining quadrants in a similar way or by locating symmetrical points on both sides of the major and minor axes and joining them by smooth curves.   ▲

NOTE *As the ellipse gets constructed within the rectangle, it is said to be inscribed.*

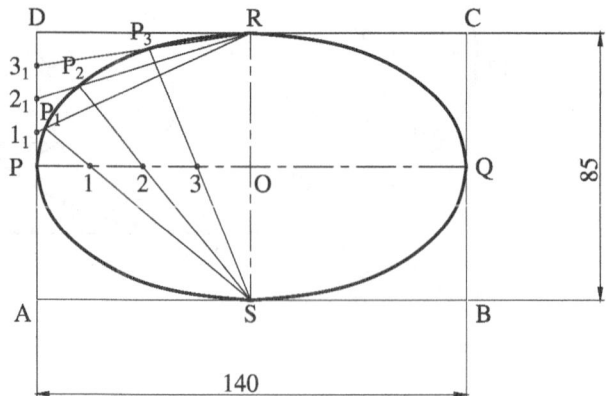

Fig. 2.17   Ellipse by rectangle or oblong method

Example 2.15

A parallelogram has sides 140 mm and 85 mm with an included angle of 65°. Inscribe an ellipse in the parallelogram. Find the major and minor axes of the ellipse.

PRINCIPLE
This is similar to inscribing an ellipse inside a rectangle, except that the sides are inclined at an angle. The two axes parallel to the sides of the parallelogram and passing through the centre are known as conjugate axes. Hence, the problem can also be referred with the conjugate axes and their included angle. In effect, when the ellipse is constructed, its major and minor axes will not be equal to the sides of the parallelogram, as the ellipse is not symmetrical in the four quadrants.

PROCEDURE (Refer Fig. 2.18)

Step 1: Draw the parallelogram ABCD of the given sides with the included angle of 65°, and mark the axes PQ and RS. Alternatively, draw the given conjugate axes PQ and RS at an angle of 65° and construct a parallelogram ABCD.

Step 2: Construct one portion of the ellipse in one quadrant, say PORD in similar lines of the rectangle method in Example 2.14. The semi-conjugate axes PO and OR (and hence PD) are divided into equal number of parts. They are joined with S and R, respectively, to get the points P_1, P_2, etc.

Step 3: Repeat the construction similarly in the other quadrants.

Step 4: A smooth ellipse emerges when these points are joined.

Step 5: The major and minor axes of the ellipse are identified as follows:

    (a)   With O as centre and OR as radius, draw a semicircle to cut the ellipse at T.

    (b)   Join the line TR and draw a line parallel to it through O to cut the ellipse at H and G, respectively. Measure HG, which is equal to the length of the minor axis of the ellipse.

    (c)   Draw EF through O and make it perpendicular to HG. Measure EF, which is equal to the length of the major axis of the ellipse.

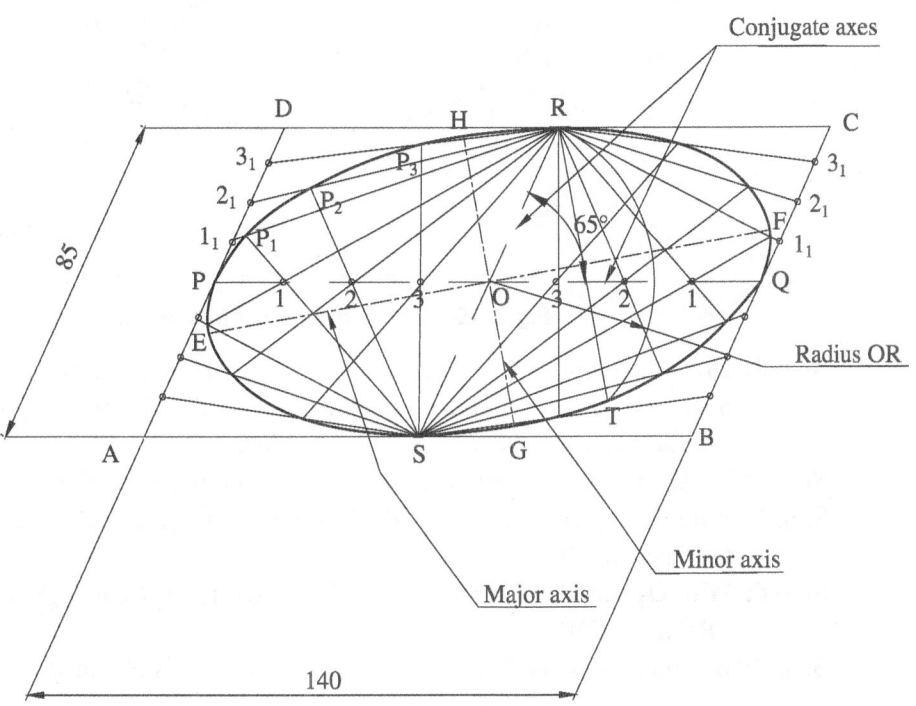

Fig. 2.18 Ellipse by parallelogram method

Example 2.16

The major and minor axes of an ellipse are 120 mm and 80 mm, respectively. Draw an ellipse using four-centre method.

PROCEDURE (Refer Fig. 2.19)

Step 1: Draw the major and minor axes PQ and RS; join PR.

Step 2: With O as centre and OP as radius, draw an arc to cut RS extended at A. Set an arc, RA, to cut PR at B.

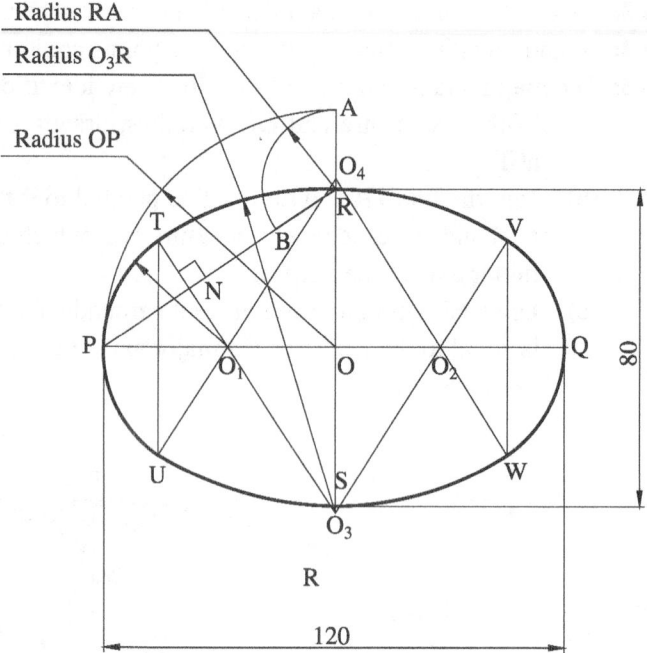

Radius RA

Radius O_3R

Radius OP

Fig. 2.19   Ellipse by four centre method

Step 3: Draw the perpendicular bisector of the line PB and extend it to meet the semi-major axis OP at O_1 and RS extended to meet at O_3. Set the points O_2 and O_4 symmetrically on OQ and on OR extended.

Step 4: O_1, O_2, O_3, and O_4 are the four centres for drawing the ellipse.

Step 5: With O_1 and O_2 as centres and radius equal to O_1P and O_2Q, draw smaller arcs UPT and WQV.

Step 6: With O_3 and O_4 as centres and radius equal to O_3R and O_4S, draw larger arcs TRV and USW.

Step 7: By completing the four arcs, the required ellipse is obtained.   ▲

Specific Construction Methods for Parabola

Example 2.17 ▲

Construct a parabola with base 60 mm and length of the axis 40 mm using rectangle method. Draw a tangent to the curve at a point 20 mm from the base. In addition, locate the focus and directrix to the parabola.

PROCEDURE   (Refer Fig. 2.20)

Step 1: This method is, in principle, similar to the rectangle method used to construct the ellipse as explained in Example 2.14, but is limited to the top adjacent quadrants.

Step 2: Mark the base AB of the parabola on the horizontal axis and the axis height CD vertically.

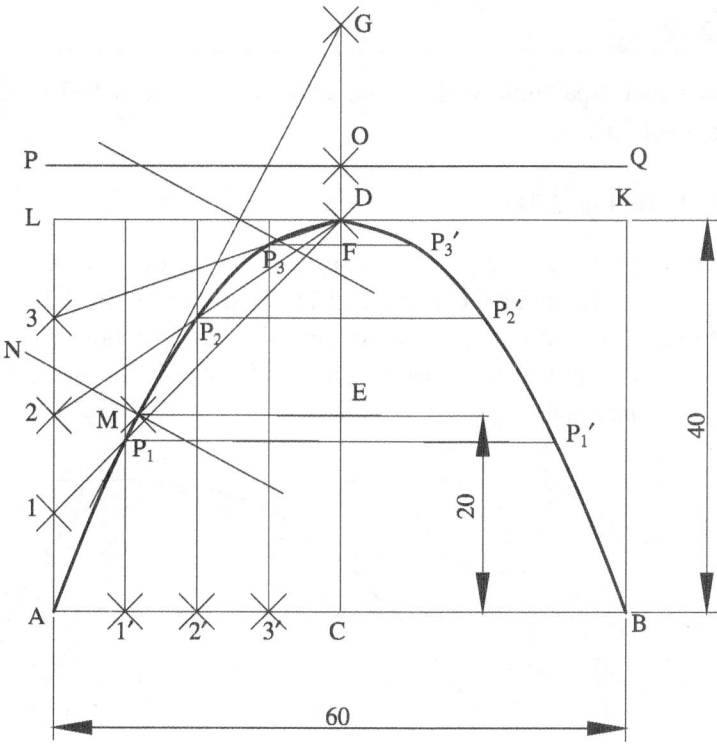

Fig. 2.20   Parabola by rectangle method

Step 3: Construct a rectangle ABKL to inscribe the parabola through the point D which becomes the vertex.

Step 4: Divide AC and AL into the equal number of parts as shown by 1', 2', and 3' and 1, 2, and 3, respectively. Join D–1, D–2, etc., and draw vertical lines parallel to CD through 1', 2', etc.

Step 5: Obtain the meeting points of these to get P_1, P_2, etc, that lie on the parabola

Step 6: Join the points A, P_1, P_2, ..., D by a smooth curve in the left half. Similarly, obtain the curve in the right half that completes the parabola.

Step 7: Locate the point M on the curve at the distance of 20 mm above the base and obtain ME perpendicular to the axis. Set the distance DG = DE on the axis extended. Join GM which gives the tangent to the curve.

Step 8: Draw a perpendicular MN to the tangent at the point 'M' and obtain the normal.

Step 9: The directrix PQ is located as follows:

    (a)   Draw a perpendicular bisector to GM which intersects the axis at F which is the focus of the parabola. Set DO equal to FD, as D is a point on the parabola and the vertex.

    (b)   Draw a line PQ perpendicular to the axis through O and obtain the directrix.

Example 2.18 ▲

Construct a parabola within a parallelogram of sides 100 × 60 mm with an included angle of 75°.

PROCEDURE (Refer Fig. 2.21)

Step 1: Construct the parallelogram ABKL with the base AB and the axis length of 60 mm at an included angle of 75° as shown in Fig. 2.21.

Step 2: The entire procedure of obtaining the parabola is similar to Example 2.17 except that the lines drawn on AB from 1′, 2′, etc., are parallel to CD and are inclined. ▲

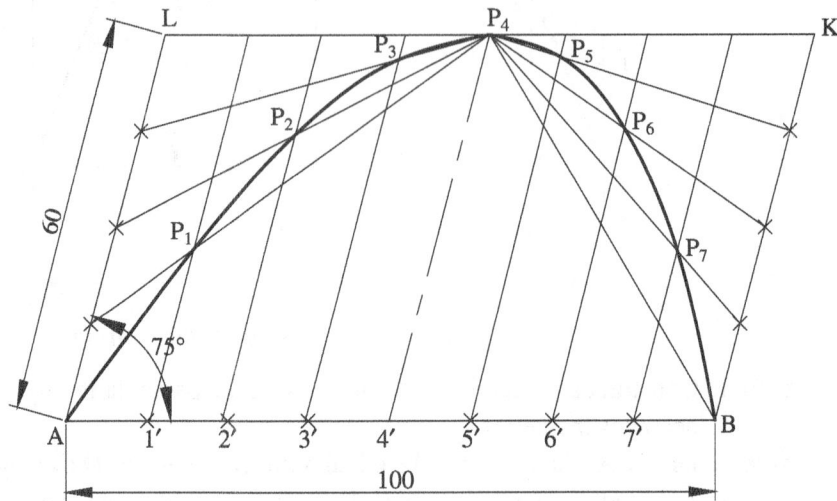

Fig. 2.21 Parabola by parallelogram

Example 2.19 ▲

A bullet is fired from the ground level at an angle 60° to the horizontal. The bullet hits the target at a point 80 m from its point of firing. Trace the path of the bullet using tangent method and by taking a suitable scale.

PROCEDURE (Refer Fig. 2.22)

Step 1: Draw the base PQ = 80 mm (to represent 80 m) and set an angle of 60° at its two ends and obtain the triangle PRQ. It can be noted that the path traced by the bullet in a parabola.

Step 2: Draw the perpendicular bisector RO and obtain its midpoint V. The line RVO becomes the axis, V is the vertex of the parabola, and VO is the height of the parabola. A rectangle as in Example 2.17 can be set up and the parabola can be drawn.

Step 3: Alternatively, the parabola can be drawn by setting up a series of tangent lines as follows:

(a)   Divide the lines RP and RQ into the same number of equal parts and number them with opposite schemes as shown by 1, 2, … on PR, and 1′, 2′, … on RQ as shown in Fig. 2.22.

(b)   Join 1–1′, 2–2′, … and draw a curve tangential to these lines and obtain the parabola desired.

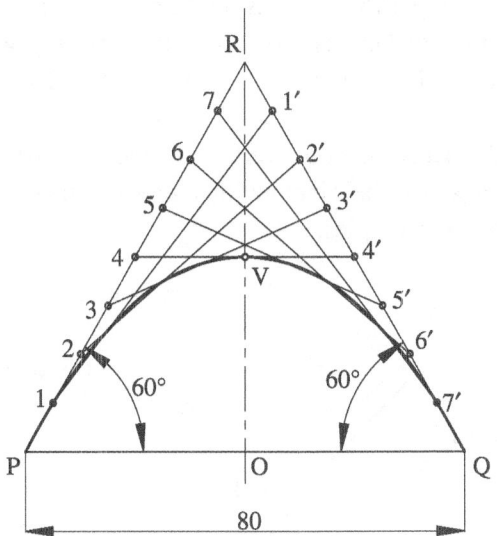

Fig. 2.22   Parabola by tangent method

Specific Construction Methods for Hyperbola

The general properties and specific parameters of the hyperbola are mentioned here and the construction methods suggested in this section make use of one or more of these properties.

Property 1   As the hyperbola has two branches on its left and right halves with two foci F_1 and F_2, the locus of any point on it can be described as the difference between its distances from the foci and is always a constant. In Fig. 2.23, if A is a point on the hyperbola, then F_2A–F_1A = constant. This constant distance is equal to the distance between the two vertices V_1 and V_2. It is known as the transverse axis of the hyperbola. This property is used when the two foci and one point on the hyperbola are given or known.

Property 2   The hyperbola is the locus of a point moving in such a way that the ratio of its distance to the focus and to that of the fixed line or directrix is always greater than 1. Here, $e = AF_1/AB > 1$. This property can be used when the eccentricity or the distances of focus and directrix are given.

Property 3 The hyperbola has got the unique presence of asymptotes, which are the two intersecting lines obtained between the left and the right curves of the hyperbola. In Fig. 2.23, the asymptotes are shown by the two lines PQ and RS intersecting at the centre O of the transverse axis. These asymptotes, when extended, approach nearer to the hyperbola curves and will become tangential to them at infinity. The asymptotes and the angle between them are also used for drawing the hyperbola curves. When the angle between the asymptotes is 90°, then the hyperbola is known as a rectangular or equilateral hyperbola. The asymptotes can be drawn by drawing a circle with O as centre and OF_1 or OF_2 as radius and by erecting two perpendiculars at the vertices and by joining the intersection points P, Q and R, S by straight lines as shown in Fig. 2.23.

Hence, the data of the transverse axis, the directrix, the foci distances, the location of the asymptotes, and their inclination angle, can be used to draw the hyperbola as shown in the following examples.

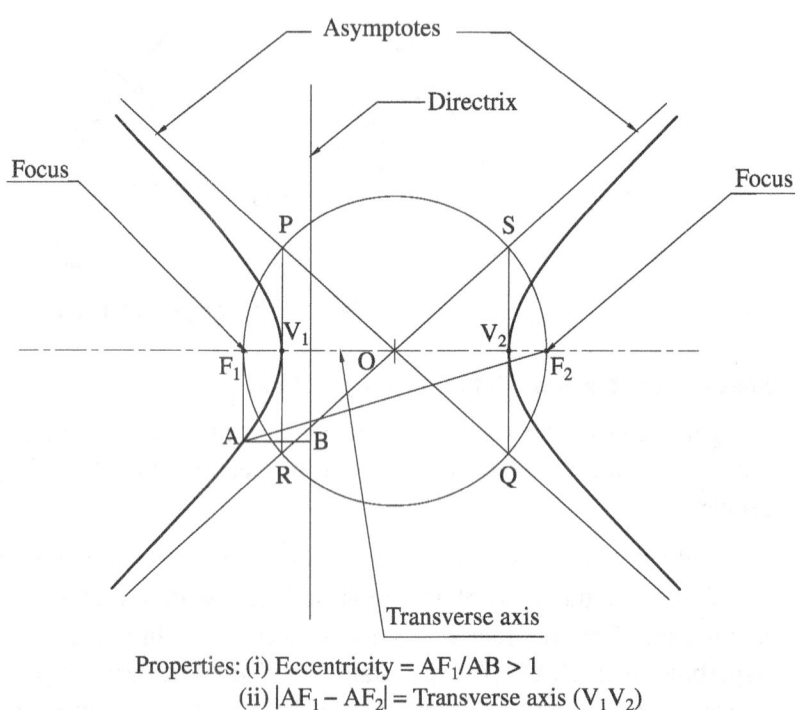

Properties: (i) Eccentricity = $AF_1/AB > 1$
(ii) $|AF_1 - AF_2|$ = Transverse axis (V_1V_2)

Fig. 2.23 Parameters of hyperbola

Example 2.20

Construct a pair of hyperbola, where the distance between the foci is 80 mm and the length of the transverse axis is 60 mm. Find the eccentricity of the curve.

PROCEDURE (Refer Fig. 2.24)

Step 1: Draw the axis line, and mark V_1V_2 as the length of the transverse axis. Mark the foci, F_1 and F_2 spaced 80 mm apart symmetrically on either side, as shown in Fig. 2.24.

Step 2: Mark a number of points 1, 2, ... arbitrarily on the axis (say on the left side).

Step 3: With F_1 as centre and V_1–1 as radius, draw arcs above and below the axis on the left-hand side.

Step 4: With F_2 as centre and V_2–1 as radius, draw arcs to cut the previously drawn arcs at P_1, P_1'.

Step 5: Repeat the same procedure with the other points on the left side and join the points P_1, P_2, ..., P_1', P_2'... to get the hyperbola on the left-hand side.

Step 6: A similar constructional procedure can be followed to construct the right-hand side hyperbola by choosing some arbitrary points on the right side of the axis.

Step 7: To find the eccentricity, the directrix is to be drawn first and the distances of any point on the hyperbola to the focus and the directrix are to be used as follows:

    (a) At F_1, erect a perpendicular to the axis to meet the hyperbola at A. Join AF_2 and bisect the angle F_1AF_2. When the angle bisector meets the transverse axis at C, erect a perpendicular to the axis, which gives the directrix. From A, draw a perpendicular to the directrix to obtain B. The eccentricity is obtained by measuring the distances AF_1 and AB and finding the ratio AF_1/AB.

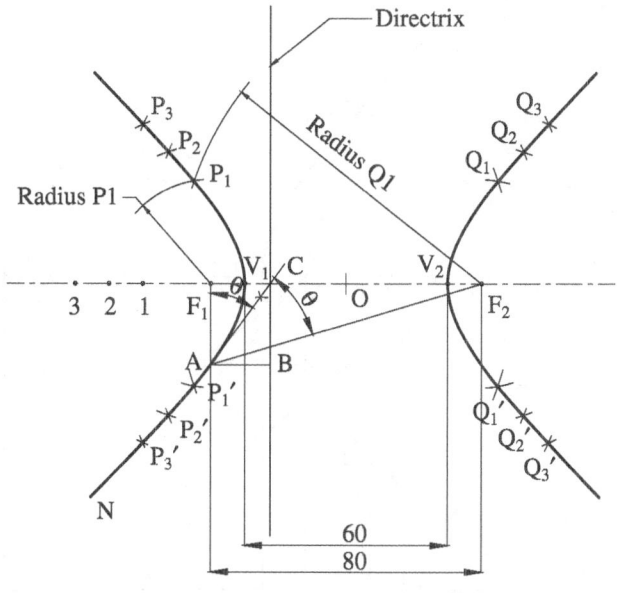

Eccentricity = AF1/AB = 1.334

Fig. 2.24 Construction of a pair of hyperbola with foci and transverse axis distances

Example 2.21   ▲

Construct a pair of hyperbola, where the distance between the foci is 80 mm and the length of the transverse axis is 60 mm. Draw a tangent to the hyperbola from a point P lying outside the curve at a distance of 40 mm from its vertices.

PROCEDURE (Refer Fig. 2.25)

Step 1: Draw a pair of hyperbola as mentioned in Example 2.20.

Step 2: Locate the point M (from which the tangent is to be drawn) by cutting the arcs 40 mm each from the points V_1 and V_2.

Step 3: Draw a circle with the transverse axis V_1V_2 as diameter.

Step 4: Connect M to the nearest focus F_1 and draw a circle with MF_1 as diameter.

Step 5: Mark the intersecting points of the two circles as A and A_1.

Step 6: Join F_1A and extend it to locate the point B such that $AB = F_1A$.

Step 7: Join F_2B and extend it to meet the hyperbola (constructed with F_1 as the focus) at the point C.

Step 8: Join the points M and C, and extend the line. MCT is the required tangent to the hyperbola from the exterior point M.

Step 9: At C, erect the line NN perpendicular to MCT. The line NN is the normal at the point of tangency.   ▲

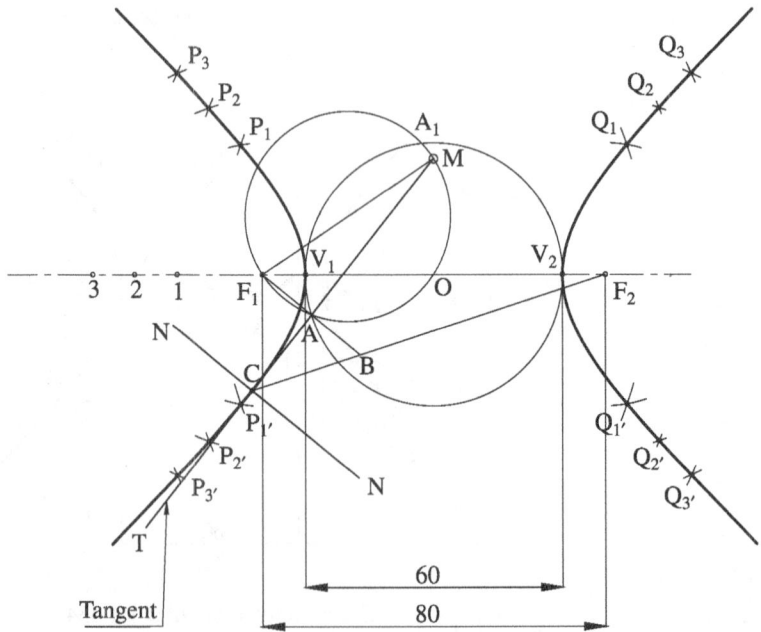

Fig. 2.25   Construction of tangent and normal to the hyperbola

Example 2.22

Draw a rectangular hyperbola with a given point P on it, at a distance 25 mm and 20 mm from the two asymptotes.

PROCEDURE (Refer Fig. 2.26)

Step 1: Draw the two asymptotes OX and OY at right angles to each other.

Step 2: Mark the point P at a distance of 25 mm from one axis and 20 mm from the other axis and draw lines A_1A_2 and B_1B_2, parallel to the asymptotes OX and OY, respectively.

Step 3: Mark a set of points 1, 2, 3, ..., 6 on A_1A_2, and join them radially with O. Find their intersection points (or their extensions) with B_1B_2 and name them as $1'$, $2'$, ..., $6'$ as shown in Fig. 2.26.

Step 4: From the points 1 and $1'$, draw lines parallel to OY and OX axes, respectively, and mark their intersection as P_1.

Step 5: Similarly, obtain P_2, P_3, etc., at the intersection of lines from 2, $2'$ and 3, $3'$, etc., and join them by a smooth curve to get the hyperbola desired. ▲

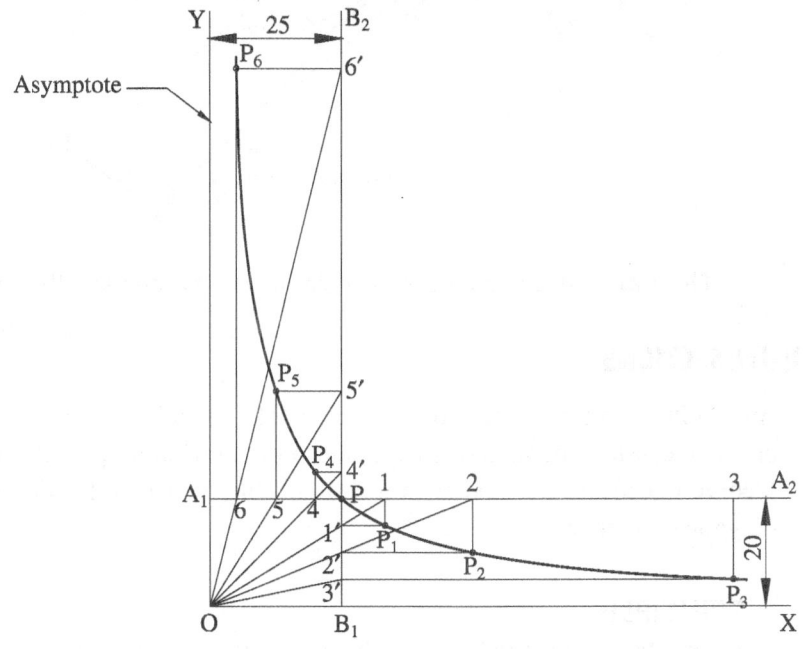

Fig. 2.26 Rectangular hyperbola

Example 2.23

The asymptotes of a hyperbola are inclined at 80° to each other. A point P on the curve is at a distance of 30 mm and 20 mm from the two asymptotes. Draw the hyperbola passing through the point P. Find the eccentricity of the curve.

PROCEDURE (Refer Fig. 2.27)

Step 1: The procedure is similar to that explained in Example 2.22, except the placement of the asymptotic lines OX and OY at an angle of 80°.

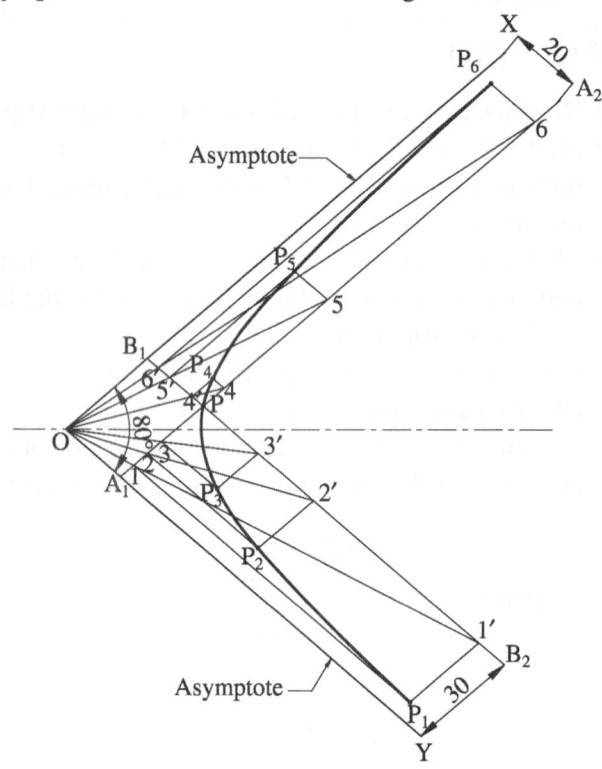

Fig. 2.27   Construction of hyperbola with asymptotes and their included angle

2.5 INVOLUTE SECTIONS

An involute is a curve traced out by the end of a thread unwound from a line, a polygon, or a circle, when the thread is kept taut in the unwinding process. It is also defined as a curve traced out by a point in a straight line, which rolls without slipping along a polygon or a circle.

PRINCIPLE

When a thread is wound over a polygon (circle) and meets its ends at a point, and one of the ends of the thread is opened to unwind, the portion of the thread unwound falls in alignment with the existing side of the polygon. This happens because the thread is kept tight during this process. The length of the portion unwound each time is equal to the length of the number of sides unwound. In the case of a circle, the thread becomes tangential to the circle at the point up to which it is unwound. The length of the thread becomes equal to the circumferential length of the portion unwound.

Example 2.24

Draw the involute of a regular hexagon of side 20 mm and mark a tangent and a normal to the curve at a distance 100 mm from the centre of the hexagon or at any point on the involute.

PROCEDURE (Refer Fig. 2.28)

Step 1: Construct the hexagon ABCDEF of sides 20 mm long, and assume a thread is wound from A to A.

Step 2: Consider the unwinding or opening of the thread from the end A in clockwise direction, the first portion being AB.

Step 3: Extend CB through B. With B as centre and BA as radius draw an arc to intersect the line CB extended at P_1.

Step 4: Similarly, extend the lines at the other corners C, D, E, and F as shown in Fig. 2.28. With C as centre and CP_1 as radius, draw an arc to cut DC extended to get P_2. Obtain the other points P_3, ..., P_6 in a similar manner, on the respective extended lines from the respective corners.

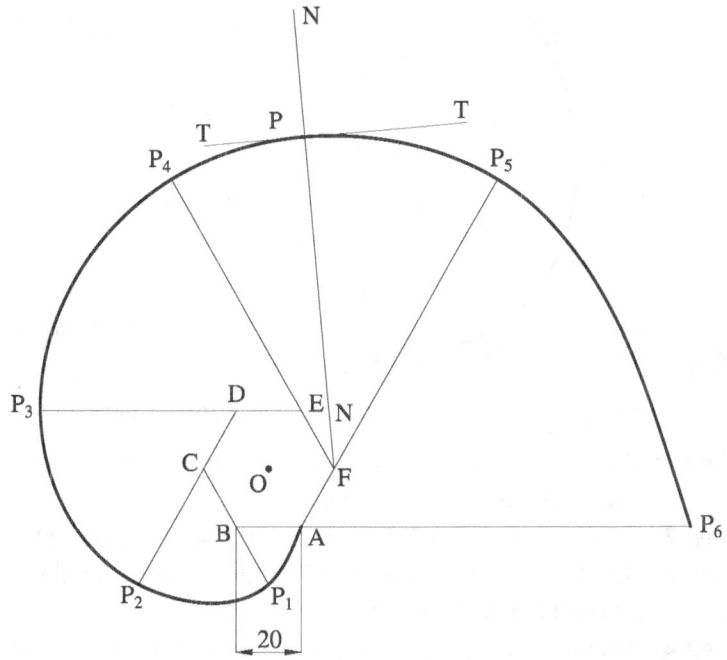

Fig. 2.28 Involute of a hexagon

Step 5: Join P_1 ... P_6 by a smooth curve, which gives the involute of the hexagon.

Step 6: Locate the point P by cutting an arc of 100 mm radius from O, the centre of the polygon.

Step 7: As the individual curves AP_1, P_1P_2, etc., have been drawn with the respective corners B, C, etc., as centres, the segment where P is located will also have the corner F as its centre. Therefore, join P and F, which becomes the normal N–N for the curve at P. Draw a perpendicular line to NN at P and obtain TPT, which is the required tangent.   ▲

Example 2.25   ◢

Draw the involute of a circle of 40 mm diameter. In addition, draw a tangent and a normal to the curve at a point 95 mm from the centre of the circle or at any point on the involute.

PROCEDURE   (Refer Fig. 2.29)

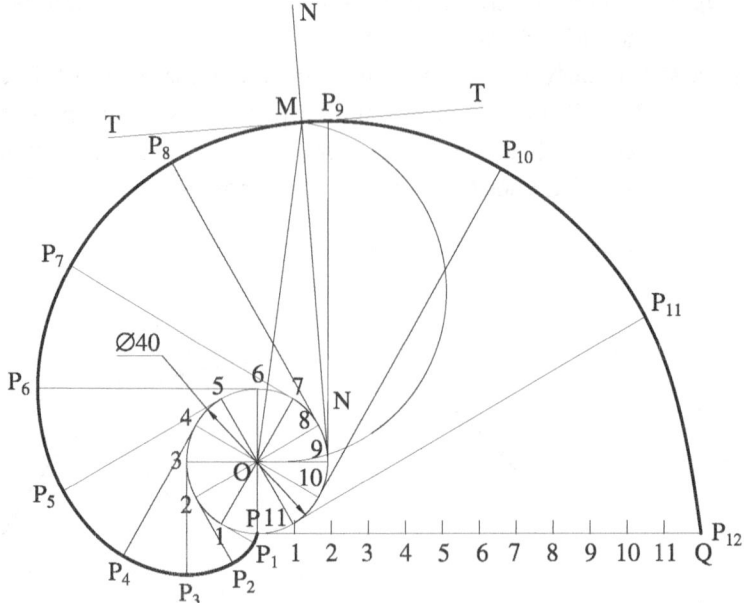

Fig. 2.29   Involute of a circle

Step 1: Draw the circle of 40 mm diameter with O as the centre, and OP as the radius with P as a generating point.

Step 2: Set a tangent PQ at P for a length equal to the circumference of the circle.

Step 3: Divide the circle and the line PQ into the same number of equal parts. Name the division points as shown.

Step 4: Assume that a thread is tied upon the circle from end to end at P. When the thread is unwound from P in the clockwise direction up to point 1, the portion of the thread from 1 to P becomes tangential at point 1, as it is kept taut. The length of that portion is equal to the arc radius, 1–P. It can be measured from the tangential line PQ from P to 1 and set it along the tangent at 1 and hence mark the point P_1.

Step 5: Similarly, the other portions pertaining to 2 to P, etc., are laid on the respective tangents at 2, 3, etc., at a radius equal to P–2, P–3, etc. Thus, the points P_2, P_3, etc., are obtained.

Step 6: Join the points P_1 ... P_{12} (Q) by a smooth curve, which gives the involute of the circle.

Step 7: To draw a tangent and normal at any point (say M), mark the point on the involute circle, join it with the centre, and draw a semicircle with that line as the diameter. Obtain its meeting point N with the circle. Join N and M, which gives the normal and draw a perpendicular at M to the line MN and obtain the tangent TMT. ▲

2.6 CYCLOIDAL CURVES

Cycloidal curves are a family of curves traced by a point on the circumference of a circle, when the circle rolls without slipping on a straight line or on another circle.

The rolling circle is known as the generating circle, and the straight line on which it rolls is known as the directing line or the base line. When the circle rolls on another circle, the latter is known as the directing or the base circle.

2.6.1 Cycloid

A cycloid is a plane curve generated by a point on the circumference of a circle, when it rolls on a fixed straight line without slipping.

> **PRINCIPLE**
> When a circle rolls on a straight line, the instantaneous positions of the point of intersection are obtained by periodically moving the centre of the circle to different positions parallel to the line, while maintaining the same height of the division points.

 **Example 2.26**

 A circle of 40 mm diameter rolls clockwise along a line for one revolution. Draw the locus of a point on the circle, which is in contact with the line. In addition, draw a tangent and a normal to the curve at a point 35 mm from the directing line or at any point on the cycloid.

PROCEDURE (Refer Fig. 2.30)

Step 1: Draw the generating circle of 40 mm diameter and mark its centre O, and divide it into 12 equal parts. Mark the points 1 ... 12 (P).

Step 2: As the circle rolls on the line, mark the coincident generating point P. Draw the generating line PA of length equal to the circumference of the circle and divide it equally as that of the rolling circle. Mark the points as $1'$ to $12'$ (Q).

Step 3: Draw the locus line of the centre, parallel to the base line, and mark the points O_1 to O_{12} (B) by erecting perpendiculars at $1'$, $2'$, etc.

Step 4: $O_1 \ldots O_{12}$ are the instantaneous positions of the centre.

Step 5: When the circle rolls forward (or clockwise), the points 1 and 1' coincide and the centre moves to O_1. At that instant, the generating point P also moves forward, maintaining the level or height of point 1 before rotation. As the generating point P also maintains the same radial distance in all its positions of movement, obtain the point P_1 by drawing an arc equal to OP from O_1 as centre on the horizontal line through 1.

Step 6: Obtain the other points $P_2, \ldots, P_{12}$, similarly with $O_2, \ldots, O_{12}$ as centres and join them by a smooth curve to get the cycloid.

Step 7: To draw a tangent and normal at any point (say M), mark the point on the cycloid and cut an arc equal to the radius of the circle to meet at O'. Drop a perpendicular on the line PA and obtain N. Join M and N and extend it a little to get the normal. Erect a perpendicular at M to the line MN and obtain the tangent TMT. ▲

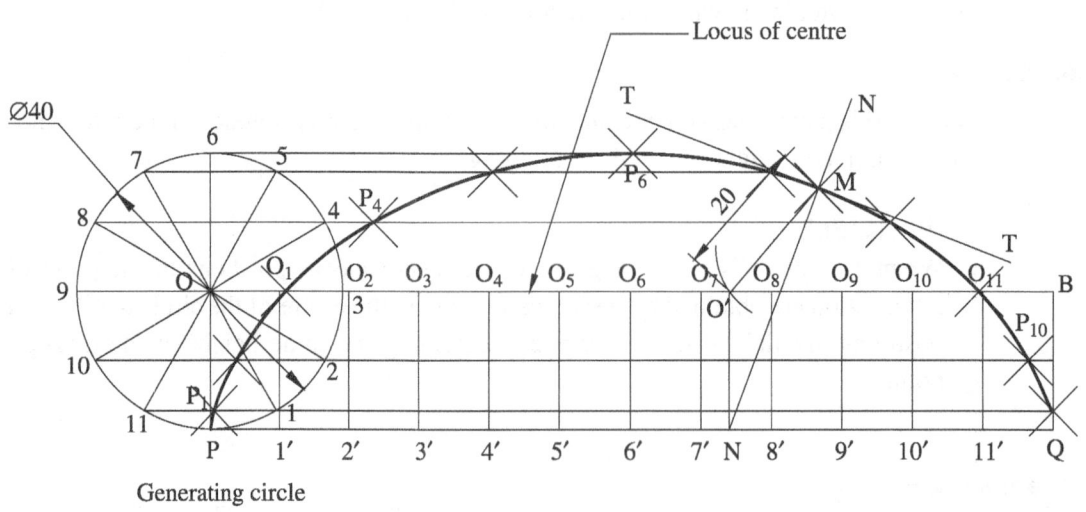

Fig. 2.30 Cycloid on a straight path

Example 2.27 ◢

A circle of 50 mm diameter rolls clockwise on a horizontal line for half a revolution, and then on a line inclined at 60° to the horizontal for another half a revolution in the same direction. Draw the curve traced by a point P on the circumference of the circle, taking the top-most point on the rolling circle as the initial position of the generating point.

PROCEDURE (Refer Fig. 2.31)

Step 1: Draw the generating circle of 50 mm diameter and mark its centre O, and divide it into 12 equal parts as shown by 1, ..., 12 (P). This is shown in Fig. 2.31. The generating point P is located at the top.

Step 2: As the circle rolls in two stages, draw the line $6P_6$ equal to half the circumference of the circle and divide it into six equal parts $1'$ to $6'$ (P_6). Mark the circle with O_6 as centre against P_6. Set the tangential line at 4 at 60° for half the circumference of the circle and mark the points $6'$ to $12'$.

Step 3: Draw the locus line of the centre, parallel to the base line, and mark the instantaneous positions of the centre O_1 to O_6, by erecting perpendiculars at $1'$ to $6'$. Extend the locus line of the centre parallel to the base line at 60° and mark the points O_7 to O_{12} (B).

Step 4: When the circle rolls forward (clockwise), the points 5 and $1'$ coincide and the centre moves to O_1. At that instant, the generating point P also moves forward, and maintains the level or the height of the point 1, before rotation. As the generating point P also maintains the same radial distance in all its positions of movement, obtain the point P_1, by drawing an arc equal to OP from O_1 as centre, on the horizontal line through 1.

Step 5: Similarly obtain the other points P_2, ..., P_6. Join all the points, including P and P_6, to get one-half of the cycloid curve.

Step 6: The other half of the cycloid curve can be obtained by locating the points P_7 to P_{12} on the lines drawn parallel to the inclined base line.

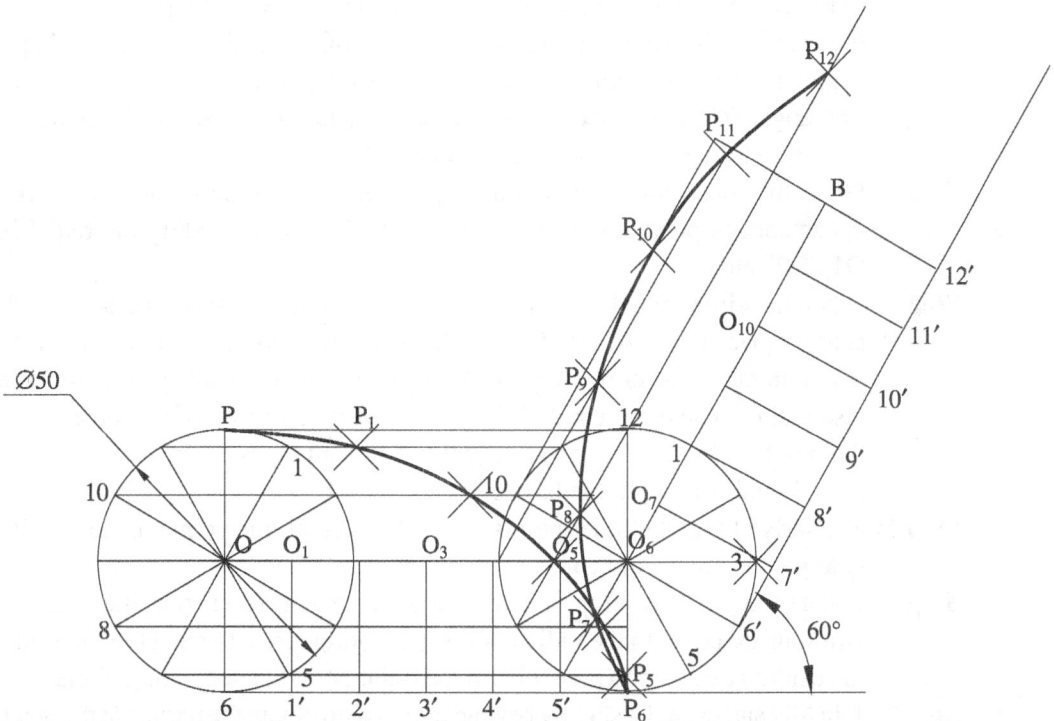

Fig. 2.31 Cycloid on straight and inclined path

2.6.2 Epicycloid

An epicycloid is a plane curve generated by a point on the circumference of a circle, when the circle rolls without slipping on the circumference of another circle.

As in the case of a cycloid, the instantaneous positions of the point of interest are obtained by periodically moving the centre of the circle to different positions parallel to the arc of the circle on which it rolls, while maintaining the same distance intended for that position from the centre of rotation. The locus of the instantaneous positions of the point of interest is obtained by joining them with a smooth curve.

Example 2.28

Draw an epicycloid of a circle of 60 mm diameter, which rolls outside on another circle whose diameter is 160 mm for one revolution. Draw a tangent and a normal at a convenient point on the curve.

PROCEDURE (Refer Fig. 2.32)

Step 1: Draw the generating circle of 60 mm diameter and mark its centre C, and divide it into 12 equal parts as shown by 1 … 12 (P) as shown in Fig. 2.32.

Step 2: As the circle rolls on another circle of 160 mm diameter, draw an arc PQ equal to the circumference of the circle. Divide that arc into 12 equal parts 1′ to 12′ (Q). This can be achieved by setting an angle POQ equal to $\theta°$ and dividing that angle into the same number of parts as the rolling circle.

The angle $\theta°$ is obtained as $(r/R) \times 360°$, where r and R are the radii of the rolling and the generating circle, respectively.

Step 3: Draw the locus line of the centre, parallel to the base circle and mark the instantaneous positions of the centre C_1 to C_{12}, by extending the radial lines O1′, O2′, etc.

Step 4: When the circle rolls forward (clockwise), the points 1 and 1′ coincide and the centre moves to C_1. At that instant, the generating point P also moves forward, maintaining the level of the point 1, before rotation. As the generating point P also maintains the same radial distance in all its positions of movement, obtain the point P_1, by drawing an arc equal to CP from C_1 as centre on the arc drawn from 1, and parallel to the base circle.

Step 5: Similarly obtain the other points P_2 … P_{11}. Join all the points, including P and Q to get the epicycloid curve.

Step 6: Locate the point M on the curve at any convenient level from the base circle. Draw an arc equal to the radius of the generating circle to meet the locus line of the centre at E. Join EO and obtain the point N on the generating circle.

Step 7: Join MN and extend it slightly beyond the curve to obtain the normal at the point M.

Step 8: Draw TMT through M and perpendicular to MN. This gives the tangent to the curve at the point M.

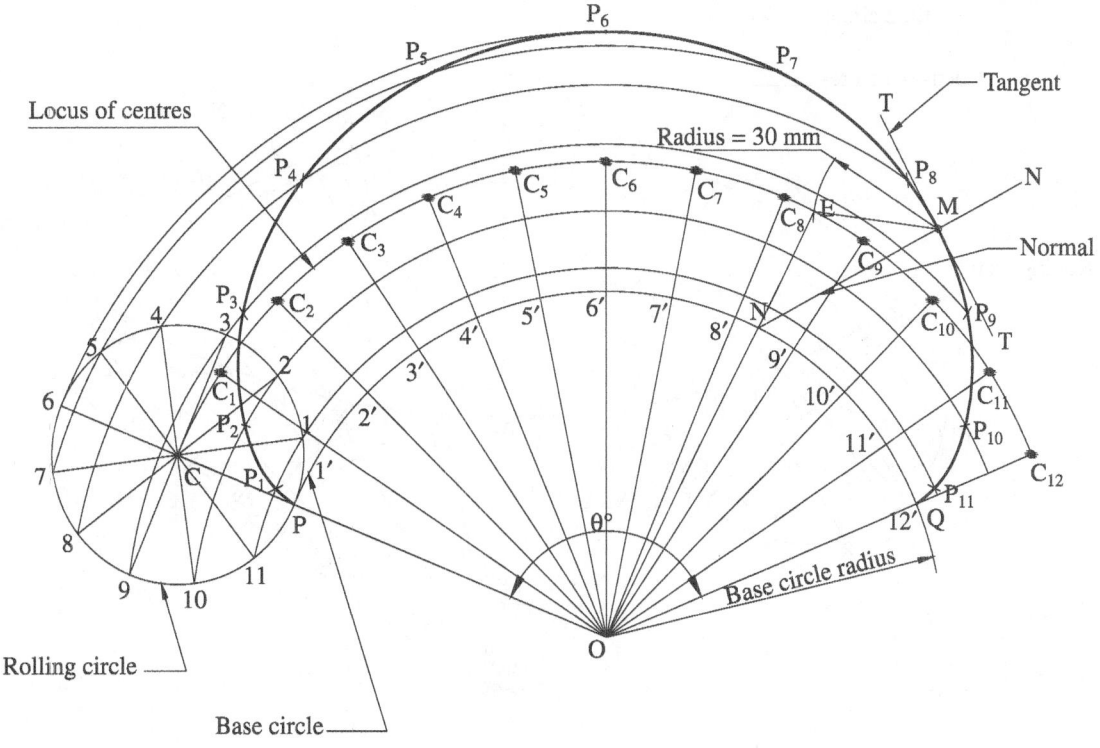

Fig. 2.32 Epicycloid

2.6.3 Hypocycloid

A hypocycloid is a plane curve generated by a point on the circumference of a circle, when it rolls without slipping on another circle and inside it.

As in the case of a cycloid, the instantaneous positions of the point of interest are obtained by periodically moving the centre of the circle to different positions parallel to the arc of the circle on which it rolls. One should maintain the same distance intended for that position from the centre of rotation. The locus of the instantaneous positions of the point of interest is obtained by joining them with a smooth curve.

Example 2.29

Draw a hypocycloid of a circle of 60 mm diameter, which rolls inside another circle whose diameter is 240 mm for one revolution. Draw a tangent and a normal at any point on the curve.

PROCEDURE (Refer Fig. 2.33)

Step 1: Draw the generating circle of 60 mm diameter and mark its centre C, and divide it into 12 equal parts as shown by 1, …, 12 (P) as shown in Fig. 2.33.

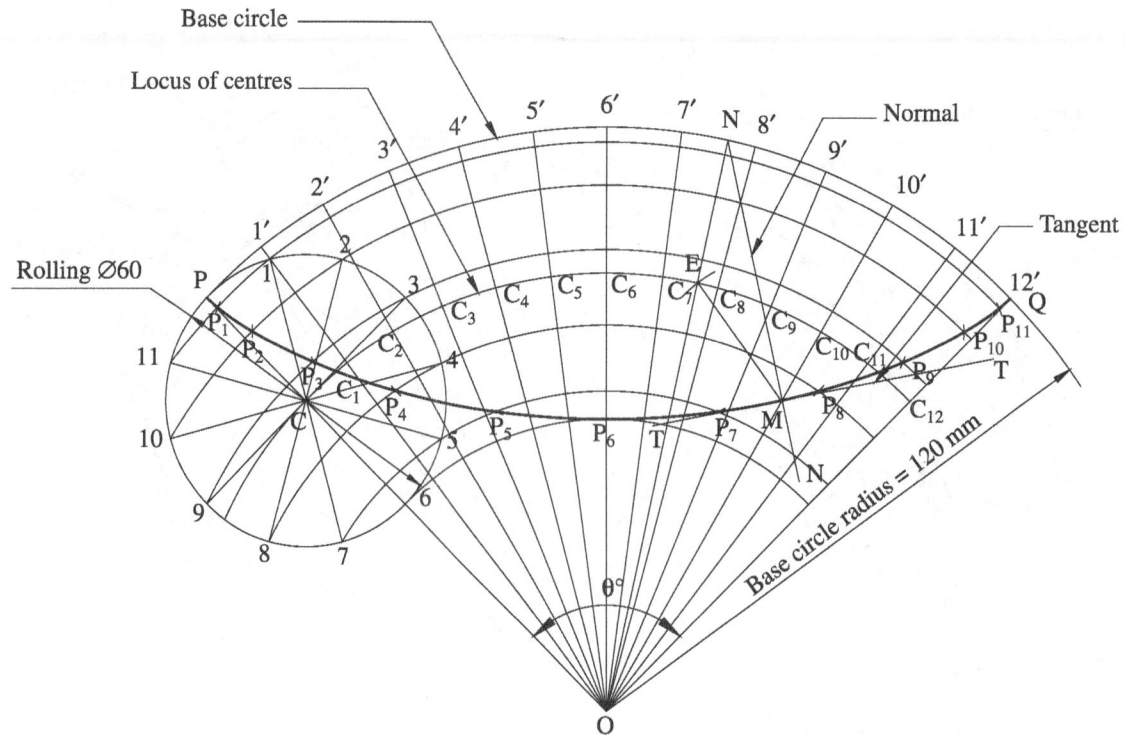

Fig. 2.33 Hypocycloid

Step 2: As the circle rolls on another circle of 240 mm diameter and inside it, draw an arc PQ equal to the circumference of the circle. Divide it into 12 equal parts, 1′ to 12′(Q). This can be achieved by setting an angle POQ equal to $\theta°$ and dividing that angle into the same number of parts as the rolling circle.

The angle $\theta°$ is obtained as $(r/R) \times 360°$, where r and R are the radii of the rolling and the generating circle, respectively.

Step 3: Draw the locus line of the centre, parallel to the base circle and mark the instantaneous positions of the centre C_1 to C_{12} on the radial lines O1′, O2′, etc.

Step 4: When the circle rolls forward (clockwise), the points 1 and 1′ coincide and the centre moves to C_1. At that instant, the generating point P also moves forward, maintaining the level of the point 1, before rotation. As the generating point P also maintains the same radial distance in all its positions of movement, obtain the point P_1, by drawing an arc equal to CP from C_1 as centre on the arc drawn from 1, and parallel to the base circle.

Step 5: Obtain the other points $P_2 \dots P_{11}$ similarly. Join all the points including P and Q to get the hypocycloid curve.

Step 6: Locate the point M on the curve at any convenient level from the base circle. Draw an arc equal to the radius of the generating circle to meet the locus line of the centre at E. Join EO and obtain the point N on the generating circle.

Step 7: Join MN and extend slightly beyond the curve to obtain the normal at the point M.

Step 8: Draw TMT through M and perpendicular to MN. This gives the tangent to the curve at the point M. ▲

2.7 LOCI OF POINTS

'Loci' is the plural of the word 'locus' and is the path traced by a point while it moves under the specified constraint either on a plane or in the space.

2.7.1 Loci of Points in Simple Cases

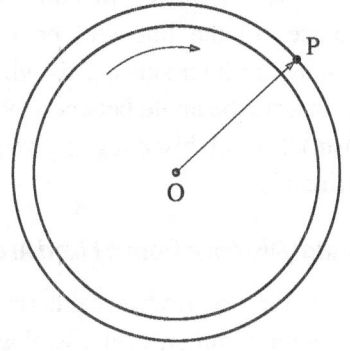

Locus of a Point Maintaining a Constant Distance from a Fixed Point

When a point moves with respect to another point such that its distance from it is a constant, its locus becomes a circle. Any point moving on the periphery of a circle is always at constant distance from the centre of the circle. This distance is known as the radius of the circle. Figure 2.34 shows that a point P is at a distance from O. When P moves to another position such that PO is constant, it traces a circle.

Fig. 2.34 Locus of point moving at constant distance from another point

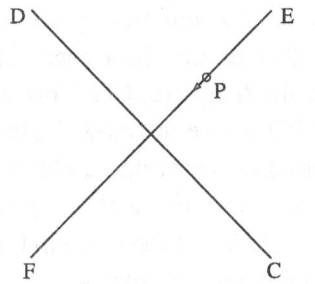

Locus of a Point Equidistant from Two Fixed Points

When a point moves such that its distance from two fixed points is a constant, then its locus is the perpendicular bisector of the line joining the two fixed points. In Fig. 2.35, the point P is equidistant from two fixed points C and D such that PC = PD. When the point P moves to E, then EC = ED, and to F, then FC = FD. This will happen only when the line FPE is the perpendicular bisector of the line joining C and D. The construction of the perpendicular bisector of a line adopts this principle.

Fig. 2.35 Locus of point moving equidistant from two fixed points

Locus of a Point Equidistant from a Fixed Straight Line

When a point moves such that its distance from a fixed straight line is constant, then its locus results in a line parallel to the fixed straight line. Figure 2.36 shows the point A that maintains the

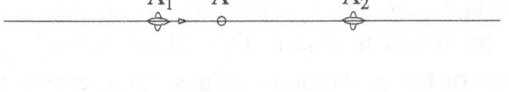

Fig. 2.36 Locus of point maintaining same distance from fixed straight line

same distance from the fixed straight line PQ, when it moves to another position. The locus becomes a straight line parallel to PQ through A. A_1 and A_2 represent two different positions. An example of an aircraft or a bird moving at a constant height will explain this case.

Locus of a Point Equidistant from Two Fixed Straight Lines

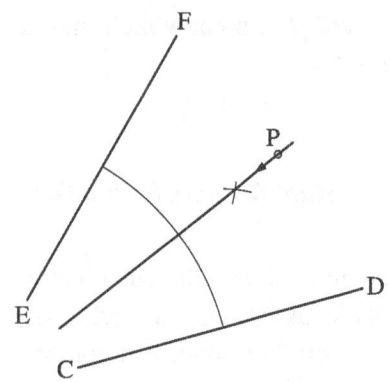

When a point moves such that it is equidistant from two fixed straight lines, then its locus is a straight line bisecting the angle between the fixed lines. In Fig. 2.37, CD and EF are the fixed straight lines. Point P moves so that its perpendicular distance from the two lines is constant. When it moves to another position, the perpendicular distances to the line will be of different lengths but will be the same on both sides. This will happen only when the locus of P bisects the angle between the two lines CD and EF. The construction of bisecting a given angle in Example 2.3 uses this principle.

Fig. 2.37 Locus of point moving between two fixed straight lines

Locus of a Point at a Constant Radial Distance from a Fixed Arc

When a point moves at a constant distance from a fixed arc, then its locus will be another arc with common centre as that of the fixed arc. The reader is advised to refer to the section on locus of a point equidistant from a fixed straight line. When a fixed line becomes a fixed arc CD as in Fig. 2.38, the point P will move parallel to the arc or will trace another arc EF whose centre also coincides with that of the fixed arc, CD. This will hold good if the point P is above or below the fixed arc. For example, in children's bicycles, a glittering object or a colourful light fixed in one of the spokes at a constant radial distance below the rim will trace the path of the rim when the wheel rotates.

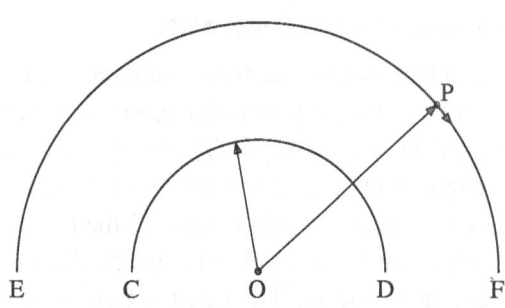

Fig. 2.38 Locus of point at constant radial distance from fixed arc

Locus of a Point Moving between a Fixed Point and a Fixed Straight Line

It was discussed in Section 2.4 that a conic is defined as the locus of a point moving in a plane such that the ratio of its distances from a fixed point and a fixed straight line is always constant. The fixed point is known as focus, whereas the fixed straight line is known as directrix. The ratio of these distances is known as the eccentricity. Based on the distances or the eccentricity values, three curves are generated—ellipse, parabola, and hyperbola. They possess eccentricity values equal to less than 1, equal to 1, and greater than 1, respectively, as shown in Fig. 2.9.

2.7.2 Locus of Point Equidistant from Fixed Point and Fixed Circle

PRINCIPLE

A basic point on the proposed locus is obtained by choosing the mid-point on the line joining the given fixed point and the fixed circle. The mid-point becomes a point on the locus as it is equidistant from the fixed point and the circle. Several points are chosen on this line and their distances from the circle are marked by arcs. The same distances are set from the fixed point, and are made to cut the arcs already drawn to yield the points that form the desired locus.

Example 2.30

Draw the locus of a point equidistant from a fixed circle of radius 60 mm and a fixed point that lies 120 mm from the centre of the fixed circle.

PROCEDURE   (Refer Fig. 2.39)

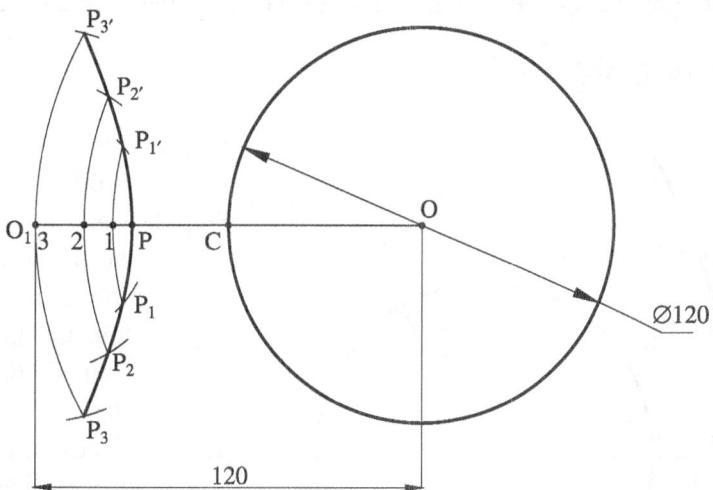

Fig. 2.39   Locus of point equidistant from fixed point and fixed circle

Step 1: Draw the given circle for a radius of 60 mm and mark its centre O.

Step 2: Mark the given point O_1 at a distance of 120 mm from O.

Step 3: Join the line OO_1 and obtain the point C, which is the meeting point of the circle.

Step 4: Choose a point P as the mid-point of the line CO_1. As point P is equidistant from O_1 and the circle at C, it lies on the locus proposed.

Step 5: Mark the points 1, 2, etc., on the line PO_1. Draw arcs through them, taking O as centre.

Step 6: Draw an arc with O_1 as centre and radius equal to C–1 which cuts the arc through 1 at points P_1 and P_1'.

Step 7: Repeat the same to obtain other points P_2 and P_2', P_3, and P_3', etc. Join P–P_1–P_2–P_3 and P–P_1'–P_2'–P_3', which is the locus of the point P. ▲

2.7.3 Locus of Point Equidistant from Fixed Straight Line and Fixed Circle

> **PRINCIPLE**
>
> A basic point on the proposed locus is obtained by choosing the mid-point on the line joining the given fixed line and the fixed circle. The mid-point becomes a point on the locus as it is equidistant from the fixed line and the circle. Several points are chosen on this line and their distances from the fixed line are set by drawing parallel lines. When the same distances are marked from the circle by circular arcs, the intersecting points form the locus desired.

Example 2.31 ▲

Draw the locus of a point equidistant from a fixed circle of radius 60 mm and a fixed straight line that lies 120 mm from the centre of the fixed circle.

PROCEDURE (Refer Fig. 2.40)

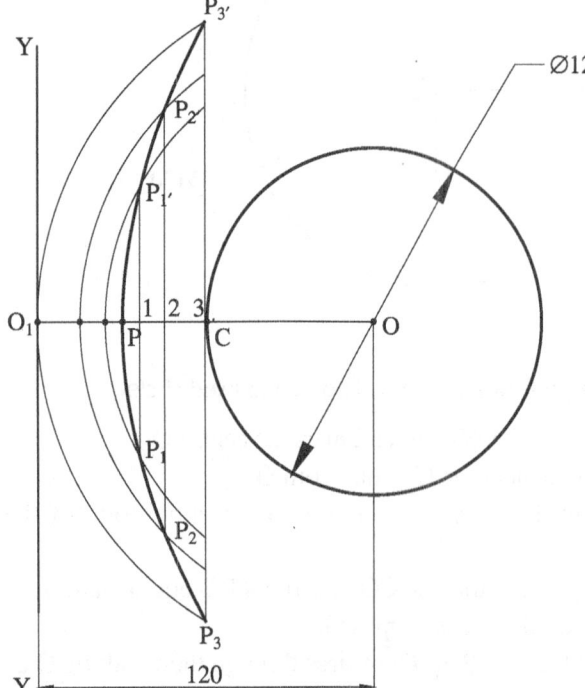

Fig. 2.40 Locus of point equidistant from fixed circle and fixed straight line

Step 1: Draw the given circle for a radius of 60 mm and mark its centre O.

Step 2: Mark the given straight line YY at the distance 120 mm from O and draw a line and mark O_1 on it.

Step 3: Join the line OO_1 and obtain the point C, the meeting point of the circle.

Step 4: Choose a point P as the mid-point of the line CO_1. As point P is equidistant from O_1 and the circle at C, it lies on the locus proposed.

Step 5: Mark the points 1, 2, etc., on the line CP and through them draw lines parallel to YY.

Step 6: With O as centre and radius equal to OC + $O_1$1, draw an arc to cut the line erected at 1, and obtain P_1 and P_1'.

Step 7: Repeat the same procedure with the points 2, 3, etc., to obtain other points P_2 and P_2', P_3, and P_3', etc. Join $P-P_1-P_2-P_3$ and $P-P_1'-P_2'-P_3'$ which is the locus of the point P.

2.7.4 Locus of Point Equidistant from Two Circles

> **PRINCIPLE**
> The same principle of drawing the locus of a point equidistant from a fixed point and a circle can be used, thereby, conceiving the fixed point as a point obtained on a circle of given radius. The mid-point of the line joining the meeting points of the two circles becomes a point on the locus. A few more points can be selected on this line and arc can be drawn from these two centres as discussed earlier. The meeting points of these arcs give the desired locus.

Example 2.32

Draw the locus of a point equidistant from a fixed circle of radius 60 mm and another fixed circle of radius 80 mm. The distance between the centres of these two circles is equal to 200 mm.

PROCEDURE   (Refer Fig. 2.41)

Step 1: Mark the two circles with O and O_1 as centres at their respective radii. Join the line OO_1 and mark the meeting points C and D of the circles.

Step 2: Locate the mid-point P on the line OO_1. By virtue of its position, P is a point on the locus.

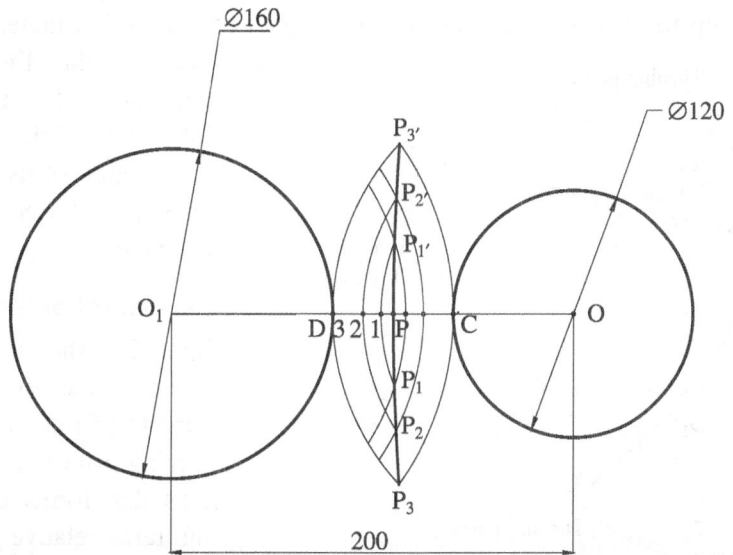

Fig. 2.41   Locus of a point which is equidistant from two fixed circles

Step 3: Mark the points 1, 2, etc., on the line DP.

Step 4: With O as centre and O–1 (OC + C1) as radius, draw an arc. As the same distance C1 has to be set from the other circle, draw an arc of radius equal to $O_1D + C1$ with O_1 as centre.

Step 5: The meeting points of the two arcs are noted as P_1 and P_1'.

Step 6: Repeat the same procedure with the points 2, 3 to obtain P_2 and P_2', P_3 and P_3', etc.

Step 7: Join $P–P_1–P_2–P_3$ and $P–P_1'–P_2'–P_3'$, which is the locus of the point P.   ▲

2.8 LOCI OF POINTS IN MECHANISMS

In engineering applications, machines are used to transform energy from one form to the other and to utilize the energy, thus obtained for doing some work. Machines consist of a number of parts, links, or bodies that have relative motion among themselves. Assembling these parts or links to achieve a particular motion is often referred to as *mechanism*. Mechanisms can result in motion transmission, power transmission, or both. If motion transmission is the only aspect involved, then the process is known as mechanism. When power and/or power and motion transmission occurs, the output, thus, obtained is known as a *machine*. These mechanisms can be achieved by coupling various elements or links in such a way that their motion occurs in a definite or regulated path. The movement of every part of the link follows a specific pattern to achieve the overall requirement. Hence, the study of the motion of different parts of a mechanism or machine becomes the fundamental requirement for an engineer. The study of these areas is often referred to as kinematics.

Since the link or any part is an assembly of various points, the kinematic study will help to understand the movement or path of the motion undertaken by the different points of that link. This section deals with the path undertaken by different links of a mechanism during one cycle of operation of its input member. This is generally referred to as loci of points in mechanisms.

2.8.1 Four-bar Mechanism

Figure 2.42 shows the four-bar mechanism, which is the simplest mechanism that consists of four links with each of them forming a turning pair (the two adjacent links that forms the pair in which one link turns relative to another link) at O_1, P, Q, and O_2. The four links may be of different lengths. According to Grashof's

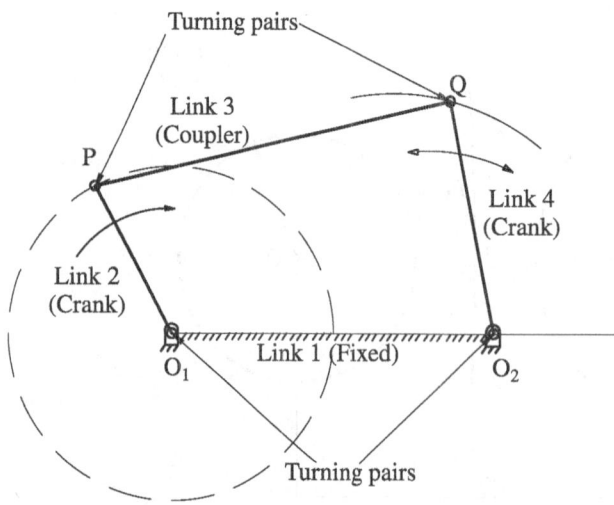

Fig. 2.42   Four-bar mechanism

law of four-bar mechanism, the sum of the shortest and the longest link lengths should not be greater than the sum of the remaining lengths of the other two links to get the continuous relative motion between any of two links of the mechanism. In the four-bar mechanism, O_1O_2—represented as link 1—is fixed and O_1P, PQ, and O_2Q are not fixed and represented as links 2, 3, and 4, respectively. Links 2 and 4 are termed as the input or driving link, and the output or the driven link cranks and link 3 is termed as the coupler. A *crank* occurs when the driving link makes a rotary motion.

When both the input and output links are of equal length, the rotary motion of the input link is transmitted through the coupler to the output link (which is driven) as rotary motion. The movement of the wheels of the locomotive engine is an example of this application.

When the input and output links are of unequal lengths, then the rotary motion given to the input link is transmitted through the coupler to the output link (which is driven), and it is known as *oscillatory motion*. This mechanism is widely used in various applications such as crank-lever mechanism, double-lever mechanism, and so on, to convert one type of motion into another in various industrial applications.

2.8.2 Single Slider Crank Mechanism

When one of the links in the four-bar chain becomes a slider, then the linkage arrangement is modified to facilitate the movement of the slider within a frame, which is usually known as cylinder. The coupler link connecting the crank and the slider is known as the connecting rod. The arrangement shown in Fig. 2.43 is a single-cylinder mechanism that consists of one sliding pair between the fixed cylinder and a slider or a piston and three turning pairs. The Pairs include the bearing O, the crank OC, and the connecting rod CD, and between the connecting rod and the slider/piston. In Fig. 2.43, OCD represents the single slider crank mechanism in which OC is a crank that rotates about the point O. It transfers the motion through the connecting rod CD to the piston or slider D, which has a reciprocating motion, whereas the connecting rod has the oscillatory motion. It is the one of the simplest mechanisms and is widely used in most internal combustion engines.

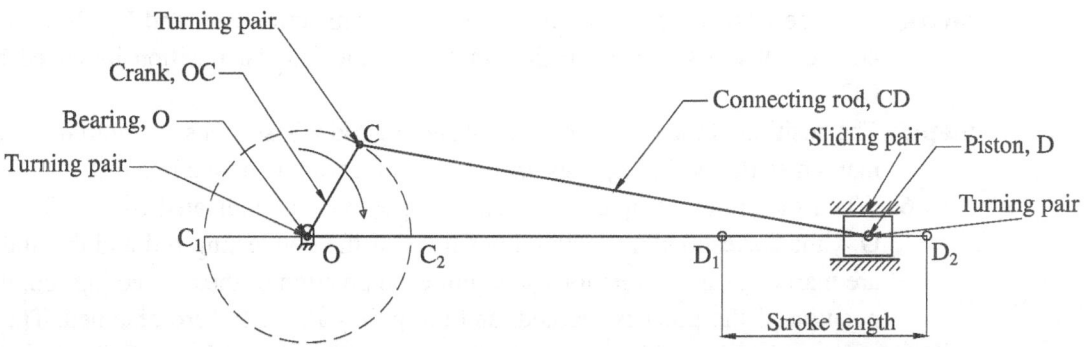

Fig. 2.43   Single slider crank mechanism

2.8.3 Problems on Single Slider Crank Mechanism

Example 2.33

 In a simple slider crank mechanism, the length of the connecting rod CD is 1000 mm, and the crank OC is 200 mm long. The slider D moves along a straight line passing through O. Draw the locus of a point P located 400 mm from point C along the connecting rod for one complete revolution of the crank OC.

> **PRINCIPLE**
> When the crank occupies an angular position, the corresponding positions of the connecting rod and slider are identified. When the position of any point (say P) on the connecting rod is marked and identified for various angular positions of the crank, the locus of the point on the connecting rod can be drawn by simply joining these instantaneous positions of the point P by a smooth curve.

PROCEDURE (Refer Fig. 2.44)

Step 1: Draw a circle with O as centre and radius equal to the length of the crank (200 mm) and divide it into 12 equal parts and mark them as 1, 2, ..., 12, which indicate the various angular positions of the crank as it rotates.

Step 2: The connecting rod and the slider position are identified and marked for one angular position of the crank. For clarity, the fifth position of the crank is indicated as C, and the position of the point P whose locus is to be drawn is marked at that instantaneous position of the connecting rod.

Step 3: When the crank occupies the angular position indicated by 5, the end of the crank is marked as C. The connecting rod position corresponding to this is marked by drawing an arc with C as centre, and radius equal to the length of the connecting rod (1000 mm) to cut the horizontal line passing through O at D_5.

Step 4: The line CD_5 indicates the position of the connecting rod, and D_5 shows the position of the slider when the crank is at the angular position indicated by OC.

Step 5: The position of the point P on the connecting rod whose locus is required is then marked as P_5, by setting a distance of 400 mm from C on the line CD_5.

Step 6: When the crank occupies other angular positions as indicated by O_1, O_2, ..., O_{12}, the corresponding inclined positions of the connecting rod and the slider are marked as in the previous step. For each position of the connecting rod, the position of the point is marked, and the points P_1 ... P_{12} are obtained. These points, when joined by a smooth curve, give the locus of the point P, when the crank completes one revolution. ▲

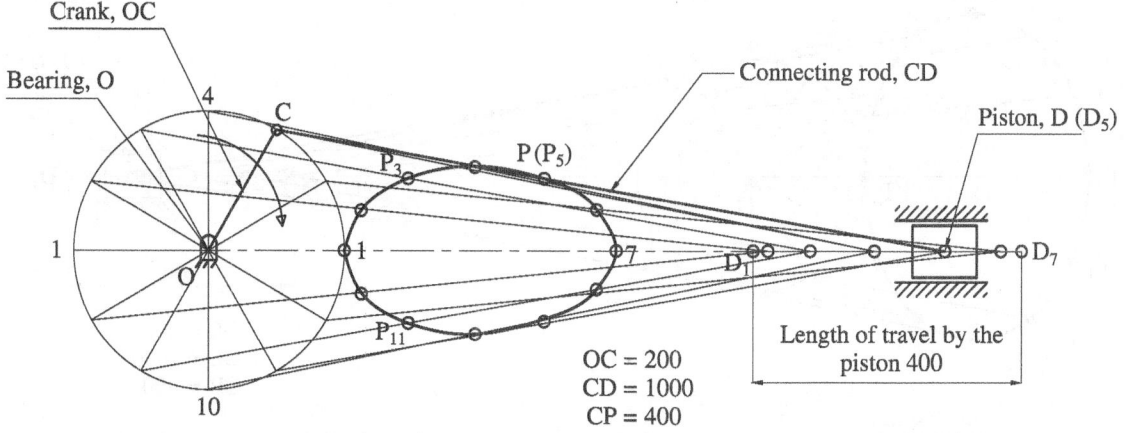

Crank, OC

Bearing, O

Connecting rod, CD

Piston, D (D$_5$)

OC = 200
CD = 1000
CP = 400

Length of travel by the piston 400

Note: All dimensions are 'mm'

Fig. 2.44 Locus of point on connecting rod of single slider crank mechanism

Example 2.34

In a simple slider crank mechanism, the length of the connecting rod CD is 1000 mm and the crank OC is 200 mm long. The slider D moves along a straight line passing through O. Draw the locus of a point P located 400 mm from point C along the left extension of the connecting rod for one complete revolution of the crank OC.

PRINCIPLE

The procedure for obtaining the locus of a point on the extended portion of the connecting rod is in similar lines to that of a point on the connecting rod. The position of the point P is marked on the extension of each instantaneous positions of the connecting rod and joined by a curve.

PROCEDURE (Refer Fig. 2.45)

Step 1: Draw the various angular positions of the crank and obtain the instantaneous positions of the connecting rod and the slider as in the previous example.

Step 2: Locate the point P on the connecting rod at a distance of 400 mm from C along its left extension.

Step 3: Obtain the various positions of P on the instantaneous connecting rod positions towards their left extensions, and join them by a smooth curve.

Step 4: As the point on the connecting rod is on the extended portion, the locus is also seen at outside.

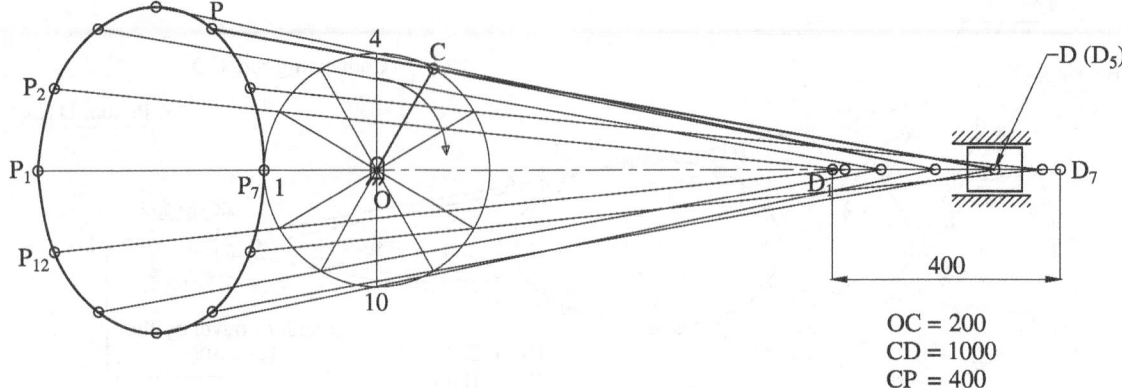

OC = 200
CD = 1000
CP = 400

400

Fig. 2.45 Locus of point on extended connecting rod of single slider crank mechanism

Example 2.35

The end C of a link CD rotates about O_1, whereas the end D slides along a straight line. The crank O_2F oscillates about the point O_2. Draw the locus of the mid-point M of the connecting link EF for one revolution of the crank O_1C. The following are the dimensions of various links of the mechanism: $O_1C = 500$ mm; CD = 1700 mm; EF = 850 mm; and $O_2F = 1300$ mm.

PRINCIPLE

The procedure for obtaining the locus of a point on a link connected to a connecting rod is similar to that of obtaining a point on the connecting rod. Initially, the instantaneous positions of the attachment point of the connecting rod and the connecting link are obtained for one complete revolution of the crank. The various positions of the connecting link are obtained from the instantaneous positions of the attachment point by duly setting the corresponding lengths of connecting link and its attached link. The instantaneous positions of the point whose locus is to be drawn are then marked on the respective connecting link positions, and joined by a smooth curve during one revolution of the driver crank.

PROCEDURE (Refer Fig. 2.46)

Step 1: Draw the various angular positions of the crank OC and obtain the instantaneous positions of the connecting rod and the slider as in the previous examples. For clarity, the fifth position of the crank is indicated as C.

Step 2: Locate the point E on the connecting rod at a distance of 1000 mm from C.

Step 3: Mark O_2 at distance of 2500 mm from O_1 and 700 mm below the axis line of the slider. With O_2 as centre, draw an arc for a length 1300 mm. From the point

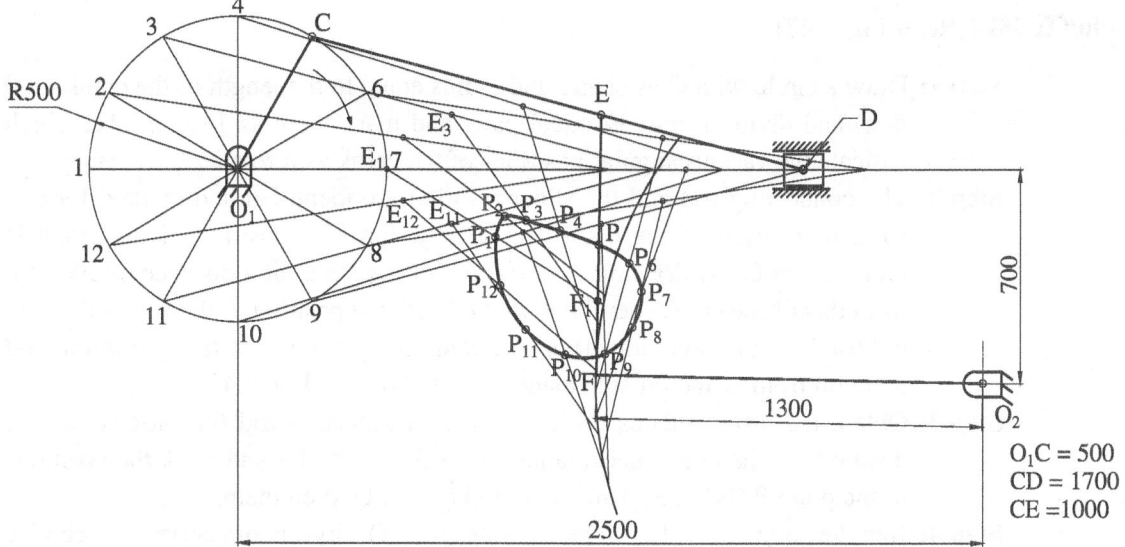

Fig. 2.46 Locus of point on attached link of slider crank mechanism

E on the connecting rod CD, draw an arc for a length of 850 mm to intersect the previous drawn arc at the point F.

Step 4: Obtain the various positions of E (E_1, E_2, ..., E_{12}) on the instantaneous connecting rod positions CD.

Step 5: From each instantaneous position of E (E_1 ... E_{12}), set an arc of length 850 mm to intersect with an arc of radius 1300 mm from O_2, and obtain the instantaneous positions of the point F (F_1 to F_{12}). Join the lines connecting E_1F_1 ... $E_{12}F_{12}$.

Step 6: Mark the various positions of the mid-point P on each instantaneous position of the lines E_1F_1 ... $E_{12}F_{12}$ and name them as P_1 ... P_{12}.

Step 7: Join the points P_1 ... P_{12} by a smooth curve that gives the locus of the point of interest in this example.

Example 2.36

In the offset slider crank mechanism, the vertical distance between the centre of rotation of the crank and the slider is 200 mm. The length of the connecting rod CD is 1000 mm, and the crank OC is 200 mm long. The slider D moves along a straight line. Draw the locus of a point P located 400 mm from point C along the connecting rod and a point Q located 500 mm from point C along the left extension of the connecting rod for one complete revolution of the crank OC.

> **PRINCIPLE**
>
> The same principle of obtaining the locus of a point on a connecting rod of a single slider crank mechanism explained in earlier problems can be applied to this case as well. The axis of the slider is not in line with the centre of the crank but is below it.

PROCEDURE   (Refer Fig. 2.47)

Step 1: Draw a circle with O as centre and radius equal to the length of the crank (200 mm) and divide it into 12 equal parts and mark them as 1, 2, ..., 12, which indicate the various angular positions of the crank as it rotates.

Step 2: The connecting rod and the slider position are identified and marked for one angular position of the crank (for clarity, the fifth position of the crank is indicated as C) by drawing the axis line of the slider at a distance of 200 mm from that of the crank rotation point O. Mark the position of the points P and Q (whose loci are to be drawn) on the connecting rod at a distance 400 mm and 500 mm from C and on either side of it as shown in Fig. 2.47.

Step 3: Obtain the instantaneous positions of the connecting rod for various angular positions of the crank as explained in earlier examples and mark the positions of the point P (P_1 ... P_{12}) and that of Q (Q_1 ... Q_{12}) on them.

Step 4: Join the points P_1 ... P_{12} and the points Q_1 ... Q_{12} by smooth curves, which give the locus of the points P and Q, respectively.

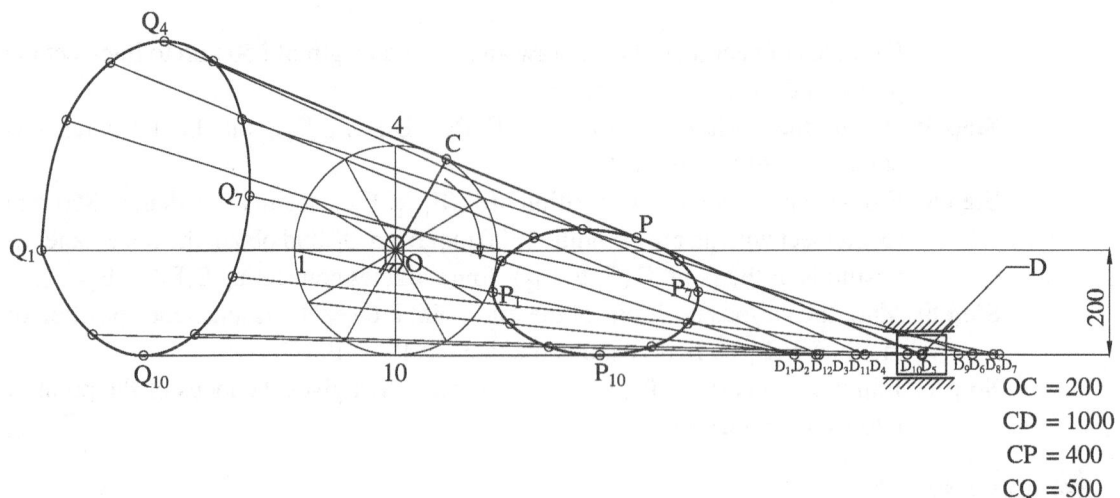

$$OC = 200$$
$$CD = 1000$$
$$CP = 400$$
$$CQ = 500$$

Fig. 2.47   Locus of points on connecting rod of offset slider crank mechanism

2.8.4 Problems on Four-bar Mechanism

Example 2.37

In a four-bar mechanism, the two cranks O_1C and O_2D of equal length of 500 mm rotate in opposite directions and are connected by the link CD of length 2000 mm. The centres O_1 and O_2 are on the same horizontal straight line and are 2000 mm apart. The point P is the left extension of link CD by 400 mm from point C. Another point Q is on the link CD by a distance of 500 mm from point D. Draw the locus of points P and Q for one complete revolution of the cranks.

PROCEDURE (Refer Fig. 2.48)

Step 1: Draw a circle with O_1 as centre and radius equal to 500 mm, divide it into 12 equal parts, and mark them as 1, 2, ..., 12. The points indicate the various angular positions of the crank as it rotates. For clarity, the fifth position of the crank is indicated as C.

Step 2: Mark the centre of the other crank as O_2 at a horizontal distance of 2000 mm from O_1. The position of the crank O_2D corresponding to the position of the crank OC can be obtained at the intersection of two arcs of 2000 mm and 500 mm from C and O_2, respectively. Thus, O_1CDO_2 describes the four-bar mechanism when the crank O_1C is at the angular position 5 and CD is the coupler link of the mechanism.

Step 3: Locate two points, P and Q (whose loci are to be drawn) on the link CD at a distance of 400 mm and 1500 mm, respectively, from C on outside and inside the link CD.

Step 4: When the crank position C occupies different angular positions such as 1, 2, etc., obtain the corresponding positions of D at the intersection of two arcs of length 2000 mm and 500 mm from each angular position and O_2. The instantaneous locations of the point D corresponding to the various crank positions indicated by 1, 2, ..., 12 are obtained as 1′, 2′, ..., 12′.

Step 5: Join the points 1 and 1′, 2 and 2′, ..., 12 and 12′, which indicate the instantaneous positions of the coupling link CD during one revolution of the crank.

Step 6: Mark the instantaneous positions of the outer point P on each of these lines as P_1 ... P_{12}. Join them by a smooth curve which gives the locus of the point P, located on the extended link.

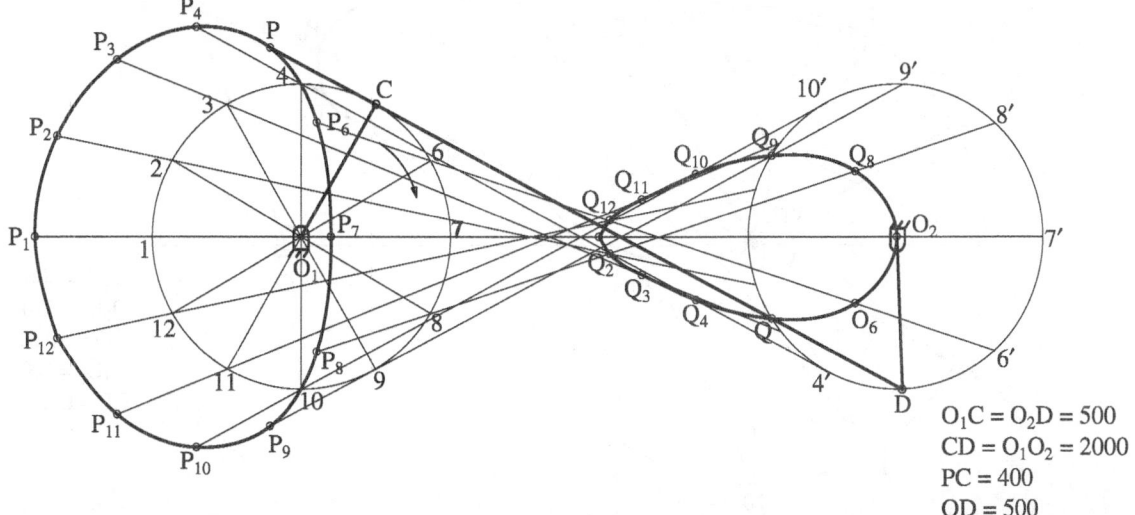

$$O_1C = O_2D = 500$$
$$CD = O_1O_2 = 2000$$
$$PC = 400$$
$$QD = 500$$

Fig. 2.48 Locus of points on connecting link of four-bar mechanism having their cranks of equal length

Step 7: Similarly, mark the instantaneous positions of the inner point Q on each of these lines as Q_1 ... Q_{12}. Join them by a smooth curve which gives the locus of the point Q located on the link CD and inside it. ▲

Example 2.38 ▲

@ The two cranks O_1C and O_2D of a four-bar mechanism are of length 600 mm and 850 mm, respectively. They are connected by a link CD of length 1250 mm. The centres of rotation O_1 and O_2 lie on the same straight line and are 1200 mm apart. The crank O_1C rotates about O_1, whereas the crank O_2D oscillates about O_2. Draw the locus of the mid-point P for one complete revolution of the crank O_1C.

PROCEDURE (Refer Fig. 2.49)

Step 1: Draw a circle with O_1 as centre, radius equal to 600 mm and divide it into 12 equal parts, and mark them as 1, 2, ..., 12. The points indicate the various angular positions of the crank as it rotates. For clarity, the fifth position of the crank is indicated as C.

Step 2: Mark the centre of the other crank as O_2 at a horizontal distance of 1200 mm, from O_1. The position of the crank O_2D corresponding to the position of the crank OC can be obtained at the intersection of two arcs of 1250 mm and 850 mm from C and O_2, respectively. Thus, O_1CDO_2 describes the four-bar mechanism when the crank O_1C is at the angular position 5, and CD is the coupler link of the mechanism.

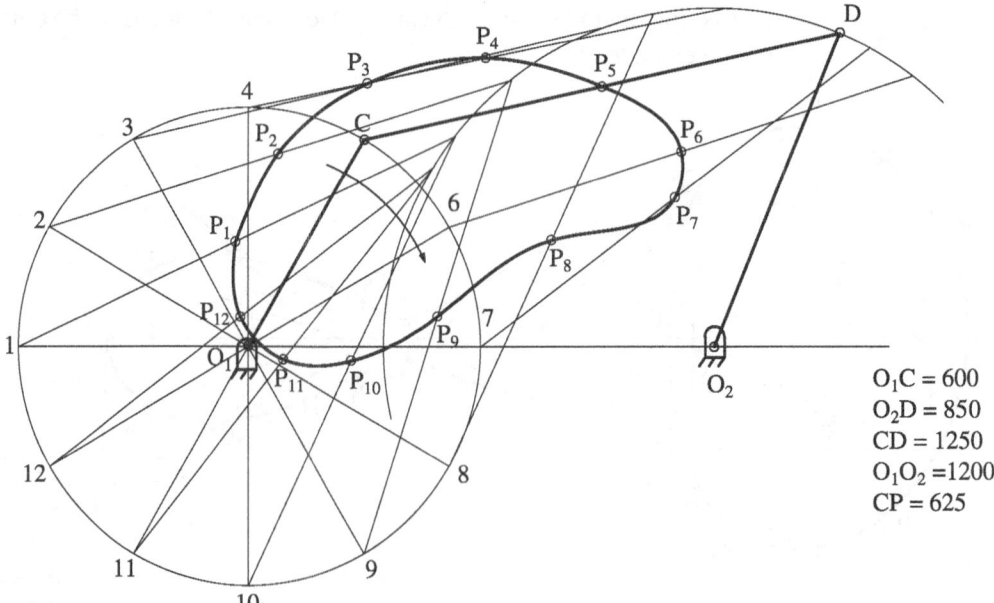

$O_1C = 600$
$O_2D = 850$
$CD = 1250$
$O_1O_2 = 1200$
$CP = 625$

Fig. 2.49 Locus of mid-point of connecting link of four bar mechanism having their cranks of different lengths

Step 3: Locate the point P (whose locus is to be drawn) on the link CD, at a distance of 625 mm from C, as shown in Fig. 2.49.

Step 4: When the crank position C occupies different angular positions such as 1, 2, etc., obtain the corresponding positions of D at the intersection of two arcs of length 1250 mm and 850 mm from each angular position and O_2. Mark the instantaneous positions of the point P as P_1 ... P_{12}, and join them by a smooth curve which gives the locus of the point P located on the coupler link CD.   ▲

2.9   ENGINEERING APPLICATIONS

Conic Sections

Circle   A circle is a very well-known commonly used mathematical curve in all domestic and engineering applications. It is one of the very basic curves known from early childhood. Circular objects in the form of rings, bangles, thin washers, thin metallic plates, compact discs, lids, domestic kitchen plates, 360° protractors, and so on, are seen in common day-to-day applications. Circular objects in the form of bicycle wheels, wheels for carts and carriages, automobile wheels, flywheels and rims for engines, and giant wheels are applications that attract us.

Ellipse   The ellipse is a common shape adopted in architecture and civil engineering. Constructions such as bridges, arches, and heritage buildings are based on the ellipse as a means to improve the aesthetics. Whenever cylindrical pipes are connected to the inclined surfaces, the resulting shape is an elliptical interface. The flanges of the pipes of automobile stuffing box covers and the lids of valve assemblies are made in elliptical designs. Elliptical gears are used in textile and printing machineries for variable speed ratios in one revolution. The elliptical trammel is one of the oldest mechanisms used to trace the ellipse. The movement of planets and satellites in elliptical orbits is a well-known scientific fact. Elliptical holes are provided in machinery base frames for adjustments in their alignments. The spokes or arms in the wheels are provided with the elliptical cross-sections. The ellipse is used as dished ends in boilers, tanks, and pressure vessels.

Parabola   Parabolic curves are extensively used in engineering and scientific applications. Light reflectors, mirrors, search lights, head lamps of motor vehicles, sound reflectors, solar concentrators, and satellite and telecommunication dishes use the principle of using their focus as the concentration point to maximize the output phenomenon. In structural and civil engineering applications that are subject to heavy bending loads, parabolic shapes are assumed to offer uniform resistance as in wall brackets, anvils, and so on. The trajectories of missiles or objects that are thrown, such as stones, balls, and bullets, are shaped as a parabola. The path of the jet of water from orifices in vertical vessels is parabolic-curved.

Hyperbola   Hyperbolic shapes are often adopted in cooling towers, flower vases, wooden handles, decorative pieces, mirrors, and so on. The well-known Boyle's law for gas expansion and compression obeys the hyperbolic path (rectangular hyperbola) when represented in indicator diagrams of the engines.

Involute-shaped curves The involute shape is adopted in the design of gear tooth profiles, as it enables easy manufacture and smooth functioning. Gear cutters also use involute profiles. The centrifugal pump casings are often involute shape as it helps to increase the pressure discharge and reduce the loss of flow due to impact.

Cycloidal curves Cycloidal curves find extensive usage in mechanical engineering applications. Like involute shapes, cycloidal shapes are also adopted in gear tooth profiles. They are also used in racks, as they are easier to cut and are the most suited, particularly, for small-sized gears. These gears are used in many precision equipments and mechanisms. Cam profiles using the cycloidal shape offer gradual and smooth lifting and lowering of followers. The path traced by a gear on another fixed gear, which is external or internal to it, is widely studied in gear assemblies. These are known as epicyclic gear trains. Large reduction ratios and compact-sized arrangements are the advantages of such gear trains. The path traced by a point on the circumference of a bicycle wheel, a coin, or rolling object on a flat surface is cycloidal in nature. The path traced by a circus man riding a motor cycle in a metallic cage in the form of a globe and the path traced by roller coasters in the amusement parks are some of the standing examples of the cycloidal curves.

Loci of points in mechanisms In mechanisms, as coupler links are used to connect the input and the output links to obtain an uninterrupted output motion for a given rotary input, the continuous movement of the coupler has to be ensured. Every point on the coupler link has to satisfy the motion requirements. In order to ascertain this technically, the locus of the desirable locations of the coupler should yield a closed curve and, hence, the locus generation draws the attention of the engineers. These are also known as the coupler curves. The design of the mechanism should yield coupler curves that are continuous to avoid the redundancy of that location. The locus generation of any point in the coupling link of the four-bar mechanism and any point in the connecting rod of a single-cylinder mechanism are some of the fundamental requirements in the mechanism design. This will also help the assembly of additional linkages in the mechanism.

RECAPITULATION

Basic Geometrical Constructions

- Regular polygon shapes are being used as cross-sections of many engineering objects. A polygon of n sides can be constructed with the length of one side and the internal angles given by $\theta = 180° - (360°/n)$. Setting the length of one side and forming an isosceles triangle with semi-internal angles, alternatively, the centre of the polygon can be located. With that as centre and the length of one identical side as radius, a circle can be drawn and the polygon can be obtained within.

Conic Sections

- When a solid cone is cut at different orientations, the cut sections obtained are known as conics. They are circle, ellipse, parabola, and hyperbola.
- Setting the orientation parallel or inclined to the base, the closed curves—circle and ellipse—are obtained, respectively.
- Setting the cutting planes parallel to the extreme generator are open-ended curves—parabola and hyperbola may be generated.
- These curves have common nomenclature terms such as eccentricity, focus or foci, vertex or vertices, axis, and directrix.
- The eccentricity, which is one of the distinguishable parameters for different conics, is defined as the ratio of the distance of a point on the curve from the focus to that of the directrix.
- The ellipse, parabola, and hyperbola will have the eccentricity less than, equal to, and greater than one, respectively. All these curves have a common procedure of construction when eccentricity is given.
- Ellipse is a closed curve with the following unique properties:
  - Any point on the ellipse moves in such a way that the sum of its distance from the two foci is always a constant and equal to the length of the major axis.
  - As the ellipse is symmetrical about the major and minor axis, the distance of the end point of the minor axis from the focus is equal to the length of semi-major axis.
  - The construction methods for ellipse make use of these properties. Some of the common methods are pin and thread method, arc of circle, and concentric circles methods.
  - As ellipse is symmetrical about the major and the minor axis, the construction of one quadrant portion can be done using rectangle or oblong methods.
  - An elliptical curve can be considered as composed of four segments or four arcs, whose individual centres can be located. This is a quick method for the construction of ellipse.
- Parabola is an open curve, but is referred with a base and a height or the transverse axis.

- Two methods—rectangle method and parallelogram method—use similar constructional features mentioned for the ellipse.
- When the base of the parabola and the angles subtended at the base are given, the tangent method, which involves a series of tangent lines constructed in both the halves, can be used.
- Hyperbola is not a closed curve and has the following unique properties:
  - As the hyperbola has two branches on its left and right halves with two foci, the locus of any point on it can be described as the difference between its distances from the foci and is always a constant.
  - The hyperbola has got the unique presence of asymptotes, which are the two intersecting lines obtained in between the left and the right halves. These asymptotes when extended approach nearer and nearer to the hyperbola curves and will become tangential to them at infinity. The asymptotes and the angle between them are also used for the construction of hyperbola curves.
  - When the angle between the asymptotes is 90°, then the hyperbola is known as rectangular or equilateral hyperbola.

Involute Sections

- The involute is a plane curve traced out by the end of a thread unwound from a line, a polygon, or a circle when the thread is kept taut in the unwinding process.
  - If one of the ends of the thread is opened to unwind, the portion of the thread unwound falls in alignment with the existing side in the case of a polygon or becomes tangential to the circle at that point. The length of the portion unwound is equal to the circumferential length involved in it. This principle is used for the construction of the involute.

Cycloidal Curves

- The cycloidal curves are the family of curves traced by a point on the circumference of a circle when the circle rolls without slipping. When the circle rolls on a straight line, the locus of a point on it is known as a cycloid. When the circle rolls on another circle and outside it, it is known as epicycloid, and when it rolls inside, it is known as hypocycloid.

- The rolling circle is known as the generating circle. The straight line or the circle on which it rolls is known as the base, the directing line, or the circle as the case may be.

Loci of Points

- The path traced by the instantaneous positions of any point in a moving member is known as locus. This is very important to ensure the proper functioning of that member in any machine or in any application.

- This has been developed from the fundamental principles of tracing the locus of a point with respect to another point, a line, and a circle when the latter is stationary.

- The procedure of drawing the locus of a point in the connecting rod of a slider crank mechanism has been addressed when the point lies within the connecting rod or on its extension. A special case where the slider axis is offset is also addressed.

- Similarly, the procedure of drawing the locus of a point in the coupler link of a four mechanism has been explained when the point lies within the coupler link or on its extension.

MULTIPLE-CHOICE QUESTIONS

2.1 The sum of internal angles of a regular polygon of n sides is equal to
   (a) $(2n-4) \times 90°$ (c) $(n-4) \times 90°$
   (b) $(2n-4) \times 180°$ (d) $360°$

2.2 A regular polygon of n sides subtends at the centre a total angle of
   (a) $180°$ (c) $n \times 90°$
   (b) $n \times 60°$ (d) $360°$

2.3 The curve traced by the locus of a point moving in a plane such that the ratio of its distances from fixed point and fixed straight line is equal to one is known as
   (a) ellipse   (c) circle
   (b) parabola   (d) hyperbola

2.4 When the eccentricity of a conic curve is greater than 1, the curve known as
   (a) parabola   (c) ellipse
   (b) hyperbola   (d) circle

2.5 The locus of a point in a conic moves in such a way that the sum of its distances from the foci is a constant. The name of the curve is
   (a) circle   (c) ellipse
   (b) parabola   (d) hyperbola

2.6 In an ellipse, the sum of the distance of any point from the two foci is equal to
   (a) minor axis
   (b) major axis
   (c) transverse axis
   (d) difference between the major and the minor axes

2.7 The length of the transverse axis in a hyperbola is equal to
   (a) the sum of the distances between a point on a curve to the foci
   (b) the product of distances between a point on a curve to the foci
   (c) the difference of the distances between a point on a curve to the foci
   (d) the ratio of the distances between a point on a curve to the foci

2.8 In a rectangular hyperbola, the angle between the two asymptotes is equal to
   (a) $180°$ (c) $90°$
   (b) $60°$ (d) $45°$

2.9 The curve traced out by the end of a thread unwound from a circle is known as
   (a) cycloid   (c) hypocycloid
   (b) epicycloid   (d) involute

2.10 When a circle rolls inside another circle, the curve traced by a point on the rolling circle is known as
   (a) cycloid   (c) hypocycloid
   (b) epicycloid   (d) involute

2.11 The locus of a point maintaining a constant distance from a fixed point is known as

(a) an ellipse   (c) a parabola

(b) a circle   (d) a straight line

2.12 The locus of a point maintaining a constant distance from a straight line is

(a) parallel to it

(b) perpendicular to it

(c) a perpendicular bisector

(d) inclined at a constant angle

2.13 The locus of a point maintaining a constant radial distance from a fixed arc is

(a) a parabola   (c) an ellipse

(b) a concentric arc (d) a circle

2.14 The locus of a point equidistant from two fixed points is

(a) parallel to the line joining the two fixed points

(b) inclined to the line joining the two fixed points

(c) perpendicular to the line joining the two fixed points

(d) perpendicular bisector to the line joining the two fixed points

2.15 The locus of a point moving between a fixed point and a fixed straight line at constant ratio is

(a) a conic   (c) a cycloid

(b) an involute   (d) a polygon

WORK PRACTICE LEVEL – 1

Problems on Geometrical Constructions

2.1 Divide a straight line of length 110 mm into nine equal parts.

2.2 Divide a straight line of length 130 mm into proportionate parts of one-half, one-third, one-fourth, and so on.

2.3 Draw an arc of radius 40 mm such that it is tangential to two lines, each of 80 mm long, when these lines are (a) inclined at 70°, (b) inclined at 100°, and (c) perpendicular to each other.

2.4 Bisect an angle of 45°.

2.5 Divide an angle of 75° into 12 equal divisions.

2.6 Construct a regular (a) pentagon, (b) hexagon, and (c) octagon, each of 40 mm side, using circumscribing circle method.

Problems on Conics

2.7 The distance between two fixed points is 110 mm. A point is moved in such a way that the sum of its distances from the two fixed points is always constant and is equal to 160 mm. Trace the locus of the point and name the curve.

2.8 Construct an ellipse when the distance of the focus from the directrix is equal to 60 mm and the eccentricity is 3/5. Draw a tangent and normal at any point on the curve.

2.9 Two points C and D are 110 mm apart. The third point E is 80 mm from C and 65 mm from D. Draw an ellipse passing through the points C, D, and E and measure the focal distance.

2.10 The major axis of an ellipse is 130 mm long and minor axis is 90 mm long. Draw one half of the ellipse by using rectangle method and the other half by using concentric circles method.

2.11 The distance of the focus from the directrix is 65 mm. Draw the locus of the point which moves such that its distance from the focus is equal to its distance from the directrix.

2.12 The vertex of a hyperbola is 70 mm from its focus. Draw two parts of the hyperbola if the eccentricity is 2.

2.13 The distance between the foci is 65 mm and the transverse axis is 25 mm. Draw the hyperbola.

2.14 The distance between the foci is 110 mm and the vertices are 20 mm from the foci. Draw the two parts of a hyperbola.

2.15 The angle between two straight lines OC and OD is equal to 80°. A point P is 45 mm from OC and 55 mm from OD. Draw a hyperbola through P, with OC and OD as asymptotes.

Simple Problems on Loci of Points

2.16 Draw the locus of a point which moves and maintains a constant distance of 20 mm from a fixed straight line AB of length 100 mm.

2.17 Draw the locus of a point which moves between two fixed straight lines which are inclined at 70° to each other.

2.18 Draw the locus of a point which is equidistant from a fixed circle of diameter 100 mm and a fixed horizontal straight line which is situated at 100 mm from the centre of the circle.

WORK PRACTICE LEVEL – 2

Problems on Involutes

2.1 A string is wound around the circumference of the regular hexagon of side 40 mm completely. Holding one end free, the string is unwound completely such that the string is always tightly stretched. Trace the path of the free end and name the curve.

2.2 An inelastic string of 200 mm is wound over a wheel of diameter 420 mm. Draw the locus of the free end of the string, when it is unwound by keeping the string taut.

2.3 Draw the curve traced by the ends of the straight line of 130 mm long which rolls without slipping on a semicircle of 90 mm diameter. One end of the line is touching the circle and the line is perpendicular to the vertical diameter of the semicircle in the beginning.

Problems on Cycloids

2.4 A circle of 60 mm diameter rolls on a straight line without slipping. Trace the locus of the point on the circumference if the circle rolls for one and quarter revolution. Name the curve. Draw a tangent and normal to the curve at a point 40 mm above the straight line and on the descending side of it.

2.5 A circle of 55 mm diameter rolls for the first three-fourth of the revolution on a horizontal line and next three-fourth revolution on a vertical line. Draw the curve traced by the point P which is initially lying at a point of contact of the horizontal line and the circle.

2.6 A circle of 50 mm diameter rolls outside on another circle of 110 mm diameter and completes one revolution. Draw to full scale the locus of a point on the circumference of the rolling circle and name the curve.

2.7 A circular wheel of 40 mm diameter rolls on another fixed wheel of 70 mm diameter and above it for one complete revolution of the rolling wheel. Draw the curve traced by a point on the circumference of the rolling wheel which is initially situated at the common contact between the wheels. Draw a tangent and normal to the curve at the point on it which is 70 mm from the centre of the fixed wheel.

2.8 The diameter of the rolling circle is 40 mm and the diameter of the base circle is 120 mm. Draw the hypocycloid traced by the rolling circle and add a tangent and normal at any point on the curve.

2.9 A circle of 50 mm diameter rolls without slipping on the inside of another circle of diameter 200 mm. Draw the path traced by the point on the periphery of the rolling circle which is situated diametrically opposite to the initial point of contact between the circles for one complete revolution in the clockwise direction. Name the curve.

WORK PRACTICE LEVEL – 3

Problems on Engineering Applications (Conics, Involutes, and Cycloids)

2.1 The distance between two straight parallel fencings of a plot is 120 m. An elliptical swimming pool is to be constructed inside the plot with major axis length of 90 m. Locate the foci and the two ends of the major axis by considering the two straight parallel fencings as directrices of the ellipse. Draw the complete ellipse using the concentric circles method and find its eccentricity (take suitable scale).

2.2 A plot is in the shape of parallelogram of 16 × 12 m, the angle between the sides being 55°. Inscribe an elliptical food court in it by taking suitable scale.

2.3 A particle 'A' moves such that the sum of its distances from two fixed points C and D, 100 mm apart, is always constant. When A is at equal distance from C and D, its distance from each one of them is measured to be 80 mm. Draw the locus of the particle.

2.4 A baseball is thrown from the top of a building terrace of 8 m height and the highest point of reach when it just crosses a coconut tree 15 m high. Draw the locus of the ball if the distance between the point of location of the building terrace and the coconut tree is 4 m (take suitable scale).

2.5 A toy rocket thrown up in the air reaches a maximum height of 50 m and travels a horizontal distance of 80 m. Trace the path of the rocket by choosing a suitable scale.

2.6 A mango in a tree which is located at 12 m above the ground is targeted by a man with the help of a stone from a position 20 m from the tree and 5 m above the ground. Draw the curve traced by the stone if the maximum position of the stone coincides with the position of mango. What is the eccentricity of this curve?

2.7 A suspension bridge has a span of 400 m and the cables dip at the centre of the span is 60 m. The cables form a parabola. Draw the shape of the cable and add a tangent to it at its dip.

2.8 The asymptotes of a hyperbola are inclined at 75° to each other. A point P on the curve is 35 mm and 45 mm from the asymptotes. Construct the curve and determine its eccentricity.

2.9 The diameter of a circle AB is 80 mm long. A string is tied tightly around the circumference of the semicircle from point A and ends at point B. The string is then unwound from point B by keeping it taut from the circle until it becomes tangent at A. Draw the curve traced by the moving end of the string.

2.10 A water fountain discharges water from the ground level at an inclination of 50° with respect to the horizontal. The water jet travels to the distance of 8 m horizontally from its discharge and then falls again on the ground. Draw the curve traced by the water jet and name the curve.

Problems on Engineering Applications (Loci of Points)

2.11 In a simple slider crank mechanism, the length of the connecting rod AB is 1200 mm and the crank OA is 250 mm long. The slider B moves along straight line passing through point O. Draw the locus of (a) the point P located 300 mm from point A on the connecting rod, (b) the point Q located at the mid of the connecting rod, and (c) the point R located 350 mm from the point A along the left extension of the connecting rod for one complete revolution of the crank OA.

2.12 Draw the locus of a point P, located 200 mm from the point F along the bottom extension of link EF. The dimensions of links are $O_1C = 400$ mm, $CD = 1600$ mm, and $CE = 900$ mm. The vertical distance between D and O_2 is equal to 650 mm. The horizontal distance between O_1 and O_2 is equal to 2400 mm and O_2F is 1200 mm (refer Fig. 2.46).

2.13 In the offset slider crank mechanism, the vertical distance between the centre of rotation of the crank and the slider is 300 mm. The length of the connecting rod is 1200 mm and the length of the crank OA is 300 mm. The slider B moves along a straight line. Draw the locus of (a) a point P located 350 mm from A along the connecting rod, (b) mid-point M of the connecting rod, and (c) a point Q located 400 mm from point A along the left extension of the connecting rod for one complete revolution of the crank OA.

2.14 Two equal cranks O_1A and O_2B as shown in Fig. 2.50 rotate in opposite direction about O_1 and O_2. They are connected by a coupler AB. Plot the locus of end D of an attachment link CD which is at 90° to the coupler AB and at its mid-point C for one complete revolution of the crank. $O_1A = O_2B$ = 250 mm; O_1O_2 = 1000 mm; AB = 950 mm; CD = 175 mm.

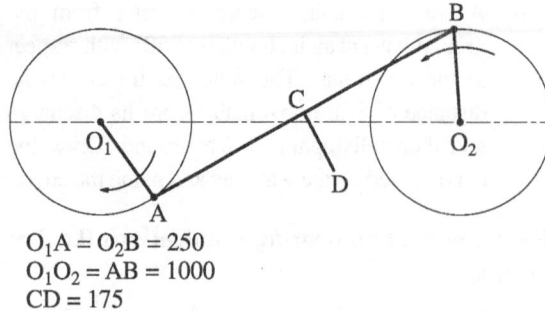

$O_1A = O_2B = 250$
$O_1O_2 = AB = 1000$
$CD = 175$

Fig. 2.50

2.15 Two cranks O_1C and O_2C are connected by a coupler CD. The end C rotates on the circumference of the circle with centre O_1 and D oscillates on an arc with O_2 as centre. Plot the locus of (a) mid-point M of the coupler CD and (b) a point A, located on left extension of 200 mm from point C of the coupler CD, for one complete revolution of the crank O_1C. $O_1C = 400$ mm,

$O_2D = 900$ mm, $CD = 1000$ mm, and $O_1O_2 = 1300$ mm (refer Fig. 2.51).

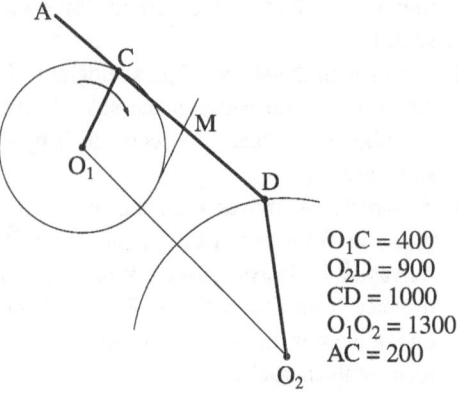

$O_1C = 400$
$O_2D = 900$
$CD = 1000$
$O_1O_2 = 1300$
$AC = 200$

Fig. 2.51

2.16 The link O_1A, 200 mm long rotates about a hinge O_1 in clockwise direction. An insect situated at 40 mm from the hinge moves towards the end A with constant velocity. Draw the locus of the insect for one-fourth revolution of the link O_1A.

Answers for Multiple-choice Questions					
2.1 (a)	2.2 (d)	2.3 (b)	2.4 (b)	2.5 (c)	2.6 (b)
2.7 (c)	2.8 (c)	2.9 (d)	2.10 (c)	2.11 (b)	2.12 (a)
2.13 (b)	2.14 (d)	2.15 (a)			

Scales

3

OBJECTIVES

This chapter will help the reader to understand the following:

- Need for scaling the dimension of an engineering object and the different types of scaling methods
- Construction of customized scales of different types based on the accuracy of the measurements needed

3.1 INTRODUCTION

Actual sizes are necessary to manufacture any object, and these sizes are to be indicated in the drawings that are drawn either manually using drawing instruments or using commercially available Computer-aided Drafting packages like AutoCad. The sizes of the drawing boards and drawing sheets were discussed in Chapter 1. It was mentioned that the maximum size of the drawing sheet used in college practice is A2, whose length and width dimensions are 594 mm and 420 mm, respectively. The working or the drawing space is decided after setting the margins or borders and the title block area. In practice, the objects have varied sizes and dimensions, which are often difficult to fit within the available drawing space. Moreover, in order to describe the object completely, its dimensions along its three directions are to be represented in the drawing, which calls for multiple views of the same object. If provision is to be made for drawing these views to the actual sizes of the objects using the drawing sheet, then the drawings can seldom be accommodated in the drawing sheet. Hence, it is important to adopt a suitable proportion between the sizes used in the drawing sheet and the sizes of the original object, and this proportion or the ratio is known as the *scale*. Therefore, the scale of a drawing is defined as the ratio of the dimension of an object as represented in the drawing sheet to the actual dimension of the object itself.

The drawings drawn of the same size of the object are called *full-scale drawings* or *full-sized drawings*. Accommodating the actual dimensions of the large-sized objects such as buildings, machines, automobiles, ships, aeroplanes, and geographical maps on the drawing sheet becomes difficult, and such objects require that all the dimensions are to be reduced proportionately to set them on the drawing sheet within the available space, by choosing a suitable reduced scale.

On the contrary, if tiny objects such as electronic chips, microsensors, miniaturized parts used in mobile phones and wrist watches, and nano-sized objects are drawn to their original sizes in the drawing sheets, then such drawings or pictures will not only be difficult to draw but will also fail to convey the features or details involved in the components. The drawings for such objects require that all the dimensions are to be increased proportionately and set on the drawing sheet, so that their features are explained clearly for manufacturing purpose. The scale adopted for this purpose is called an enlarged scale where the sizes of the objects conform comfortably to the drawing space available in the drawing sheet.

3.2 SCALING METHODOLOGIES

The scale of a drawing can be mathematically represented by *representative fraction* (RF), which is defined as the ratio of the drawing size to the respective actual size of an object, both expressed in the same units.

$$\text{Thus, RF} = \frac{\text{Drawing size}}{\text{Actual size}}$$

3.2.1 Full-size Scale

In this scale, as the sizes of an object represented on the drawing sheet are the same as the respective actual dimensions of the object, the RF value is equal to one. The normally available scales correspond to this ratio only.

$$\text{For this scaling, RF} = 1$$

For example, if an object is 5 cm in size and is represented on the drawing sheet as a line with length 5 cm, then the RF value is expressed as follows:

$$\text{RF} = \frac{\text{Drawing size}}{\text{Actual size}} = \frac{5 \text{ cm}}{5 \text{ cm}} = \frac{1}{1}$$

Hence, the scale is represented in the ratio of 1:1 in the drawing sheet.

3.2.2 Reduced Scale

In this scale, the sizes of an object represented on the drawing sheet are less than the respective actual dimensions of the object, such that RF < 1.

For example, if a tunnel of length 1 km is represented in a map or in the drawing sheet as a 2-cm line, then the drawing size is 2 cm and the object size is 1 km. After setting the object size to the units adopted in the drawing sheet, the object size becomes 1,00,000 cm.

$$\text{Therefore, RF} = \frac{\text{Drawing size}}{\text{Actual size}} = \frac{2 \text{ cm}}{1 \text{ km}} = \frac{2 \text{ cm}}{1 \times 1000 \times 100 \text{ cm}} = \frac{1}{50,000}$$

Hence, the scale is represented as the ratio of 1:50,000 in the drawing sheet. This means that one unit of the drawing sheet refers to 50,000 units of the object.

3.2.3 Enlarged Scale

In this scale, the sizes of an object represented on the drawing sheet are greater than the respective actual dimensions of the object, such that RF > 1.

For example, if an object is of length 5 μm and is referred in the drawing sheet as a 5 cm long line, then the drawing size is 5 cm and the object size is 5 μm. After setting the object size to the units adopted in the drawing sheet, the object size becomes 5×10^{-4} cm.

$$\text{Therefore, RF} = \frac{\text{Drawing size}}{\text{Actual size}} = \frac{5 \text{ cm}}{5 \text{ μm}} = \frac{5 \text{ cm}}{5 \times 10^{-6} \times 100 \text{ cm}} = \frac{10,000}{1}$$

Hence, the scale is represented as the ratio of 10,000:1 in the drawing sheet. This means that one unit of the drawing sheet refers to 1/10,000 unit of the object.

3.3 CUSTOMIZED SCALES

3.3.1 Need

A set of standard scales explained in Table 1.2 of Chapter 1 ranging from M1 (consisting of full-size scale of 1:1 and a reduced scale of 1:2 on either side) to M8 (consisting of reduced scales of 1:1000 and 1:2000 on either side), are used to mark the measurements in engineering drawings. These scales adopt standard RF values as marked on them, which are universally followed in many engineering applications. Generally, these scales are of 30 cm in length and are made of cardboard material. However, if the dimensions are to be scaled and expressed in a manner other than these standard RF values and are desired to be measured with two- or three-digit decimal place accuracies, then customized scales can be constructed and used as explained in Sections 3.3.2 and 3.3.3. Occasionally, the same scale can be constructed to measure two different units with the specific RF value, and the customized scales can be used for this purpose also. The customized scales are constructed on the drawing sheet.

3.3.2 Parameters for Construction

To construct any customized scale, the following parameters are to be considered:

1. Representative fraction (RF)

2. The required units in which the measurements are to be made
3. The maximum length that is to be measured with the proposed scale

3.3.3 General Constructional Procedure

In this section, the basic and common procedure used for the construction of all types of scales is discussed with the requisite changes in the corresponding categories.

Calculation of the Length of the Scale

In the first step, the length of the scale is calculated as follows:

Length of the scale = RF × Maximum measurement to be made using the scale

> **NOTE** *In the drawing sheet, the length of the scale is generally represented in millimetre or centimetre as they can be conveniently measured and marked with any normal scale available. It is advisable to convert the quantity to be expressed to the same units adopted in the drawing sheet. When the length of the scale obtained is in decimals, it is rounded off to the next whole number value.*

Setting the Scale Length and Marking its Subdivisions

1. Set the length of the scale, hereinafter called the main scale, and divide it into convenient number of equal divisions.
2. The first unit of the main scale can be further subdivided into a number of equal divisions as indicated by the relationship of the main scale and its subunits. The main scale division points are marked towards the right and that of subdivided scales are marked towards the left as shown in Fig. 3.1.

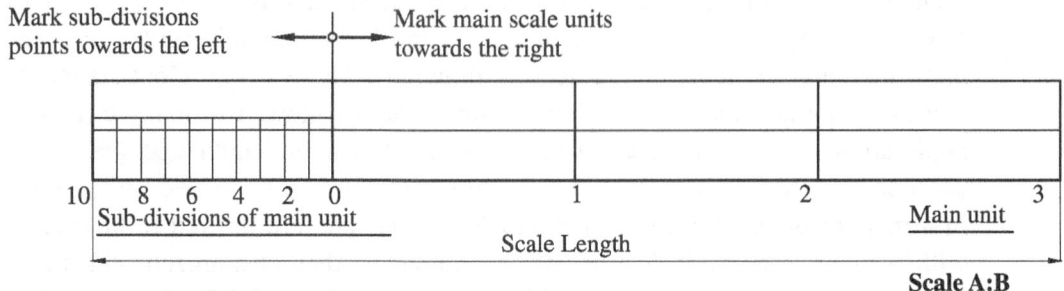

Fig. 3.1 Features of a scale

3. The subdivision points will measure the dimension of the main scale to the first decimal or first unit accuracy.
4. If the scale is required to measure up to the second decimal accuracy, then the subdivision points marked in the main scale are further subdivided into its smaller parts as indicated by their units.

5. The units of the main scale and their subdivisions can correspond to the metric or imperial system of units. The reader is advised to become conversant with the corresponding system of units, their subunits, and their interrelationships for making the required number of division and subdivision points.

6. Instead of marking the division points on a straight line, a rectangle of small width (about 3 mm) is constructed to represent the scale. Furthermore, the alternate divisions and subdivisions are darkened for better clarity and appeal.

3.4 TYPES OF SCALES

This section deals with the following three important types of scales:

1. Plain scale
2. Diagonal scale
3. Vernier scale

These scales are often used in various engineering applications.

3.4.1 Plain Scales

A plain scale can be used to measure only *two consecutive units* such as a metre (m) and a decimetre (dm), a decimetre and a centimetre (cm), or a centimetre and a millimetre (mm) in a metric system. When imperial system units are to be used in the scale, their consecutive units will be yard and foot or foot and inches. The plain scales are intended for measurements up to first decimal accuracy. The construction procedure is as per the general guidelines explained in Section 3.3 and will be explained in detail in Examples 3.1–3.3.

Figure 3.1 shows a plain scale, and the procedure to construct the plain scale is followed on similar lines as mentioned in Section 3.3. The units of measurement are indicated on the main scale and on the subdivision scale towards right and left ends, respectively. All information such as RF value, length of the scale, and printing of division numbers are represented on the main scale and subdivision scale.

Example 3.1

Construct a plain scale to read decimetre and centimetre, and the scale should be long enough to measure 8 dm with $RF = \dfrac{1}{5}$. Mark the distances 5.2 and 7.6 dm on the scale.

Given data $RF = \dfrac{1}{5}$

Units to be measured: centimetres and decimetres
Maximum measurement of the scale = 8 dm or 80 cm

PROCEDURE (Refer Fig. 3.2)

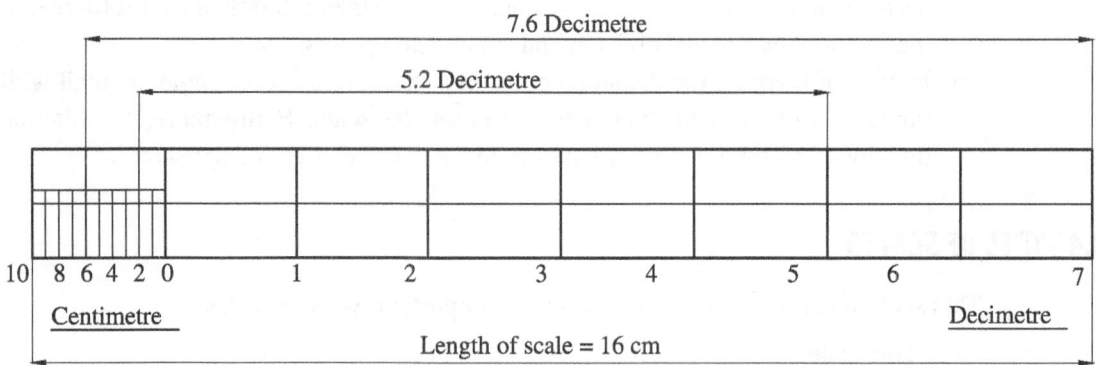

Fig. 3.2

Step 1: Calculate the length of the scale in small units as 16 cm using the following formula:

Length of scale = RF × Maximum measurement in centimetre

$$= \frac{1}{5} \times (8 \times 10 \text{ cm}) = 16 \text{ cm}$$

Step 2: Draw a horizontal line of length 16 cm and divide it into eight equal main parts, each representing 1 dm. Mark 0 at the end of the first main part and 0 to 7 for the other parts towards the right.

Step 3: Subdivide the first main division into 10 equal parts, which represents the small units of decimetre, that is, the centimetre. Mark the subdivision points 0 to 10 towards the left as shown in Fig. 3.2, and complete the scale arrangement.

Step 4: To mark a distance of 5.2 dm on the scale, mark 5 main divisions and 2 small divisions, both from 0 but towards the right and the left, respectively.

Step 5: Similarly, mark 7.6 dm also on the scale. ▲

Example 3.2 ▲

Construct a plain scale to read centimetre and millimetre to measure 5 cm with RF = 2. Show on it a distance of 3.6 cm.

Given data   RF = 2

Units to be measured: centimetres and millimetres
Maximum measurement of the scale = 5 cm

PROCEDURE (Refer Fig. 3.3)

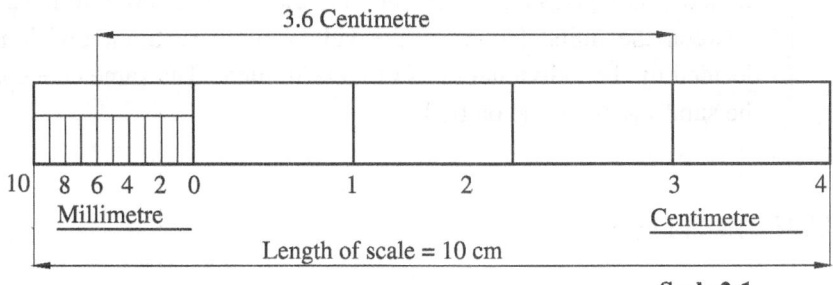

3.6 Centimetre

10 8 6 4 2 0   1   2   3   4
Millimetre   Centimetre
Length of scale = 10 cm

Scale 2:1

Fig. 3.3

Step 1: Calculate the length of the scale as 10 cm using the following formula:

Length of scale = RF × Maximum measurement in centimetre

$$= 2 \times 5 = 10 \text{ cm}$$

Step 2: Although RF > 1, the construction procedure is similar to that explained in Example 3.1, and hence is not repeated here.

Step 3: Marking the required distance of 3.6 cm is done by taking 3 units in the main division and 6 subunits, both measured from 0 but to the right and left, respectively. ▲

Example 3.3 ▲

An automobile is running at a speed of 50 kmph. Construct a plain scale to read the minimum of a kilometre and a minute. The scale should measure up to a maximum of 60 km. The RF of the scale is $\dfrac{1}{5,00,000}$. Show the distance covered by the automobile in 42 min on the scale.

Given data $RF = \dfrac{1}{5,00,000}$

Units to be measured: centimetres and millimetres in one scale and minutes and its subparts in the other scale

Maximum measurement on the kilometre scale: 60 km

> ### PRINCIPLE
> Since a plain scale is to be constructed to measure two different units of measurement, that is, kilometre and minute, they can be constructed on the top and bottom edges, respectively, following the same procedure outlined in Example 3.1.
>
> The length of the scale required is decided based on one of the units of measurement (km) and is adopted for the other, since a one-to-one correspondence exists through

the speed relation. As the length of the scale is the same for both, the number of main division points are to be retained the same for both, to have correspondence between the units. However, the values of these main division points will be decided by the individual units of measurement. The same concept holds good for the subdivision points on both.

PROCEDURE (Refer Fig. 3.4)

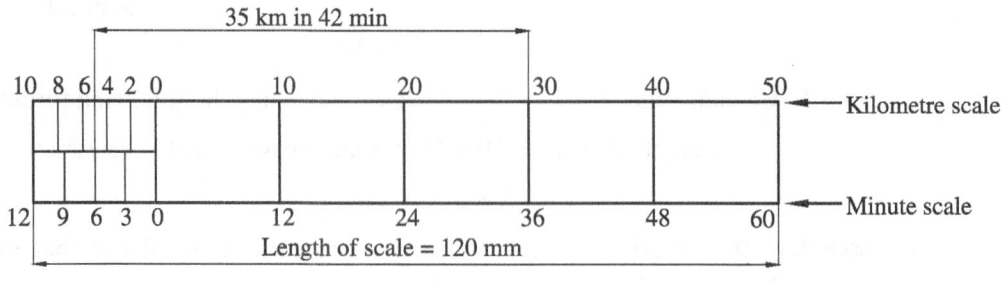

Fig. 3.4

Detailed Working of the Respective Scales

Length of the Scales

Step 1: The length of the kilometre scale is decided by the maximum kilometre value to be measured and the RF using the following formula:

$$\text{Length of kilometre scale} = \frac{1}{5,00,000} \times (60 \times 1000 \times 100 \times 10 \text{ mm})$$

$$= 120 \text{ mm}$$

Step 2: Length of the minute scale also equals that of the kilometre scale, that is, 120 mm, as both are connected by the speed relation.

Recording Maximum Units of Measurement in the Respective Scales

Step 3: The kilometre scale corresponds to a maximum of 60 km as per the requirement of the problem.

Step 4: The maximum quantity for the minute scale is decided by the speed relation. As per the speed relation of 50 kmph, 50 km is covered in one hour or in 60 min. The corresponding time to cover the maximum distance of 60 km mentioned in the kilometre scale will be $\frac{60}{50} \times 60 = 72$ min .

Therefore, the minute scale corresponds to a maximum time of 72 min.

Marking of Division Points in Scales

Step 5: As both the scales are of the same length, they also should bear the same number of main divisions, and the maximum values mentioned in the respective scales should be amenable for that.

Step 6: It can be noted that if both the maximum values are not amenable for a common division, then these values can be revised till a common division becomes feasible, and the length corresponding to the revised quantities should be treated as the revised length for further usage.

Step 7: In this example, the maximum kilometre (60) and the maximum time in minutes (72) are amenable for division by 6. Then, each main division point in the kilometre scale corresponds to 10 km and that in the minute scale corresponds to 12 min.

Step 8: As in the previous examples, mark the main division points in the kilometre scale as 10, …, 50 starting from 0 and towards the right and divide the first left unit into five subdivision points.

Step 9: Mark the corresponding main division points in the minute scale as 12, 24, …, 60 starting from 0 and towards the right and divide its first left unit into four subdivision points as 0, 3,…,12 as shown in Fig. 3.4.

Step 10: This completes the scale arrangement for both the kilometre and minute units.

Marking of Distance Corresponding to a Particular Time

To mark the distance covered by the automobile in 42 min, select the 42 min on the minute scale (3 main division points to the right and 2.5 subdivision points to the left) and project these two measured ends on to the kilometre scale, totalling to 35 km. ▲

3.4.2 Diagonal Scales

Customized plain scales are constructed to measure the lengths in two consecutive units or in the same unit to the accuracy of the first decimal place. If a customized scale is required to measure to the accuracy of the second decimal place or in three consecutive units, then diagonal scales are to be prepared. For example, the lengths of three successive units such as metre, decimetre, and centimetre or metre correct to two decimal places, like, 6.57 m and 7.83 m, can be measured using diagonal scales. Diagonal scale is an extension of the plain scale of two consecutive units to three consecutive units. In the diagonal scale, the first main unit of the plain scale, which was subdivided into a number of equal parts, is again divided to obtain its next consecutive unit. This can be done by dividing the second smaller unit into the third smallest unit by constructing diagonals to the rectangles drawn on each subdivision of the main unit. The constructional procedure of the diagonal scale is given in the following subsection.

Basic Constructional Procedure

The following procedure explains the construction of a diagonal to one of the rectangles that are drawn over each subdivision of the main unit as shown in Fig. 3.5.

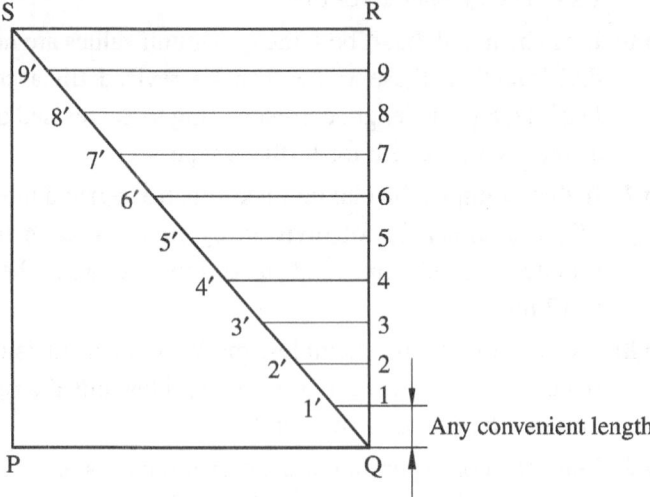

Fig. 3.5   Basic constructional arrangement of diagonal scale

Step 1: Draw a rectangle PQRS in which PQ represents one of the subdivisions of the main unit.

Step 2: Erect perpendiculars at the points P and Q for a convenient length. Divide the perpendicular line drawn from the point Q into the required equal number of parts as per the ratio between this subdivision and its next unit. Let us assume these division points as 10 and mark the points as 1, 2, ..., 9, 10 (R).

Step 3: Complete the rectangle PQRS and draw the diagonal joining the points Q and S. Draw the horizontal lines through the points 1, 2, ... to intersect the diagonal at 1', 2', ..., respectively.

Step 4: The horizontal intercepts 11', 22', 33', ..., etc., are progressively greater in length by 0.1RS or 0.1PQ. Hence, the diagonal line serves to represent the division of the second unit to obtain the third consecutive unit in the increasing order. For example, the eighth division of the third consecutive unit (line 88') can be shown to be 0.8 times the subdivision unit as follows.

From the similar triangles, Q88' and QRS,

$$\frac{Q8}{QR} = \frac{88'}{RS}$$

Since $Q8 = \dfrac{8}{10} \times QR$ or 0.8 QR, the line 88' = 0.8 RS or 0.8 PQ

The distance 88′ represents the fraction of the subdivision length PQ. Similarly, the lengths 11′, 22′, etc., represent their proportionate fraction of the subdivision PQ, and hence measure to the next decimal accuracy.

Example 3.4

Construct a diagonal scale of RF 1:25, which is capable of measuring a maximum length of 4 m, and also mark a distance of 3.28 m on it.

PROCEDURE (Refer Fig. 3.6)

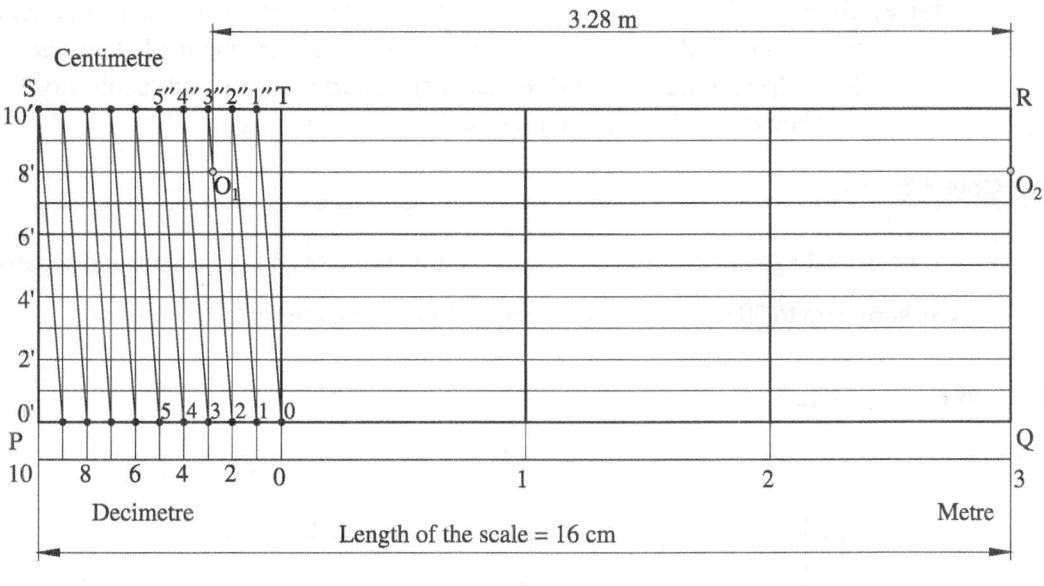

Scale = 1:25

Fig. 3.6

Step 1: Calculate the length of the scale in centimetre units, as 16 cm, using the following formula:

Length of scale = RF × Maximum measurement in centimetre

$$= \frac{1}{25} \times (4 \times 100 \text{ cm}) = 16 \text{ cm}$$

Step 2: Draw a horizontal line PQ of length 16 cm and divide it into four equal main parts, each representing 1 m. Mark 0 at the end of the first main part and 0 to 3 for the other parts towards the right.

Step 3: Subdivide the first left main division into 10 equal parts, which represent the immediate small units of metre, that is, the decimetre. Mark the subdivision points 0 to 10 towards the left as shown in Fig. 3.6.

Step 4: Erect a vertical line at P for any convenient length, divide it into 10 equal parts (as the next consecutive unit is a division of 10), mark the points as 0′, 1′, 2′, ..., 9′, 10′(S) and draw horizontal lines through them.

Step 5: Draw vertical lines through each main scale division and complete the rectangle PQRS.

Step 6: Erect perpendiculars from the points 0, 1, 2, ..., 10 of the first subdivision unit of the main scale division, and mark subdivision points of the main scale on ST as 1″, 2″, 3″, ..., 10″.

Step 7: Draw the diagonals of the subdivisions of the rectangles by connecting the opposite corner points 0–1″, 1–2″, 2–3″, etc. This completes the diagonal scale arrangement.

Step 8: To mark a distance of 3.28 m on the scale, select three main scale units and first two small divisions, both from 0 but towards the right and the left, respectively. Mark the eighth intercept O_1 on the diagonal from the second subdivision point. The horizontal line O_1O_2 represents 3.28 m on the scale. ▲

Example 3.5 ▲

Construct a diagonal scale of $RF = \dfrac{1}{4}$ to show centimetre and millimetre and capable of measuring up to 50 cm. Mark a distance of 36.4 cm on the scale.

PROCEDURE (Refer Fig. 3.7)

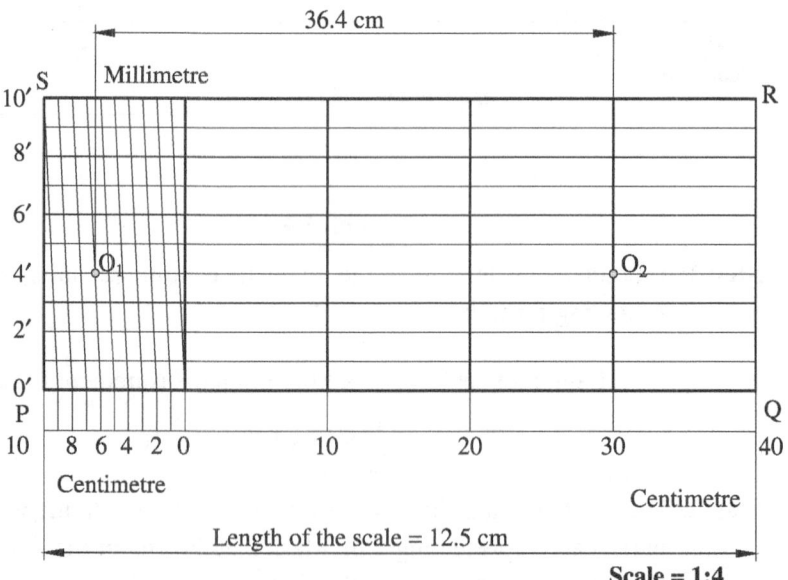

Fig. 3.7

Step 1: Calculate the length of the scale in centimetre units, as 12.5 cm, using the following formula:

Length of scale = RF × Maximum measurement in centimetre

$$= \frac{1}{4} \times (50 \text{ cm}) = 12.5 \text{ cm}$$

Step 2: Draw a horizontal line PQ of length 12.5 cm and divide it into five equal main parts, each representing 10 cm. Mark 0 at the end of the first main part and 0 to 40 for the other parts towards the right.

Step 3: Subdivide the first left main division into 10 equal parts, which represents 1 cm. Mark the subdivision points 0 to 10 towards the left as shown in Fig. 3.7.

Step 4: Locate the diagonal scale point O_1 as mentioned in Example 3.4 and mark the distance of 36.4 cm on the scale.

Example 3.6

The area of a land in rectangular shape is equal to 0.64 hectare and is represented on the map by a similar rectangular shape of 4 sq. cm area. Calculate the RF of the scale of the map. Based on the RF value, construct a diagonal scale to read up to a minimum of 1 m in the map. The required maximum scale to be measured is 600 m. Also, mark 486 m in the scale.

PROCEDURE (Refer Fig. 3.8)

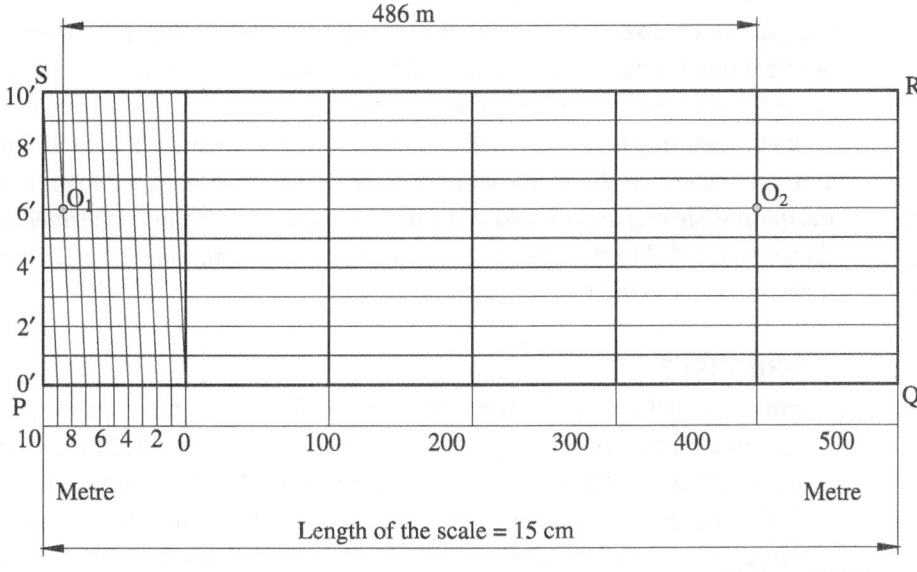

Fig. 3.8

Determination of Representative Fraction

The given area of 0.64 hectare (i.e., 0.64 × 10,000 sq. m) is represented in the map by 4 sq. cm, or 4 sq. cm on the drawing represents 6,40,00,000 sq. cm actual land.

Therefore, 1 sq. cm on the drawing represents 1,60,00,000 sq. cm land, or 1 cm on the drawing represents = $\sqrt{1,60,00,000}$ sq. cm or 4000 cm actual land.

$$\text{Therefore, RF} = \frac{\text{Drawing size}}{\text{Actual size}} = \frac{1}{4000}$$

Determination of Length of Scale and Construction of Diagonal Scale

Step 1: From the RF calculated earlier, the length of the scale in centimetre units is obtained as 15 cm using the following formula:

$$\text{Length of scale} = \text{RF} \times \text{Maximum measurement in centimetre}$$

$$= \frac{1}{4000} \times (600 \times 100 \text{ cm}) = 15 \text{ cm}$$

Step 2: This line is divided and further subdivided as per the diagonal arrangements explained in Examples 3.4 and 3.5. The complete constructional arrangements are shown in Fig. 3.8 and the distance of 486 m has been marked as shown by the line O_1O_2 on the scale. ▲

3.4.3 Vernier Scales

A Vernier is a short supplementary scale constructed to support the plain scale. Here, it is known as the main scale or primary scale to measure the lengths exact to any customized fraction of smaller division on the main scale. In 1630, a French scientist Pierre Vernier invented this scale, and hence, it is named after him as the Vernier scale. It is very compact in comparison with the diagonal scales and widely used in many engineering applications for the measurement of length and depth of the object. The main scale reads up to two consecutive units, and the Vernier scale supplements the main scale by further measuring the third smallest consecutive unit, which is the fraction of the second consecutive unit of the main scale. Each smallest division of the main scale is termed as the *main scale division* or msd and each division of the Vernier is termed as *Vernier scale division* or vsd. The use of the Vernier scale is very common for length and cross-section measurements in the Physics laboratory.

PRINCIPLE

For constructing the Vernier scale, the main scale or the plain scale has to be constructed first as per the procedure given in Section 3.4.1. Then, the Vernier to be constructed should be equal to the length of $n - 1$ or $n + 1$ times the msd (where n is the number of divisions taken on the main scale). This length is divided into the same n number of equal parts and each division is known as 1 vsd. Therefore, 1 vsd measures $(1 - 1/n)$ msd or $(1 + 1/n)$ msd as in the previous cases.

The difference between one msd and one vsd is known as the least count (LC) and that gives the least or the smallest measurement that is possible with the Vernier scale. Therefore, the LC becomes $\left(\dfrac{1}{n}\right) \times \text{msd}$ in both the cases.

Types of Vernier Scale

There are two types of Vernier scales; one is the forward or direct Vernier and another is the backward or retrograde Vernier.

Forward Vernier   In this scale, the length of each Vernier scale division is smaller than that of the main scale division, and the divisions on the Vernier are numbered in the same direction as that of the main scale.

Backward Vernier   In this scale, the length of each Vernier scale division is greater than that of the main scale division, and the divisions on the Vernier are numbered in the direction opposite to that of the main scale.

Basic Constructional Procedure

Step 1: The main scale is constructed based on the RF value and the maximum length to be measured, as explained in Section 3.4.1, and is divided into suitable number of parts.

Step 2: Each part of the main scale is subdivided into msd and the value of one msd is noted.

Step 3: The given or the desired LC is expressed as the fraction of msd, say $1/n$, and the value of n is calculated.

Step 4: If a forward Vernier is to be constructed, then one vsd should be smaller than one msd by an amount equal to the LC, such that

$$1 \text{ vsd} = 1 \text{ msd} - \text{LC or } 1 \text{ msd} - (1/n) \text{ msd or } (1 - 1/n) \text{ msd}$$

Therefore, $n \times \text{vsd} = (n - 1) \times \text{msd}$

Hence, the length of the forward Vernier is calculated as equal to $(n - 1) \times \text{msd}$, it is marked from the left end of the main scale, divided into n equal parts, and numbered in the forward direction or towards the right as shown in Fig. 3.9.

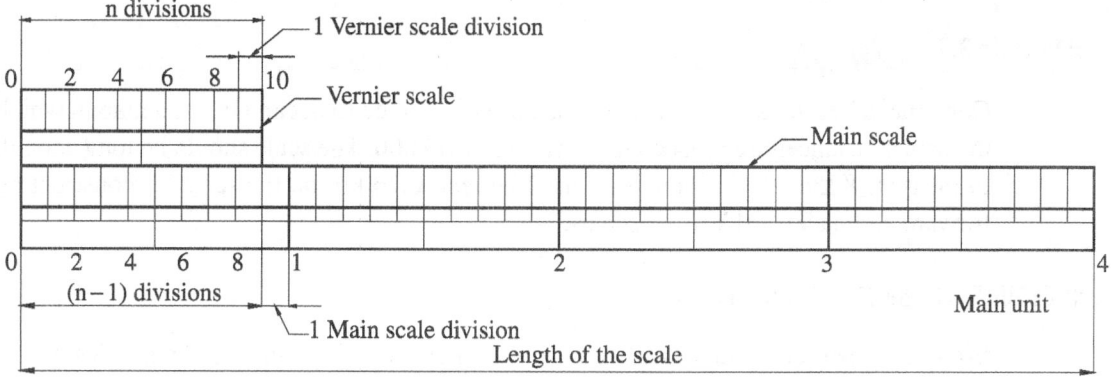

Fig. 3.9   Forward Vernier

Step 5: If a backward Vernier is to be constructed, then one vsd should be greater than one msd by an amount equal to the LC, such that

$$1 \text{ vsd} = 1 \text{ msd} + \text{LC or } 1 \text{ msd} + (1/n) \text{ msd or } (1 + 1/n) \text{ msd}$$

Therefore, $n \times \text{vsd} = (n + 1) \times \text{msd}$

Hence, the length of the backward Vernier is calculated as equal to $(n + 1) \times \text{msd}$, it is marked from the end of the first part of the main scale, divided into n equal parts, and numbered in the backward direction or towards the left as shown in Fig. 3.10.

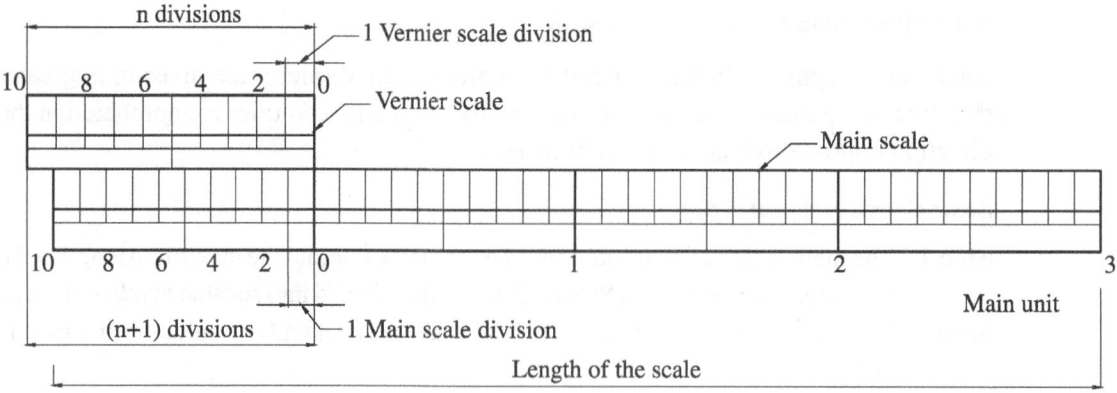

Fig. 3.10 Backward Vernier

Step 6: To mark any required measurements on the scale, the measurement is split into two parts, one on the main scale and another on the Vernier scale, such that the sum of their lengths indicates the required distance. The length on the Vernier scale is decided by the length of one vsd and the number of such divisions required. The number of divisions required on the Vernier is decided such that the product of 1 vsd and the number of Vernier divisions, when subtracted from the total length to be marked, enables that to be made in the main scale, including its first main part. ▲

Example 3.7 ▲

Construct a Vernier scale to read the distances corrected to decametre on a map in which the actual distances are reduced in the ratio of 1:50,000. The scale should be long enough to measure 8 km. Mark a length of 4.82 km and 0.76 km on the scale by constructing forward and backward Vernier scales.

PROCEDURE (Refer Figs 3.11a and b)

Step 1: Construct the main scale in a similar manner as explained in Section 3.4.1. The length of the scale is calculated using the formula,

Length of the scale = RF × Maximum measurement in centimetre

$$= \frac{1}{50,000} \times (8 \times 1000 \times 100 \text{ cm})$$

$$= 16 \text{ cm}$$

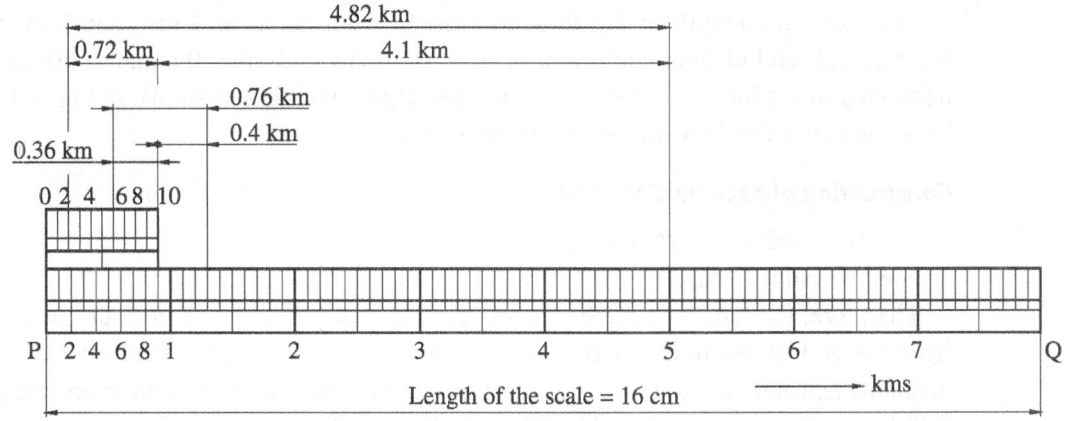

(a) By forward Vernier

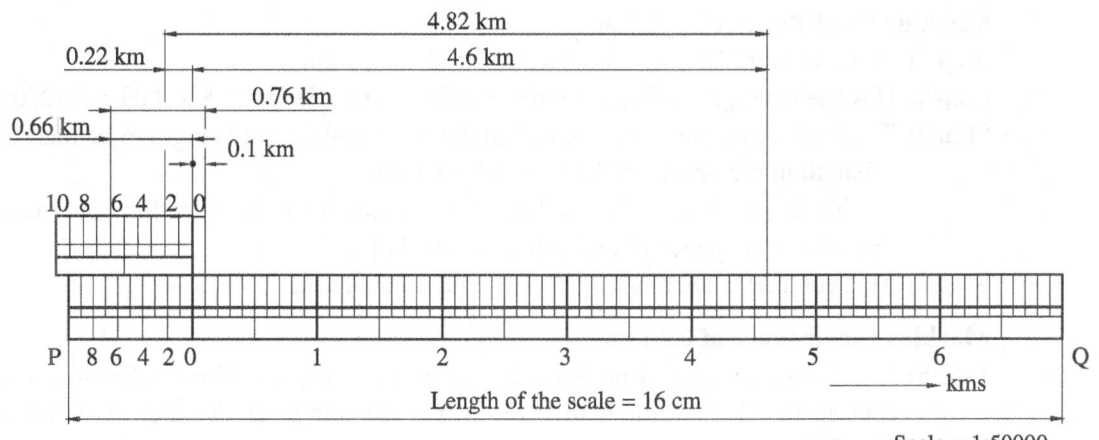

(b) By backward Vernier

Fig. 3.11

Step 2: Draw a line PQ of length 16 cm to represent the main scale, divide it into 8 equal parts, and mark them as 0, 1, 2, ..., 8 (Q), which represent the units of kilometre from the left towards the right as shown in Fig. 3.11(a).

Step 3: Divide each main scale division into 10 subdivisions (a kilometre and a decametre bear a ratio of 1:10), and mark the first main scale subdivision as 0, 1, 2, ..., 10 from the left and towards the right such that 1 msd = 0.1 km.

Step 4: As the length to be measured is of two-decimal accuracy, LC = 0.01 km.

Step 5: Using the relation LC = $(1/n)$ × msd and substituting the LC and msd values, n is obtained as 10.

Construction of Forward Vernier

As $n = 10$, 1 vsd = $(1 - 1/10)$ msd or $(9/10)$ msd

That is, 10 vsd = 9 msd

Therefore, the length of the forward Vernier is set equal to 9 msd and is marked from the left end of the main scale. Then, it is subdivided into 10 equal divisions and numbered in the forward direction, each representing 0.09 km as shown in Fig. 3.11(a). This completes the forward Vernier arrangement.

Construction of Backward Vernier

As $n = 10$, 1 vsd = $(1 + 1/10)$ msd or $(11/10)$ msd
That is, 10 vsd = 11 msd

Therefore, the length of the backward Vernier is set equal to 11 msd and is marked from the end of the first part of the main scale. Then, it is subdivided into 10 equal divisions and numbered towards the left in the backward direction, each representing 0.11 km as shown in Fig. 3.11(b). This completes the backward Vernier arrangement.

Marking Required Distances in Forward Vernier Scale

Marking the distance of 4.82 km

Step 1: In the forward Vernier constructed, 1 vsd = 0.09 km
Step 2: If we select eight divisions on the Vernier, then its length = $8 \times 0.09 = 0.72$ km
Step 3: The balance length to be marked on the main scale = total length to be marked
– length in the Vernier = $4.82 - 0.72 = 4.1$ km

This is possible to be marked in the main scale by selecting four main divisions and one small division as shown in Fig. 3.11(a).
Step 4: This completes the marking of the required distance.

Marking the distance of 0.76 km

Step 5: In a similar manner of marking the earlier distance, four Vernier divisions and four subdivisions in the main scale will enable the given distance as shown in Fig. 3.11(a).

Marking Required Distances in Backward Vernier Scale

Marking the distance of 4.82 km

Step 1: In the backward Vernier constructed, 1 vsd = 0.11 km
Step 2: If we select two divisions on the Vernier, then its length = $2 \times 0.11 = 0.22$ km
Step 3: The balance length to be marked on the main scale = total length to be marked
– length in the Vernier

$$= 4.82 - 0.22 = 4.6 \text{ km}$$

This is possible to be marked in the main scale by selecting four main parts and six small divisions as shown in Fig. 3.11(b).
Step 4: This completes the marking of the required distance.

Marking the distance of 0.76 km

In a similar manner of marking the earlier distance, six Vernier divisions and one subdivision in the main scale will enable the given distance as shown in Fig. 3.11(b). ▲

Example 3.8

@ Construct a Vernier scale with an RF value of 2:1 to give readings of 1/10th of a millimetre measured up to 60 mm and mark the following points on the scale: (a) 38.4 mm (b) 20.7 mm.

PROCEDURE   (Refer Fig. 3.12)

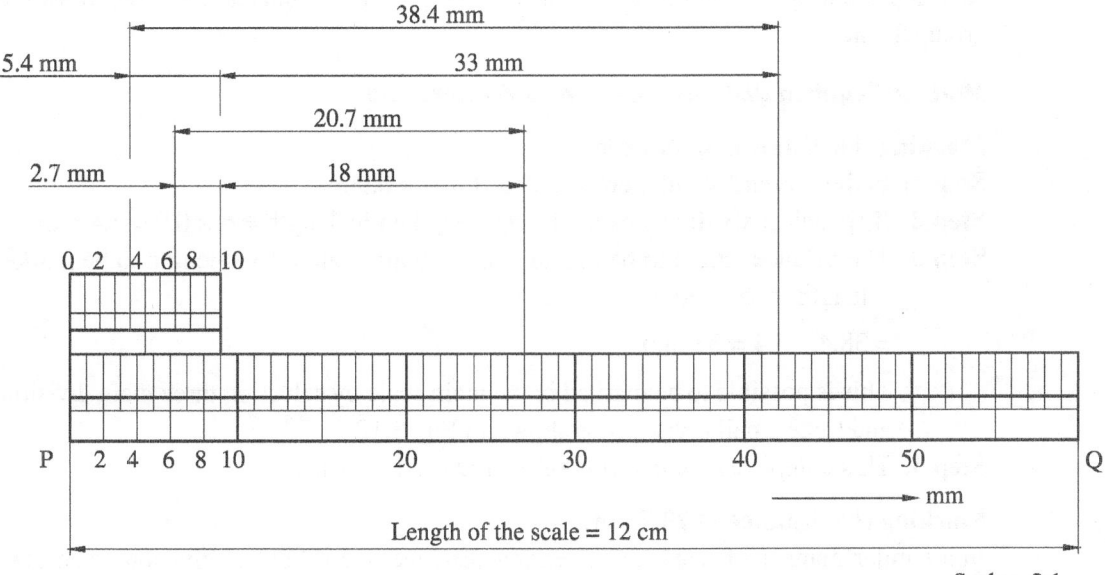

Fig. 3.12

Step 1: Construct the main scale in a similar manner as explained in Plain Scales of Section 3.4.1.

The length of the scale is calculated using the formula,

Length of the scale = RF × Maximum measurement in centimetre

$$= 2 \times (60/10) \text{ cm} = 12 \text{ cm}$$

Step 2: Draw a line PQ of length 12 cm to represent the main scale, divide it into 6 equal parts, and mark them as 0, 10, 20, ..., 60 (Q), which represent the units of millimetre from the left towards the right as shown in Fig. 3.12.

Step 3: Divide each main scale division into 10 subdivisions and mark the first main scale subdivision as 0, 1, 2, ..., 10 from the left towards the right such that,

1 msd = 1 mm

Step 4: As the length to be measured is of single-decimal accuracy, LC = 0.1 mm.

Step 5: Using the relation LC = (1/n) × msd and substituting the LC and msd values, n is obtained as 10.

Construction of Forward Vernier

As no specific type of Vernier scale arrangement is asked, a forward Vernier has been explained here.

As $n = 10$, 1 vsd = $(1 - 1/10)$ msd or $(9/10)$ msd

That is, 10 vsd = 9 msd

Therefore, the length of the forward Vernier is set equal to 9 msd and is marked from the left end of the main scale. Then, it is subdivided into 10 equal divisions, each representing 0.9 mm as shown in Fig. 3.12. This completes the forward Vernier arrangement.

Marking Required Distances in Forward Vernier Scale

Marking the distance of 38.4 mm

Step 1: In the forward Vernier constructed, 1 vsd = 0.9 mm

Step 2: If we select six divisions on the Vernier, then its length = $6 \times 0.9 = 5.4$ mm

Step 3: The balance length to be marked on the main scale = total length to be marked − length in the Vernier

= 38.4 − 5.4 = 33 mm

This is possible to be marked in the main scale by selecting three main divisions and three small divisions as shown in Fig. 3.12.

Step 4: This completes the marking of the required distance.

Marking the distance of 20.7 mm

In a similar manner of marking the earlier distance, three Vernier divisions, one main division, and eight subdivisions in the main scale will enable the given distance as shown in Fig. 3.12.　▲

Example 3.9　▲

Construct a Vernier scale having a LC of 0.004 cm. The scale should be long enough to measure up to 6 cm. Show the following distances on the scale: (a) 3.28 cm (b) 2.36 cm. Select a scale with RF = 3:1.

PROCEDURE　(Refer Fig. 3.13)

Step 1: Construct the main scale in a similar manner as explained in Plain Scales of Section 3.4.1.

The length of the scale is calculated using the formula,

Length of the scale = RF × Maximum measurement in centimetre = 3×6 cm

= 18 cm

Step 2: Draw a line PQ of length 18 cm to represent the main scale, divide it into six equal parts, and mark them as 0, 1, 2,..., 6 (Q), which represent the units of centimetre from the left towards the right as shown in Fig. 3.13.

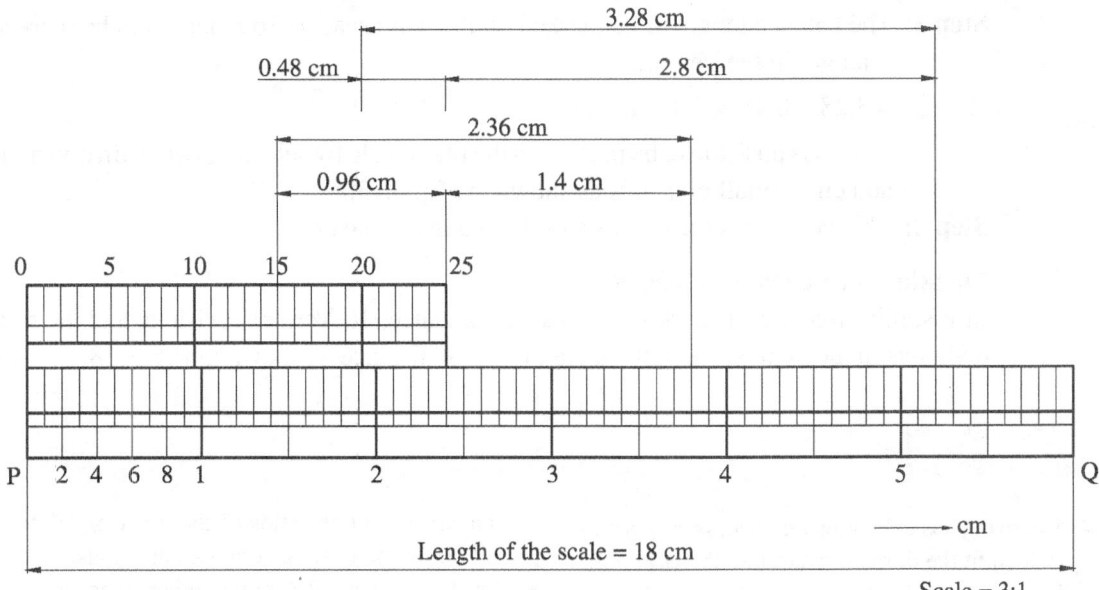

Forward Vernier

Fig. 3.13

Step 3: Divide each main scale division into 10 subdivisions and mark the first main scale subdivision as 0, 1, 2,…, 10 from the left towards the right, such that, 1 msd = 1 mm.

Step 4: The LC is mentioned as LC = 0.004 cm or 0.04 mm.

Step 5: Using the relation LC = (1/n) × msd and substituting the LC and msd values, n is obtained as 25.

Construction of Forward Vernier

As no specific type of Vernier scale arrangement is asked, a forward Vernier has been explained here.

As $n = 25$, 1 vsd = (1 – 1/10) msd or (9/10) msd

That is, 25 vsd = 24 msd

Therefore, the length of the forward Vernier is set equal to 24 msd and is marked from the left end of the main scale. Then, it is subdivided into 25 equal divisions, each representing (24/25) = 0.96 mm or 0.096 cm as shown in Fig. 3.13. This completes the forward Vernier arrangement.

Marking the Required Distances in the Forward Vernier Scale

Marking the distance of 3.28 cm

Step 1: In the forward Vernier constructed, 1 vsd = 0.096 cm

Step 2: If we select five divisions on the Vernier, then its length = 5 × 0.096 = 0.48 cm.

Step 3: The balance length to be marked on the main scale = Total length to be marked – length in the Vernier

= 3.28 – 0.48 = 2.8 cm.

This is possible to be marked in the main scale by selecting two main divisions and eight small divisions as shown in Fig. 3.13.

Step 4: This completes the marking of the required distance.

Marking the distance of 2.36 cm

In a similar manner of marking the earlier distance, 10 Vernier divisions and 14 subdivisions in the main scale will enable the given distance as shown in Fig. 3.13.   ▲

RECAPITULATION

- A drawing has to be prepared necessarily with a scale to fit the dimensions of the object into the drawing space available in the drawing sheet. It is expressed as the ratio of the drawing size of an object to the actual size of it, and is known as the 'representative fraction', commonly abbreviated as RF.
- When the dimensions of an object are such that the drawings can be prepared to the actual dimensions of the object itself in either hard or soft form, then the drawings are known as 'full-scale' drawings, which have RF value equal to one.
- When the drawings are to be made to represent various large sized engineering applications such as survey maps, machines, and building plans on the drawing sheets, certain proportionate reduction in all the dimensions is required and this scale is called 'reduced scale', which has RF value less than one.
- In some of the applications, when the component dimensions are very small such as electronic chips, nano products, sensors, and wrist watch components, the dimensions in the drawing sheet are to be proportionately increased to represent them, and this type of scaling of dimension is known as 'enlarged scale', which has RF value greater than one.
- A set of cardboard scales are available ranging from M1 to M8 sizes and are widely used with

standard reduction ratios of the order of 1:2 to 1:2000 for many engineering applications.
- However, when RF ratios other than those mentioned earlier are required, then customized scales can be prepared on the drawing sheet. Three types of scales are popular, namely the plain scale, the diagonal scale, and the Vernier scale. All these scales bear the same concept of construction, differing slightly in their arrangement due to the need of placing up to the first or second decimal requirements.
- *Plain and diagonal scales*: A plain scale is used to measure only two consecutive units in the same system of measurement or will read up to a single decimal place only, whereas a diagonal scale is used to represent three consecutive units or up to two decimal accuracies.
- *Vernier scale*: A Vernier is a short supplementary scale constructed to support the plain scale. It is known as main scale to measure the lengths exact to any customized fraction of smaller division on the main scale. The difference between one main scale division and one Vernier scale division is called the least count (LC), and the Vernier scales can be designed as per the desired LC. These scales are compact in size compared with the diagonal scales and are widely used in many engineering measurements.
- There are two types of Vernier scales—forward or direct Vernier and backward or retrograde Vernier. In the forward Vernier scale, the length

of each Vernier scale division is smaller than that of the main scale division, and the divisions on the Vernier are numbered in the same direction as that of the main scale. In the backward Vernier scale, the length of each Vernier scale division is greater than that of the main scale division, and the divisions on the Vernier are numbered in the direction opposite to that of the main scale.

MULTIPLE-CHOICE QUESTIONS

3.1 The representative fraction (RF) is the ratio of
   (a) the drawing size to the actual size of the object
   (b) the drawing size to the bench mark size of an object
   (c) the drawing size to the standard size of an object
   (d) the actual size of an object to the drawing size

3.2 If the RF < 1, then the scale is termed as the
   (a) enlarged scale   (c) full scale
   (b) reduced scale   (d) M1 scale

3.3 The length of a customized scale is the product of
   (a) RF and minimum measurement of the scale
   (b) RF and maximum measurement of the scale
   (c) RF and the average measurement of the scale
   (d) The drawing size and maximum measurement of the scale

3.4 In a scale, 10 cm is drawn on the drawing sheet to represent an actual length of 10 mm. The RF value is
   (a) 5:1   (c) 10:1
   (b) 1:1   (d) 1:10

3.5 In a scale, a line of length 2 cm is drawn on the drawing sheet to represent an actual length of 1 m. The RF value is
   (a) 50:1   (c) 1:1
   (b) 1:50   (d) 1:2

3.6 The customized plain scales are used to measure
   (a) two consecutive units
   (b) three consecutive units
   (c) one unit
   (d) fraction of a main scale division

3.7 The customized diagonal scales are used to measure
   (a) two consecutive units
   (b) three consecutive units
   (c) one unit
   (d) fraction of a main scale division

3.8 The LC in a Vernier scale is
   (a) 1 vsd − 1 msd   (c) equal to 1 msd
   (b) 1 msd − 1 vsd   (d) equal to 1 vsd

3.9 In a forward Vernier scale, 1 vsd is
   (a) equal to 1 msd   (c) less than 1 msd
   (b) greater than 1 msd (d) always equal to 1

3.10 In a backward Vernier scale, 1 vsd is
   (a) equal to 1 msd   (c) less than 1 msd
   (b) greater than 1 msd (d) always equal to 1

WORK PRACTICE LEVEL − 1

Problems on Plain Scales

3.1 The RF of a scale is 1/25. Construct a scale to measure up to 4 m. Mark a distance of 2.7 m on the scale.

3.2 Construct a scale to measure centimetres and millimetres and read up to 12 cm by considering the RF equal to 1:2. Mark the lengths of 6.8 cm and 9.9 cm on the scale.

3.3 The actual distance between two places is 1.5 km apart, which is marked as a 5 cm line on a survey map. Construct a scale to read 4.7 km. What is the RF of the scale?

3.4 The distance between two different bus stops is 1.8 km and that is shown by a line of 9 cm in a drawing. What is the RF of the scale? Construct a scale to read up to 2.6 km and show on the scale a distance of 1.7 km.

3.5 Construct a plain scale to read metres and decimetres. The scale should be long enough to measure 8 m with RF = 1/4. Show the distance of 5.9 m and 6.4 m on the scale.

3.6 An automobile is running at a speed of 60 kmph. Construct a plain scale to read a minimum of a kilometre and a minute. The scale should measure up to a maximum of 64 km. The RF of the scale is 1/4,00,000. Show on the scale the distance covered by the automobile in 42 min.

Problems on Diagonal Scales

3.7 Draw a diagonal scale to read metre, decimetre, and centimetre when RF = 1/25. Show a distance of 5.67 m on the scale.

3.8 Construct a diagonal scale with RF = 1/40 to read metre, decimetre, and centimetre. Show a distance of 5.76 m on the scale.

3.9 Construct a diagonal scale with RF = 7/200 showing the divisions of 0.01 m and capable of measuring 7 m. Mark a distance of 5.87 m on the scale.

3.10 Construct a diagonal scale with RF = 1/40 showing metres and decimetres and capable of measuring 10 m. Show a distance of 9.46 m on the scale.

3.11 The area of a square-shaped land is equal to 0.6561 hectare, which is represented on the map by a similar square shape of 9 sq. cm. Calculate the RF of the map. Based on the RF value, construct a diagonal scale to read up to a maximum of one meter in the map. The required maximum scale to be measured is 700 m. Show a dimension of 549 m in the scale.

WORK PRACTICE LEVEL – 2

Problems on Vernier Scales

3.1 Construct a Vernier scale of RF = 1/30 to read centimetre and up to a maximum length of 5 m. Mark the dimensions 3.57 m and 4.64 m on the scale.

3.2 Construct a Vernier scale of RF = 1/40 to read metre, decimetre, and centimetre. Mark the dimensions 5.57 m and 2.84 m on the scale.

3.3 Construct a Vernier scale to read distances corrected upto the second decimal of a kilometre unit on a map in which the actual distances are reduced in the ratio of 1:40,000. The scale should be long enough to measure 10 km. Mark the dimension 8.59 km and 2.27 km on the scale. Use forward Vernier.

3.4 Construct a Vernier scale to read distances corrected to the second decimal of a kilometre unit on a map in which the actual distances are reduced in the ratio of 1:80,000. The scale should be long enough to measure 10 km. Mark the dimensions 8.76 km and 2.98 km on the scale. Use backward Vernier.

3.5 Construct a forward Vernier scale with a least count of 4 mm. The RF of the scale is given as 1/4. The scale should be long enough to measure 50 cm. Mark 27.4 cm and 39.6 cm using the scale.

3.6 The distance between Chennai to Hyderabad is 640 km and is represented on a map as 12.8 cm. Draw a Vernier scale long enough to measure 800 km and correct up to kilometre. Mark a distance of 766 km using the scale.

Answers for Multiple-choice Questions

3.1 (a)	3.2 (b)	3.3 (b)	3.4 (c)	3.5 (b)	3.6 (a)
3.7 (b)	3.8 (b)	3.9 (c)	3.10 (b)		

Visualization Concepts and Freehand Sketching

4

OBJECTIVES

This chapter will help the reader to understand the following:

- Importance of pictorial sketches of objects and the application of visualization principles that help to observe the forms and shapes from them
- Various types of pictorial sketches and their limitations on preciseness in certain dimensions and the need for multiple views to represent the different surfaces of three-dimensional objects and their dimensions
- Choice of freehand sketches as a rapid means to represent multiple views and convey ideas with focus on communication
- Step-by-step development procedure to obtain multiple views when various features such as slopes, cuts, slots, and holes are added to a basic rectangular block
- Familiarization of hidden edges and surfaces and their representation
- Co-ordination of details in different views
- Sketching multiple views of common products/gadgets used in domestic environment and in engineering applications

4.1 VISUALIZATION CONCEPTS

The pictorial representations of the objects are adopted universally, as they are not bound by any formal language (vernacular), region, or special skills. The details communicated through the pictures are easily understandable, since they create a visual impact on the viewer. As a result, the communication through pictures continues to play a major role than other means of communications that have limitations in one or many number of ways. Understanding the pictures of an object beyond its general level of appearance creates an awareness of its form and shape which are vital to produce or manufacture it. Conceiving the form and shape of an object amounts to capturing its dimensions and the geometrical features in the mind and translating them through suitable pictorial representations. This chapter stresses the importance of the visualization or observation capabilities of the viewer as the prime requirement to extract the details from any pictorial sketches, and deals with the systematic ways of developing them.

4.2 PICTORIAL PROJECTIONS

Since any physical object is three-dimensional in nature, its form and shape can be completely understood only when all the three dimensions are available in the pictorial sketch (or sketches) meant for that. One of the simplest and most convenient forms of such a representation is through the arrangement of a single picture. Three major types of pictorial sketches or projections are used to convey the appearance of the object. They are, namely axonometric projection, oblique projection, and the perspective projection as shown in Fig. 4.1.

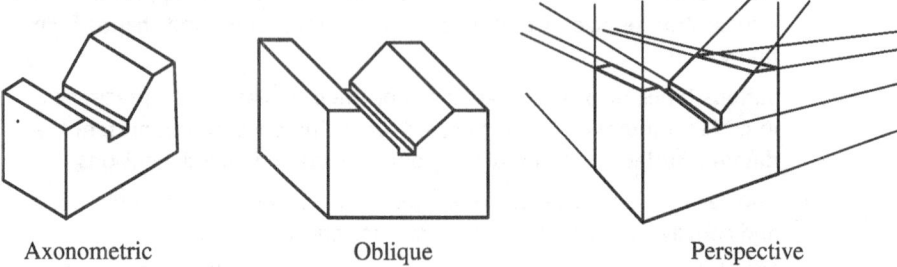

Axonometric   Oblique   Perspective

Fig. 4.1   Axonometric, oblique, and perspective projection

Axonometric projection is the representation of an object in a single picture, when the object is turned and tilted so that its three adjacent surfaces (faces) and its three sides are visualized by the viewer. It can be further subclassified into the following:

1. Isometric, where all the three sides of the object are equally inclined
2. Dimetric, where two of the three sides are equally inclined
3. Trimetric, where all the three sides of the object are inclined differently

In oblique projection, one face of the object is kept parallel to the viewer, or two of the axes are visually perpendicular to each other, while the third axis recedes at a convenient angle.

Perspective projection, on the other hand, is a single pictorial drawing, representing all the three sides of an object such that they have a tendency to converge at a point. This corresponds to the normal visualization process of physical objects, when we see them through our eyes. Though perspective projection conveys the natural view of an object, it does not reveal the exact size and shape in all the directions and hence is not fit for manufacture. However, this projection is generally used by engineers where aesthetic appeal and looks matter to attract the customers, as done in architectural engineering or advertising and marketing.

Hence, in all the pictorial projections, one or the other edges are foreshortened and hence do not represent the true dimensions or size completely. Such pictorial sketches are not fit for manufacturing purposes, but become useful where the focus is on communication and appearance of an idea and not on the exactness of dimensions.

4.3 MULTI-VIEW PROJECTIONS AND LAYOUT OF VIEWS

To extract the dimensions and features of the object in all the three directions exactly, each surface has to be visualized individually, and its shape and form are to be presented. The images of the individual surfaces are called views, and hence, a three-dimensional object can be explained with multi-view drawings or projections rather than a single pictorial arrangement.

Consider a three-dimensional rectangular block with corners A to H as shown in Fig. 4.2. When the object is viewed along the direction of the arrow, the front surface is identified and with respect to that the other surfaces can be classified. The following are the six surfaces of the object.

1. Front surface is identified as ABCD
2. Rear surface is identified as EFGH
3. Top surface is identified as EFBA
4. Bottom surface is identified as HGCD
5. Right-hand surface is identified as BFGC
6. Left-hand surface is identified as EADH

The view of the block from the front is known as the front view, while the view from the top is known as the top view. The views from the left and the right are known as the respective side views.

While representing the views, the reader is advised to adhere to the following conventions:

1. Front view is represented by the lower case letters of the corners with single prime
2. Top view is represented by the lower case letters of the corners only
3. Side views are represented by the lower case letters of the corners with double prime

It can be observed that the front view represents the dimensions of length and height of the object, while the top view represents the dimensions of length and width of the object. Since the length dimensions are common for both these views, these views are laid one below the other. In practice, in the beginning, let us represent the front view and lay the top view directly below that having the length dimension as common to both the views. (This will be clarified later in detail, when we discuss the chapters on 'Orthographic Projections' and 'Projection of Points')

Similarly, since the side surfaces are perpendicular to the front/rear and the top surfaces, the views from the sides indicate the dimensions other than the common length dimension, that is, the height and the width dimensions. The views from side are always laid by the side of the front view, on the right and the left, respectively, since the height dimension is common to both the views. While sketching the side view, the width dimensions are taken from the top view and the height dimensions are taken from the front view. The right-side view will be placed to the left of the front view, while the left-side view will be placed on its right, with the height dimensions in common in both the views. (More clarifications will be given in the chapters on 'Orthographic Projections' and 'Projection of Points'.)

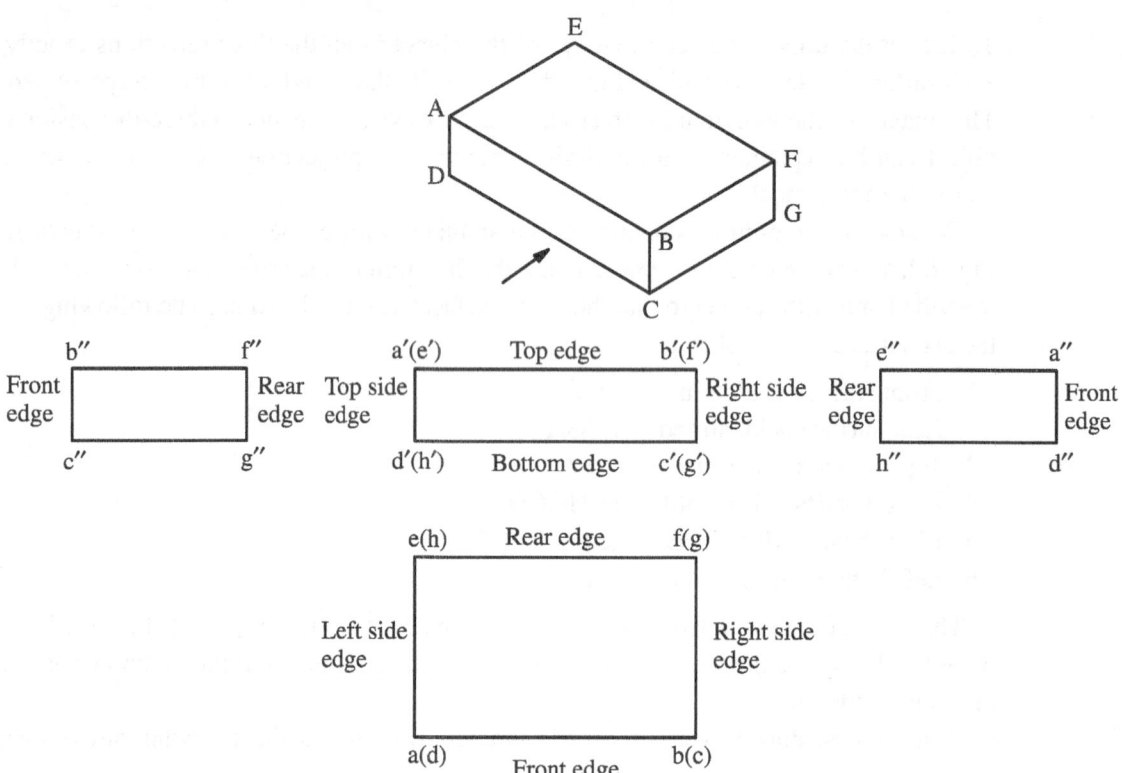

Fig. 4.2   Layout of views of the three-dimensional block

For the three-dimensional rectangular block shown in Fig. 4.2, the contour joining the various corners of the front surface, are shown as, a′, b′, c′, and d′ in the front view. The corners of the rear surface that coincide with the front corners are shown as e′, f′, g′, and h′ and are marked in brackets in the front view.

The top view is laid directly below the front view and is shown as e, f, b, and a. The corresponding corners of the bottom surface that coincide with the corners of the top surface are shown as h, g, c and d and are marked in brackets.

The right-hand side view is represented as b″, f″, g″, and c″ and is marked on the left side of the front view. Similarly, the left-side surface, corresponding to e″, a″, d″, and h″ is marked on the right-hand side of the front view. The aforementioned views can also be referred with the appropriate edges that lie on the respective surfaces. Figure 4.2 explains all the views and the edges connected with them.

4.4 FREEHAND SKETCHING PROCEDURE

Engineers are often required to prepare the multi-view drawings by freehand sketches without drafting tools such as scales, drafters, and protractors to convey the technical ideas in meetings with their colleagues, draftsmen, technicians, etc. This serves as the first level of idea exchange and quicker means of transformation of ideas of the details of the objects visualized in the mind. To acquire the skill in freehand sketching, one should practise drawing straight lines, circles, arcs, and ellipse by freehand to fairly resemble the accurately drawn ones with the instruments. Dimensions of the object are to be made by mental judgment and to the proportions approximately.

4.4.1 Sketching a Straight Line

To sketch a straight line, the pencil should be held between the fingers in a comfortable position. Horizontal lines are sketched from the left to the right, while vertical lines are sketched from the top to the bottom, from a suitable starting point. Inclined lines can be sketched in the upward or downward direction, depending on the starting point and the nature of the inclination. Figure 4.3 shows the above category of lines with arrowheads indicating the movement of the pencil from the respective starting points.

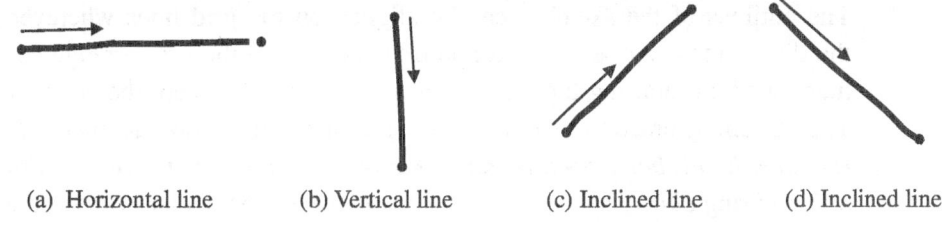

(a) Horizontal line   (b) Vertical line   (c) Inclined line   (d) Inclined line

Fig. 4.3   Sketching a straight line

4.4.2 Sketching a Circle

A circle can be drawn by sketching a vertical and a horizontal line, both intersecting at a point. With this point as centre and radius equal to radius of the circle, marked on the above lines, a circle can be sketched as shown in Fig. 4.4(a). For sketching a finer circle, a set of radial lines can be used as shown in Fig. 4.4(b).

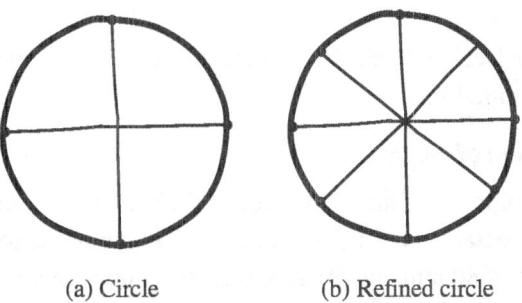

(a) Circle   (b) Refined circle

Fig. 4.4   Sketching a circle

4.4.3 Sketching an Ellipse

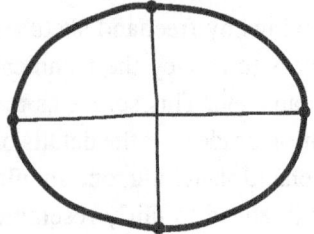

A similar procedure as outlined in sketching a circle can be followed. Semimajor axis distances can be marked on either sides of a line (horizontal line), while semiminor axis distances can be used on the other line (vertical line), and all these points can be sequentially joined by hand to form an ellipse as shown in Fig. 4.5.

Fig. 4.5   Sketching an ellipse

4.5 FREEHAND SKETCHING OF MULITIPLE VIEWS OF OBJECTS

This section deals with the freehand sketching of the multiple views of three-dimensional objects. In all the examples, a solid model of the object will be shown, followed by a pictorial sketch, indicating the various edges of the block and its dimensions. The arrow shown in the pictorial sketch is considered to be the direction of the view for the front surface, and the top and side surfaces are classified with respect to that. The laying out of the views is planned as discussed in Section 4.3 on grid sheets for a few examples. The outlines of the sketches can be aligned on the grid lines wherever possible and the dimensions can be selected approximately from the grid arrangement. As already mentioned, no drafting tools or scales are to be used, except the pencil and the eraser. The sketching practice mentioned in Section 4.4 has to be adopted. *However, in the examples in this book, the lines and other features are shown in regular drawing pattern, only to bring clarity to the student or the learner.* The reader is advised to use the grid sheet arrangements for the other examples too. After sufficient practice, even the grid arrangement can be dispensed with and the sketches can be made by free hand! The examples are planned with a systematic knowledge building by the inclusion of different features to a base solid as described in the following sections.

4.5.1 Multiple Views of Simple Rectangular Block

Example 4.1(a) ◢

Sketch by freehand, the front, top, and the side views of a simple rectangular block shown in Fig. 4.6.

Observation of Views

Since the block is of uniform length, width, and thickness, the front and the rear surfaces appear as rectangles of 80 mm and 20 mm, while the top and bottom surfaces appear as rectangles of 80 mm and 50 mm. Both the side views also appear as rectangles of 50 mm and 20 mm size. Hence, the front view is a rectangle of size 80×20 mm, the top view 80×50 mm, and the side views are of 50×20 mm.

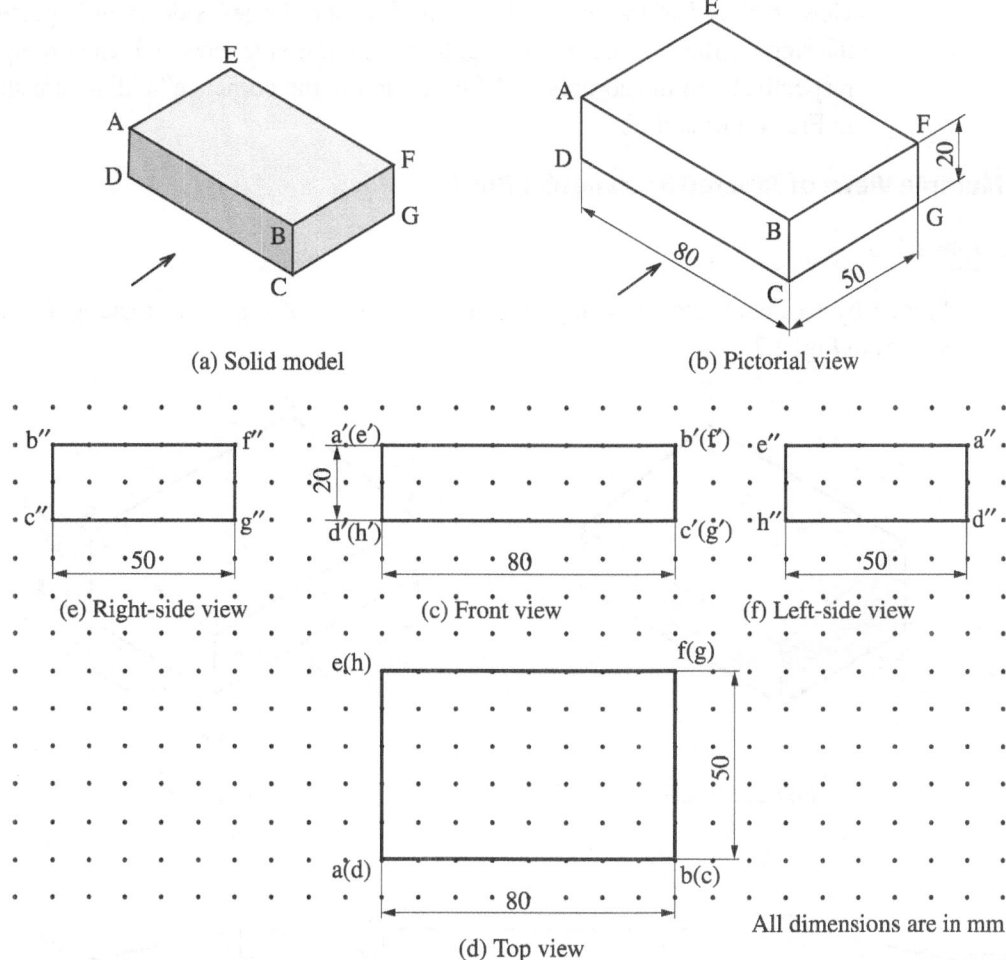

(a) Solid model

(b) Pictorial view

(e) Right-side view

(c) Front view

(f) Left-side view

(d) Top view

All dimensions are in mm

Fig. 4.6 Multiple views of a simple rectangular block

SKETCHING PROCEDURE

Step 1: Sketch a rectangle of length 80 mm and height 20 mm and mark the front surface corner points as a′, b′, c′, and d′. In addition, the coincident rear surface corner points are marked as e′, f′, g′, and h′ in brackets. This completes the front view as shown in Fig. 4.6(c).

Step 2: Sketch another rectangle of length 80 mm and width 50 mm, directly below the front view at any convenient distance. The top surface corner points are marked as e, f, b, and a, while their coincident bottom surface corner points are marked in brackets as h, g, c, and d. The complete the top view is shown in Fig. 4.6(d).

Step 3: Sketch two rectangles of size 50 mm × 20 mm (height) on either side of the front view, to get the side views. As discussed in Section 4.3, the right-side view is placed to the left of the front view and the left-side view is placed to the right of the front view. The right- and the left-side views which correspond, respectively, to the corners b″ f″ g″ c″ and to the corners e″a″d″h″ are shown in Fig. 4.6(e) and (f). ▲

4.5.2 Multiple Views of Tapered Rectangular Block

Example 4.1(b) ▲

Sketch by freehand, the front, top, and the side views of a tapered rectangular block shown in Fig. 4.7.

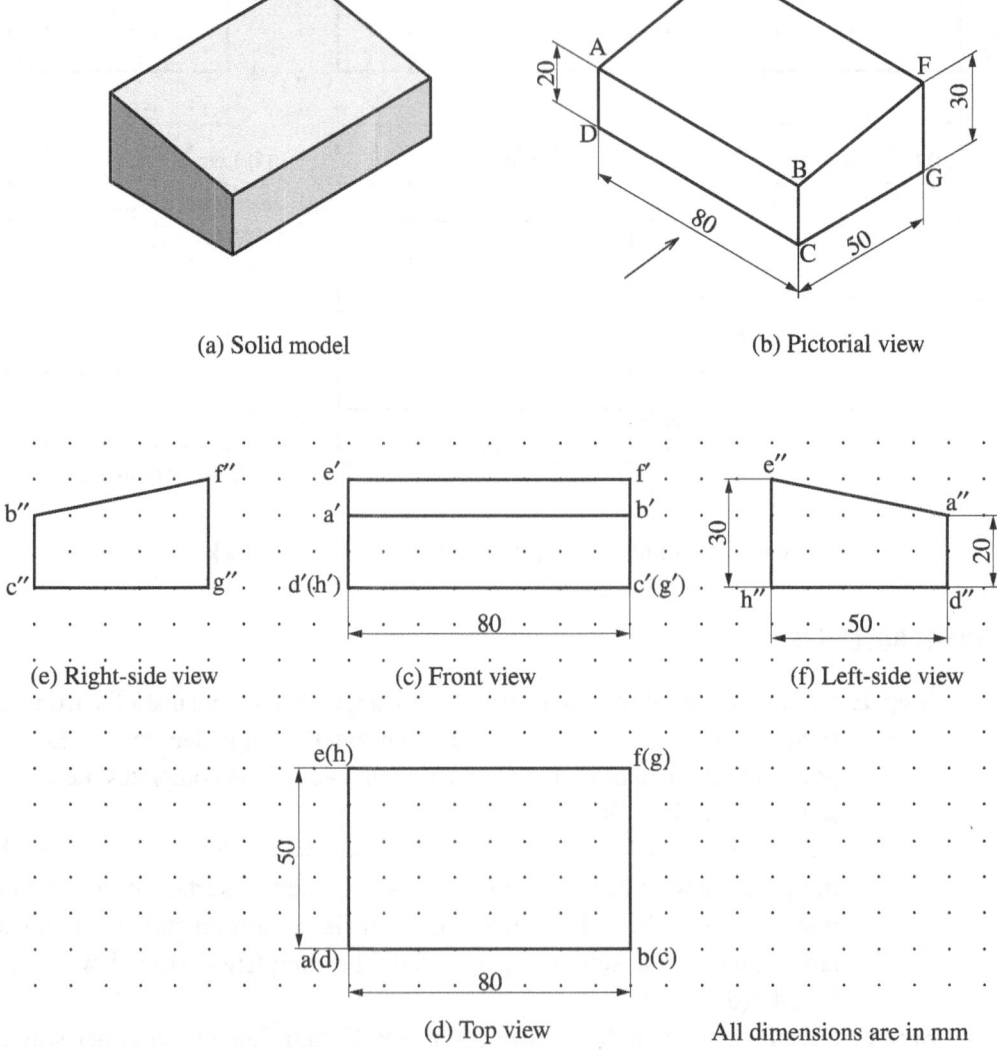

(a) Solid model   (b) Pictorial view

(e) Right-side view   (c) Front view   (f) Left-side view

(d) Top view   All dimensions are in mm

Fig. 4.7 Multiple views of a tapered rectangular block

Observation of Views

Multi-views of a tapered rectangular block are similar to that of a straight rectangular block explained in Example 4.3(a), except marking the slope that happens from the rear surface to the front surface. Hence, the front view is composed of two rectangles, one corresponding to the rear surface and the other representing the front surface. The rectangles have the same base length of 80 mm but with different heights. However, the top view remains as one rectangle, since the top and the bottom surfaces have the same size. In the side view, the taper or the slope of the surface is observed as the line joining the rear edge and the front edge. Note that though the inclined surface is visualized in the front view, the actual slope of the surface is available only in the side views.

SKETCHING PROCEDURE

Step 1: Refer the views discussed in Example 4.1(a) and in Fig. 4.6.

Step 2: Sketch a rectangle of length 80 mm and height 20 mm and mark the front surface corner points a′, b′, c′, and d′. In addition, draw an another rectangle on the same base for a height of 30 mm and mark the coincident rear surface corner points e′, f′, g′, and h′. The points g′ and h′ are marked in brackets as they coincide with the front corner points c′ and d′. This completes the front view as shown in Fig. 4.7(c).

Step 3: Sketch another rectangle of length 80 mm and width 50 mm, directly below the front view at any convenient distance. The top surface corner points e, f, b, and a and their coincident bottom corner points h, g, c, and d marked in brackets completes the top view as shown in Fig. 4.7(d).

Step 4: Sketch the line f″ g″ of height 30 mm to represent the rear edge and the line c″ b″ of height 20 mm to represent the front edge, at a distance of 50 mm, being the width of the block. The right-side view of the object b″f″g″c″ is shown in Fig. 4.7(e). Similarly, construct the mirror image e″a″d″h″ to get the left-side view as shown in Fig. 4.7(f).

Note that though the inclined surface is visualized in the front view, the actual slope of the surface is available only in the side views.   ▲

4.5.3 Mulitple Views of Rectangular Block with Cut-outs and Additional Features

In this section, the realization of multiple views of a rectangular block will be discussed, when some features such as slots or cut-outs are made on it or when some features are built over it. The front, top, and the side views are drawn for the basic rectangular block as discussed in Section 4.5.1 and additions or deletions are carried out in the respective views according to the nature of the feature introduced. The example shown in Fig. 4.6 is retained as the basic rectangular block and its corresponding views are considered as reference views and discussions or changes in the relevant views will be made as per the forthcoming exercises and the features introduced there on. The corners of the basic rectangular block are indicated with alphabets—A, B, C, D, …., whereas the corners of the feature discussed are referred in numbers 1, 2, 3, 4, …. Since the observation of views and the freehand sketches are elaborated for the basic rectangular block in Example 4.1(a), they will not be repeated in this section again. The observation of views along with the additional features and their effect on the respective views will be discussed. Since this section deals with each feature in turn, the numbering will be made for the current feature discussed, while the views of the previous features will be retained with no numbering scheme, to avoid repetition and also to bring more clarity on the feature of interest in that example.

Rectangular Block with a Rectangular Cut-out in the Front Surface

Example 4.2(a)

Sketch by freehand, the front, top, and the side views of a rectangular block with a rectangular cut-out in the front surface shown in Fig. 4.8.

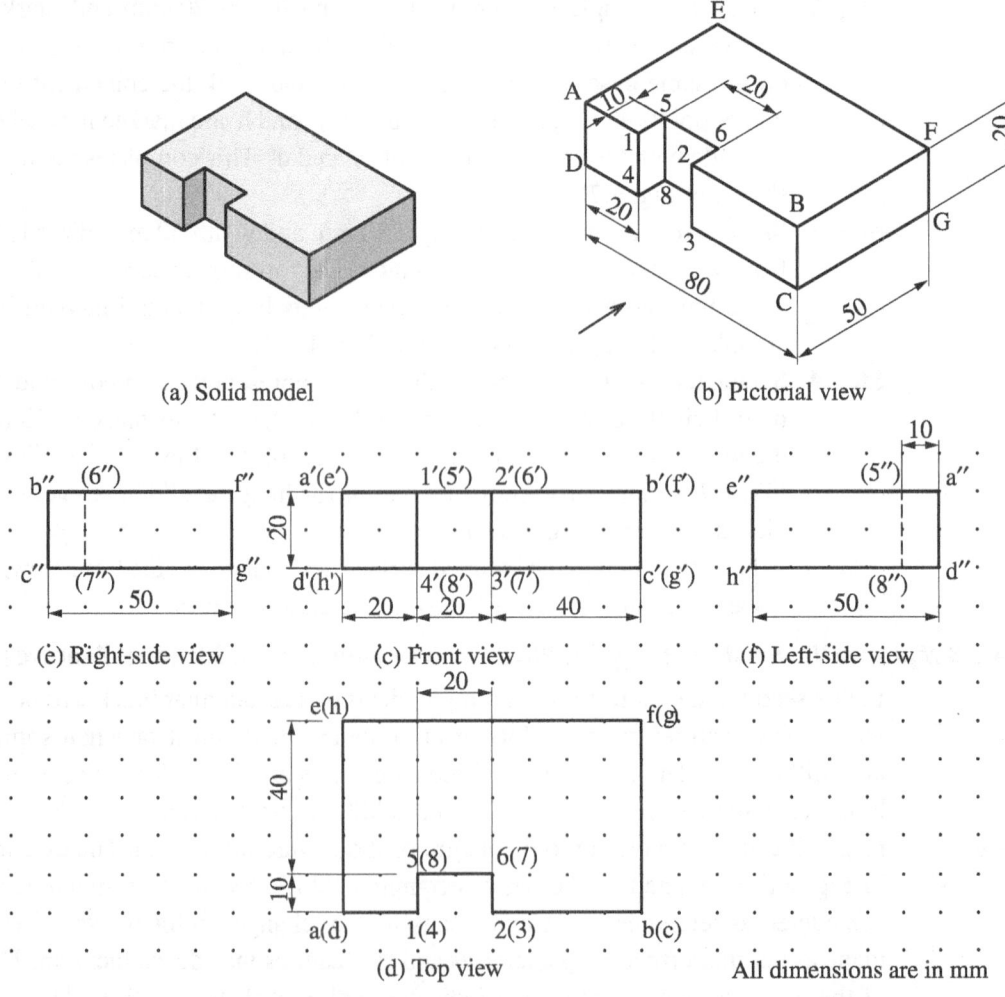

(a) Solid model

(b) Pictorial view

(e) Right-side view

(c) Front view

(f) Left-side view

(d) Top view

All dimensions are in mm

Fig. 4.8 Rectangular block with a front cut-out

Observation of Views

Since the cut-out made an opening in the front surface, two lines corresponding to the front edges of the cut-out will also appear in the front view. The top view does not appear as a full rectangle but with a discontinuity in the front edge. Both the right-hand side and the left-hand side surfaces appear as rectangles at their surface level and the outline of the side views do not change from that of the basic rectangular block. However, since the cut-out happens in the inside surface of the block, the inner edges of the cut-out are marked in dotted lines.

SKETCHING PROCEDURE

Step 1: Refer the views discussed in Example 4.1(a) and in Fig. 4.6.

Step 2: Since the cut-out begins at a distance of 20 mm from the front edge a′d′, mark a line 1′4′ in the front view. Similarly, draw a line 2′3′ at a distance of 40 mm from a′d′, to represent the end of the cut-out. This completes the front view as shown in Fig. 4.8(c).

Step 3: The line 'ab' is not drawn as a continuous line as shown in Fig. 4.8(c) but as a line a1 for a length of 20 mm, a perpendicular line 15 for a length of 10 mm, a horizontal line 56 of length 20 mm, a perpendicular line 62 for 10 mm length and finally as a line 2b for the remaining length as shown in Fig. 4.8(d). The complete top view is marked as 'ea1562bf'.

Step 4: Sketch two rectangles of size 50 mm × 20 mm (height) on either side of the front view, to get the side views and mark dotted lines at a distance of 10 mm from the front edges as shown in Fig. 4.8(e) and (f).   ▲

Rectangular Block with a Step Cut at the Corner

Example 4.2(b)

Sketch by free hand the front, top, and the side views of the rectangular block shown Fig. 4.9 with a step cut at the corner.

Observation of Views

This object has an additional feature in the form of a cut in the left corner for a depth of 10 mm, resulting in the removal of material in the front-, top-, and left-side surfaces. Hence, squares of 10 mm size will appear in all the views around the corner A, which is removed now. Accordingly, the corner A will not appear in the views.

SKETCHING PROCEDURE

Step 1: Refer the views discussed in Example 4.2(a) and in Fig. 4.8.

Step 2: Add a square of size 10 mm around the top left corner in the front view and mark the visible edges 5′2′, 2′3′, 3′4′, and 4′5′ to get the front view as shown in Fig. 4.9(c). Note the corner a′ is not marked as it is removed due to the cut.

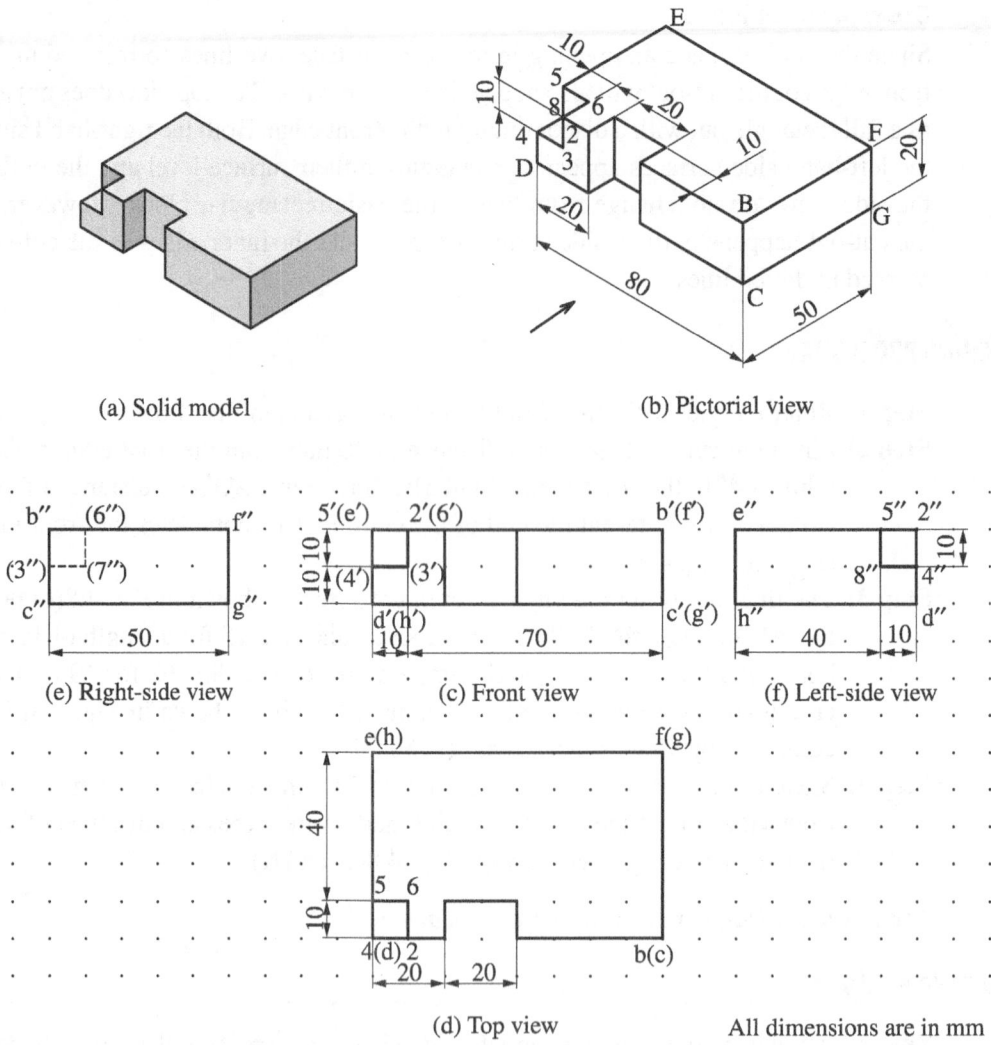

(a) Solid model

(b) Pictorial view

(e) Right-side view

(c) Front view

(f) Left-side view

(d) Top view

All dimensions are in mm

Fig. 4.9 Rectangular block with a corner cut-out

Step 3: Add a square of size 10 mm around the bottom left corner in the top view and mark the visible edges 56, 62, 24, and 45 to get the top view as shown in Fig. 4.9(d). Note the corner 'a' is not marked as it is removed due to the cut.

Step 4: Similarly, draw two squares of size 10 mm each at the top corner a″ of the front edge. Hence, in the left-side view, a square of 10 mm size appears around the top right corner. However, since the right-side surface does not experience the cut on its face but only at the rear surface, the square is marked in dotted line. The right-hand and the left-hand side views are shown in Figs 4.9(e) and (f). ▲

Rectangular Block with a Slope Cut in One of the Corners

Example 4.2(c)

Sketch by free hand the front, top, and side views of the rectangular block shown in Fig. 4.10, which has a slope cut in one of the corners.

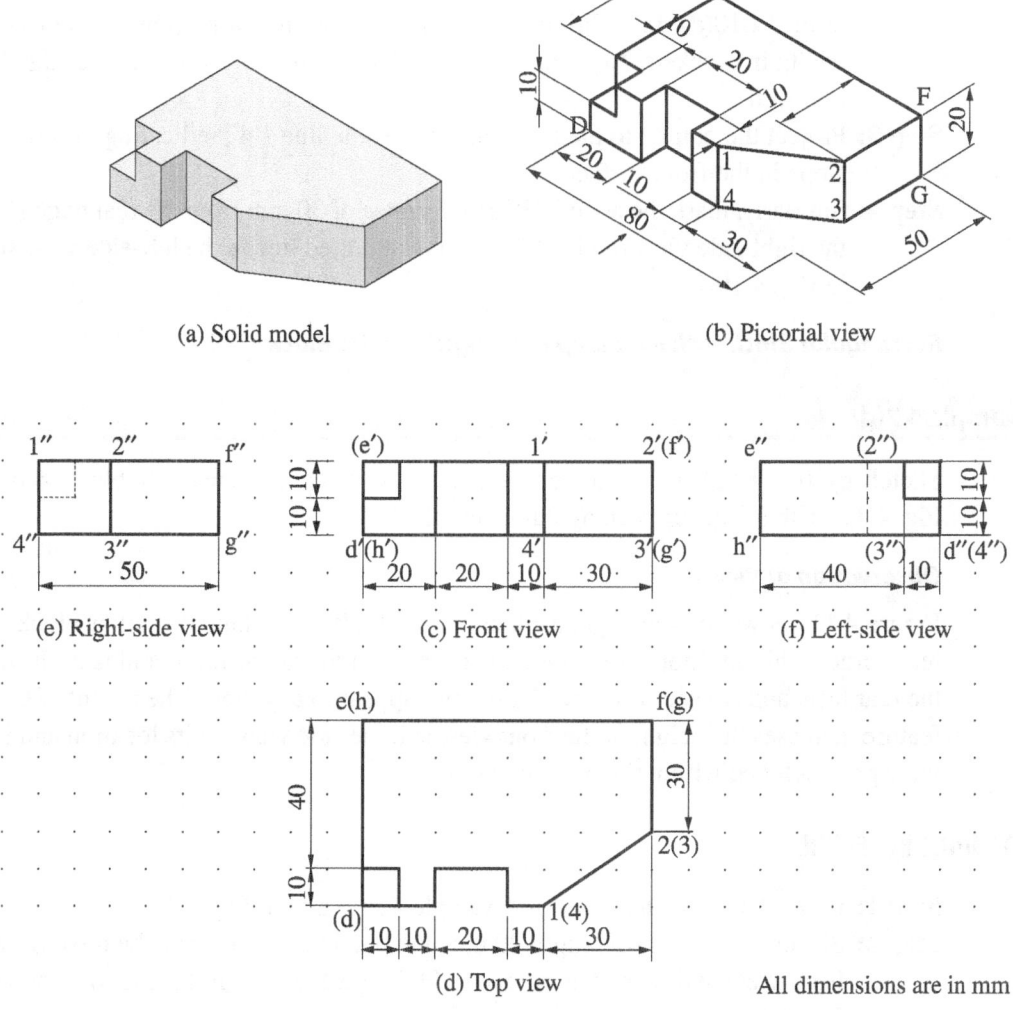

(a) Solid model   (b) Pictorial view

(e) Right-side view   (c) Front view   (f) Left-side view

(d) Top view   All dimensions are in mm

Fig. 4.10   Rectangular block with a slope cut

Observation of Views

The slope or the chamfer made has removed the right front edge BC, and hence, it does not appear in the views. The top view is not a rectangle but with the sloping line on the right, with edge BC removed. Since the slope starts in the front surface at 14, a line to this effect will appear within the front view.

SKETCHING PROCEDURE

Step 1: Refer the views discussed in Example 4.2(b) and in Fig. 4.9.

Step 2: As the slope starts at point 1 in the front surface at a distance of 30 mm from the edge BC, mark it in the top view on the line 'db' at a distance of 30 mm from b. Similarly, as the slope ends at 2 on the side surface, mark it at a distance of 30 mm from the rear corner f, in the top view. The top view is marked as 'ed12f' in Fig. 4.10(d). In addition, the coincident bottom corner points are marked as hd43g in brackets. Note that the points b and c do not appear as the edge BC is removed.

Step 3: Project the point 1 to the front view to get the line 1'4', indicating the change of slope in the front surface.

Step 4: Similarly, mark the edge 2"3" at a distance of 30 mm from the rear edge f"g" in the right-side view in Fig. 4.10(e), and in dotted line in the left-side view shown in Fig. 4.10(f).   ▲

Rectangular Block with a Built up Rectangular Pillar Block

Example 4.2(d) ▲

Sketch by free hand the front, top, and side views of a rectangular block shown in Fig. 4.11, with a built-up rectangular pillar block.

Observation of Views

Figure 4.11 shows the rectangular block with a built-up rectangular pillar block at the rear corner. This additional feature made is around the rear corner F and is in flush with the rear face, and hence, the corner F does not appear in any view. The height of the new feature increases the height of the front view and the side views at its location and its top view gets included within the base rectangle.

SKETCHING PROCEDURE

Step 1: Refer the views discussed in Example 4.2(c) and in Fig. 4.10.

Step 2: Sketch a rectangle of length 20 mm and height 30 mm above the top edge in the front view and mark the points 1'2'3'4' and 5'6'7'8' (in brackets) as shown in Fig. 4.11(c). Remove the corner f' as discussed.

Step 3: Mark a rectangle of 20 mm length and 10 mm width around the corner g to get the top view as shown in Fig. 4.11(d).

Step 4: Add a rectangle of height 30 mm and width 10 mm in the side view, at the rear edge through the corner g", to get the right-side view as shown in Fig. 4.11(e).

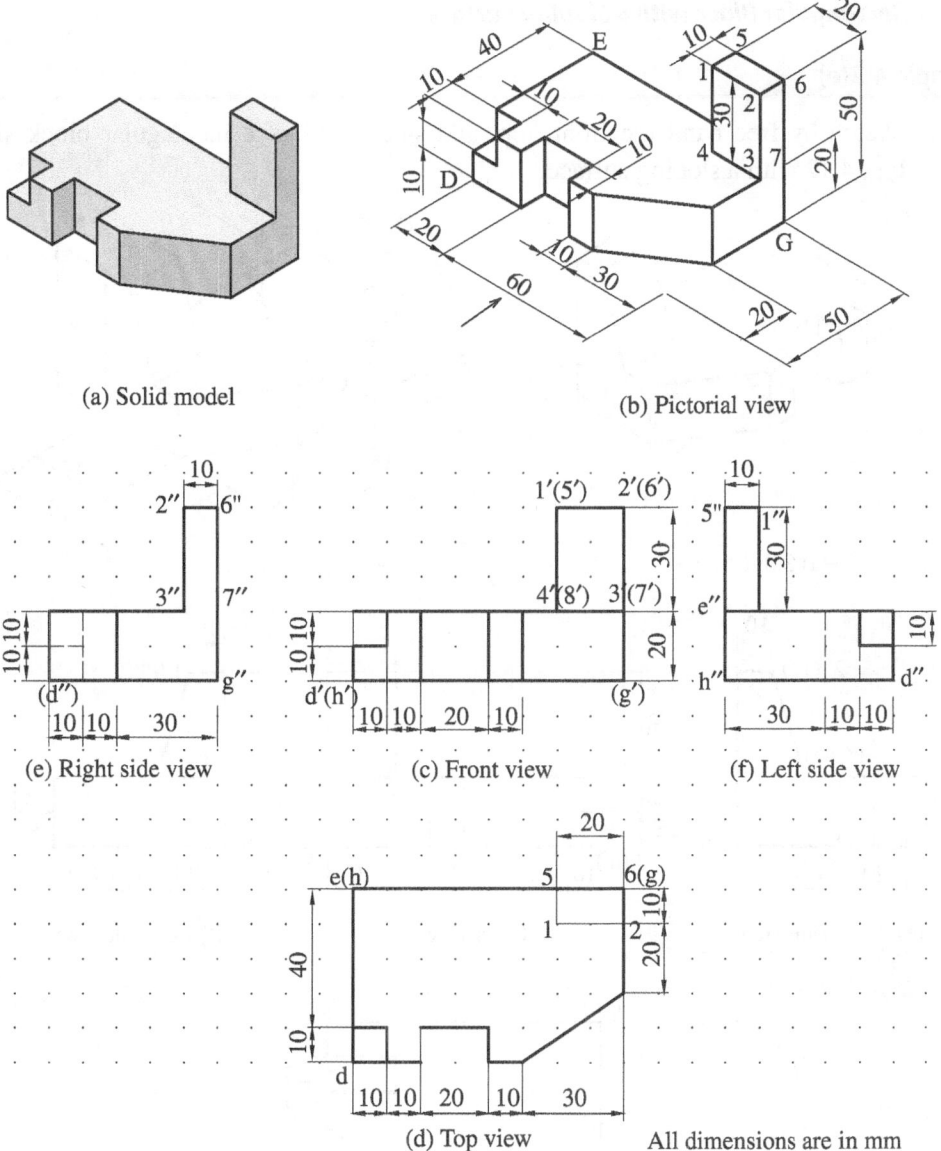

(a) Solid model

(b) Pictorial view

(e) Right side view

(c) Front view

(f) Left side view

(d) Top view   All dimensions are in mm

Fig. 4.11   Rectangular block with a built-up block

Mark the points 2″, 6″, and 3″. Note that the corner 7″ and f″ do not appear as the side surfaces are in flush with each other. Note also the side view appears L shaped with no joining line in between.

Step 5: Make a similar construction for the other side view and indicate the separation line of the rectangle as in Fig. 4.11(f).

Rectangular Block with a Sloping Surface

Example 4.2(e)

Sketch by free hand the front, top, and side views of a rectangular block shown in Fig. 4.12 with a sloping surface.

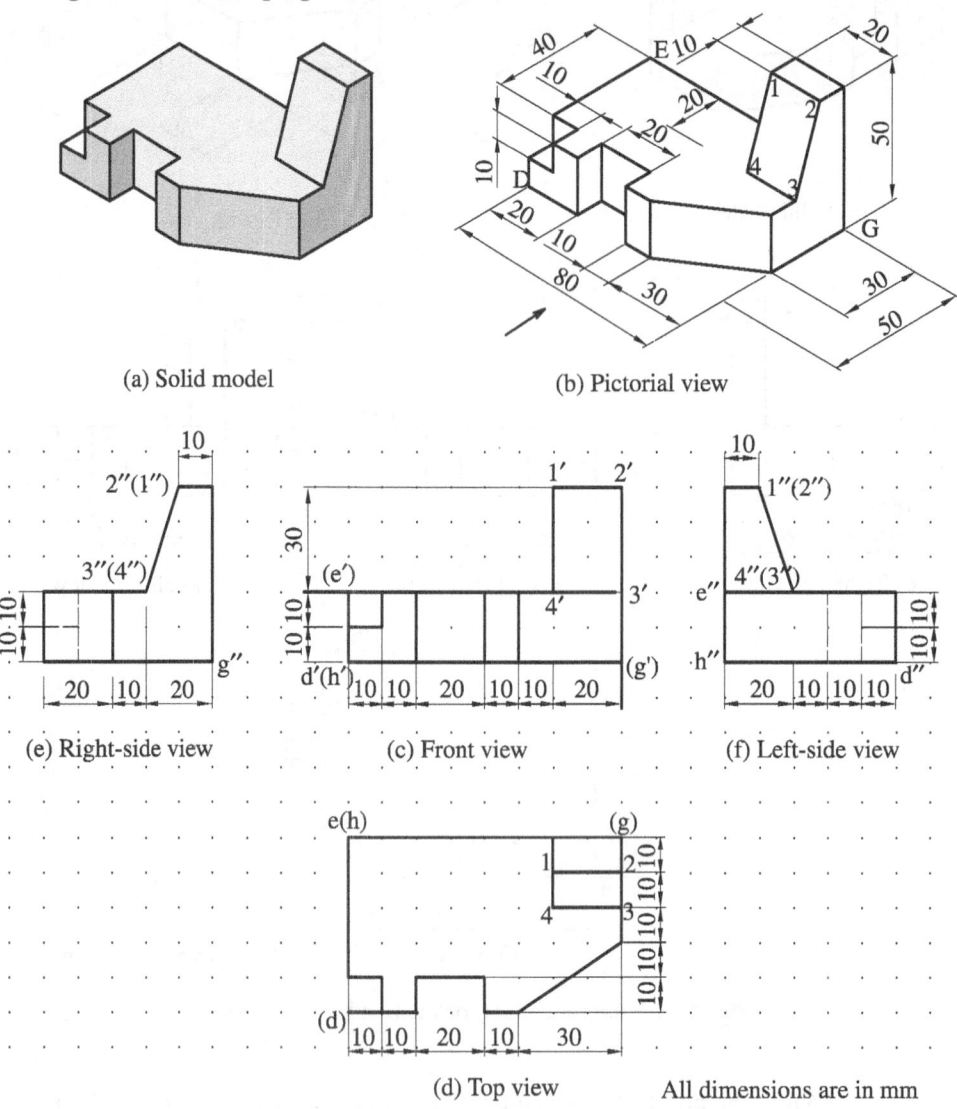

(a) Solid model

(b) Pictorial view

(e) Right-side view

(c) Front view

(f) Left-side view

(d) Top view   All dimensions are in mm

Fig. 4.12   Rectangular block with a built-up sloping surfaces

Observation of Views

Since the sloping surface is added in the front face of the pillar block, it does not change the front view explained in the Example 4.2(d) but for a small additional rectangle in the top view. As discussed in earlier examples, the side views register the slope.

SKETCHING PROCEDURE

Step 1: Refer the views discussed in Example 4.2(d) and in Fig. 4.11.

Step 2: The front view remains same as shown in Example 4.2(d) (refer Fig. 4.11).

Step 3: Add in the top view an additional rectangle 1, 2, 3, 4 of length 20 mm and width 10mm to the existing one in the top corner to get the top view as shown in Fig. 4.12(d).

Step 4: Mark the sloping lines in the side views above the top surface of the base block as shown in Fig. 4.12(e) and (f). ▲

Rectangular Block with a Centre Block

Example 4.2(f)

Sketch by free hand the front, top, and side view of the rectangular block shown in Fig. 4.13 with a centre rectangular block.

Observation of Views

A centre rectangular block is raised on the base rectangular block. This feature is similar to the one erected at the corner and discussed in Fig. 4.12. Owing to this feature, rectangles of 10 × 40 mm, 10 × 20 mm, and 20 × 40 mm appear, respectively, in the front, top, and the side views.

SKETCHING PROCEDURE

Step 1: Refer the views discussed in Example 4.2(e) and in Fig. 4.12.

Step 2: Erect a rectangle of 10 mm length and 40 mm height above the top surface in the front view, by positioning the 40 mm edge at a distance of 35 mm from the edge through g′, to get the front view as shown in Fig. 4.13(c).

Step 3: Add a rectangle of 10 mm length and 20 mm width in the top view by positioning the smaller edge at a distance of 15 mm from the front edge (lower horizontal line) and exactly below the front view of the new feature available in the front view. This gets the top view as shown in Fig. 4.13(d).

Step 4: Sketch a rectangle of 20 mm width and 40 mm height in the side views (refer Fig. 4.13e and f). In the left-hand side view, it is shown as a full rectangle, while in the right-hand side view it is shown partially, as the sloping block is visible before that. ▲

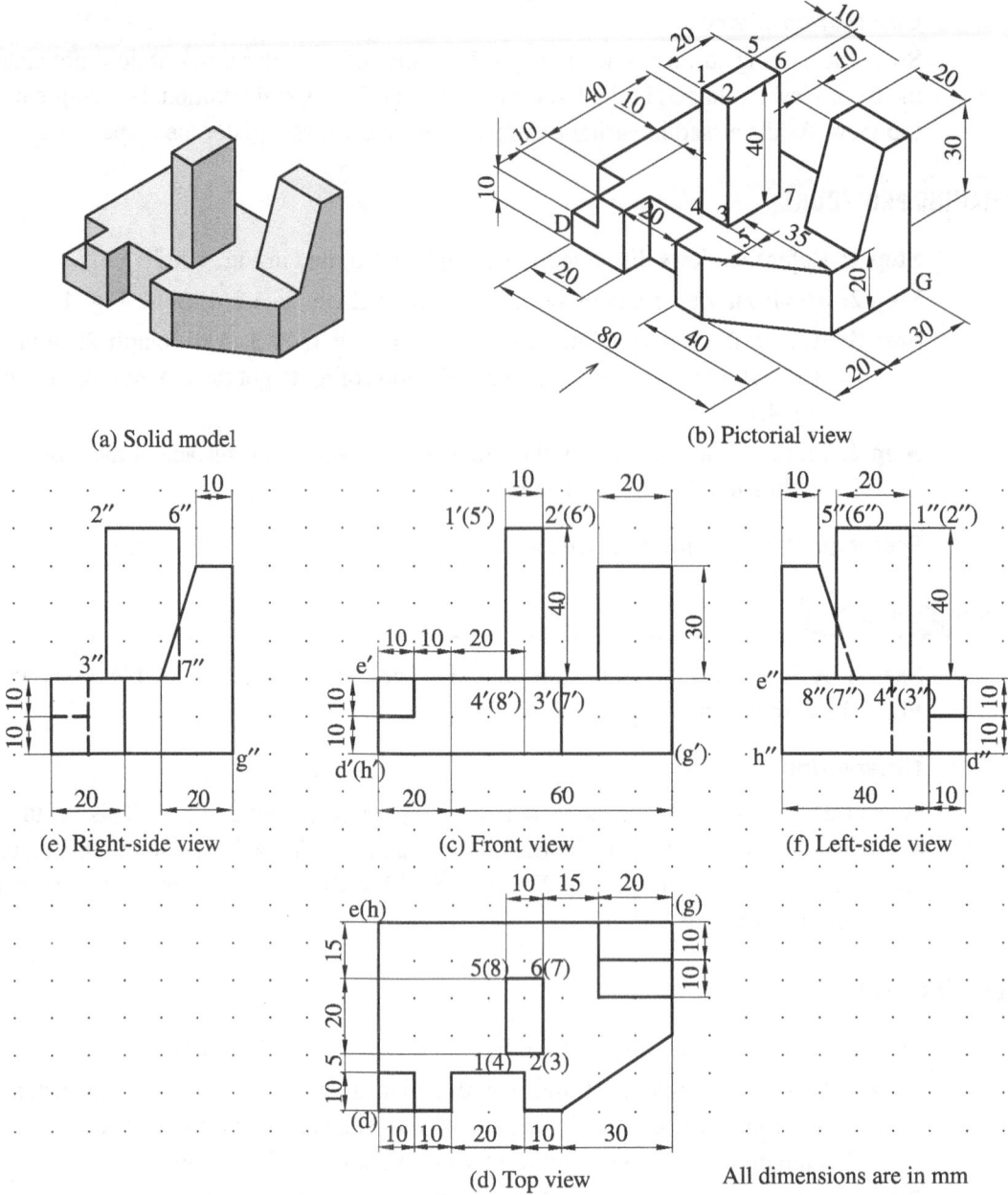

(a) Solid model

(b) Pictorial view

(e) Right-side view

(c) Front view

(f) Left-side view

(d) Top view   All dimensions are in mm

Fig. 4.13   Rectangular block with an additional centre block

Rectangular Block with Rectangular Slot and a Circular Hole

Example 4.3(a)

Sketch by free hand the front, top, and the side views of the basic rectangular block with a circular hole in the centre as shown in Fig. 4.14.

Observation of Views

The rectangular slot and the circular hole are made from the top surface of the basic rectangular block. They are made throughout the depth of the rectangular block.

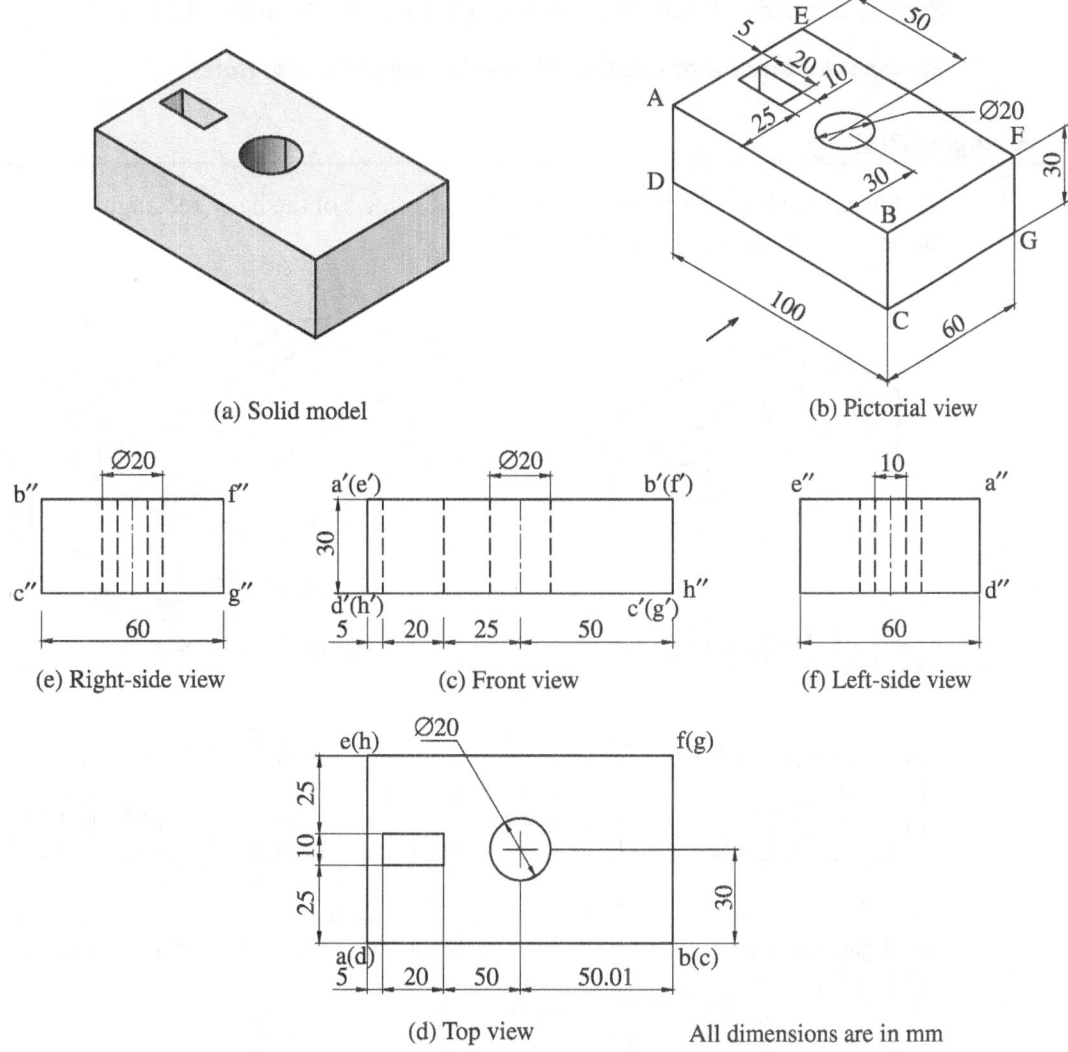

(a) Solid model

(b) Pictorial view

(e) Right-side view

(c) Front view

(f) Left-side view

(d) Top view   All dimensions are in mm

Fig. 4.14   Rectangular block with a central circular hole

Generally, when no mention is made about the depth of the slot/hole, it is considered to be made throughout the depth of the block. Therefore, none of the boundary edges of these features are visible in the views except their cross-sections in the top view.

SKETCHING PROCEDURE

Step 1: Refer the views of the simple rectangular block as in Example 4.1(a).

Step 2: Add a rectangle and a circle in the top view as per the dimensions and the orientations (refer Fig. 4.14d).

Step 3: Obtain dotted lines in the front view (refer Fig. 4.14c) corresponding to the slot and the circular hole made in the top view. The axis line of the cylindrical hole is marked from the centre of the circle in the top view.

Step 4: Similarly, sketch the right- and left-side views as in Fig. 4.14e and f.   ▲

Rectangular Block with Elliptical Slot and Multiple Circular Holes

Example 4.3(b) ▲

Sketch by free hand the front, top, and the side views of the basic rectangular block with the set of slots and holes as shown in Fig. 4.15.

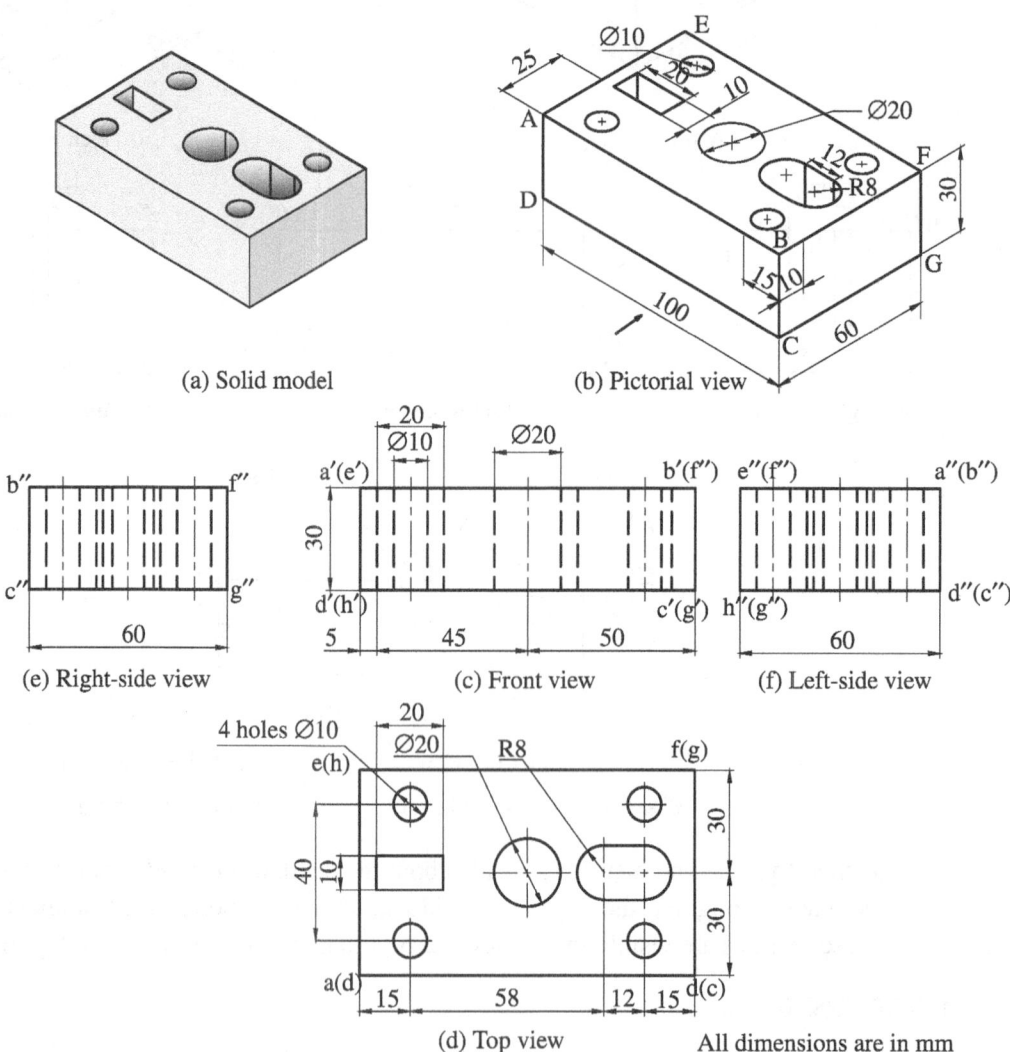

(a) Solid model   (b) Pictorial view

(e) Right-side view   (c) Front view   (f) Left-side view

4 holes Ø10

(d) Top view   All dimensions are in mm

Fig. 4.15   Rectangular block with a set of slots and circular holes

Observation of Views

The current features of interest are an elliptical slot and a set of holes on the top surface and extending throughout the depth of the block. As given in Example 4.3(a), their cross-section appear in the top view as per their location scheme, with dotted lines representing the curved edges in the other views.

SKETCHING PROCEDURE

Step 1: Refer the views discussed in Example 4.3(a) and in Fig. 4.14.

Step 2: Mark the centres of various holes and elliptical slot in the top view at their respective locations from the edges and draw circles/slot contours corresponding to their diameters/widths as shown in Fig. 4.15(d).

Step 3: Sketch the axis lines corresponding to their centres, in the front view and draw dotted lines corresponding to their boundaries as shown in Fig. 4.15(c).

Step 4: Similarly sketch the right- and left-side views as in Fig. 4.15(e) and (f).   ▲

Rectangular Block with Two Curved Surfaces

Example 4.4(a) ▲

Sketch by free hand the front, top, and side views of the rectangular block with two curved surfaces as shown in Fig. 4.16.

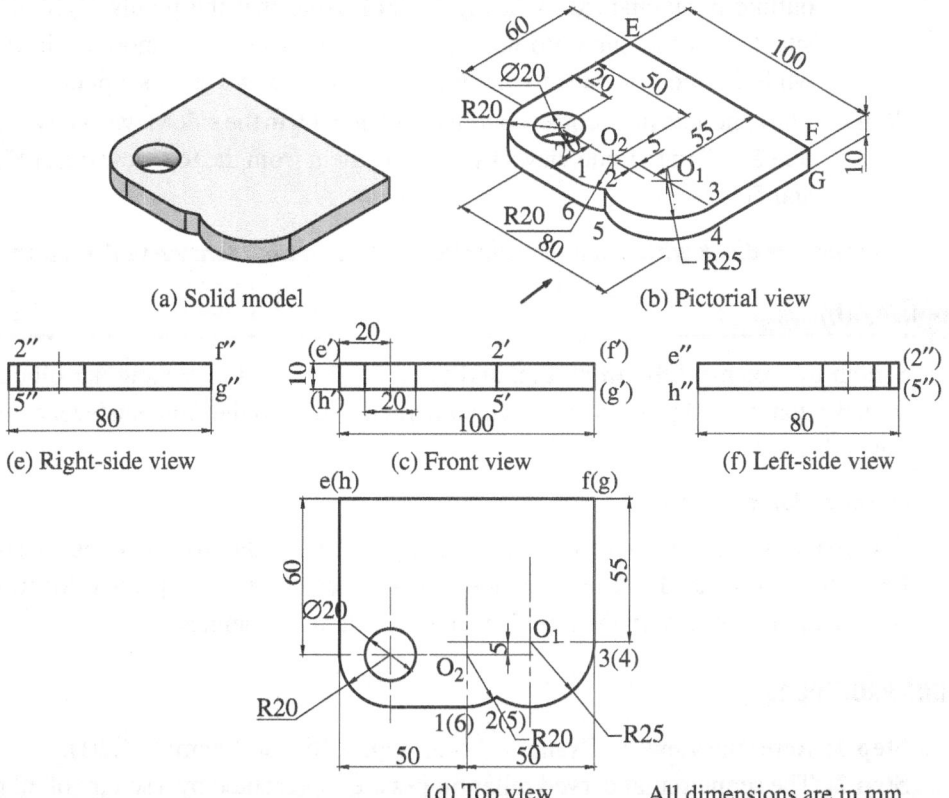

(a) Solid model

(b) Pictorial view

(e) Right-side view

(c) Front view

(f) Left-side view

(d) Top view   All dimensions are in mm

Fig. 4.16   Rectangular block with two curved surfaces

Observation of Views

All the curved surfaces are provided in the front surface. As discussed earlier, both the curves appear in the top view, while the other views are realized with their boundaries. The curves are drawn after locating their centres.

SKETCHING PROCEDURE

Step 1: Sketch the outline views of the full basic rectangular block as in Example 4.1(a) and in Fig. 4.6. Mark an arc of 20 mm radius at the bottom left corner in the top view region after marking its centre from the front and left-side edge. Locate the centre O_1 at a distance of 25 mm from ab and bf and draw an arc of 25 mm radius to get the point 3 on the edge bf as shown in Fig. 4.16(d).

Step 2: Similarly, locate the centre O_2 at its respective distances from 'ab' and 'ea' and make an arc of radius 20 mm to get the point 1 on the front edge. Both the arcs meet at the point 2 to get the curved surface 1-2-3. Mark the coincident points 6, 5, and 4 in brackets. Remove the corners a, b, and c and their coincident points.

Step 3: Mark a line 2′-5′ in the front view, corresponding to the points 2 and 5 from the top view. Though the front corners a′, b′, and c′ are removed, the front view outline is maintained as in Fig 4.16(c). Note that the points 1′, 3′, 4′, and 6′ situated on the curve do not appear as the curves join smoothly in the faces, while 2′5′ corresponds to the intersection of the two curves, appears.

Step 4: Mark the axis lines corresponding to O_1 and O_2 in the side views. Draw the dotted line 2″-5″ in the side views by locating them from the top view (refer Fig. 4.16e and f).

Rectangular Block with a Semicircular Pillar Block in the Transverse Direction

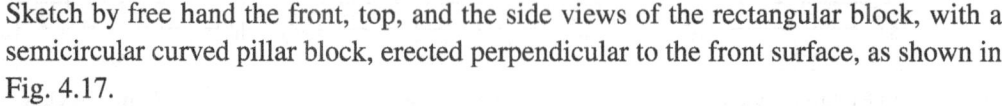

Example 4.4(b)

Sketch by free hand the front, top, and the side views of the rectangular block, with a semicircular curved pillar block, erected perpendicular to the front surface, as shown in Fig. 4.17.

Observation of Views

The features that are added in this example are similar to the ones discussed in Example 4.2(f) and the front and side views appear at their respective locations with semicircular contours. In the top view, they appear as rectangles.

SKETCHING PROCEDURE

Step 1: Refer the views in Example 4.4(a), Fig. 4.16, and Example 4.2(f).

Step 2: The semicircular curved pillar blocks are represented by a square of 30 mm size and a semicircle built over it, in the front view (Fig. 4.17c) and in the side views (Figs 4.17e and 4.17f).

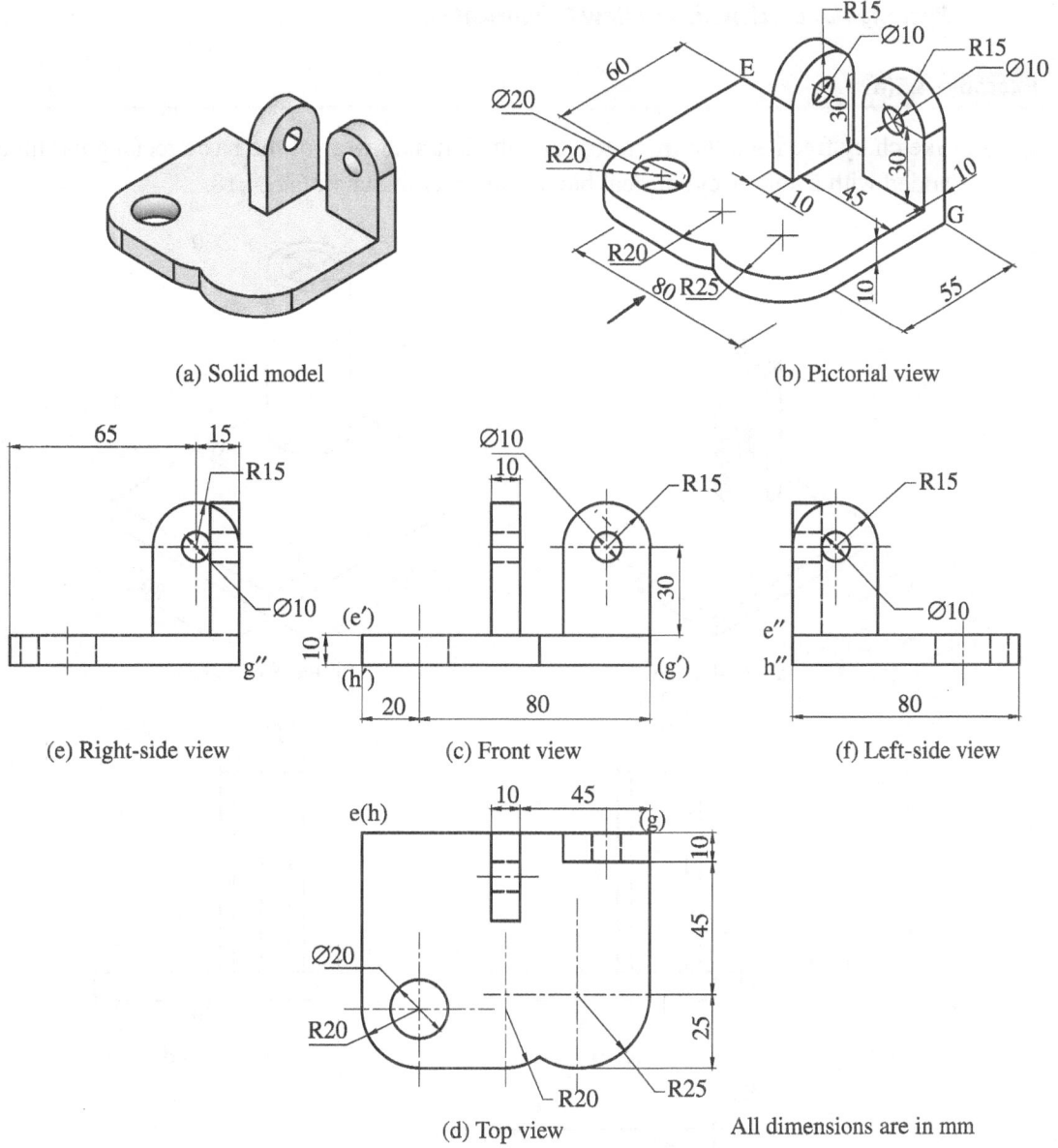

(a) Solid model

(b) Pictorial view

(e) Right-side view

(c) Front view

(f) Left-side view

(d) Top view   All dimensions are in mm

Fig. 4.17 Rectangular block with semicircular pillar blocks

Step 3: Two rectangles of size 10 mm × 45 mm are placed in perpendicular directions in the top view (Fig. 4.17d).

Step 4: The circular holes are also marked in the respective locations.

4.5.4 Multiple Views of Simple Rectangular Block with Cylindrical and Conical Blocks

In continuation with the discussions in the Section 4.5.5, dealing with the curved surfaces and the arcs, extension works are carried out in this section with cylindrical and conical-shaped blocks added to the basic rectangular block explained earlier.

Rectangular Block with a Hollow Cylindrical Block

Example 4.5(a) ▶

Sketch by free hand the front, top, and the left-side view of the basic rectangular block added with a hollow cylindrical block over it as shown in Fig. 4.18.

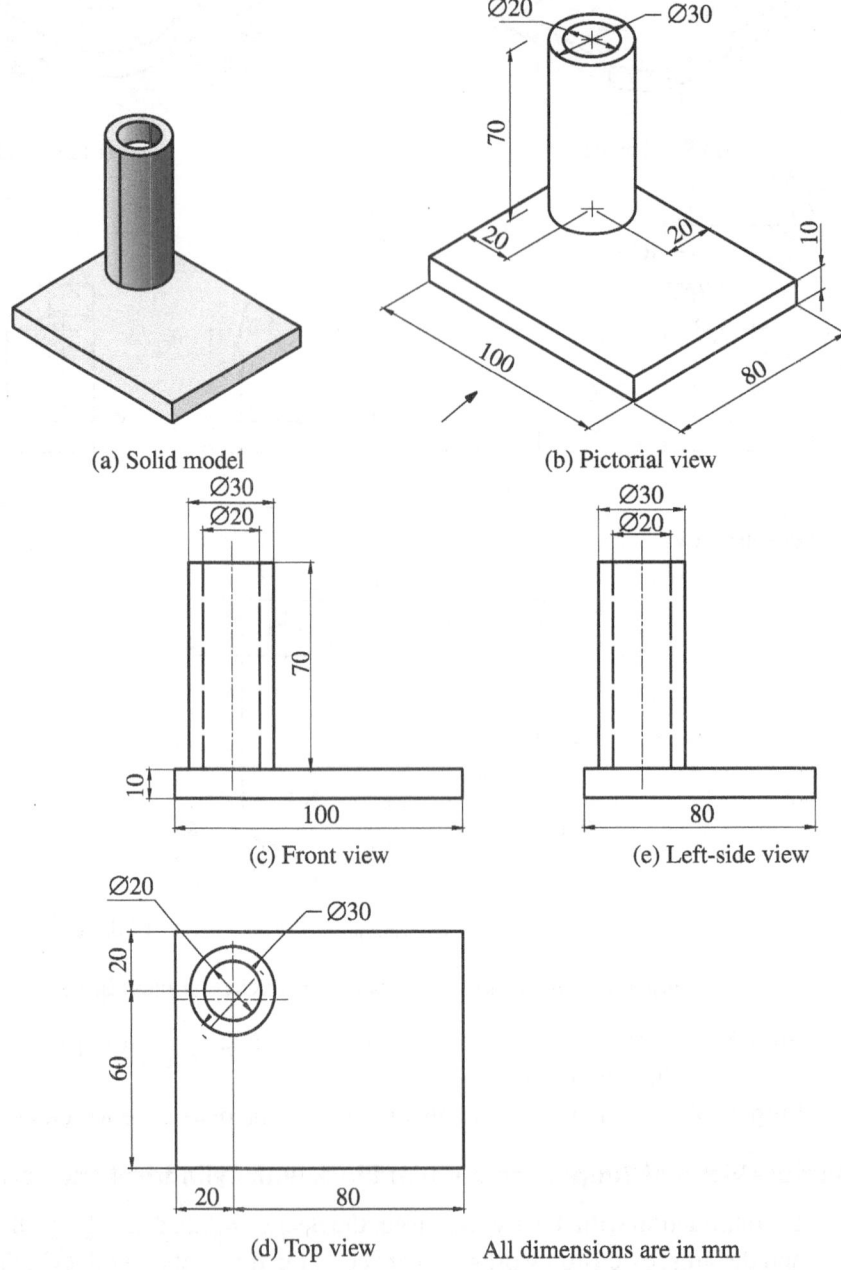

(a) Solid model

(b) Pictorial view

(c) Front view

(e) Left-side view

(d) Top view

All dimensions are in mm

Fig. 4.18 Rectangular block with an additional hollow cylindrical block

Observation of Views

Since the hollow cylindrical block is raised over the top surface of the rectangular block, its circular cross section is seen as two concentric circles in the top view, while the other views maintain their outlines as rectangles. The inner surfaces are marked in dotted lines.

SKETCHING PROCEDURE

Step 1: Sketch the views of the basic rectangular block.

Step 2: Mark two concentric circles of outer and inner diameter of 30 mm and 20 mm, after locating the centre at a distance of 20 mm from the rear and the left-side edges as shown in Fig. 4.18(d).

Step 3: Mark the axis line in the front view and draw a rectangle of length 30 mm and height 70 mm. Add dotted lines inside the rectangle as shown in Fig. 4.18(c).

Step 4: Similarly obtain the left-side view of the cylindrical block as shown in Fig. 4.18(e). ▲

Rectangular Block with a Truncated Conical Block

Example 4.5(b) ◢

Sketch by free hand the front, top, and the side views of the basic rectangular block added with a truncated conical block over it, as shown in Fig. 4.19.

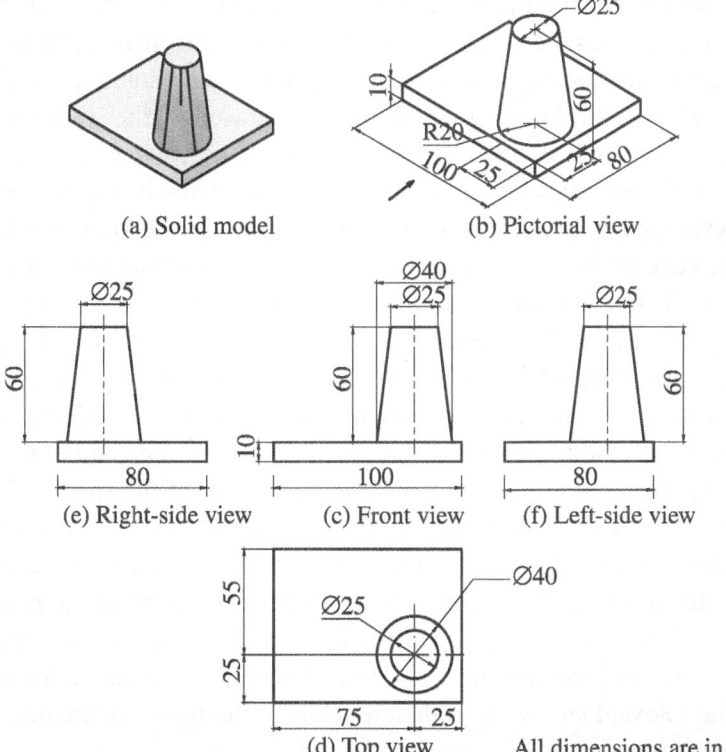

(a) Solid model

(b) Pictorial view

(e) Right-side view

(c) Front view

(f) Left-side view

(d) Top view   All dimensions are in mm

Fig. 4.19   Rectangular block with truncated conical block

Observation of Views

The feature added is a truncated conical block of base diameter 40 mm, top diameter 25 mm, and height 60 mm, built over the top surface of the basic rectangular block. While the top view shows the cross section of the top and the bottom surfaces of this feature, the other views register the out lines, which are sloping lines.

SKETCHING PROCEDURE

Step 1: Sketch the views of the basic rectangular block.

Step 2: Sketch two concentric circles corresponding to the top and bottom diameters of 40 mm and 25 mm, after locating the centre at a distance of 25 mm from the front and the right-side edges as shown in Fig. 4.19(d).

Step 3: Mark the axis line in the front view and add two parallel lines of 25 mm and 40 mm length at a vertical distance of 60 mm and join their ends with sloping lines as shown in Fig. 4.19(c).

Step 4: Similarly obtain the side views of the conical block as shown in Fig. 4.19(e) and (f). ▲

4.5.5 Mulitiple Views of Various Three-dimensional Objects

In Sections 4.5.1–4.5.4, freehand sketching of the multiple views of a basic three-dimensional rectangular block was discussed, when many features in the form of cut-outs, slots, and holes were made on it. The views were also explained when the cuts were resulting in sloping or inclined edges or cut-outs made in the form of circular arcs, single, or multiple in nature. With this knowledge, the students can familiarize themselves with their experiences on birthday cakes, when they are cut into small pieces of different shapes for distribution to their friends. Further, such knowledge can be developed from one's own home experiences in the kitchen, when they engage in cutting vegetables and fruits, partly to help their parents and partly to enjoy their own share! This topic can be understood clearly and confidently only, when one exposes himself/herself unassumingly to such experiences. In this section, we will discuss the sketching of views of three-dimensional objects, when multiple features are added in combinations, to get the desired shape of the object. The fundamental approach that we adopted earlier will be kept as the basis and explanations will be given as and when required to understand the impact of the specific feature in all the three views. In order to increase the confidence levels and understanding abilities, the students are advised to refer the pictorial view of the object (and not the solid model picture). The solid model pictures are given only for initial practice and not as a support for ever. As discussed earlier, looking in the direction of the arrow decides the front view and the top view is decided when we go above the object and look down. The relevant side view is decided depending upon which side of the object is available for view.

L-shaped Block with a Rectangular Cut-out

Example 4.6

Sketch by free hand the front view, top view, and the side views of the L-shaped object shown in Fig. 4.20.

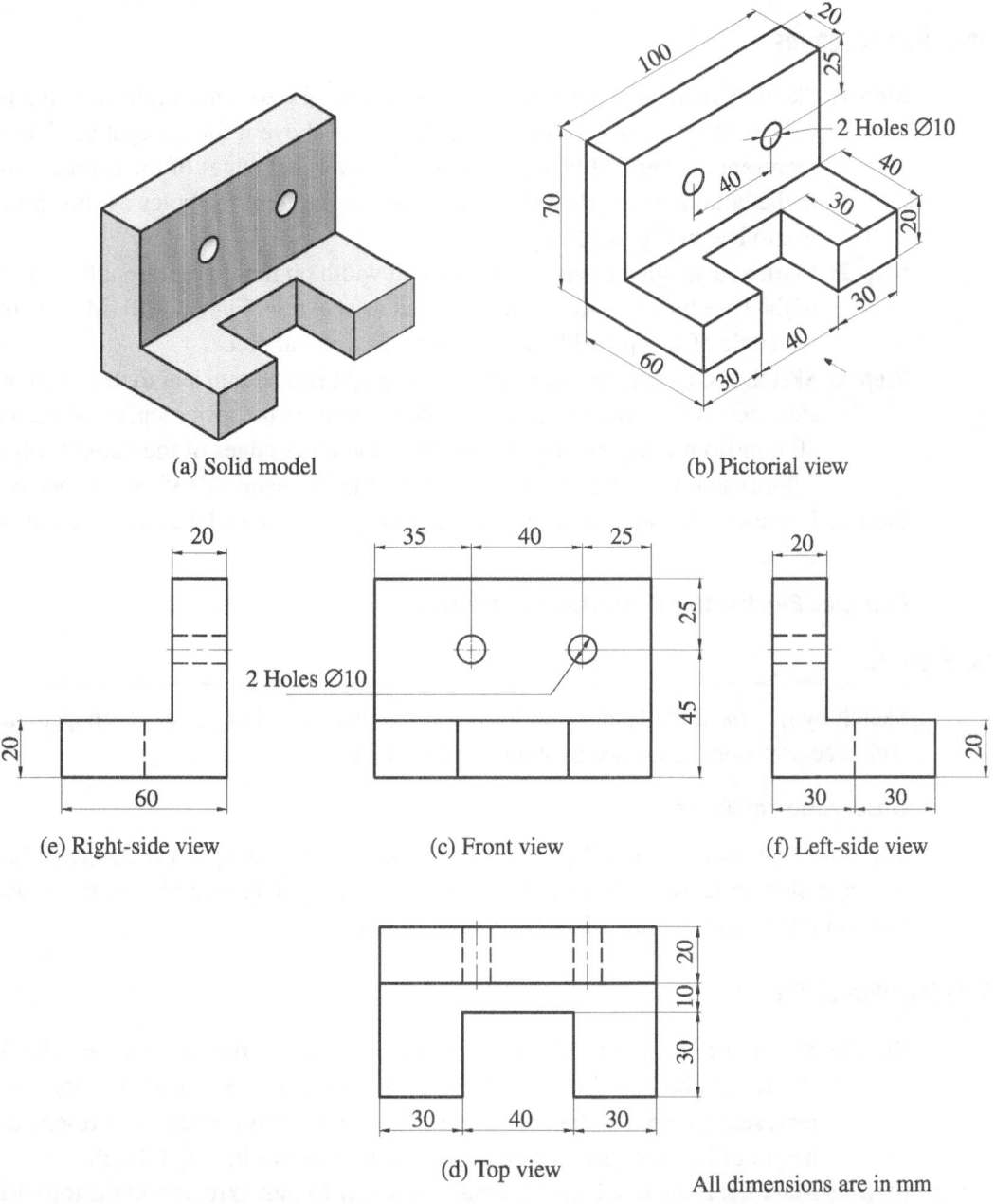

(a) Solid model

(b) Pictorial view

2 Holes Ø10

(e) Right-side view

(c) Front view

(f) Left-side view

2 Holes Ø10

(d) Top view

All dimensions are in mm

Fig. 4.20 L-shaped block with a rectangular cut-out

Observation of Views

Figure 4.20 shows an L-shaped block, with a vertical and horizontal portion. A central rectangular cut-out is provided in the horizontal block and two holes are drilled in the vertical block. The front, top, and the side views of the block comprise of respective views of the two portions.

SKETCHING PROCEDURE

Step 1: Sketch a rectangle of length 100 mm and height 20 mm to represent the front view of the base block and another rectangle above it for a height of 50 mm to represent the vertical block. Represent the vertical edges of the central cut-out in the bottom rectangle and mark circles to represent the holes as shown in the front view in Fig. 4.20(c).

Step 2: Mark a rectangle of length 100 mm and width 60 mm to represent the top view of the base block, with a central cut-out as shown in Fig. 4.20(d). Mark another rectangle of 20 mm width to represent the vertical block.

Step 3: Sketch two L-shaped figures of 70 mm height and 60 mm length to represent the side views of the two portions. Add dotted lines in the lower leg, at a distance of 40 mm from the front edge to represent the inner edges of the cut-out. Figures 4.20(e) and (f) shows the right-hand and the left-hand side views, respectively.

Step 4: Represent the holes as circles in the front view and as dotted lines in the other views. ▲

L-shaped Block with a Central Sloping Surface

Example 4.7 ▲

Sketch by free hand the front view, top view, and the side views of the L-shaped object with a central sloping surface as shown in Fig. 4.21.

Observation of Views

The object has two L–shaped pieces, connected in between with a rectangular block, having a sloping surface. Connect this with any sloping steps available in the house or Public Park, imagining the side walls to be L shaped.

SKETCHING PROCEDURE

Step 1: Sketch a rectangle of length 60 mm and height 15 mm to represent the front surface of the base block and add two squares of 15 mm size at its ends to represent the vertical legs. The rear edge of the middle block is represented at a height of 25 mm. The complete front view is shown in Fig. 4.21(c).

Step 2: Mark a rectangle of length 60 mm and width 15 mm to represent the top surface of the vertical portion and add two rectangles at its ends to mark the top view of the L-shaped legs as shown Fig. 4.21(d).

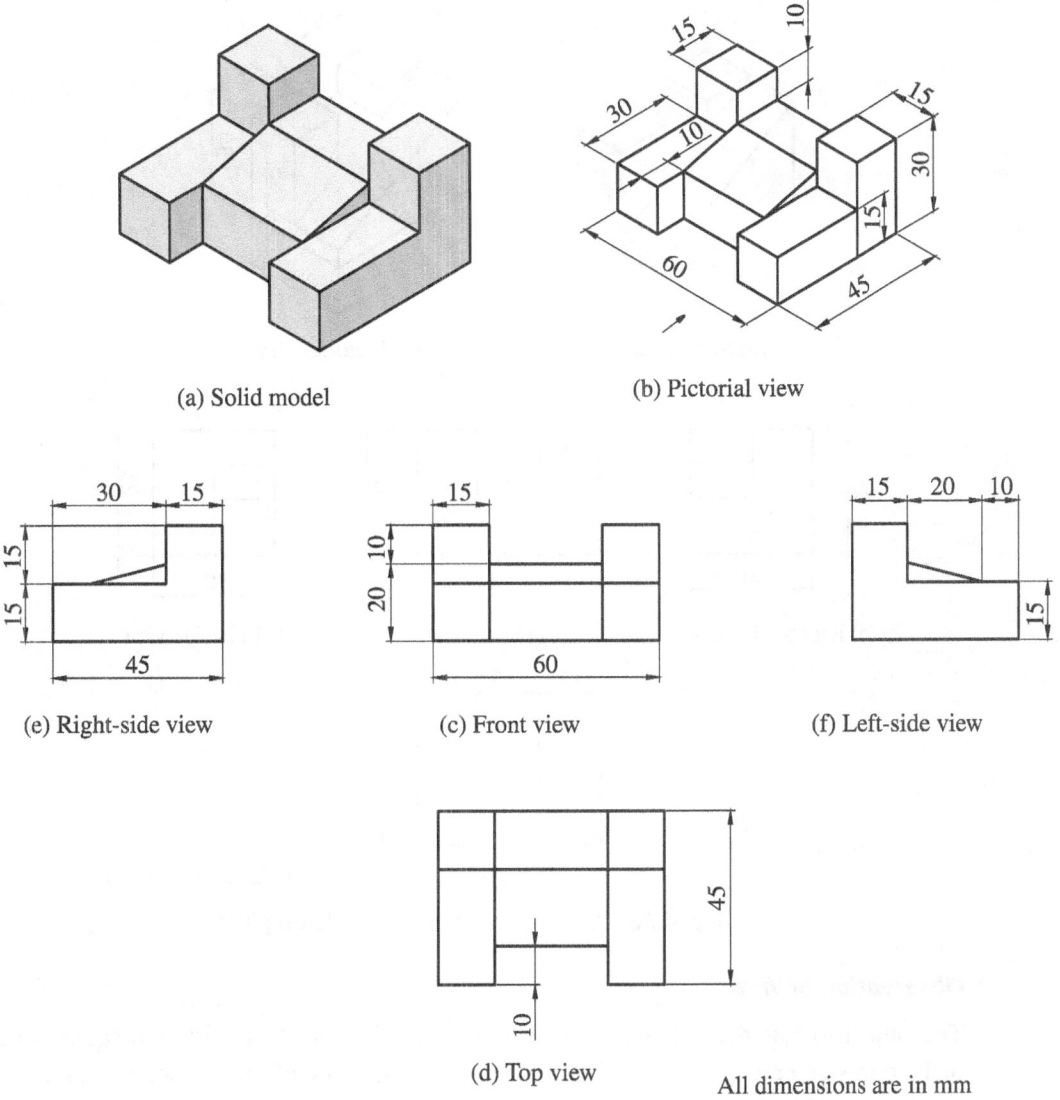

(a) Solid model

(b) Pictorial view

(e) Right-side view

(c) Front view

(f) Left-side view

(d) Top view   All dimensions are in mm

Fig. 4.21   L-shaped block with a central sloping surface

Step 3: Sketch two L-shaped figures by the side of the front view for an overall width of 45 mm to represent the side views. The beginning of the slope is obtained corresponding to the top edge from the front view as shown in Figs 4.21(e) and (f).

Block with Sloping Surface and Add on Pillars

Example 4.8

Sketch by free hand the front view, top view, and the side views of the block with a sloping surface and add on pillars as shown in Fig. 4.22.

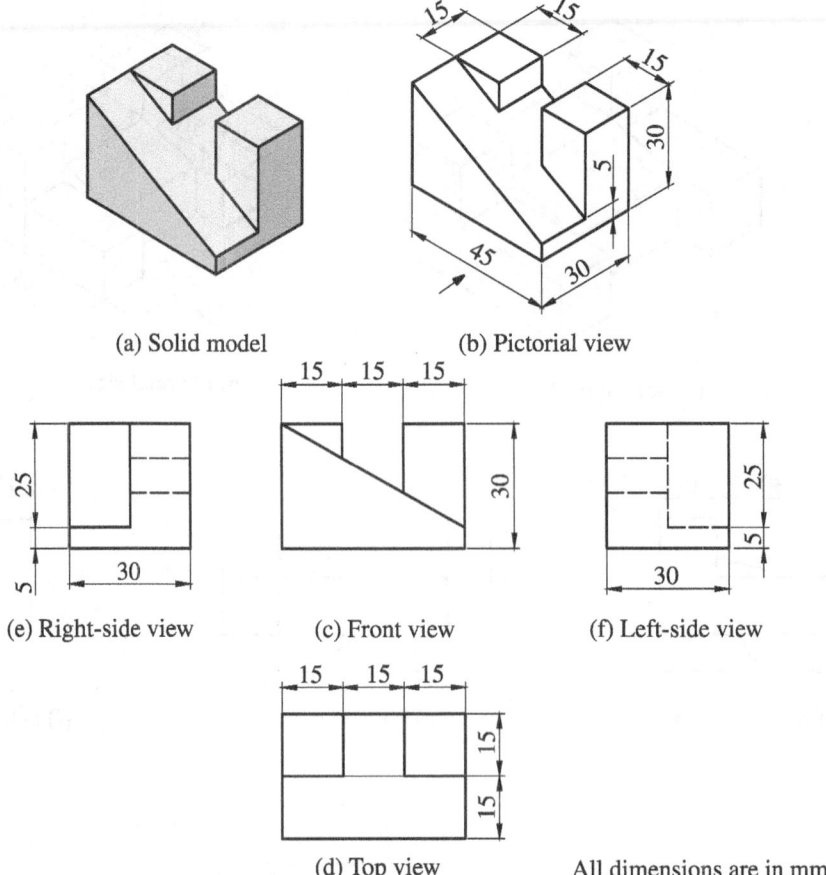

(a) Solid model   (b) Pictorial view

(e) Right-side view   (c) Front view   (f) Left-side view

(d) Top view   All dimensions are in mm

Fig. 4.22 Block with a slope and add on pillars

Observation of Views

The object in Fig. 4.22 shows a block with an inclined surface with two square pillars of 15 mm size erected over it. The inclination or the slope of the surface is seen in front view and so the overall height of the pillars.

SKETCHING PROCEDURE

Step 1: Sketch front view with the inclined surface as shown in Fig. 4.22(c) and add two rectangles of 15 mm length at an overall height of 30 mm to represent the square pillars.

Step 2: Mark a rectangle of 45 mm length and 30 mm width, with two squares of 15 mm size contained within, at the rear corners as shown in the top view in Fig. 4.22(d).

Step 3: Mark a square of 30 mm sides corresponding to the width and height of the base unit in the side view and project all the edges of the vertical pillars in dotted lines.

Step 4: The transition edge of the inclined surface and the vertical surface of the base block is indicated by a thick line in the right-side view, while it appears dotted in the left-side view as shown in Figs 4.22(e) and (f), respectively.   ▲

Block with Multiple V-cuts

Example 4.9   ▲

@ Sketch by free hand the front view, top view, and the left-side view of the block with multiple V-cuts as shown in Fig. 4.23.

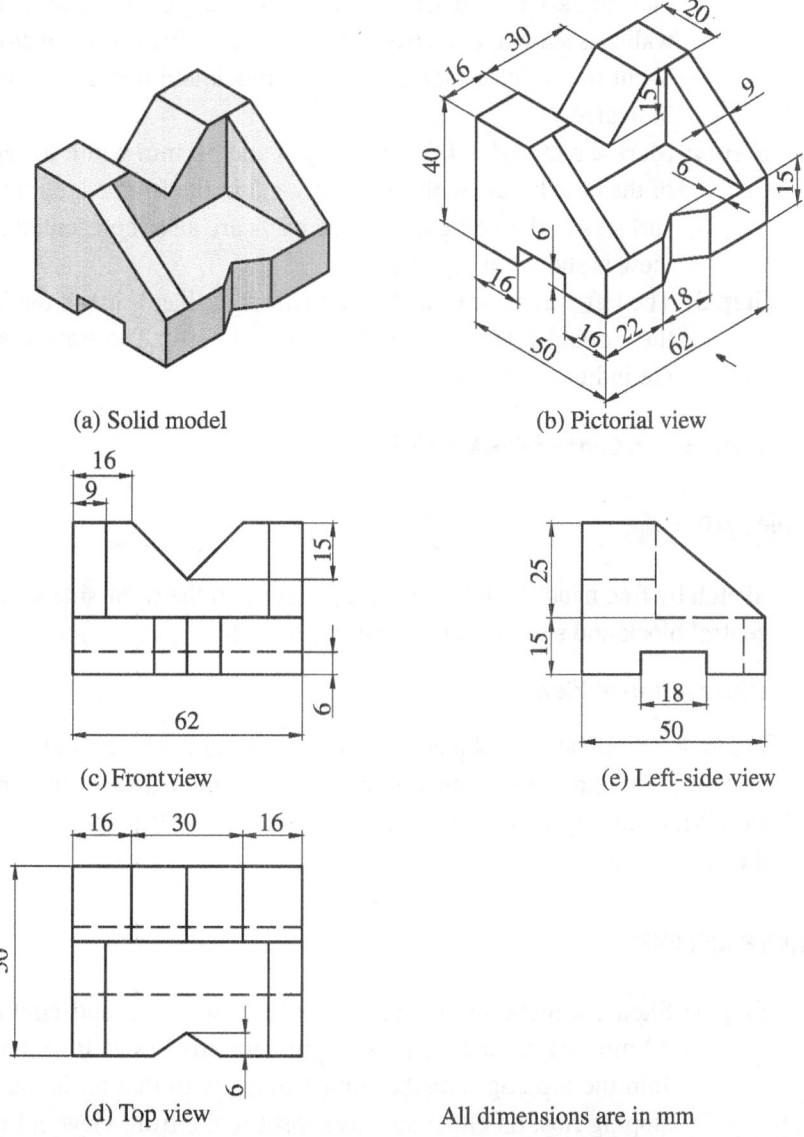

(a) Solid model   (b) Pictorial view

(c) Front view

(e) Left-side view

(d) Top view   All dimensions are in mm

Fig. 4.23   Block with V-cuts

Observation of Views

The object has a horizontal portion with a V-cut in the vertical face, while the vertical portion also has got a V-cut, the slope of which is visible in the front view. Further, the bottom surface has got a rectangular slot running throughout its length and the edges of it are visible only in the side view.

SKETCHING PROCEDURE

Step 1: Sketch a rectangle of 62 mm length and 15 mm height to represent the front view of the base block and erect two rectangles at its ends to represent the sloping walls, as shown in the front view in Fig. 4.23(c). The edges corresponding to the V-cut in the front face of the base block and that at the vertical block are also indicated.

Step 2: Mark a rectangle of 62 mm length and 50 mm width to represent the top view of the base block, with a central V-cut in the front edge for 9 mm width. The top surfaces of the vertical and side walls are also represented and the complete top view is shown in Fig. 4.23(d).

Step 3: The left-side view is projected to the right-hand side of the front view, indicating the sloping surface as shown in Fig. 4.23(e). The transition edges of the walls are indicated in dotted lines. ▲

Block with a Central Block with Side Ribs

Example 4.10 ▲

Sketch by free hand the front view, top view, and the right-side view of an object with a central block and side ribs as shown in Fig. 4.24.

Observation of Views

Figure 4.24 indicates an object with a base rectangular block and a square block centrally built above it and joined with sloping ribs. The inclination of the ribs will be seen in the front view. In a symmetric arrangement, two ribs will be placed, each on either side of the centre block.

SKETCHING PROCEDURE

Step 1: Sketch a rectangle to represent the base block and another rectangle of size 40 mm length and 30 mm height centrally above it to mark the square block. Join the top edges of the square block with that at the base block to mark the sloping ribs, on either side as shown in the front view in Fig. 4.24(c).

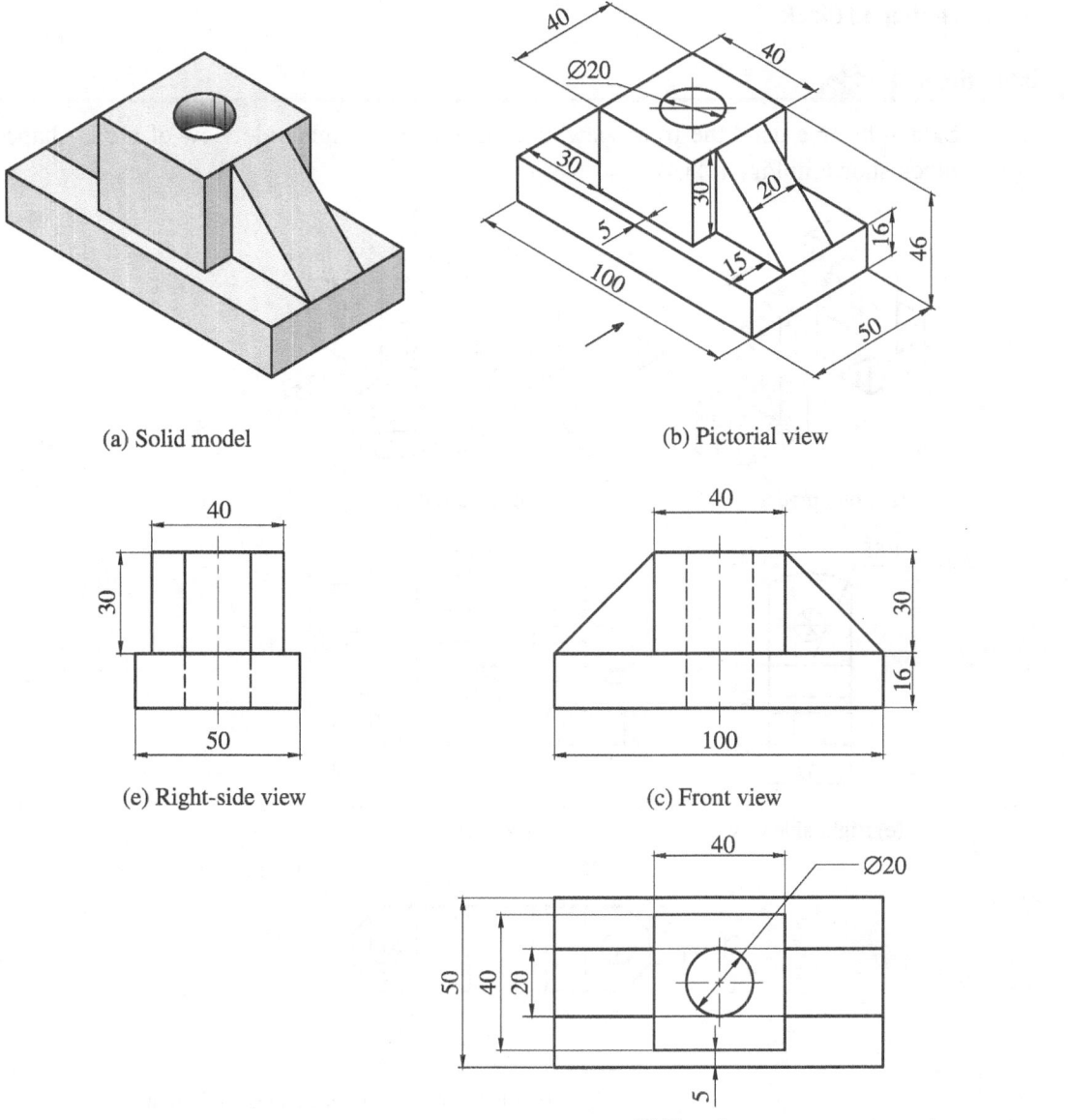

(a) Solid model

(b) Pictorial view

(e) Right-side view

(c) Front view

(d) Top view   All dimensions are in mm

Fig. 4.24   Block with a central block with side ribs

Step 2: Sketch a rectangle of length 100 mm and width 50 mm and mark a square of 40 mm sides in the centre, to represent the top view of the base and the centre block. Mark the hole and the side edges to indicate the ribs as shown in the top view in Fig. 4.24(d).

Step 3: Project the side view of the base unit and the centre block along with the ribs as shown in right-side view in Fig. 4.24(e). Represent the hole in dotted lines. ▲

H-shaped Block

Example 4.11

Sketch by free hand the front view, top view, and the right-side view of the H-shaped block shown in Fig. 4.25.

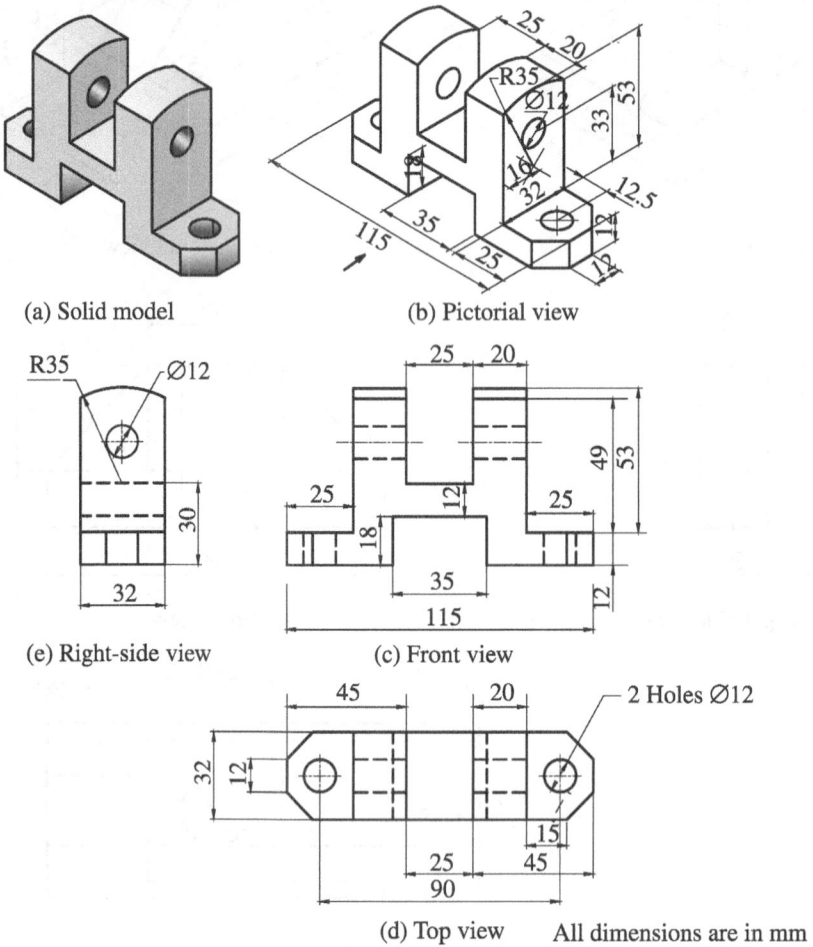

(a) Solid model   (b) Pictorial view

(e) Right-side view   (c) Front view

(d) Top view   All dimensions are in mm

Fig. 4.25 H-shaped block

Observation of Views

Figure 4.25 shows H-shaped block with a symmetrical base. The vertical walls of the H-shaped block have got a curved surface at the top and the curves are visible in the side views.

SKETCHING PROCEDURE

Step 1: Mark the H-shaped arrangement as shown in the front view in Fig. 4.25(c) by noting the heights of different units. Project the circular holes in dotted lines.

Step 2: Sketch a rectangle to represent the top view of the base unit and mark the chamfers symmetrically. Project the vertical walls and all the circular holes in the top view as shown in Fig. 4.25(d).

Step 3: Sketch a rectangle to represent the side view of the vertical walls. The top surface is indicated by an arc of 35 mm radius as shown in Fig. 4.25(e).

Step 4: The top point and the end points of the arc as obtained in the side view are projected to the front view to get the height of the front surface.

Cylinder with Rectangular Block Seated on its Curved Surface

Example 4.12

Sketch by free hand the front view, top view, and the right-side view of the cylindrical block with a rectangular block seated as shown in Fig. 4.26.

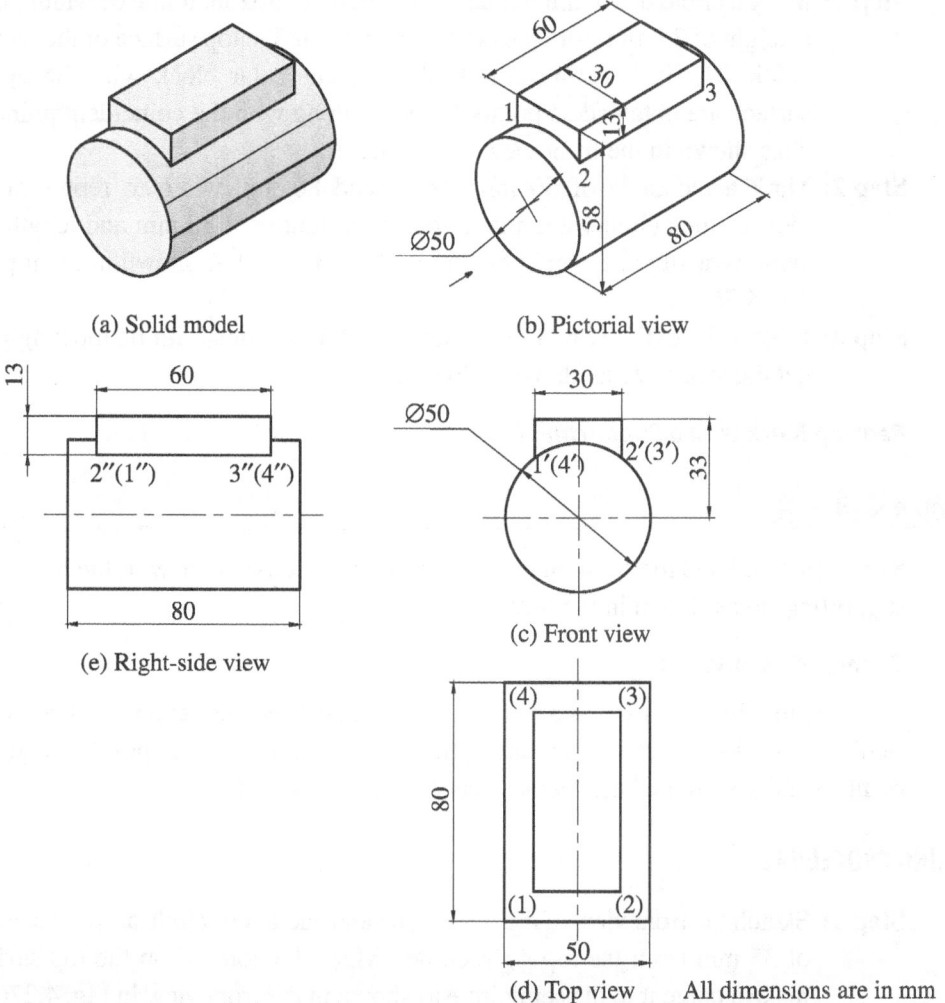

(a) Solid model

(b) Pictorial view

(c) Front view

(d) Top view   All dimensions are in mm

(e) Right-side view

Fig. 4.26   Cylinder with a rectangular block

Observation of Views

A solid cylinder with rectangular-shaped block is placed on its cylindrical surface. This can resemble a seat or saddle provided on the top of an elephant or horse or any other animal. The bottom surface of the block seated on the cylinder suits the contour of the cylindrical surface. In mechanical engineering applications, this arrangement is depicted as a shaft and a key. When the width of the top rectangular block is equal to the diameter of the cylinder, it will perfectly merge with the cylindrical surface, leaving no impression of its joining. When the width is less than the diameter of the cylinder, the vertical surface of the block and the cylindrical surface will leave an impression in the form of a line.

SKETCHING PROCEDURE

Step 1: Mark a circle of 50 mm diameter and mark a horizontal line of width 30 mm, at a height of 33 mm from the centre to represent the top surface of the rectangular block. The intersecting points of the rectangular block with the cylindrical surface are obtained as points 1′ and 2′ along with the coincident points 3′ and 4′ as shown in the front view in Fig. 4.26(c).

Step 2: Mark a rectangle of 50 mm length and 80 mm width to represent the top view of the cylinder and mark another rectangle of 30 mm and length 60 mm, symmetrically and obtain the points 1, 2, 3, and 4 as shown in the top view in Fig. 4.26(d).

Step 3: Project the cylinder and the rectangular block along with its meeting points to get the side view as shown in Fig. 4.26(e).   ▲

Bearing Block with a Supporting Rib

Example 4.13

Sketch by free hand the front view, top view, and the left-side view of the bearing with a supporting rib as shown in Fig. 4.27.

Observation of Views

A rectangular block is used as a base block to support a semicircular block above it. The semicircular block is connected tangentially with a sloping rib at points 2 and 3 and at points 1 and 4 with the base block as seen from Fig. 4.27(b).

SKETCHING PROCEDURE

Step 1: Sketch the front view of the base block and add a semicircle above it at a distance of 35 mm from the top right corner. Mark the point 1′ on the top surface and make a tangent to the semicircle as shown in the front view in Fig. 4.27(c) to get the point 2′. Mark the coincident points 3′ and 4′.

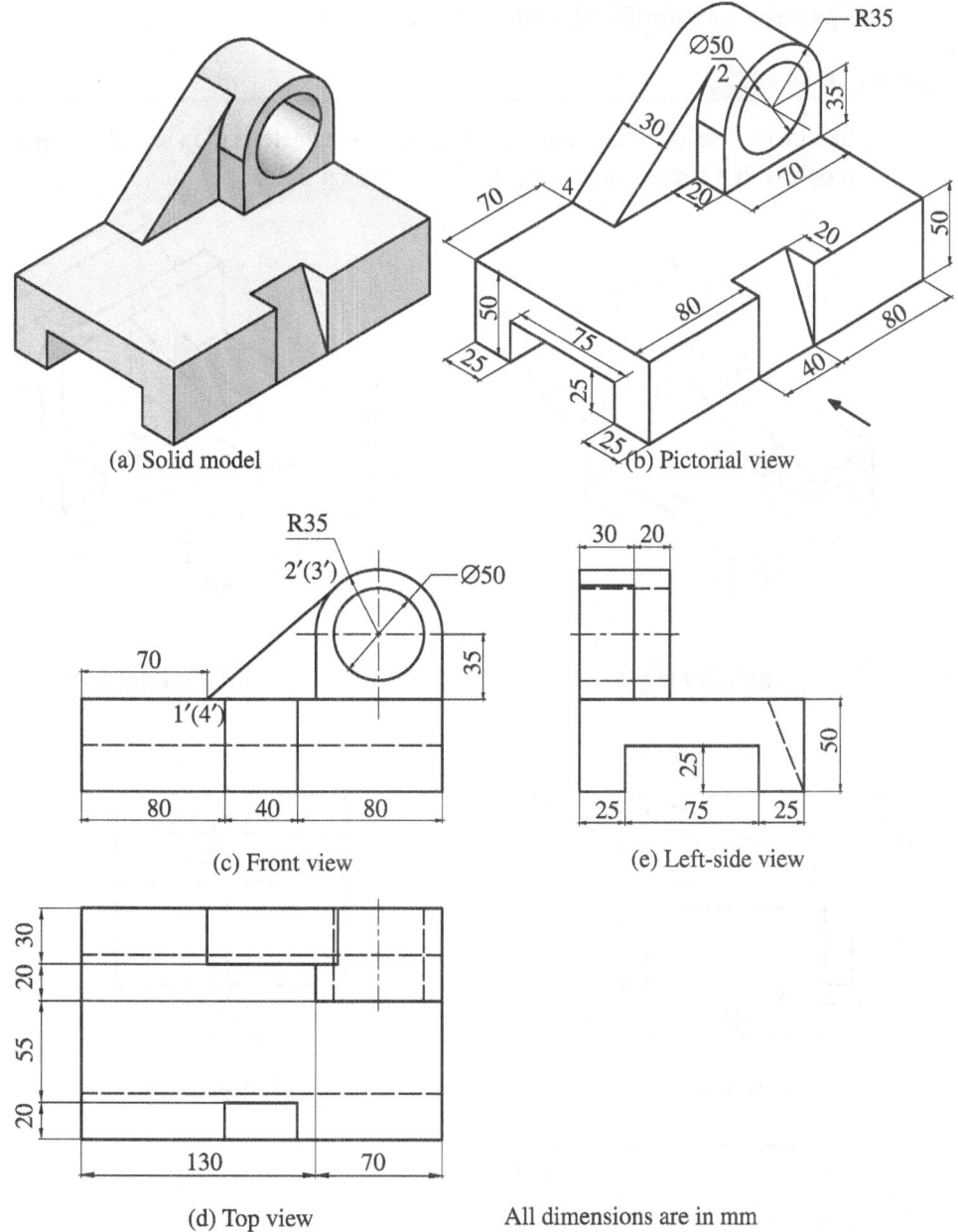

(a) Solid model

(b) Pictorial view

(c) Front view

(e) Left-side view

(d) Top view

All dimensions are in mm

Fig. 4.27 Bearing block with a supporting rib

Step 2: The top view of the base block is drawn with the central cut-out in the front edge. Project the semicircular block and the rib as shown in Fig. 4.27(d). Mark the circular hole and the rectangular slot in dotted lines.

Step 3: An L-shaped arrangement indicating the base block and the bearing block is shown in the left-side view in Fig. 4.27(e). Project the end points of the rib. The slope of the central cut-out is marked at a distance of 20 mm.

Open Semicircular Block with a Support Frame

Example 4.14

Sketch by free hand the front view, top view, and the left side view of the open semicircular block with a supporting frame as shown in Fig. 4.28.

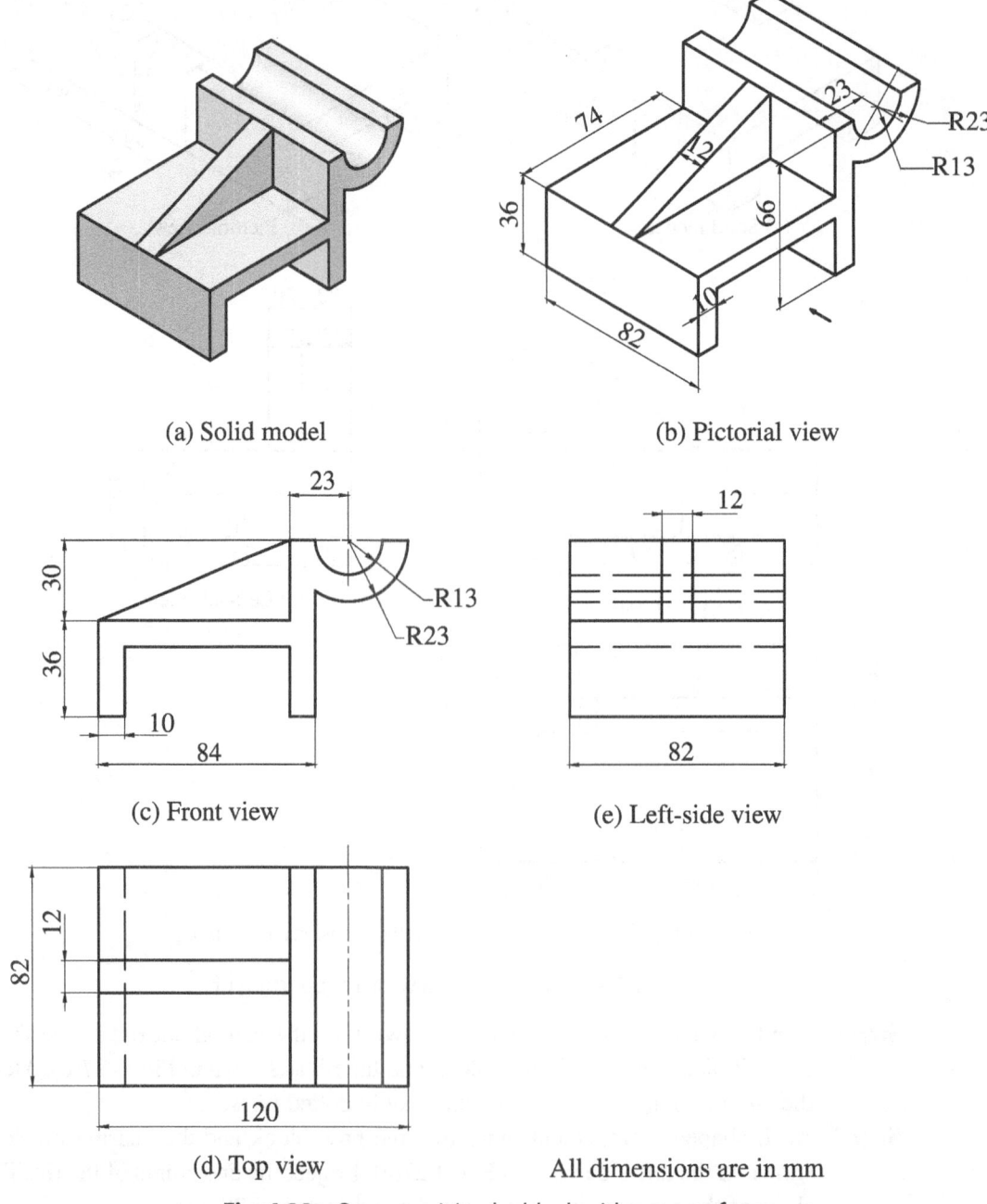

(a) Solid model

(b) Pictorial view

(c) Front view

(e) Left-side view

(d) Top view   All dimensions are in mm

Fig. 4.28   Open semicircular block with support frame

Observation of Views

Figure 4.28 shows a semicircular open block fixed at the right-hand side of a rectangular frame with a central rib. Since the sloping rib joins with the vertical surface, the joining line is seen on both the horizontal and the vertical surfaces.

SKETCHING PROCEDURE

Step 1: Sketch the contour of the rectangular block with the cut-out as shown in Fig. 4.28(c). Locate the centre of semicircular open block at a distance of 23 mm and at a height of 66 mm and mark two concentric semicircles at radii of 13 mm and 23 mm. Join the top right- and left-hand corner by a straight line to represent the inclined rib.

Step 2: Mark a rectangle of length 120 mm and width 82 mm to represent the overall top view of the block as shown in Fig. 4.28(d). Mark the outlines of the semicircular block and the central rib.

Step 3: Mark a rectangle of 82 mm and height 66 mm to represent the side view as shown in Fig. 4.28(e). Mark the outlines of the semicircular block and the central rib. ▲

4.6 ENGINEERING APPLICATIONS

A systematic development achieved by the addition of different features in the form of arcs, circles, holes rectangular cut-outs, slopes, etc., to a basic rectangular block was discussed in Sections 4.5.1–4.5.6. In Section 4.5.7, a combination of these features in three-dimensional objects or components was discussed. In this section, the pictorial views of the objects, which are used in practical or in some specific engineering applications, are discussed. The products/appliances or gadgets, which are used in domestic purposes, are shown in pictorial views, along with their multiple views. Large-sized objects, such as a car, an aeroplane, and house, are also explained with their views. Electrical appliances and electronic gadgets, such as electrical motors, circuit boards, and computer monitors, are also discussed with their multi-views. In order to familiarize these products, these devices are drawn without any specific dimension. It is hoped that this will inspire the students to practise the freehand sketching of various objects that are used in day-to-day life, without bothering about the intricacies of the dimensions or their accuracy. And it will certainly increase the confidence and the pride, since the objects are real life examples.

The multiple views of a bowl, a flower vase, a bicycle crank, and a domestic mixie are shown in Figs 4.29–4.32.

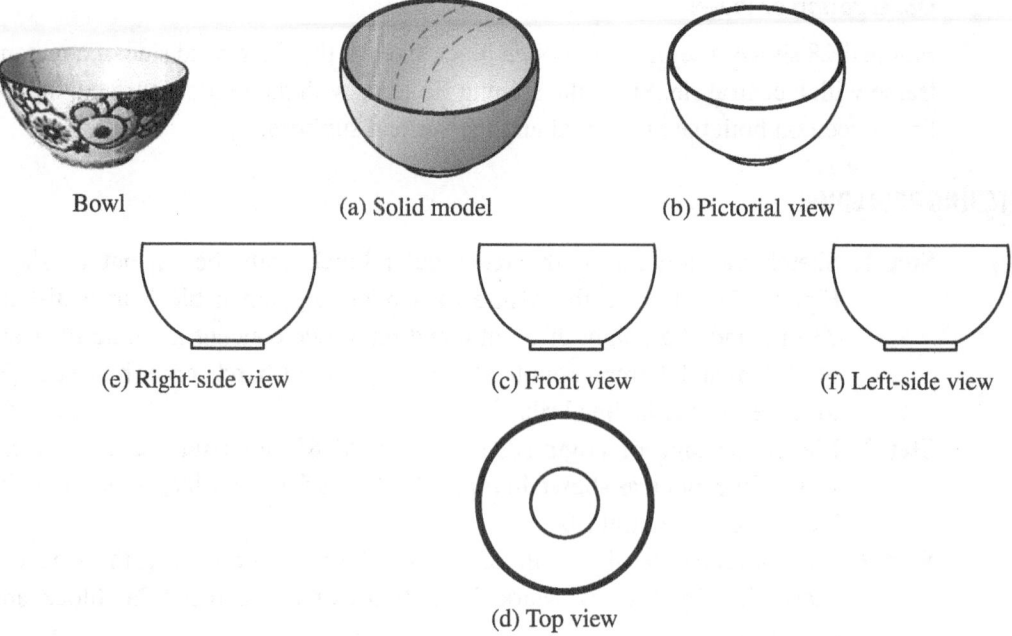

Bowl   (a) Solid model   (b) Pictorial view

(e) Right-side view   (c) Front view   (f) Left-side view

(d) Top view

Fig. 4.29   Multiple views of bowl

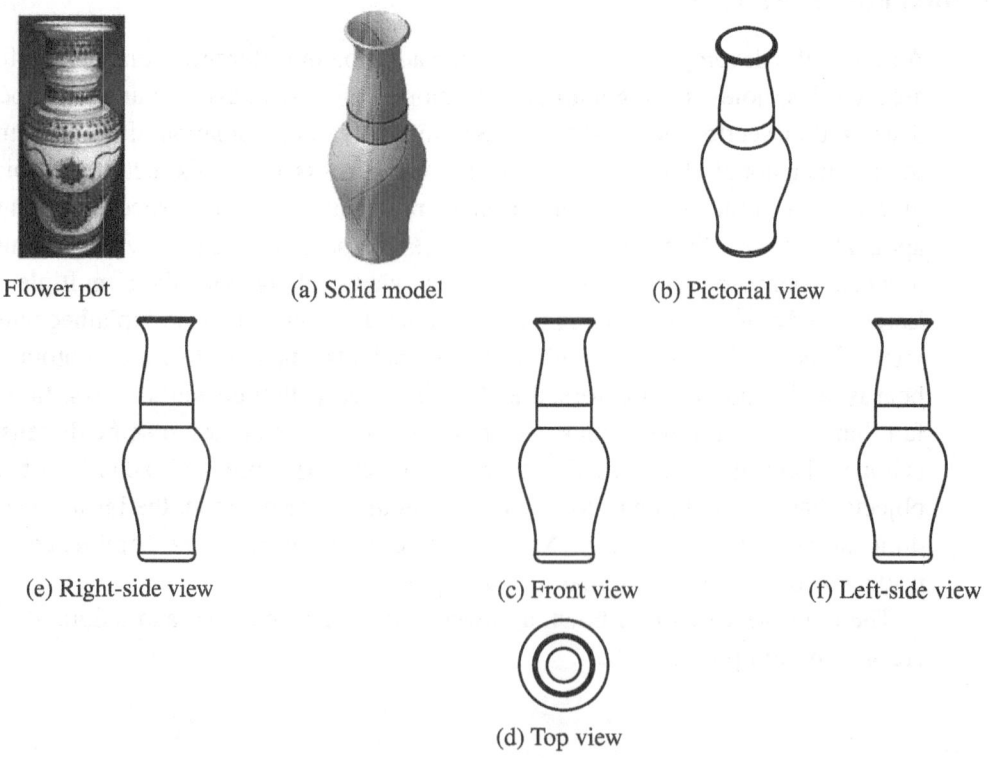

Flower pot   (a) Solid model   (b) Pictorial view

(e) Right-side view   (c) Front view   (f) Left-side view

(d) Top view

Fig. 4.30   Multiple views of a flower pot

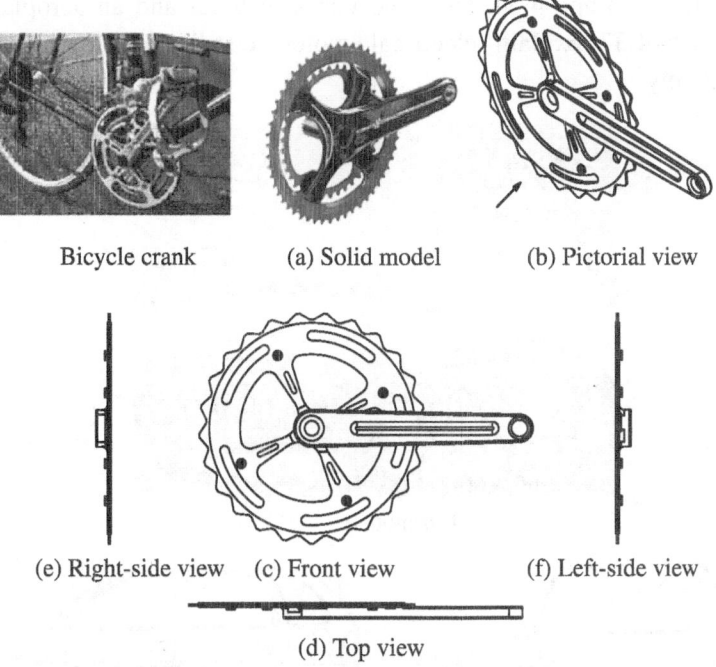

Bicycle crank   (a) Solid model   (b) Pictorial view

(e) Right-side view   (c) Front view   (f) Left-side view

(d) Top view

Fig. 4.31   Multiple views of a bicycle crank

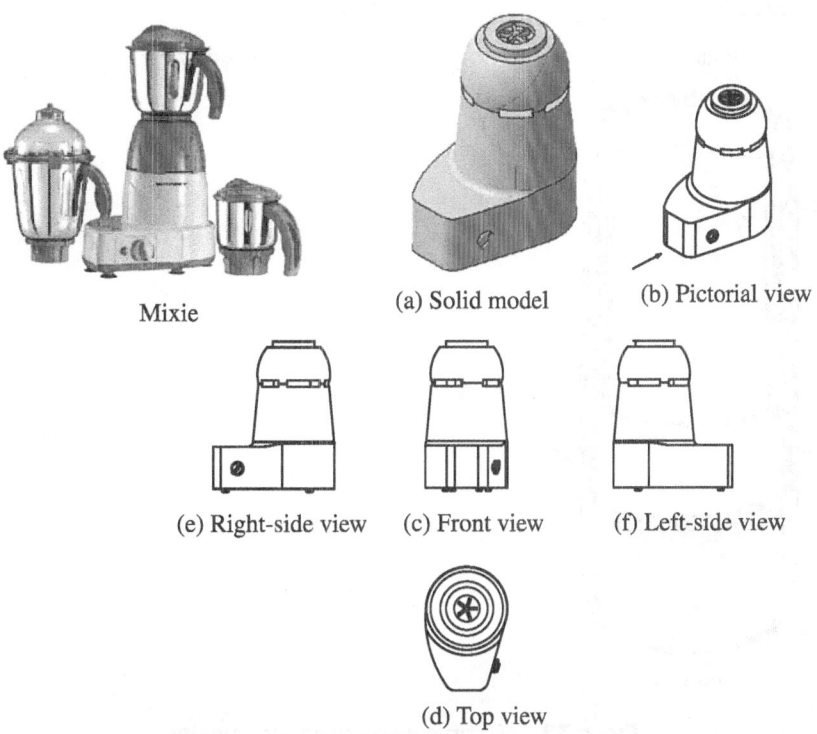

Mixie   (a) Solid model   (b) Pictorial view

(e) Right-side view   (c) Front view   (f) Left-side view

(d) Top view

Fig. 4.32   Multiple views of a domestic mixie

In Figs 4.33 and 4.34, the three views of a car and an aeroplane are shown, and Figs 4.35–4.37 explain electrical motor, circuit board, and computer monitor, respectively.

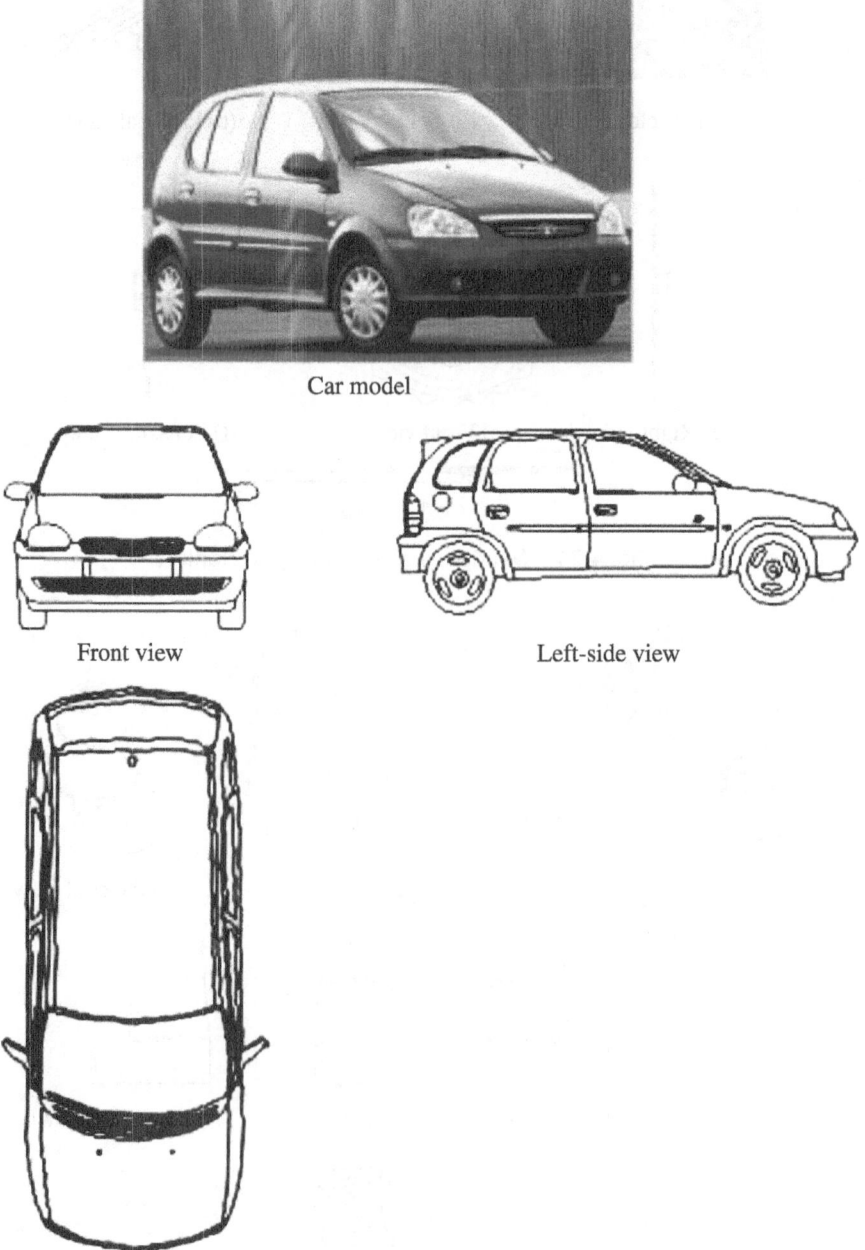

Car model

Front view   Left-side view

Top view

Fig. 4.33   Multiple views of a four wheeler

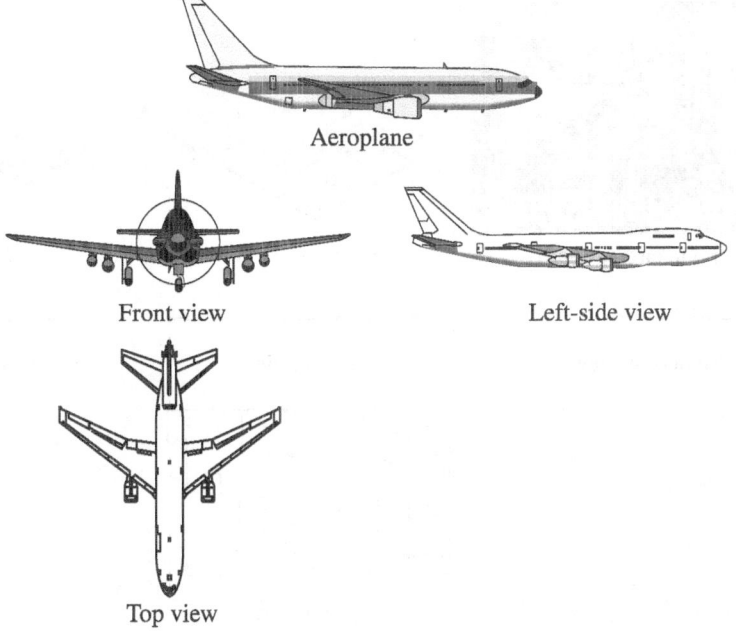

Fig. 4.34   Multiple views of an aeroplane

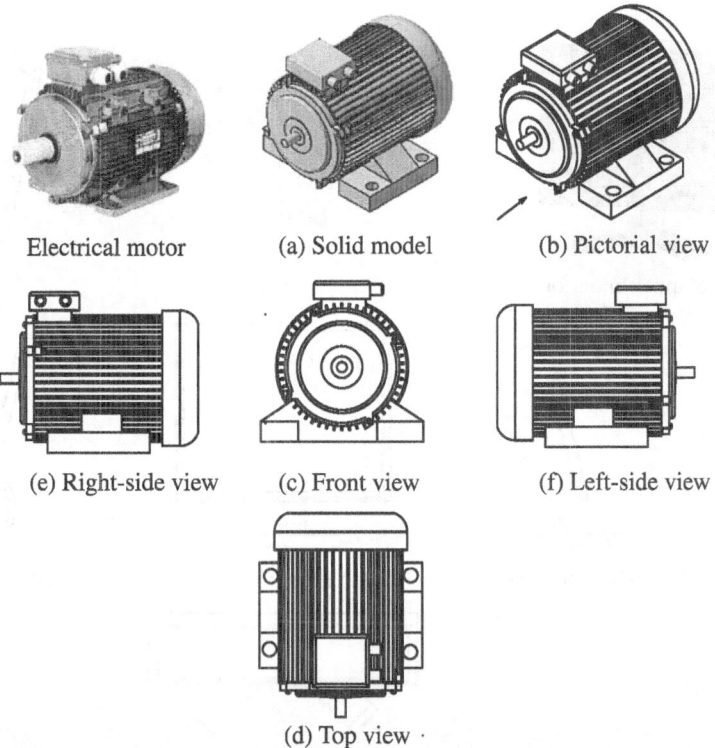

Fig. 4.35   Multiple views of an electrical motor

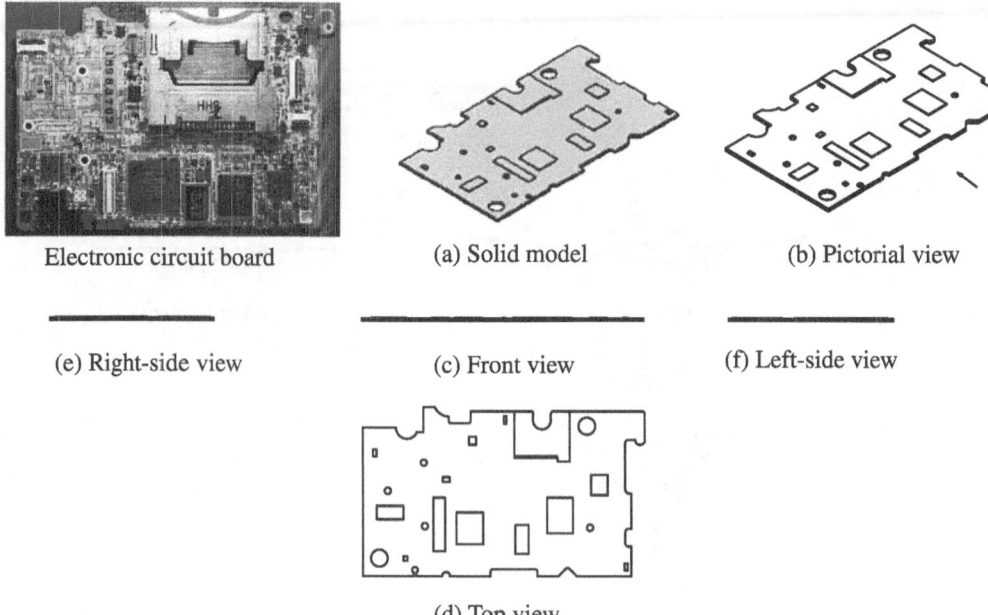

Electronic circuit board   (a) Solid model   (b) Pictorial view

(e) Right-side view   (c) Front view   (f) Left-side view

(d) Top view

Fig. 4.36   Multiple views of an electronic circuit board

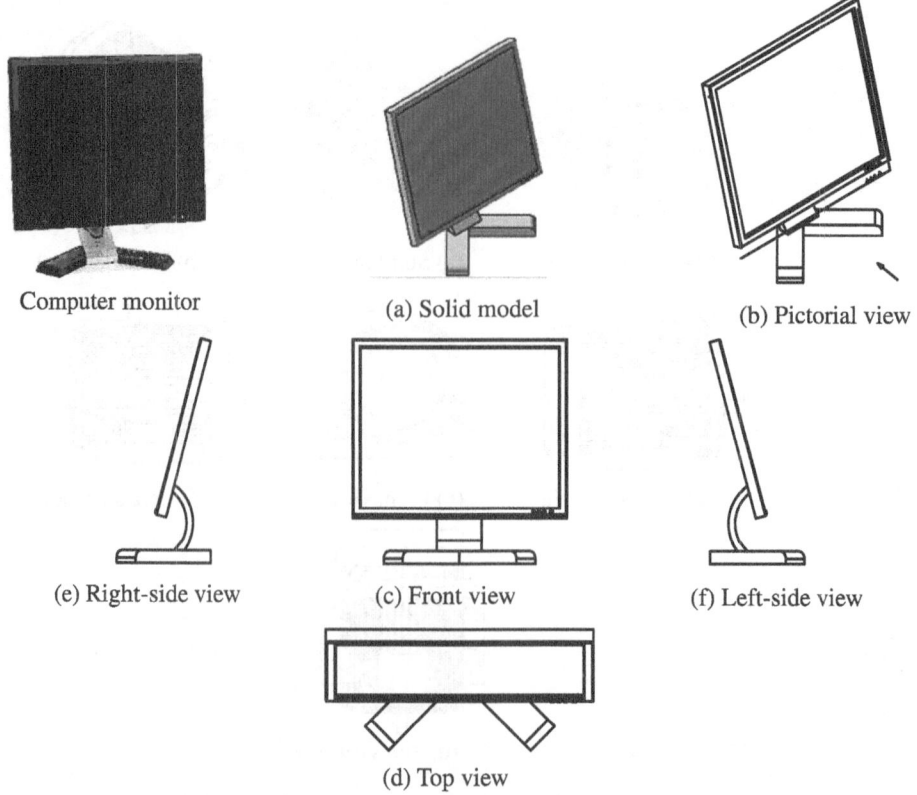

Computer monitor   (a) Solid model   (b) Pictorial view

(e) Right-side view   (c) Front view   (f) Left-side view

(d) Top view

Fig. 4.37   Multiple views of a computer monitor

The three views of a house are shown in Fig. 4.38.

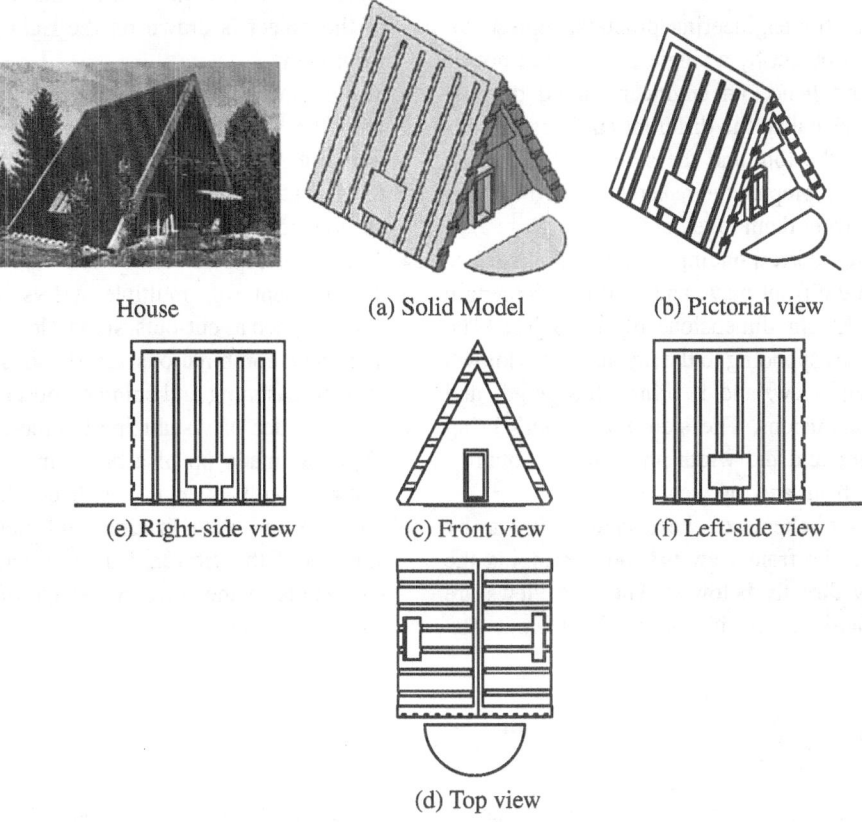

House   (a) Solid Model   (b) Pictorial view

(e) Right-side view   (c) Front view   (f) Left-side view

(d) Top view

Fig. 4.38   Multiple views of a house

The objects explained are sample pieces only, and the students are advised to look at various domestic objects such as kitchen accessories, showpieces, electrical home appliances, refrigerators, television sets, cell phones, digital cameras, etc. and develop skills to get their different views.

RECAPITULATION

- The pictorial representation of objects is used as the common mode of communication and is universally adopted for the simple reason that it is not bound by any formal language (vernacular) or special skills, but is easily understandable and creates a visual impact on the viewer. The pictorial sketches should be capable of conveying information about all the three-dimensions of an object.

- Three major types of pictorial projections, namely axonometric projection, oblique projection, and perspective projection, are used to convey the appearance of the object. Although all these pictorial projections show the three sides of an object in a single picture, one or the other edges are foreshortened, losing their true dimensions or size, thereby making them unsuitable for engineering purpose related to manufacture and physical realization of products.

- Therefore for engineering practice, representation of an object by a single picture will not be useful and hence an engineer should develop skills to visualize the different surfaces such as the front, the top, and the side surfaces of the object, and prepare their sketches and meaningfully interpret them together.
- The sketch corresponding to the frontal surface is called the front view, and it brings the length and the height dimensions of the object. The sketch corresponding to the top surface is known as the top view, and it brings the length and width dimensions. The side appearances bring the height and the width and can be obtained from the front and top view.
- The layout scheme of these views is made by sketching the front view first and arranging the top view directly below it. The side views are always laid by the side of the front view. The right-side appearance of the object is laid to the left of the front view, whereas the left-side view of the object is drawn on the right side of the front view.
- The engineer should familiarize himself/herself with the sketching practice by free hand first (without using instruments), as the focus is to visualize a picture or an object and develop abilities to read the details from it.
- Typical examples that show the systematic development of multiple views of various features such as cut-outs, slope, slots, arcs, holes, and their combinations are demonstrated with logical reasoning and common observations.
- A novel idea of visualizing a three-dimensional object as made up of a base unit and multiple features built over it has been discussed to improve the visualization and understanding abilities of the viewer. This will help to extract the details of the form and shape of the object with more ease.

MULTIPLE-CHOICE QUESTIONS

4.1 Study the pictorial views of the objects shown in the exercises A1 to A4 and choose the correct views (A, B, or C)

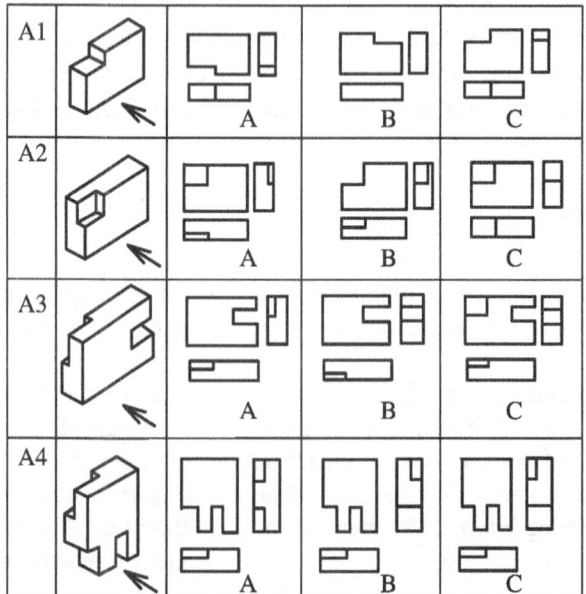

4.2 Study the pictorial views of the objects shown in the exercises B1 to B4 and decide which multi-view drawings (A, B, or C) best describes the correct minimum number of views.

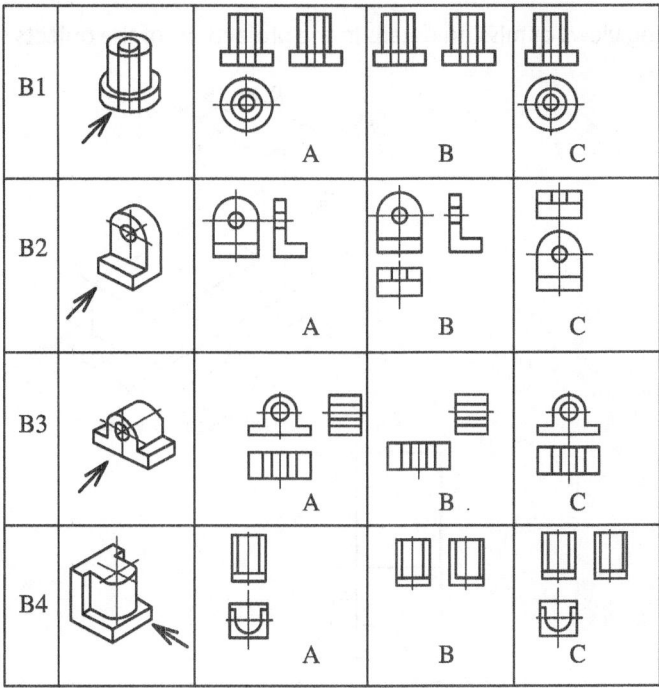

4.3 Study the objects shown in the exercises C1 to C4 and match their views.

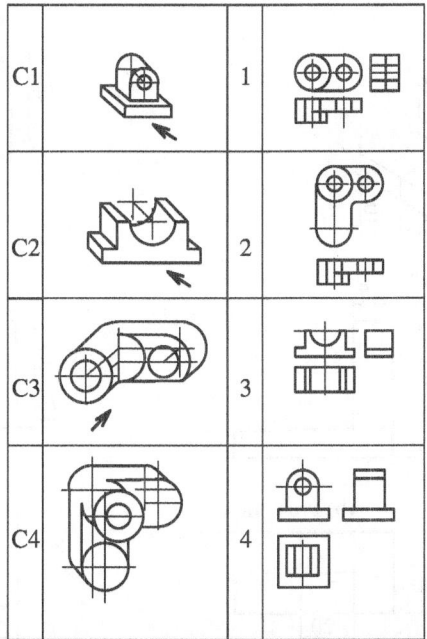

Sketch the missing views or missing details in the projections of the objects shown.

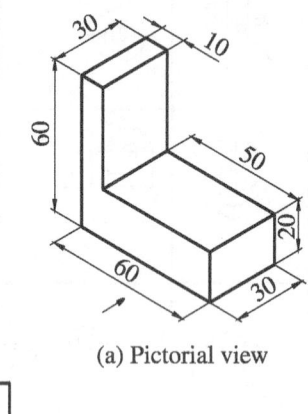

(a) Pictorial view

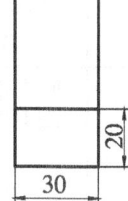

(c) Right-side view   (b) Front view

Fig. 4.39

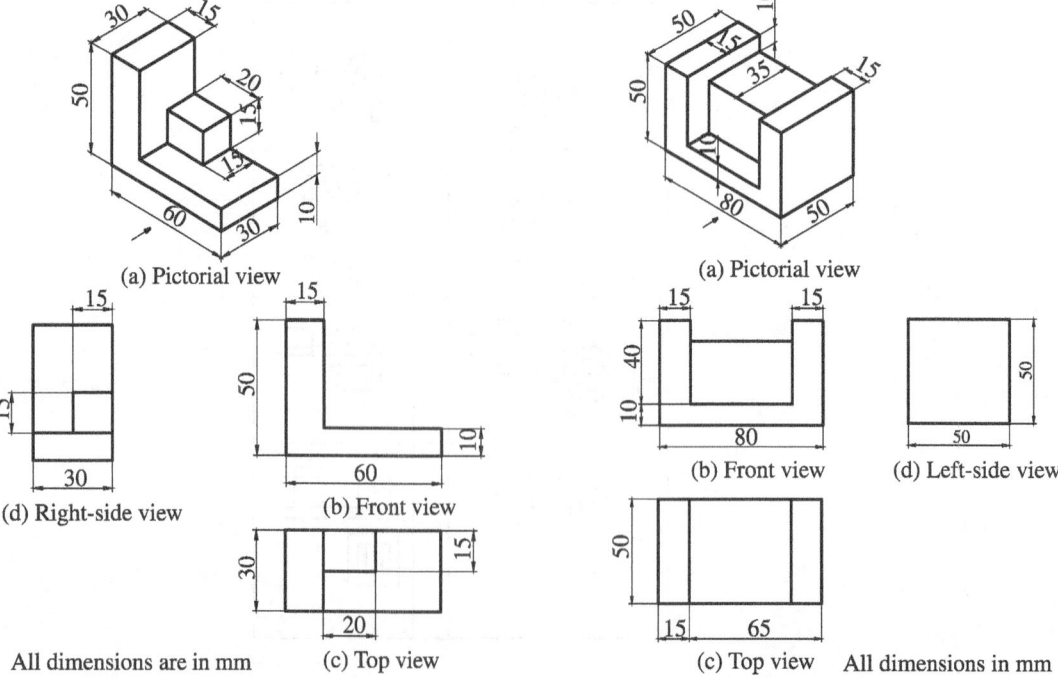

Fig. 4.40

Fig. 4.41

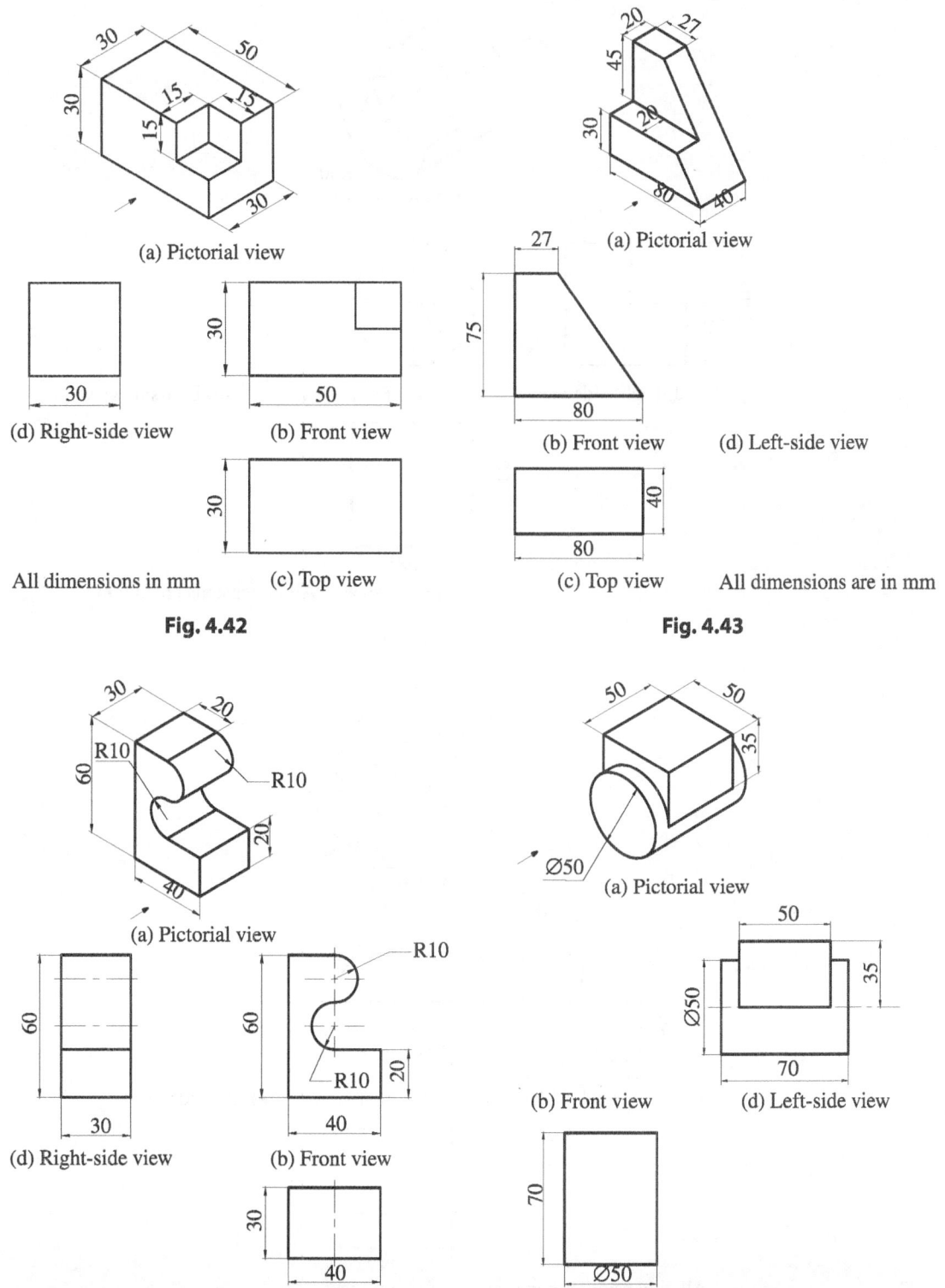

(a) Pictorial view

(d) Right-side view

(b) Front view

All dimensions in mm

(c) Top view

Fig. 4.42

(a) Pictorial view

(b) Front view

(d) Left-side view

(c) Top view

All dimensions are in mm

Fig. 4.43

(a) Pictorial view

(d) Right-side view

(b) Front view

All dimensions are in mm

(c) Top view

Fig. 4.44

(a) Pictorial view

(b) Front view

(d) Left-side view

(c) Top view

All dimensions are in mm

Fig. 4.45

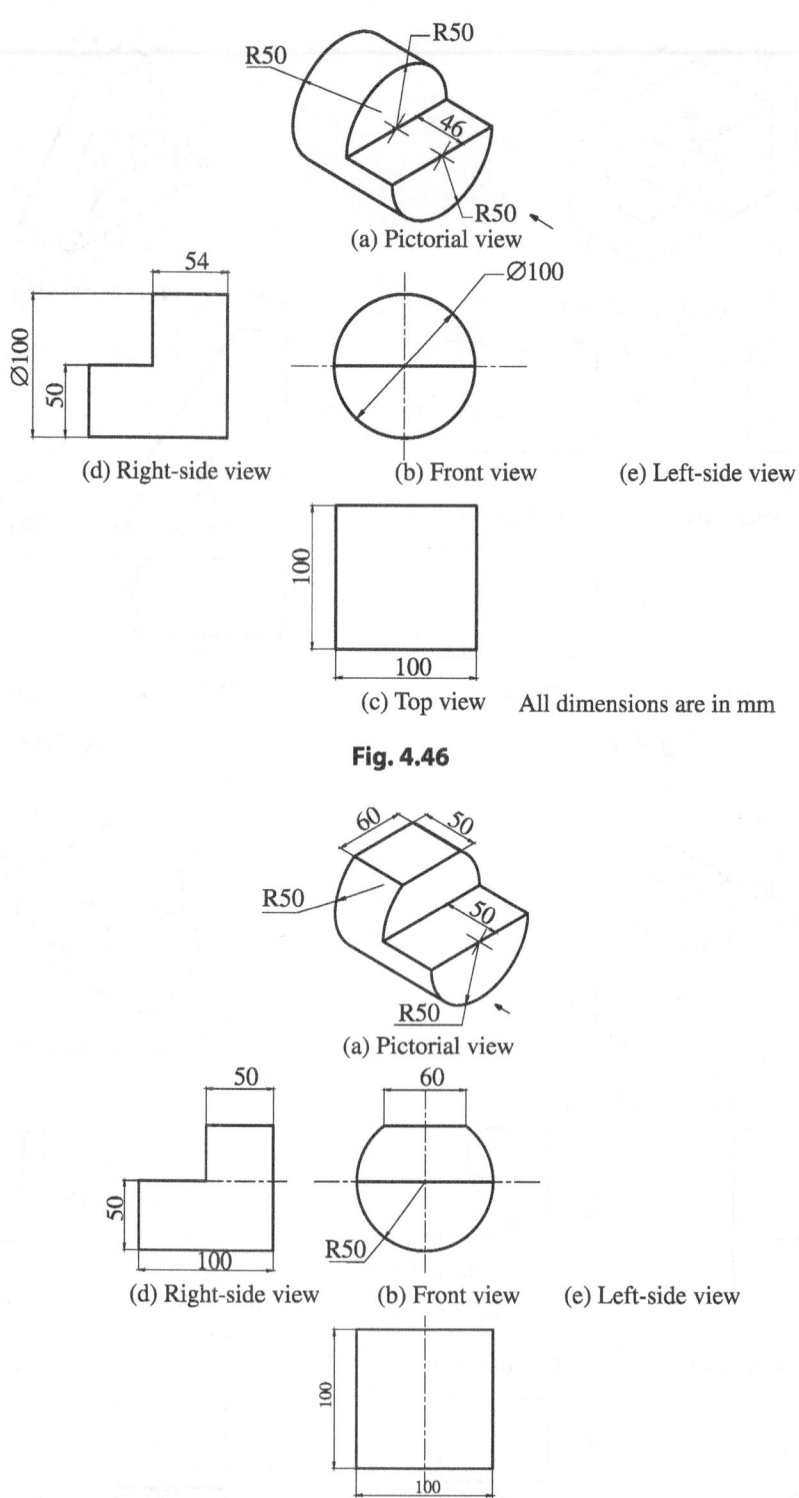

R50
R50
46
R50
(a) Pictorial view

54
Ø100
50
(d) Right-side view

Ø100
(b) Front view

(e) Left-side view

100
100
(c) Top view   All dimensions are in mm

Fig. 4.46

60   50
R50
50
R50
(a) Pictorial view

50
50
100
(d) Right-side view

60
R50
(b) Front view

(e) Left-side view

100
100
(c) Top View

All dimensions are in mm

Fig. 4.47

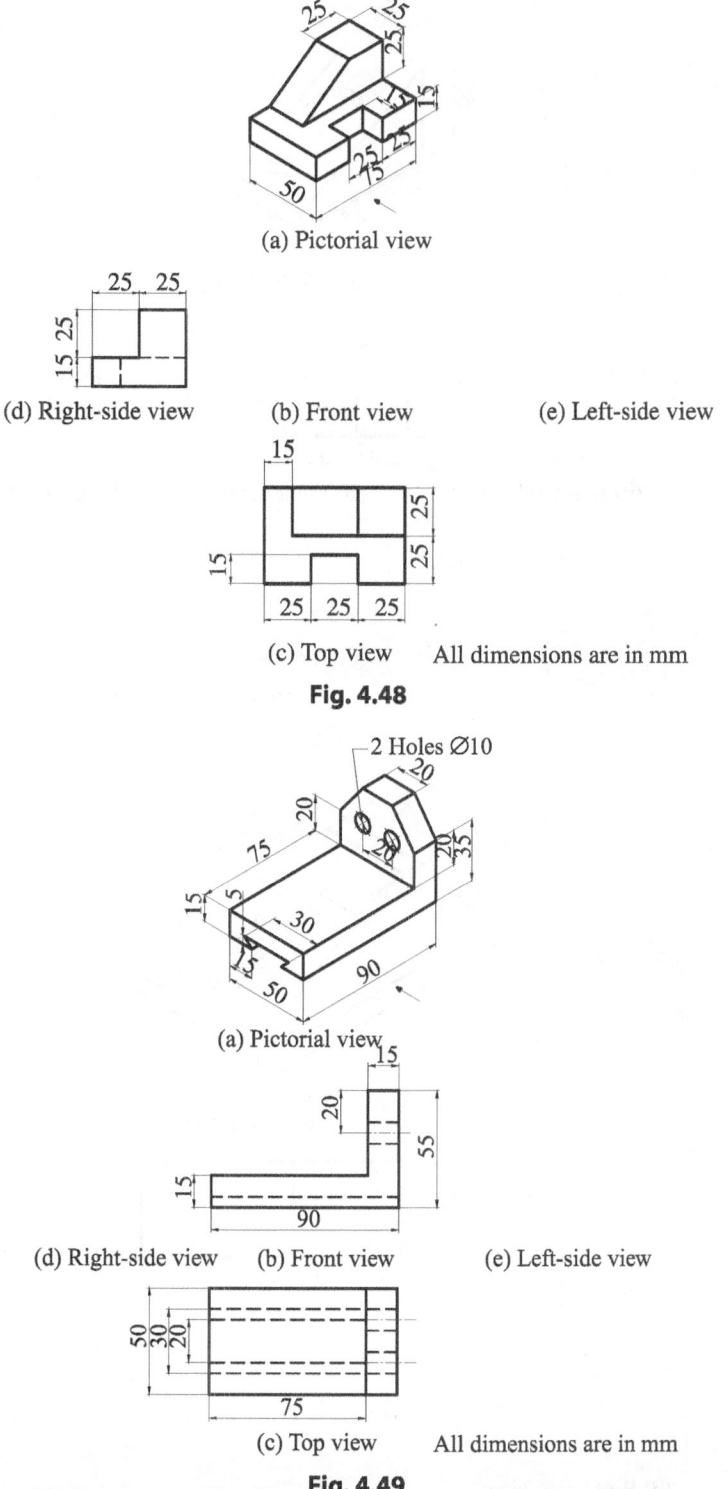

(a) Pictorial view

(d) Right-side view   (b) Front view   (e) Left-side view

(c) Top view   All dimensions are in mm

Fig. 4.48

2 Holes Ø10

(a) Pictorial view

(d) Right-side view   (b) Front view   (e) Left-side view

(c) Top view   All dimensions are in mm

Fig. 4.49

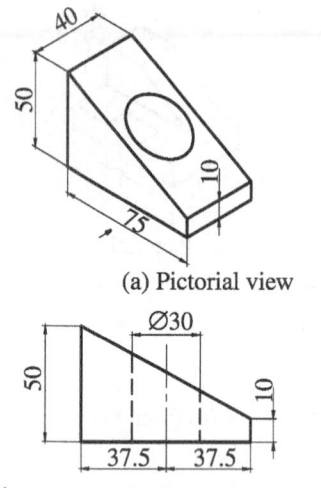

(a) Pictorial view

(d) Right-side view   (b) Front view   (e) Left-side view

(c) Top view

All dimensions in mm

Fig. 4.50

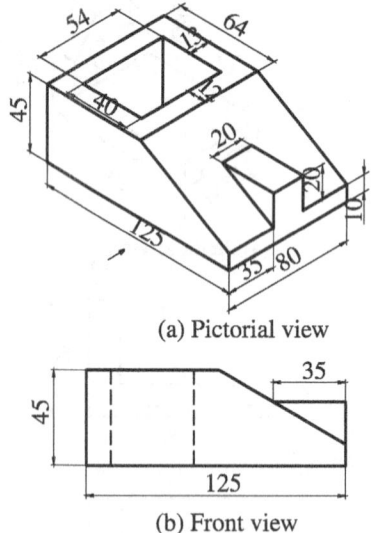

(a) Pictorial view

(b) Front view

(d) Right-side view   (c) Top view   All dimensions in mm

Fig. 4.51

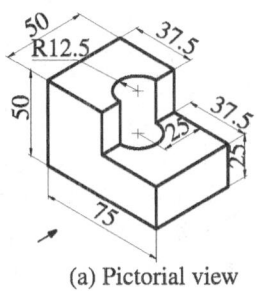

(a) Pictorial view

(d) Right-side view   (b) Front view   (e) Left-side view

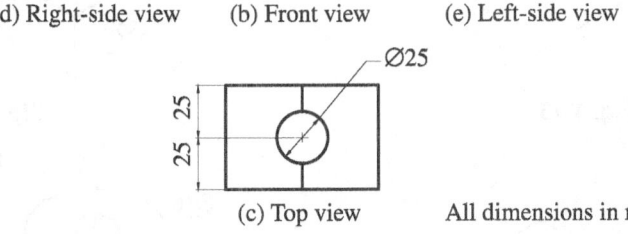

Ø25

(c) Top view   All dimensions in mm

Fig. 4.52

Observe along the direction of the arrow and sketch by free hand, the front, the top, and the relevant side views of the objects shown.

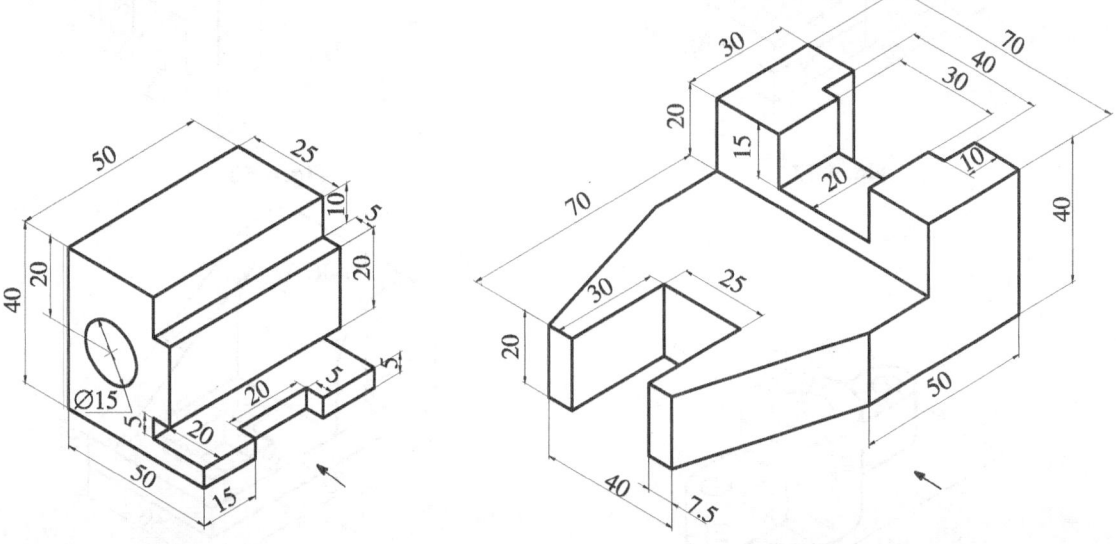

Fig. 4.53   **Fig. 4.54**

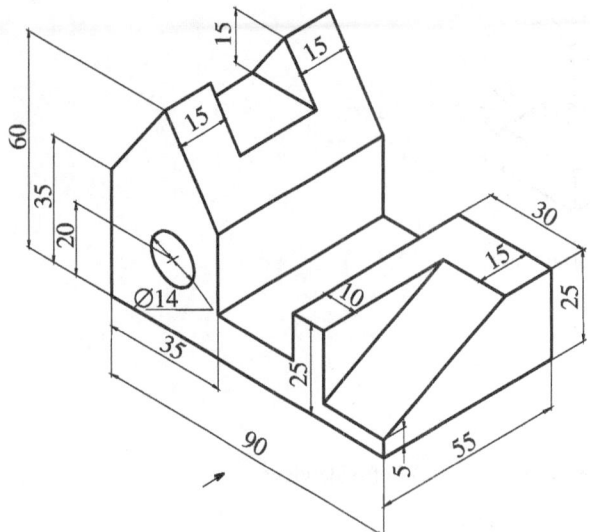

Fig. 4.55

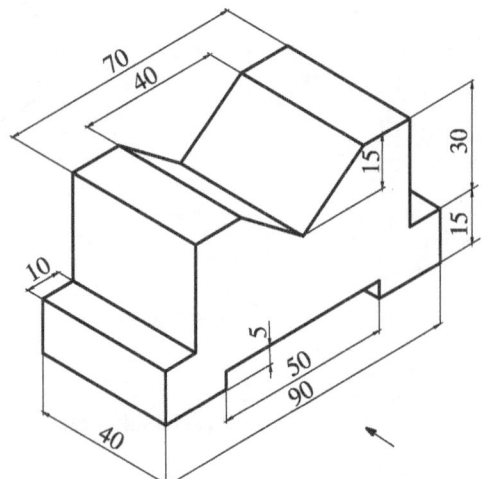

Fig. 4.56

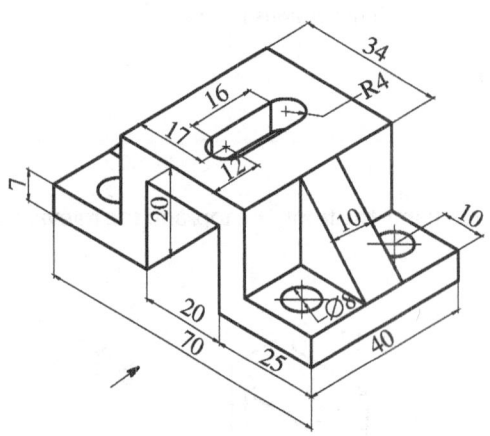

Fig. 4.57

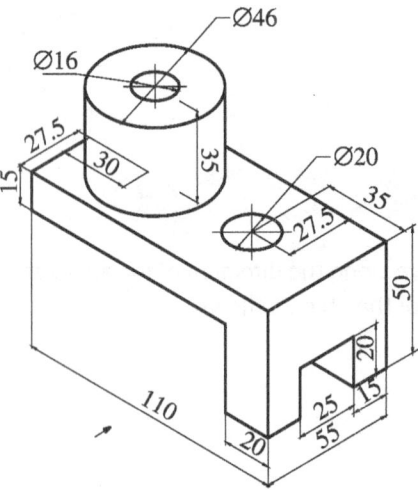

Fig. 4.58

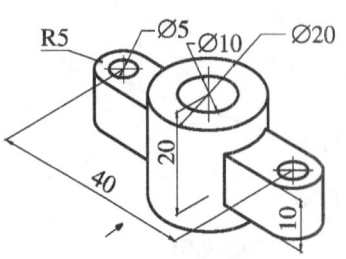

Fig. 4.59

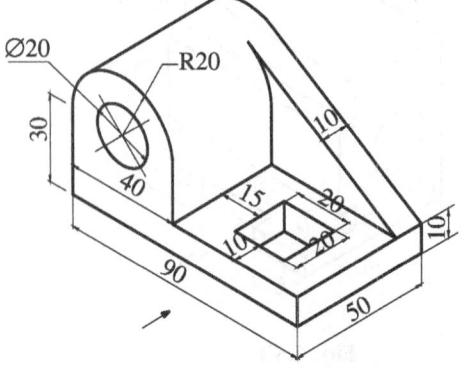

Fig. 4.60

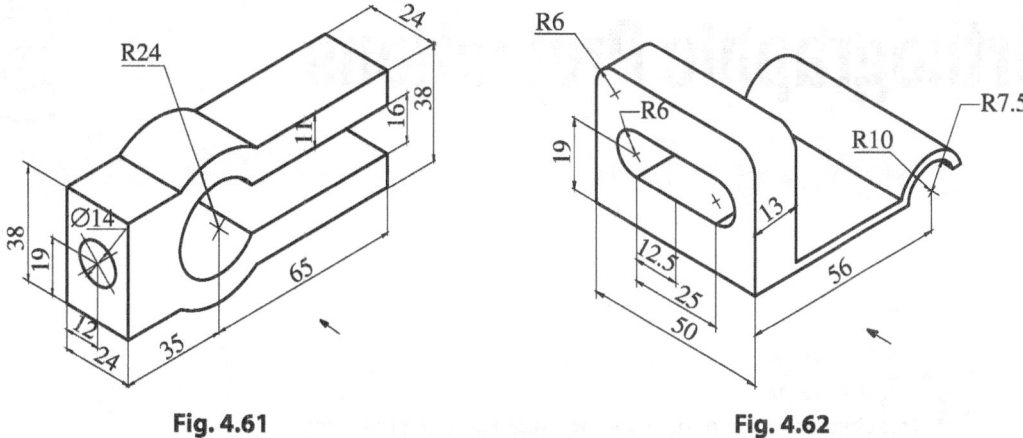

Fig. 4.61   **Fig. 4.62**

Answers for Multiple-choice Questions

4.1 A1 (C) A2 (A) A3 (A) A4 (C)   4.2 B1 (C) B2 (A) B3 (C) B4 (A)

4.3 C1 matches with 4; C2 matches with 3; C3 matches with 1; C4 matches with 2

Orthographic Projections

5

OBJECTIVES

This chapter will help the reader to understand the following:

- Adopt the projection of three-dimensional object orthogonally on a set of vertical and horizontal planes and obtain the appearances or the views of the frontal and the top surfaces, which explain the dimensions along the three axes

- Extend the orthogonal projection on a set of perpendicular planes known as 'profile planes' or 'side planes', where the appearances or the views of the right- and left-side surfaces of the object are revealed

- Device a suitable layout scheme, where all the views described earlier are combined for a meaningful interpretation

- Understand the impact of the views of the object when it is placed in different quadrants, which are obtained from the imaginary extension of the planes of projection

- Identify and familiarize the orthographic projection methods suitable for engineering practice

- Exposure to the standards practised in different countries for these drawings and adopt the one recommended by Bureau of Indian Standards (BIS), as a beginner

5.1 INTRODUCTION

In Chapter 4, it was explained that a three-dimensional object can be represented by the views of its different surfaces. Each view is represented by two dimensions of the three-dimensional object at a time and by suitably arranging them as the front, top, and side views, all the three dimensions can be expressed. Each view is obtained by the projection of surfaces on a set of planes, so that its image is obtained on these planes. When the projectors are parallel to each other and also perpendicular to the plane, the projection formed is known as *orthographic projection*. This chapter explains the basic principles of orthographic projection, the principal planes upon which the object is projected, and the views that are generated on them.

5.2 GENERAL PRINCIPLES OF ORTHOGRAPHIC PROJECTIONS

The orthographic projection of an object is the way of visualizing different surfaces of an object on mutually perpendicular or orthogonal planes, which are known as the principal planes. The object is observed from infinite distance and the rays of the vision are parallel among themselves and are perpendicular or orthogonal to the reference planes. The resulting images obtained on the reference planes are known as orthographic views. The rays that pass through various corners of the object and meet the reference planes are known as projectors. Since the projectors are parallel and pass through the various points on the surface of the object, they represent the dimensions of the surface of the object. If the surface of the object is parallel to one of the reference planes of projection, the projection of the surface on that plane to which it is parallel will have its true and exact dimensions of the surface. Similarly, if the other surfaces are also viewed, all the three dimensions of the object are obtained without any distortion or reduction in their sizes. This is the basic principle of orthographic projections. Figure 5.1 shows the basic elements, such as the observer, the object, the projection plane, and the parallel projectors that are involved in the orthographic projection.

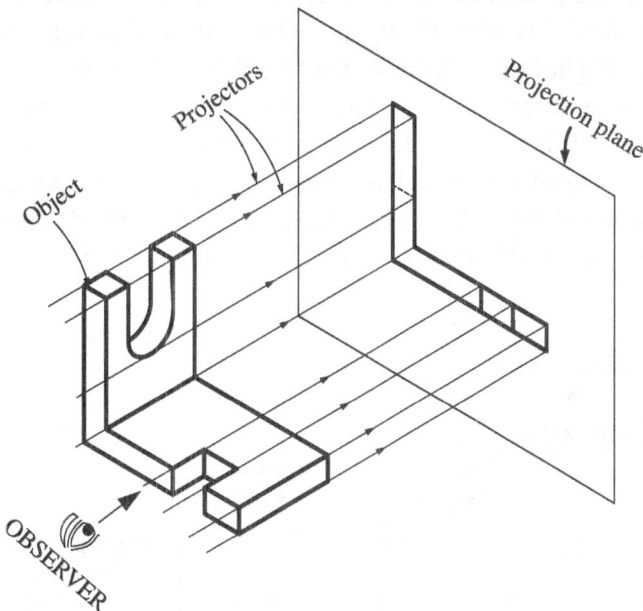

Fig. 5.1   Basic elements of orthographic projection

5.2.1 Principal Planes

The imaginary plane that is placed vertically or in front of an observer is called 'vertical plane (VP)' and a corresponding perpendicular plane placed horizontally is called 'horizontal plane (HP)'. In general, these two planes are known as the principal planes and their intersecting line is called horizontal reference line or reference line. Figure 5.2 shows the arrangement of the principal planes, such as the VP and the HP.

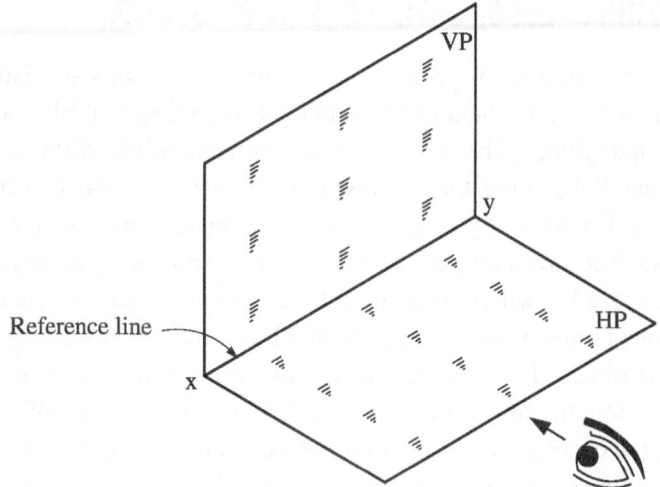

Fig. 5.2   Arrangement of VP and HP

A plane placed at 90° to the HP and the VP is known as the 'profile plane (PP)' or 'side plane'. The PP at the right-hand side of the observer, who is facing the VP, is known as the 'Right Profile Plane' or the 'Right-hand Side Plane'. Similarly, the PP at the left-hand side of the observer, who is facing the VP, is known as the 'Left Profile Plane' or the 'Left-hand-Side Plane'. Figure 5.3(a) shows the arrangement of right-hand-side plane along with the principal planes and Fig. 5.3(b) shows the arrangement of left-hand-side plane along with the principal planes. The vertical plane (VP), the horizontal plane (HP), and the side planes intersect along lines, which are known as vertical reference lines.

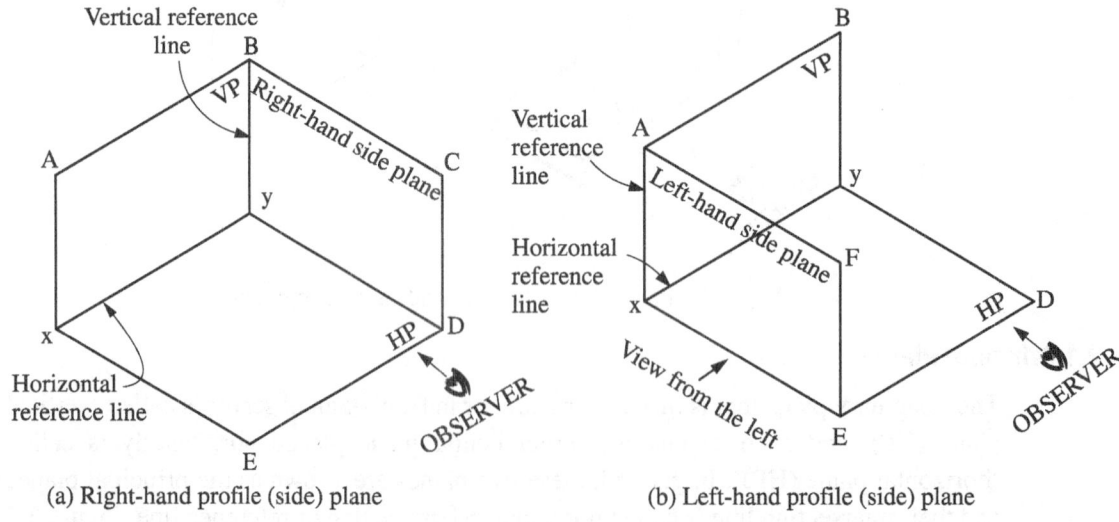

(a) Right-hand profile (side) plane   (b) Left-hand profile (side) plane

Fig. 5.3   Principal and profile planes

5.2.2 Classification of Views

The projection or view that is obtained on the VP is known as the front view or the view from the front (or the elevation), while the view from the top is obtained on the HP and is known as the top view (or the plan). The views from the right and the left, which are projected on the respective side planes, are known as the right-side view and the left-side view.

5.2.3 Layout of Views

The views that are obtained on the VP, HP, and the PPs are suitably laid out to provide the complete information of the front, top, and side appearances of the object. Generally, the VP is kept stationary and HP or the PPs are rotated about their intersecting lines till they become in flush with the VP. To make the views distinctly different and clear, these planes are rotated at 90° such that the planes 'open out'. The reason for keeping the VP stationary and rotating the other planes is due to the need of bringing all the planes in flush with VP, so that the views can be explained or displayed easily in one mode. Further, when the observer views the objects, the frontal appearance is the conventional view that is observed first and the other views are added to that to build up the whole object. Furthermore, all the medium through which we explain to others are all vertical in nature, such as walls or screens, class room boards, projection screens in any cinema house, and computer monitor screens etc., on which visuals or pictures or videos are displayed. For obvious reasons, such explanations cannot be made on horizontal medium such as floor due to the limitations of visibility by many people at a time. The views are laid with respect to the reference line or the intersecting line of these planes with the VP. The reference line indicating the intersection of VP and HP is used to distinguish the front view and the top view. The intersecting line of the VP and the PPs is used to distinguish the side views. For the object shown in Fig. 5.4, the front view obtained in the direction of the arrow is laid above the reference line and the top view is laid below the reference line (xy), when the HP is rotated as discussed earlier. These views are laid directly one below the other. The front view gives the distance or the height at which the object is placed from HP, while the top view gives the distance of the object from the VP. These distances are, reflected above and below the reference line respectively, as shown in Fig. 5.4.

Figure 5.4(a) shows the arrangement of the object and the reference planes and Fig. 5.4(b) shows the front view and the top of the object along with the projector lines.

Similarly, the side views are placed by the side of the front view, with the reference line or the intersecting line of the VP and the PPs, in between. The side views convey the distance of the object from the HP as indicated from the horizontal reference line and the distance of the object from the VP as shown from the vertical reference line. The left-side view of the object falls to the right of the front view and the vertical reference line and the right-side view of the object falls to the left of the front view and the vertical reference line, when the side planes or PPs are 'opened out' about their vertical reference lines x_1y_1 and x_2y_2 respectively as shown in Figs 5.5 and 5.6. Figure 5.5(a) shows the arrangement of the reference planes, the right-hand side plane and the object.

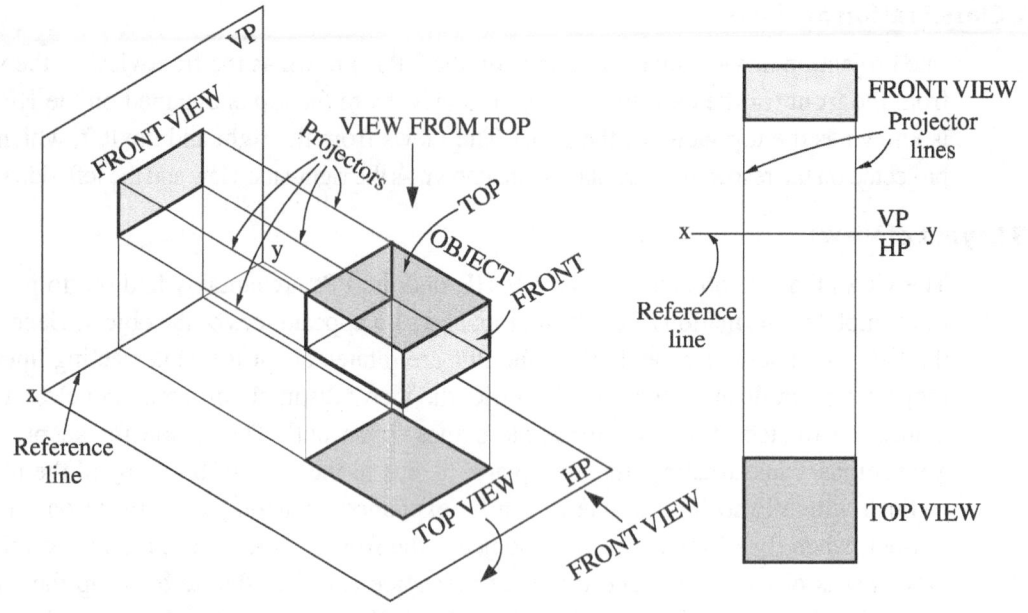

(a) Arrangement of object and the principal planes   (b) Arrangement of views

Fig. 5.4 Principal planes and the object along with the frontal and the top appearances

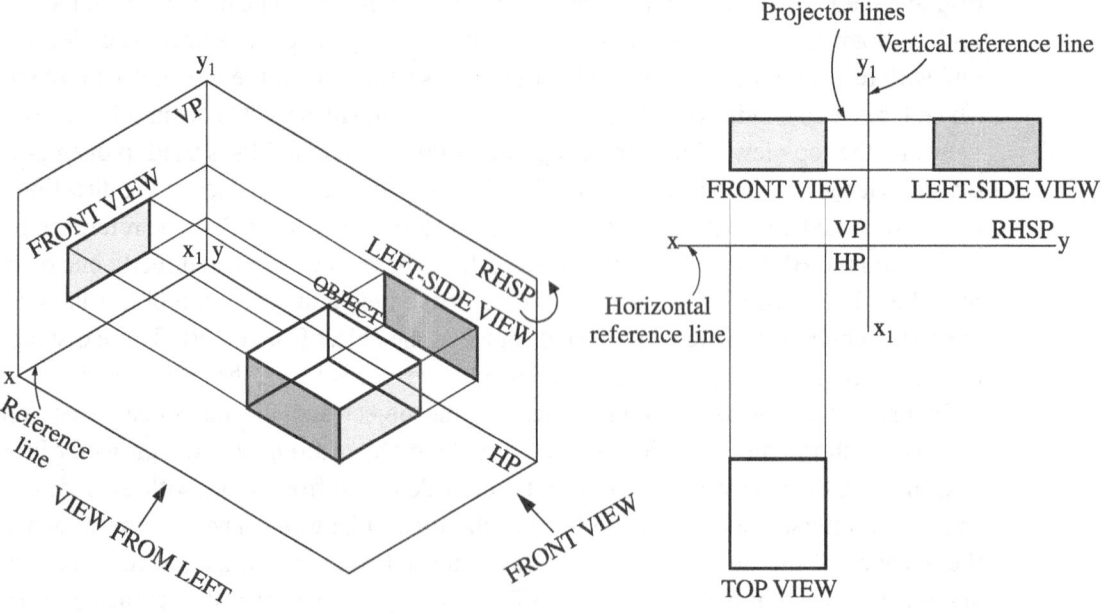

(a) Arrangement of object, principal planes, and left-side plane   (b) Arrangement of views

Fig. 5.5 Principal planes, right-hand side plane, and the object along with its views

Figure 5.5(b) shows the front view and the left-side view of the object along with the projector lines and the reference lines.

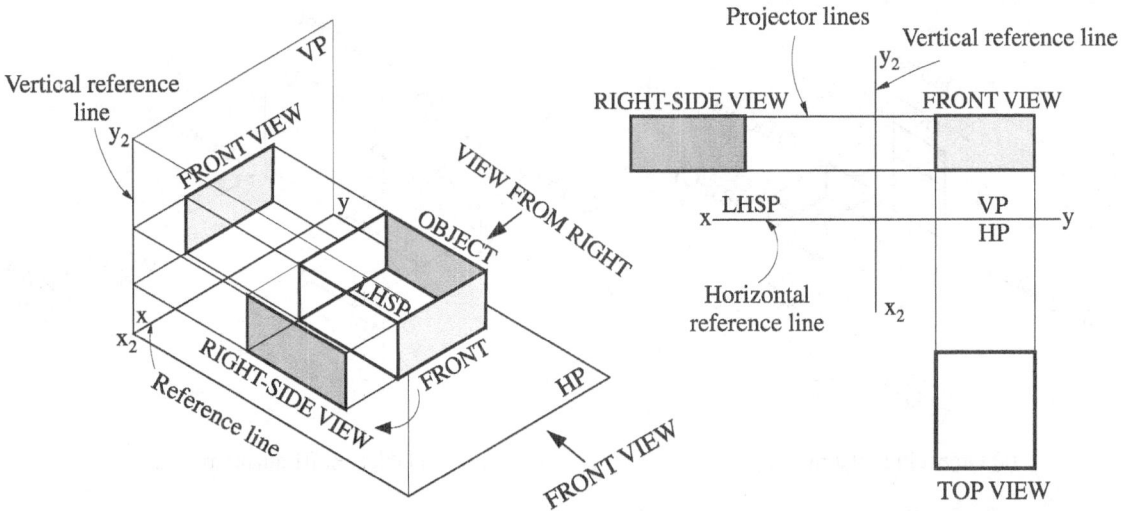

(a) Arrangement of the object, principal planes, and right-side plane   (b) Arrangement of views

Fig. 5.6 Principal planes, left-hand side planes, and the object along with its views

Figure 5.6(a) shows the arrangement of the reference planes, the left hand side plane and the object. Figure 5.6(b) shows the front view and the right-side view of the object along with the projector lines and the reference lines.

5.3 ORTHOGRAPHIC PROJECTIONS WITH RESPECT TO DIFFERENT QUADRANTS

When a set of vertical and horizontal planes discussed in Section 5.2.1 is extended beyond the line of intersection, it forms four quadrants as shown in Fig. 5.7. The object can be positioned in any one of the quadrants and its projection on the VP and the HP can be discussed along similar lines as in Section 5.2.1. The observer or the viewer stands in front of the VP for all the cases, and the direction of viewing is indicated by an arrow. The planes are assumed to be transparent. As discussed before, the front view is obtained when the object is projected orthogonally onto the VP, and its top view is obtained when the object is viewed from the above and projected to the HP. As discussed in Section 5.2.1, after the projections were individually obtained on the VP and the HP, the VP is held stationary and the HP is rotated 90° about the intersecting line in the clockwise direction and gets opened out. The views are thus classified with respect to the intersecting line or the reference line. The discussions made in Section 5.2.1 pertain to the positioning of the object in the first quadrant as shown in Fig. 5.7(a). The object can be located in any quadrant and its corresponding projections can be drawn. The position of any object with reference to a particular quadrant can be described as follows:

1. When the object is above HP and in front of VP, it is said to be in first quadrant.
2. When the object is above HP and behind VP, it is said to be in second quadrant.
3. When the object is below HP and behind VP, it is said to be in third quadrant.
4. When the object is below HP and in front of VP, it is said to be in fourth quadrant.

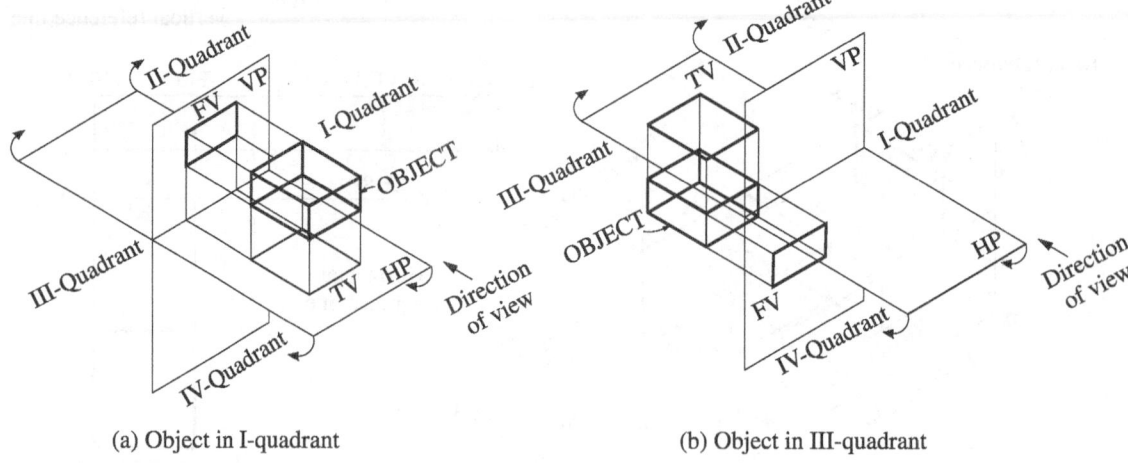

(a) Object in I-quadrant   (b) Object in III-quadrant

Fig. 5.7   Arrangement of principal planes and different quadrants

After obtaining the projections of the object and getting the respective views (front view and top view) in these cases, they can be arranged with respect to the intersecting line, keeping the VP stationary and rotating the HP as explained earlier. It should be noted that the sense of rotation of HP adopted in the case of the first quadrant is maintained for all the other quadrants to get a relative comparison of the placement of views.

As mentioned in Section 5.2.1, the distance of the object from HP will be reflected in the placement of the front view from the reference line, while that from VP will decide the position of the top view from the reference line. In the first-quadrant location, since the object is placed above the HP, the front view is located above the reference line. Logically, if the object is located below the HP as in the cases of third and fourth quadrants, the front view will be below the reference line. When the object is in front of the VP, the top view is drawn below the reference line, in the case of first-quadrant location. Therefore, logically, if the object is behind the VP, the top view should be positioned above the reference line in the second- and third-quadrant locations. The position of the object in different quadrants and the relative placement of the views are shown in the Table 5.1 and Fig. 5.8.

Table 5.1 indicates that the front view and the top view are brought out on either side of the reference line in first and third quadrant locations, while the views lie on the same side of the reference line in second and fourth quadrant locations. If the distances from HP and VP become equal, then the views will overlap and lose the sanctity of clarity of dimensions in second and fourth quadrant location and hence becomes useless for practice. That is why the practice of placing the object in first or third quadrant locations and getting the corresponding views is prevalent in engineering practice. The corresponding projections obtained by placing the object in first and third quadrant locations are, known as first-angle and third-angle projections respectively and will be dealt with in the subsequent sections in detail.

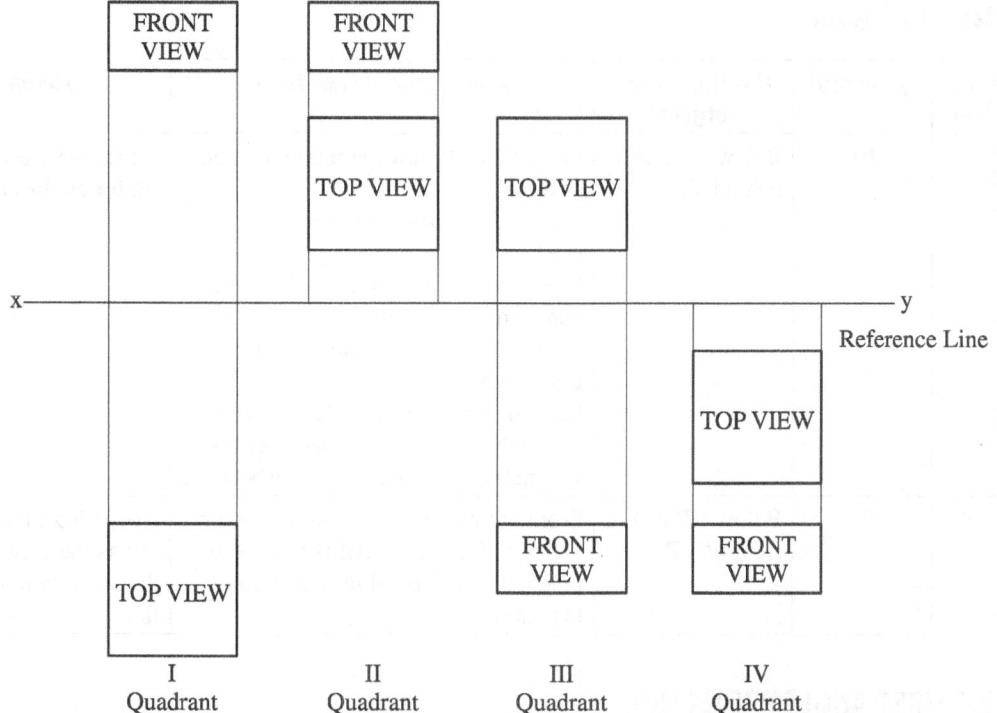

Fig. 5.8 Positioning of views of the object when placed in different quadrants

Table 5.1 Position of the object in different quadrants and the relative placement of views

S. no.	Quadrant	Position of the object	Placement of the views	Remarks
1.	I	Above HP and in front of VP	Front view: Above the horizontal reference line (xy line) Top view: Below the horizontal reference line (xy line) Right-side view: Placed to the left-hand side of the front view and above xy line, with the vertical reference line in between Left-side view: Placed to the right-hand side of the front view and above xy line, with the vertical reference line in between.	First-angle projection – Fit for engineering use
2.	II	Above HP and behind VP	Front view, top view, right-side view and left-side views : All the views are above the horizontal reference line (xy line)	Not fit for engineering use as the views are on the same side of reference line

(Contd)

Table 5.1 (*Contd*)

S. no.	Quadrant	Position of the object	Placement of the views	Remarks
3.	III	Below HP and behind VP	Front view: Below the reference line (xy line) Top view: Above the reference line (xy line) Right-side view: Placed to the right-hand side of the front view and below xy line, with the vertical reference line in between Left-side view: Placed to the left-hand side of the front view and below xy line, with the vertical reference line in between	Third-angle projection – Fit for engineering use
4.	IV	Below HP and in front of VP	Front view, top view, right-side view and left-side views: All the views are below the horizontal reference line (xy line)	Not fit for engineering use as the views are on the same side of reference line

5.4 FIRST-ANGLE PROJECTION

As discussed earlier in Section 5.2, when the object is placed above the HP and in front of the VP and the projections are obtained corresponding to the case of first quadrant location, it is known as the first-angle projection. While obtaining the projections or the views, the object lies between the observer and the plane of projection.

Figure 5.9 shows the location of the object, the plane of projection, and the observer as per the first-angle projection method. The HP can be compared to the ground, and

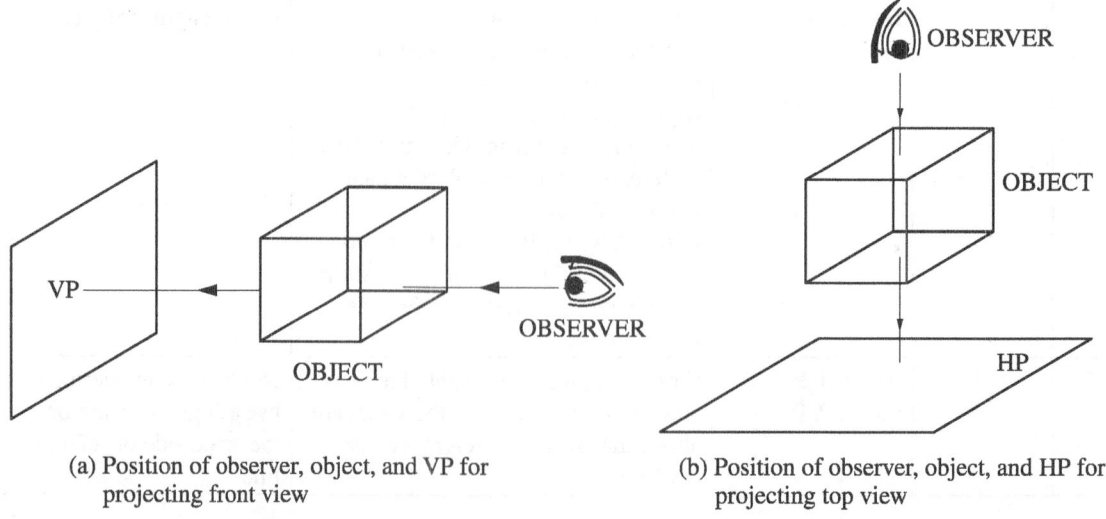

(a) Position of observer, object, and VP for projecting front view

(b) Position of observer, object, and HP for projecting top view

Fig. 5.9 Location of the object, principal planes, and the observer as per first-angle projection

the VP can be imagined as a vertical screen or wall behind the object. Since the object is above the HP, the front view is drawn above the reference line to the extent of its distance from the HP. The top view is drawn below the reference line depending on its distance from the VP. Both the views lie directly one below the other and are joined by the projector lines that are drawn as thin lines. The projector lines are perpendicular to the reference line, which is generally marked by the letters 'xy'.

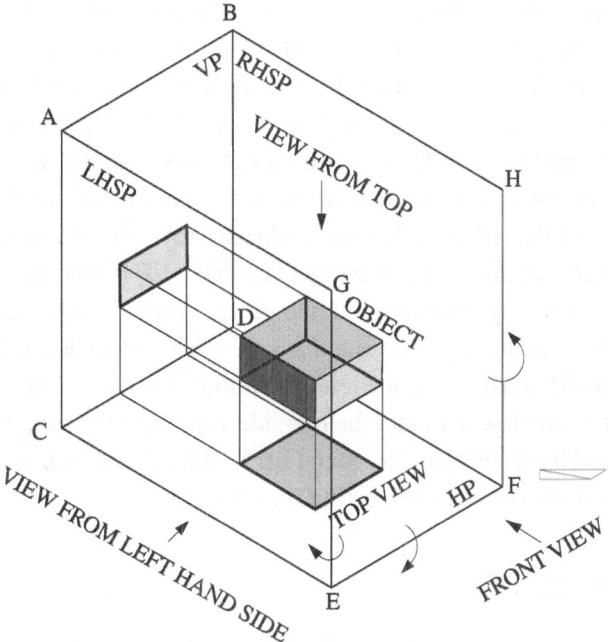

(a) Arrangement of principal and profile planes and the object

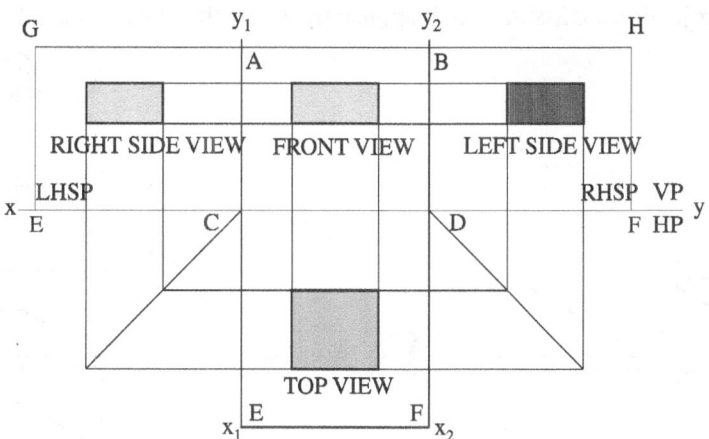

(b) Arrangement of views as per first angle projection

Fig. 5.10   Principal and PPs and views of the object in first-angle projection

Figure 5.10(a) shows the arrangement of VP, HP, and PPs along with the object. The VP is shown by ABDC, and the HP is shown by CDFE. The right profile is shown by

BHFD and the left PP is shown by AGEC. Figure 5.10(b) shows the layout of these planes when the HP, and the PPs are opened out along with the intersecting reference lines. The front view and the top view of the object are shown above and below the reference line xy.

When the object is viewed from the right-hand side of the observer and projected orthogonally to the PP situated at the left, it produces the right-hand side view of the object. When the left PP (indicated by AGEC) which contains the side view is rotated in the 'opening mode' and made to coincide with the vertical plane, the front view and the right-side view fall adjacent to each other. The intersecting line of the VP and the left PP lies in between the front and the side views as shown in Fig. 5.10(b). The above intersecting line is known as the vertical intersecting line and is denoted as 'x_1y_1'. Hence, the right-side view of the object falls to the left of the front view with the line x_1y_1 in between. The side views convey the distance of the object from the HP and the VP. Since in this case, the object is above the HP, the side view is placed above the horizontal reference line xy by the same amount. The distance of the object from the VP is marked from the vertical reference line x_1y_1 horizontally towards the left to get the complete right-side view of the object. The vertical reference line x_1y_1 is located from the front view depending on the distance of the object's left-side surface and the left PP. Similar constructional arrangement is made to the right of the front view to obtain the left-side view of the object, when projected to the right PP (indicated by BHFD) as shown in Fig. 5.10(b). It can be noted that the left-side view is marked from the vertical intersecting line 'x_2y_2'.

5.5 THIRD-ANGLE PROJECTION

In third-angle projection, a three-dimensional object is considered to lie in the third angle, that is, placed behind the VP and below the HP as shown in Fig. 5.7(b). The planes of projection are assumed to be transparent, and they lie between the object and the observer.

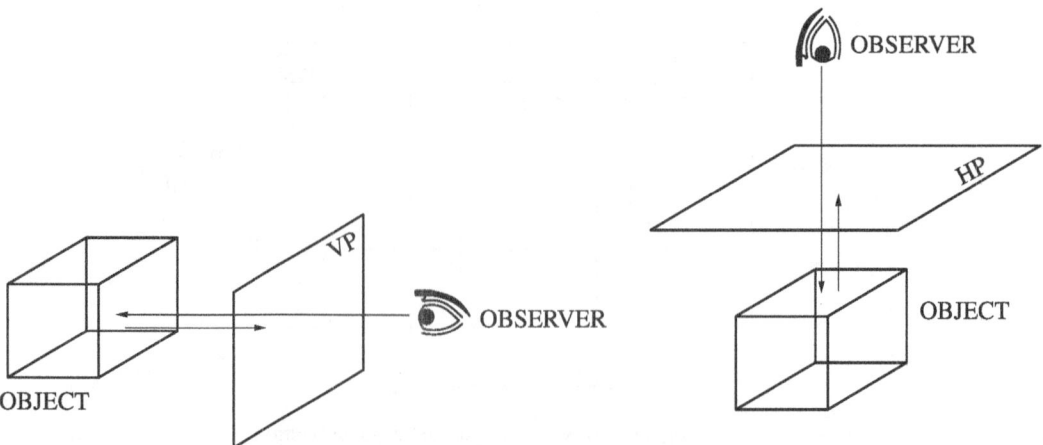

(a) Position of observer, VP, and object for projecting front view in third angle projection

(b) Position of observer, HP, and object for projecting top view in third angle projection

Fig. 5.11 Location of the object, principal planes, and the observer in third-angle projection

Figure 5.11 shows the location of the object, the plane of projection, and the observer as per the third-angle projection method. The VP can be imagined as a vertical transparent screen before any object, and the HP can be considered as the roof below which the object lies. Since the object is below the HP, the front view appears below the reference line depending on the distance from the HP. The top view, when projected to the HP, gets rotated with it, as the HP gets opened out and is placed above the reference line, depending on the distance from the VP. As described previously, both the views lie directly one below the other with the projector lines joining them. The projectors are perpendicular to the reference line xy. Figure 5.12(a) shows the arrangement of VP,

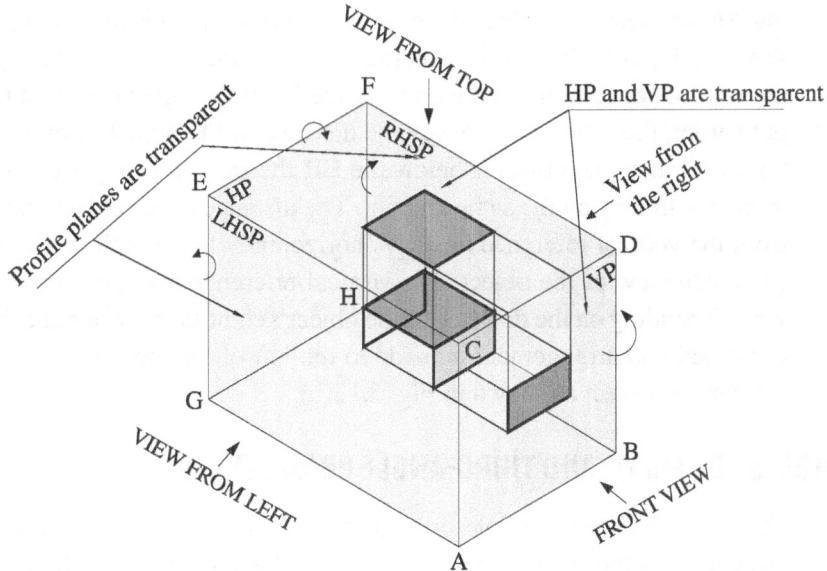

(a) Arrangement of principal and profile planes and the object

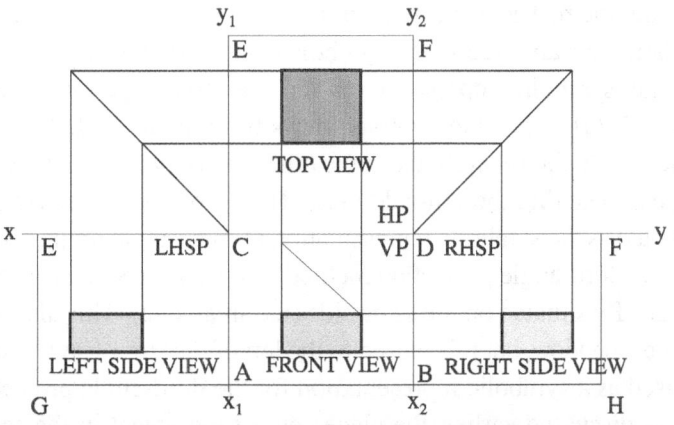

(b) Arrangement of the views as per third angle projections

Fig. 5.12 Principal, PPs, and views of the object as per third-angle projections

HP, and the PPs along with the object. The VP is shown by ABDC, and the HP is shown by CDFE. The right PP is shown by BHFD and the left PP is shown by AGEC. Figure 5.12(b) shows the layout of these planes, when the HP and the PPs are opened out along with the intersecting reference lines. The front view and the top view of the object are shown below and above the reference line xy.

When the object is viewed from the right-hand side of the observer, an imaginary profile plane FDBH is used to obtain the right-side view. When that PP is rotated 90° in the opening mode and made to coincide with the VP, both the surfaces are brought in the same vertical plane. The side view of the object that is available in the PP also gets turned and placed by the side of the front view and to the right of it. The vertical intersecting line of the PP and the VP denoted as x_2y_2 falls in between these two views as shown in Fig 5.12(b). In this case, the right-side view of the object appears to the right of the front view, contrary to the one obtained in first-angle projection method. As pointed out earlier, the side views convey the distance of the object from the HP and the VP. In this case, since the object is below the HP, the side view is placed below the horizontal reference line xy by the same amount. The distance of the object from the VP is marked from the vertical reference line x_2y_2 horizontally towards the right to get the complete right-side view of the object. The vertical reference line x_2y_2 is located from the front view depending on the distance of the object's right side surface and the right PP. Similar constructional arrangement is made to the left of the front view to obtain the left-side view of the object as shown in Fig. 5.12(b).

5.6 SYMBOLS FOR FIRST- AND THIRD-ANGLE PROJECTIONS

The drawings with multiple orthographic views obtained by the first- and third-angle projection methods are represented at the footnote or at the title blocks of those drawings by suitable symbols. The engineers understand the views and the method of projection used to get them by these symbols only, as no text or formal language is written to explain them. The projections of the frustum of a cone that has two different diameters at their ends are used as the symbols for the drawings.

The symbolic representation for the first-angle projection method is shown in Fig. 5.13(a). When the observer views from the left side (the small end) and project the object on to the PP available at the other end (larger diameter end), two concentric circles appear, denoting the left-side view. This is placed to the right of the front view and is adopted as the symbolic representation for first-angle projection as shown in Fig. 5.13(b).

The third-angle projection method also uses the same principle but gets the side view on to a PP situated on the same side (the small end). This also gets two concentric circles in the side view, but it is placed to the left of the front view as shown in Fig. 5.13(d). This is used as a symbolic representation for the third-angle projection method.

As discussed earlier, the placement of the object in the second and fourth quadrant locations do not distinguish the front and the top views clearly as they overlap and hence are not practicable, and no symbolic representations are attached to them.

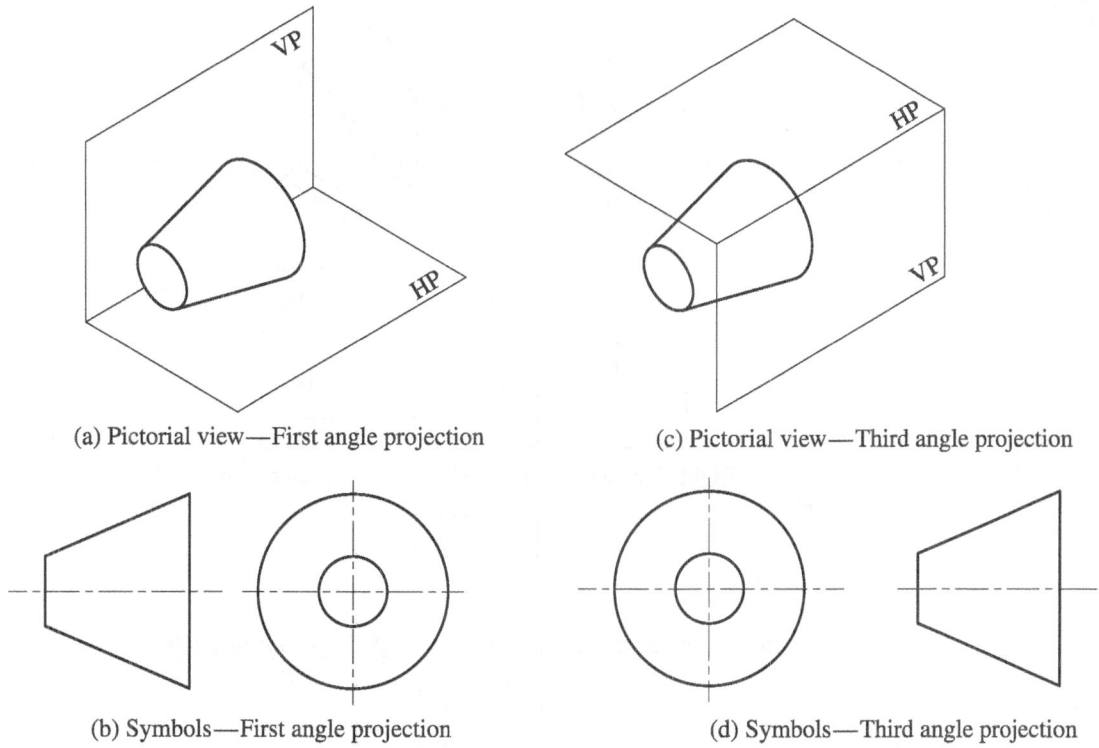

(a) Pictorial view—First angle projection   (c) Pictorial view—Third angle projection

(b) Symbols—First angle projection   (d) Symbols—Third angle projection

Fig. 5.13   Pictorial views and symbols for first- and third-angle projections

5.7 SALIENT FEATURES OF FIRST- AND THIRD-ANGLE PROJECTIONS

Since both the first- and third-angle projection methods equally convey all the details of the object and are adopted in engineering practice, the engineer should be fairly conversant with both methods. The details of the views are the same in both methods but only their layout schemes are different. The front view remains the same in both methods. While the top view is placed below the front view in the first-angle projection method, it appears above the front view in the third-angle projection method. The side views appear on the sides of the front view only in both methods. In the first-angle projection method, the view from the right-hand surface is placed to the left of the front view and that of the left-hand side surface is arranged to the right of the front view. The third-angle projection method, however, maintains the views of the respective surfaces on the same sides of the front view. The reference lines are drawn to distinguish these views, only for initial practice and once the learner gets the complete grip of the positions of the object and the arrangement of the views, reference lines are not mandatory. The industrial and technical drawings do not use the reference lines. However, in this book, the usage of the reference lines is maintained.

Figures 5.14(a) and (b) show the solid model and the corresponding line model of a three-dimensional block in a pictorial view. The direction of view of the arrow 'A'

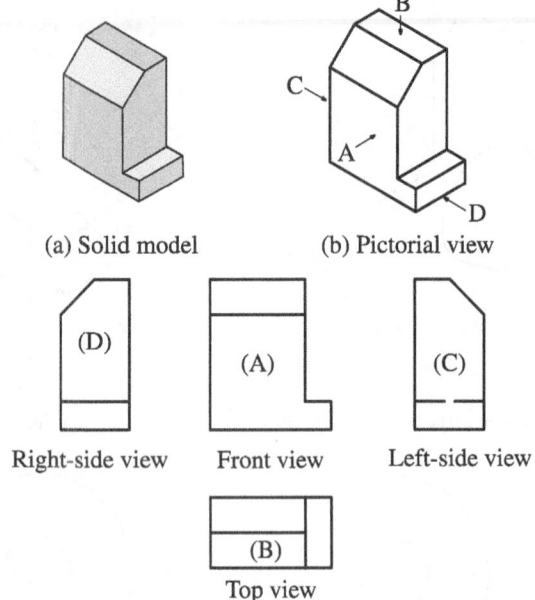

(a) Solid model   (b) Pictorial view

(c) Representation of views by first-angle projection methods

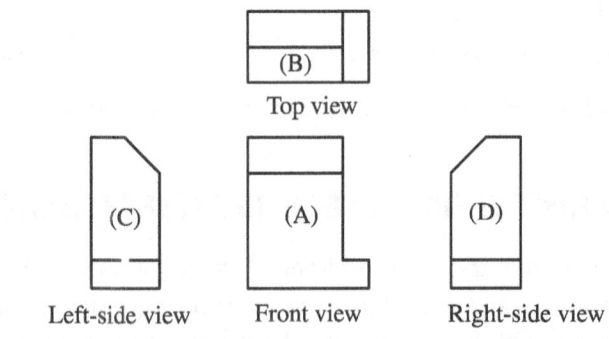

(d) Representation of views by third-angle projection methods

Fig. 5.14   Representation of an object by first- and third-angle projection methods

represents the front surface. The top, the left-, and the right-side surfaces are shown by the arrows 'B, C, and D', respectively. The front, top, and side views as obtained by the first- and third-angle projection methods are shown in Figs 5.14(c) and (d).

Table 5.2 summarizes the distinct features of both the methods.

Table 5.2   Distinct features of the first- and third-angle projections

S. no.	Feature	First-angle projection	Third-angle projection
1.	Position of the object	The object is placed above the HP and in front of VP in the first quadrant.	The object is placed below the HP and behind the VP in the third quadrant.

(Contd)

Table 5.2   (*Contd*)

S. no.	Feature	First-angle projection	Third-angle projection
2.	Location of the object, reference planes and the observer	The object is located in between the observer and the plane of projection (HP, VP, or PP). Observer → Object → Plane of projection	The plane of projection (HP, VP, or PP) is located in between the observer and the object. Observer → Plane of projection → Object
3.	Nature of the planes of projection	The planes of projections (HP, VP, or PP) are assumed to be non-transparent	The planes of projections (HP, VP, or PP) are assumed to be transparent
4.	Layout of views	Front view: above the reference line Top view: below the reference line Right-side view: placed to the left of the front view and above reference line Left-side view: placed to the right of the front view and above reference line	Front view: below the reference line Top view: above the reference line Right-side view: placed to the right of the front view and below the reference line Left-side view: placed to the left of the front view and below the reference line
5.	Symbolic representation in the drawing		

5.8 BIS CODE OF PRACTICE

The Bureau of Indian Standards (BIS) follows the British Standard practice, which uses the preparation and presentation of the multiple views by the first-angle projection method. United States and other European countries follow the practice of the third-angle projection method in their drawings. In the present-day world, where projects are handled irrespective of the affiliations of the countries, an engineer should become conversant to understand all the standards of practice.

In our country, Indian Standards Institution (ISI), the standardizing body, had recommended the first-angle projection method in 1955 in its standard, "IS: 696-Code of Practice for General Engineering Drawing". After a brief change over to the third-angle projection method, BIS once again reverted to the use of the first-angle projection method only and has issued Special Publications SP: 2003—Engineering Drawing Practice for schools and colleges to be followed in this regard.

In this book, the first-angle projection method has been adopted in general, and the usage of the third-angle projection method is cited for additional interest for the purpose to visualize the object through the transparent HP and VP in certain examples.

- Representation of a three-dimensional object by a single pictorial projection, although widely used, has an inherent disadvantage of foreshortening one or the other dimensions of the object. This distortion of the dimensions happen as we try to represent an object in three-dimensional space on to a two-dimensional medium such as the wall, board, screen, paper, or drawing sheet, where only two dimensions can be reflected.

- Generation of multiple two-dimensional pictures or views representing the various surfaces of a three-dimensional object is mandatory to understand all the three dimensions of the object completely and correctly.

- Principle of orthographic projection offers a means of generating the two-dimensional views of the different surfaces of the three-dimensional object by projecting it on to mutually perpendicular or orthogonal reference planes.

- Imaginary vertical plane in front of the observer is called VP and the corresponding perpendicular plane placed horizontally is known as the HP. These are called the principal planes. Two planes arranged side-by-side and placed at 90° to the VP and the HP is known as the profile planes or the side planes. The object is kept amidst these planes and its projections or views are obtained on them.

- The view that is obtained on the VP is called front view and the view that is obtained on the HP is called top view. The side views are obtained when the object is projected on to the PPs.

- The front view and the top view are laid from the reference line (intersecting line of VP and HP) depending on the distances of the object from these planes. The front view is placed above the reference line and the top view directly below it. The side views on the right and left sides are arranged by the sides of the front view, with the vertical reference line in between. The right-hand side view of the object is placed on to the left of the front view, whereas the left-hand side view of the object is placed to the right of the front view. Positioning of the object and the placement scheme thus discussed is called as the first-angle projection.

- Positioning of the object behind the vertical plane and below the horizontal plane results in the interchange of views in the layout and such reference scheme is called third-angle projection.

- These projections are of engineering interest since the views do not overlap and convey the details correctly. The salient features of these methods are summarized in Table 5.2.

- The drawings confirming to these projections are referred by symbols as recommended in the National and International Standards.

5.1. Orthographic projection means
   (a) projecting an object such that the projectors are parallel to each other and 30° to the plane of projection.
   (b) projecting an object such that the projectors are parallel to each other and 120° to the plane of projection
   (c) projecting an object such that the projectors are parallel to each other and 180° to the plane of projection
   (d) projecting an object such that the projectors are parallel to each other and 90° to the plane of projection

5.2 Principal planes refer to _____.
   (a) VP and HP
   (b) auxiliary planes
   (c) central plane and ground plane
   (d) auxiliary ground plane and picture plane

5.3 The profile plane is a plane which is
   (a) 45° to both HP and VP
   (b) 90° to HP and inclined VP
   (c) 90° to both HP and VP
   (d) 90° to VP and inclined to HP

5.4 The reference line denotes
   (a) the front edge of the VP
   (b) the top edge of the VP
   (c) the intersecting line between HP and VP
   (d) a line drawn 30° to the horizontal line

5.5 As per the the first-angle projection, the appearance of the right-side surface of the object is drawn
   (a) on the right side of the front view
   (b) on the left side of the front view
   (c) on the top of the front view
   (d) below the front view

5.6 Which of the following positions of the object favours the use of practical drawings?
   (a) object placed in the first quadrant
   (b) object placed in the fourth quadrant
   (c) object placed in the second quadrant
   (d) object placed partly in the first quadrant and partly in the second quadrant

5.7 Choose the right choice of layout for the first-angle projection

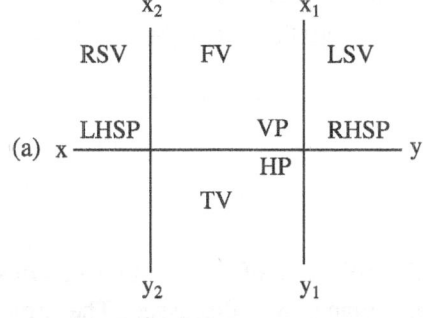

5.8 Choose the right option that yields the front view as per the first-angle projection.

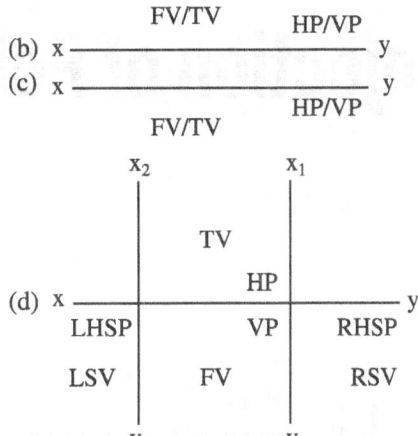

5.9 Choose the right option that yeilds the top view as per the first-angle projection.

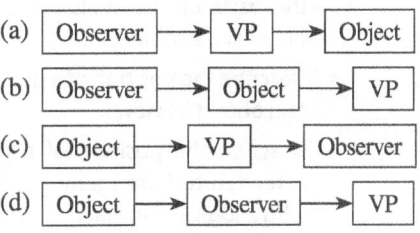

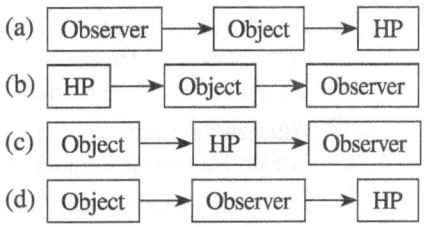

5.10 In the third-angle projection
   (a) HP and VP are assumed transparent
   (b) HP and VP are assumed to be non-transparent
   (c) observer is to be stationed behind VP and behind the object
   (d) observer is to be stationed below HP and below the object

Answers for Multiple-choice Questions

5.1 (d)	5.2 (a)	5.3 (c)	5.4 (c)	5.5 (b)	5.6 (a)
5.7 (a)	5.8 (b)	5.9 (a)	5.10 (a)		

Projection of Points

6

OBJECTIVES

This chapter will help the reader to understand the following:

- Appreciate the principle of conceiving the idea that a point is an infinitesimal object with no dimensions to describe its shape but forms the basis of the evolution process for various one-, two-, and three-dimensional objects
- Describe the position of a point with respect to all the planes of projection and obtain its views
- Describe the position of a point when the planes of projections are extended to form many quadrants and obtain the corresponding views with a logical approach
- Extend the idea of projection of one point to two or many points situated in space in the same quadrant and obtain their views
- Visualize the network of many points in space resulting in a straight line or a plane figure and to understand their respective views, when such points are located in different orientations
- Visualize the spatial arrangement of points from their projections/views, as a kind of reverse engineering practice
- Expose the reader to some of the day-to-day practical applications
- Finally, building exercises to understand chapters on projection of straight lines, planes, and solids

6.1 INTRODUCTION

In Chapter 5, the basic principle of orthographic projection of an object located in space between the vertical, horizontal, and the profile planes was discussed. The projectors emanating from the object from various points on its boundaries meet the projection planes orthogonally or perpendicularly and create an image or view on the respective projection plane. The projection on the vertical plane (VP) is termed as the front view (FV) and that on the horizontal plane (HP) is termed as the top view (TV). The projection on the profile planes (PP) gives the side views, depending on to which side it is placed.

It was also emphasized that the front view and the top view of the object are marked from the horizontal reference line (the intersecting line of VP and HP), depending on their distances from the HP and the VP respectively. The side views of the object are represented by the side of the front view, with the combined distances from HP and VP but separated by a vertical reference line, placed suitably on the respective side of the front view. This chapter deals with the orthographic projection of points, since a three-dimensional object can be obtained from many points, as in the case of a matter evolved from molecules or the atoms. While the arrangement of many points leads to the formation of a straight line, the arrangement of straight lines together forms a plane surface. A three-dimensional object can be realized by arranging its planar surfaces, and hence the projection of a point forms a mandatory basis for the other chapters discussed in this book.

6.2 PROJECTION OF A POINT AND ITS BASICS

The object under consideration is a point that does not have any dimension for itself to describe its shape, and hence, its projections do not involve any dimensions to describe its views. However, the positions of the point are described by the distances with respect to the reference planes and hence are used to mark the views. Figure 6.1(a) shows the arrangement of the VP and HP and a point (object).

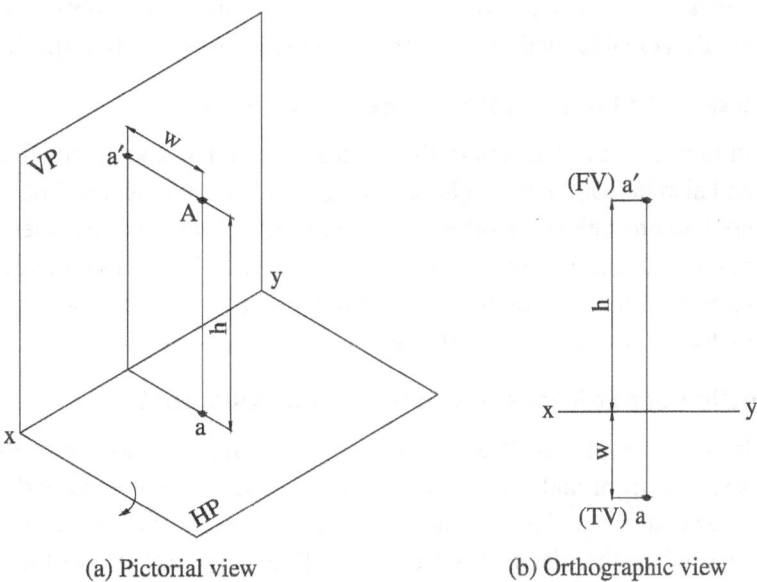

(a) Pictorial view   (b) Orthographic view

Fig. 6.1   Orthographic projection of a point in the first quadrant

The position or location of a Point 'A' located in the space and bounded by these planes can be described as above the HP and in front of the VP. Let these distances be denoted as 'h' and 'w', respectively. Let the respective views of the Point A on the VP and HP are denoted by a' and a. The views that are obtained on these individual planes can be brought on to a common vertical surface such as VP, by keeping the VP stationary and rotating the HP to align with the surface of the VP, as explained in Section 5.2 in Chapter 5. This can be achieved by opening HP as shown in Fig. 6.1(a). The corresponding views obtained are as shown in Fig. 6.1(b) and are marked from the intersecting lines, depending on the distances from the projection planes. While the front view appears above the reference line xy at a distance h, the top view appears directly below that at a distance w as shown in Fig. 6.1(b). The projector lines from the point A to the projection planes are also indicated by the line joining a' and a. Based on the principle of the orthographic projection, this projector line is perpendicular to the reference line xy as shown in Fig. 6.1(b).

6.3 PROJECTION OF A POINT IN DIFFERENT QUADRANTS

The reference planes shown in Fig. 6.1 can be extended to form four quadrants as discussed in Section 5.3 and in Fig. 5.7 in Chapter 5. The point (object) can be positioned in any of these quadrants, and its corresponding orthographic projections can be obtained. The reader is advised to refer Table 5.1 in Chapter 5. It is also reminded that the HP is rotated and brought on to the same plane as VP. As a convention and also as a means of comparison of the position of the object in different quadrants and their corresponding views, the same sense of rotation adopted in the first-quadrant position for the HP and the PP is maintained, while discussing the position in other quadrants.

6.3.1 Projection of a Point Located in the First Quadrant

In this case, the position of the point is referred as above the HP and in front of the VP and also in front of the right or left-hand-side profile planes. The front and the top views are laid one below the other with the reference line xy in between. The distance h of the point from the HP appears above the reference line, while the distance w from the VP appears below the reference line. Figure 6.1(a) shows the position of the Point A in the pictorial view, and Fig. 6.1(b) shows its orthographic views.

6.3.2 Projection of a Point Located in the Second Quadrant

In this case, the position of the point is represented as above the HP and behind the VP. The front and the top views are obtained on the same side of the reference line xy and above it. The distance h of the point from the HP appears above the reference line, while the distance w from the VP also appears above the reference line. Figure 6.2(a) shows the position of the Point A in the pictorial view and Fig. 6.2(b) shows its orthographic views.

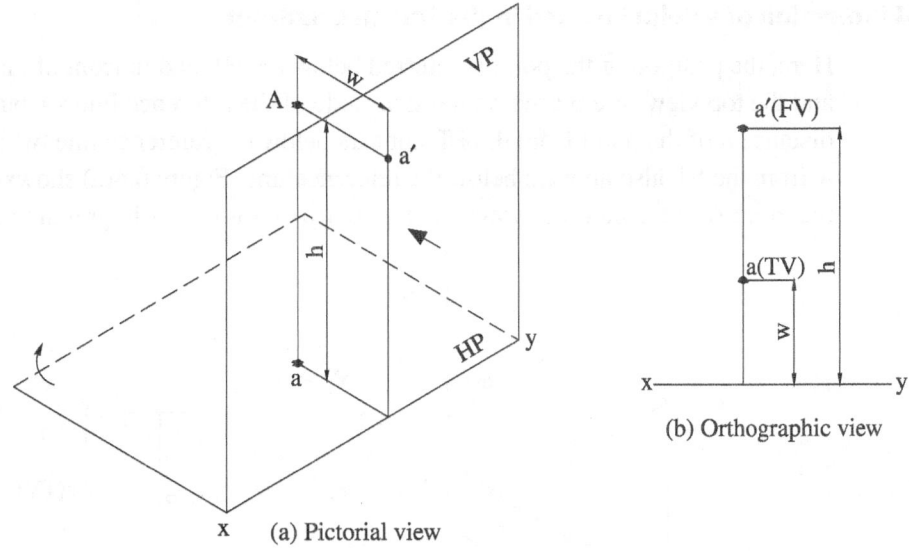

Fig. 6.2 Orthographic projection of a point in the second quadrant

6.3.3 Projection of a Point Located in the Third Quadrant

The position of the point in this case is referred as below the HP and behind the VP. The front and the top views are positioned one below the other with the reference line xy in between. The distance h of the point from the HP appears below the reference line, while the distance w from the VP appears above the reference line. Figure 6.3(a) shows the position of the Point A in the pictorial view and Fig. 6.3(b) shows its orthographic views.

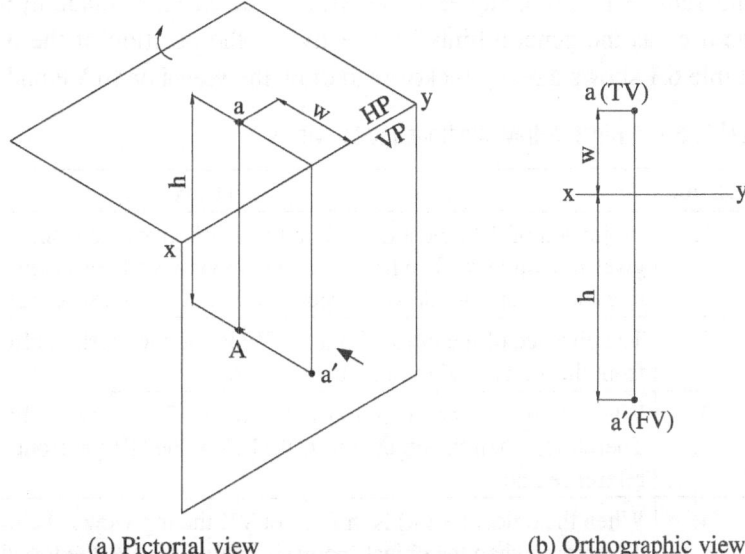

(a) Pictorial view   (b) Orthographic view

Fig. 6.3 Orthographic projection of a point in the third quadrant

6.3.4 Projection of a Point Located in the Fourth Quadrant

Here, the position of the point is referred below the HP and in front of the VP. The front and the top views are placed on the same side of the reference line xy, but below it. The distance h of the point from the HP appears below the reference line, while the distance w from the VP also appears below the reference line. Figure 6.4(a) shows the position of the Point A in the pictorial view and Fig. 6.4(b) shows its orthographic views.

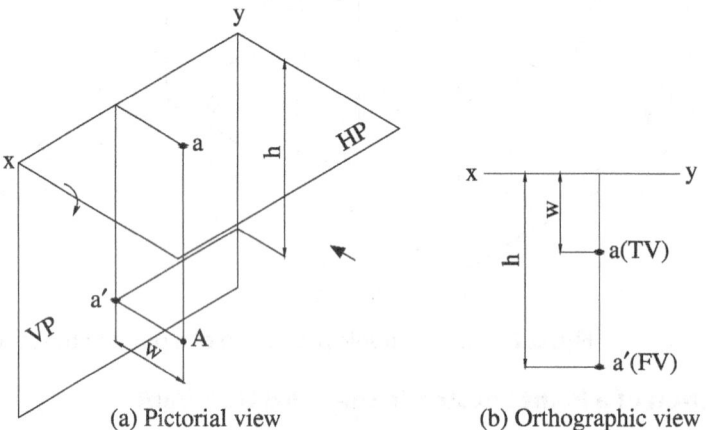

(a) Pictorial view (b) Orthographic view

Fig. 6.4 Orthographic projection of a point in the fourth quadrant

6.3.5 Summary

Projection of a Point on to VP and HP

The projection of an object on to VP and HP can be summarized in conjunction with the Table 5.1 from Chapter 5 and with the discussions made in Section 6.3. These can be used as the general hints irrespective of the position of the object in any quadrant. Table 6.1 shows a ready reckoner to draw the views on to VP and HP.

Table 6.1 Hints to draw the front and top views

S. no.	Hints
1.	Projection of the object (point) on to the VP gives the front view and that to the HP gives the top view. The front and the top view will always lie on the same vertical line (projector line) drawn perpendicular to the reference line.
2.	The distance of the object from the HP is used to mark the front view, while that from the VP is used to mark the top view.
3.	When the object (point) is above the HP, the front view is above the reference line. Therefore, when the object (point) is below the HP, the front view is below the reference line.
4.	When the object (point) is in front of VP, the top view is below the reference line. Therefore, when the object (point) is behind the VP, the top view is above the reference line.

6.3.6 Illustrative Examples

Example 6.1

Draw the projections of the following points on to VP and HP.

Point A 25 mm Above HP and 20 mm in Front of VP

Inference

1. The point is above the HP, and hence, the front view is above the reference line.
2. The point is in front of the VP, and hence, the top view is below the reference line

PROCEDURE (Refer Fig. 6.5a)

Step 1: Draw the horizontal reference line xy.

Step 2: From any point on the xy line, draw a perpendicular line and mark a', 25 mm above and a, 20 mm below the reference line.

Step 3: Draw the projector line joining a' and a and dimension the views as indicated.

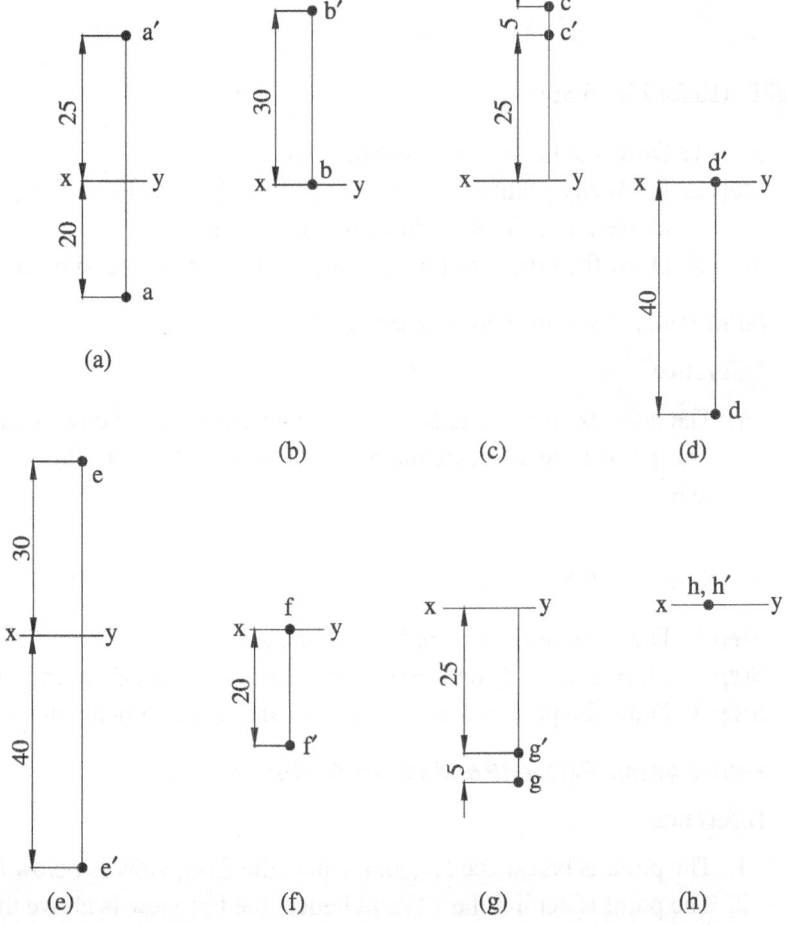

Fig. 6.5   Projection of a point in different quadrants

Point B on VP and 30 mm Above HP

Inference

1. The point is above the HP, and hence, the front view is above the reference line.
2. The point is on the VP, and hence, the top view is on the reference line.

PROCEDURE (Refer Fig. 6.5b)

Step 1: Draw the horizontal reference line xy.
Step 2: Mark a point b on the reference line and mark b', 30 mm above it.
Step 3: Draw the projector line joining b' and b and dimension the views as indicated.

Point C 30 mm Behind VP and 25 mm Above HP

Inference.

1. The point is above the HP, and hence, the front view is above the reference line
2. The point is behind the VP, and hence, the top view is also above the reference line.

PROCEDURE (Refer Fig. 6.5c)

Step 1: Draw the horizontal reference line xy.
Step 2: From any point on the xy line, draw a perpendicular line and mark c', 25 mm above and c, 30 mm above the reference line.
Step 3: Draw the projector line joining c' and c and dimension the views as indicated.

Point D on HP and 40 mm in Front of VP

Inference

1. The point is on HP, and hence, the front view is on the reference line.
2. The point is in front of the VP, and hence, the top view is below the reference line.

PROCEDURE (Refer Fig. 6.5d)

Step 1: Draw the horizontal reference line xy.
Step 2: Mark a point d' on the reference line and mark d, 40 mm below it.
Step 3: Draw the projector line joining d' and d and dimension the views as indicated.

Point E 40 mm Below HP and 30 mm Behind VP

Inference

1. The point is below the HP, and hence, the front view is below the reference line.
2. The point is behind the VP, and hence, the top view is above the reference line.

PROCEDURE (Refer Fig. 6.5e)

Step 1: Draw the horizontal reference line xy.

Step 2: From any point on the xy line, draw a perpendicular line and mark e′, 40 mm below and e, 30 mm above the reference line.

Step 3: Draw the projector line joining e′ and e and dimension the views as indicated.

Point F on VP and 20 mm Below HP

Inference

1. The point is below the HP, and hence, the front view is below the reference line.
2. The point is on the VP, and hence, the top view is on the reference line.

PROCEDURE (Refer Fig. 6.5f)

Step 1: Draw the horizontal reference line xy.

Step 2: Mark a point f on the reference line and mark f′, 20 mm below it.

Step 3: Draw the projector line joining f′ and f and dimension the views as indicated.

Point G 25 mm Below HP and 30 mm in Front of VP

Inference

1. The point is below the HP, and hence, the front view is below the reference line.
2. The point is in front of the VP, and hence, the top view is also below the reference line.

PROCEDURE (Refer Fig. 6.5g)

Step 1: Draw the horizontal reference line xy.

Step 2: From any point on the xy line, draw a perpendicular line and mark g′, 25 mm below and g, also 30 mm below the reference line.

Step 3: Draw the projector line joining g′ and g and dimension the views as indicated.

Point H at the Intersection of VP and HP

Inference

Since the point is at the intersection of VP and HP, its views are also on the intersecting line (reference line).

PROCEDURE (Refer Fig. 6.5h)

Step 1: Draw the horizontal reference line xy.

Step 2: Mark any point on the xy line, and it represents both the views.

Example 6.2

The projections of the following points are as indicated in Fig. 6.6. Define the positions of these points with respect to the planes of projection and mention the quadrants in which each point is situated.

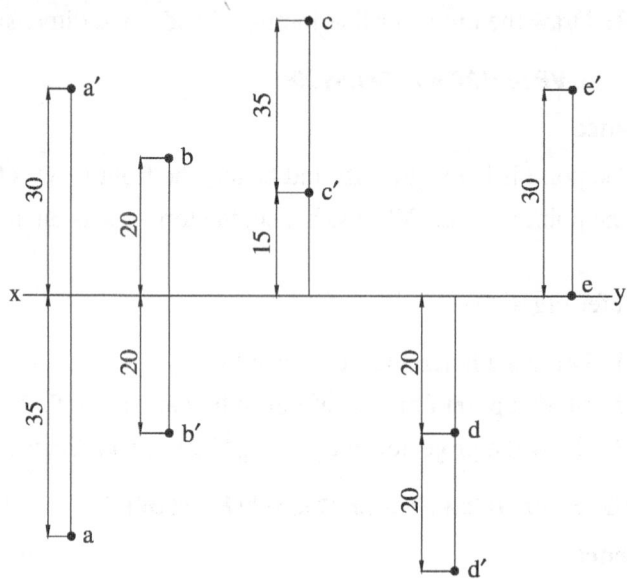

Fig. 6.6   Orthographic views of a point in different quadrants

Point A

1. Since a′ is above the reference line, the Point A is above the HP.
2. Since a is below the reference line, the Point A is in front of the VP.
3. Since the position of the object (Point A) is above the HP and in front of the VP, it is situated in the first quadrant.

Point B

1. Since b′ is below the reference line, the Point B is below the HP.
2. Since b is above the reference line, the Point B is behind the VP.
3. Since the position of the object (Point B) is below the HP and behind the VP, it is situated in the third quadrant.

Point C

1. Since c′ is above the reference line, the Point C is above the HP.
2. Since c is above the reference line, the Point C is behind the VP.
3. Since the position of the object (Point C) is above the HP and behind the VP, it is situated in the second quadrant.

Point D

1. Since d′ is below the reference line, the Point D is below the HP.
2. Since d is also below the reference line, the Point D is in front of the VP.
3. Since the position of the object (Point D) is below the HP and in front of the VP, it is situated in the fourth quadrant.

Point E

1. Since e′ is above the reference line, the Point E is above the HP.
2. Since e is on the reference line, the Point E is on the VP.
3. Since the position of the object (Point E) is above the HP and on the VP, it is situated in the common plane of the first and second quadrants. ▲

Projection of a Point on to VP, HP, and PP

When the object (point) is projected to the PP, in addition to VP and HP, the following are to be noted in addition to the discussions made in the earlier section and in Table 6.1. The reader is advised to refer Table 6.2 shown below as a ready reckoner to draw the side views of an object. Figures 6.7(a) and (b) show constructional procedure to obtain left-side view of a Point A projected on to the right profile plane in pictorial view and its orthographic views, respectively, in addition to the front view and the top view in first-angle projection.

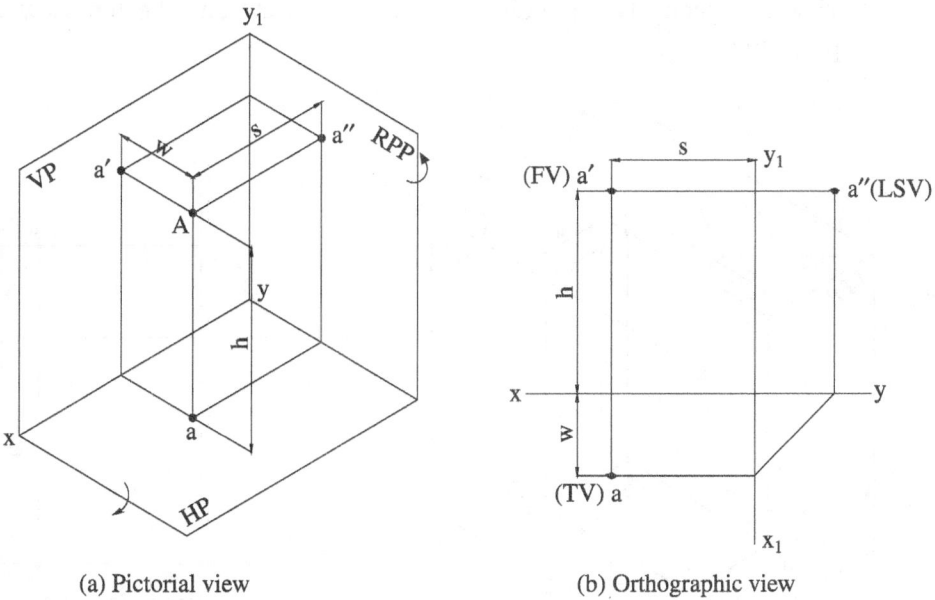

(a) Pictorial view   (b) Orthographic view

Fig. 6.7 Orthographic Projection of a point in the first Quadrant

Figures 6.8(a) and (b) show constructional procedure to obtain left-side view of a Point A projected on to the left profile plane in pictorial view and its orthographic views, respectively, in addition to the front view and the top view in second-angle projection.

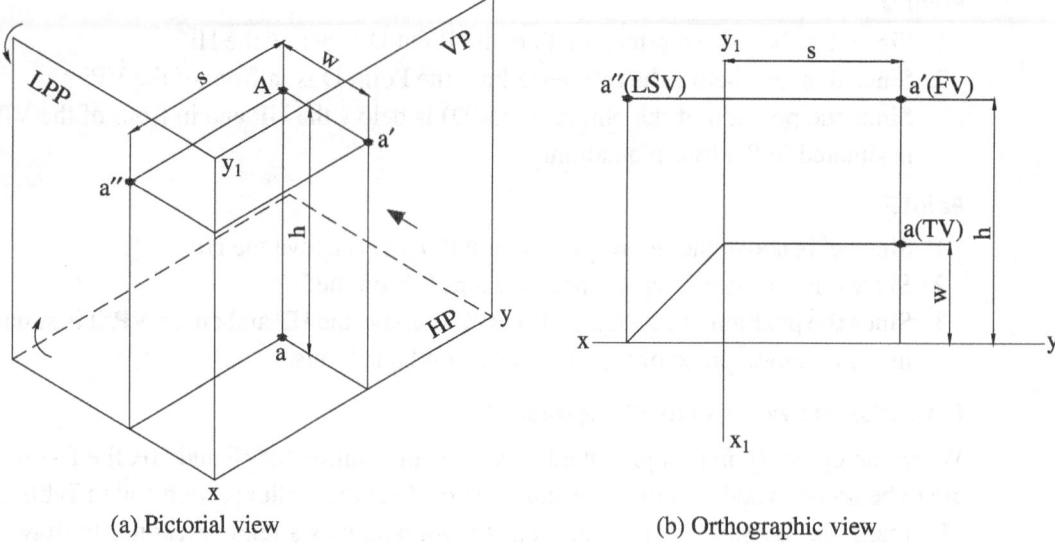

(a) Pictorial view   (b) Orthographic view

Fig. 6.8 Orthographic projection of a point in the second quadrant

Figures 6.9(a) and (b) show constructional procedure to obtain left-side view of a Point A projected on to the left profile plane in pictorial view and its orthographic views, respectively, in addition to the front view and the top view in third-angle projection.

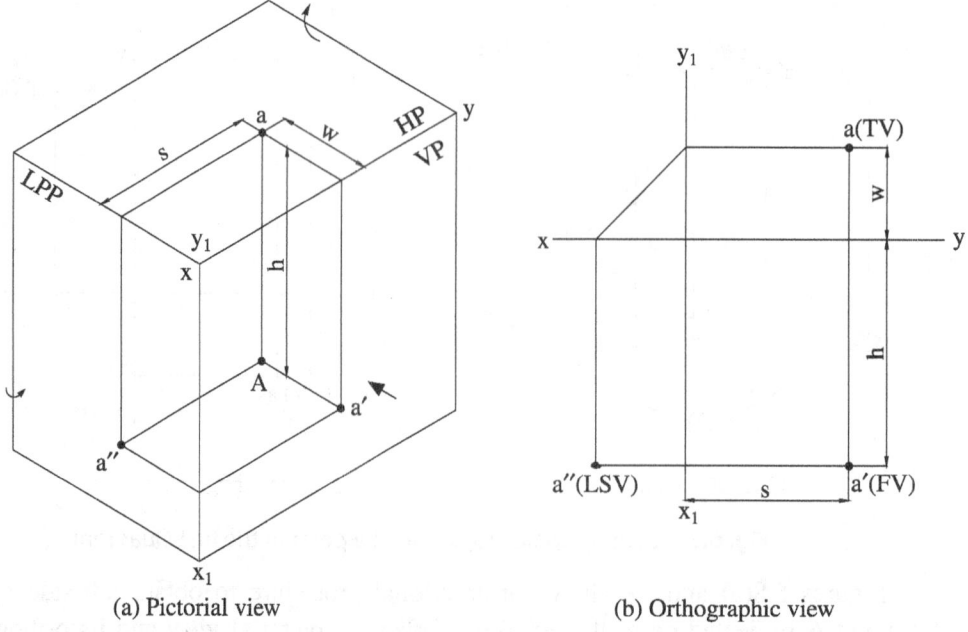

(a) Pictorial view   (b) Orthographic view

Fig. 6.9 Orthographic projection of a point in the third quadrant

Figures 6.10(a) and (b) show constructional procedure to obtain left-side view of a Point A projected on to the right profile plane in pictorial view and its orthographic views, respectively, in addition to the front view and the top view in fourth-angle projection.

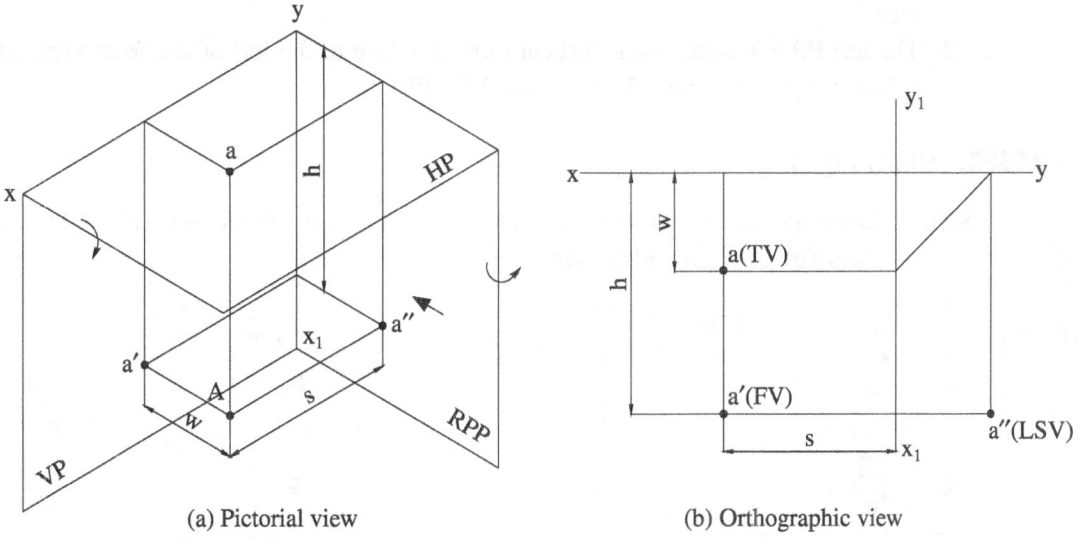

(a) Pictorial view   (b) Orthographic view

Fig. 6.10   Orthographic projection of a point in the fourth quadrant

Table 6.2   Hints to draw the side views

S. no.	Hints
1.	The projection on to the profile plane (PP) gives the side view of the object.
2.	The side view is positioned always by the side of the front view by drawing a vertical reference line (on to the right or the left as the case may be) at a distance equal to that of the object and the PP.
3.	The front view is extended horizontally to meet the vertical reference line and the side view is obtained on the other side of it, at a distance equal to that of the object and the VP. (This can also be obtained by extending the top view to meet the vertical reference line, drawing a 45° line or a quarter circle to meet the horizontal reference line, and then drawing a perpendicular line to meet the front view extended.)
4.	The front view and the side view always lie on a horizontal line but on either side of the vertical reference line.

Example 6.3

Draw the projections of the following points on to VP, HP, and PP.

NOTE   *Since the procedure for obtaining the projections of the point on to VP and HP was explained in detail in the aforementioned section and also in Example 6.1, it will not be repeated here. The procedure for obtaining the projection on to PP alone will be elaborated here.*

Point P 25 mm Above HP, 20 mm in Front of VP, and 20 mm in Front of Left PP

Inference

1. Since the point is in front of the left PP, it is projected on to it to get the right-side view.
2. The left PP is marked by a vertical reference line to the left of the front view, at a distance equal to that of the object and the PP.

PROCEDURE (Refer Fig. 6.11a)

Step 1: Draw the horizontal reference line xy and locate the front view (p′) and the top view (p) as explained earlier.

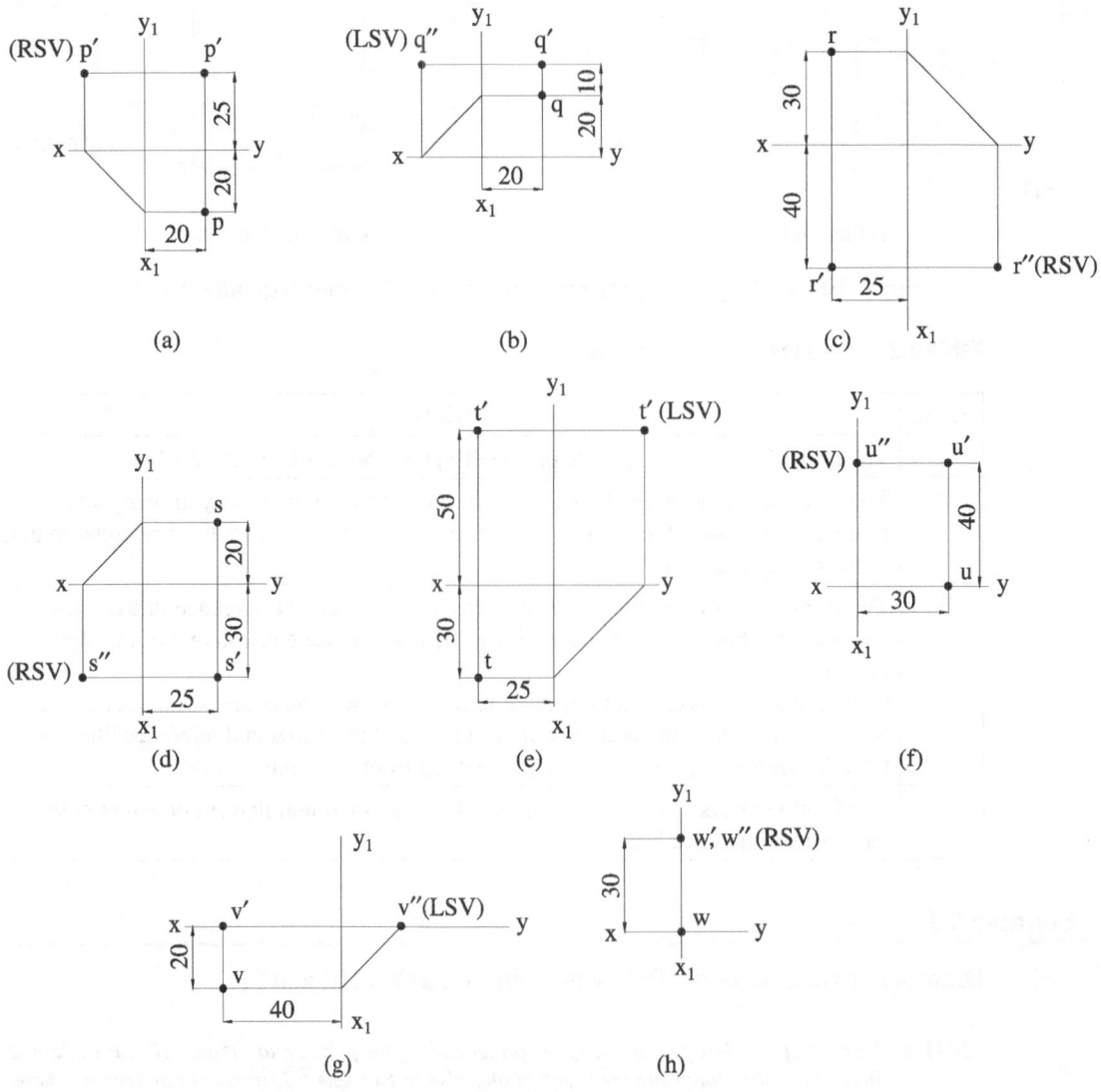

Fig. 6.11   Projection of a point on to VP, HP, and PP

Step 2: Place the vertical reference line x_1y_1 at a distance of 20 mm from p′ and to the left of it.

Step 3: Extend the front view horizontally on to the other side of the vertical reference line and mark the distance of the object from the VP, that is, 20 mm to get the side view p″.

Step 4: Draw the projector lines joining p′ and p, p′ and p″ and the vertical line from p″ and dimension the views as indicated.

Point Q 20 mm Behind VP, 30 mm Above HP, and 20 mm Behind Left PP

Inference

1. Since the point is behind the left PP, it is projected on to it to get the left-side view.
2. The left PP is marked by a vertical reference line, to the left of the front view, at a distance equal to that of the object and the PP.

PROCEDURE (Refer Fig. 6.11b)

Step 1: Draw the horizontal reference line xy and locate the front view (q′) and the top view (q) as explained earlier.

Step 2: Place the vertical reference line x_1y_1 at a distance of 20 mm from q′ and to the left of it.

Step 3: Extend the front view horizontally on to the other side of the vertical reference line and mark the distance of the object from the VP, that is, 20 mm to get the side view q″.

Step 4: Draw the projector lines joining q′ and q, q′ and q″ and the vertical line from q″ and dimension the views as indicated.

Point R 30 mm Behind VP, 40 mm Below HP, and 25 mm Behind Right PP

Inference

1. Since the point is behind the right PP, it is projected on to it to get the right-side view.
2. The right PP is marked by a vertical reference line, to the right of the front view, at a distance equal to that of the object and the PP.

PROCEDURE (Refer Fig. 6.11c)

Step 1: Draw the horizontal reference line xy and locate the front view (r′) and the top view (r) as explained earlier.

Step 2: Place the vertical reference line x_1y_1 at a distance of 25 mm from r′ and to the right of it.

Step 3: Extend the front view horizontally on to the other side of the vertical reference line and mark the distance of the object from the VP, that is, 30 mm to get the side view r″.

Step 4: Draw the projector lines joining r′ and r, r′ and r″ and the vertical line from r″ and dimension the views as indicated.

Point S 30 mm Below HP, 20 mm Behind VP, and 25 mm in Front of Left PP

Inference

1. Since the point is in front of the left PP, it is projected on to it to get the right-side view.
2. The left PP is marked by a vertical reference line, to the left of the front view, at a distance equal to that of the object and the PP.

PROCEDURE   (Refer Fig. 6.11d)

Step 1: Draw the horizontal reference line xy and locate the front view (s′) and the top view (s) as explained earlier.

Step 2: Place the vertical reference line x_1y_1 at a distance of 25 mm from s′ and to the left of it.

Step 3: Extend the front view horizontally on to the other side of the vertical reference line and mark the distance of the object from the VP, that is, 20 mm to get the side view s″.

Step 4: Draw the projector lines joining s′ and s, s′ and s″ and the vertical line from s″ and dimension the views as indicated.

Point T 50 mm Above HP, 30 mm in Front of VP, and 25 mm in Front of Right PP

Inference

1. Since the point is in front of the right PP, it is projected on to it to get the left-side view.
2. The right PP is marked by a vertical reference line, to the right of the front view, at a distance equal to that of the object and the PP.

PROCEDURE   (Refer Fig. 6.11e)

Step 1: Draw the horizontal reference line xy and locate the front view (t′) and the top view (t) as explained earlier.

Step 2: Place the vertical reference line x_1y_1 at a distance of 25 mm from t′ and to the right of it.

Step 3: Extend the front view horizontally on to the other side of the vertical reference line and mark the distance of the object from the VP, that is, 30 mm to get the side view t″.

Step 4: Draw the projector lines joining t′ and t, t′ and t″ and the vertical line from t″ and dimension the views as indicated.

Point U on VP, 40 mm Above HP, and 30 mm in Front of Left PP

Inference

1. Since the point is in front of the left PP, it is projected on to it to obtain the right-side view.

2. The left PP is marked by a vertical reference line, to the left of the front view, at a distance equal to that of the object and the PP.

PROCEDURE (Refer Fig. 6.11f)

Step 1: Draw the horizontal reference line xy and locate the front view (u′) and the top view (u) as explained earlier.

Step 2: Place the vertical reference line x_1y_1 at a distance of 30 mm from u′ and to the left of it.

Step 3: Extend the front view horizontally on to the other side of the vertical reference line and mark the distance of the object from the VP. Since that distance is zero, the side view u″ falls on the vertical reference line itself.

Step 4: Draw the projector lines joining u′ and u, u′ and u″ and the vertical line from u″ and dimension the views as indicated.

Point V on HP, 20 mm in Front of VP, and 40 mm in Front of Right PP

Inference

1. Since the point is in front of the right PP, it is projected on to it to obtain the left-side view.
2. The right PP is marked by a vertical reference line, to the right of the front view, at a distance equal to that of the object and the PP.

PROCEDURE (Refer Fig. 6.11g)

Step 1: Draw the horizontal reference line xy and locate the front view (v′) and the top view (v) as explained earlier.

Step 2: Place the vertical reference line x_1y_1 at a distance of 40 mm from v′ and to the right of it.

Step 3: Extend the front view horizontally on to the other side of the vertical reference line and mark the distance of the object from the VP, that is, 20 mm to get the side view v″. Since the point is on the HP, extension of v′ happens on the reference line xy itself.

Step 4: Draw the projector lines joining v′ and v, v′ and v″ and dimension the views as indicated.

Point W 30 mm Above HP, on VP, and on Left PP

Inference

1. Since the point is on the left PP, its projection, that is, the right-side view falls on the PP itself.
2. The left PP is marked by a vertical reference line, to the left of the front view, at a distance equal to that of the object and the PP.

PROCEDURE (Refer Fig. 6.11h)

Step 1: Draw the horizontal reference line xy and locate the front view w' and the top view w as explained earlier.

Step 2: The vertical reference line x_1y_1 is placed on the projector line w'w itself, as the distance of PP is zero.

Step 3: Extension of the front view horizontally to the other side of the vertical reference line happens on the same line and the intersection point is the view w".

Step 4: Draw the projector lines joining w', w, and w", and dimension the views as indicated. ▲

Example 6.4 ▲

The projections of the following points are as indicated in Fig. 6.12. Define the positions of these points with respect to the planes of projection.

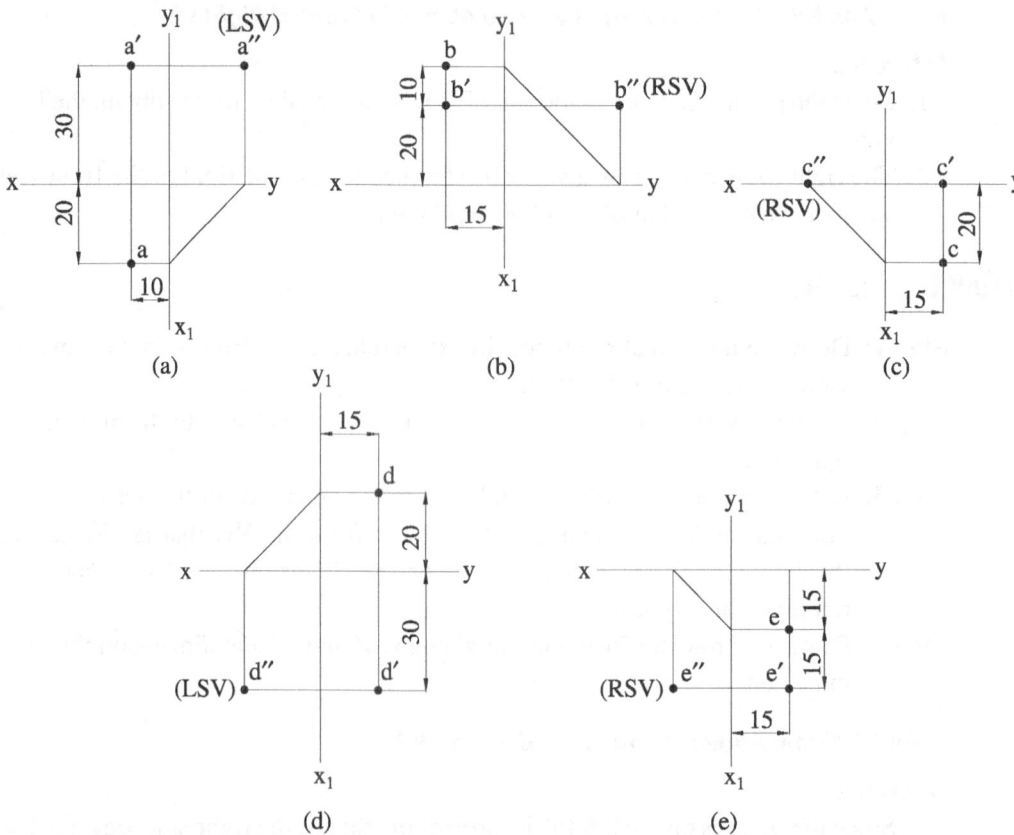

Fig. 6.12 Orthographic views of a point on to VP, HP, and PP

Point A

1. Since a' is 30 mm above the reference line, the Point A is 30 mm above the HP.
2. Since a is 20 mm below the reference line, the Point A is 20 mm in front of the VP.

3. Since the vertical reference line is 10 mm to the right of the front view and a" is the left-side view, the Point A 10 mm in front of the right profile plane.

Point B

1. Since b' is 20 mm above the reference line, the Point B is 20 mm above the HP.
2. Since b is 30 mm above the reference line, the Point B is 30 mm behind the VP.
3. Since the vertical reference line is 15 mm to the right of the front view and b" is the right-side view, the Point B 15 mm behind the right profile plane.

Point C

1. Since c' is on the reference line, the Point C is on the HP.
2. Since c is 20 mm below the reference line, the Point C is 20 mm in front of the VP.
3. Since the vertical reference line is 15 mm to the left of the front view and c" is the right-side view, the Point C 15 mm in front of the left profile plane.

Point D

1. Since d' is 30 mm below the reference line, the Point D is 30 mm below the HP.
2. Since d is 20 mm above the reference line, the Point D is 20 mm behind the VP.
3. Since the vertical reference line is 15 mm to the left of the front view and d" is the left-side view, the Point D 15 mm behind the left profile plane.

Point E

1. Since e' is 30 mm below the reference line, the Point E is 30 mm below the HP.
2. Since e is 15 mm below the reference line, the Point E is 20 mm in front of the VP.
3. Since the vertical reference line is 15 mm to the left of the front view and e" is the right-side view, the Point E 15 mm in front of the left profile plane. ▲

Miscellaneous Examples

Example 6.5 ▲

Draw the projections of the Point P lying 30 mm above HP and situated in the first quadrant, if its shortest distance from the line of intersection of VP and HP is 40 mm. Determine the distance of the point from the VP.

Inference

1. The line joining the point in space and the line of intersection of VP and HP will be seen in the side view.
2. Since the front view and the side view of the point appear at the same height, project the side view to get the other views

PROCEDURE (Refer Fig. 6.13)

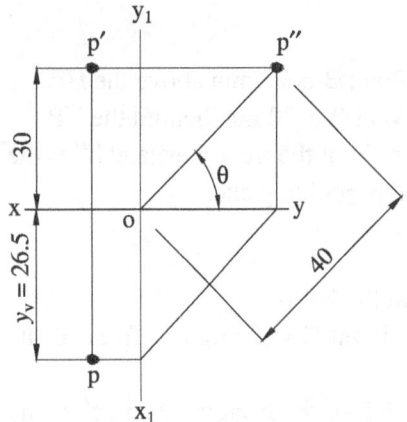

Fig. 6.13 Problem using shortest distance in space

Step 1: Draw the horizontal reference line xy and the vertical reference line x_1y_1 at any convenient point 'O'. The point O indicates the side view of the intersecting line.

Step 2: Since the height of the Point P is the same in the front view and in the side view, draw a horizontal line at a distance of 30 mm from the horizontal reference line.

Step 3: Draw an arc of length 40 mm to indicate the shortest distance of the Point P from 'O'. When it cuts the horizontal line drawn at 30 mm height, it gives the side view p″ of the Point P.

Step 4: At any convenient point on the horizontal reference line, draw a perpendicular projector line.

Step 5: Extend the side view p″ to meet this line and obtain p′.

Step 6: Measure the distance of p″ from the vertical reference line (in the horizontal direction) and mark from the xy line to get 'p' on the projector line.

Step 7: The distance of the Point P from the VP is obtained by measuring the distance of its top view, p, from the reference line. This is indicated by y_v measuring 26.5 mm. ▲

Example 6.6 ▲

A thread is connected between the two diagonally opposite corners of a room and it appears as a straight line of length 50 mm on the side plane at an angle of 45° to the horizontal reference line. If the top corner is equidistant from the walls and the ground, determine its distance using the geometric construction of views.

Inference

1. The front wall, the floor, and the side wall of a room resemble the VP, the HP, and the PP, respectively.
2. In the side view, the thread (connecting the bottom corner and the top corner) appears as a straight line.
3. Reversal of the side view gives the other views and the distances involved.

PROCEDURE (Refer Fig. 6.14)

Step 1: Draw the horizontal reference line xy and the vertical reference line x_1y_1 at any convenient point 'O_1'. The Point O_1 indicates the side view of the intersecting line and also the bottom corner of the room.

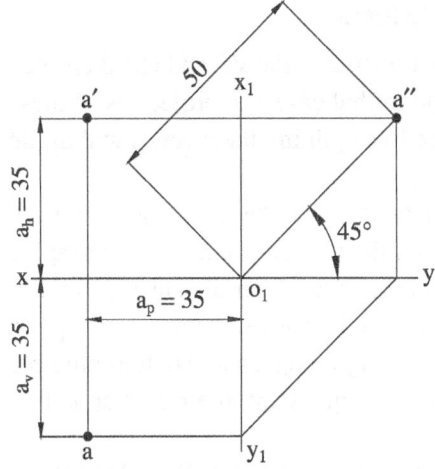

Fig. 6.14 Problem equidistant from HP, VP, and PP

Step 2: Set at O_1 a distance of 50 mm at an angle of 45°, being the side view of the thread. Mark the side view of the point as, a″.

Step 3: Measure the distance of a″ from the xy line and x_1y_1 line. These distances represent the location of the front view, a′ and the top view, a. They are found to be 35 mm and are equal as the Point A is equidistant from the HP and the VP.

Step 4: Draw the projector line on the other side of x_1y_1 and at a distance of 35 mm from it, since the Point A is also at the same distance from the PP.

Step 5: Complete the views and dimension them as indicated. ▲

6.4 PROJECTION OF POINTS LOCATED IN THE SAME QUADRANT

In the sections discussed earlier, the projection of a point in space was discussed. If two or more points are located in space, their projections can be obtained in similar ways. The knowledge of the projection of many points in space becomes vital to understand the objects such as a straight line, a plane figure, or even a solid, whose shapes are formed or built with points. These objects are termed as one-dimensional, two–dimensional, and three-dimensional objects, respectively. In this section, the basic principle of projection of two and three points (resulting in a straight line and in a triangular plane) will be discussed, although the forthcoming chapters will explain in detail. As discussed earlier, while projecting a point, the imaginary lines known as the projectors emanate from the point and meet the projection planes orthogonally to create the views on VP, HP, and PP. These projector lines are indicated in the views by the lines joining the respective views. For example, the projector lines of the Point A are indicated in Fig. 6.1(b) as the lines joining the front view a′ and the top view 'a'. Logically, if another Point B is also located in space, its projector lines can be recognized by the lines joining its front view b′ and its top view b. The projector lines of the points in space can be situated in a common single plane or they can be located in different planes. In the former case, the projector lines joining the views will lie on a single line, while in the latter case, they will get separated. The lines joining a′a and that joining b′b will lie together in a single plane case, while they will get separated in the other case. Depending on the position of the points in any quadrant, their views can be discussed. The readers are advised to refer Table 5.1 and Table 6.1 to familiarize the position of the points and the layout of their views.

If both the points are located in the same quadrant, their respective views will also lie on the same side of the reference line as mentioned in Table 5.1 in Chapter 5.

6.4.1 Projection of Points Whose Projectors Lie in a Single Plane

If any two points in space lie on a single plane perpendicular to the VP and HP, then such a plane will become parallel to the profile plane and in that case, the projectors of these points also lie on a single plane. In other words, the lines joining their views will all lie on a single line.

Figure 6.15(a) shows the pictorial arrangement of the two Points A and B, whose projectors lie in a single plane. Figure 6.15(b) shows their corresponding orthographic views. The front views of the Points A and B are marked with their distances h_1 and h_2 (being the distances from the HP) and their top views are located at distances of w_1 and w_2 (being the distances from the VP). The projector lines joining a'a and that joining b'b lie on a single line. Since both the points lie in the same quadrant, their corresponding views lie on the same side of the reference line.

Figure 6.16(a) shows the pictorial arrangement of the three points A, B, and C, whose projectors lie in a single plane. Figure 6.16(b) shows their corresponding orthographic views. The front views of the Points A, B, and C are marked with their distances h_1, h_2, and h_3 and their top views are located at distances of w_1, w_2, and w_3. The projector lines joining a'a, b'b, and c'c all lie on a single line. Since all these points are located in the same quadrant, their corresponding views also lie on the same side of the reference line.

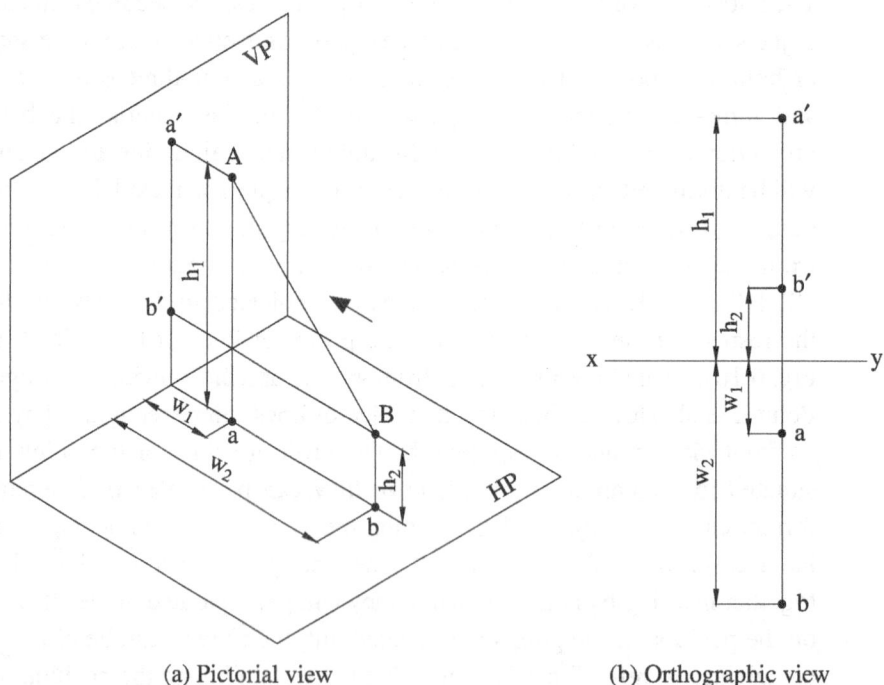

(a) Pictorial view   (b) Orthographic view

Fig. 6.15 Projection of two points in the same quadrant with projectors lying in a single plane

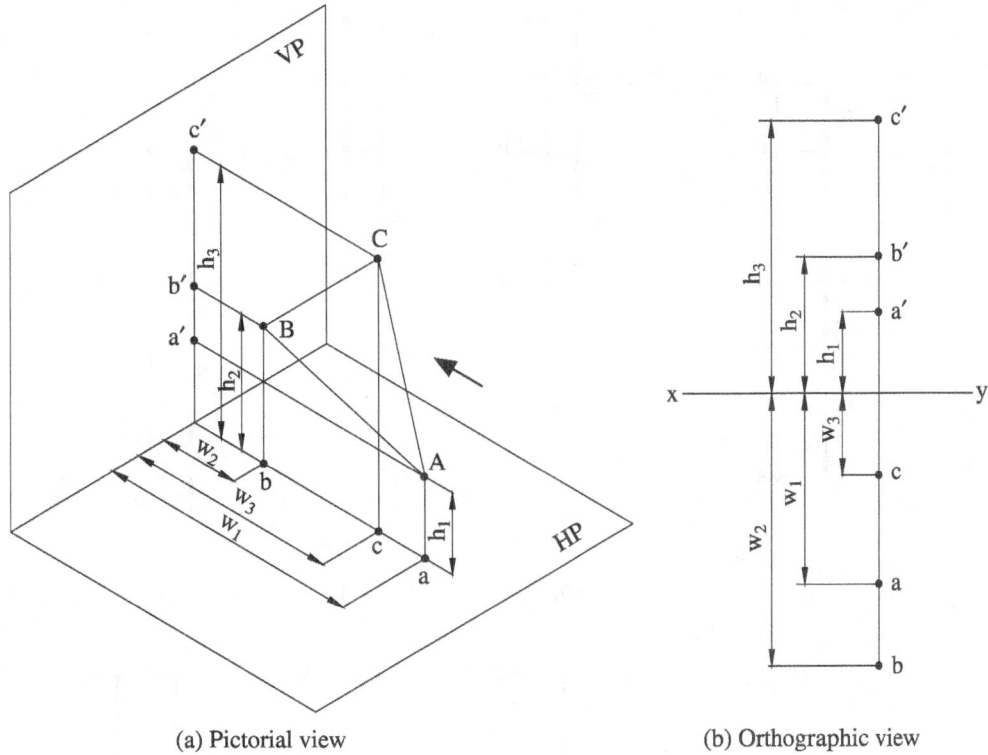

(a) Pictorial view   (b) Orthographic view

Fig. 6.16   Projection of three points in the same quadarnt, with projectors lying in a single plane

6.4.2 Illustrative Examples

Example 6.7

Draw the projections of the line joining the following points as per their locations, if their projectors lie on a single plane.

Point A 30 mm Above HP and 30 mm in Front of VP and Point B 20 mm Above HP and 10 mm in Front of VP

Inference

1. Since both the points are above the HP, their front views lie above the reference line.
2. As both the points are in front of the VP, their top views lie below the reference line.
3. All these views lie on the same line, as the projectors lie in a single plane.

PROCEDURE   (Refer Fig. 6.17a)

Step 1: Draw the horizontal reference line xy.
Step 2: From any point on the xy line, draw a perpendicular line and mark a′, 30 mm above and a, 30 mm below the reference line.

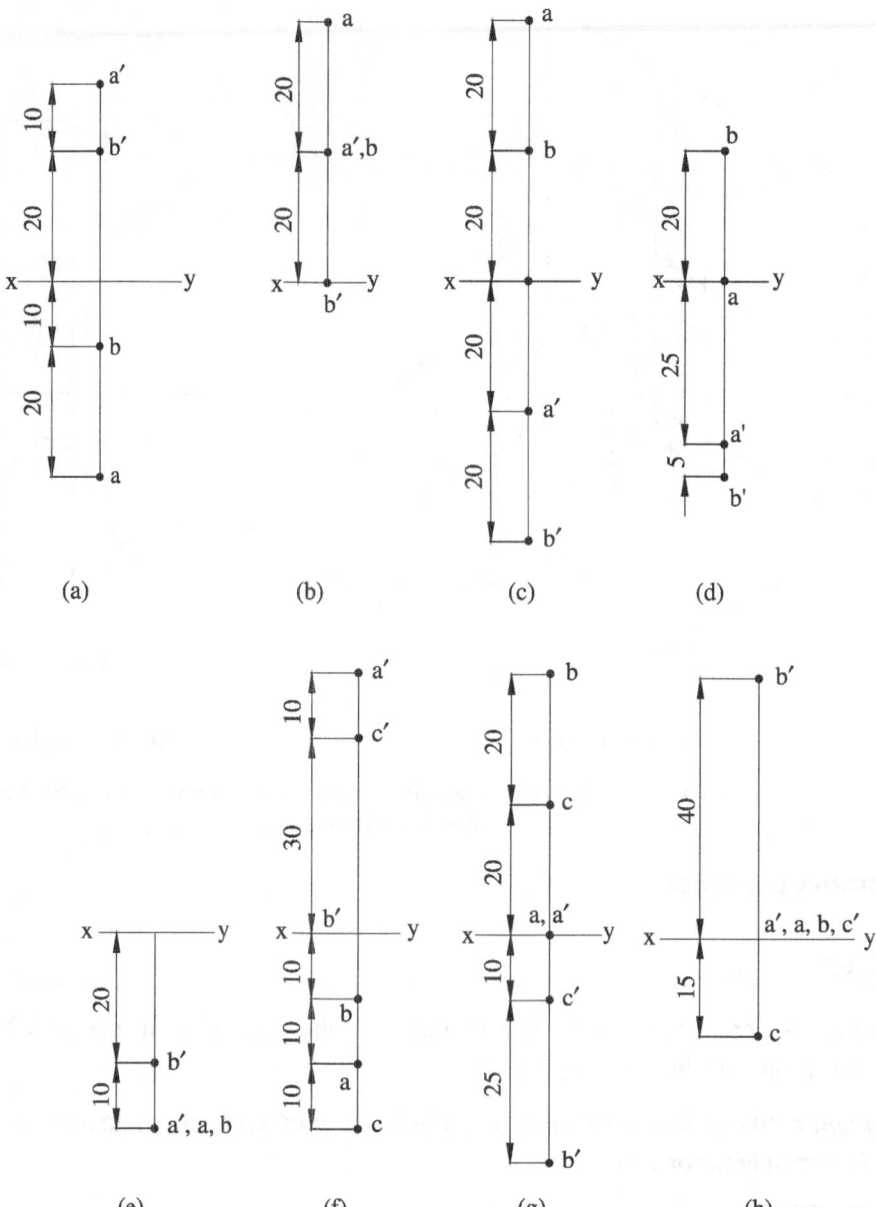

Fig. 6.17   Projection of points located in the same quadrant with projectors lying in a single plane

Step 3: Mark on the same projector line, b′, 20 mm above and b, 10 mm below the reference line.

Step 4: Join the views and dimension them as indicated.

Point A 40 mm Behind VP and 20 mm Above HP and Point B 20 mm Behind VP and on HP
Inference

1. Since one point is on HP and the other above the HP, their front views lie on and above the reference line, respectively.

2. As both the points are behind the VP, their top views lie above the reference line.
3. All these views lie on the same line, as the projectors lie in a single plane.

PROCEDURE   (Refer Fig. 6.17b)

Step 1: Draw the horizontal reference line xy.

Step 2: From any point on the xy line, draw a perpendicular line and mark a' 20 mm above and a, 40 mm above the reference line.

Step 3: Mark on the same projector line b' on the reference line and b, 20 mm above the reference line.

Step 4: Join the views and dimension them as indicated.

Point A 20 mm Below HP and 40 mm Behind VP and Point B 40 mm Below HP and 20 mm Behind VP

Inference

1. Since both the points are below the HP, their front views lie below the reference line.
2. As both the points are behind the VP, their top views lie above the reference line.
3. All these views lie on the same line, as the projectors lie in a single plane.

PROCEDURE   (Refer Fig. 6.17c)

Step 1: Draw the horizontal reference line xy.

Step 2: From any point on the xy line, draw a perpendicular line and mark a', 20 mm below and a, 40 mm above the reference line.

Step 3: Mark on the same projector line, b', 40 mm below and b, 20 mm above the reference line.

Step 4: Join the views and dimension them as indicated.

Point A 25 mm Below HP and on VP and Point B 30 mm Below HP and 20 mm Behind VP

Inference

1. Since both the points are below the HP, their front views lie below the reference line.
2. As one point is on the VP and another behind it, their top views lie on and above the reference line, respectively.
3. All these views lie on the same line, as the projectors lie in a single plane.

PROCEDURE   (Refer Fig. 6.17d)

Step 1: Draw the horizontal reference line xy.

Step 2: From any point on the xy line, draw a perpendicular line and mark a', 25 mm below and a, on the reference line, respectively.

Step 3: Mark on the same projector line, b', 30 mm below and b, 20 mm above the reference line.

Step 4: Join the views and dimension them as indicated.

Point A 30 mm Below HP and 30 mm in Front of VP and Point B 20 mm Below HP and 30 mm in Front of VP

Inference

1. Since both the points are below the HP, their front views lie below the reference line.
2. As both the points are in front of the VP, their top views also lie below the reference line.
3. All these views lie on the same line, as the projectors lie in a single plane.

PROCEDURE (Refer Fig. 6.17e)

Step 1: Draw the horizontal reference line xy.

Step 2: From any point on the xy line, draw a perpendicular line and mark a' and a, both 30 mm below the reference line.

Step 3: Mark on the same projector line, b', 20 mm below and b, also 30 mm below the reference line. Note that a', a, and b all lie on the same point.

Step 4: Join the views and dimension them as indicated.

Point A 40 mm Above HP and 20 mm in Front of VP, Point B on HP and 10 mm in Front of VP, and Point C 30 mm Above HP and 30 mm in Front of VP

Inference

1. Since two points are above the HP, their front views lie above the reference line.
2. As all the points are in front of the VP, their top views lie below the reference line.
3. All these views lie on the same line, as the projectors lie in a single plane.

PROCEDURE (Refer Fig. 6.17f)

Step 1: Draw the horizontal reference line xy.

Step 2: From any point on the xy line, draw a perpendicular line and mark a', 40 mm above and a, 20 mm below the reference line.

Step 3: Mark on the same projector line b' and c' on and 30 mm above the reference line, respectively. Mark also b and c, 10 mm and 30 mm below the reference line.

Step 4: Join the views and dimension them as indicated.

Point A on HP and on VP, Point B 35 mm Below HP and 40 mm Behind VP, and Point C 10 mm Below HP and 20 mm Behind VP

Inference

1. Since two points are below the HP, their front views lie below the reference line.
2. As two points are behind the VP, their top views lie above the reference line.
3. Points lying on the reference plane will have their views on the reference line itself.
4. All these views lie on the same line, as the projectors lie in a single plane.

PROCEDURE (Refer Fig. 6.17g)

Step 1: Draw the horizontal reference line xy.

Step 2: From any point on the xy line, draw a perpendicular line, and mark a′ on the reference line and b′ and c′ 35 mm and 10 mm below the reference line, respectively.

Step 3: Mark on the same projector line, a, on the reference line and b and c 40 mm and 20 mm above the reference line, respectively.

Step 4: Join the views and dimension them as indicated.

Point A on HP and on VP, Point B 40 mm Above HP and on VP, and Point C on HP and 15 mm in Front of VP

Inference

1. Points on the reference planes will have their views on the reference line.
2. Points above the HP will have their front views above the reference line.
3. Points in front of the VP will have their top views below the reference line.
4. All these views lie on the same line as the projectors lie in a single plane.

PROCEDURE (Refer Fig. 6.17h)

Step 1: Draw the horizontal reference line xy.

Step 2: From any point on the xy line, draw a perpendicular line, and mark a′ and c′ on the reference line and b′ 40 mm above it.

Step 3: Mark on the same projector line, a, b on the reference line and c, 15 mm below it.

Step 4: Join the views and dimension them as indicated.   ▲

6.4.3 Projection of Points Whose Projectors Lie in Different Planes

As discussed earlier, if the two points lie in space such that projectors are separated, then the projector lines joining their views also will be separated.

Figure 6.18(a) shows the pictorial arrangement of the two Points A and B, whose projectors are separated at a distance 'p'. Figure 6.18(b) shows their corresponding orthographic views. The front views of the Points A and B are marked with their distances h_1 and h_2 and their top views are located at distances of w_1 and w_2. The projector lines joining a′ and a and that joining b′ and b lie separated by a distance p, which is generally known as 'the projector distance'. Since the shortest distance joining the Points A and B in the space results in a straight line, the lines a′ and b′ and that joining a and b, respectively, are known as the front view and the top view of the line AB. As both the points lie on the same quadrant, their corresponding views also lie on the same side of the reference line.

Figure 6.19(a) shows the pictorial arrangement of the three Points A, B, and C, whose projectors are separated. The projectors of A and B are separated by a distance of p_1

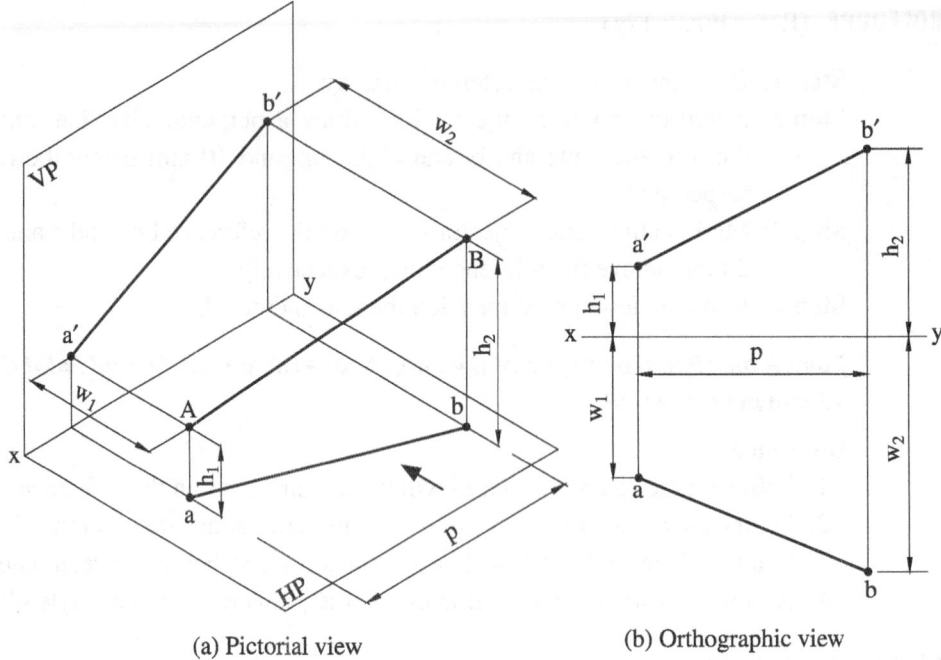

(a) Pictorial view   (b) Orthographic view

Fig. 6.18   Projection of two points in the same quadrant with projectors lying in different planes

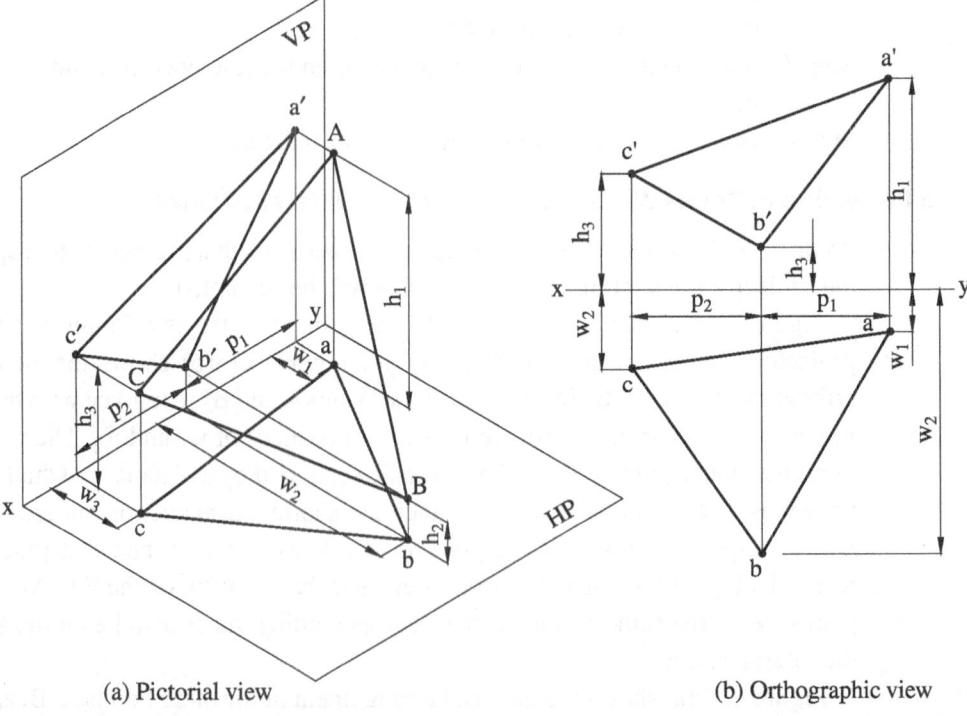

(a) Pictorial view   (b) Orthographic view

Fig. 6.19   Projection of three points located in the same quadrant with projectors lying in different planes

and that of B and C are at a distance of p_2. Figure 6.19(b) shows their corresponding orthographic views. The front views of the Points A, B, and C are marked with their distances h_1, h_2, and h_3, and their top views are located at distances of w_1, w_2, and w_3. The projector lines joining a′ and a, b′ and b, and c′ and c, are separated as shown in Fig. 6.19(b). Since all these points are situated in the same quadrant, their corresponding views also lie on the same side of the reference line. Since the imaginary lines joining the Points A, B, and C in the space result in a plane, the lines a′, b′, and c′ and that joining a, b, and c, respectively, are known as the front view and the top view of the plane ABC.

6.4.4 Illustrative Examples

Example 6.8

Draw the projections of the line joining the following points as per their locations.

> **NOTE** *The problems discussed in Example 6.7 are followed here but the points of projector lines are separated as shown below. Since the views are not going to change but for their placement on the respective projector lines, the inference note is not repeated here but the procedure for drawing the views are explained.*

Point A 30 mm Above HP and 30 mm in Front of VP and Point B 20 mm Above HP and 10 mm in Front of VP

The distance between the projectors is 60 mm apart.

PROCEDURE (Refer Fig. 6.20a)

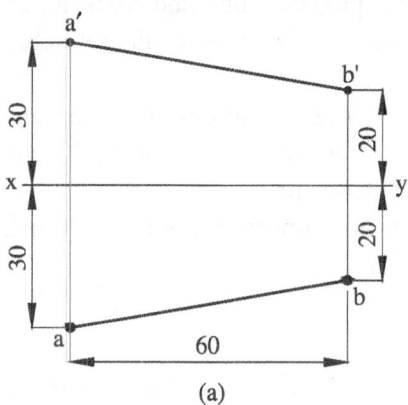

(a)

Fig. 6.20 Projection of points located in the same quadrant with the projector placed in different planes (*Contd*)

Step 1: Draw the horizontal reference line xy.

Step 2: From any point on the xy line, draw a perpendicular projector line and mark on it, a′, 30 mm above and a, 30 mm below the reference line.

Step 3: Draw another projector line, 60 mm apart and mark on it, b′, 20 mm above and b, 10 mm below the reference line.

Step 4: Join the views and dimension them as indicated.

Point A 40 mm Behind VP and 20 mm Above HP and Point B 20 mm Behind VP and on HP

The distance between the projectors is 60 mm apart.

PROCEDURE   (Refer Fig. 6.20b)

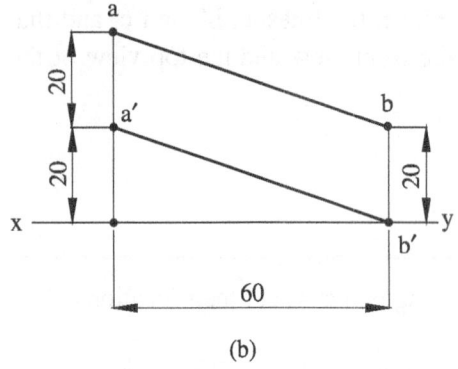

(b)

Step 1: Draw the horizontal reference line xy.
Step 2: From any point on the xy line, draw a perpendicular projector line and mark a′ 20 mm above and a, 40 mm above the reference line.
Step 3: Draw another projector line, 60 mm apart and mark on it, b′, on the reference line and b, 20 mm above the reference line.
Step 4: Join the views and dimension them as indicated.

Fig. 6.20   *(Contd)*

Point A 20 mm Below HP and 40 mm Behind VP and Point B 40 mm Below HP and 20 mm Behind VP

The distance between the projectors is 60 mm apart.

PROCEDURE   (Refer Fig. 6.20c)

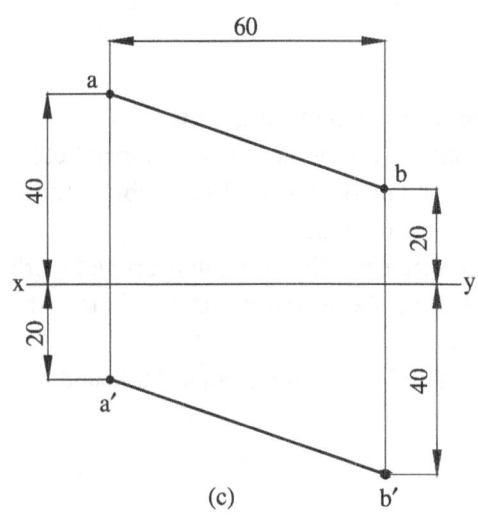

(c)

Step 1: Draw the horizontal reference line xy.
Step 2: From any point on the xy line, draw a perpendicular projector line and mark a′, 20 mm below and a, 40 mm above the reference line.
Step 3: Draw another projector line, 60 mm apart and mark on it, b′, 40 mm below and b, 20 mm above the reference line.
Step 4: Join the views and dimension them as indicated.

Fig. 6.20   *(Contd)*

Point A 25 mm Below HP and on VP and Point B 30 mm Below HP and 20 mm Behind VP

The distance between the projectors is 60 mm apart.

PROCEDURE (Refer Fig. 6.20d)

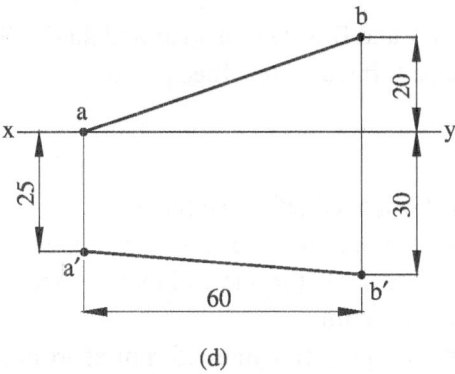

(d)

Fig. 6.20 (Contd)

Step 1: Draw the horizontal reference line xy.

Step 2: From any point on the xy line, draw a perpendicular line and mark a', 25 mm below and a, on the reference line, respectively.

Step 3: Draw another projector line, 60 mm apart and mark on it, b', 30 mm below and b, 20 mm above the reference line.

Step 4: Join the views and dimension them as indicated.

Point A 30 mm Below HP and 30 mm in Front of VP and Point B 20 mm Below HP and 30 mm in Front of VP

The distance between the projectors is 60 mm apart.

PROCEDURE (Refer Fig. 6.20e)

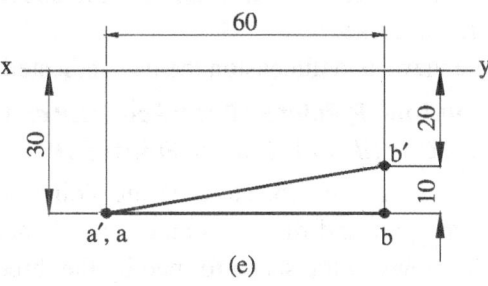

(e)

Fig. 6.20 (Contd)

Step 1: Draw the horizontal reference line xy.

Step 2: From any point on the xy line, draw a perpendicular line and mark a' and a, both, 30 mm below the reference line. Note that a' and a lie on the same point.

Step 3: Draw another projector line, 60 mm apart and mark on it, b', 20 mm below and b, also, 30 mm below the reference line.

Step 4: Join the views and dimension them as indicated.

Point A 40 mm Above HP and 20 mm in Front of VP, Point B on HP and 10 mm in Front of VP, and Point C 30 mm Above HP and 30 mm in Front of VP

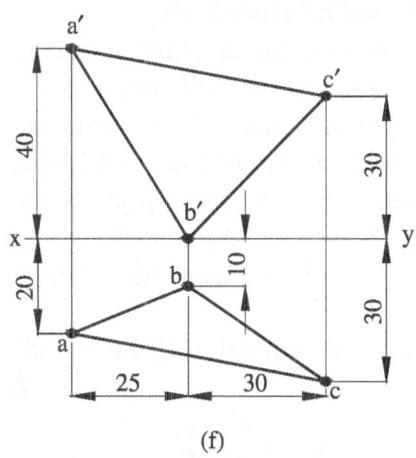

(f)

Fig. 6.20 (Contd)

The distance between the projectors of the Points A and B is 25 mm apart and that of B and C is 30 mm apart. Show the view of the shape formed by the three points.

PROCEDURE (Refer Fig. 6.20f)

Step 1: Draw the horizontal reference line xy.

Step 2: From any point on the xy line, draw a perpendicular projector line and mark a', 40 mm above and a, 20 mm below the reference line.

Step 3: Draw another projector line, 25 mm apart and mark on it, b' and b, on and 30 mm below the reference line, respectively. Draw another projector line from this at a distance of 30 mm and mark on it, c' and c, 30 mm above and 30 mm below the reference line.

Step 4: Join the views and dimension them as indicated

Point A on HP and on VP, Point B 35 mm Below HP and 40 mm Behind VP, and Point C 10 mm Below HP and 20 mm Behind VP

The distance between the projectors of the Points A and B is 25 mm apart and that of B and C is 30 mm apart. Show the views of the shape formed by the three points.

PROCEDURE (Refer Fig. 6.20g)

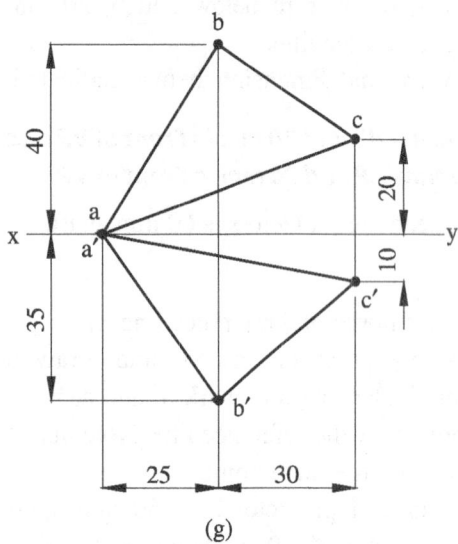

(g)

Fig. 6.20 (Contd)

Step 1: Draw the horizontal reference line xy.

Step 2: From any point on the xy line, draw a perpendicular projector line and mark a′ and a, on the reference line.

Step 3: Draw another projector line, 25 mm apart and mark on it, b′ and b, 35 mm below and 40 mm above the reference line. Draw another projector line from this at a distance of 30 mm and mark on it, c′ and c, 10 mm below and 20 mm above the reference line.

Step 4: Join the views and dimension them as indicated.

Point A on HP and on VP, Point B 40 mm Above HP and on VP, and Point C on HP and 15 mm in Front of VP

The distance between the projectors of the Points A and B is 25 mm apart and that of B and C is 30 mm apart. Show the views of the shape formed by the three points.

PROCEDURE (Refer Fig. 6.20h)

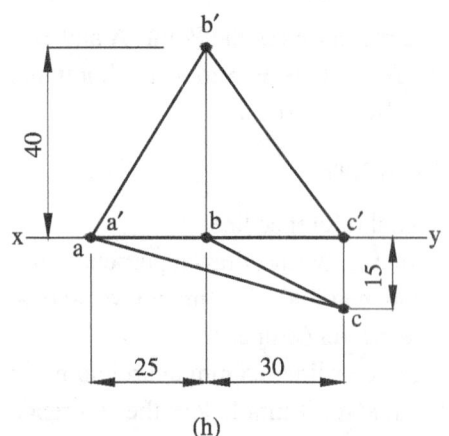

(h)

Fig. 6.20 (Contd)

Step 1: Draw the horizontal reference line xy.

Step 2: From any point on the xy line, draw a perpendicular line and mark a′ and a, on the reference line.

Step 3: Draw another projector line, 25 mm apart and mark on it, b′ 40 mm above and b on the reference line, respectively. Draw another projector line from this at a distance of 30 mm and mark on it, c′ and c, on and 15 mm below the reference line.

Step 4: Join the views and dimension them as indicated. ▲

- The imagination of the object as a point serves as the basis of projection of all types of objects.
- The projection of the point on to the planes of projection yields the respective views, which are also points and are marked with their positional references.
- The front view of the point is positioned depending on its distance from the HP, while the top view marked with that of the VP. The front and the top views are laid on the same vertical projector line but separated by a reference line. The details are shown in Table 6.1.
- The side view of the point is marked by the side of the front view, to its right or left, depending on the distance of the point from the respective side or profile plane (PP). A vertical reference line is drawn to effect this, and the side view is located on the other side of the vertical reference line, depending on the distance of the

point from the VP but at the same height as that of the front view. The side view and the front view lie on the same horizontal projector line. The details are shown in Table 6.2.
- The projection of two or many points in space situated in the same quadrant can be obtained in similar lines as that of a single point, with due care for their positions and the corresponding placement of views. The details are shown in Table 6.1.
- The spatial connection of two points results in a straight line, while that of three or many points results in the shape of a plane figure. By joining the respective views of the points, the front view or the top view of the shape can be realized.
- The extensional idea of the projection of many points in space, results in a stronger foundation to understand chapters on projection of straight lines, planes and solids.

MULTIPLE-CHOICE QUESTIONS

6.1 Choose the right option from the following for the orthographic projections of a Point A situated in the I-quadrant.

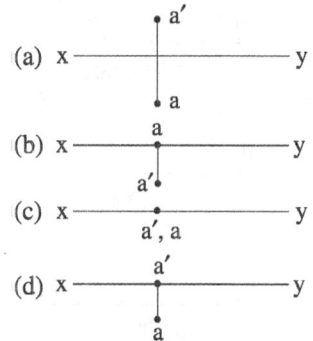

6.2 Choose the right option for the orthographic projections of a Point A situated in the third quadrant from the following:

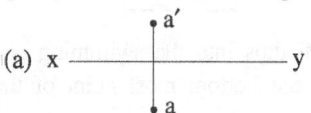

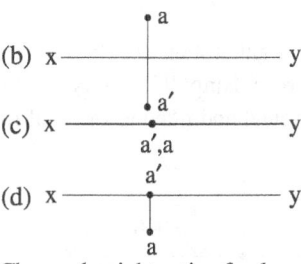

6.3 Choose the right option for the orthographic projections of a Point A located in HP and in VP and 10 mm in front of right-side plane from the following:

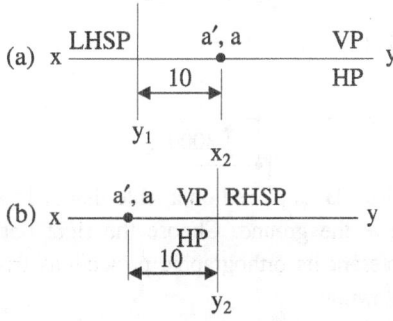

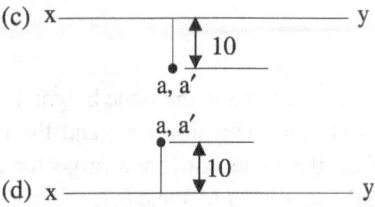

(c)

(d)

6.4 Choose the right option for the orthographic projections of a Point A located in HP and in VP and 10 mm in front of left-side plane from the following:

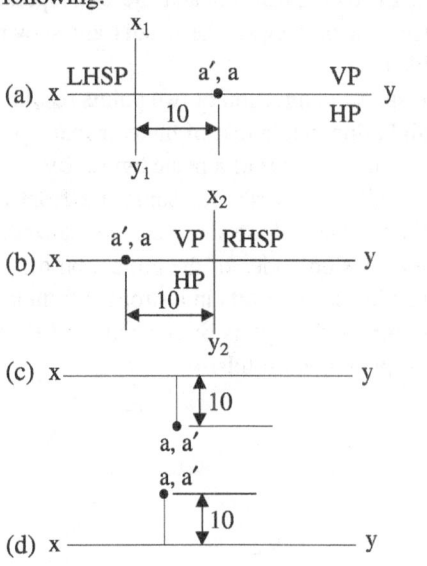

(a)

(b)

(c)

(d)

6.5 Choose the right option for the orthographic projections of a lamp 'L' that is located 3 m above the ground and fitted to the wall from the following:

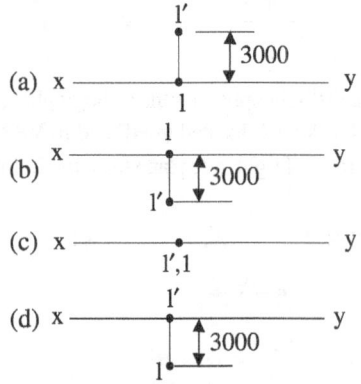

(a)

(b)

(c)

(d)

6.6 A ball 'B' is placed on a table that is 1500 mm above the ground. Choose the right option to represent its orthographic projections from the following:

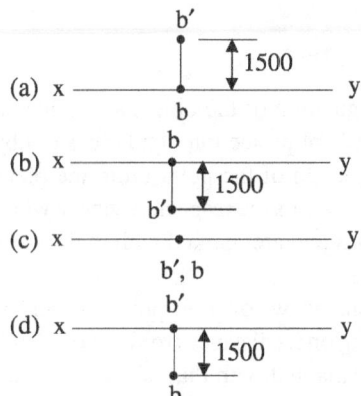

(a)

(b)

(c)

(d)

6.7 For the Point A, the top view 50 mm above 'xy' line and the front view is 25 mm above the top view. State the quadrant in which the point is situated.
   (a) First quadrant   (c) Third quadrant
   (b) Second quadrant   (d) Fourth quadrant

6.8 For the Point A, the projections coincide with each other at 50 mm above 'xy' line. State the quadrant in which the point is situated.
   (a) First quadrant   (c) Third quadrant
   (b) Second quadrant   (d) Fourth quadrant

6.9 Two particles 'A' and 'B' are moving horizontally on a ground, one behind each other, from the same destination. 'A' moves faster than B and after 5 seconds 'A' is 500 m ahead of 'B'. Choose the right option to represent their orthographic projections.

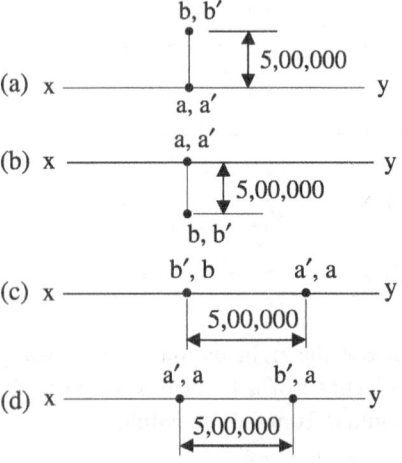

(a)

(b)

(c)

(d)

6.10 A man 'M' dips into the swimming pool and reaches to the bottom most point of the pool which is 12 m from the ground level. Represent the position of the man, neglecting his height.

(a)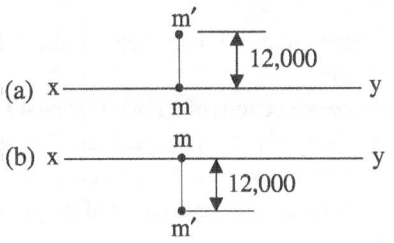

(b) x ——————— m ——————— y
⤒ 12,000
m′

(c) x ——————— m′, m ——————— y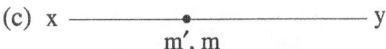

(d)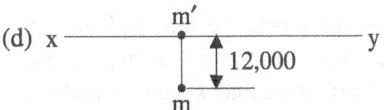

Note: All dimensions are in mm.

6.1 State the quadrants in which the following points are located:
   (a) The projections of a Point A coincide with each other at 50 mm below the xy line
   (b) The front view of the Point B is 40 mm below xy line and top view is 50 mm above the xy line
   (c) The front view of the Point C is 30 mm below xy line and top view is 40 mm below the front view.

6.2 Draw the projections of the following points on the same reference line, keeping the projectors 30 mm apart.
   (a) P, 30 mm above HP and 20 mm in front of VP
   (b) Q, 20 mm above HP and on VP
   (c) R, on HP and 30 mm in front of VP
   (d) S, is 40 mm above HP and 20 mm behind VP

   (e) T, on HP and 40 mm behind VP
   (f) V, is 40 mm below HP and 50 mm in front of VP
   (g) W, on both HP and VP

6.3 The orthographic projections of various points are shown in Fig. 6.21. State the position of each point with respect to the reference planes.

6.4 A Point P is situated 40 mm in front of VP, 30 mm above HP and 20 mm in front of the right-side plane. Draw its projections.

6.5 A Point Q is situated on VP and 30 mm below HP and 20 mm in front of the left-side plane. Draw its projections.

6.6 Draw the top, front, and the side views of a Point P, situated 50 mm in front of VP, 40 mm above HP, and 30 mm from both the right- and the left-side planes.

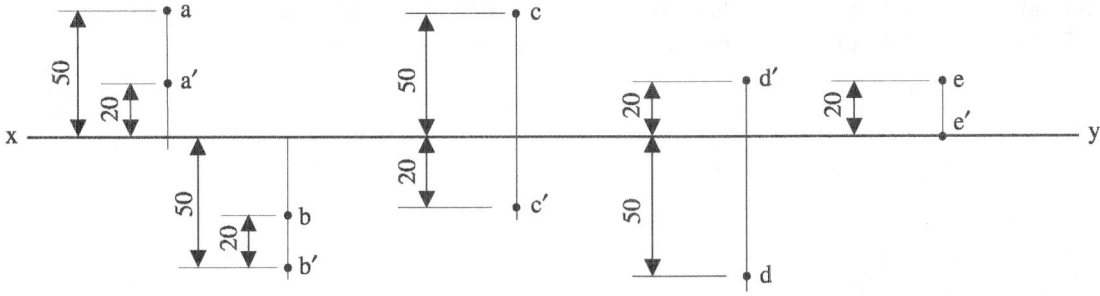

Note: All dimensions are in mm

Fig. 6.21

6.1 Draw the front, the top, and the appropriate side view for the following points, on the same reference line, keeping suitable distances between each case.

(a) A Point A on HP, 50 mm in front of VP and 30 mm in front of the left-side plane.

(b) A Point B is on VP, 40 mm above HP and 30 mm in front of the right-side plane.

(c) A Point C is 30 mm above HP, 40 mm in front of VP, 40 mm in front of right-side plane and 30 mm behind the left-side plane.

(d) A Point D is on HP, 50 mm behind VP and 30 mm behind the left-side plane.

(e) A Point E is on VP, 40 mm above HP and 30 mm behind the right-side plane.

(f) A Point F is 30 mm above HP, 40 mm behind VP, 40 mm behind the right-side plane, and 30 mm in front of the left-side plane.

(g) A Point G is on VP, 40 mm below HP and 30 mm behind the right-side plane.

(h) A Point H is 30 mm below HP, 40 mm behind VP, 40 mm behind the right-side plane, and 30 mm in front of the left-side plane.

(i) A Point I is 30 mm below HP, 40 mm in front of VP, 40 mm in front of right-side plane, and 30 mm behind the left-side plane.

6.2 Draw the projections of a Point P lying 40 mm in front of VP and in the first quadrant, if its shortest distance from the line of intersection of HP and VP is 50 mm. Find the distance of the point from the HP.

6.3 Draw the projections of a Point Q lying in the first quadrant such that its shortest distance from the point of intersection of HP, VP, and right-side plane is 75 mm, and it is equidistant from the three planes of projections. Also, find the position of the point with reference to the HP and the VP.

6.4 A point 30 mm below 'xy' line is the front view of two Points P and Q. The Point P is situated 40 mm in front of VP and Point Q 50 mm behind VP. Draw the projections of the two points and state their positions with the reference planes and the quadrants in which they lie if their projections lie on the same plane.

6.5 Point A is 40 mm below HP and 30 mm in front of VP and Point B is 30 mm below HP and 40 mm in front of VP. Draw the views of the straight line connecting these points in space, if

(a) their projectors lie on the same plane

(b) their projectors are 40 mm apart

Answers for Multiple-choice Questions

6.1 (a)	6.2 (b)	6.3 (b)	6.4 (a)	6.5 (a)	6.6 (a)
6.7 (b)	6.8 (b)	6.9 (c)	6.10 (b)		

Projection of Lines

7

OBJECTIVES

This chapter will help the reader to understand the following:

- Straight line which is the locus of a point that moves linearly describing its shape in one dimension and forms the basis of the evolution process for various two- and three-dimensional objects

- Position of the line with respect to all the planes of projection and obtain its views

- Extend the method of projection of the two end points of the line situated in space in the same or in different quadrants and obtain their projections/ views

- Visualize the spatial arrangement of two end points of the line from their projections or views, as a kind of reverse engineering practice

- Obtain the traces of the line on the reference planes and solve the problems involved with the traces and the projections

- Expose the reader to practical applications using the principles of straight lines and lay the foundations for other chapters on the projection of planes and solids

7.1 INTRODUCTION

A straight line is defined as the locus of a point that moves linearly. In other words, it is defined as the shortest distance measured between any two end points. In the chapters discussed earlier, the basic principle of orthographic projection of an object, which is placed in the space between principal planes, was analysed. Various points on the boundary of each side of an object were projected orthogonally or perpendicularly to obtain their front, top, and the side views on the vertical plane (VP), horizontal plane (HP), and profile planes (PP), respectively. In this chapter, the projection of a straight line is explained by projecting its two ends, that is, the end points, on the respective planes of projection and joining them appropriately.

The location of the straight line can be described with respect to the principal planes by specifying the distances of its two end points from the HP, VP, and PP. In addition, it can be described by its position from the reference planes, for example in front or behind the VP, PP, and above or below the HP. With respect to the principal planes, the spatial position of the line can also be described as follows:

(i) Line parallel to both the reference planes (HP and VP) (refer Fig. 7.1a).
(ii) Line parallel to one of the reference planes and perpendicular to the other (refer Fig. 7.1b). The position of the line as per (i) and (ii) is known as simple position and its projections are discussed in Section 7.2.
(iii) Line parallel to one of the reference planes and inclined to the other (refer Figs 7.1c and 7.1d).This position of the line is referred to as simple inclined position and is discussed in Section 7.3.
(iv) Line inclined to both the reference planes (refer Fig. 7.2). The position of this line is referred to as line in inclined position and is discussed in Section 7.4.

As discussed earlier, the distances, inclinations, and other details of the straight line pertaining to the HP will be visualized in the front view, while those connected with the VP will be marked in the top view.

7.1.1 Terminology

The following terminologies are often used in the projection of straight lines.

True length (TL)   It is the physical length of the line in space, measured along the line's own direction or orientation. It is not generally visible in the projections. However, when the line is parallel to one reference plane, the length of the view formed on that plane is equal to the true length of the line.

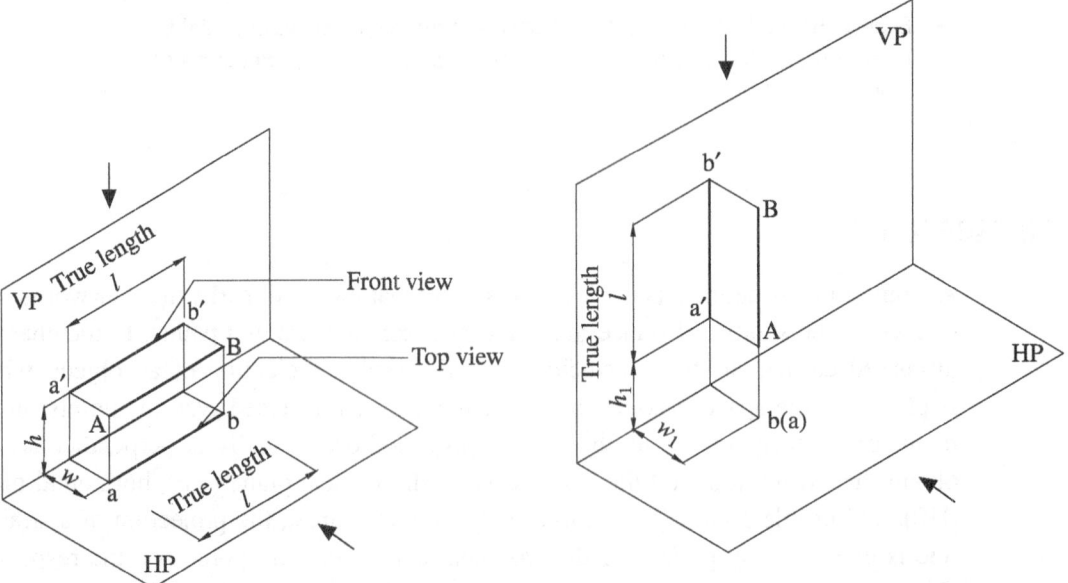

(a) Line parallel to both the planes

(b) Line parallel to VP and perpendicular to HP

Fig. 7.1   Different positions of a straight line (*Contd*)

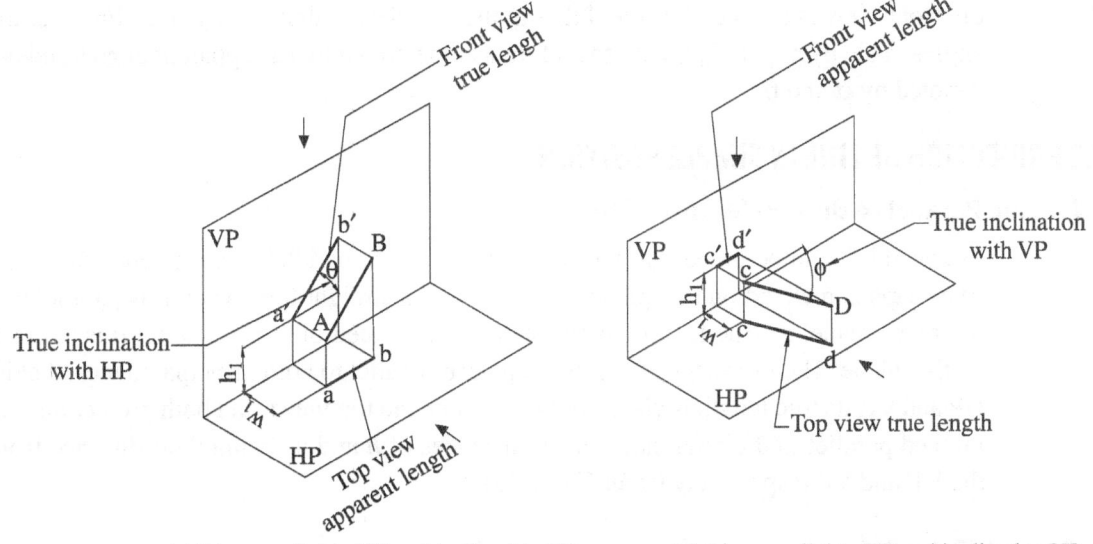

(c) Line parallel to VP and inclined to HP   (d) Line parallel to HP and inclined to VP

Fig. 7.1 (*Contd*)

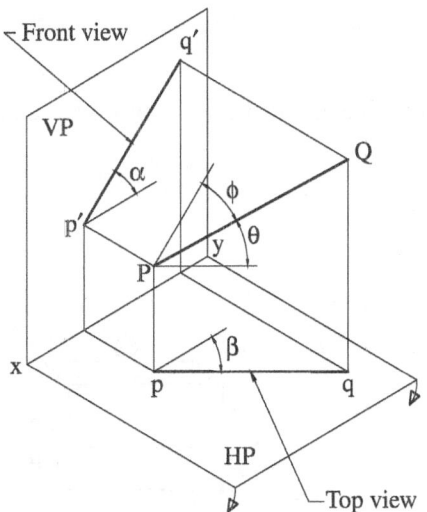

Fig. 7.2 Line inclined to both
the reference planes

In Figs 7.1 and 7.2, the length of AB is the true length of the line. In Fig 7.1(a), both the projections represent the true length; in Figs 7.1(b) and (c), the front view a'b' gives the true length; and in Fig 7.1(d), the top view gives the true length.

True inclination It is the physical inclination of the line in space with the reference planes (refer Fig. 7.2). In this book, the angle made by the line with the HP is denoted by θ and the angle made with VP is denoted by ϕ. Similar to the true length, these angles are not visible in the projections. However, when a line is parallel to a reference plane and inclined to the other, it is seen along with the projection that gives the true length. In Fig. 7.1(c), the angle (θ) made by a'b' with the reference line and that (ϕ) made by cd with the reference line in Fig 7.1(d) represent the true inclination of the line with the HP and VP respectively.

Apparent length It is the length of the projection of a line, when the line is not parallel to any reference plane and is always shorter than the true length. Both the projections in Fig. 7.2 represent the apparent length, while ab in Fig. 7.1(c) and c'd' in Fig. 7.1(d) give the apparent lengths.

Apparent angles The inclinations of the projections that represent the apparent lengths are known as apparent angles and these are always more than the true angles of inclinations of the line. In this book, the apparent angle made by the projection in the front view is denoted by α and that in the top view is denoted by β. In Fig. 7.2, the angles made by p'q' and pq with the reference line are known as apparent angles and are denoted by α and β.

7.2 PROJECTION OF LINE IN SIMPLE POSITIONS

7.2.1 Line Parallel to Both Reference Planes

In general, it can be noted that when an object is kept parallel to a reference plane, its true shape is realized on that plane. In the case of a straight line, when it is parallel to a reference plane, its true length (as the shape is described only by its length) is realized on that plane. Therefore, when the line is placed parallel to both principal planes, that is, HP and VP, its true length is visualized in its front and top views and both projections are marked parallel to the reference line, at distances of h and w, being their distance from the HP and VP respectively (refer Fig. 7.1a).

Example 7.1

Draw the projections of the following straight lines of 60 mm length as per their respective position:

Line AB, 30 mm above HP and 20 mm in front of VP

Figures 7.3(a) and (b) show the spatial position of the line and its projections on HP and VP. Figure 7.3(c) shows its orthographic views.

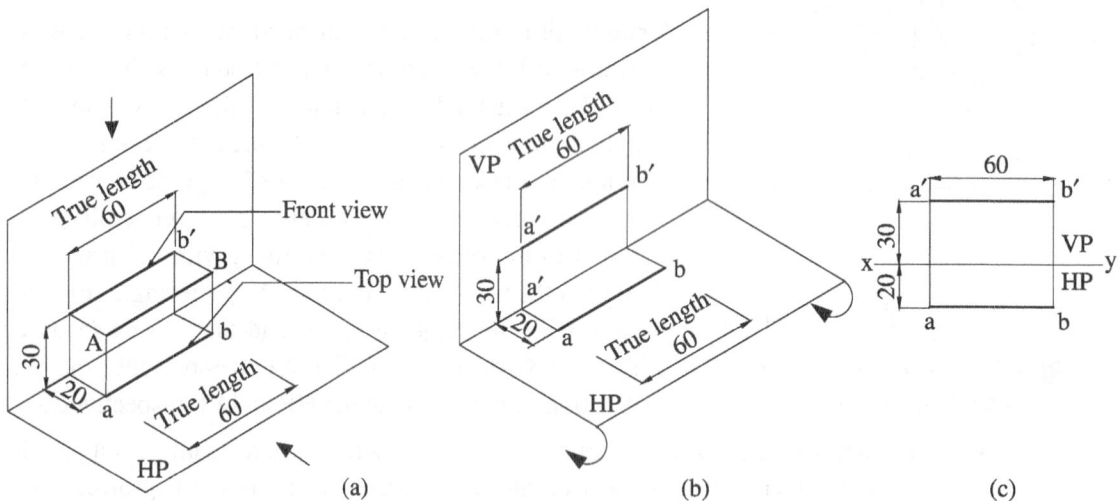

(a)   (b)   (c)

Fig. 7.3   Line parallel to both the planes

Inference

The front and the top view represent the true length, and both are parallel to the reference line.

PROCEDURE (Refer Fig. 7.3c)

Step 1: Draw the horizontal reference line 'xy'.

Step 2: Draw a horizontal line with a true length of 60 mm at a distance of 30 mm above the 'xy' line to mark the front view as a'b'.

Step 3: Draw the projectors through the points a' and b' and draw a line at a distance of 20 mm below the reference line to mark the top view 'ab'.

Line Parallel to One Reference Plane and Contained on the Other

When a line is contained by one reference plane or otherwise lying on that plane, it is parallel to that reference plane and the distance between the plane and the line is zero. Therefore, its projection on the other perpendicular reference plane falls on the reference line. If a plane is on the VP, its top view is on the reference line, and if it is on HP, its front view is on the reference line. The relevant points discussed in the earlier problems are also valid here.

Line CD, 30 mm above HP and on VP

Figure 7.4(a) shows the spatial position of the line and its projections on HP and VP. Figure 7.4(b) shows its orthographic views.

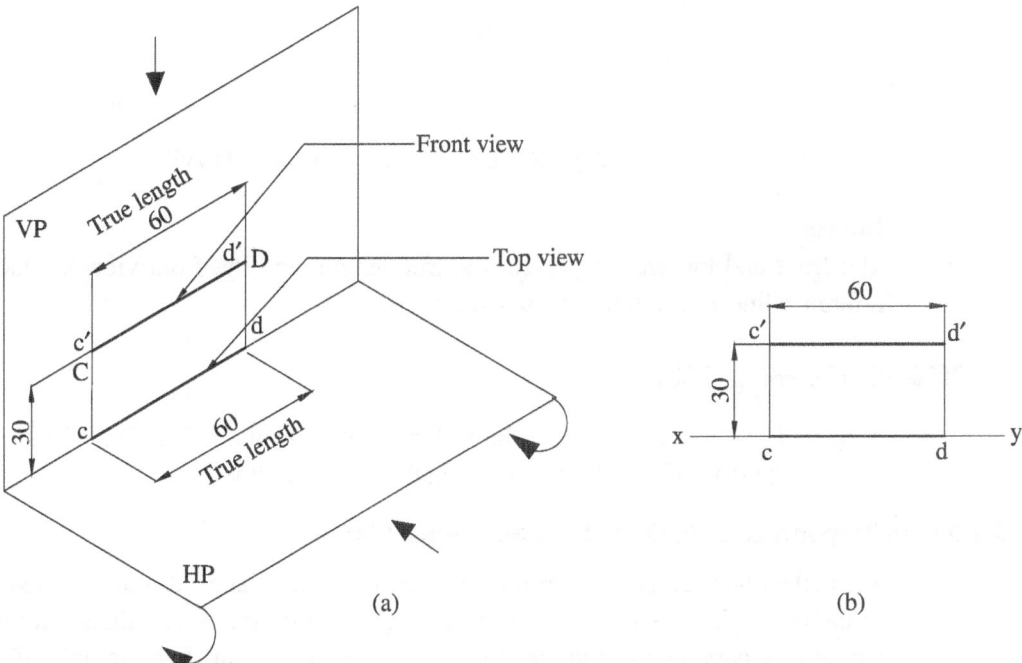

Fig. 7.4 Line on the VP and parallel to HP

Inference

The front and top view represent the true length, and the top view is placed on the reference line, as the line is contained on VP.

PROCEDURE (Refer Fig. 7.4b)

> **Step 1:** The construction is similar to the problem explained in Fig. 7.3(c), except the cd is on the reference line.

Line EF, 30 mm in front of VP and on HP

Figure 7.5(a) shows the spatial position of the line and its projections on HP and VP. Figure 7.5(b) shows the orthographic views.

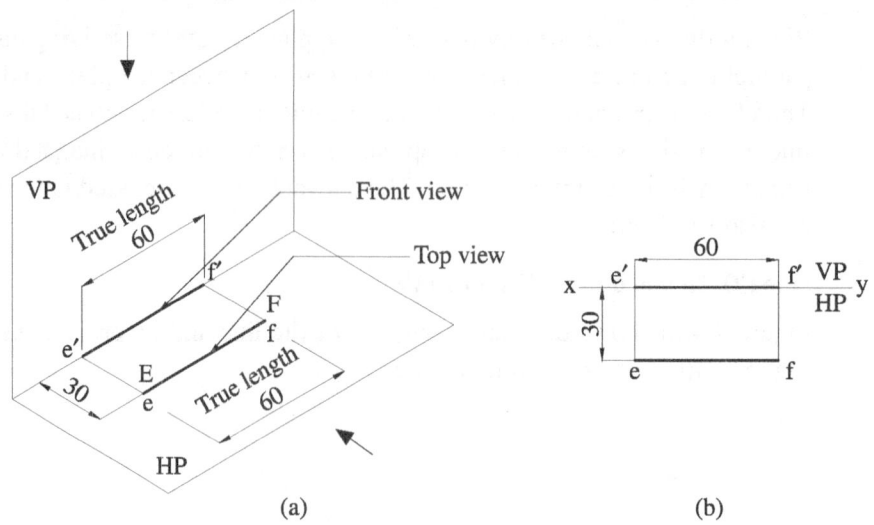

(a)   (b)

Fig. 7.5 Line is on HP and parallel to VP

Inference

The front and top view represent the true length, and the front view is placed on the reference line, as the line is visualized on HP.

PROCEDURE (Refer Fig. 7.5b)

> **Step 1:** The construction is similar to the problem explained in Fig. 7.3(c), except that the front view e'f' is marked on the reference line.   ▲

7.2.2 Line Perpendicular to One of the Reference Planes

When the line is perpendicular to one of the principal planes, it is automatically parallel to the other plane, and hence, the true length on the line is visualized on the plane to which it is parallel and the other view appears as a point. For example, if the line is perpendicular to HP, it is parallel to VP, and hence, the true length is visualized in its front view on VP and the top view is observed as a point on HP. Similarly, if the line is perpendicular to VP, it is parallel to HP, and hence, the true length is visualized in its top view on HP and the front view is seen as a point on VP. Refer Fig. 7.1(b).

Example 7.2

Draw the projections of the following straight lines of length 60 mm as per their respective positions.

Line CD Perpendicular to VP and on HP and End C 20 mm in Front of VP

Figure 7.6(a) shows the spatial position of the line and its projections on HP and VP. Figure 7.6(b) shows the orthographic projections.

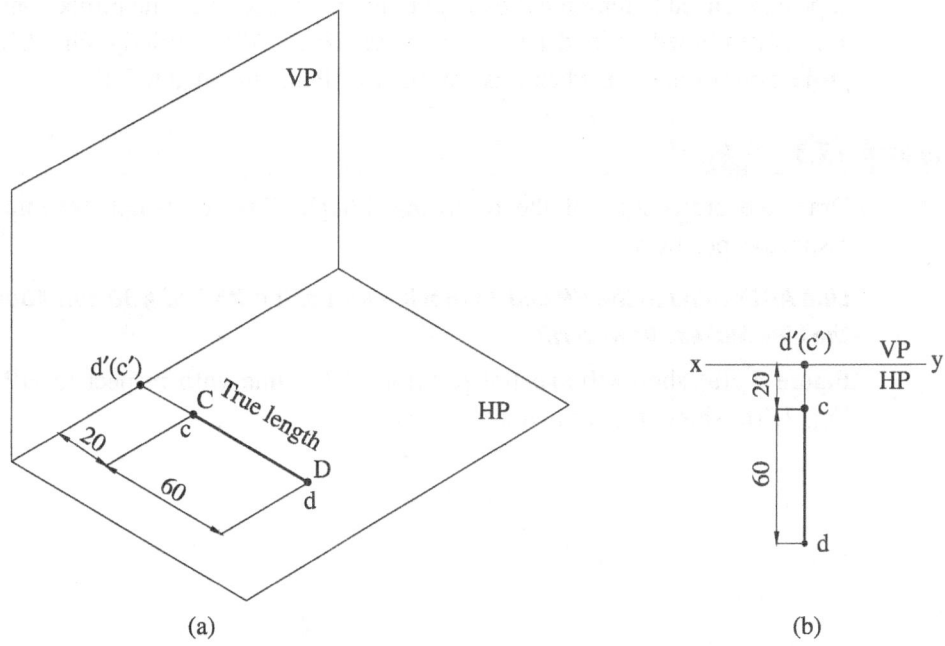

(a)   (b)

Fig. 7.6   Line on HP and perpendicular to VP

Inference

The top view gives the true length of the line as it is on HP and the front view appears as a point on the reference line as it is perpendicular to VP.

PROCEDURE (Refer Fig. 7.6b)

Step 1: Draw the horizontal reference line 'xy'.

Step 2: Mark the top view of the end 'C' at a distance of 20 mm below the 'xy' line and draw a vertical line of length 60 mm (TL) to get the top view as 'cd'.

Step 3: Draw the projector through the points 'd' and 'c' on to the 'xy' line and mark the front view as point d' (c'). The point d' denotes the front view of the end 'D' of the line which is visible and (c') denotes the front view of the other end 'C' of the line which is behind it and is not visible.

7.3 PROJECTION OF A LINE IN SIMPLE INCLINED POSITIONS

When the line is inclined to one of the principal planes, that is, HP or VP and parallel to other plane, it is said to be in a simple inclined position. Then, its true length and inclination are seen on the plane to which it is parallel and the projection on the other plane appears parallel to the reference line but reduced in length.

For example, in Fig. 7.1(c), the line (AB) is inclined to HP and parallel to VP. Then the front view represents the true length and the HP inclination angle (θ) and its top view appears reduced in length but parallel to the reference line. The projections of the end A are marked based on its distances from the HP and VP. Similarly, Fig. 7.1(d) shows the projections of the line when it is parallel to HP and inclined to VP.

Example 7.3

Draw the projections of the following straight lines of length 60 mm as per their respective positions.

Line AB Parallel to the VP and 25 mm in Front of it With End A 30 mm Above HP; the Line Makes 30° with HP

Figure 7.7(a) shows the spatial position of the line with respect to HP and VP and Fig. 7.7(b) shows the orthographic views.

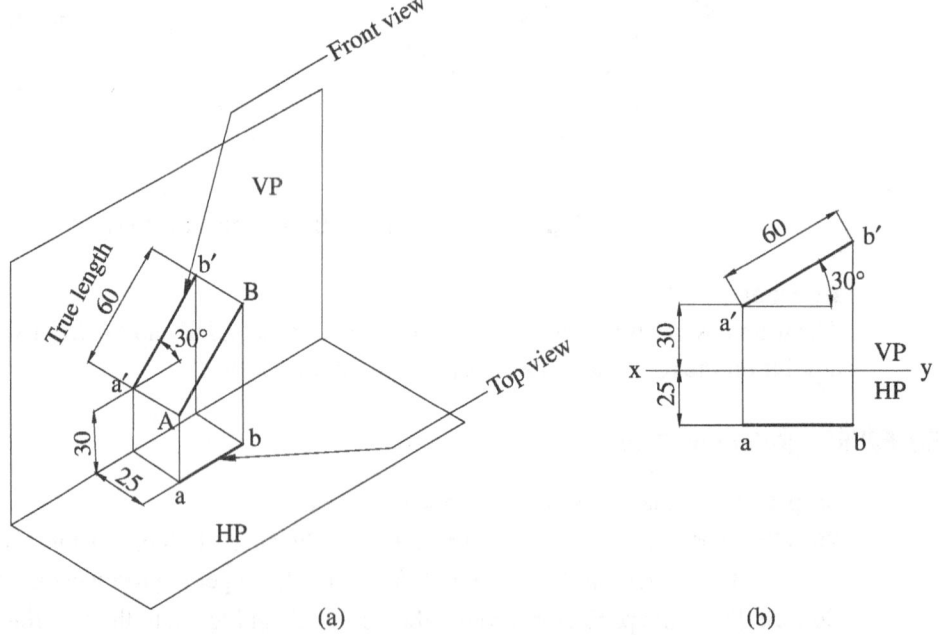

(a)   (b)

Fig. 7.7   Line parallel to VP and inclined to HP

Inference

Since the line is parallel to VP and inclined to HP, the front view represents the true length and the HP inclination angle. The top view is parallel to the reference line.

PROCEDURE (Refer Fig. 7.7b)

Step 1: Draw the horizontal reference line 'xy'.

Step 2: Mark the front view of the end A at a distance of 30 mm above the 'xy' line as a'. Set the HP inclination (30°) and mark the true length and get a'b'. Note that the inclination set at a' is such that the end b' is obtained above the reference line, as the line is in the first quadrant.

Step 3: Draw the projector through the points a' and b' to a distance of 25 mm below the 'xy' line and mark the top view as 'ab' parallel to the reference line.

Line AB is on the VP and Makes 30° with the HP; the End A is 30 mm Above HP

Figure 7.8(a) shows the projection of the line on HP and VP and Fig. 7.8(b) shows the orthographic views.

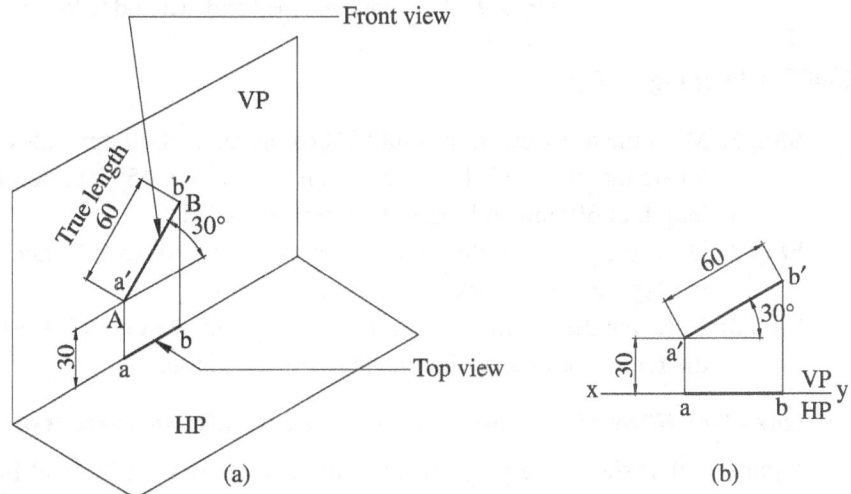

Fig. 7.8   Line on VP and inclined to HP

Inference

Since the line is visualized on VP and inclined to HP, the front view represents the true length and the HP inclination angle. The top view is on the reference line.

PROCEDURE (Refer Fig. 7.8b)

Step 1: The construction is similar to the problem explained in Fig. 7.7(b), except that the top view ab is on the reference line.

Line CD 25 mm Above the HP and Makes 45° with VP; the End C is 20 mm in Front of VP

Figure 7.9(a) shows the spatial position of the line and its projections on HP and VP and Fig. 7.9(b) shows the orthographic views.

Inference

Since the line is parallel to HP and inclined to VP, the top view represents the true length and the VP inclination angle. The front view is parallel to the reference line.

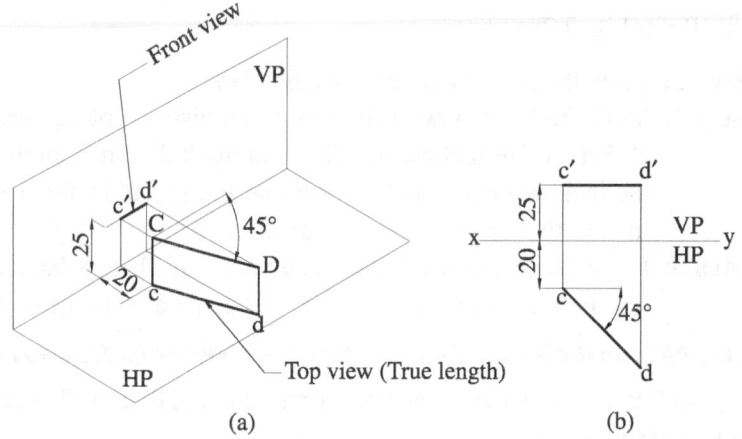

Fig. 7.9   Line parallel to HP and inclined to VP

PROCEDURE   (Refer Fig. 7.9b)

Step 1: Mark the top view of the end 'C' at a distance of 20 mm below the 'xy' line and locate the point 'c'. From the point 'c', draw a 45° inclined line with the true length of 60 mm and mark the top view as 'cd'.

Step 2: Draw the projector through the points 'c' and 'd' to a distance of 25 mm above the 'xy' line and mark the front view as c'd'.

Step 3: Note that the inclination set at 'a' is such that the end 'b' is also obtained below the reference line, as the line is in first quadrant.

Line CD on HP and Makes 30° with VP; the End C is 20 mm in Front of VP

Figure 7.10(a) shows the projections of the line on HP and VP and Fig. 7.10(b) shows the orthographic views.

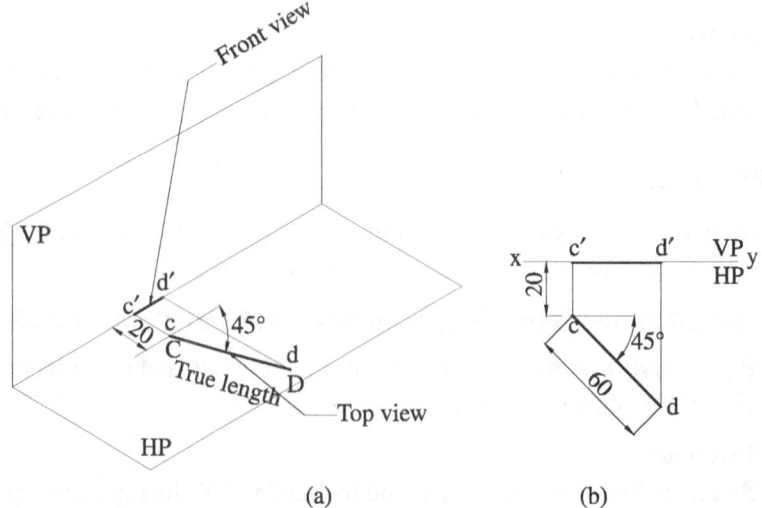

Fig. 7.10   Line on HP and inclined to VP

Inference

Since the line is contained on HP and inclined to VP, the top view represents the true length and the VP inclination angle. The front view is on the reference line.

PROCEDURE (Refer Fig. 7.10b)

Step 1: The construction is similar to the problem explained in Fig. 7.9(b), except that the front view c'd' is on the reference line. ▲

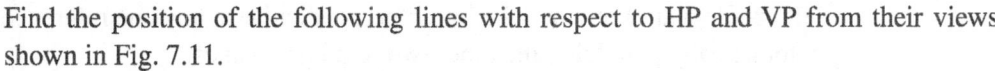

Example 7.4 ▲

Find the position of the following lines with respect to HP and VP from their views shown in Fig. 7.11.

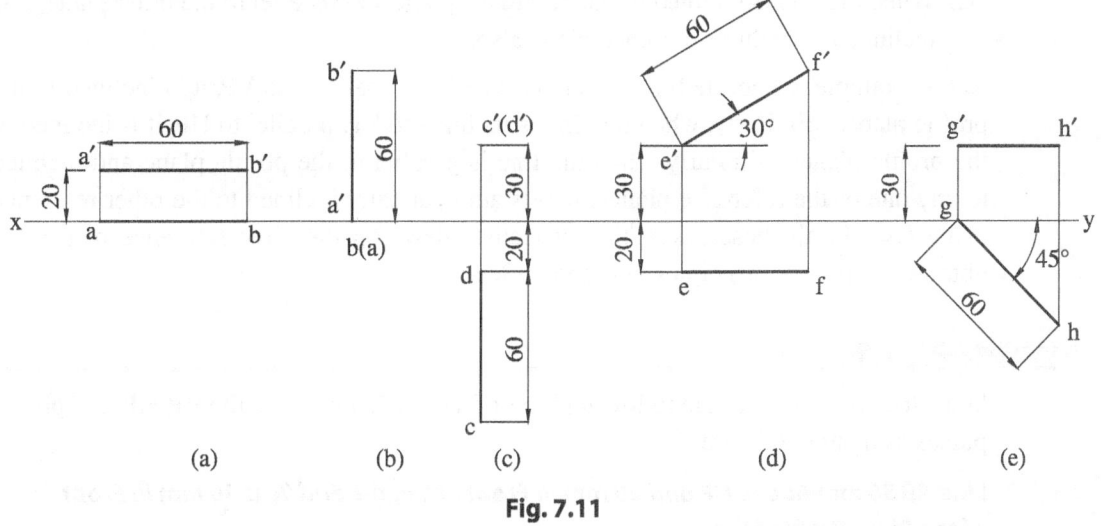

Fig. 7.11

(a) Line AB is 20 mm above HP and on the VP.

(b) Line AB is perpendicular to HP and on VP and the end A is on the HP.

(c) Line CD is parallel to HP and 30 mm above it and perpendicular to VP and the end D is 20 mm in front of VP.

(d) Line EF is parallel to VP, inclined at 30° to HP, 20 mm in front of VP and the end E is 30 mm above HP.

(e) Line GH is parallel to and 30 mm above the HP, inclined at 45° to VP and the end G is on VP. ▲

7.4 PROJECTIONS OF LINES ON THE THREE REFERENCE PLANES

In the sections discussed earlier, the projections of the lines are referred with respect to only two principal planes, namely HP and VP. In this section, the projections of the lines are referred with respect to the HP, the VP, and their right and left profile planes.

The method of projections of the line on to the profile planes (right and left) is similar to the projections of lines on the VP, because the right or left profile plane is only an another vertical plane, not in front of the observer but only to the sides of the observer. It can be recalled that the projection on to the right profile plane will yield the left-side view while that on the left profile plane will yield the right-side view. Both the side views are placed on the sides of the front view with a vertical reference line in between. The following observations are valid, when the line is placed between three reference planes:

1. When the line is parallel to any two reference planes (HP and VP), it is automatically perpendicular to the third reference plane, that is, the profile plane (the right and the left profile plane) or when the line is perpendicular to any one reference plane, it is automatically parallel to the other two reference planes.
2. When the line is inclined to one reference plane and parallel to the other plane, it is inclined to the third reference plane also.

For example, when the line is inclined to HP and parallel to VP, it is inclined to the profile planes. Similarly, when the line is inclined to VP, parallel to HP, it is inclined to the profile planes. Similarly when the line is parallel to the profile plane and inclined to any one of the reference planes, it gets automatically inclined to the other reference plane also. In all these cases, the respective views on the three reference planes are obtained as per the principles outlined in Section 7.3.

Example 7.5

Draw the projections of the following lines of length 70 mm on to the VP, HP, and profile planes as mentioned here.

Line AB 30 mm Above HP and 20 mm in Front of VP; the End 'A' is 30 mm in Front of the Right Profile Plane

Figures 7.12(a) and (b) show the spatial position of the line with respect to HP, VP, and right profile plane.

Inference

Since the line is viewed from the front and projected to VP to get the front view, it has to be viewed in the same manner from the left and projected onto the right profile plane to get the left-side view, which appears as a dot.

PROCEDURE (Refer Fig. 7.12c)

Step 1: After drawing the front view (b'a') and the top view (ba), draw a vertical reference line x_1y_1 to the right of the front view at a distance of 30 mm from a'.

Step 2: Extend top view to meet the vertical projector and set an angle of 45° or an arc, to cut the horizontal reference line and draw a projector at that point.

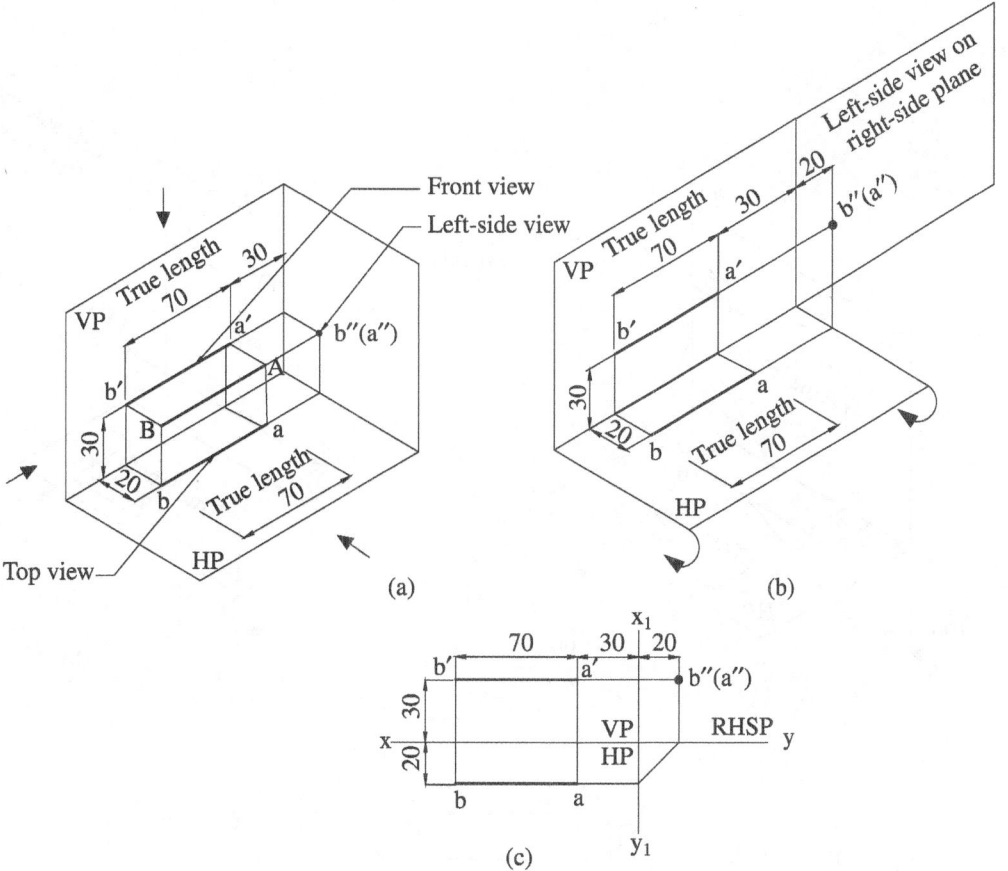

Fig. 7.12 Line in front of right profile plane and parallel to both HP and VP

Step 3: Extend the front view to meet this projector line and locate the point of
intersection as b″ (a″). The point b″ denotes the left-side view of the end B of
the line which is visible and (a″) denotes the left-side view of the other end A of
the line which is hidden.

Line CD 20 mm in Front of VP and on HP; the End Point 'D' is 25 mm in Front of the Right Profile Plane

Figure 7.13(a) and (b) shows the spatial position of the line with respect to HP, VP, and
right profile plane.

Inference
Same as in Example 7.5(i) except the front view is on the reference line due the line
contained on HP.

PROCEDURE (Refer Fig. 7.13c)

Step 1: Same as in Example 7.5(i) except the front view and the left-side view are on
the reference line.

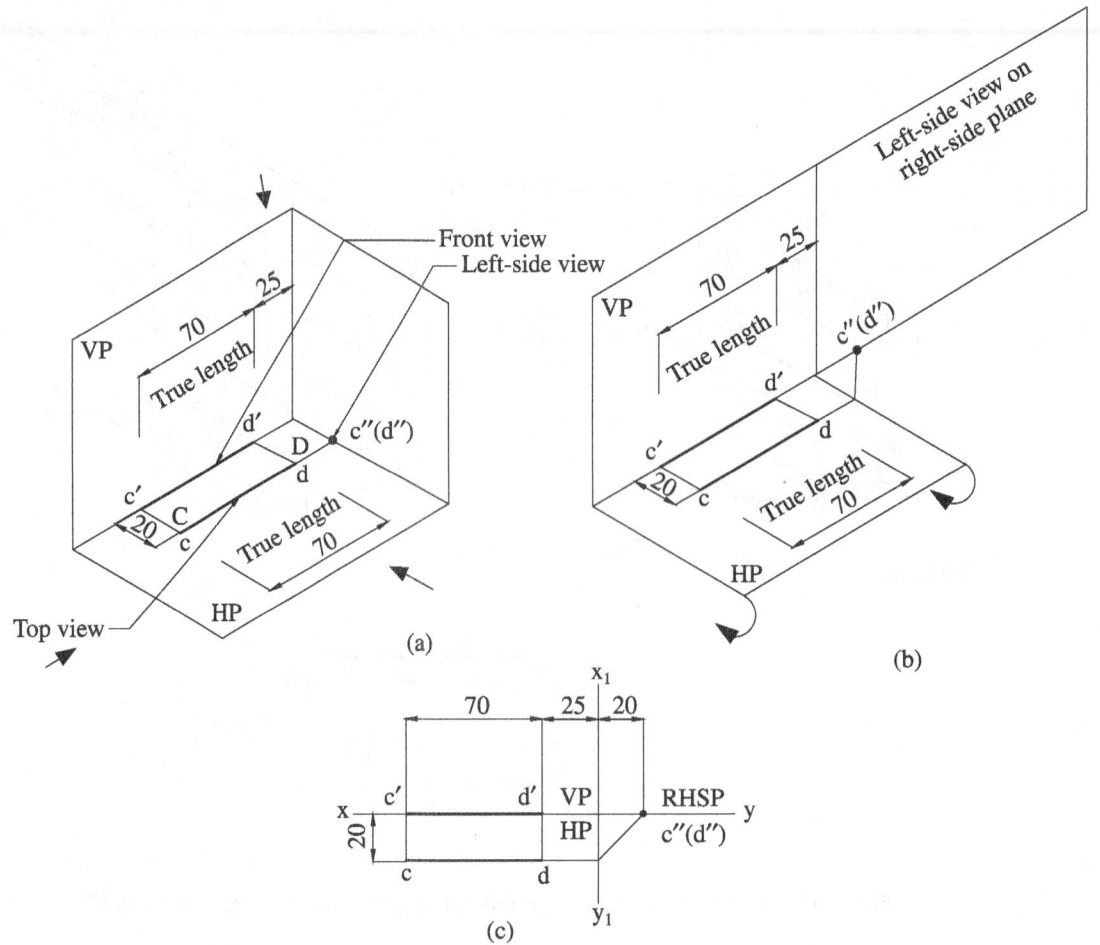

Fig. 7.13 Line in front of right profile plane, on HP, and parallel to VP

Line AB 20 mm above HP and on VP; the end A is on left profile plane

Figure 7.14(a) and (b) shows the spatial position of the line with respect to HP, VP, and left profile plane.

Inference

Since the line is viewed from the front and projected onto VP to get the front view, it has to be viewed from the right and projected to the left profile plane to get the right-side view, which appears as a dot on the vertical reference line.

PROCEDURE (Refer Fig. 7.14c)

Step 1: Locate the front and top view of the line as problems discussed earlier.

Step 2: Since the end A is on the left profile plane, the right-side view appears on the vertical reference line at 20 mm above HP as a point.

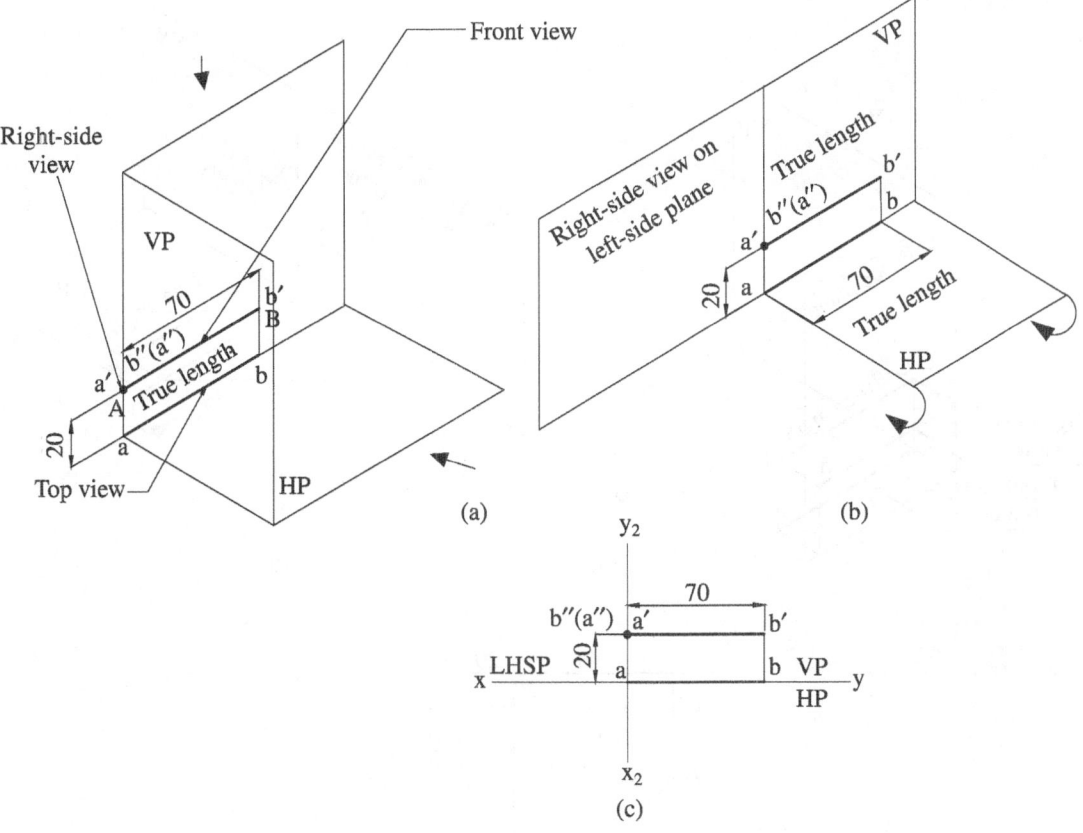

Fig. 7.14 Line on VP, parallel to HP, and one end on left profile plane

Line CD 30 mm Above HP and Perpendicular to the VP and 20 mm in Front of the Left Profile Plane; the End C 20 mm in Front of VP

Figure 7.15(a) and (b) shows the spatial position of the line with respect to HP, VP, and left profile plane.

Inference

1. Since the line is parallel to HP and LPP, the top view and the right-side view represent the true length and are parallel to the concerned reference lines.
2. Since the line is perpendicular to VP, it appears as a point in the front view. Refer to Fig. 7.14 for positioning the side view.

PROCEDURE (Refer Fig. 7.15c)

Step 1: Locate the front and top view of the line as problems discussed earlier.

Step 2: Obtain the right-side view c″d″ from the front and top views as shown in the Fig. 7.15(c).

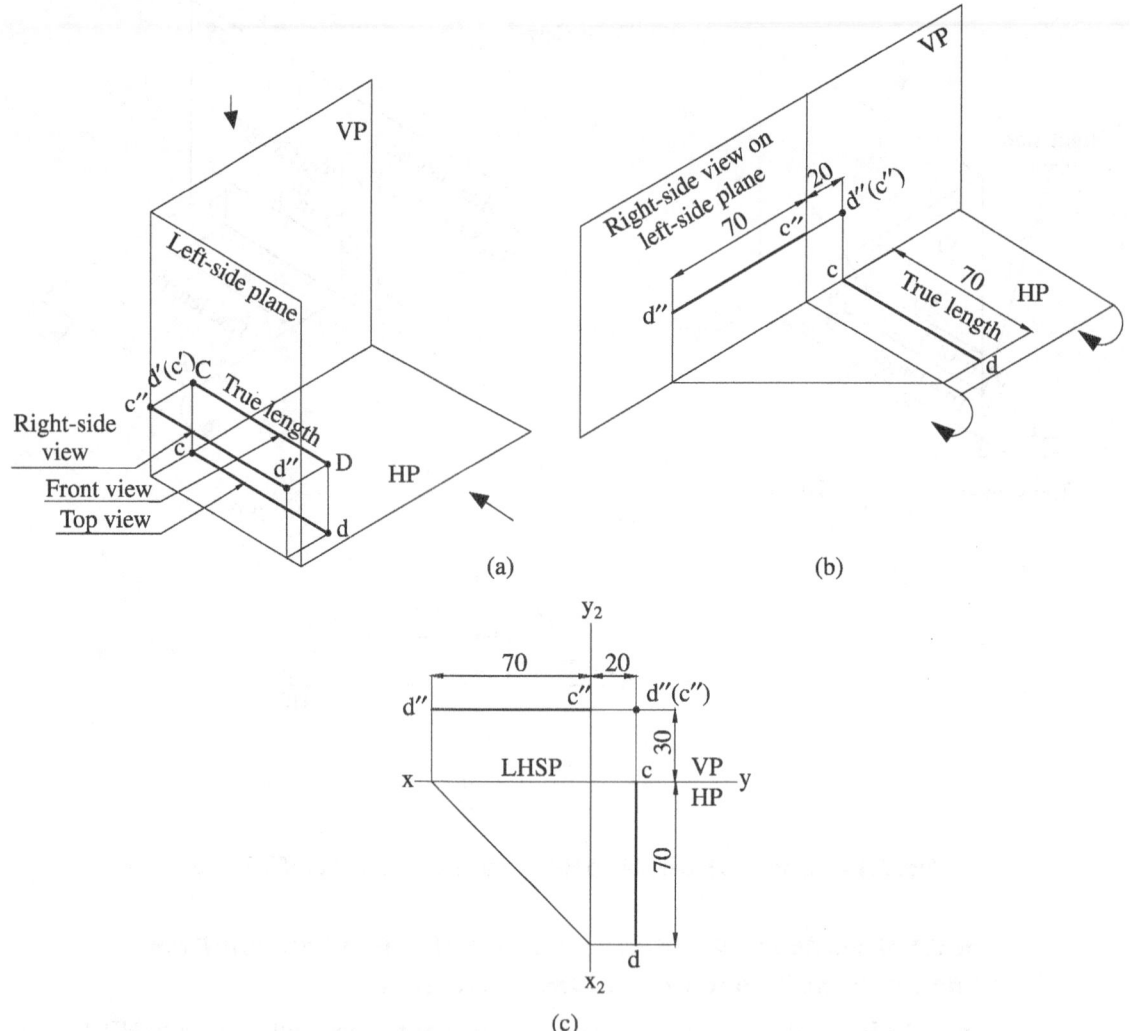

(a)

(b)

(c)

Fig. 7.15   Line parallel to HP, LPP, and perpendicular to VP

Line CD 30 mm Above HP and Makes 45° with the VP; the End C 25 mm in Front of VP and 20 mm in Front of Left Profile Plane

Figure 7.16(a) and (b) shows the spatial position of the line with respect to HP, VP, and left profile plane.

Inference

1. Since the line is parallel to HP and inclined to VP, the top view represents the true length and VP angle, while the front view is parallel to the reference line.
2. Since the line is parallel to HP and inclined to LPP, the right-side view is parallel to reference line. Refer to Fig. 7.14 for positioning the side view.

PROCEDURE (Refer Fig. 7.16b and c)

Step 1: Locate the front and top view of the line as problems discussed earlier.

Step 2: Obtain the right-side view c″d″ from the front and top views as shown in the figure.

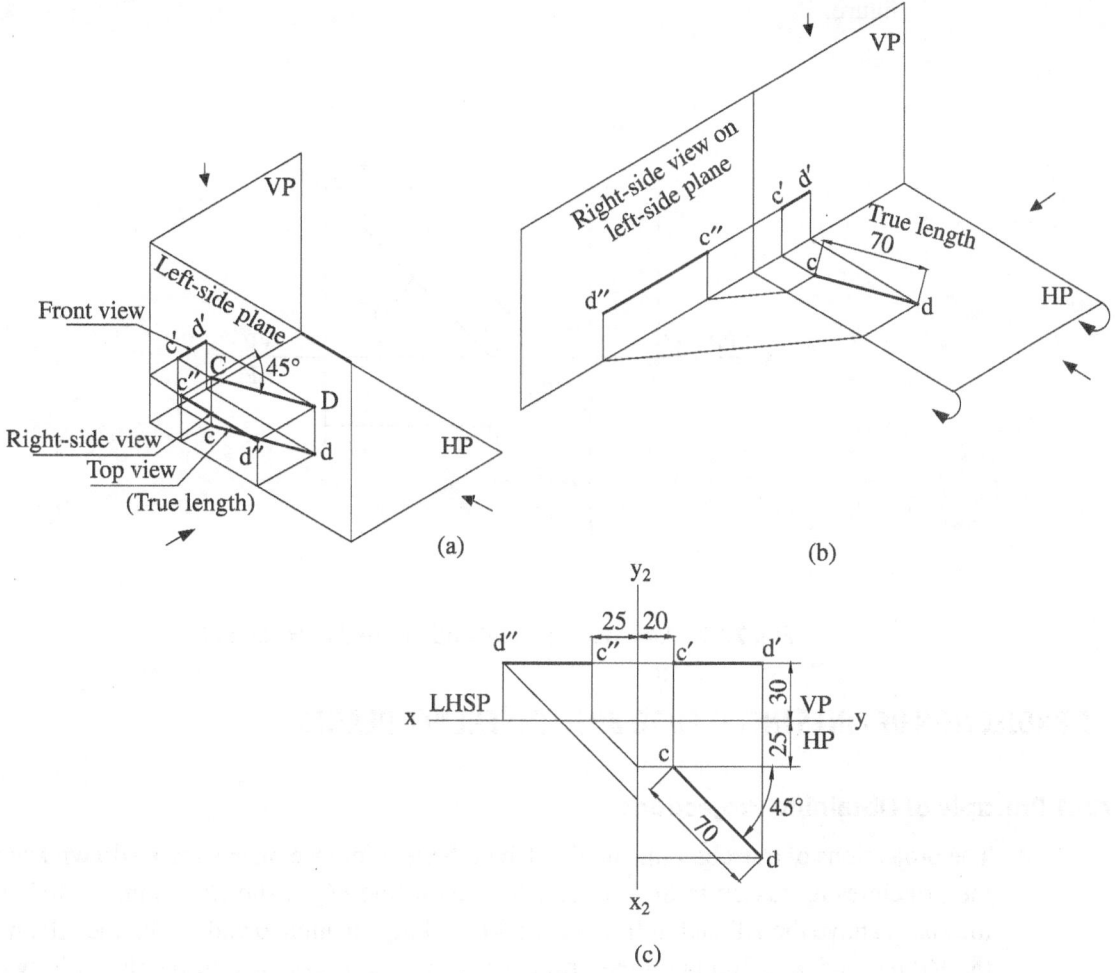

Fig. 7.16 Line parallel to HP and inclined to VP and PP

Line CD 20 mm in Front of VP and Makes 45° with the HP; the End 'D' is 30 mm in Front of Left Profile Plane

Inference

1. Since the line is parallel to VP and inclined to HP, the front view represents the true length and HP angle, while the top view is parallel to the reference line.

2. Since the line is parallel to VP and inclined to LPP, the right side view is parallel to the vertical reference line. Refer to the inference (Fig. 7.14) for details on the side view.

PROCEDURE (Refer Fig. 7.17)

Step 1: Locate the front and top view of the line as problems discussed earlier.

Step 2: Obtain the right-side view c"d" from the front and top views as shown in the figure.

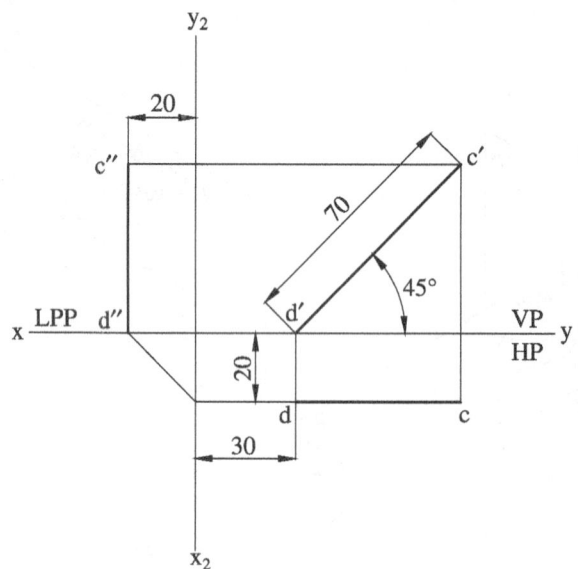

Fig. 7.17 Line parallel to VP and inclined to HP and PP

7.5 PROJECTION OF LINES INCLINED TO BOTH REFERENCE PLANES

7.5.1 Principle of Obtaining Projections

The projections of a straight line inclined to both the reference planes can be drawn from the principles discussed in Section 7.3. A straight line PQ is shown in Fig. 7.18(a) in the space above the HP and in front of the VP making an angle θ and ϕ with the HP and the VP respectively. To obtain the projections, the line is assumed to be placed in two stages, their corresponding projections are drawn and are later combined suitably as per the following scheme.

(i) In the first stage, the line is assumed to be parallel to one reference plane (say, VP) and inclined to the other (HP) at an angle θ and this is shown as PQ_1. Its projections are readily obtained as $p'q_1'$ and pq_1, as discussed in Section 7.3. Then the front view $p'q_1'$ represents the true length and the HP inclination, while the top view pq_1 is parallel to the reference line and shorter than the true length.

(ii) Since the line has to be additionally inclined to VP, it has to be tilted from this position PQ_1 to PQ. However, the inclination of the line with the HP (and so, the distances of its ends from the HP) due to (i) should not be disturbed or changed, during this process of movement of the line.

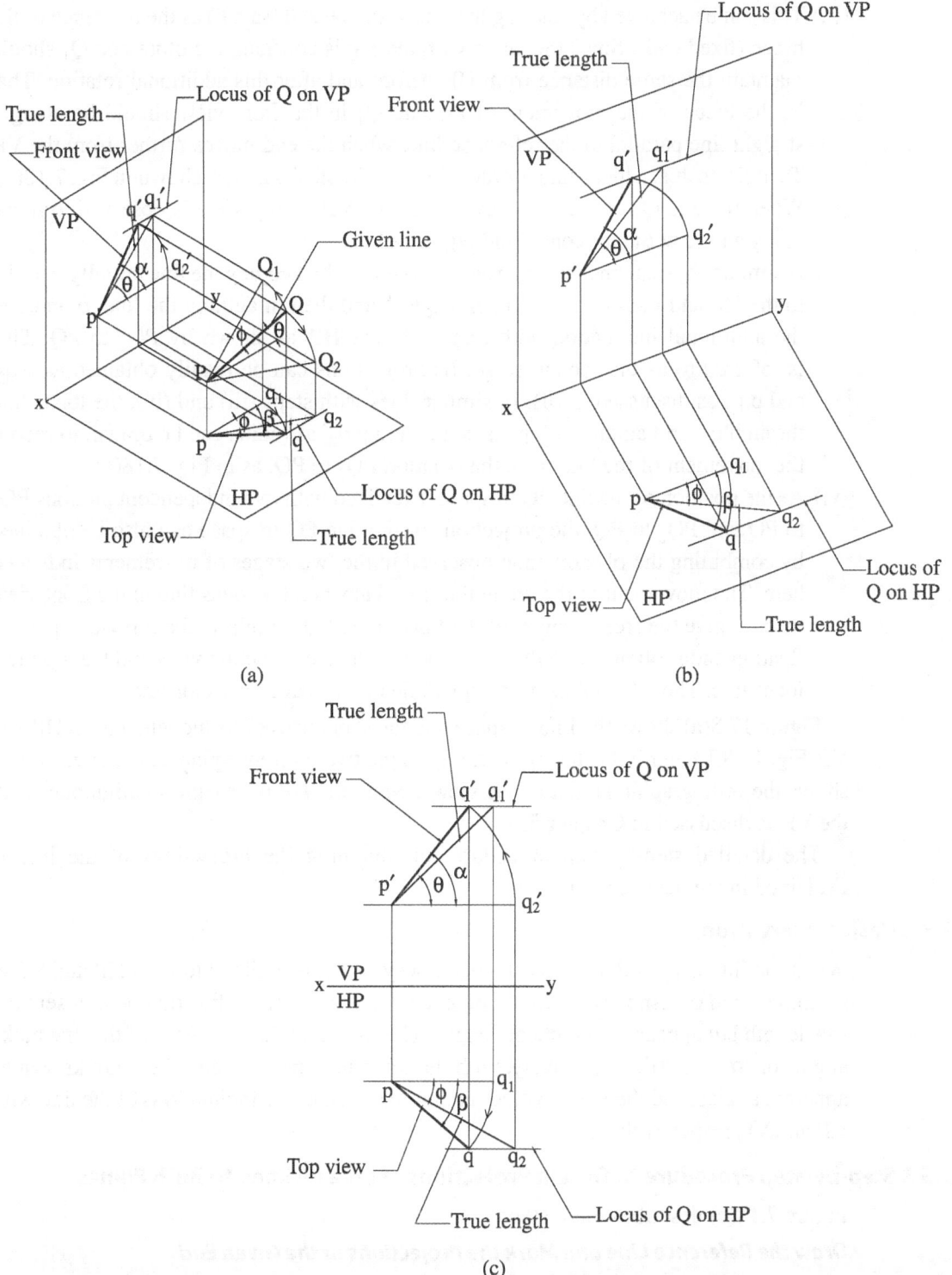

Fig. 7.18 Projection of a line inclined to both the planes

(iii) This can be achieved by rotating the line with one end (say, P) as the reference or the hinge (fixed end). Since the angle with the HP is constant, the other end Q_1 should maintain the same distance from HP, before and after this additional rotation. That is, the locus of the movement of the end Q_1 in the front view should be along a straight line parallel to the reference line, when the end moves further from the VP. To indicate this, a locus line is drawn from q_1' in the front view shown in Fig. 7.18(a).

(iv) When the end Q_1 of the line moves from the VP, its top view is observed to move along an arc with p as centre and pq_1 as radius.

(v) A similar physical phenomenon can be observed by keeping the line initially parallel to the HP and inclined to VP at an angle ϕ and then imagining the line to undergo the additional inclination with respect to the HP, as shown by PQ_2 to PQ. This is referred to as the second stage. Its projections can be readily obtained as $p'q_2'$ and pq_2, as discussed in (ii). In similar lines with steps (iii) and (iv), the locus line through q_2 and an arc with p' as centre and $p'q_2'$ as radius can be drawn, to record the movement of the line from the position PQ_2 to PQ, as in Fig. 7.18(a).

(vi) As the position of the line PQ in space is realized with two independent motions PQ_1 to PQ and PQ_2 to PQ, the projections of the line PQ in space can also be obtained by combining the phenomenon observed in the two stages of movements indicated here. The movement of the arc in the front view and the locus line in the front view meet to give the front view q' of the floating end Q. Similarly the top view q of the floating end is obtained by the interaction of the arc in the top view and the top view locus line. Then the points q' and q will lie on the same projector line.

Figure 17.8(a) shows the line in space and the sequences of its movements on HP and VP, Fig. 17.8(b) highlights the projections on the two reference planes, and Fig. 17.8(c) shows the orthographic layout of the views, when the HP is brought in alignment with the VP as discussed in Chapter 5.

The detailed step-by-step procedure for obtaining the projections of the line is explained in the next section.

7.5.2 Basic Observations

When the line is placed spatially in such a way that it is inclined to both HP and VP at an angle θ and ϕ, respectively, the front and the top views of the line do not represent the true length but appear with reduced length. The front and the top views of the line make angles of 'α' and 'β', respectively, with the reference line 'xy' and these are known as apparent angles, and the whole values are more than the true inclinations of the line with HP and VP, respectively.

7.5.3 Step-by-step Procedure to Obtain Projections of Line Inclined to Both Planes

Figure 7.18(c) describes these steps.

Draw the Reference Line and Mark the Projections of the Given End

This is done by drawing the projections of the given end 'P' of line PQ on VP and HP as p' and p, respectively, according to its position with respect to HP and VP.

Draw the Projections of the Line by Keeping it Parallel to One Reference Plane and Inclined to the Other

Draw the projections of the line when the line is kept parallel to VP and inclined to HP at an angle θ. This is achieved by drawing a line equal to the true length, set at the given HP angle, from the front view of the given point and marking its corresponding top view parallel to the reference line. Draw a line $p'q_1'$ equal to the true length of the line, set at the HP inclination angle θ from the point p' in the front view. It can be noted that the HP angle is set at p_1' such that q_1' is also located above the reference line being the requirement, as both the ends are assumed to be in the first quadrant. Its corresponding projection in the top view is obtained as pq_1, being parallel to the reference line as the line is assumed parallel to VP.

Draw the Projections of the Line with the Other Inclination

Draw the projections of the line parallel to the HP and inclined to VP at an angle φ such that q_2 is also located below the reference line, since both the ends are in the I quadrant. Repeat the previous step by setting the true length pq_2 in the top view with the VP inclination angle and obtain its parallel projection $p'q_2'$ as shown in Fig. 7.18(c).

Draw the Locus Lines (Parallel to the Reference Line) through the Floating Ends of the True Lengths and Identify Them

This is obtained by drawing horizontal locus lines (parallel to the reference line) through q_1' and q_2 in the front view and in the top view. These locus lines can be identified and named as the 'locus of Q on VP' and 'locus of Q on HP' as shown in Fig. 7.18(c). This identification is necessary to fix the final projections of the end Q properly.

Obtain the Final Projections of the Line by Superimposing the Individual Projections

These are obtained by selecting the parallel projections, identifying them, and associating them with their corresponding locus lines. For example, the final front view of the line $p'q'$ is obtained by selecting the front view parallel projection, $p'q_2'$ and cutting an arc on the front view locus line through q_1' with point p' as centre. Similarly, the final top view of the line 'pq' is obtained by selecting the top view parallel projection, pq_1 and cutting an arc on the top view locus line through 'q_2' with point 'p' as centre. It can be noticed that q_1 and q lie on the same vertical projector line due to orthographic projection principle. Alternatively, this step can be exercised by simply drawing a projector line from q' (obtained earlier) to meet the locus line in the top view as q' and q have to lie on the same projector. The user can obtain 'q' as explained earlier and draw a projector line to meet the front view locus line to get q'.

7.5.4 Parameters Governing the Projections of a Line

From Fig. 7.18(c), it is clear that the problems related to the projections of straight line involve the following parameters or data:

1. True length of the line (TL)
2. Position of the first end point (or) last end point (or) the mid-point (or) any other point on the straight line
3. Angle of inclination of the line with the HP (θ)
4. Angle of inclination of the line with the VP (ϕ)
5. Distance between the end projectors
6. Length of the front view (FV)
7. Length of the top view (TV)
8. Angle of inclination of the front view with 'xy' line or the apparent angle of inclination with HP (α)
9. Angle of inclination of the top view with 'xy' line or the apparent angle of inclination with VP (β)

In general, any five of these data can be used to solve a problem on straight lines. Appropriate combinations of these parameters constitute a variety of problems that will be discussed in the forthcoming examples. It can also be noted that if a straight line problem does not consist of five data parameters directly or indirectly, it cannot be solved. If more than five data parameters are mentioned, the user is expected to distinguish their dependency and use only the independent parameters.

Example 7.6

 A line AB of length 70 mm is inclined at 30° with HP and 45° with VP. The end A is 20 mm above HP and 30 mm in front of VP. Draw the projections of the line and find its apparent inclinations with respect to the principal planes.

Given data   Position of the end A of the line, true length, and angles of inclination with respect to HP and VP

PROCEDURE (Refer Fig. 7.19)

Step 1: Draw the horizontal reference line 'xy' and mark the given end 'A'. Draw a' at a distance of 20 mm above the 'xy' line and a 30 mm below 'xy' line.

Step 2: Draw the projections of the line when the line is parallel to VP and inclined at an angle θ = 30° with HP. Set a'b$_1$' of length equal to the true length of the line 70 mm inclined at an angle 30° to the 'xy' line and above it. Draw the respective parallel projection in the top view as ab$_1$.

Step 3: Draw the projections of the line when the line is parallel to HP and inclined at an angle ϕ = 45° with VP. Set 'ab$_2$' of length equal to the true length of the line 70 mm inclined at an angle 45° to the 'xy' line and below it. Draw the respective parallel projection in the front view as a'b$_2$'.

Step 4: Draw the horizontal locus lines through points b$_1$' and b$_2$ and name them as shown in Fig. 7.19.

Step 5: Identify the parallel projections and associate them with the corresponding locus lines. With a' as centre a'b$_2$' as radius draw an arc to cut the locus line in the front view as the parallel projection a'b₂' belongs to the front view family. This gives the front view a'b'.

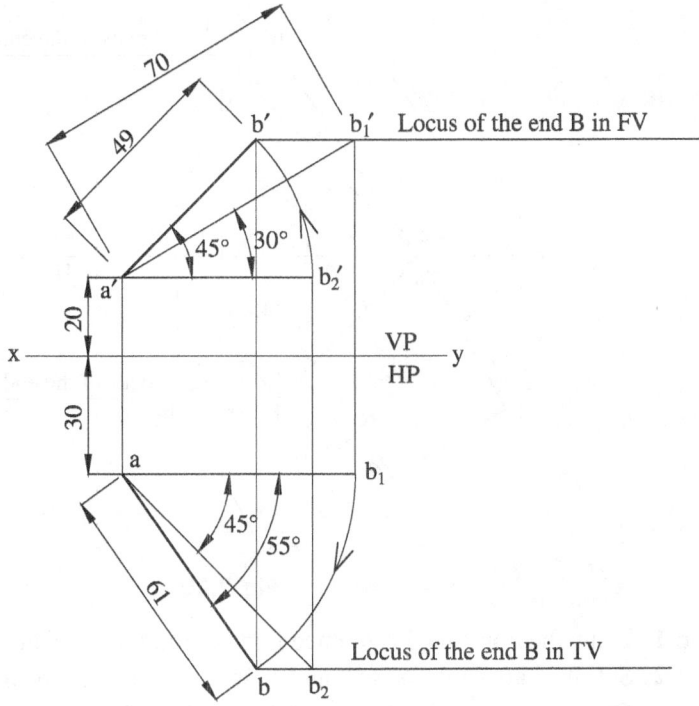

Fig. 7.19

Step 6: Project the point b' to meet the horizontal locus line on HP and obtain b. Alternatively, this can be obtained with 'a' as centre and 'ab₁' as radius and by drawing an arc to cut the locus line in the top view.

Step 7: Join a'b' and measure its length and angle which are denoted as length of the front view (FV) = 49 mm and apparent inclination of the front view with the 'xy' line as α = 45°. Similarly, join 'ab' and measure its length and angle which are denoted as length of the top view (TV) = 61 mm and apparent inclination of the top view with the 'xy' line as 'β = 55°'. It can be observed that the lengths of the projections are less than true length of the line, while their apparent angles of inclination are more than respective true angles of inclinations.

Example 7.7

A line AB of length 70 mm has its end A at the intersection of both the reference planes. The line is inclined at 45° with HP and 30° with VP. Draw the projections of the line and find its apparent inclinations with respect to the principal planes.

Given data Position of the end A of the line, true length, and angles of inclination with respect to HP and VP

PROCEDURE (Refer Fig. 7.20)

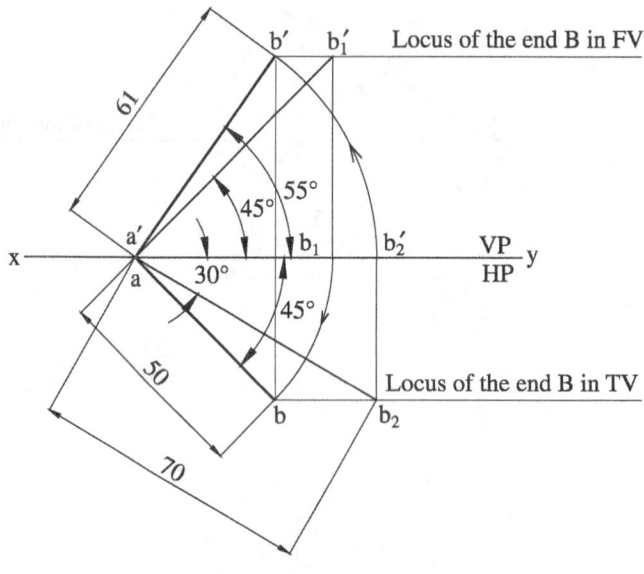

Fig. 7.20

Step 1: Draw the horizontal reference line 'xy' and mark a′ and a on it.

Step 2: Set at a′ an angle of 45° and mark the true length of 70 mm and obtain b_1'. Obtain its parallel projection 'ab_1' on the reference line.

Step 3: Similarly, set at 'a' an angle of 30° and mark the true length of 70 mm and obtain 'b_2'. Obtain its parallel projection $a'b_2'$ on the reference line.

Step 4: Draw the locus lines at b_1' and b_2 and name as shown in Fig. 7.20.

Step 5: With a′ as centre and $a'b_2'$ as radius cut an arc on the front view locus line and obtain b′. Project b′ on to the other locus line and obtain b.

Step 6: Obtain the projections by joining a′b′ and 'ab'. Measure the respective included angles $\alpha = 55°$ and $\beta = 45°$. ▲

7.5.5 Obtaining True Length/True Inclinations from the Projections

Reversal or Rotating Line Method

> **PRINCIPLE**
>
> 1. The complete procedure of obtaining the projections of a straight line was explained in Section 7.5.3, when the position of one end, the length, and the inclinations of the line with respect to HP and VP were given. Figure 7.18(a) explained the sequential movements of the line located in space to obtain the projections and Fig.7.18(c) gave the complete arrangement as per the ortho-graphic projection layout.
>
> 2. In Fig. 7.18(a), the line PQ in space when moved parallel to VP gave its true length ($p'q_1'$) in the front view along with the HP inclination angle (θ),

while its other view 'pq₁' became the parallel projection on the HP. The latter became the top view of the line, when the line made its additional inclination with respect to VP.

3. The principle of the reversal of this sequence of operation can be used to obtain the true length, HP inclination angle when the top view of the line is given.

4. In other words in Fig. 7.18(c), when the top view 'pq' of the line is laid parallel to the reference line as pq_1 and projected to the front view domain, then the length of the line and the HP inclination angle can be obtained as noted from $p'q_1'$. Similarly, when the front view of the line $p'q'$ is laid parallel to the reference line as $p'q_2'$ and projected to the top view domain, then the length and the VP inclination angle can be obtained as shown by pq_2.

5. Since this method has adopted the reversal procedure of the various operations used to get the projections of the line, this is known as a reversal method, and since the line is rotated in all these positions, this method is also known as rotating line method.

PROCEDURE (Refer Fig. 7.21)

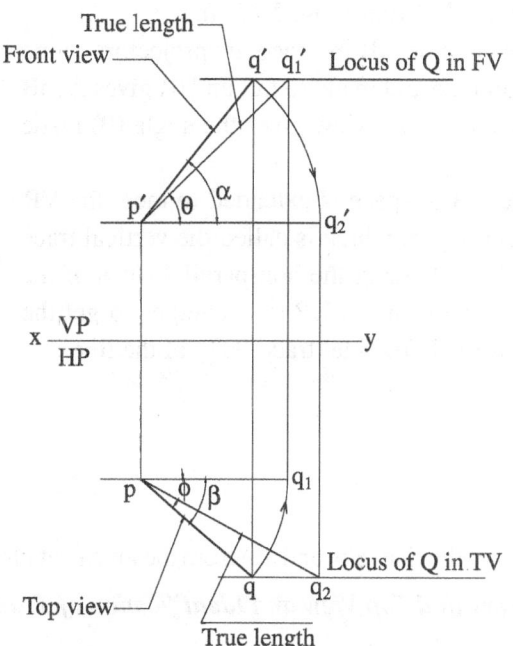

Fig. 7.21 Finding true length and true inclinations from the projections of a line inclined to both the planes—rotating line method

Step 1: Locate the given end and draw the projections of the line as per the details and mark the locus lines in the VP and HP locations.

Draw $p'q'$ and pq as per the details of the problem and draw locus lines as shown in Fig. 7.21.

Step 2: Mark any one projection parallel to the reference line, through the given end and project it to the other domain area.

Make $p'q'$ parallel to the reference line as shown by the arc and obtain $p'q_2'$.

Step 3: When the floating end projector meets the locus line in the other domain area, the line joining the meeting point and the fixed end gives the true length and the angle of inclination in that domain area.

Project q_2' to meet the locus line in the top view and obtain q_2. Join pq_2. This gives the true length and the angle subtended with the reference line gives the inclination of the line with VP (ϕ).

Step 4: Repeat the same set of operations with the other projection of the line and obtain the true length and the inclination of the line with the other reference plane.

Make 'pq' parallel to the reference line as shown by the arc and obtain pq_1. Project q_1 to meet the locus line in the front view and obtain q_1'. Join $p'q_1'$. This gives the true length and the angle subtended with the reference line gives the inclination of the line with HP (θ).

Trapezoidal Plane Method

PRINCIPLE

This method involves considering the line in space and its projections on VP and HP and rotating the trapezoidal planes formed there in about the projections, to get the true length and the inclinations of the line.

The reader is advised to recall the procedure in Section 7.5.1 for obtaining the projections of a line inclined to VP and HP. In Fig. 7.22(a), the line AB is in space and its projections a'b' and ab as obtained on the reference planes are shown. Since the projectors are perpendicular to the reference planes, the original line AB in space and two projectors emanating from its ends A and B and the front view a'b' form a trapezoidal plane. This plane when assumed to be rotated about the front view a'b' and made to fall on VP, forms a trapezium Aa'b'B with four sides as mentioned earlier. The included angle between the line and the front view in the trapezium gives the angle (ϕ) made by the line with the VP. Refer Fig. 7.22(b).

Similarly, the trapezium consisting of the line AB in space, its projectors to the HP and the top view ab, when rotated about ab and made to fall on HP, gives AabB and the included angle between the line and its top view gives the angle (θ) made by the line with the HP. Refer Fig. 7.22(b).

It can be also be noted when the line AB in space is extended to meet the VP, the trapezoidal plane converges to a point on VP, which is called the vertical trace (VT) of the line, and this can obtained by extending the non-parallel sides of the trapezium located on VP. Similarly, the trapezium on HP to converges to get the meeting point of the line in HP, known as the horizontal trace (HT) of the line.

PROCEDURE (Refer Fig. 7.22c)

Construction of Orthographic Views

Locate the given end and draw the projections of the line (a'b' and ab) from the given details.

Construction of the Trapezium in Front View and Top View and Identification of True Length and Angles

Consider any one projection and set perpendiculars at its ends and mark the distances of these ends from the other view. The resulting line gives the true length and the included angle between the line and concerned projection gives the angle made by the line with the other reference plane.

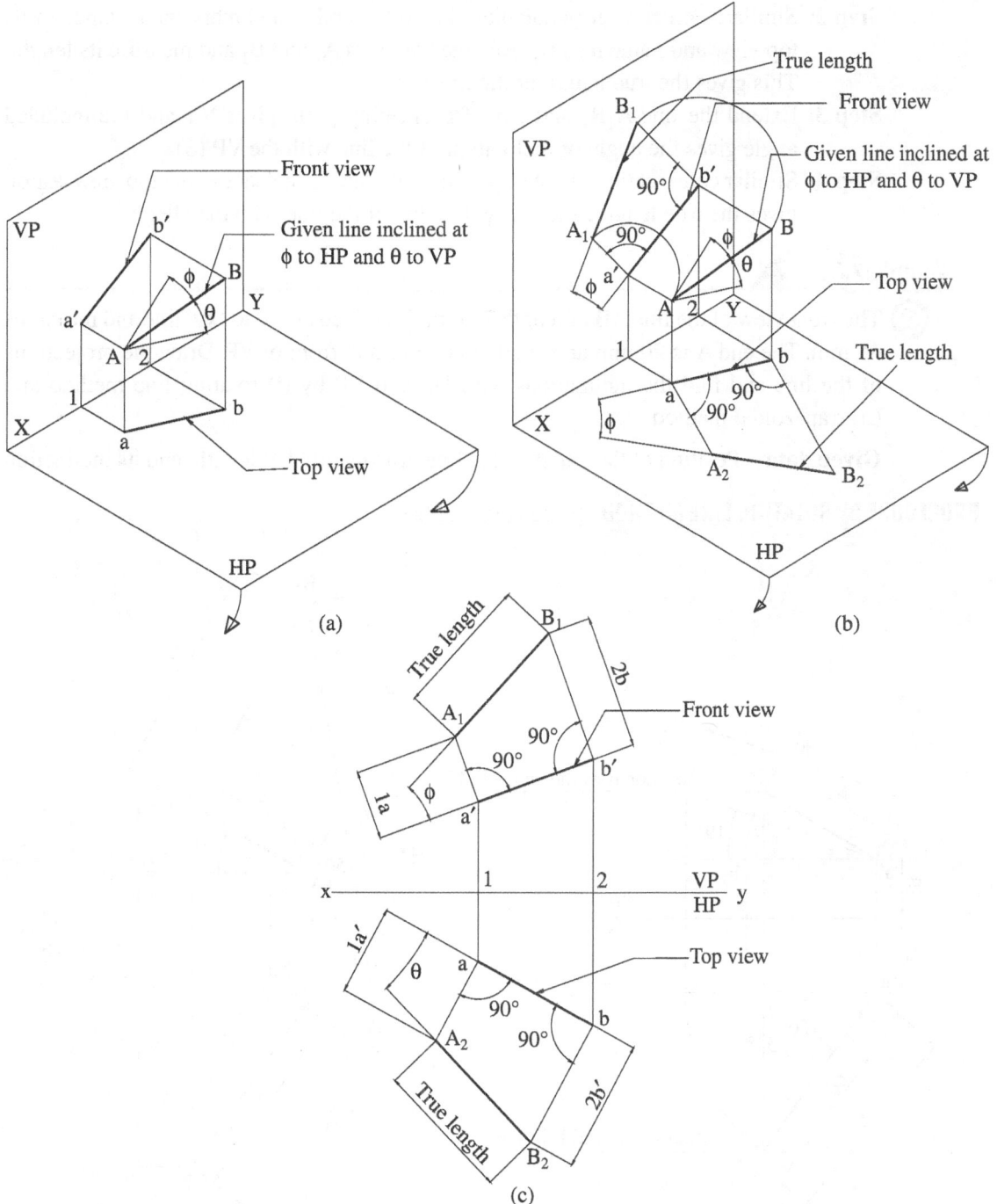

Fig. 7.22 Trapezium method of obtaining true length and true inclinations of the line

Step 1: For example, consider the front view a′b′ in Fig. 7.22(c). Erect the perpendicular at its one end a′ and mark the distance of its top view (i.e. the distance between the reference line and top view a) and name it as A₁.

Step 2: Similarly, erect a per pendicular at its other end b′ and mark the distance of its top view and name it as B_1. Join the two ends A_1 and B_1 and measure its length. This gives the true length of the line.

Step 3: Extend the line A_1B_1 and a′b′. The meeting point gives VT and the included angle gives the angle of inclination of the line with the VP (ϕ).

Step 4: Similar construction of the trapezium with 'ab' as the base in the top view region gives the true length, and the inclination of the line with the HP.

Example 7.8

@ The front view of the line AB of length 70 mm is inclined at 30° to 'xy' line and measures 45 mm. The end A is 20 mm above HP and 25 mm in front of VP. Draw the projections of the line and find the inclinations with HP and VP by (i) rotating line method and (ii) trapezoidal method.

Given data Position of the end A of the line, true length, FV length, and its inclination

PROCEDURE BY ROTATING LINE METHOD (Refer Fig. 7.23a)

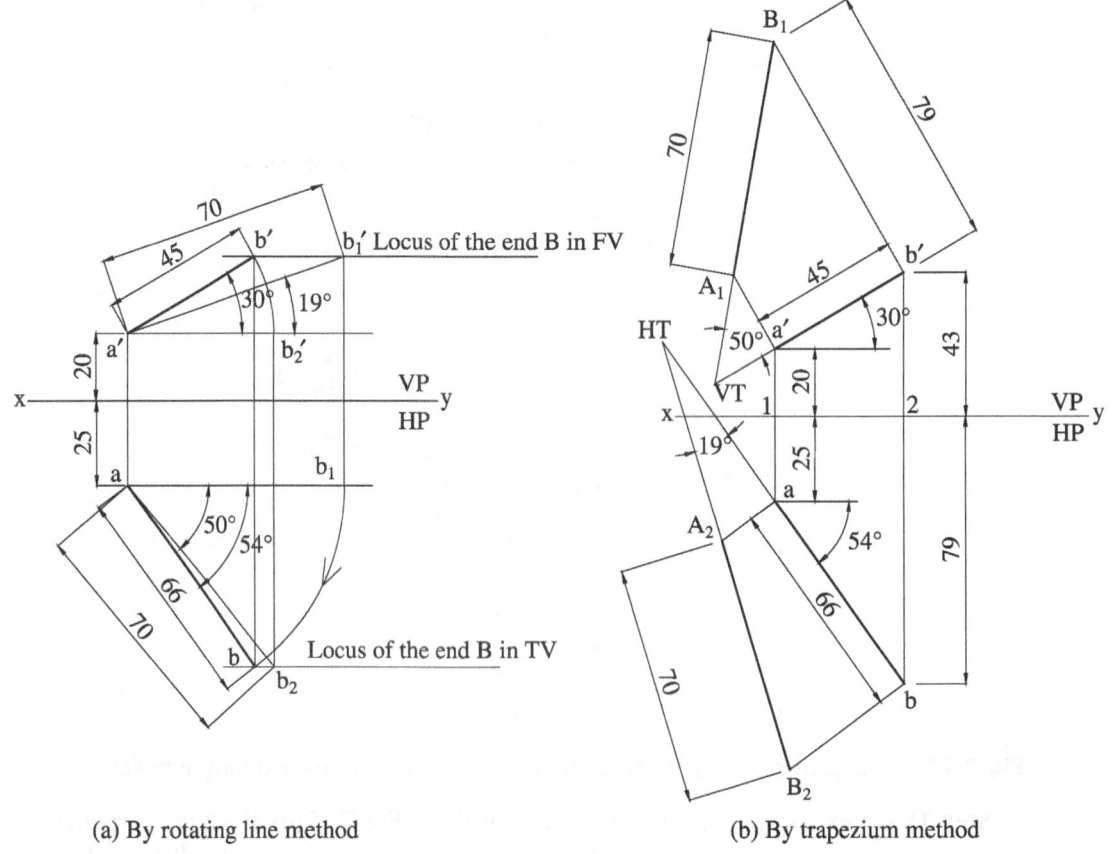

(a) By rotating line method   (b) By trapezium method

Fig. 7.23

Step 1: Draw the reference line and mark the views of the end point 'A' of the line as shown in the Fig. 7.23.

Step 2: At a', set the front view a'b' of length 45 mm at 30° to the reference line and make it parallel to the reference line as shown by a'b$_2$' by drawing an arc. Draw a projector line through b$_2$'. With 'a' as centre and true length as radius cut an arc on the projector line to get b$_2$.

Step 3: Draw horizontal locus lines at b' and b$_2$ and name them as indicated.

Step 4: As ab$_2$ indicates the true length in the top view domain, the angle subtended by it with the reference line indicates the angle with VP ($\phi = 50°$). With a' as centre and the true length as radius draw an arc on the front view locus line and get b$_1$'. The angle subtended by a'b$_1$' gives the angle with the HP ($\theta = 19°$).

Step 5: Project b' to meet the locus line in the top view and get 'b'. The line 'ab' gives the top view that measures 66 mm and the apparent angle $\beta = 54°$. Alternatively an arc can be drawn from b$_1$ to get b.

PROCEDURE BY TRAPEZOIDAL METHOD (Refer Fig. 7.23b)

Step 1: Draw the reference line and mark the views of the end point 'A'. At a', set the front view a'b' for a given length of 45 mm at 30° as indicated in Fig. 7.23.

Step 2: Erect perpendiculars at a' and b'. Set the distance of end A at a' and mark as A$_1$. With A$_1$ as centre and true length as radius cut an arc on the perpendicular at b' and get B$_1$. Extend A$_1$B$_1$ and a'b' to meet at VT. The included angle gives the inclination of the line with VP ($\phi = 50°$).

Step 3: Draw the projector line at b' and mark the distance b'B$_1$ (79 mm measured from Fig. 7.23(b) from the reference line and get 'b'. The line 'ab' gives the top view.

Step 4: Erect perpendiculars at 'a' and 'b' and mark the distances of a' and b' to get A$_2$ B$_2$ as shown in Fig. 7.23(b). Extend A$_2$B$_2$ and 'ab' to meet at HT. The included angle gives the inclination of the line with HP ($\theta = 19°$). ▲

Example 7.9   ◤

The top view of the line CD of length 65 mm is inclined at 45° to 'xy' line and measures 40 mm. The end D is 50 mm above HP and 60 mm in front of VP. Draw the projections of the line and find the inclinations with HP and VP by (i) rotating line method and (ii) trapezoidal method.

Given data Position of the end D of the line, true length, TV length, and its inclination

PROCEDURE BY ROTATING LINE METHOD (Refer Fig. 7.24a)

Step 1: Draw the reference line and mark the projections of the end D and draw the top view dc as per the given details.

Step 2: Make the top view 'dc' parallel to reference line and obtain the angle made with HP in similar lines as discussed in Example 7.8.

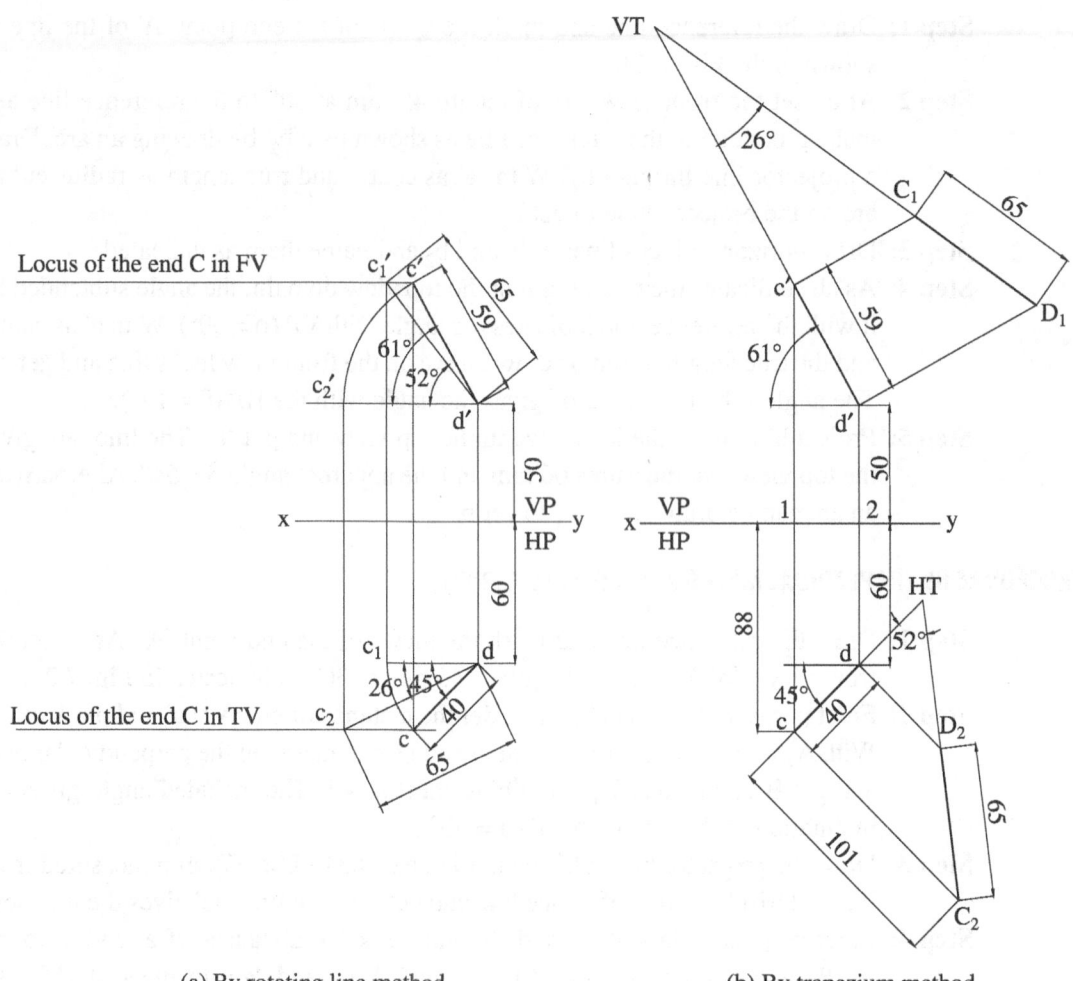

(a) By rotating line method　　　　　　(b) By trapezium method

Fig. 7.24

Step 3: Draw locus lines at c and c_1' and obtain the angle with VP in similar lines as discussed in the problem earlier.

Step 4: Measure the views and apparent angles.

PROCEDURE BY TRAPEZIUM METHOD　(Refer Fig. 7.24b)

Step 1: Draw the reference line and mark the projections of the end D and draw the top view 'dc' as per the given details.

Step 2: Construct the trapezium at 'dc' in similar lines as discussed in the previous problem and obtain D_2C_2. Extend D_2C_2 and 'dc' to get the HT and the included angle with the HP ($\theta = 52°$).

Step 3: Obtain the front view $d'c'$ in similar lines as discussed in the previous problem and obtain D_1C_1. Extend D_1C_1 and $d'c'$ to get the VT and the included angle with the VP ($\phi = 26°$). ▲

Example 7.10

A line CD of length 65 mm has its end 'C' 25 mm above HP and 20 mm in front of VP. The other end 'D' is 45 mm above HP and 55 mm in front of VP. Draw the projections of the line and find its inclinations with respect to the principal planes.

Given data   Position of the two ends and true length

PROCEDURE BY ROTATING LINE METHOD   (Refer Fig. 7.25)

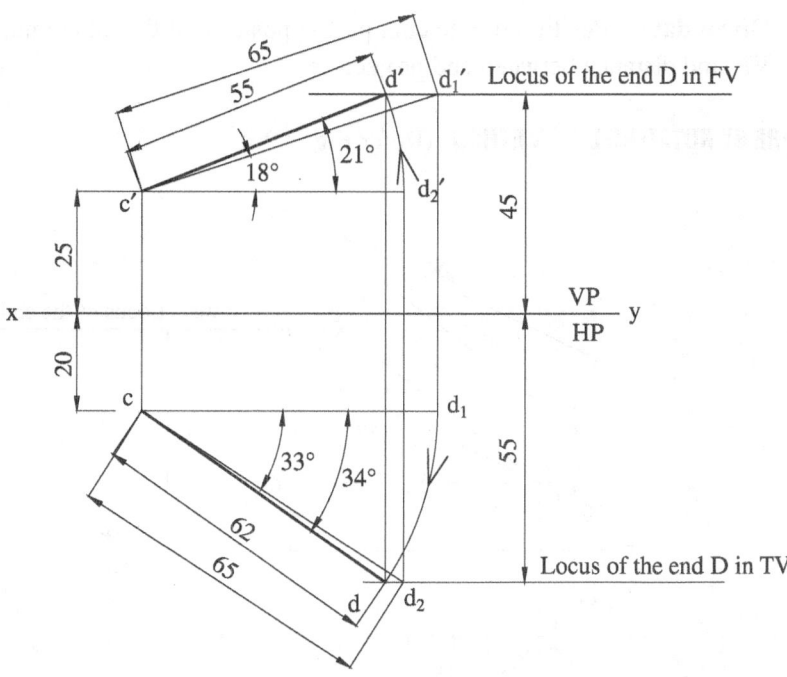

Fig. 7.25

Step 1: Draw the reference line and mark the projections c′ and c of the end point 'C'.

Step 2: Draw a horizontal line 45 mm above and another 55 mm below the 'xy' line to indicate the locus lines of the end D in the front and top views.

Step 3: With c′ and c as centres, draw arcs equal to the true length of the line and cut the respective locus lines and get points d_1' and d_2.

Step 4: Since $c'd_1'$ is the true length, the angle subtended by it with reference line gives the angle of inclination of the line with the HP ($\theta = 18°$). Similarly, cd_2 gives the angle of inclination with the VP ($\phi = 33°$).

Step 5: Obtain the parallel projections $c'd_2'$ and cd_1 corresponding to the true lengths.

Step 6: With c' and c as centres and $c'd_2'$ and cd_1 as radii, cut arcs on the respective locus lines and obtain d′ and d as shown in Fig. 7.25, and d′ and d should lie on the same projector.

Step 7: Join c'd' and cd to get the projections of the line and measure apparent inclination angles α = 21° and β = 34°.   ▲

Example 7.11   ▲

A line CD has its one end C 20 mm above HP and 25 mm in front of VP. The other end 'D' is 65 mm above HP and the line is inclined at 45° to VP. The distance between the end projectors of the line which is measured parallel to the line of intersection of HP and VP is 60 mm. Draw the projections of the line and find its inclination with HP.

Given data   Position of one end, partial position of the other end, inclination with the VP, and distance between end projectors

PROCEDURE BY ROTATING LINE METHOD   (Refer Fig. 7.26)

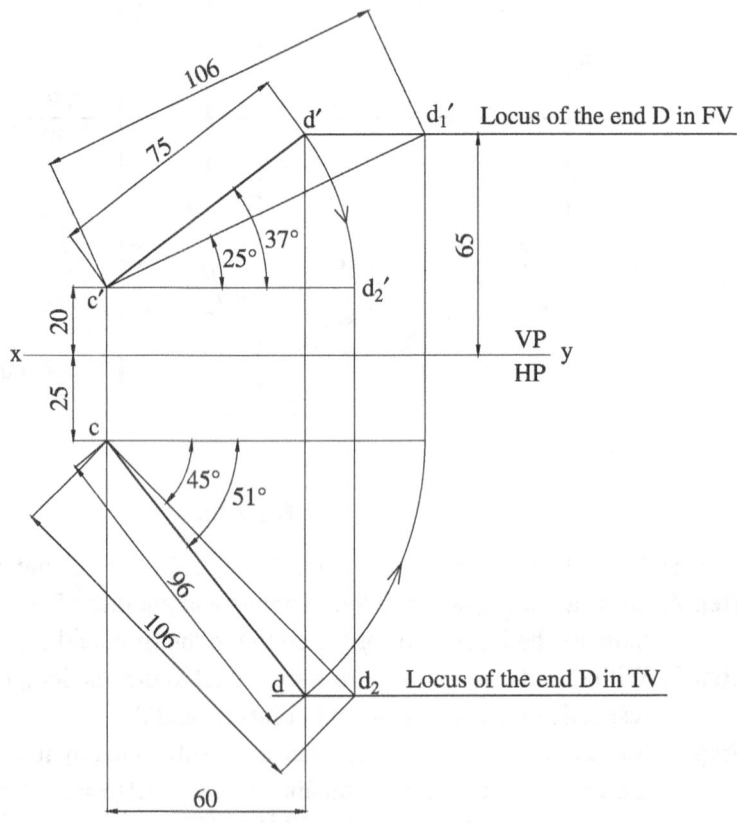

Fig. 7.26

Step 1: Draw the reference line and mark the projections c' and c of the end point 'C'.
Step 2: Draw a horizontal line at a distance of 65 mm above 'xy' line to mark the locus of the end D in front view.

Step 3: Set the projector of end D at a distance of 60 mm along the reference line and mark d′ on the locus line in the front view. Make c′d′ parallel to the reference line by drawing an arc and draw a projector through d_2'.

Step 4: Set the VP angle ($\phi = 45°$) at 'c' and obtain 'd_2' on the projector drawn.

Step 5: The line cd_2 gives the true length of the line. Draw a locus line through d_2 and obtain the point d on the projector line of d′.

Step 6: Draw the parallel projection cd_1 by drawing an arc from 'cd'. Project d_1 to the front view locus line and obtain d_1' and measure the included HP angle ($\theta = 25°$).

Step 7: Measure the apparent angles of inclination with HP and VP as 37° and 51°.

Example 7.12

The left profile view of a line CD 70 mm long makes 30° to 'xy' line and measures 55 mm. The end C is 20 mm above HP and 25 mm in front of VP and is nearer to it. Draw the projections of the line.

Given data   Position of one end, length and angle of left-side view, and length of the line

PROCEDURE BY ROTATING LINE METHOD   (Refer Fig. 7.27)

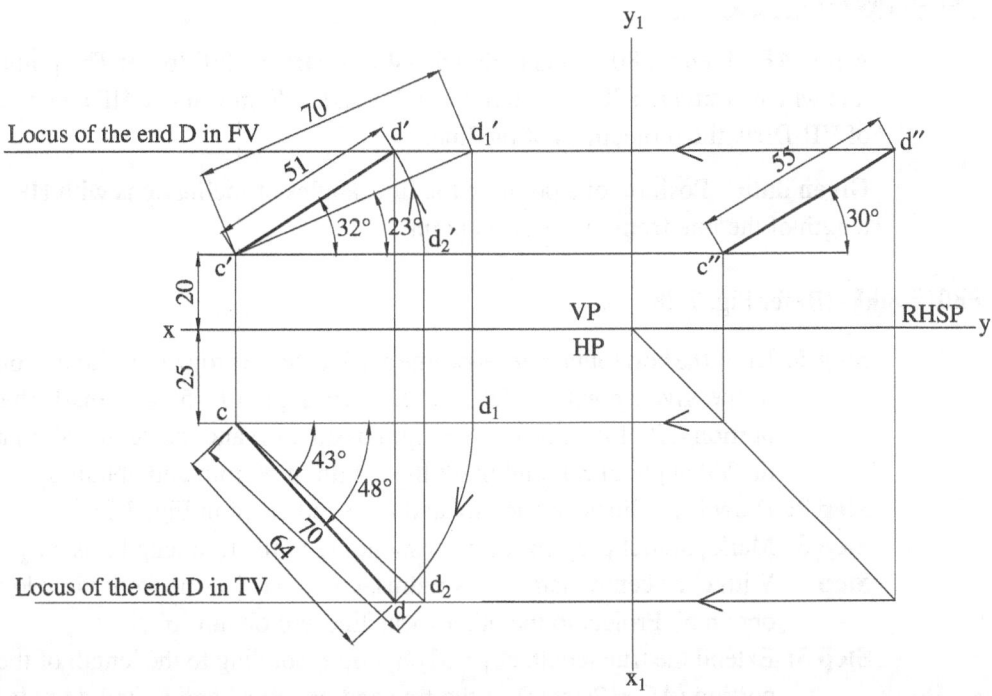

Fig. 7.27

Step 1: Draw the reference line and mark the projections c' and c of the end point 'C' of the line.

Step 2: Draw at any convenient distance a vertical reference line x_1y_1 on the right side to separate VP and RHSP. Draw the side view c'' of the end C and set 30° and obtain d'' for a length of 55 mm.

Step 3: Extend d'' to get the locus lines corresponding to the end D in the front view and top view domains as indicated in the Fig. 7.27.

Step 4: With c' and c as centres and lengths equal to the true length of the line, draw arcs to meet the respective locus lines and get d_1' and d_2. The angles made by $c'd_1'$ and cd_2 give the inclination of the line with HP and VP ($\theta = 23°$ and $\phi = 43°$).

Step 5: Draw parallel projections corresponding to the true lengths and obtain d' and d on the locus lines as indicated in Fig. 7.27.

Step 6: Join $c'd'$ and 'cd' to complete the projections and measure the apparent angles of inclination with HP and VP as 32° and 48°. ▲

7.6 SPECIFIC CASES OF STRAIGHT LINES

7.6.1 Line with Any Point as Reference Instead of the End Points

Example 7.13 ▲
_____

A line AB of length 80 mm is inclined at 45° to HP and 30° to VP. The point C is on the line which is situated 20 mm from the end A and is 30 mm above HP and 40 mm in front of VP. Draw the projections of the line.

Given data Position of a point on the line, angles of inclinations with HP and VP, and length of the line from the referred point

PROCEDURE (Refer Fig. 7.28)

Step 1: Draw the horizontal reference line and locate the projection c' and c corresponding to the given point C. Set at c' the HP angle of 45° and mark the length the portion CB of the line (i.e. a length of 60 mm) and locate b_1'. Similarly, at c, set the VP angle of 30° and mark the length of 60 mm and obtain b_2.

Step 2: Draw locus lines through b_1' and b_2 as indicated in Fig. 7.28.

Step 3: Mark parallel projections corresponding to the true lengths as $c'b_2'$ and cb_1.

Step 4: With c' as centre and cb_2' as radius, cut an arc on the front view locus line and obtain b'. Project to the other locus line and obtain 'b'.

Step 5: Extend the true length cb_1' and cb_2 corresponding to the length of the remaining portion (AC = 20 mm) of the line and obtain a_1' and a_2 and draw locus lines as shown in Fig. 7.28.

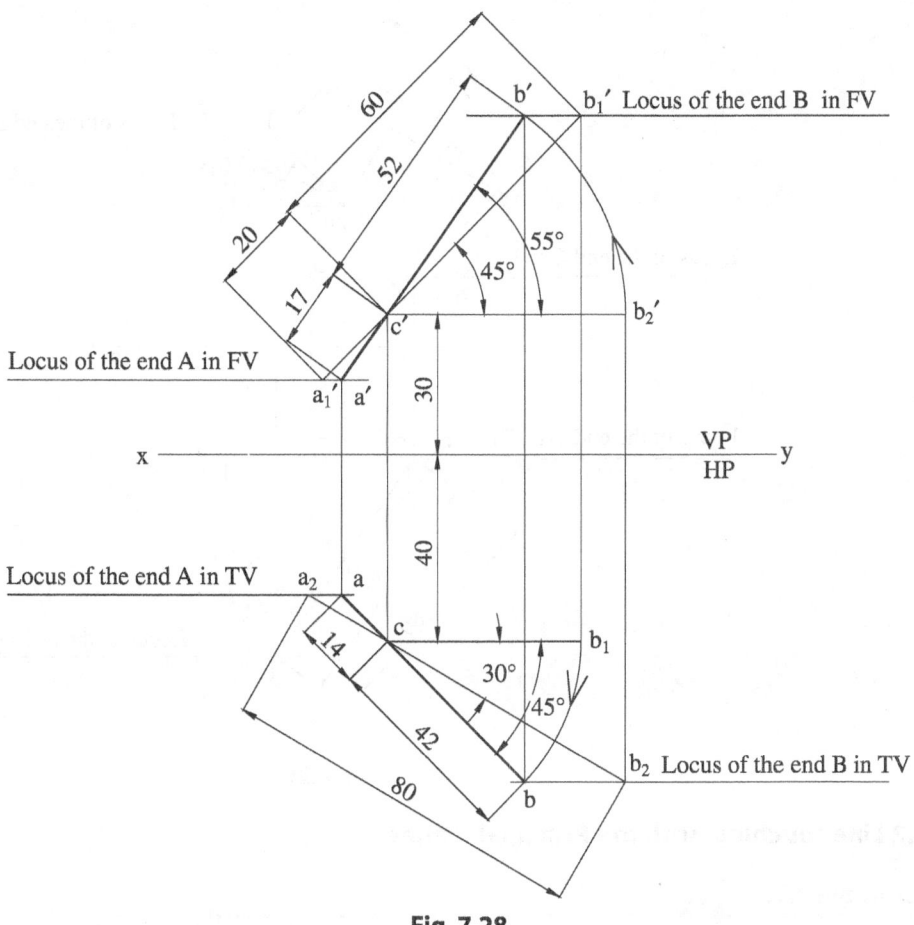

Fig. 7.28

Step 6: Join c′b′ and cb and extend them to meet the corresponding locus lines at a′₁ and a₂ respectively and obtain a′and a that lie on the same projector.

Step 7: The lines a′b′ and ab are the projections of the line and measure their angles as 55° and 45° in the front and top views, respectively. ▲

Example 7.14

A line CD of length 80 mm is inclined at 30° to HP and 45° to VP. The midpoint of the line 'M' is 40 mm above HP and 60 mm in front of VP. Draw the projections of the line.

Given data   Position of a point (midpoint) on the line, angles of inclinations with HP and VP, and length of the line from the referred point

PROCEDURE   (Refer Fig. 7.29)

Step 1: Similar to the problem discussed earlier, except that the portions of the true lengths extended or equal on either side, since M is the midpoint of the line. ▲

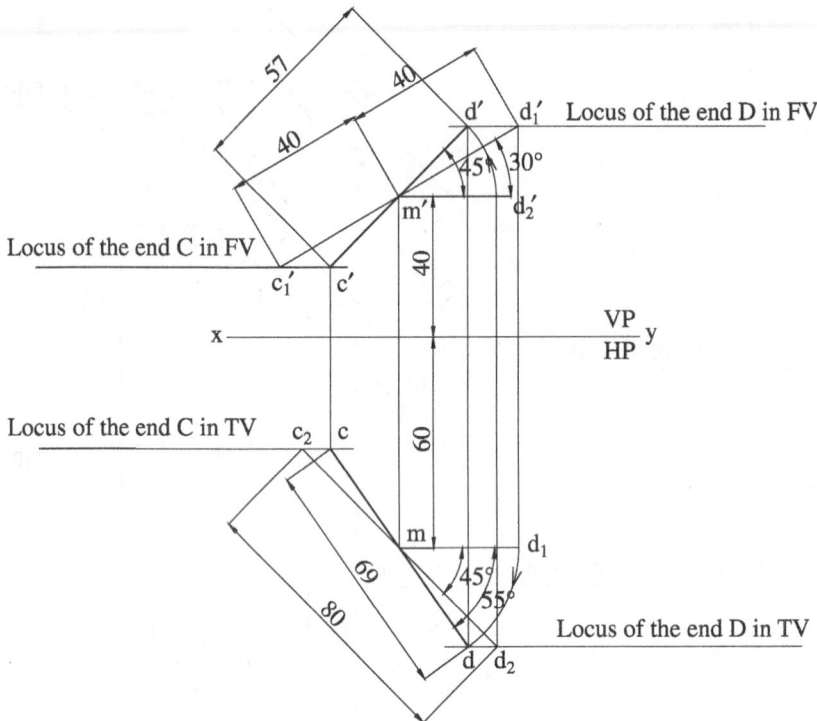

Fig. 7.29

7.6.2 Line Touching Both the Principal Planes

Example 7.15

A line AB of length 80 mm is inclined at 30° VP and 60° to HP. The end A is on HP and end B is on VP. Draw the projections of the line.

Given data Position of one end on HP, position of other end on VP, angles of inclinations with HP and VP, and length of the line

PROCEDURE (Refer Fig. 7.30)

> **NOTE** *Since one end of the line is in HP and the other end is in VP (instead the same end referred from both planes as in problems discussed earlier), the general procedure mentioned in Section 7.5 is followed, but the stages indicated there in are done separately from first principles and later combined.*

Drawing the Projections of the Line Inclined to HP and Parallel to VP (Stage 1)

Draw the horizontal reference line and mark a' on it and set the true length (80 mm) and the HP inclination (60°) and obtain $b_1{}'$. Its parallel projection indicated by ab_1 measures 40 mm along the reference line, but this cannot be located in the top view, as the distance of end A from VP is not given.

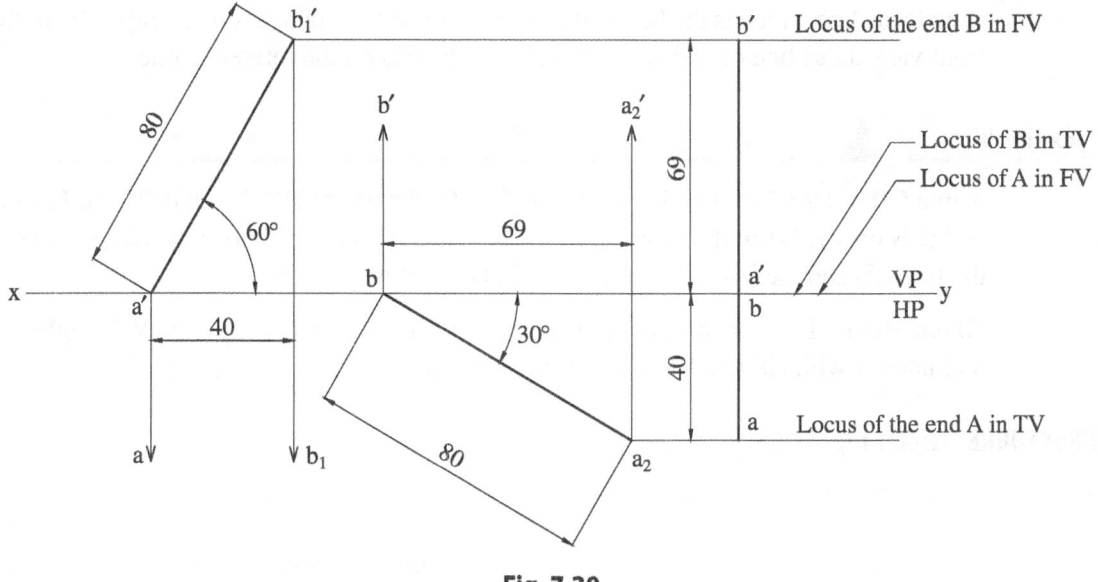

Fig. 7.30

Drawing the Projections of the Line Inclined to VP and Parallel to HP (Stage 2)

Similarly, mark the other end B on VP as b is on the reference line at any convenient distance and set the true length (80 mm) and the VP inclination (30°) and obtain a_2. Its parallel projection indicated by $b'a_2'$ measures 69 mm along the reference line, but this cannot be located in the front view, as the distance of end A from HP is not given.

Drawing Locus Lines at the Ends of the True Lengths

It can be noted that the locus lines are to be drawn parallel to the reference lines at both the ends, unlike the previous problems where one end is referred.

Since a′ and b are on the reference lines, the reference line itself indicates their locus lines. Draw locus lines from b_1' and a_2 and name them as indicated.

Combining the Projections of the Individual Stages and Obtaining the Final Views (Stage 3)

This can be done at a reasonable spacing from the previous stage. Since end A is on HP, mark a′ arbitrarily on the reference line. With a′ as centre and the front view parallel projection as radius (69 mm corresponding to $b'a_2'$ in the second stage), draw an arc to cut the front view locus line of end B to get b′. Project b′ to the top view locus line of the end B, which is the reference line itself and obtain b on it. Project a′ to the top view locus line of the end A, which is below, and obtain a on it. As a check a′ and a, b′ and b should lie on the respective projectors.

Note that instead of marking a′ and following the earlier mentioned procedure, 'b' can be marked on the reference line (as the end B is on VP) and the top view parallel projection corresponding to ab_1 (40 mm length in the first stage) can be cut on the top view locus line which is the below the reference line to obtain 'a' on it. Project 'a' to the front view locus line of end A and obtain a', which is on the reference line. ▲

Example 7.16 ▲

A line CD of length 80 mm is inclined at 30° VP and 45° to HP. The end C is on HP and end D is on VP. Draw the projections of the line and find the shortest distance between the line CD and the line of intersection of planes of projection.

Given data Position of one end on HP, position of other end on VP, angles of inclinations with HP and VP, and length of the line

PROCEDURE (Refer Fig. 7.31)

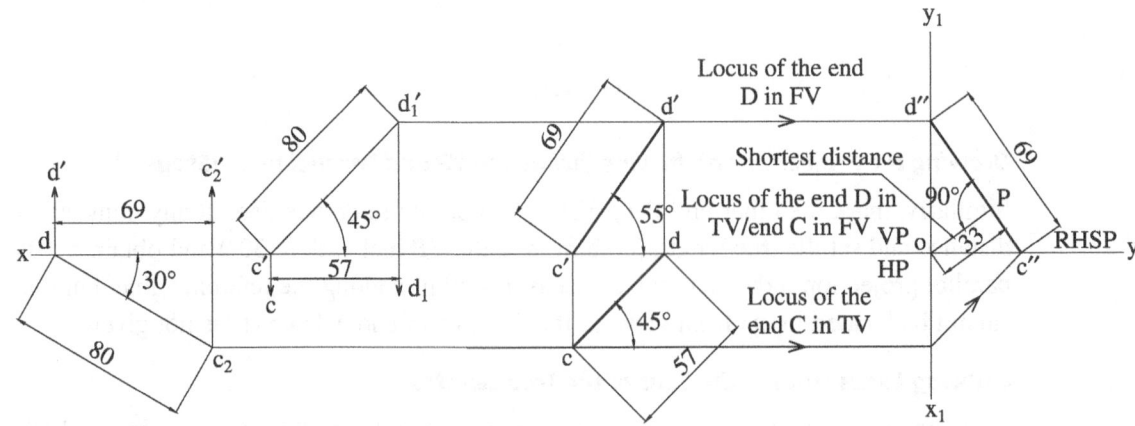

Fig. 7.31

Step 1: The projections are obtained in similar lines as discussed in the previous example. Since the sum of angle of inclinations with HP and VP is not 90°, the projections do not lie on a single vertical line as in previous problem.

Step 2: After obtaining the projections c'd' and cd, a side view c"d" of the line is drawn as shown in Fig. 7.31.

Step 3: The shortest distance between the line and the point of intersection of the planes of projection is obtained by drawing a perpendicular OP to c"d", from O which is the point of intersection of VP and HP in the side view. ▲

7.6.3 Line Parallel to the Profile Planes

Example 7.17 ▲

A line AB of length 80 mm parallel to the profile plane is inclined at 40° VP. The end A is 100 mm above HP and 30 mm in front of VP. Draw the projections.

Given data   Position of one end from VP and HP, angle of inclination with VP, and length of the line

Line parallel to profile plane

> **NOTE**   *Since the line is parallel to PP, it is inclined to VP and HP and the sum of the angles is equal to 90°. Therefore, the inclination of line with HP is 50°.*
>
> *Since the line parallel to PP, the side view gives the true length and the VP and HP angles.*

PROCEDURE   (Refer Fig. 7.32)

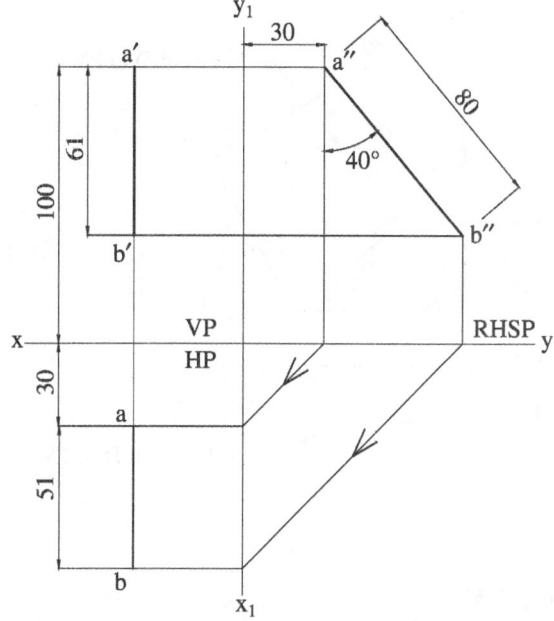

Fig. 7.32

Step 1: Draw the horizontal reference line 'xy' and mark a' and a 100 mm above 30 mm below it, respectively.

Step 2: Draw a vertical reference line x_1y_1 on the right side arbitrarily. Mark right-hand side plane—RHSP. Obtain the side view of end A by extending 'a' to meet the vertical reference line and arranging the 45° line and a vertical projector. The front view when extended meets this projector to give a''.

Step 3: Since the line is parallel to the side plane, the side view gives the true length and the true angles. Therefore, at a'', set VP angle 40° and mark the true length and get b''.

Step 4: Draw the locus lines in the front and top view by drawing horizontal from a'' and b'' as shown in Fig. 7.32.

Step 5: Since the line is parallel to PP, the front and top view are perpendicular to the reference line and lie on the same projector. Obtain a'b' and 'ab' as shown in Fig. 7.32.   ▲

7.6.4 Lines with Ends in Different Quadrants

Example 7.18   ▲

A line CD of length 70 mm has its one end C 20 mm below HP and 25 mm behind VP. The other end D is in first quadrant and the line is inclined at 30° to VP and 45° to HP. Draw the projections of the line.

Given data   Position of one end is behind VP and below HP, angle of inclination with HP and VP, and length of the line. The other end is in first quadrant.

NOTE *Since the ends are in different quadrants, after positioning one end, the inclination angles are to be set in such way that the other end satisfies its requirement.*

PROCEDURE (Refer Fig. 7.33)

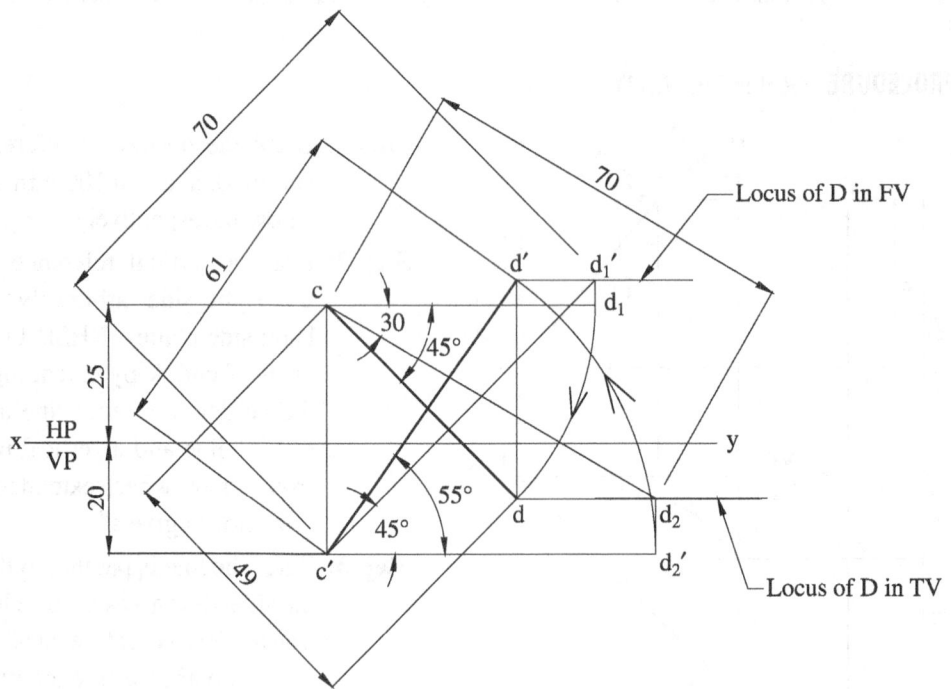

Fig. 7.33

Step 1: Draw the horizontal reference line 'xy' and mark c' and c at a distance 20 mm below and 25 mm above 'xy' line, respectively.

Step 2: Set at c', the HP angle ($\theta = 45°$) and mark the true length 70 mm to get d_1' on the other side of the reference line as end D is in the first quadrant and obtain its parallel projection cd_1. Similarly, set at c, the VP angle ($\phi = 30°$) and mark the true length 70 mm to get d_2 below the reference line as end D is in first quadrant and obtain its parallel projection $c'd_2'$.

Step 3: Draw locus lines of the floating end D, that is, through d_1' and d_2 in the front and top views.

Step 4: As cited in the general procedure, consider one parallel projection say $c'd_2'$. With c' as centre and $c'd_2'$ as radius, draw an arc to cut the front view locus line of D (which is lying above the reference line) and get d'.

Step 5: Project d' to meet the locus line in the top view and get 'd'.

Step 6: Join c'd', 'cd', and the projector line d'd and complete the projections.   ▲

Example 7.19

A line AB of length 70 mm has its one end A 25 mm above HP and 25 mm in front of VP. The other end B is in second quadrant and the line is inclined at 45° to HP and 30° to VP. Draw the projections of the line.

Given data   Position of one end is above HP and in front of VP, angle of inclination with VP and HP, length of the line. The other end is in second quadrant.

NOTE   *Since the ends are in different quadrants, after positioning one end, the inclination angles are to be set in such way that the other end satisfies its requirement.*

PROCEDURE   (Refer Fig. 7.34)

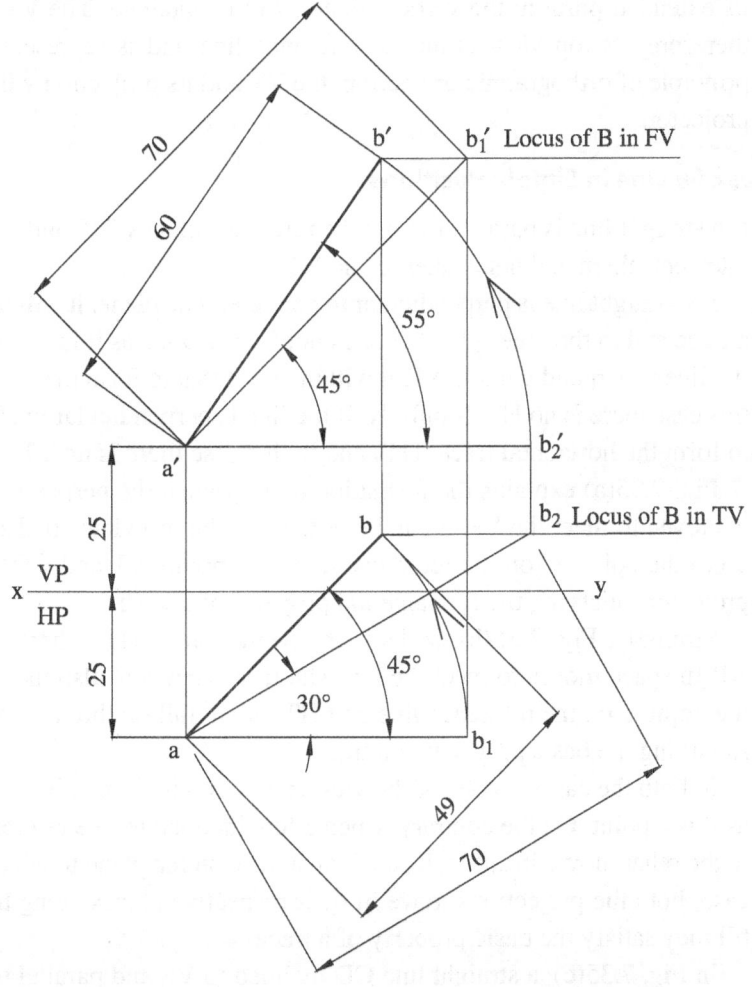

Fig. 7.34

Step 1: Locate a' and a as given. Set HP angle and VP angle to get b' and b above the reference line, since end B is in second quadrant for which both front view and top view are above the reference line. Other constructional procedures are in similar lines as that of the previous problem. ▲

7.7 TRACES OF A LINE

The trace of a line is a point at which the line intersects a reference plane or meets it imaginarily when extended from either of its ends.

When the line or its extension from one of the ends intersects the HP or its extended portion, the horizontal trace (HT) is obtained. The HT is a point on HP, and therefore, its front view is on the reference line and is represented as h'. As per the principle of orthographic projection, the HT and its projection h' lie on the same vertical projector.

Similarly, when the line or its extension from one of the ends intersects the VP or its extended portion, the vertical trace (VT) is obtained. The VT is a point on VP, and therefore, its top view is on the reference line and is represented as 'v'. As per the principle of orthographic projection, the VT and its projection v lie on the same vertical projector.

7.7.1 Traces of a Line in Simple Positions

If a straight line is parallel to both the reference planes (HP and VP), then the line never intersects them and hence there is no HT or VT.

If a straight line is perpendicular to one reference plane, it will meet that plane to form a trace and in the other plane, it does not form a trace as it is parallel to it. For example, if a line is perpendicular to VP, it will intersect that to form the vertical trace (VT) and in this case there is no HT. Similarly, if the line is perpendicular to HP, it will intersect that to form the horizontal trace (HT) and in this case there is no VT.

Fig. 7.35(a) explains the formation of VT, when the perpendicular line AB in space is moved to meet the VP. It can be noted that the top view ab also moves along the HP up to the point 'v' on the reference line. The points VT and v fall on the same vertical projector satisfying the fundamental property of a trace.

Similarly, Fig. 7.35(b) explains the formation of HT, where the perpendicular line AB in space moves to meet the HP. The front view a'b' also moves along the VP up to the point h' on the reference line and HT and h' fall on the same vertical projector line, satisfying the basic property of a trace.

In both the cases, as one of the views moves on its respective plane, the other doesn't, as it is a point. On the contrary, when a line inclined to one reference plane and parallel to the other moves in space, it can form a trace on the plane to which it is inclined. In this case, both the projections move in their respective planes along their own orientations, till they satisfy the basic property of a trace.

In Fig. 7.35(c), a straight line CD inclined to VP and parallel to HP is shown. While the line in space moves to meet the VP and forms the VT, the top view and the front view move in their respective planes in their own orientations and satisfy the fundamental property of a trace. In other words, in Fig. 7.35(c), when the top view of the line is extended to meet the reference line at v and a perpendicular is erected at v and when the front view is extended to meet this perpendicular, the VT results. This is explained in Fig. 7.36 (a) for a straight line of 65 mm length inclined at 30° to the VP.

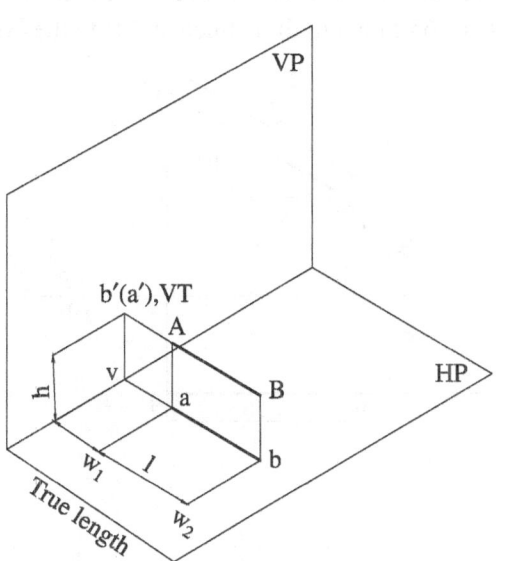

(a) Line parallel to HP and perpendicular to HP (No HT)

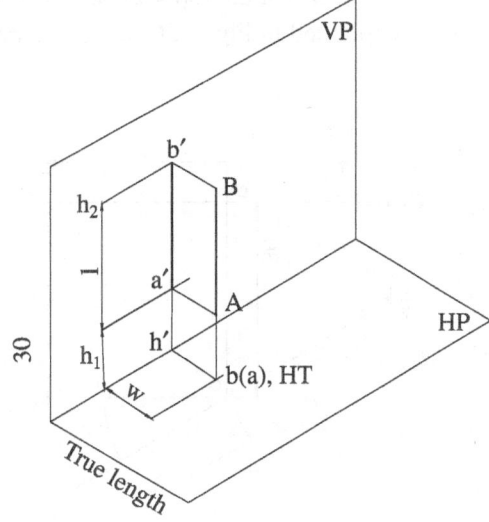

(b) Line parallel to VP and perpendicular to HP (No VT)

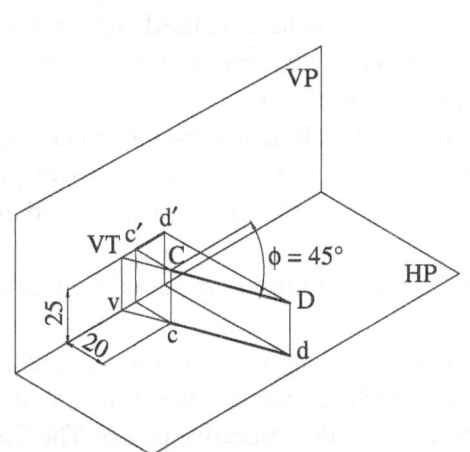

(c) Line Parallel to HP and inclined to VP (No HT)

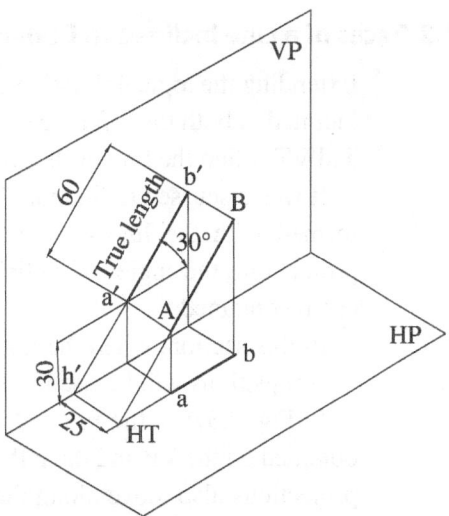

(d) Line parallel to VP and inclined to HP (No VT)

Fig. 7.35 Location of traces for the line perpendicular/inclined to one plane and parallel to another plane

Similarly, in Fig. 7.35(d), a straight line AB inclined to HP and parallel to VP is shown. While the line in space moves to meet the HP and forms the HT, the top view and the front view move in their respective planes in their own orientations and satisfy the fundamental property of a trace. In other words, in Fig. 7.35(d), when the front view of the line is extended to meet the reference line at h' and a perpendicular is erected at h' and when the top view is extended to meet this perpendicular, the HT results. This is explained in Fig. 7.36 (b) for a straight line of 65 mm length inclined at 30° to the HP.

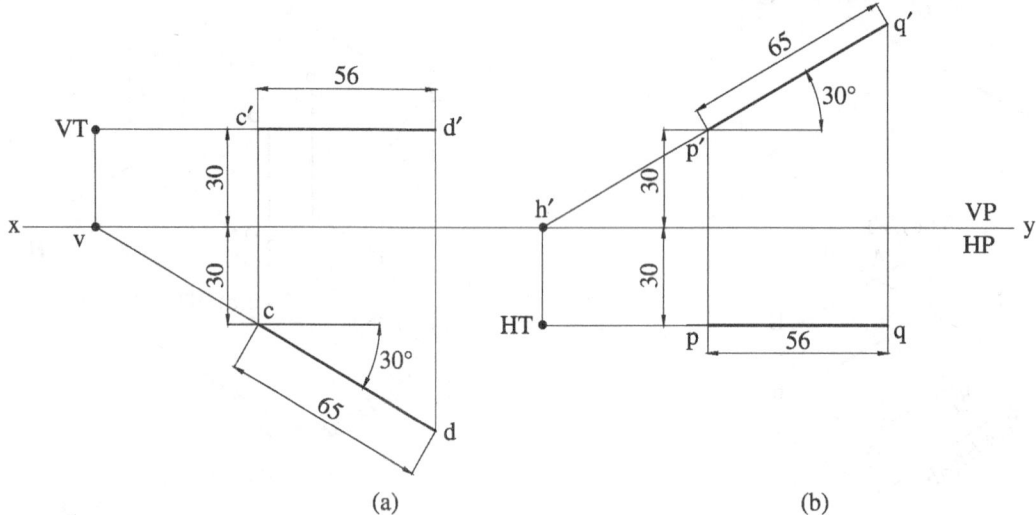

(a)   (b)

Fig. 7.36 Location of traces for the line inclined to one plane and parallel to another

7.7.2 Traces of a Line Inclined to Both the Reference Planes

Extending the logical thinking of Section 7.7.1, it can be concluded that if a straight line inclined to both the reference planes (HP and VP), will form two traces, namely, the HT and VT, when the line or its extension meets these planes.

It was discussed in Section 7.5.5 and in Fig. 7.23(b) that when the spatial trapeziums formed with the line and its projections are aligned with the respective planes of projections, the angles of inclinations of the line with the reference planes and the traces can be obtained.

In this section, when the line in space is moved to form the traces, the movement of the projections will be used to locate them.

In Fig. 7.37(a), the line PQ in space is shown along with its projections p'q' and pq obtained on the VP and the HP. As discussed earlier, when the line in space moves, these projections also move along their orientations on the respective planes. The front view p'q' moves on the VP up to h' on the reference line and the top view moves up to HT, where the line meets the HP. The trace (HT) and its projection on the reference line (h') lie on the same projector and this satisfies the fundamental property of a trace. Similar observations can be made while getting the VT and its projection, v.

Figure 7.37(b) shows the aforementioned movement scheme in the orthographic layout arrangement, when the HP is rotated and brought in alignment with the VP, as discussed in Chapter 5. As discussed in Section 7.7.1, it can be observed from Fig. 7.37(b) that when the front view of the line is extended to meet the reference line at h′, a perpendicular is erected at h′, and when the top view is extended to meet this perpendicular, the HT can be obtained.

Similarly, when the top view of the line is extended to meet the reference line at v, a perpendicular is erected at v, and when the front view is extended to meet this perpendicular, the VT results.

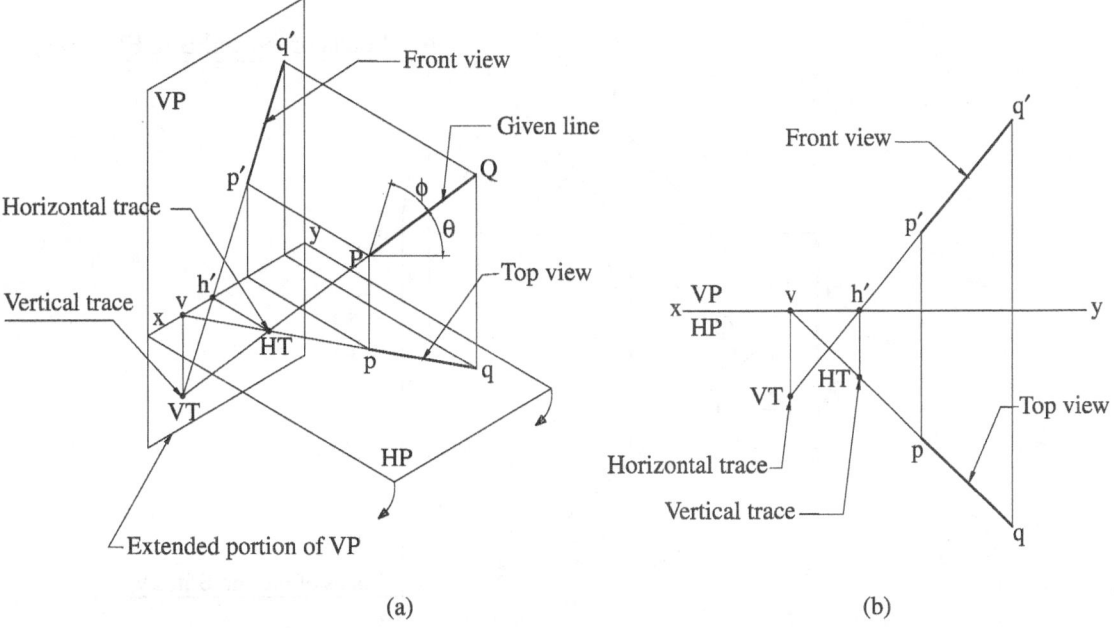

(a)   (b)

Fig. 7.37   Location of traces for the line inclined to both planes

It can also be observed from Fig. 7.37(b) that while the VT, h′ (the front view of HT), and the front view of the straight line lie on one single line, the HT, v (the top view of VT), and the top view of the straight line also lie together on an another single line.

All these observations and the general principles outlined in the trapezium method can be used in combination to solve problems on straight lines involving traces. A few problems are discussed in the forthcoming examples.

Example 7.20

A line AB of length 65 mm is inclined at 40° to HP and 30° to VP. The end A is 25 mm above HP and 30 mm in front of VP. Draw the projections of the line and locate its traces.

Given data   Position of one end, angles of inclination with VP and HP, and length of the line

PROCEDURE   (Refer Fig. 7.38)

Step 1: The reader is advised to refer Section 7.5.1 and Example 7.6 for drawing the projections of the line.

Step 2: To locate the HT, extend the front view a'b' to meet the 'xy' line at h' and erect a perpendicular and allow the top view to meet this perpendicular.

Step 3: Similarly, to locate the VT, extend the top view 'ab' to meet the 'xy' line at 'v' and erect a perpendicular and allow the front view to meet this perpendicular.

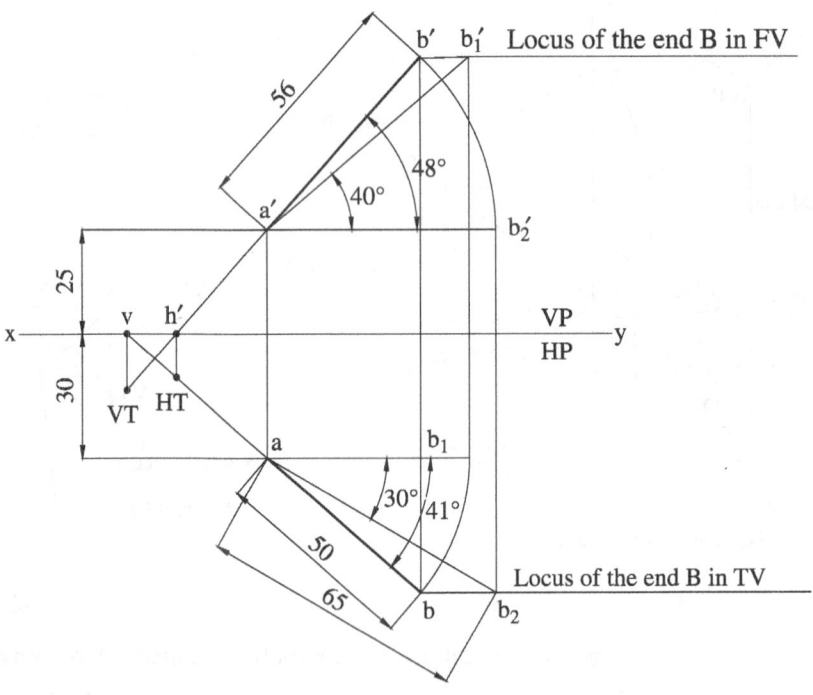

Fig. 7.38

Example 7.21

A line CD of length 65 mm is inclined at 45° to HP and 30° to VP. The end D is 50 mm above HP and 45 mm in front of VP. Draw the projections of the line and locate its traces.

Given data   Position of one end, angles of inclination with VP and HP, and length of the line

PROCEDURE   (Refer Fig. 7.39)

Step 1: The reader is advised to refer Section 7.6.1 and Example 7.6 for drawing the projections of the line.

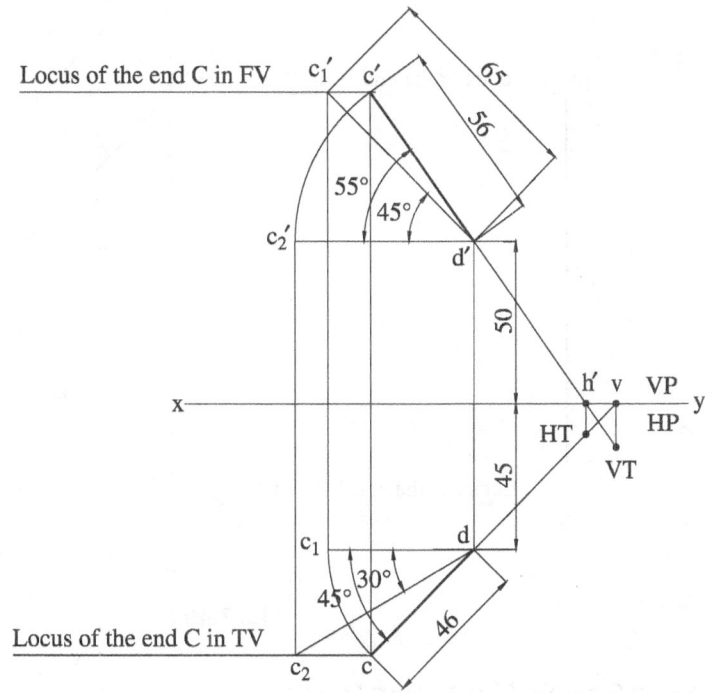

Fig. 7.39

Step 2: To locate the HT, extend the front view c′d′ to meet the 'xy' line at h′ and erect a perpendicular and allow the top view to meet this perpendicular.

Step 3: Similarly, to locate the VT, extend the top view 'cd' to meet the 'xy' line at 'v' and erect a perpendicular and allow the front view to meet this perpendicular.

Example 7.22

A line AB of length 65 mm is inclined at 45° to HP and parallel to the profile planes.
    The end A is on HP and end B is on VP. Draw the projections of the line and locate its traces.

Given data   Position of one end on HP, another end on VP, angles of inclination with VP and HP, and length of the line

PROCEDURE (Refer Fig. 7.40)

Step 1: The reader is advised to refer Section 7.6.1 and Example 7.17 for drawing the projections of the line.

Step 2: It can be observed from the side view that a″ also represents the HT as end A is on HP. Similarly, b″ also represents VT as end B is on VP. They are also marked in the front and top views for reference.

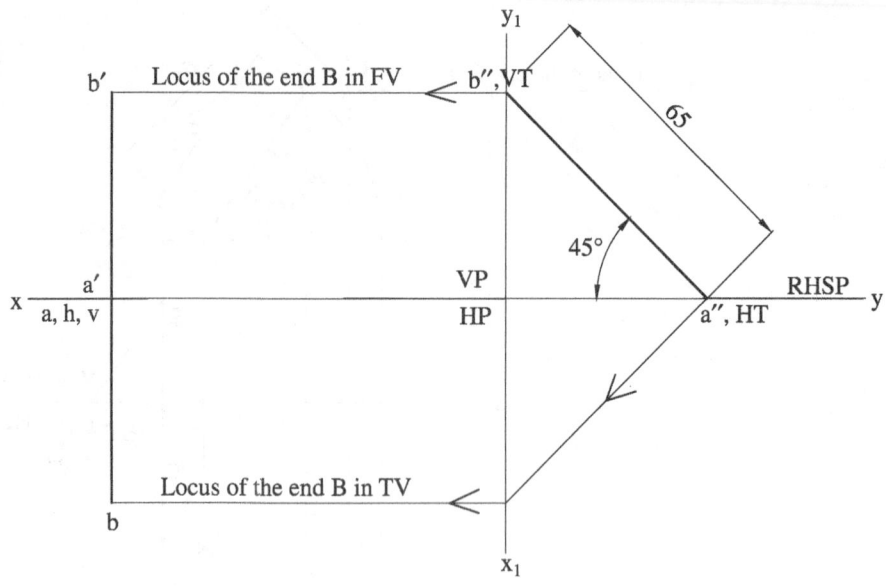

Fig. 7.40

7.7.3 Problems on Straight Lines using Traces

Example 7.23

The front view of the line PQ is measured as 65 mm and is inclined at 45° to 'xy' line, the end P being 20 mm above HP and 20 mm in front of VP. The VT is 20 mm below HP. Draw the projections of the line and find its true length and the inclinations with the reference planes. Also locate HT.

Given data   Position of one end, position of HT, front view length, and its inclination

PROCEDURE   (Refer Fig. 7.41)

Step 1: Draw the reference line and mark the mark the projections of the given end P.

Step 2: Set at p′ the front view p′q′ at an angle of 45° for a length of 65 mm.

Step 3: Draw the locus of VT at a distance 20 mm below the 'xy' line. Since VT, h′, and front view should lie on same line, extend q′p′ to cut the locus line of VT and obtain VT and v. The intersection of the front view with the reference line gives h′ and drop a projector through it.

Step 4: Since v, HT, and the top view of the line should lie together, join v and p and when the projector through h′ intersects this, HT is located.

Step 5: Draw a vertical projector through the point q′ to meet the extended line joining the points HT, v, and p at point q.

Step 6: Draw locus lines at q′ and q.

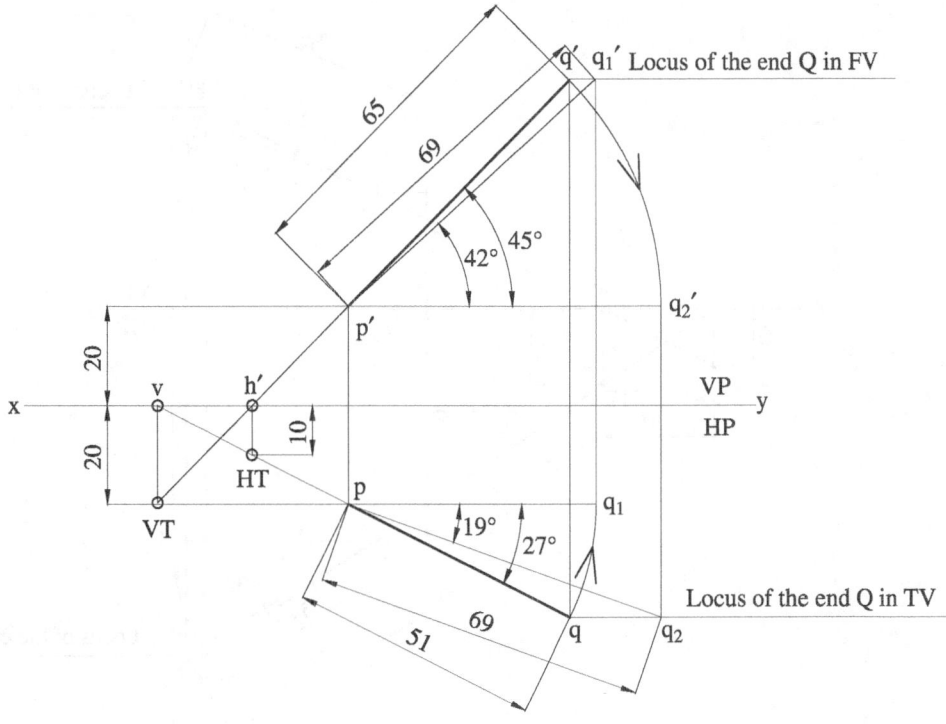

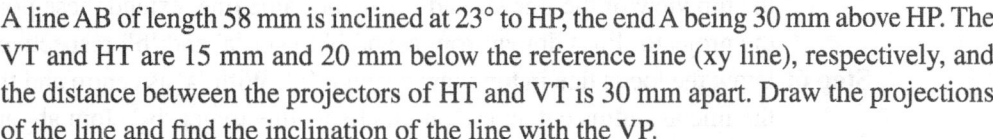

Fig. 7.41

Step 7: The projections p'q' that are made parallel and projected on to the locus line in the top view give q₂. Join pq₂ and measure its length and the angle subtended to get the true length of the line and the VP angle as 69 mm and 19°, respectively.

Step 8: Repeat similar operation by making 'pq' parallel and obtain the true length and the HP inclination as indicated by p'q₁'. The inclination with HP is found to be 42°. ▲

Example 7.24 ▲

A line AB of length 58 mm is inclined at 23° to HP, the end A being 30 mm above HP. The VT and HT are 15 mm and 20 mm below the reference line (xy line), respectively, and the distance between the projectors of HT and VT is 30 mm apart. Draw the projections of the line and find the inclination of the line with the VP.

Given data Position of one end from HP, position of traces, projector distance of traces, angle of inclination with HP, and length of the line

PROCEDURE (Refer Fig. 7.42)

Step 1: Draw the reference line and mark VT 15 mm below it and locate its projection v (top view of the VT) on 'xy' line.

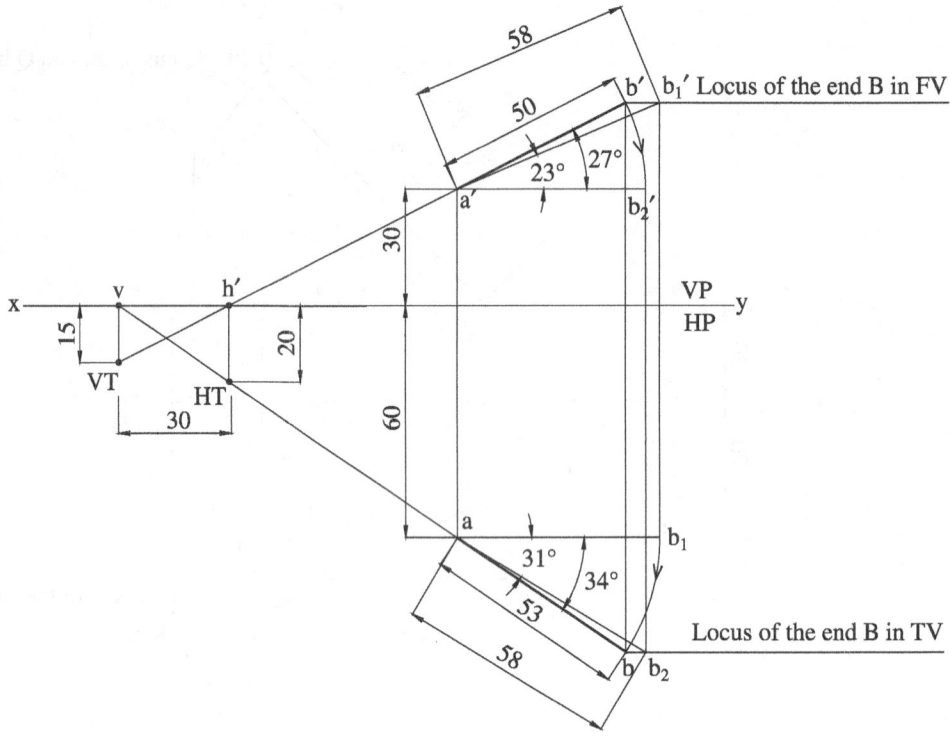

Fig. 7.42

Step 2: Draw the projector of HT at a distance of 30 mm from v along the reference line and locate HT at 20 mm below and h' on the reference line.

Step 3: Join VT and h' and extend the line and locate a' on this line at a distance of 30 mm above the 'xy' line. Set at a', the HP angle 23° and mark the true length of the line to get b₁' and draw the locus line in the front view.

Step 4: Since VT, h', and the front view of the line should lie on the same line, extend these points to meet locus line and get b'.

Step 5: Draw vertical projector lines through the points a' and b'. Since the HT, v, and the top view of the line should lie on the same line, extend these points to meet the projector lines drawn from a' and b' to get 'a' and 'b', respectively.

Step 6: Draw the locus line in top view through 'b'. With 'a' as centre and the length of the line as radius cut an arc on this locus line to get 'b₂'. Join ab₂ and measure the included angle with the VP which is equal to 31°.

Example 7.25

The line joining HT and v is 10 mm long and HT is in front of VP. The top view of the line PQ measures 65 mm and is inclined at 45° with 'xy' line and the end P is 20 mm above HP and 25 mm in front of VP. Draw the projections of the line and find its true length and true angles of inclinations.

Given data   Position of one end, position of HT, top view length, and its inclination

PROCEDURE (Refer Fig. 7.43)

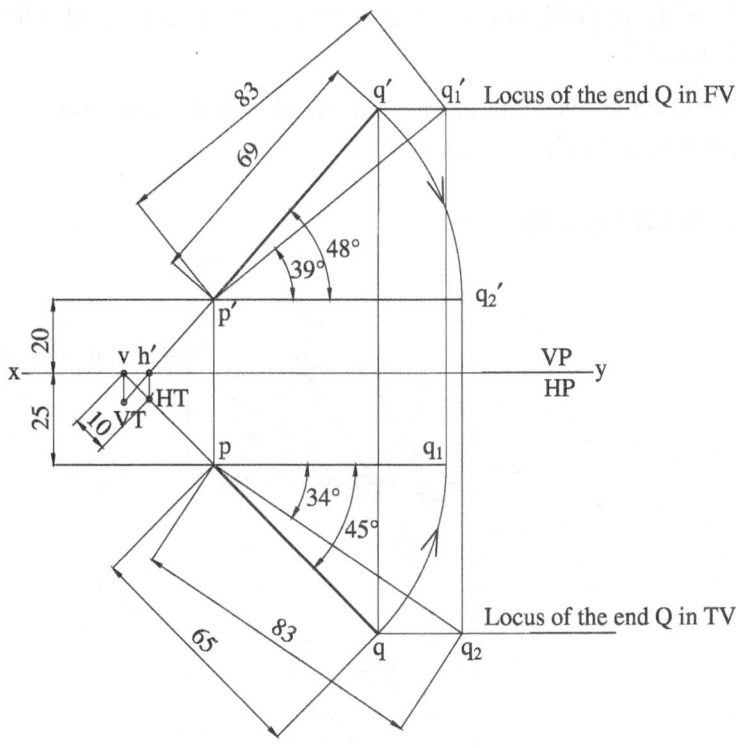

Fig. 7.43

Step 1: Draw the reference line 'xy' and mark the projections of the given end point P.

Step 2: Set at 'p' the top view 'pq' at an angle of 45° for a length of 65 mm.

Step 3: Extend the top view 'pq' to meet the 'xy' line and mark the intersecting point 'v'. Since the HT, v, and the top view of the line should lie together HT can be marked on the line joining v, p, and q at a distance of 10 mm from v. Project HT to get h'.

Step 4: Join h'p' and extend on both sides. Since VT, h', and front view of the line should lie together, the projectors from v and q, respectively, yield VT and q'.

Step 5: Draw locus lines through q' and q.

Step 6: By making the top view of the line 'pq' parallel to the reference line and drawing a projector to meet the locus line in front view, q_1' is obtained. Join $p'q_1'$ and measure its length and the subtended HP angle as 69 mm and 39°, respectively.

Step 7: By making the front view of the line p'q' parallel to the reference line and drawing a projector to meet the locus line in top view, q_2 is obtained. Join pq_2 and measure its length and the subtended VP angle as 69 mm and 34°, respectively.

Example 7.26

A line AB of length 65 mm is inclined at 30° to HP, the end A being 30 mm above HP and 25 mm in front of VP. The line joining the front view of the end A and h′ is 35 mm. Draw the projections of the line and locate the traces and also find the inclination of the line with VP.

Given data   Position of one end, position of the projection of HT, and true length of the line, and its inclination with HP

PROCEDURE (Refer Fig. 7.44)

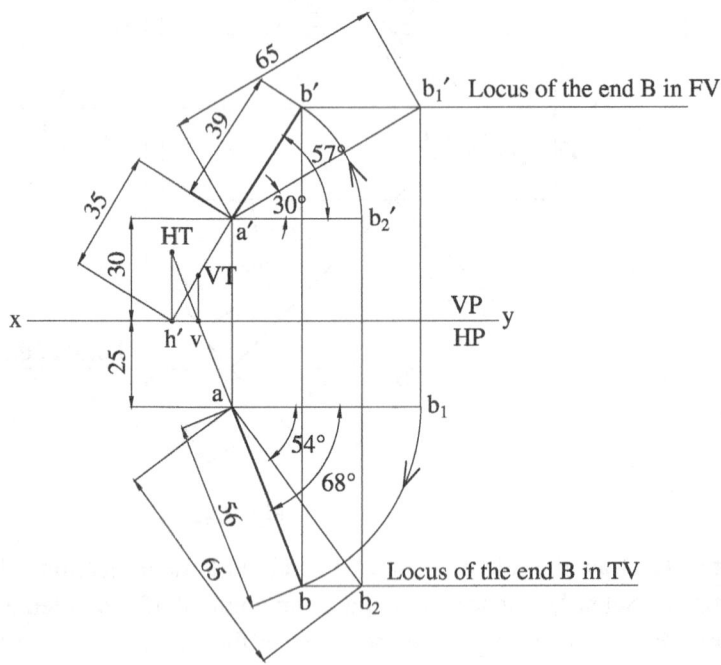

Fig. 7.44

Step 1: Draw the horizontal reference line and mark the projections of the given end A.

Step 2: Set at a′ the true length of 65 mm at the HP angle of 30° and obtain b₁′ and draw the locus line.

Step 3: Since the distance of a′ and h′ is given as 35 mm, and h′ has to be on the reference line, draw an arc from a′ for a length of 35 mm to cut the 'xy' line and get h′.

Step 4: Join h′ and a′ and extend it to meet the locus line in the front view and obtain b′ and draw the projector line.

Step 5: Obtain the parallel projection ab₁ corresponding to the true length a′b₁′. With 'a' as centre and 'ab₁' as radius, draw an arc to meet the projector line through b′ and obtain b.

Step 6: Extend the top view 'ab' to meet the perpendicular at h' and get HT. When this line meets the reference line, 'v' is obtained.

Step 7: Erect a perpendicular at 'v' and when a'b' intersects it, VT is obtained.

Step 8: Since the projections a'b' and ab are known the inclination of the line with VP can be found as discussed in the previous problems and the VP angle is found to be 54°.

7.8 ENGINEERING APPLICATIONS

Simple cases of the projection of lines, when the line parallel to one plane and perpendicular or inclined to another plane and the projections of the line with reference to profile planes in additions to the principal planes, were discussed in earlier sections. It was also discussed about the projection of the line when the line is inclined to both planes and locating the traces. In this section, problems related to some of engineering applications that can be related to the projection of line when the line is inclined to both planes are discussed which can help the learner to understand the importance of this topic with reference to real problems.

Example 7.27

A rope that forms a straight line AB has its one end A tied to a hook on the ground and 25 mm in front of a vertical wall. The other end B is tied to a pole which is 65 mm in front of vertical wall and the rope is inclined at 30° to both the vertical wall and the ground. Draw the projections of the line joining the rope and find the length of the rope.

Given data Position of one end, position of other end with respect to VP, and angles of inclination with VP and HP

PROCEDURE (Refer Fig. 7.45)

Step 1: Draw the horizontal reference line and mark a' and a, which represent the projections of the end A attached to the hook on the ground.

Step 2: Draw a locus line 65 mm below the reference line to represent the end B of the rope with respect to the vertical wall.

Step 3: Set at a, the VP angle of 30° to meet the locus line of B in the top view and obtain b_2. Join ab_2 and measure the true length of the rope AB (80 mm).

Step 4: Set at a', the HP angle of 30° and mark the true length obtained above to get b_1' and draw a locus line through that.

Step 5: Obtain the parallel projection of any true length say $a'b_1'$ as ab_1 and cut an arc on the locus line in the top view and obtain b. Project b to get b' on the front view locus line.

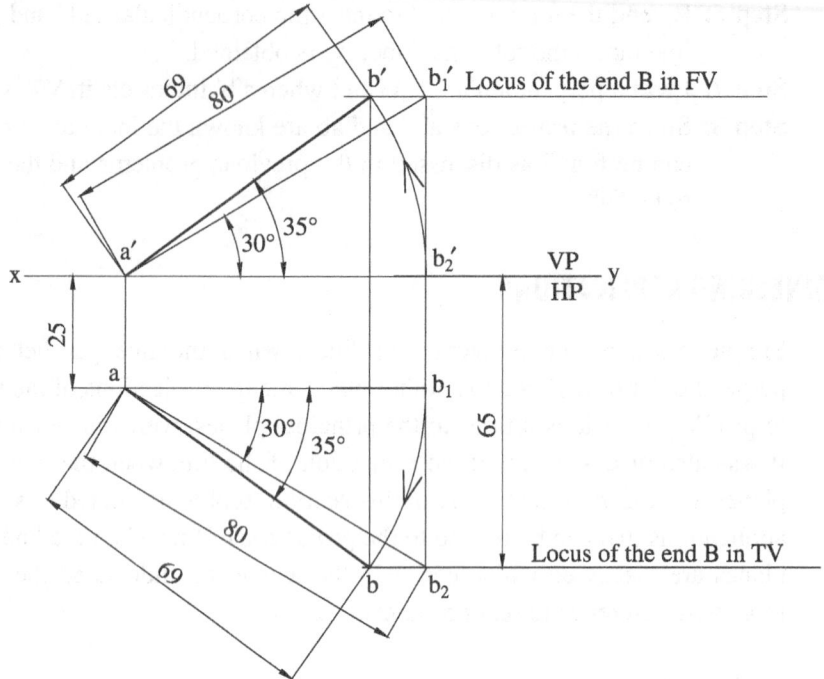

Fig. 7.45

Step 6: a′b′ and ab give the projections of the rope AB.

Example 7.28

A flag mast is held in vertical position by means of three ropes OP, OQ, and OR all tied to the mast at point O, 4 m above the ground. The ends P, Q, R all lying on the ground make 120° with each other, in the top view. OP is vertical and has the length of 2 m, OQ = 3.5 m and OR = 5 m, all representing the distances in the top view. Draw the projections of the ropes and the mast and find the length of each rope and their inclinations with the ground. Neglect the diameter of mast.

Given data   Projections of three ropes (OP, OQ, and OR)

PROCEDURE   (Refer Fig. 7.46)

Step 1: Draw the horizontal reference line 'xy' and mark vertical line to indicate the flag mast and mark a point o′ at a height of 4 m.

Step 2: Obtain the projection of o′ in the top view as o at a convenient distance and mark the three radial lines at angle 120° to each other to represent the top view of the three ropes tied to the mast.

Step 3: Mark on the radial line, the distances of 2 m, 3.5 m, and 5 m and obtain the points p, q, and r, respectively, to indicate the ropes ties to the ground. Project these points to the reference line and obtain p′, q′, and r′.

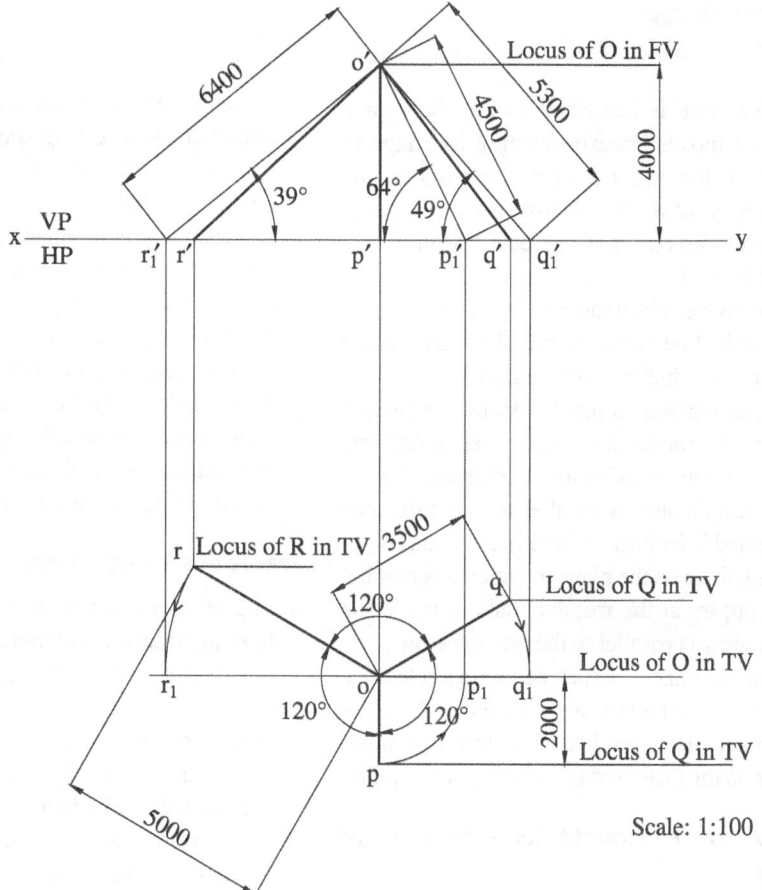

Fig. 7.46

Step 4: oq, oq' indicate the top and front views of the rope OQ. Its length and inclination can be found by method indicated in Section 7.5.

Step 5: Draw the horizontal locus lines through points q and q' for floating point Q. As per the general procedure, make line 'oq' parallel to 'xy' line and project it to the locus line of q' to get q_1'. Join o'q, which gives the true length (measured equal to 5.3 m). Measure the included angle (49°), which gives the inclination of the rope OQ with the ground.

Step 6: In similar lines, the projections op, or can be made parallel to the reference line and projected to meet the corresponding locus lines in the front view and the length of the ropes OP and OR and their inclinations with the ground can be obtained.

- A straight line is described as the locus of a point that moves linearly forming its shape in one dimension and becomes the basis of the evolution process for various two-dimensional and three-dimensional objects, which are derived from it.
- The following observations of the projection of a straight line in simple position serve as the basis for other inclined orientations:
  - If a straight line is parallel to both principal planes, its true length is visualized in both the views and are parallel to the reference line.
  - If a straight line is parallel to one reference plane and is inclined to the other, its true length is visualized on the plane to which it is parallel but appears at the true inclination, while the other view is parallel to the reference line.
  - If a straight line is perpendicular to a reference plane, it is automatically parallel to the other reference plane and hence the true length appears on the latter and the other view is a point.

Projection of a Straight Line in Inclined Positions

- When a straight line is inclined to both reference planes, none of the final views indicate the true length but appear shorter and make apparent inclinations that are more than the true inclinations.
- The projections are obtained by independently considering the two inclinations and superimposing the individual projections suitably. This method is known as rotating line method.
- The projections of the line are discussed when the line is touching both the principal planes, parallel to the profile plane and when its ends are in different quadrants.

Obtaining the Parameters of the Straight Line from their Projections

- When the projections are given, the parameters such as the true length of the line and angles

of inclinations with the reference planes can be obtained by reversing the procedure that was adopted to draw the projections. This is called reversal method.
- Alternatively, the imagination of the line in space, its projectors and the projections on the reference planes involves two trapeziums in space. By rotating these trapeziums about their respective projections on to the reference planes help to find the parameters of the line. This method is called trapezium method. This method also helps to find the traces of the line by extending the trapeziums to meet at triangles.

Traces of a Straight Line

- The trace of a line is a point at which the line intersects a reference plane or meets it imaginarily when extended from either of its ends.
- When the line or its extension intersects the horizontal plane (HP), the horizontal trace (HT) is obtained. A HT is necessarily a point on HP, and therefore, its front view is on the reference line and is represented as h'.
- Similarly, when the line or its extension intersects the vertical plane (VP), the vertical trace (VT) is obtained. A VT is necessarily a point on VP, and therefore, its top view is on the reference line and is represented as v.
- When the line is parallel to HP and VP, it never intersects these planes, and hence, there is no HT and VT.
- Since the traces are the points that lie on the views of a straight line or their extension, problems of straight lines involved with traces can also be solved in similar lines of other problems.
- The estimation of the length, orientation of the typical engineering applications such as the guy ropes tied to the mast, ropes connected between the walls, and the solid diagonal in a three-dimensional space can be solved by using the principles outlined earlier.

7.1 True length of the line can be seen in top view when the line is
(a) parallel to PP, VP, and perpendicular to HP
(b) inclined to HP and parallel to VP
(c) parallel to VP and perpendicular to HP
(d) inclined to VP and parallel to HP

7.2 True length of the line can be seen in front view when the line is
(a) inclined to both HP, VP, and parallel to profile plane
(b) inclined to HP and parallel to VP
(c) inclined to VP and parallel to HP
(d) equally inclined to HP and profile plane

7.3 Both FV and TV lies on the line and perpendicular to xy when
(a) one end touching the HP, inclined to it and parallel to VP
(b) one end touching the HP, another end touching the VP and line inclined to HP, VP, and PP
(c) line inclined to both HP and VP and parallel to profile plane
(d) line in the HP and inclined to VP

7.4 When the line is in the VP and inclined to HP, its top view is
(a) apparent on xy line and parallel to it
(b) true length on xy line and parallel to it
(c) seen as a point and on and on xy line
(d) apparent and perpendicular to xy line

7.5 The length of FV and TV of the line when the line is inclined to both the planes are_____ than the TL of the line
(a) greater   (c) equal
(b) less   (d) 2 times greater

7.6 The inclinations of FV and TV of the line with xy when the line is inclined to both the planes are_____ than the true inclination with HP and VP, respectively.
(a) greater   (c) equal
(b) less   (d) 2 times greater

7.7 HT is
(a) the intersection of the extended line with HP
(b) the intersection of the extended line with VP
(c) intersection of the perpendicular drawn from VT with xy line
(d) intersection of the perpendicular drawn from v (projection of VT) with extension of top view

7.8 VT is
(a) the intersection of the extended line with VP
(b) the intersection of the extended line with HP
(c) intersection of the perpendicular drawn from HT with xy line
(d) intersection of the perpendicular drawn from h' (projection of HT) with extension of front view

7.9 As per the rules of traces of line, FV, _____ and _____lie on the same line.
(a) v and HT   (c) H' and HT
(b) VT and h'   (d) v and VT

7.10 As per the rules of traces of line, TV, ____ and _____lie on the same line.
(a) v and HT   (c) h' and HT
(b) VT and h'   (d) v and VT

Projections of the Line Parallel to Both the Reference Planes

7.1 Draw the projections of the following straight lines of length 70 mm on to the same VP and HP. Adopt suitable distances between the projections in each case.

(a) Line AB on both HP and VP
(b) Line AB 30 mm in front of VP and on HP
(c) Line AB 40 mm above HP on VP and parallel to both the principal planes.
(d) Line CD 30 mm above HP and 30 mm in front of VP

Projections of the Line Perpendicular to One of the Reference Planes

7.2 Draw the projections of the following straight lines of length 70 mm on to the same VP and HP.
   (a) Line EF perpendicular to HP, parallel to VP and 30 mm in front of it. The end E is 20 mm above HP.
   (b) Line AB perpendicular to VP, parallel to HP and 40 mm above it. The end A is 20 mm in front of VP
   (c) Line CD on HP and perpendicular to VP. The end C is on VP.

Projections of the Line Inclined to One of the Reference Planes

7.3 Draw the projections of the following straight lines of length 70 mm on to the same VP and HP.
   (a) Line AB is 30 mm above HP and parallel to it, the end A being 30 mm in front of VP and the line is inclined at 30° to VP.
   (b) The line AB is parallel to VP and is inclined at 30° to HP, the end A being 30 mm in front of VP and on HP.
   (c) Line CD parallel to VP and 25 mm in front of it and is inclined at 45° to HP. The end C is 20 mm above HP.
   (d) Line CD on VP and is inclined at 30° to HP, the end C being 20 mm above HP.

7.4 State the position of the straight lines with reference to VP and HP of their projections are as in Fig. 7.47.

Projections of the Lines on all the Three Planes

7.5 Draw the projections of the following straight lines of length 65 mm on to the same VP, HP, and profile planes.
   (a) Line PQ, 30 mm above HP, 20 mm in front of VP, and the end P being 100 mm in front of the right profile plane.
   (b) Line MN, 30 mm above HP, on VP, and the end M being 30 mm in front of the left PP.
   (c) Line PQ, 30 mm above HP, on VP, and the end P being on the left PP.

7.6 Draw the projections of the following straight lines of length 65 mm on to the same VP, HP, and profile planes.
   (a) Line MN, 30 mm in front of VP, 20 mm in front of the right PP, and the end M being 25 mm above HP.
   (b) Line PQ is on the left PP and 30 mm above HP, the end P being on VP.
   (c) Line MN, 30 mm in front of VP, is inclined at 30° to HP, the end M being 25 mm above HP and 20 mm in front of the right PP.
   (d) Line PQ, on VP and is inclined at 30° to HP, the end P being 25 mm above HP and 30 mm in front of the left PP.
   (e) Line MN, 20 mm above HP and is inclined at 30° to VP, the end M being 20 mm in front of VP and 100 mm in front of the right PP.

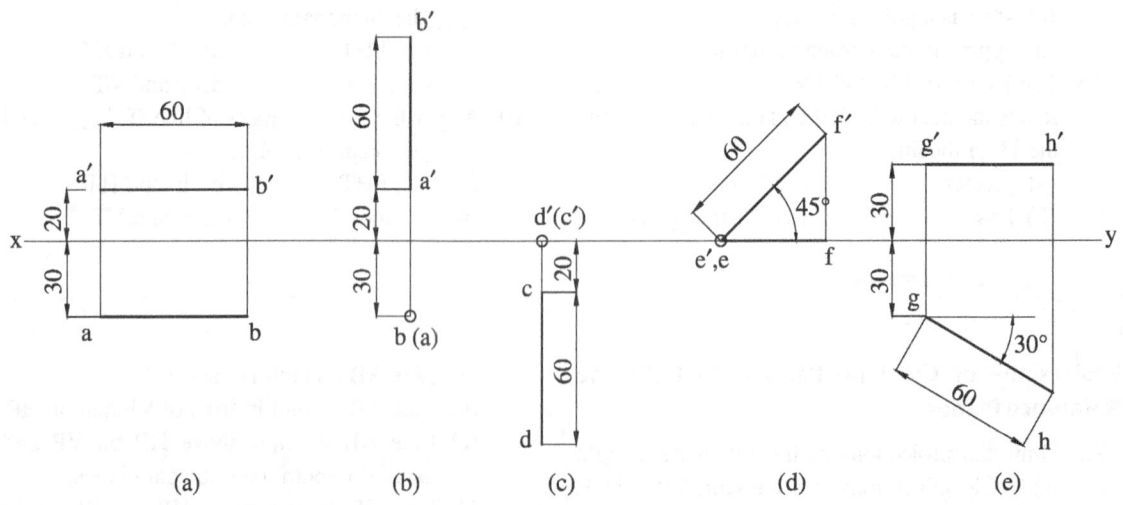

(a)   (b)   (c)   (d)   (e)

Fig. 7.47

Projection of the Line when Reference of One End is Given

7.1 A line PQ of length 65 mm is inclined at 45° with HP and 30° with VP. The end P is 20 mm above HP and 30 mm in front of VP. Draw the projections of the line and find its apparent inclinations with respect to principal planes.

7.2 A line PQ of length 65 mm is inclined at 45° with HP and 30° with VP. The end P is on HP and 30 mm in front of VP. Draw the projections of the line and find its apparent inclinations with respect to principal planes.

7.3 A line PQ of length 65 mm is inclined at 45° with HP and 30° with VP. The end P is on VP and 20 mm above HP. Draw the projections of the line and find its apparent inclinations with respect to principal planes.

7.4 A line PQ of length 65 mm is inclined at 45° with HP and 30° with VP. The end P is at the intersection of reference planes. Draw the projections of the line and find its apparent inclinations with respect to principal planes.

7.5 A line PQ of length 65 mm is inclined at 30° with HP and 45° with VP. The end Q is 50 mm above HP and 70 mm in front of VP. Draw the projections of the line and find its apparent inclinations with respect to principal planes.

Projections of the Line when Reference of Any Point on the Line is Given

7.6 A line AB of length 70 mm is inclined at 45° to HP and 30° to VP. The midpoint M is 45 mm above HP and 65 mm in front of VP. Draw the projections of the line.

7.7 A line CD of length 70 mm is inclined at 30° to HP and 45° to VP. A point E on line is at a distance 30 mm from D and is 55 mm above HP and 60 mm in front of VP. Draw the projections of the line.

Projections of the Line when Reference of End Points is Given

7.8 A line PQ has its one end P 25 mm above HP and 30 mm in front of VP. The other end Q is 70 mm in front of VP and the line is inclined at 30° to VP. The distance between the end projectors of the line which is measured parallel to the line of intersection of HP and VP is 50 mm. Draw the projections of the line and find the inclination with the HP and true length of the line.

7.9 A line AB has its end A 25 mm above HP and 20 mm in front of VP. The other end B is 60 mm in front of VP and line is inclined at 45° to HP. The distance between the end projectors of the line which is measured parallel to the line of intersection is 65 mm. Draw the projections of the line and find its inclinations with VP and true length of the line.

7.10 A line PQ of length 65 mm, the end P is 20 mm above HP and 25 mm in front of VP. The other end Q is 50 mm above HP and 65 mm in front of VP. Draw the projections of the line and find its inclination with respect to principal planes.

7.11 A line AB of length 70 mm and the end B is 60 mm above HP and 50 mm in front of VP. The other end A is on HP and 20 mm in front of VP. Draw the projections of the line and find its apparent inclinations with xy line.

Projections of the Line when Ends are in Different Quadrants

7.12 A line AB, 100 mm long, has its end A 15 mm above HP and 10 mm in front of VP. The end B is in the third quadrant. The line is inclined at 45° to HP and 30° to VP. Draw its projections.

7.13 A line AB, 100 mm long, has its end A 15 mm above HP and 10 mm in front of VP. The end B is in the second quadrant. The line is inclined at 45° to HP and 30° to VP. Draw its projections.

7.14 A line AB, 100 mm long, has its end A 15 mm above HP and 10 mm in front of VP. The end B is in the fourth quadrant. The line is inclined at 45° to HP and 30° to VP. Draw its projections.

Line with One End Touching HP and Other End Touching VP

7.15 A line CD of length 75 mm is inclined at 40° to HP and 50° to VP. The end C is on HP and end

D on VP. Draw the projections and find shortest distance between the line CD and the line of intersection of planes of projection.

Lines Parallel to the Profile Plane

7.16   A line CD of length 70 mm is inclined at 40° with VP and parallel to profile planes. The end C is on HP and end D is on VP. Draw the projections of the line.

7.17   A line CD of length 75 mm parallel to the profile plane is inclined at 45° to VP and 45° to HP. The end C is 25 mm above HP and 30 mm in front of VP. Draw the projections of the line.

Lines Involving Side Views

7.18   The right profile view of a line CD of length 65 mm makes 45° to XY line and measures 50 mm. The top view of the line is measured as 55 mm. The end C is 30 mm above HP and 25 mm in front of VP and is nearer to it. Draw the projections of the line.

To Find True Length and Inclinations of the Line from its Projections

7.19   The top view of line PQ of length 70 mm is inclined at 30° to xy line and measures 50 mm. The end P is 25 mm above HP and 20 mm in front of VP. Draw the projections of the line and find the inclinations with HP and VP by using (a) rotating line method and (b) trapezoidal method.

7.20   The front view of a line AB measuring 125 mm long, is 75 mm, and its top view is 100 mm long. The end B is 30 mm from the both planes. Draw the projections of the line and find its inclinations with reference planes.

7.21   A straight line CD has its end point C 10 mm in front of VP and 15 mm above HP. The line is inclined at 45° to VP and its top view measures 40 mm. Draw the projection of the line CD if it is 50 mm long and is in the first quadrant.

7.22   The top view of the line PQ is 60 mm long and makes 60° with xy line. The end Q is 10 mm in front of VP and 25 mm above HP. The difference between the distance of P and Q above HP is 45 mm. Draw the projection of the line and find its true inclinations and true length.

7.23   The top and front views of a line are inclined at 35° and 50° to the xy line respectively. One end of the line is touching both the reference planes. The other end is 50 mm above HP and 30 mm in front of VP. Draw the projection of the line and find its true length and true inclinations.

7.24   The top view of a 75 mm long line CD, measures 50 mm. The mid-point of the line is 50 mm in front of VP and 75 mm above the HP. The end D is 30 mm in front of the VP. Draw the projection of the line and find its inclination with the HP and VP.

7.25   The line MN measures 120 mm. Its top and front views measure 80 mm and 96 mm respectively. A point P on the line, dividing it in the ratio of 1:2, i.e MP:PN = 1:2 is at distance of 80 mm from both the reference planes. Draw the projection of the line and determine its inclinations with the reference plane.

Problems Involving Traces

7.26   Draw the projections of the following straight lines of length 70 mm on to the same VP and HP and locate its traces.
   (a)   Line AB, 25 mm above HP, the end A is 30 mm in front of VP, and the line is parallel to HP and perpendicular to VP.
   (b)   Line CD, on HP, the end C is 30 mm in front of VP, and the line is perpendicular to VP.
   (c)   Line AB, 25 mm is above HP is inclined at 45° to VP. The end A is 25 mm in front of VP.
   (d)   Line AB, 25 mm in front of VP is inclined at 45° to HP. The end A is 25 mm above HP.

7.27   A line CD of length 65 mm is inclined at 45° to HP and 40° with VP. The end A is 20 mm above HP and 25 mm in front of VP. Draw the projections of the line and locate its traces.

7.28   The top view of the line CD measures 60 mm and is inclined at 60° to xy line, the end C is 15 mm in front of VP. The VT is 25 mm above xy line. The distance between the projectors of HT and VT is 15 mm. Draw the projections of the line and find the true length and true inclination of the line with HP and VP, and locate the HT.

7.29   The line joining VT (which is above xy line) and h′ is 20 mm. One end of the line A is 25 mm above HP and 15 mm in front of VP. The front view of the line measures 60 mm and makes 45° with xy line. Draw the projections of the line, find the true length of the line and its inclinations with the reference planes, and locate HT.

7.30 The projectors through the traces of a line are 110 mm apart and those through the ends are 50 mm apart. The ends are 50 mm apart. The end A is 20 mm above HP. The top and front views of the line make 30° and 40° with the reference line. Draw the projections of the line and find the true length and inclinations of the line.

7.31 A line AB has its ends A and B, 45 mm and 25 mm in front of VP respectively. The end projectors of the line AB when measured parallel to the reference line are 50 mm apart. The HT of the line is 10 mm in front of VP. The line AB is inclined at 35° to the HP. Draw the projection of the line AB and locate the VT. Find the distance of the VT of the line from the HP and the inclination of the line with VP.

7.32 The end points A and B of a straight line AB are respectively in the HP and in the VP. The HT of the line 25 mm behind VP and VT is 40 mm below HP. Draw the projection of the line, if it is 60 mm long. Find angles of the inclinations of AB with reference planes.

Engineering Applications

7.33 A rope which forms a straight line MN has one end M tied to a hook on the ground and is 30 mm in front of a vertical wall. The other end N is tied on to a pole which is 75 mm above the ground and the line is inclined at 30° to the ground. The distance between the end projectors of the line which is measured parallel to the line of intersection of the ground and vertical wall is 55 mm. Draw the projections of the line and find the inclinations with VP and true length of MN.

7.34 An electrical bulb is hanging vertically 1 m below centre of the ceiling of a hall of size 5 m × 6 m and height 4 m. Find graphically the distance between the bulb and the floor corners of the hall.

7.35 An equilateral triangle PQR of side 70 mm is so placed that side PQ is parallel to HP and inclined 45° to the VP. The difference in height between R and P is 30 mm. Draw the projection of the triangle and determine the inclination of the lines QR and RP to the reference planes.

7.36 The vertices P, Q, and R of a triangle PQR are 30 mm, 20 mm and 50 mm above the HP respectively. They are 40 mm, 30 mm and 60 mm in front of VP respectively. The distance between the extreme projections from P and R is 50 mm. The projector through Q is 30 mm from that through P, between extreme ones. Find the true shape of the triangle PQR. Determine the true length of the sides.

Answers for Multiple-choice Questions					
7.1 (d)	7.2 (b)	7.3 (c)	7.4 (a)	7.5 (b)	7.6 (a)
7.7 (a)	7.8 (a)	7.9 (b)	7.10 (a)		

Projection of Planes

8

OBJECTIVES

This chapter will help the reader to understand the following:

- Surfaces of the objects, which are bounded by straight lines, curves, or in combinations, and their resulting appearances while viewed from different directions
- Orthographic projections of various simple plane surfaces such as triangle, square, rectangle, and regular polygonal (pentagon and hexagon) surfaces, which are bounded by straight lines, in simple and inclined positions
- Orthographic projections of plane surfaces with curved boundaries such as circle, semicircle in simple and inclined positions.
- Physical objects conforming to these shapes in various engineering applications
- Methods of obtaining the true shapes of the objects from their projections, as a reverse engineering exercise
- Composition of various shapes resulting in a three-dimensional solid

8.1 INTRODUCTION

A plane surface is defined as a surface with only two dimensions, that is, the length and the breadth with negligible thickness. The shape of the plane surface is bounded by straight lines, curves or their combinations and can be made up of regular or irregular plane figures. A regular plane consists of polygons of equal sides or with equal internal angles or both, while irregular planes have the sides and internal angles unequal. The irregular plane surface may sometimes have curved shapes or boundaries approximated as closely connected straight lines which are unequal.

All three-dimensional regular objects are bounded with regular plane surfaces. These objects, when viewed from the front, top, and the sides to obtain their orthographic projections, form views that are plane surfaces.

In engineering applications, objects that are very thin or having negligible thickness compared with the length and the breadth, are often used as the bounding surfaces. The surfaces of the cartoon boxes made of paper or cardboard or plywood, the surfaces of the computer, television cabinets, thin panels of wood or sheet metals used in the almirahs, racks, or storage cabinets, doors and panels of automobiles, ships and space vehicles, projection screens, banners, walls and roofs, partition panels used in buildings, doors and windows, coins, compact disks etc. are all treated as plane surfaces, while drawing their orthographic projections. This chapter will deal with the projection when planes are placed in the first quadrant or as per first-angle projection.

8.1.1 Position of Plane Surface

A plane surface may be positioned in space or in between two imaginary reference planes in any one of the following positions:

1. Plane surface parallel to one reference plane and perpendicular to the other
2. Plane surface inclined to one reference plane and perpendicular to the other
3. Plane surface inclined to both reference planes

The position of a plane as per (1) will be dealt in Section 8.2 and (2) and (3) in Section 8.3.

8.1.2 Projections of Plane Surface—The Basics

The orthographic projections of a plane surface can be drawn with respect to its position in the reference planes.

When a plane surface is positioned parallel to any reference plane, the projection or view on that plane will reveal the true shape of its surface, since all its sides or edges are parallel to the reference plane.

When a plane surface is positioned perpendicular to a reference plane, its projection on that plane will be a straight line. If the plane surface is also held parallel to the other reference plane, then this straight line will appear parallel to the reference line, and if it is inclined to the other reference plane, the straight line view also will be inclined to the reference line depending on the inclination of the object with the reference plane.

Logically, when a plane surface is held inclined to both the reference planes, the object is not parallel and not perpendicular either. Therefore, its views on both reference planes will not be straight lines but will be composed of shapes of reduced sizes instead of the true shape of the object.

8.2 PROJECTION OF PLANE SURFACES IN SIMPLE POSITIONS

8.2.1 Planes Held in between Two Reference Planes (HP and VP)

When a plane surface is parallel to one reference plane (and hence perpendicular to the other reference plane), it is said to be placed in simple position. As discussed in Section 8.1, its true shape will be visualized on the plane to which it is parallel, and the other view will be a straight line parallel to the reference line.

Surface Parallel to One of the Principal Planes and Perpendicular to the Other

In this projection, the true shape of the surface is visualized on the reference plane to which it is parallel and the other view will be a straight line parallel to the reference line, on the principal plane to which the object is perpendicular.

Plane Surface Parallel to VP and Perpendicular to HP

Figure 8.1(a) shows a square plane—PQRS—parallel to VP and perpendicular to HP.

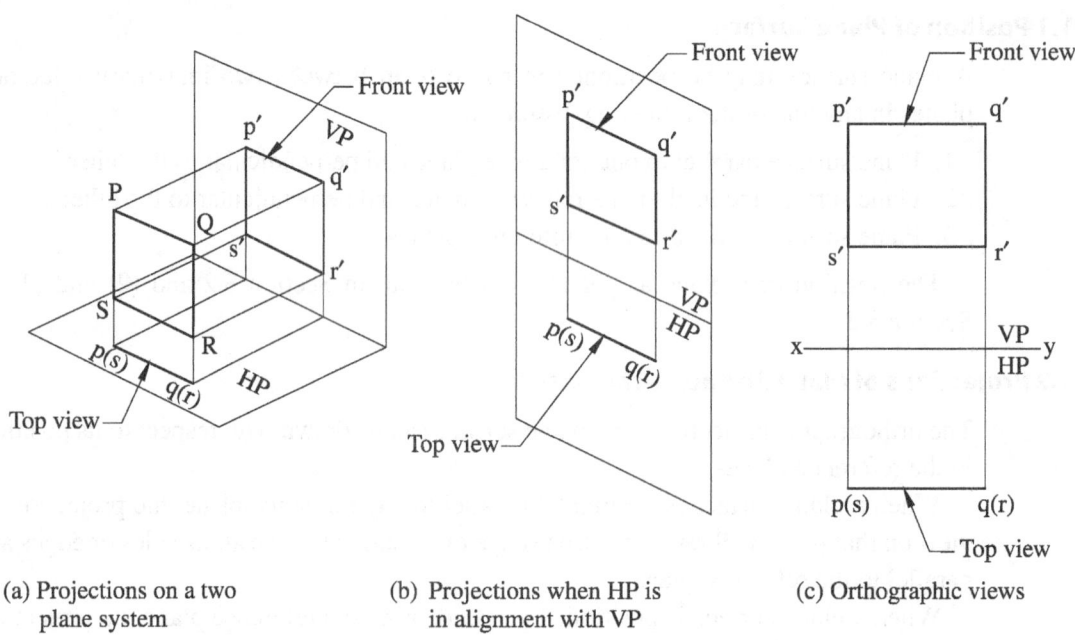

(a) Projections on a two plane system

(b) Projections when HP is in alignment with VP

(c) Orthographic views

Fig. 8.1 Plane surface parallel to VP and perpendicular to HP

The true shape of the plane can be visualized in the front view, which is represented as p′q′r′s′ on VP. The top view of the plane is a straight line, which is represented as p(s)–q(r) parallel to the reference line on HP. These are shown in Fig. 8.1(b) when HP is rotated and brought in alignment with VP, and Fig. 8.1(c) shows the orthographic projections.

Plane Surface Parallel to HP and Perpendicular to VP

Figure 8.2(a) shows a square plane—PQRS—parallel to HP and perpendicular to VP.

The true shape of the plane can be visualized in the top view, which is represented as 'pqrs' on HP. And the front view is observed as a straight line represented as p′(q′)–s′(r′) on VP parallel to the reference line. These are shown in Fig. 8.2(b), when HP is rotated and brought in alignment with VP, and Fig. 8.2(c) shows the orthographic projections.

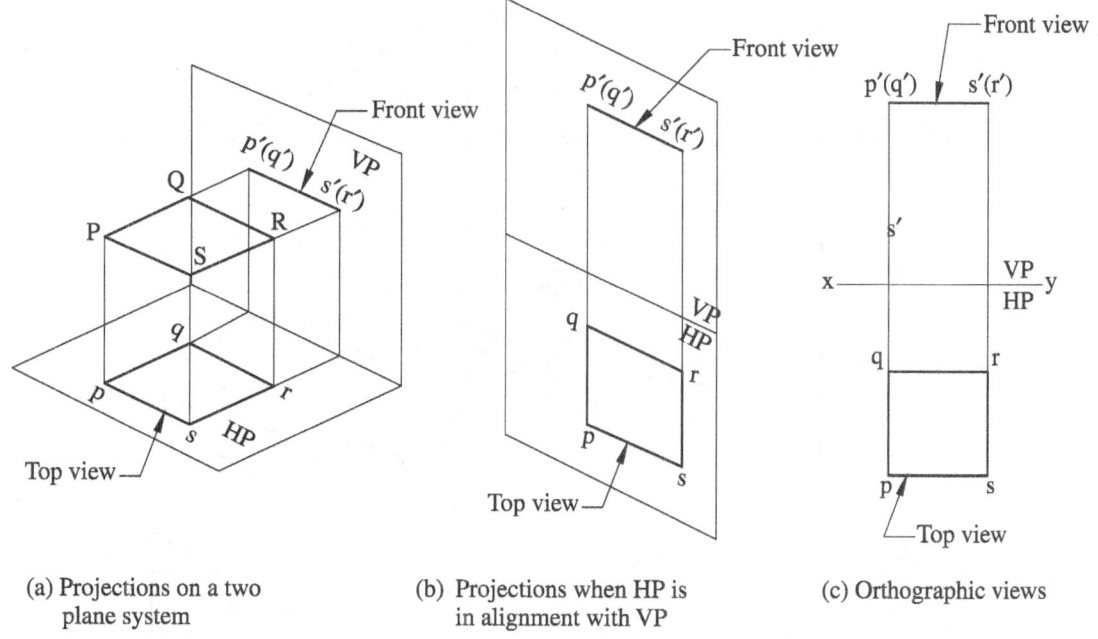

(a) Projections on a two
plane system

(b) Projections when HP is
in alignment with VP

(c) Orthographic views

Fig. 8.2   Plane surface parallel to HP and perpendicular to VP

8.2.2 Planes Held in between Three Reference Planes (HP, VP, and PP)

When a plane surface is held parallel to one reference plane, it is automatically perpendicular to the other two reference planes. As discussed in Section 8.2.1, its true shape will be visualized on the plane to which it is parallel and the views on the other two reference planes will be straight lines parallel to the respective reference lines.

Plane Surface Parallel to HP and Perpendicular to VP and RPP

Figure 8.3(a) shows a pentagonal plane—ABCDE—parallel to HP and perpendicular to VP and right profile plane (RPP).

The true shape of the plane can be visualized in the top view, which is represented as 'abcde' on HP. The front view represented as b′(a′)–c′(e′)–d′ on VP and the left-hand-side view a″–(e″)–(d″)–b″–c″ on RPP are straight lines parallel to the horizontal and vertical reference lines, respectively, as shown in Fig. 8.3(b), when HP and PP are brought in alignment with the VP. Figure 8.3(c) shows the orthographic projections.

Plane Surface Parallel to VP and Perpendicular to HP and RPP

Figure 8.4(a) shows a hexagonal plane—ABCDEF—parallel to HP and perpendicular to VP and right profile plane (RPP).

The true shape of the plane can be visualized in the front view, which is represented as a′b′c′d′e′f′ on VP. The top view represented as a(b)–f(c)–e(d) on HP and the left-hand-side view represented as f″–a″(e″)–b″(d″)–c″ on RPP are straight lines parallel to

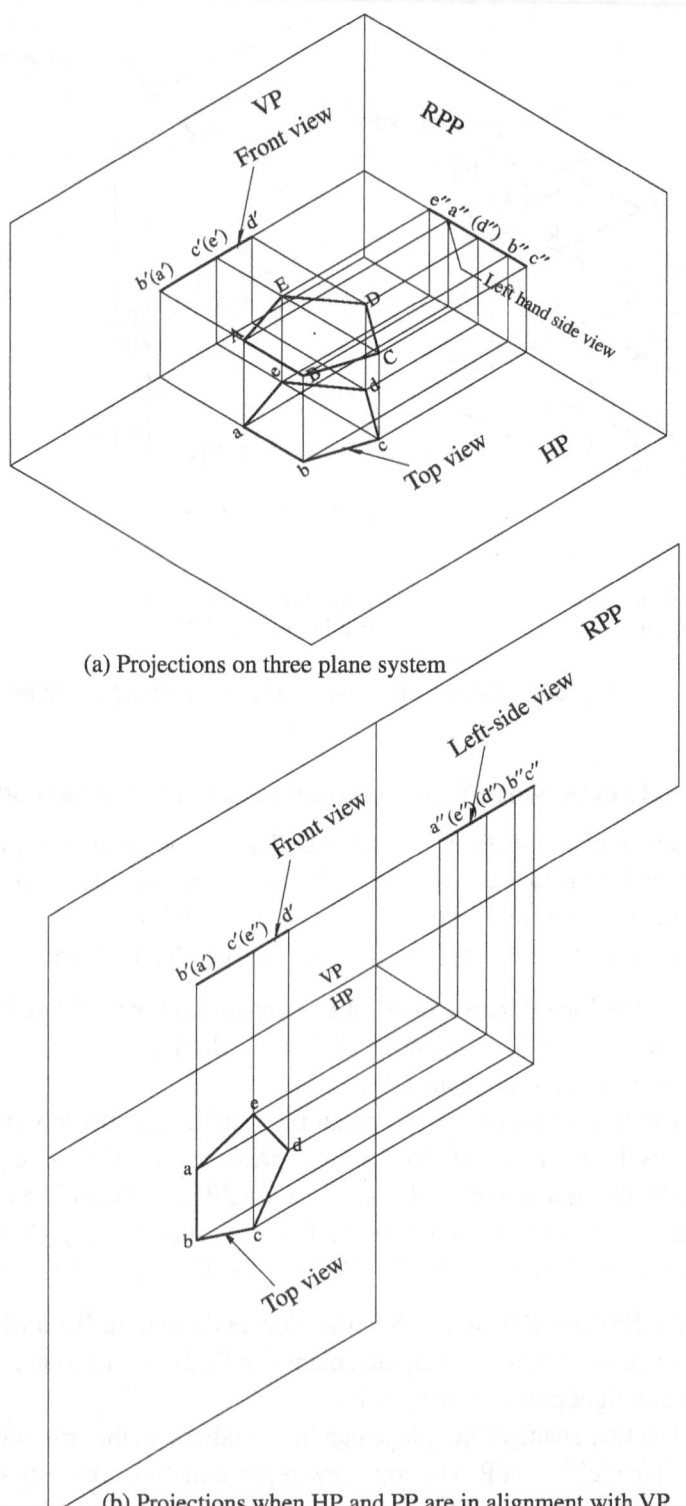

(a) Projections on three plane system

(b) Projections when HP and PP are in alignment with VP

Fig. 8.3 Pentagonal plane parallel to HP and perpendicular to VP and RPP (*Contd*)

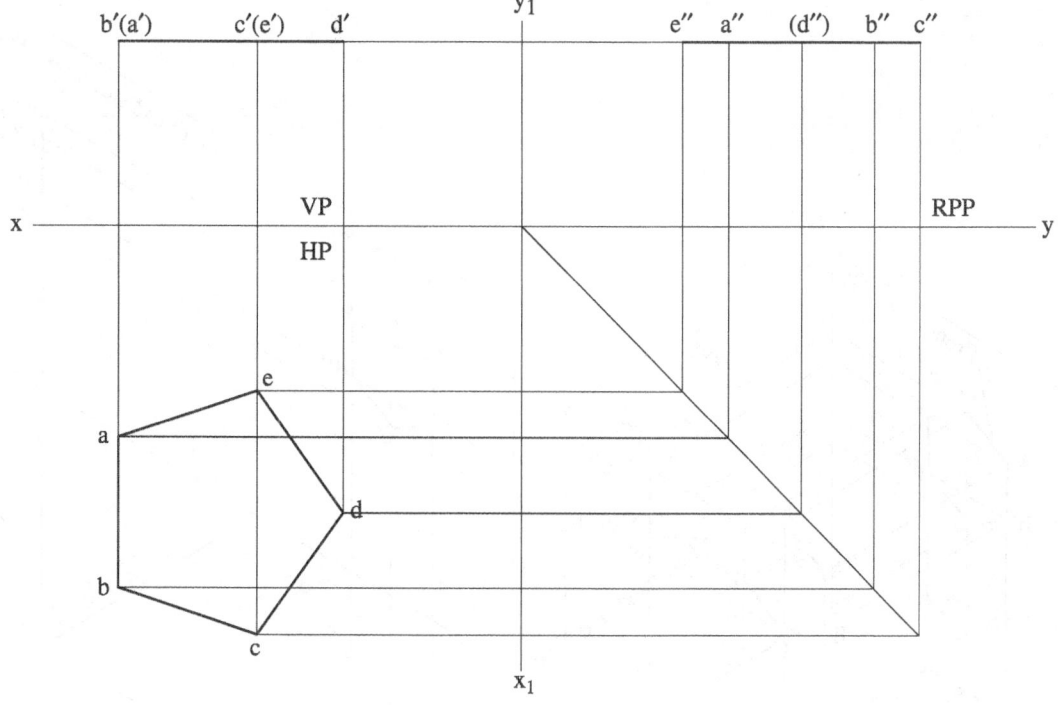

(c) Orthographic views

Fig. 8.3 (*Contd*)

the horizontal and vertical reference lines, respectively, as shown in Fig. 8.4(b), when HP and PP are brought in alignment with the VP. Figure 8.4(c) shows the orthographic projections.

8.3 PROJECTION OF PLANES IN INCLINED POSITIONS

When the object (the plane) is held in a two reference plane system (HP and VP), the resulting projections can be classified based on the following categories:

1. Plane surface inclined to one reference plane and perpendicular to the other
2. Plane surface inclined to both reference planes

8.3.1 Basic Principle

In Section 8.2, the projection of plane surface was analysed, when it is parallel to one reference plane and perpendicular to other. This is considered as the basis to draw the projection of planes in inclined positions. Initially, the object (the plane) will be held parallel to the reference plane to which it is to be inclined, and the projections will be drawn by following the procedure as discussed in Section 8.2, and then, the views are altered suitably, when the plane surfaces gets inclined.

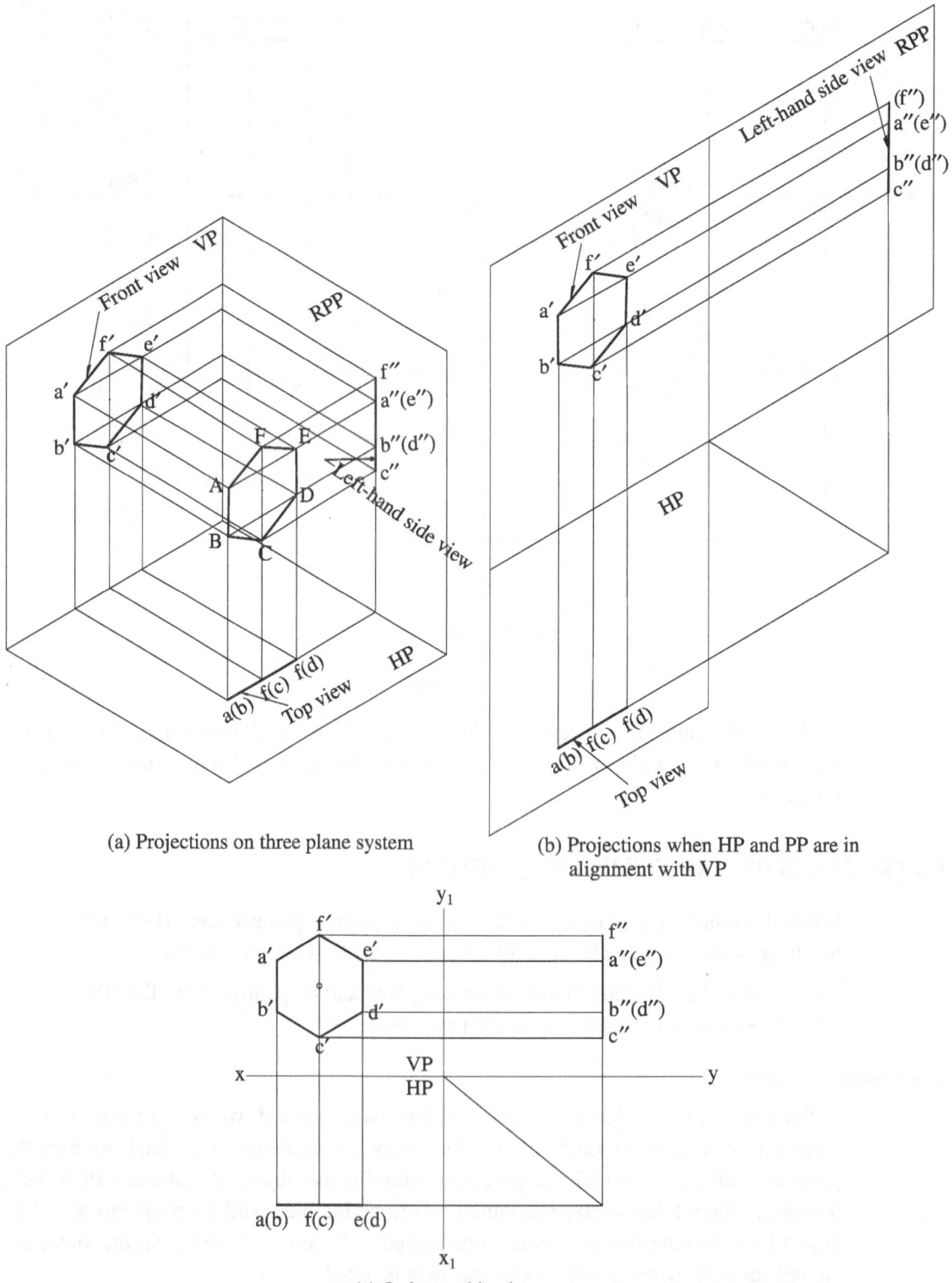

(a) Projections on three plane system

(b) Projections when HP and PP are in alignment with VP

(c) Orthographic views

Fig. 8.4   Hexagonal plane parallel to VP and perpendicular to HP and RPP

8.3.2 Plane Surface Inclined to one Reference Plane and Perpendicular to the Other

PRINCIPLE

It was observed that when the surface of a plane is parallel to one reference plane, it is perpendicular to the other reference plane, and hence, its true shape is visualized on that plane to which it is parallel, and the other view is a straight line parallel to the reference line. When the plane surface gets inclined, its perpendicularity with other reference plane remains unaltered, and hence, the straight line view remains unaltered in shape, but occupies an inclined orientation with the reference line according to the inclination desired for the plane. The view corresponding to this inclined orientation gives a reduced shape (instead of its true shape), since the plane surface is inclined to the reference plane of interest. This can be obtained by drawing locus lines from various corners/points of the plane surface, as they maintain their respective distances from the reference plane during this tilting process.

The tilting process of the plane surface can be addressed in one of the following two ways:

1. The tilting of the plane about an edge or a side
2. The tilting of the plane about a corner

When the plane surface gets tilted or inclined to a reference plane, all parts of the plane (the sides and the corners) move away from that reference plane except the side or the corner about which it is tilted. In other words, that edge/side/corner retains the same position before and after tilting. Since the plane surface has to maintain its perpendicularity with the other reference plane also, the side that retains its position before and after tilting has to be perpendicular to the reference line, so that the straight line view travels along an arc with this side as reference point and occupies the inclined position without any change in length. This principle is followed when the tilting of the plane surface occurs about an edge or side.

On the contrary, if the tilting of plane surface occurs about a corner, the required tilting of the straight line view to a new orientation can be achieved, along an arc only when the diagonal through that corner of reference is positioned parallel to the reference line while constructing the true shape. This principle is adopted when the tilting of the plane surface occurs about a corner. The word diagonal is valid in the case of polygons with even number of sides only. For the polygons with odd number of sides (e.g., triangle, pentagon, etc.), the word diagonal has to be interpreted as the line joining the centre of the shape and the corner of interest. In the former case of polygon, the tilting of the straight line view can be done with either of its ends, while in the latter case, the tilting has to be done corresponding to that corner of interest only.

It can be noted that any other tilting methodology of the plane will not satisfy the aforementioned principle and follow the movement of the straight line view along an arc and hence is not recommended to the reader, for drawing the views of the plane in inclined positions, at this stage.

PROCEDURE

The following are the sequences of operations when a plane surface is tilted about an edge:

Step 1: Keep the plane surface parallel to the reference plane to which it is to be inclined later and draw the true shape with one side or edge (about which it is to be tilted) perpendicular to the reference line and get the other view as a straight line parallel to the reference line. This is referred as initial or first-stage projections.

Step 2: Tilt the non-polygonal straight line view about the point corresponding to the perpendicular edge in the form of an arc to the required inclination and obtain its corresponding projection. This completes the second–stage projections or final projections.

The following are the sequences of operations when a plane surface is tilted about a corner:

Step 1: Keep the plane surface parallel to the reference plane to which it is to be inclined later and draw the true shape such that the line joining the corner (about which it is to be tilted) and the centre of the polygon (sometimes the diagonal through that corner) is laid parallel to the reference line and get the other view as a straight line parallel to the reference line. This is referred as initial or first-stage projections.

Step 2: Tilt the non-polygonal straight line view about the point of reference in the form of an arc to the required inclination and obtain its corresponding projection. This completes the second-stage projections or final projections

The forthcoming examples narrate the detailed procedure of drawing the projection of a plane inclined to one reference plane and perpendicular to the other. It can be noted that if a plane is resting on an edge/side or a corner, it amounts to getting tilted about an edge/side or corner respectively in the problems and the initial positions and the tilting scheme can be followed as mentioned earlier.

Example 8.1

A regular pentagonal lamina of 30 mm sides has one of its sides on VP. Draw its projections when the plane is vertical and inclined at 45° to VP.

Hint

1. Since the surface of plane is inclined to VP, position it parallel to VP in the first stage. The true shape is visualized in the front view and the polygon (pentagon) appears in the front view.

2. Since the plane has one side on VP and surface perpendicular to HP, position the pentagon with one side perpendicular to the reference line and tilt the straight line view about this edge.

PROCEDURE (Refer Fig. 8.5)

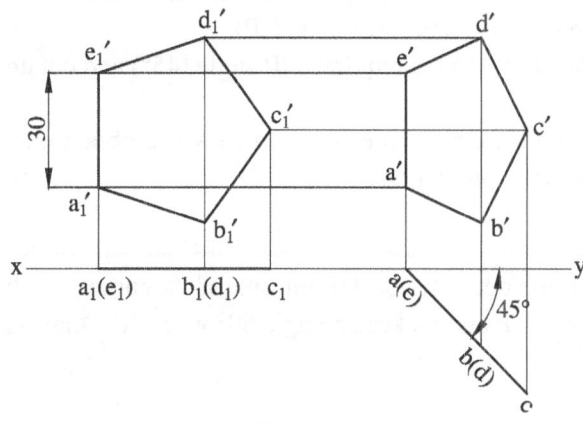

Fig. 8.5

Step 1: Draw the initial projections with the pentagon in the front view with one side $a_1'\ e_1'$ perpendicular to the reference line as shown by $a_1'\ ...\ e_1'$ and obtain its top view as a straight line on the reference line (since side is on VP).

Step 2: Tilt the straight line view about $a_1(e_1)$ to the required VP angle (45°) and obtain second-stage top view $a\ ...\ e$.

Step 3: Project this to meet the locus lines from the previous front view and obtain final front view as $a'\ ...\ e'$ and complete the projections. ▲

Example 8.2

A regular hexagonal lamina of 30 mm sides is resting on one of its sides on HP. Draw its projections when its surface is inclined at 45° to HP and the side on HP makes 90° with VP.

Hint

1. Since the surface of plane is inclined to HP, position it parallel to HP in the first stage. The true shape is visualized in the top view and the polygon (hexagon) appears in the top view.

2. Since the plane is resting on one of its sides and perpendicular to VP, position the hexagon with one side perpendicular to the reference line and tilt the straight line shape about this edge.

PROCEDURE (Refer Fig. 8.6)

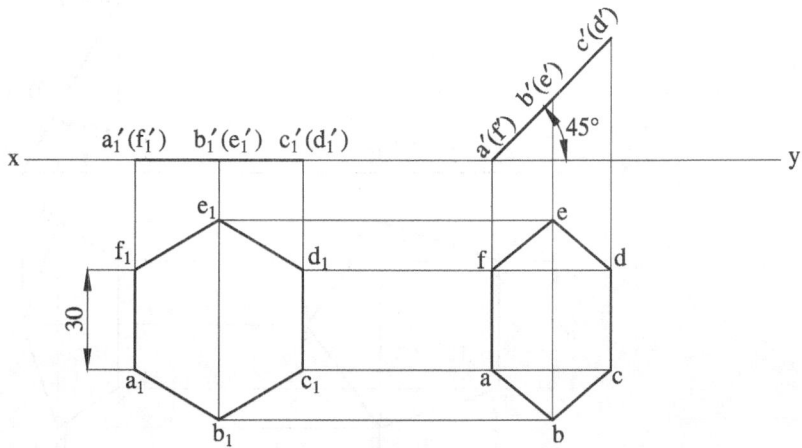

Fig. 8.6

Step 1: Draw the initial projections with the hexagon in the top view with one side a_1f_1 perpendicular to the reference line as shown by $a_1 \ldots f_1$ and obtain its front view as a straight line on the reference line (since side is on HP).

Step 2: Tilt the straight line view about $a_1'(f_1')$ to the required HP angle (45°) and obtain second-stage front view $a' \ldots f'$.

Step 3: Project this to meet the locus lines from the previous top view and obtain final top view as $a \ldots f$ and complete the projections. ▲

Example 8.3 ◤

A regular pentagonal lamina of 30 mm sides rests on HP on one of its corners such that the plane surface is perpendicular to VP and makes an angle 60° with HP. Draw its projections.

Hint

1. Since the surface of plane is inclined to HP, position it parallel to HP initially. The true shape is visualized in the top view and the polygon (pentagon) appears in the top view.
2. Since the plane is resting on a corner on HP and surface perpendicular to VP, position the pentagon such that the line joining the centre and the corner is parallel to the reference line and tilt the straight line shape about this corner.

NOTE *In the case of a pentagon shape, even if it is laid with the diagonal parallel to the reference line as mentioned here, the figure will always have one edge perpendicular to the reference line only. That is, for the edge tilting problem or the corner tilting problem, the initial projections are the same. However, the tilting of the straight line view has to be done carefully. For edge tilting problem, the straight line end corresponding to the edge has to be used, whereas for corner tilting problem, the straight line end corresponding to that corner has to be used. If done otherwise, the entire problem will be incorrect.*

PROCEDURE (Refer Fig. 8.7)

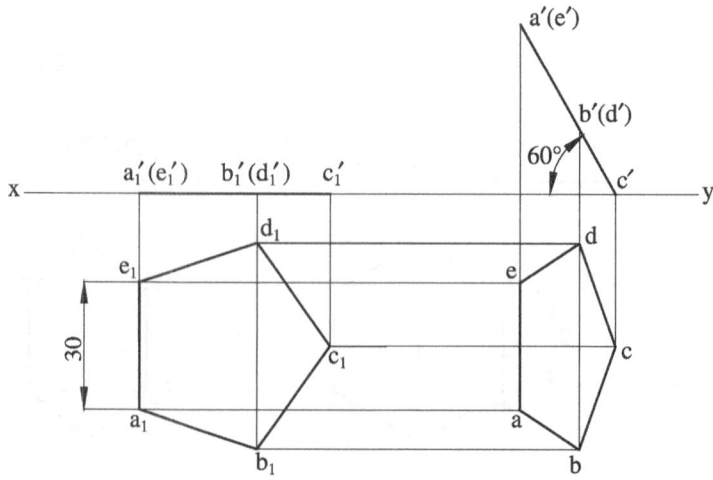

Fig. 8.7

Step 1: Draw the initial projections with the pentagon in the top view with the line joining the centre and the corner c_1 is parallel to the reference line as shown by $a_1 \ldots e_1$ and obtain its front view as a straight line on the reference line (since corner is on HP).

Step 2: Tilt the straight line view about c_1' to the required HP angle (60°) and obtain second-stage front view $a' \ldots e'$.

Step 3: Project this to meet the locus lines from the previous top view and obtain final top view as $a \ldots e$ and complete the projections.   ▲

Example 8.4   ▲

A square lamina of 30 mm sides has one of its corners on VP. Draw its projections when its surface is vertical and makes an angle 30° with the VP.

Hint   Since the surface of plane is inclined to VP, position it parallel to VP initially. The true shape is visualized in the front view and the polygon (square) appears in the front view with its diagonal parallel to the reference line.

PROCEDURE   (Refer Fig. 8.8)

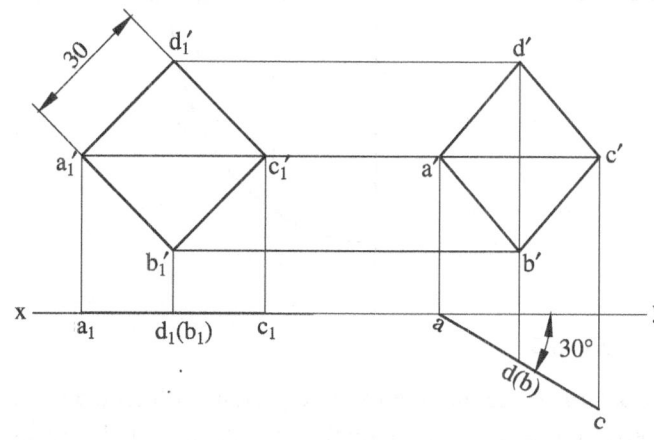

Fig. 8.8

Step 1: Draw the initial projections with the square in the front view as $a_1' \ldots d_1'$ with its diagonal $a_1'c_1'$ parallel to the reference line and obtain its top view as a straight line on the reference line (since the corner is on VP).

Step 2: Tilt the straight line view about a_1 to the required VP angle (30°) and obtain second-stage top view $a \ldots c$.

Step 3: Project this to meet the locus lines from the previous front view and obtain final front view as $a' \ldots d'$ and complete the projections.   ▲

Example 8.5   ▲

An 180° plastic protractor has its flat end on HP and perpendicular to VP. Draw its projections when its surface makes 30° with the HP. The flat end of the protractor measures 60 mm, and its thickness is negligible.

Hint   Since the surface of plane is inclined to HP, position it parallel to HP initially. The true shape (semicircle) appears in the top view with its flat end perpendicular to the reference line.

PROCEDURE (Refer Fig. 8.9)

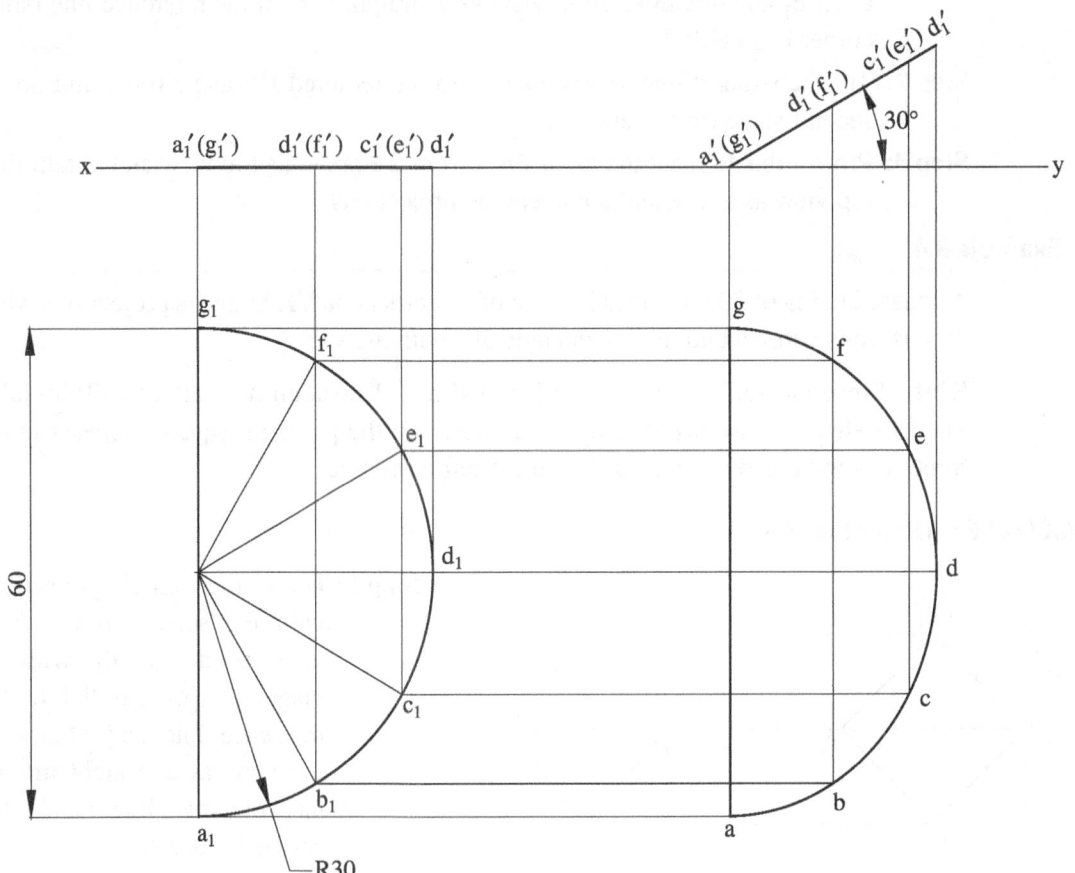

Fig. 8.9

Step 1: Draw the initial projections of the semicircle in the top view with its diameter perpendicular to the reference line. Divide the semicircle into six equal parts as shown by a_1 ... g_1 and obtain its front view as a straight line on the reference line (since the diametrical edge is on HP).

Step 2: Tilt the straight line view about $a_1'(g_1')$ to the required HP angle (30°) and obtain second-stage front view a' ... g'.

Step 3: Project this to meet the locus lines from the previous top view and obtain final top view as a ... g and complete the projections. ▲

Example 8.6 ◢

A thin hexagonal plate of sides 30 mm is lying on VP on one of its corners. Draw its projections, when the surface makes 45° with VP and the diagonal is parallel to HP.

Hint

1. Since the surface of plane is inclined to VP, position it parallel to VP initially. The true shape is visualized in the front view and the polygon (hexagon) appears in the front view.
2. Since the plane is resting on a corner on VP and surface perpendicular to HP, position the hexagon with the diagonal through that corner parallel to the reference line and tilt the straight line shape about this corner.

PROCEDURE (Refer Fig. 8.10)

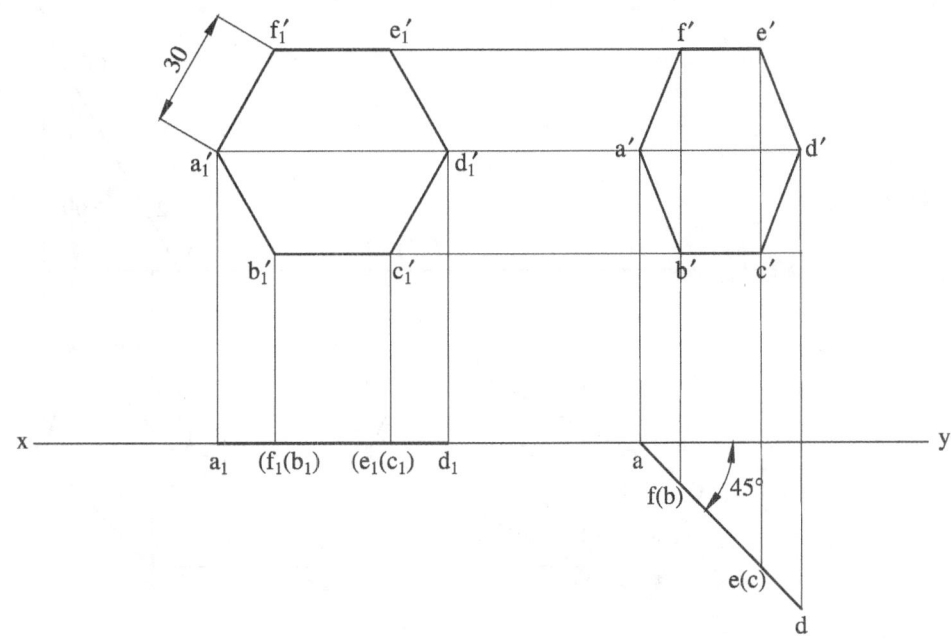

Fig. 8.10

Step 1: Draw the initial projections with the hexagon in the front view with corner a_1' and the diagonal through that is parallel to the reference line as shown by a_1' ... f_1' and obtain its top view as a straight line on the reference line (since corner is on VP).

Step 2: Tilt the straight line view about a_1 to the required VP angle (45°) and obtain second-stage top view a ... f.

Step 3: Project this to meet the locus lines from the previous front view and obtain final front view as a' ... f' and complete the projections.

Example 8.7

A thin rectangular card board lamina has one of its corners on the HP and the surface makes 60° with the HP. Draw its projections, when the diagonal passing through that corner on HP is parallel to VP. The size of the lamina is 50×25 mm.

Hint

1. Since the surface of the plane is inclined to HP, position it parallel to HP initially. The true shape is visualized in the top view and the polygon (rectangle) appears in the top view.
2. Since the plane is resting on a corner on VP and the diagonal parallel to VP, position the rectangle with the diagonal through that corner parallel to the reference line and tilt the straight line shape about this corner.

PROCEDURE (Refer Fig. 8.11)

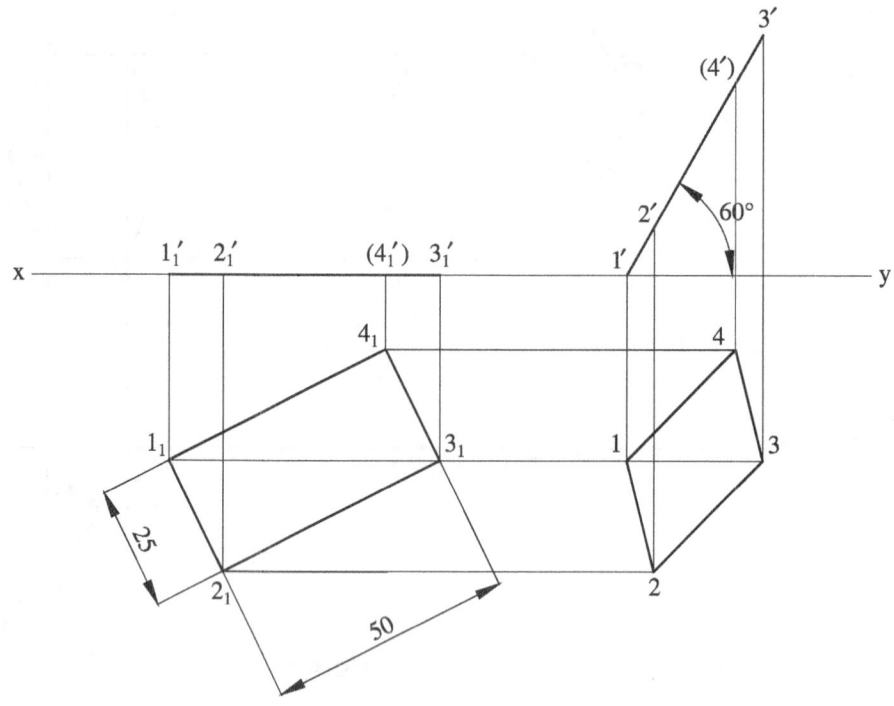

Fig. 8.11

Step 1: Draw the initial projections with the rectangle in the front view with corner '$1_1'$' and the diagonal through that corner is parallel to the reference line as shown by $1_1 \ldots 4_1$ and obtain its front view as a straight line on the reference line (since corner is on HP).

Step 2: Tilt the straight line view about $1_1'$ to the required HP angle (60°) and obtain second-stage front view $1' \ldots 4'$.

Step 3: Project this to meet the locus lines from the previous top view and obtain final top view as $1 \ldots 4$ and complete the projections. ▲

Example 8.8

A circular lamina of 60 mm diameter rests on HP such that the surface of the lamina is inclined at 30° HP and perpendicular to VP. Obtain its projection.

Hint Since the surface of plane is inclined to HP, position it parallel to HP initially. The true shape is visualized in the top view and the polygon (circle) appears in the top view with its diameter parallel to the reference line.

PROCEDURE (Refer Fig. 8.12)

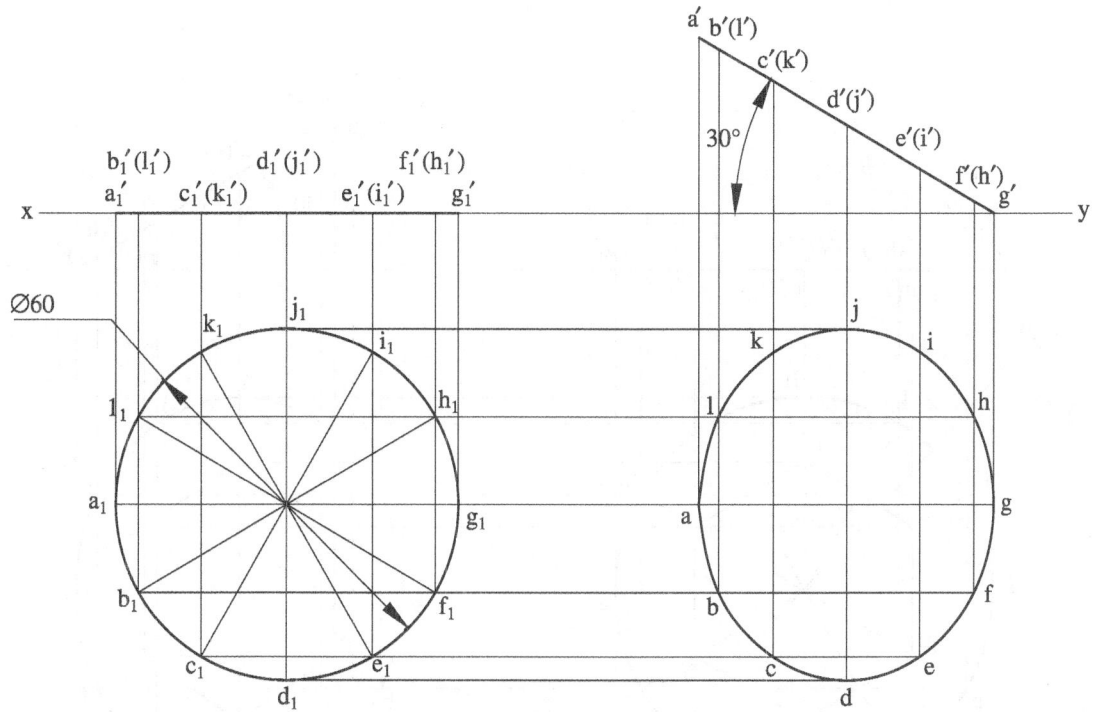

Fig. 8.12

Step 1: Draw the initial projections with the circle in the top view with the diameter parallel to the reference line and divide into 12 equal parts as shown by $a_1 \ldots l_1$ and obtain its front view as a straight line on the reference line (since one end of the diameter is on HP).

Step 2: Tilt the straight line view about g_1' to the required HP angle (30°) and obtain second-stage front view $a' \ldots l'$.

Step 3: Project this to meet the locus lines from the previous top view and obtain final top view as $a \ldots l$ and complete the projections. ▲

Example 8.9 ▲

A circular plate of 60 mm diameter has a hexagonal hole of 20 mm sides, centrally punched. Draw the projections of the lamina when its surface is inclined at 30 to HP and with two parallel sides of the hexagonal hole parallel to HP and perpendicular to VP.

Hint Since the surface of plane is inclined to HP, position it parallel to HP initially. The true shape is visualized in the top view and the polygon (circle with hexagonal hole) appears in the top view, with the diameter parallel to the reference line and two parallel sides of the hexagon perpendicular to the reference line.

PROCEDURE (Refer Fig. 8.13)

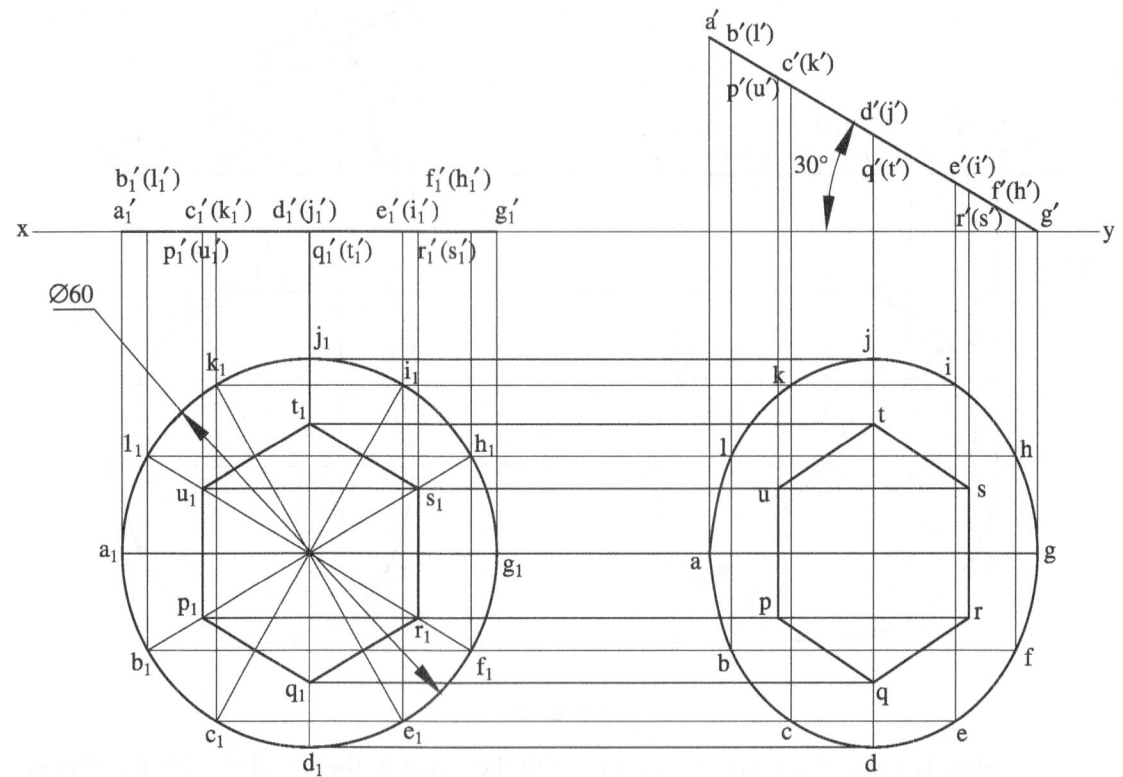

Fig. 8.13

Step 1: Draw the initial projections with the circle in the top view with the diameter parallel to the reference line and divide into 12 equal parts as shown by $a_1 \ldots l_1$ and draw the concentric hexagon $p_1 \ldots u_1$ such that the sides p_1u_1 and r_1s_1 are perpendicular to the reference line.

Step 2: Obtain its front view as a straight line on the reference line (since one end of the diameter is on HP).

Step 3: Tilt the straight line view about g_1' to the required HP angle (30°) and obtain second-stage front view $a' \ldots l'$.

Step 4: Project this to meet the locus lines from the previous top view and obtain final top view as $a \ldots l$ with the corners of the concentric hexagon $p \ldots u$ and complete the projections.

8.3.3 Projection of Planes Inclined to Both Reference Planes by Tilting Object Method

PRINCIPLE

In Section 8.3.2, the procedure for obtaining the projections of a plane surface when it is tilted about one of its sides or corners and surface inclined to one reference plane and perpendicular to the other was discussed. That relevant side was kept perpendicular or that corner diagonal was kept parallel to the reference line to ensure that the plane surface remains perpendicular to the other reference plane. When this side or corner diagonal also gets inclined to the reference plane, the plane surface is no more perpendicular to that reference plane and it is said to be inclined to both the reference planes.

In order to achieve this status, the tilting is done in three stages, the first two as discussed in Section 8.3.2 and the third-stage tilting with the side or the corner diagonal also made inclined to the reference line as per the requirement of the problem.

When the side or corner diagonal is tilted to make this additional inclination, the whole plane has to rotate about this edge or the corner as the surface inclination should not be altered. This means that the reduced shape obtained in the second stage is redrawn or rotated such that the edge or the corner diagonal makes the required inclination with the reference line and a new projection corresponding to that is obtained. This completes the third–stage projections or final projections when the plane surface is inclined to both the reference planes.

It can be noted that when the side or the corner diagonal gets inclined with the reference plane, the plane surface does not change its inclination with the other plane. Therefore, various points/corners of the plane maintain same distance before and after the tilt, and this fact is used by drawing the locus lines from the previous view to get the required final projections.

PROCEDURE

The stage-by-stage sequential procedure of drawing the projection of a plane surface inclined to both reference planes is explained in Table 8.1.

Table 8.1   Procedure to draw the projections stage by stage

Stage	Plane (object) when tilted about an edge	Plane (object) when tilted about a corner
1	Draw the true shape of the plane in the appropriate view with one edge perpendicular to the reference line and obtain a parallel line in the other view	Draw the true shape of the plane in the appropriate view with the diagonal passing through that corner, parallel to the reference line and obtain a parallel line in the other view

(Contd)

Table 8.1   (*Contd*)

Stage	Plane (object) when tilted about an edge	Plane (object) when tilted about a corner
2	Tilt the straight line view to the required inclination of the surface and obtain a reduced shape in the other view	Tilt the straight line view to the required inclination of the surface and obtain a reduced shape in the other view
3	Redraw the reduced shape such that the concerned edge (which is kept perpendicular) is set to the required inclination and obtain the corresponding projection	Redraw the reduced shape such that the diagonal (which is kept parallel) is set to the required inclination and obtain the corresponding projection

The forthcoming examples narrate the detailed procedure of drawing the projection of a plane inclined to both reference planes.

Example 8.10

A regular pentagonal lamina of 30 mm sides touches VP with one of its sides. Draw its projections when the surface is inclined at 45° to VP and the side which touches VP makes 30° to HP.

Hint

> **NOTE**   *Decide the initial projections based on the surface inclination.*

Since the surface of plane is inclined to VP, position it parallel to VP initially. The pentagon appears in the front view and placed it with one side perpendicular to the reference line.

PROCEDURE   (Refer Fig. 8.14)

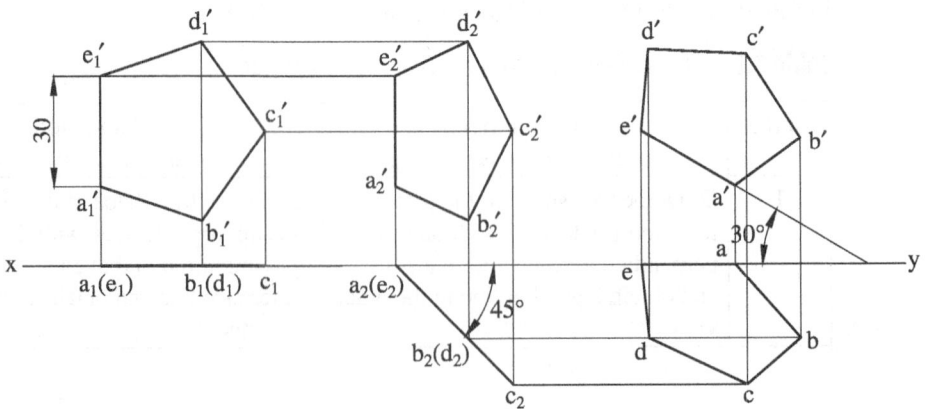

Fig. 8.14

Step 1: Complete the initial projections and surface inclination as shown in stage 1 and stage 2 in Example 8.1.

Step 2: Redraw the reduced shape a_2' ... e_2' such that the edge $a_2'e_2'$ makes 30° with reference line and draw vertical projectors.

Step 3: Draw locus line from the second-stage top view to meet the projectors and get final top view a ... e.

Example 8.11

A regular pentagonal lamina of 20 mm side rests on HP on one of its sides such that the plane surface makes an angle 45° with HP. Draw its projections when the side on HP makes an angle of 30° with the VP.

Hint

> **NOTE**   *Decide the initial projections based on the surface inclination.*

Since the surface of plane is inclined to HP, position it parallel to HP initially. The pentagon appears in the top view and placed it with one side perpendicular to the reference line.

PROCEDURE   (Refer Fig. 8.15)

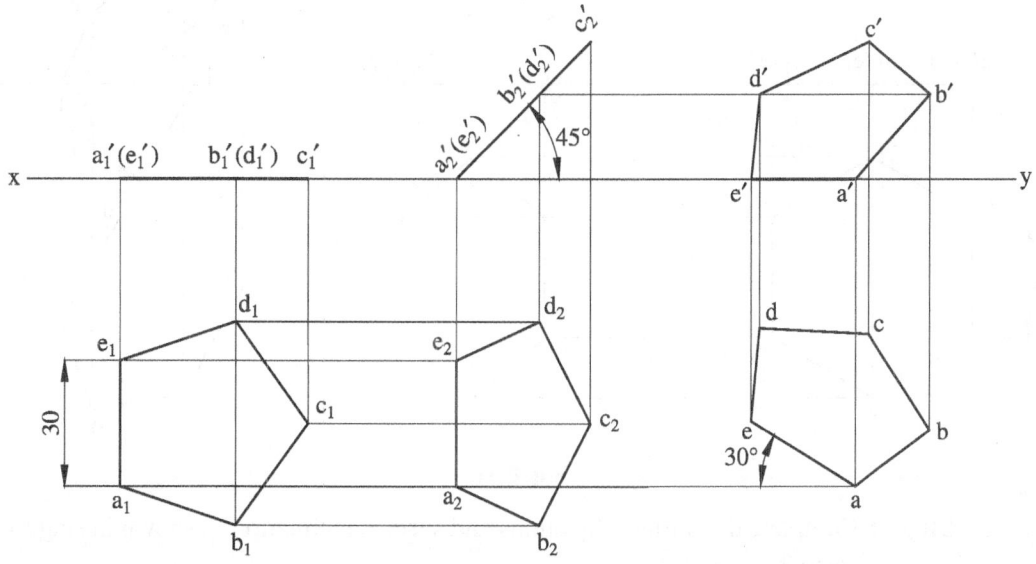

Fig. 8.15

Step 1: Complete the initial projections and surface inclination as in stage 1 and stage 2.

Step 2: Redraw the reduced shape $a_2 \ldots e_2$ such that the edge a_2e_2 makes 30° with reference line and draw vertical projectors.

Step 3: Draw locus line from the second-stage front view to meet the projectors and get final front view $a' \ldots e'$. ▲

Example 8.12 ▲

A regular hexagonal lamina of 30 mm sides rests on HP on one of its sides. The side which is on HP makes 60° to the VP and the surface of the lamina is inclined to HP at 45°. Draw the front view and top view of the lamina in its final position.

Hint

> **NOTE** *Decide the initial projections based on the surface inclination*

Since the surface of plane is inclined to HP, position it parallel to HP initially. The hexagon appears in the top view and placed it with one side perpendicular to the reference line.

PROCEDURE (Refer Fig. 8.16)

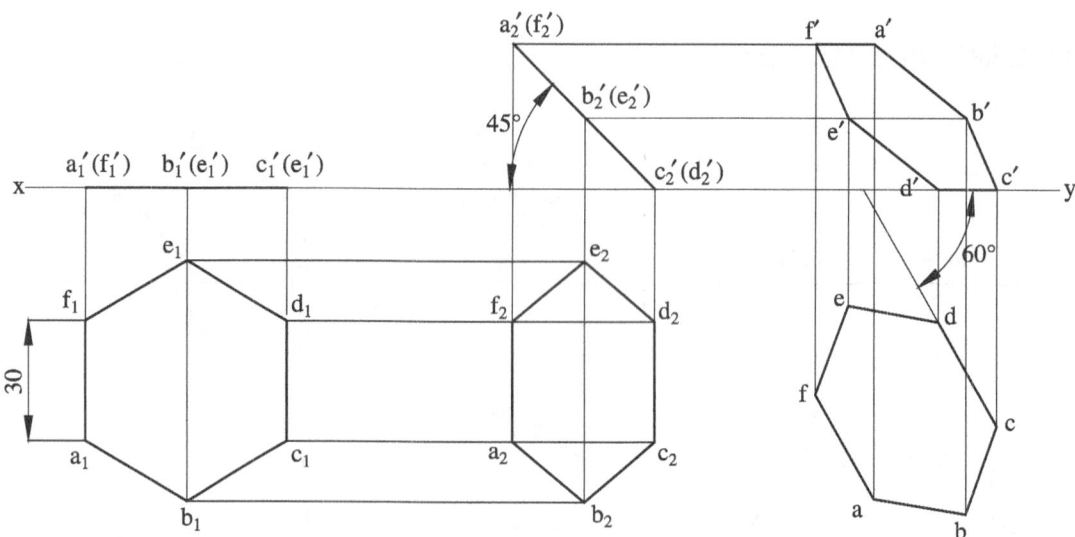

Fig. 8.16

Step 1: Complete the initial projections and surface inclination as shown in stage 1 and stage 2.

Step 2: Redraw the reduced shape $a_2 \ldots f_2$ such that the edge c_2d_2 makes 60° with reference line and draw vertical projectors.

Step 3: Draw locus line from the second-stage front view to meet the projectors and get final front view $a' \ldots f'$. ▲

Example 8.13

@ A regular pentagonal lamina of 30 mm base edges rests on one of its corners on HP. Draw its projections when the surface of the plate makes 60° with HP and the top view of the diagonal passing through that corner on HP makes 45° with the reference line.

Hint

> **NOTE**   *Decide the initial projections based on the surface inclination*

Since the surface of plane is inclined to HP, position it parallel to HP initially. The pentagon appears in the top view and placed it with the line joining the centre and the corner is parallel to the reference line.

PROCEDURE   (Refer Fig. 8.17)

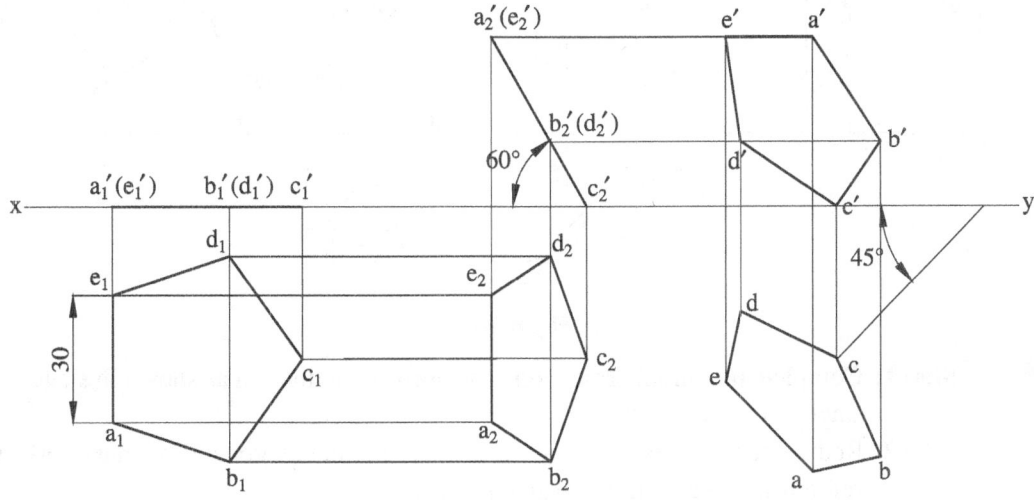

Fig. 8.17

Step 1: Complete the initial projections and surface inclination as in stage 1 and stage 2 in Example 8.3.

Step 2: Redraw the reduced shape $a_2 \ldots e_2$ such that the centre line through c_2 makes 45° with reference line and draw vertical projectors.

Step 3: Draw locus line from the second-stage front view to meet the projectors and get final front view $a' \ldots e'$.

Example 8.14

A thin hexagonal plate of sides 30 mm is touching VP on one of its corners. Draw its projections when the surface makes 45° with VP and the front view of the diagonal passing through the corner resting on VP, makes 30° with the reference line.

Hint

> ⚙ **NOTE** *Decide the initial projections based on the surface inclination*

Since the surface of plane is inclined to VP, position it parallel to VP initially. The hexagon appears in the front view and placed it with one diagonal parallel to the reference line.

PROCEDURE (Refer Fig. 8.18)

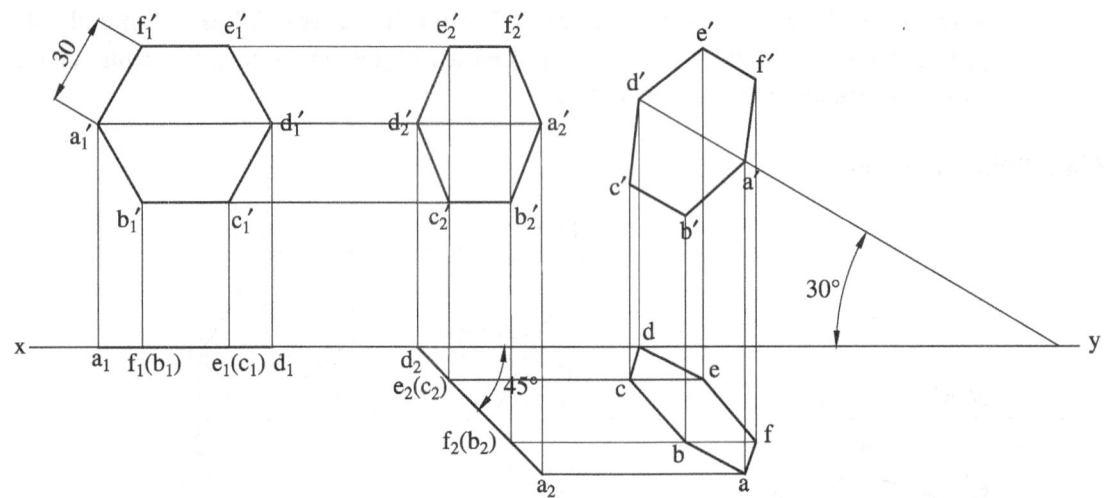

Fig. 8.18

Step 1: Complete the initial projections and surface inclination as shown in stage 1 and stage 2 in Example 8.6.

Step 2: Redraw the reduced shape $a_2' \ldots f_2'$ such that the diagonal $d_2' a_2'$ makes 30° with reference line and draw vertical projectors.

Step 3: Draw locus line from the second-stage top view to meet the projectors and get final top view a … f. ▲

Example 8.15 ▲

A thin hexagonal plate of sides 30 mm is touching VP on one of its corners. Draw its projections when the surface makes 45° with VP and the diagonal through the corner on VP, makes 30° with the HP.

Hint

> ⚙ **NOTE** *Decide the initial projections based on the surface inclination.*

Since the surface of plane is inclined to VP, position it parallel to VP initially. The hexagon appears in the front view and positioned it with one diagonal parallel to the reference line.

PROCEDURE (Refer Fig. 8.19)

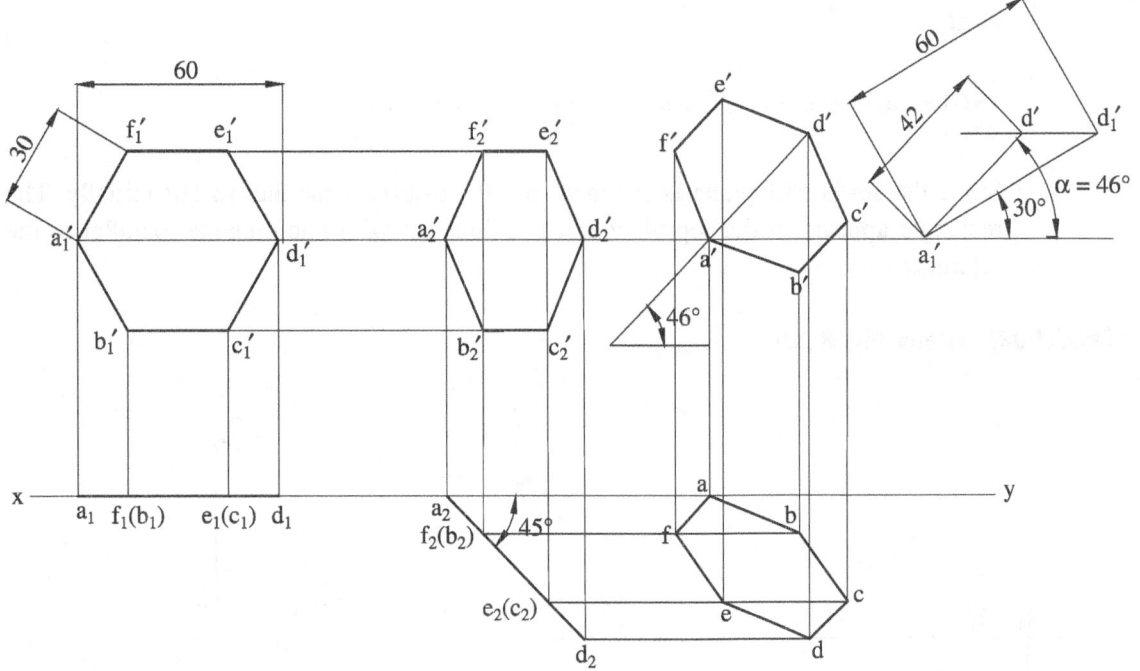

Fig. 8.19

Step 1: Complete the initial projections and surface inclination as shown in stage 1 and stage 2 in Example 8.14.

Step 2: Since the diagonal through $a_1'd_1'$ makes an angle of 30° with the HP, the front view of the diagonal will make an angle more than that (refer Chapter 7 Section 7.5.3). To this effect, set up a diagram separately indicating the true angle of inclination of the diagonal with the HP (30°) and mark the true length of the diagonal ($a_1'd_1'$) on it. Draw the horizontal locus line at d_1'. With a_1' as centre and the front view corresponding to VP inclination obtained in the second stage ($a_2'd_2'$) as radius, draw an arc to cut the locus line at d'. Join $a_1'd'$ and measure the apparent angle $\alpha = 46°$.

Step 3: Redraw the reduced shape $a_2' \dots f_2'$ such that the diagonal $a_2'd_2'$ makes 46° with reference line and draw vertical projectors.

Step 4: Draw locus line from the second-stage top view to meet the projectors and get final top view a … f.

Example 8.16

A thin rectangular card board lamina has one of its corners on the HP and the surface makes 60° with the HP. Draw its projections, when the top view of the diagonal passing through the corner on HP, makes 45° with the reference line. The size of the lamina is 50 × 25 mm.

Hint

> **NOTE**   *Decide the initial projections based on the surface inclination*

Since the surface of plane is inclined to HP, position it parallel to HP initially. The rectangle appears in the top view and positioned it with one diagonal parallel to the reference line.

PROCEDURE   (Refer Fig. 8.20)

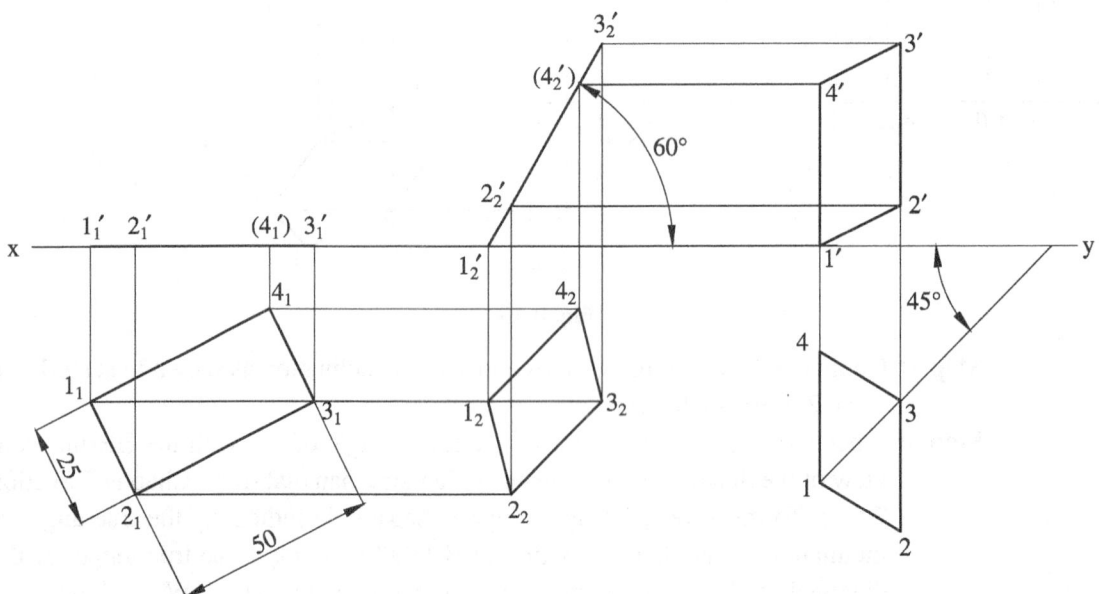

Fig. 8.20

Step 1: Complete the initial projections and surface inclination as shown in stage 1 and stage 2 in Example 8.7.

Step 2: Redraw the reduced shape $1_2 \dots 4_2$ such that the diagonal $1_2 3_2$ makes 45° with reference line and draw vertical projectors.

Step 3: Draw locus line from the second-stage front view to meet the projectors and get final front view $1' \dots 4'$.

Example 8.17

A regular circular lamina of 60 mm diameter rests on HP such that the surface of the lamina is inclined at 30° HP. Obtain the projections when the top view of the diameter passing through the point on HP makes 45° with the reference line.

Hint

NOTE *Decide the initial projections based on the surface inclination*

Since the surface of plane is inclined to HP, position it parallel to HP initially. The circle appears in the top view.

PROCEDURE (Refer Fig. 8.21)

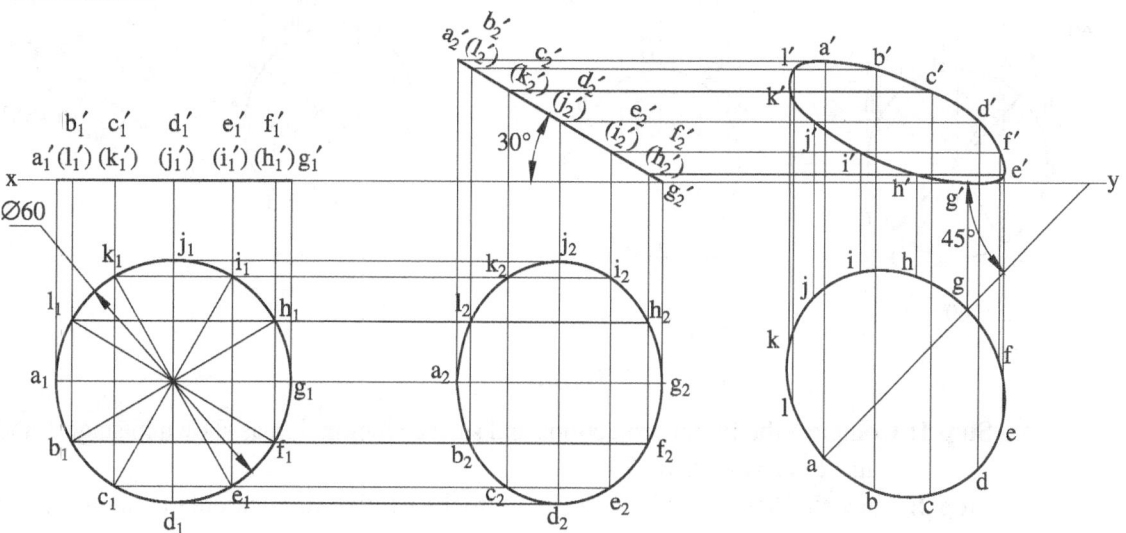

Fig. 8.21

Step 1: Complete the initial projections and surface inclination as shown in stage 1 and stage 2 in Example 8.8.

Step 2: Redraw the reduced shape $a_2 \ldots l_2$ such that the top view of the diameter $(a_2 g_2)$ makes 45° with reference line and draw vertical projectors.

Step 3: Draw locus line from the second-stage front view to meet the projectors and get final front view $a' \ldots l'$.

Example 8.18

A regular circular lamina of 60 mm diameter rests on HP such that the surface of the lamina is inclined at 30° HP. Obtain the projections when the diameter passing through the point on HP makes 30° to VP.

Hint

NOTE *Decide the initial projections based on the surface inclination*

Since the surface of plane is inclined to HP, position it parallel to HP initially. The circle appears in the top view.

PROCEDURE (Refer Fig. 8.22)

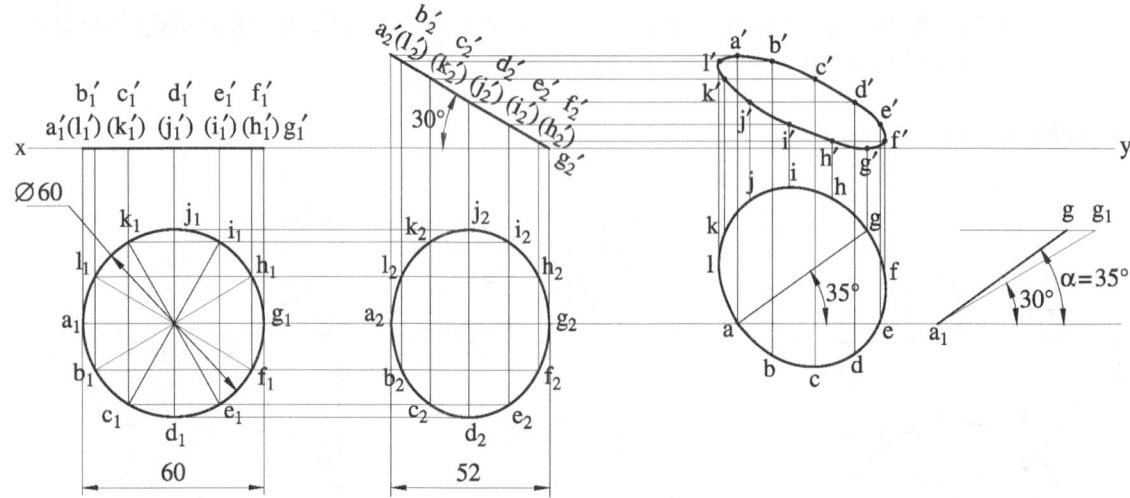

Fig. 8.22

Step 1: Complete the initial projections and surface inclination as shown in stage 1 and stage 2 in Example 8.17.

Step 2: Since the diameter is inclined at 30° to the VP, find its apparent inclination angle as indicated in Example 8.15 and it amounts to 35°.

Step 3: Redraw the reduced shape $a_2 \ldots l_2$ such that the top view of the diameter (a_2g_2) makes 35° with reference line and draw vertical projectors.

Step 4: Draw locus line from the second-stage front view to meet the projectors and get final front view $a' \ldots l'$. ▲

8.4 FINDING THE TRUE SHAPE OF THE OBJECT FROM ITS PROJECTIONS

As discussed in Section 8.3, when a plane surface is parallel to a reference plane, its true shape appears on that plane and its other view is a straight line parallel to a reference line separating the views. When the plane surface is inclined, the true shape gets reduced, but the straight line view maintains the same dimension, however, in an inclined orientation. If the projections of the inclined surface (the reduced shape and the straight line in the inclined orientation) are given, one can find the true shape of the object by following reversal procedure of the sequences adopted. If the straight line view in the inclined orientation is made parallel to the reference line, then the extended shape obtained therein should reveal the true shape. This is used as the basic principle to find the true shape of a plane surface, when its projections are known. The reader can derive this knowledge from the applications such as doors and windows, which are inclined to the wall in the open position, reveal their true shape when brought in alignment with the wall. These are explained in the forthcoming examples.

It can also be noted that when the plane surface gets tilted about one of its sides or edges, the length of that side or edge remains the same before and after tilting, and this is identified in the projection by the perpendicular side or edge to the reference line.

Example 8.19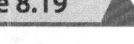

A rectangular lamina when tilted at an angle of 30° to the horizontal, about its 400 mm long edges appears as a square of 400 mm sides in the top view. Find the true shape of the lamina.

Hint Since the tilting of the lamina occurs about one of its edges, draw the tilted shape with that side perpendicular to the reference line and obtain the other view as a straight line in inclined orientation.

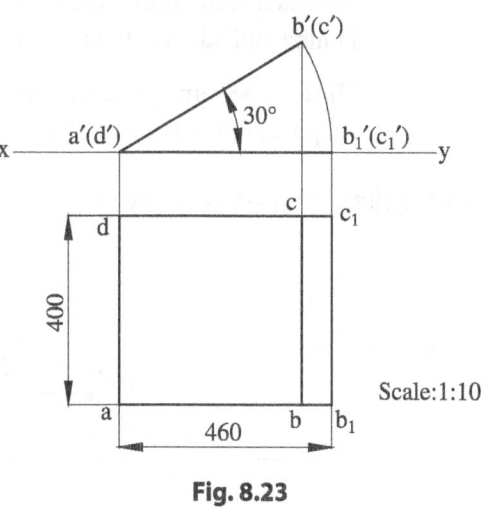

Fig. 8.23

PROCEDURE (Refer Fig. 8.23)

Step 1: Draw the top view as a square of 400 mm sides by keeping one side perpendicular to the reference line and draw its corresponding front view as a straight line inclined at 30° as shown by $a' \ldots d'$.

Step 2: Tilt the inclined straight line to the horizontal orientation about $a'(d')$ and obtain its corresponding projection ab_1c_1d which gives the true shape.

Example 8.20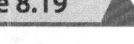

A five-sided planar lamina when tilted about one of its corners, appears as a regular pentagon of 22 mm sides in the front view. The inclination of its surface with the vertical plane is 45°. The surface is vertical in the tilted position. Find the true shape of the lamina.

Hint Since the tilting of the lamina occurs about a corner, draw the tilted shape with the corner and the diagonal through that parallel to the reference line and obtain the other view as a straight line in inclined orientation.

PROCEDURE (Refer Fig. 8.24)

Step 1: Draw the tilted shape, that is, the regular pentagon $a' \ldots e'$ with the diagonal through a' parallel to the reference line and draw its corresponding top view as a straight line inclined at 45° as shown by $a \ldots e$.

Step 2: Tilt the inclined straight line to the horizontal orientation about 'a' and obtain its corresponding projection $a'b_1'c_1'd_1'e_1'$ which gives the true shape. ▲

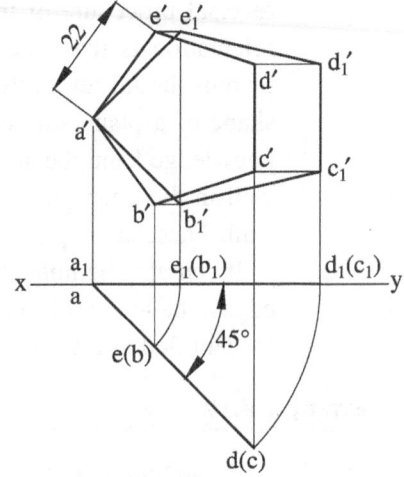

Fig. 8.24

Example 8.21 ▲

The top view of a six-sided planar lamina appears as a regular hexagon of 20 mm sides. The front view is a straight line inclined at 40° with the reference line. Find the true shape of the lamina and add a side view for the above arrangement.

Hint The same principle indicated in Example 8.19 applies except with the change of shape and orientation angle.

PROCEDURE (Refer Fig. 8.25)

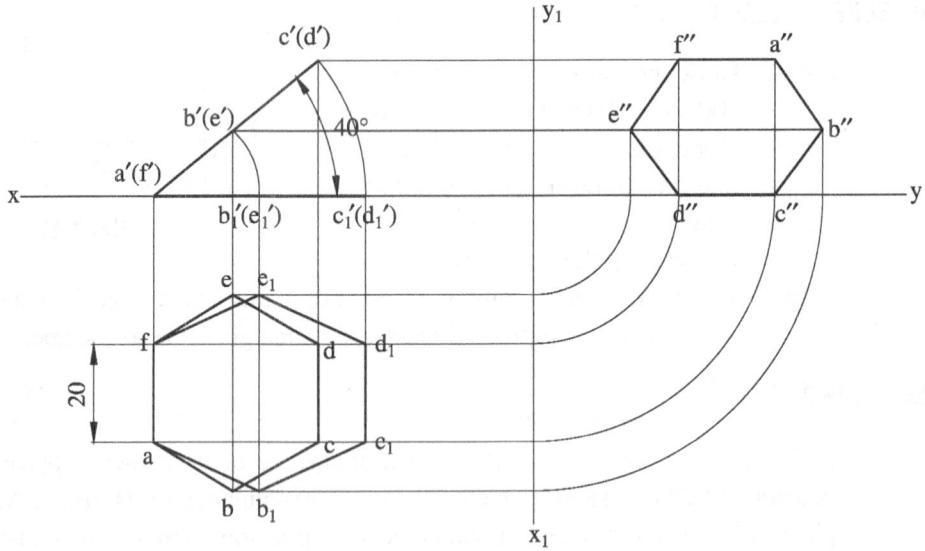

Fig. 8.25

Step 1: The reader is advised to follow the procedure indicated in Example 8.19.

Step 2: After drawing the views, a side view is also added as shown in Fig. 8.25. ▲

Example 8.22

The surface of the square lamina makes 45° with the HP when tilted about one of its corners such that the diagonal passing through this corner is parallel to VP. In this position, the lamina appears as rhombus of sides 26 mm. Find the true shape of the lamina.

Hint Since the tilting of the lamina occurs about a corner, draw the tilted shape in the top view with the corner and the diagonal through that parallel to the reference line and obtain the other view as a straight line in inclined orientation.

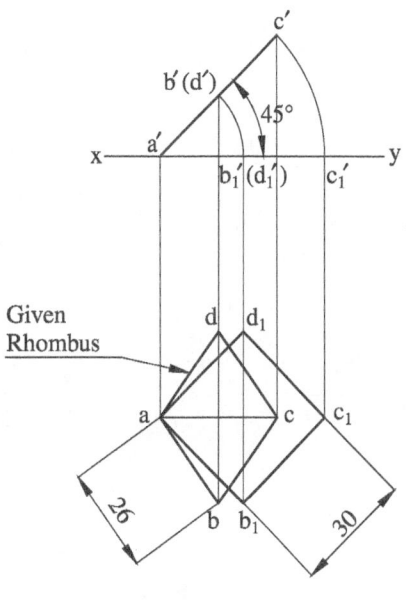

Fig. 8.26

PROCEDURE (Refer Fig. 8.26)

Step 1: Draw the tilted shape, that is, the rhombus a–b–c–d with the diagonal through 'a' parallel to the reference line and draw its corresponding front view as a straight line inclined at 45° as shown by a'–$b'(d')$–c'.

Step 2: Tilt the inclined straight line to the horizontal orientation about a' and obtain its corresponding projection ab_1 c_1 d_1 which gives the true shape as square of side. ▲

Example 8.23

A circular lamina appears as an ellipse with major and minor axes as 140 mm and 85 mm respectively, when tilted about the HP. Find the diameter of the lamina and its angle of inclination with the HP.

Hint Since the tilting of the lamina occurs about a point, draw the tilted shape as an ellipse in the top view with the point and the major axis through that point parallel to the reference line and obtain the other view as a straight line in inclined orientation.

PROCEDURE (Refer Fig. 8.27)

Step 1: Draw the tilted shape, the ellipse of minor axis pq and major axis rs, using any method mentioned in Chapter 2, such that the minor axis is parallel to the reference line in the top view. Obtain the corresponding front view p'–$r'(s')$–q', by cutting an arc of length equal to the major axis on the projector through q. Measure the inclination of this line and get the HP angle of the lamina.

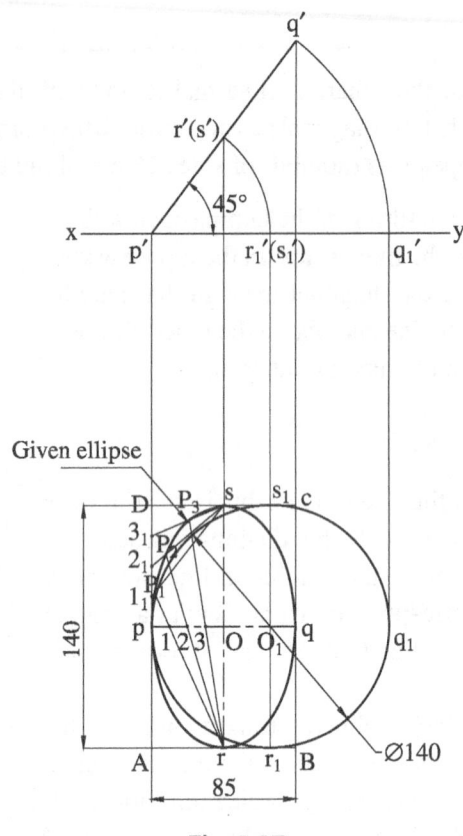

Fig. 8.27

Step 2: Tilt the inclined straight line to the horizontal orientation about p′ and obtain its corresponding projection $pr_1 q_1 s_1$ which gives the true shape as a circle of diameter equal to 140 mm. ▲

8.5 ENGINEERING APPLICATIONS

Example 8.24

The front and top view of a window ventilator appears as a rectangle of size 100 × 52 cm and 100 × 30 cm, respectively. The ventilator is inclined at 30° to the wall. Find the inclination of the ventilator with the wall. Also find true shape of the ventilator.

Hint   Ventilator is a rectangular plane fixed in the wall and is generally hinged with its longer side at the top and when opened yields rectangles in front and top views and a straight line inclined to the wall in the side view. The common dimension in the views reveals the edge or side about which it is tilted.

PROCEDURE (Refer Fig. 8.28)

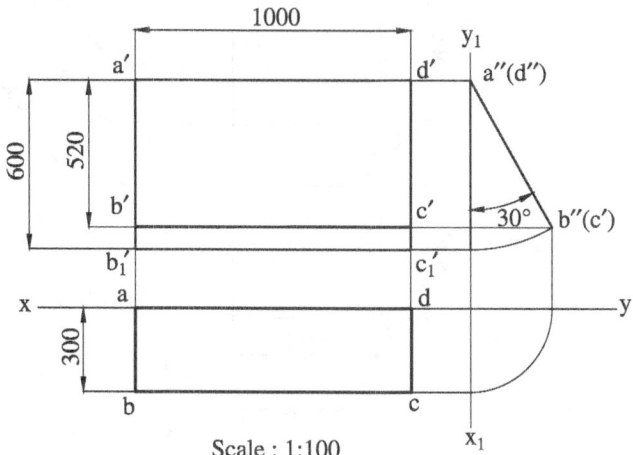

Scale : 1:100

Fig. 8.28

Step 1: Draw two rectangles as per the given dimensions in the front and top views with the common dimension 100 cm as its length.

Step 2: Add a side view corresponding to the above by keeping the 100 cm dimension as the hinge on the vertical reference line, which indicates the wall. Obtain the angle of inclination in the side view which is the inclination of the ventilator with the wall.

Step 3: Tilt the side view in alignment with the vertical reference line and find its corresponding front view $a'b_1'c_1'd'$ which gives the true shape of the ventilator. ▲

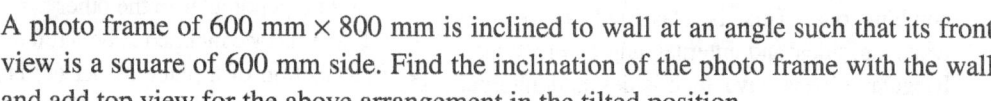

Example 8.25

A photo frame of 600 mm × 800 mm is inclined to wall at an angle such that its front view is a square of 600 mm side. Find the inclination of the photo frame with the wall and add top view for the above arrangement in the tilted position.

Hint Photo frames are usually held on the wall at its lower end and kept inclined for better visibility. It can be noted that the true dimensions of the photo frame can be visualized from the length of the lower end and the length in the side view.

PROCEDURE (Refer Fig. 8.29)

Step 1: Draw horizontal and the vertical reference lines. The vertical reference line corresponds to the wall in the side view.

Step 2: Draw the front view a' ... d' corresponding to the square of given dimensions. Draw the locus lines to the side view domain.

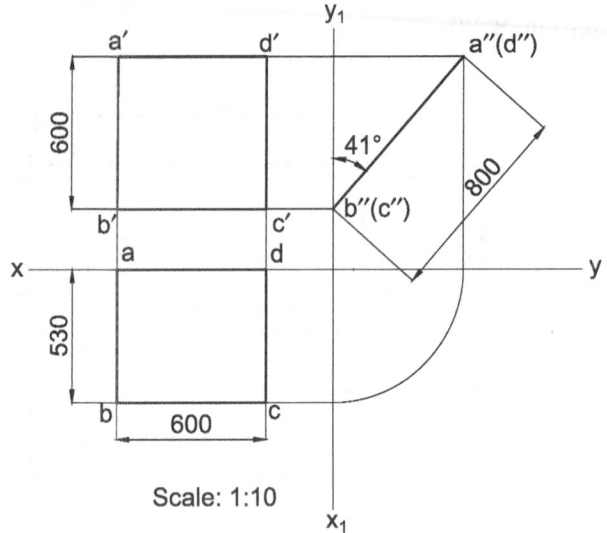

Fig. 8.29

Step 3: The bottom edge b'c' represents the edge about which the photo frame is tilted. With b"(c") as centre and the true length of one side 800 mm as the radius, cut an arc on the locus line and obtain a"(d"). The inclination made by this line gives the true inclination of the photo frame with the wall.

Step 4: Add a corresponding top view abcd as shown in Fig. ▲

Definition

- A plane surface is a two-dimensional surface bounded by straight lines, curves, or their combinations. This is classified as regular, when its sides and internal angles are equal. Irregular planes have their sides and internal angles unequal. Curved shapes or boundaries also constitute irregular planes.
- The physical objects that are very thin or having negligible thicknesses are also classified as planes and their projections are drawn in similar lines as that of any other plane surface.

Position of Plane Surface

- A plane surface may be positioned in space or in between two imaginary reference planes in any one of the following positions:

  - Plane surface parallel to one reference plane and perpendicular to the other
  - Plane surface inclined to one reference plane and perpendicular to the other
  - Plane surface inclined to both reference planes
- The position of a plane as per (1) is known as simple positions and (2) and (3) are referred as inclined positions.

Projections of Plane Surface—The Basics

- When a plane surface is held parallel to one reference plane, the projection or view on that plane will reveal the true shape of its surface, since all its sides or edges are parallel to the reference planes.
- When a plane surface is held perpendicular to a reference plane, its projection on that plane will

be a straight line. If the plane surface is also held parallel to the other reference plane, then this straight line will appear parallel to the reference line, and if it is inclined to the other reference plane, the straight line view also will be inclined to the reference line depending on the inclination of the object with the reference plane.

- Logically, when a plane surface is held inclined to both reference planes, the object is not parallel and not perpendicular either. Therefore, its views on both reference planes will not be straight lines but will be composed of shapes of reduced sizes instead of the true shape.

Projection of Plane Surface in Simple Positions

- When a plane surface is parallel to one reference plane (and hence perpendicular to the other reference plane), its true shape will be visualized on the plane to which it is parallel and the other view will be a straight line parallel to the reference line.

Projection of Planes in Inclined Positions

Projection of a Plane Inclined to One Reference Plane

- The surface of a planar object gets inclined to one reference plane, when it is tilted about one of its edges or one of its corners such that its surface is perpendicular to the other reference plane. The projections are drawn in two stages.
- In the first stage, the plane surface is kept parallel to the reference plane to which it is to be inclined later and its projections are drawn similar to that in simple positions. This will enable the true shape of the plane in one view and a straight line view in the other.

- If the object is tilted about an edge, the true shape of the plane surface is drawn with that edge perpendicular to the reference line, and if it is tilted about a corner, the true shape is drawn with its diagonal parallel to the reference line.
- In the second stage, the straight line view (non-polygonal view) is then tilted to the required inclination of the surface and its corresponding projection (a reduced shape) is obtained from that of the first stage.

Projection of a Plane Inclined to Both the Reference Planes

- When the plane surface is inclined to both reference planes, the edge about which it is tilted or the diagonal containing the corner also gets inclined over and above the positional reference mentioned in the previous case. This requires drawing one more set of projections after the second stage mentioned earlier.
- The reduced shape obtained in the second stage is further redrawn about the edge or the corner containing the diagonal kept as reference earlier and the corresponding projections are obtained.

To Find the True Shape of the Object from Its Projections

- It is known that when the plane surface is inclined, the true shape gets reduced, but the straight line view maintains the same dimension in an inclined orientation. If the projections of the inclined surface (the reduced shape and the straight line in the inclined orientation) are given, one can find the true shape of the object by following reversal procedure of the sequences adopted.

MULTIPLE-CHOICE QUESTIONS

8.1 The top view of a rectangular lamina parallel to VP and perpendicular to HP will appear as a
(a) straight line   (c) rectangle
(b) square   (d) point

8.2 The surface of a rectangular photo frame makes 30° with the wall. The true shape of the photo frame can be obtained from
(a) the side view   (c) the top view
(b) the front view   (d) cannot be obtained

8.3 A pentagonal plane rests on HP on one of its corners and its surface is inclined to HP and perpendicular to VP. Which of the following positions are suitable as its initial projections?

(a) Plane parallel to HP with the edge opposite to the corner (on which it rests) parallel to xy line

(b) Plane parallel to VP with the edge opposite to the corner (on which it rests) parallel to 'xy' line

(c) Plane parallel to HP with the edge opposite to the corner (on which it rests) perpendicular to 'xy' line

(d) Plane parallel to VP with the edge opposite to the corner (on which it rests) perpendicular to 'xy' line

8.4 The front view of a tent appears as an equilateral triangle of side 10 m. The height of the tent is

(a) 5 m   (c) 10 m
(b) 11.18 m   (d) 15 m

8.5 A semicircular lamina is resting on HP on its diametrical edge and the surface is parallel to VP. Which of the following views gives the true shape of the lamina?

(a) Top view   (c) Left-side view
(b) Right-side view (d) Front view

8.6 A circular surface is perpendicular to the VP rests on a point on its circumference on HP and the surface inclined to it. Its top view will appear as

(a) a circle   (c) a straight line
(b) an ellipse   (d) a point

8.7 A thin hexagonal plate has one of its edges touching the HP and the corresponding opposite edge touching the VP. Its side view appears as

(a) a straight line intersecting horizontal and vertical reference lines

(b) a straight line perpendicular to the horizontal reference line

(c) a straight line parallel to the horizontal reference line

(d) a hexagon

8.8 A pentagonal plane is perpendicular to both HP and VP. Its true shape can be visualized in

(a) the top view   (c) the front view
(b) the side view   (d) none of the views

8.9 An artistic design is to be made on the surface of a vertical wall. Which of the views will explain the artistic design?

(a) the top view   (c) the front view
(b) the side view   (d) none of the views

8.10 The details pertaining to the wall thickness, the cross-sectional details of pillars, and the size of the room sizes will appear in which of the following view?

(a) the top view   (c) the front view
(b) the side view   (d) exploded view

WORK PRACTICE LEVEL – 1

Plane Surface Inclined to One Reference Plane and Perpendicular to the Other

8.1 A square lamina of 40 mm sides has one of its corners on HP. Draw its projections when its surface is perpendicular to VP and 45° to HP.

8.2 A regular pentagonal lamina of 40 mm sides has one of its sides on HP. Draw its projections when the plane is perpendicular to VP and inclined at 30° to HP.

8.3 A regular hexagonal lamina of 40 mm sides is resting on one of its sides on VP. Draw its projections when its surface is inclined at 45° to VP and perpendicular to HP and the side on VP.

8.4 A semicircular plate has its flat end on VP and perpendicular to HP. Draw its projections when its surface makes 45° with the VP. The flat end of the plate measures 70 mm and its thickness is negligible.

8.5 A thin hexagonal plate of sides 40 mm is lying on VP on one of its corners. Draw its projections when the surface makes 30° with VP and the diagonal is parallel to HP.

8.6 A circular lamina of 60 mm diameter rests on VP such that the surface of the lamina is inclined at 30° VP and perpendicular to HP. Obtain its projection.

8.7 A pentagonal lamina of 40 mm sides has a circular hole of 40 mm diameter, centrally punched. Draw the projections of the lamina when its surface is inclined at 30° to VP and with one of the sides of the pentagon parallel to VP and perpendicular to HP.

WORK PRACTICE LEVEL – 2

Projection of Planes Inclined to Both Principal Planes

Planes Resting on an Edge

8.1 A square lamina of 40 mm sides rests on one of its side on HP. The lamina makes 60° to HP and the side on which it rests makes 30° to VP. Draw its projections.

8.2 A thin pentagonal plate with edge of 40 mm length is resting on HP on one of its edges. This edge is inclined at 45° to VP and the plate surface makes 30° with HP. Draw its projections.

8.3 A hexagonal lamina of 40 mm sides rests on one of its sides on VP. The surface of the lamina makes 30° with VP. Draw its projections, when the side which rests on VP, is inclined at 45° to HP.

Planes Resting on a Corner

8.4 An equilateral triangular lamina of 40 mm sides rests on one of its corners on HP such that the median passing through that corner on which it rests, is inclined at 30° to HP and 45° to VP. Draw its projections.

8.5 A thin lamina ABCD of 40 mm sides has one of its corners touching VP. The surface makes 45° with VP. Draw its projection when the front view of the diagonal passing through the corner on VP, makes an angle of 30° to reference line.

8.6 A pentagonal lamina of sides 40 mm is resting on the VP on one of its corners, so that the surface makes an angle of 45° with VP, draw the front view and top view of the pentagon, if its surface is vertical.

8.7 A hexagonal plane figure of side 40 mm is touching the VP on one of its corners, with its surface making an angle of 30° with the VP. Draw the projection of the plane figure, when

(a) The front view of the diagonal passing through that corner is inclined at 40° to the reference line

(b) The diagonal passing through that corner is inclined at 40° to the HP

8.8 A circular plate of diameter 70 mm has the end P of the diameter PQ in the HP and the plate is inclined at 40° to HP. Draw its projection when

(a) The diameter PQ appears to be inclined at 50° to VP, in the top view

(b) The diameter PQ makes 50° with VP

8.9 A circular plate of 70 mm in diameter has a hexagonal hole of 20 mm sides, centrally punched. Draw the projections of the lamina resting on HP, with its surface inclined at 30° to HP and the diameter through the point on which its rests on HP, is inclined at 50° to VP. Any two parallel sides of the hexagonal hole are perpendicular to the diameter of the circular plate, passing through the point on which it rests. Draw the projections.

Planes Touching Both the Reference Planes

8.10 A regular hexagonal plate of 40 mm sides has one corner touching VP and opposite corner touching HP. The plate is inclined at 60° to HP and 30° to VP. Draw the projections of the plate if the thickness is negligible.

8.11 A pentagonal lamina of side 40 mm is resting on one of its sides on HP with the corner opposite to the edge, touching VP, this side is parallel to VP and the corner touches VP at a height 20 mm above HP. Draw the projections of the plane, and determine the inclination of the plane with the coordinate planes and the distance at which the side is from VP.

Engineering Applications

8.12 A rectangular lamina of 60 mm × 40 mm, appears as a square of 40 mm sides, when tilted about one of its edges and inclined to HP. Find the inclination of its surface with HP.

8.13 A rectangular ventilator measuring 1 m × 6 m is hinged on one of its longer edges on a vertical wall. The ventilator is opened through an angle of 30° to the wall. Draw the projection of the ventilator.

8.14 The top view of a circular lamina of diameter 60 mm resting on HP is an ellipse of major axis 60 mm and minor axis 40 mm. find the inclination of the lamina with HP. Draw the front view when the major axis of the ellipse in the top view is horizontal.

8.15 A rhombus of diagonals 120 mm and 60 mm long appears as a square of 60 mm sides in the front view. Find line inclination of the rhombus plane with VP.

8.16 A plane object appears as a regular pentagon of 40 mm sides in the top view, when tilted about one of its edges. The front view of the object is a straight line inclined at 60° to the reference line. Find the true shape of the planar object.

Answers for Multiple-choice Questions

8.1 (a)	8.2 (a)	8.3 (c)	8.4 (b)	8.5 (d)	8.6 (b)
8.7 (a)	8.8 (b)	8.9 (c)	8.10 (a)		

Projection of Solids

9

OBJECTIVES

This chapter will help the reader to understand the following:

- General configuration of a three-dimensional solid bounded by plane surfaces, whose coordinates are spread in three directions

- Classification schemes of solids such as polyhedra in general (prisms and pyramids in particular) and solids obtained by revolution of plane surfaces such as cylinders and cones, and their related nomenclature

- Principles of orthographic projections of right and regular solids in simple positions, when their axes are perpendicular to one reference plane and parallel to the other

- Orthographic projection of the aforementioned solids explained by the change of position method or tilting solid method, when their axes are inclined to one reference plane and parallel to the other

- Need for additional planes of projection or auxiliary planes, the concept of projecting the objects on to them and obtaining the relevant views

- Extension of views by the tilting solid method and by the auxiliary projection method, when the axis of the solid is inclined to both the reference planes

- Importance of different orientations of the solids and their respective views, as a knowledge tool to cut/section or combine/assemble at any orientation

- Engineering products/assemblies that involve the integrated knowledge of the projections discussed earlier

9.1 INTRODUCTION

A solid is a three-dimensional object having length, breadth, and height or thickness as its three dimensions and bounded by plane or curved surfaces. As discussed in Chapter 4, a three-dimensional solid requires three orthographic views, namely front view, top view, and side view (right or left), to describe all its dimensions. If geometrical similarities exist, two views may become sufficient. In certain cases where the details along any other orientation of the solid are required, auxiliary or additional views can be drawn by creating additional or auxiliary planes, apart from the conventional reference planes.

In this chapter, solids whose geometry and shapes are regular and simple are discussed, although real-life solid objects have curved or irregular shapes. The knowledge of obtaining the orthographic views of a solid will enable the reader systematically to combine different shapes and sizes and to visualize an engineering product or a machine part, which is made up of the assembly of different shapes.

9.2 CLASSIFICATION OF SOLIDS AND THEIR SHAPES

9.2.1 Pictorial Views of Solids

Solids are generally classified into the following two groups:

1. Polyhedra, which are obtained by the arrangement of polygon shapes.
2. Solids of revolution, which are obtained by revolving a polygon about one of its sides.

Polyhedra

A polyhedron is a three-dimensional solid bounded by flat faces or plane surfaces and straight edges. The word 'Polyhedron' comes from the classical Greek word—'poly' means 'many' and 'hedron' means 'base, seat, or face'. The plural word for polyhedron is 'polyhedra', and they are often named by the number of faces.

Polyhedron Formed with Similar Plane Surfaces

Polyhedron is regular, when it is formed with equal and regular plane surfaces or polygons. The shapes of ployhedra are adopted in physics to represent the structure of various atoms and ions, in chemistry to represent various molecular bonds, in biology to represent many viruses and genes. While certain natural crystals exhibit some of these shapes, these are intentionally adopted in sports and toys as dices. In engineering applications, they are used for space filling with honeycomb arrangement of structures. In fact, the structural arrangement of geodesic dome was designed by the famous innovator Buckminster Fuller with regular polyhedra. The construction of large volume spherical and hemispherical structures is enabled by only the arrangement of regular polyhedron. Some of the commonly used regular polyhedra shapes are discussed in the following sections, though many combinations of these shapes are derived from them.

Polyhedron Formed with Triangular Plane Surfaces

The regular polyhedra formed with equilateral triangular faces can be categorized as follows:

Tetrahedron   A tetrahedron is a polyhedron bounded by a flat polygon base and three equilateral triangular faces connecting the base to a common point called the vertex or the apex. Its base is an equilateral triangle, and hence, this solid is known as a triangular pyramid. It has four triangular faces (including the base), six edges, and four vertices. These shapes are used to represent the covalent bonds of molecules in chemistry. Figure 9.1(a) shows the pictorial representation of a tetrahedron.

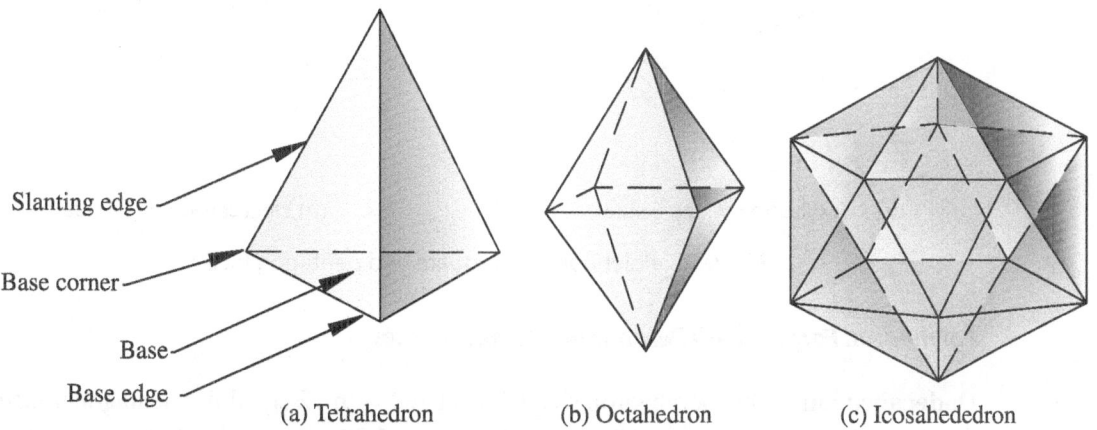

Slanting edge

Base corner

Base

Base edge

(a) Tetrahedron   (b) Octahedron   (c) Icosahededron

Fig. 9.1   Polyhedra with triangular faces

Octahedron   An octahedron is a three-dimensional shape realized with eight equal equilateral triangles as the faces, four of which meet at each vertex (2 vertices) independently. It has eight triangular faces, 12 edges, and two vertices. These shapes are observed in natural crystals of diamond, alum, and fluorite, which are also used in space-filling arrangements. Figure 9.1(b) shows the pictorial representation of an octahedron.

Icosahedron   An icosahedron is a polyhedron bounded by 20 equal equilateral triangles as the faces, 30 edges, and 12 vertices. These shapes are used in biology to represent many viruses, for example herpes virus that has icosahedral shells. They are also used in toys and games as a dice. The famous innovator Buckminster Fuller designed a world map in the form of an icosahedron known as Fuller projection, whose maximum distortion is only 2%. Figure 9.1(c) shows the pictorial representation of an icosahedron.

Polyhedron Formed with Square Plane Surfaces

Hexahedron or Cube   A hexahedron or cube is a regular polyhedron formed with six equal square faces, facets, or sides. One of the faces can be used as a base, while its parallel face above forms the top surface and the other faces form lateral surfaces. It can be cut into six identical square pyramids. It is one of the most common shapes to be imagined and produced which are very often used as a dice in games. Figure 9.2(a) shows the pictorial representation of a hexahedron.

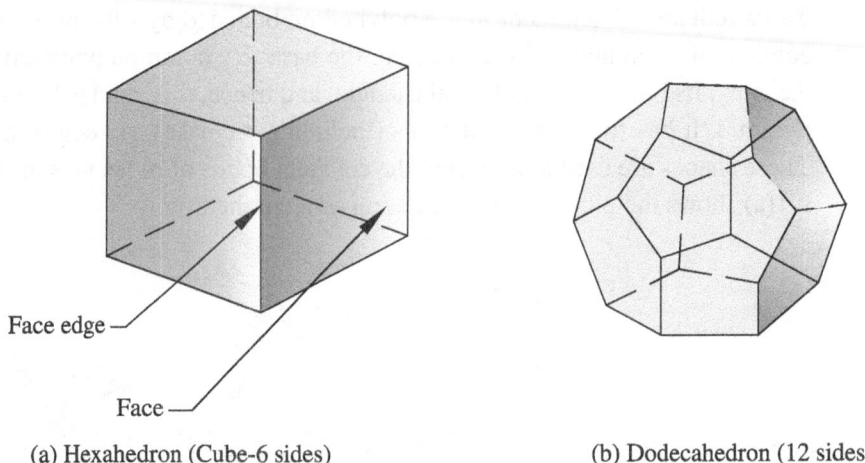

Face edge

Face

(a) Hexahedron (Cube-6 sides)　　　　　　(b) Dodecahedron (12 sides)

Fig. 9.2　Polyhedra with square and pentagon faces

Polyhedron Formed with Pentagonal Plane Surfaces

Dodecahedron　This polyhedron shape is realized with 12 equal flat pentagonal faces, three meeting at each vertex. This has 30 edges and 20 vertices. Such shapes occur in crystals called 'pyrites'. The crystals such as garnet and diamond exhibit dodecahedral habit. These shapes are used in visual arts, in sports as dice, and also in philosophy. Aristotle postulated heaven with these shapes! The pentagon shape is used to model the spherical shapes such as football and volley ball surfaces by suitably arranging them. Figure 9.2(b) shows the pictorial representation of a dodecahedron.

Polyhedron Formed with Dissimilar Plane Surfaces

A polyhedron can be realized with a polygon shape for its base and top (if available) and with another polygon shape for its bounding lateral surfaces. Such polyhedra are also called 'regular'. In many practical engineering applications, two popular polyhedra under this category play an important role and are discussed as follows.

Prism　A prism is a polyhedron having two equal and similar polygonal surfaces for its base and the top that are parallel to each other and connected by lateral surfaces known as the faces, which is made up of an another polygonal shape. In general, the lateral surfaces are either parallelograms or rectangular plane surfaces. The imaginary line connecting the centre of the base and the top is known as the axis of the solid. The prisms are named according to the shape of their bases, such as triangular, square, rectangular, pentagonal, and hexagonal.

　　When the axis of the solid is perpendicular to the base and the top that are regular polygons and the lateral faces are rectangles, then such solids are called right and regular prisms. Figure 9.3 shows the pictorial views of various types of regular and right prisms.

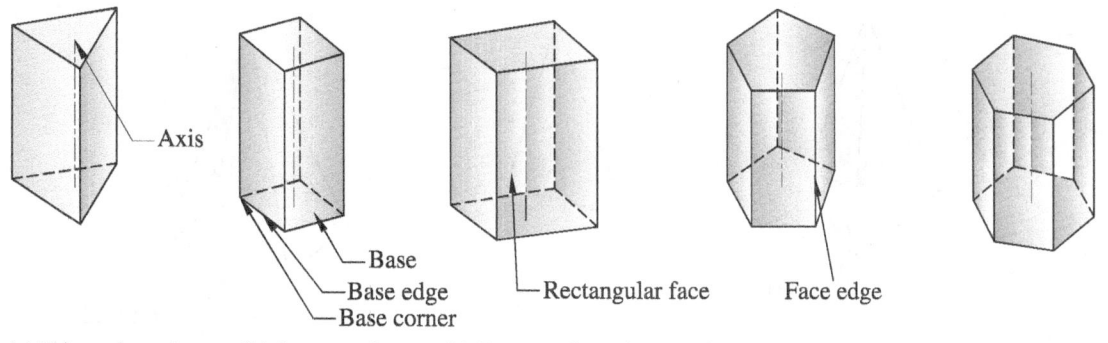

(a) Triangular prism   (b) Square prism   (c) Rectangular prism   (d) Pentagonal prism   (e) Hexagonal prism

Fig. 9.3   Regular prisms

When the axis of the solid is inclined to the base and the top and the lateral faces are parallelograms, such solids are called oblique prisms.

Pyramid   A pyramid is a polyhedron having its base as a polygonal surface and a number of triangular faces as its lateral surfaces, connecting the base to a common meeting point, known as the apex or the vertex. In general, the lateral surfaces are triangular plane surfaces and are slanting as they converge to a point. The imaginary line connecting the centre of the base and apex is called axis of the solid. The pyramids are named according to the shape of their bases such as triangular, square, rectangular, pentagonal, and hexagonal.

When the axis of the solid is perpendicular to the base that is made of a regular polygon and the lateral faces are isosceles triangles, such solids are called the right and regular pyramids. Figure 9.4 shows the pictorial representation of various types of regular and right pyramids.

Solids of Revolution

These are solids that are obtained by the revolution of a polygonal plane surface about one of its sides, which remains fixed. That edge of the polygon that resolves and facilitates the generation of the lateral surface is known as 'the generator'. The base, which is obtained during the revolution, is a closed curve, that is, a circle. As discussed earlier, when the axis of the solid is perpendicular to the base, it is called a right circular (since the base is a circle) solid, and when the axis is inclined to the base, it is known as an oblique solid.

Cylinder   A right circular cylinder is a solid obtained by the revolution of a rectangle, about one of its sides, which remains fixed. It has two circular ends or the base and the top that are parallel to each other. As mentioned earlier, the axis of the cylinder is an imaginary line connecting the centre of the base and that of the top and makes 90° with the base and the top ends. Figure 9.5(a) shows the pictorial representation of a cylinder.

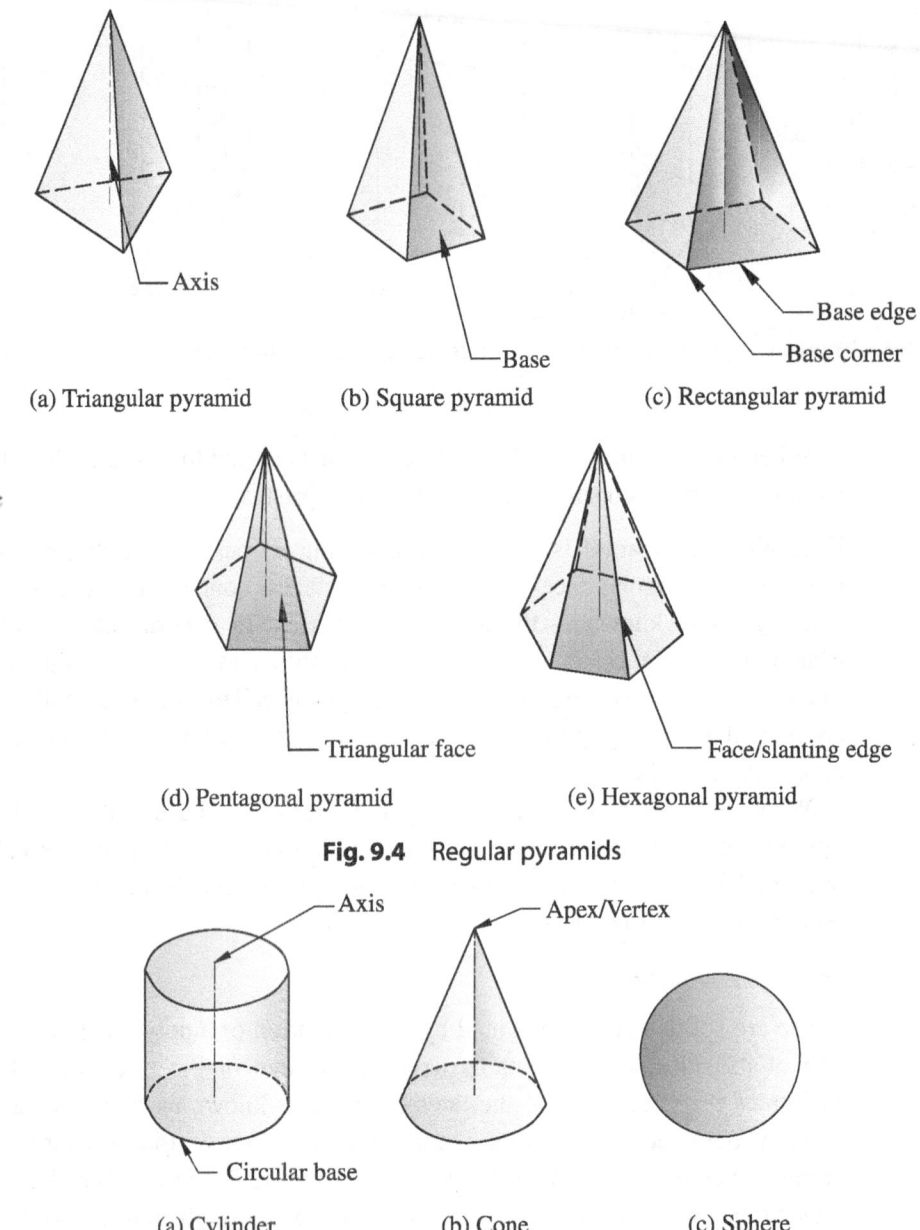

(a) Triangular pyramid   (b) Square pyramid   (c) Rectangular pyramid

(d) Pentagonal pyramid   (e) Hexagonal pyramid

Fig. 9.4   Regular pyramids

(a) Cylinder   (b) Cone   (c) Sphere

Fig. 9.5   Solids of revolution

Cone   A right circular cone is a solid formed by the revolution of a right-angled triangle about one of its perpendicular sides, which remains fixed. It has one circular end or the base and has an apex or vertex with which the base joins, while the sloping generators form the curved lateral face. As mentioned earlier, the axis of the cone is an imaginary line connecting the centre of the base and the vertex and makes 90° with the base. Figure 9.5(b) shows the pictorial representation of a cone.

Sphere   Sphere is a solid formed by the revolution of a semicircle about its diameter, which remains fixed and forms the imaginary axis. The midpoint of the diameter is known as centre of the sphere, and all points on the surface are equidistant from it. Figure 9.5(c) shows the pictorial representation of a sphere.

Oblique Solids

When the axis of the solid is inclined to the base, in its natural status during the formation of the solid itself, it is known as oblique solid. Therefore, in such solids, the axis and the base (the top if it exists) are not perpendicular to each other but inclined. Hence, the polygons pertaining to the base and the top (if applicable) are not regular and do not have sides whose lengths are equal.

In the case of oblique prisms, since the axis of the solid is inclined to the base and the top, the lateral faces are parallelograms. Figure 9.6(a) shows the pictorial representation of an oblique prism.

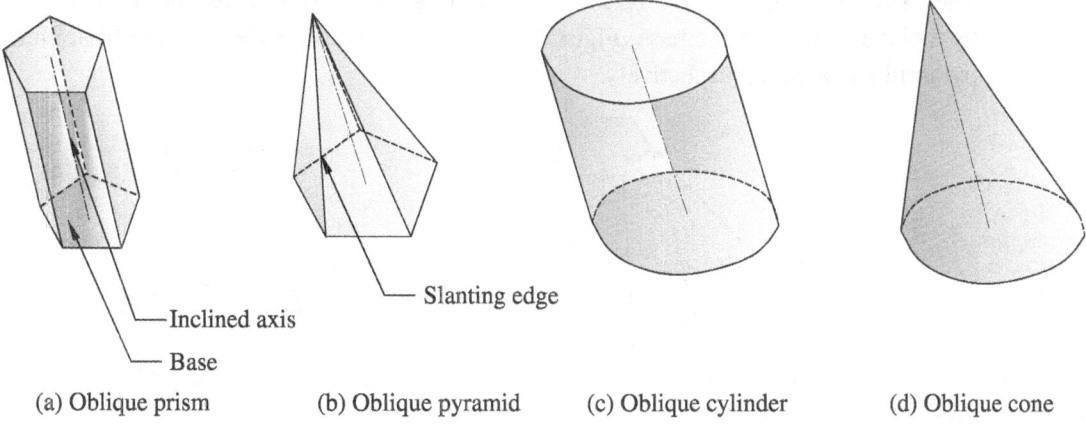

Inclined axis

Base

Slanting edge

(a) Oblique prism   (b) Oblique pyramid   (c) Oblique cylinder   (d) Oblique cone

Fig. 9.6   Oblique solids

In the case of an oblique pyramid, since the axis of the solid is inclined to the base, the lateral faces are triangles of different dimensions. Figure 9.6(b) shows the pictorial view of an oblique pyramid.

Similarly, the axes of the oblique cylinder and oblique pyramid are also inclined to the base, and the base cross-section is also not circular but elliptical. Figure 9.6(c) and (d) shows the pictorial views of these solids.

In this chapter, the projections of oblique solids are not discussed, as they are beyond the scope of this book.

Truncated Solids

When a solid is cut by a plane inclined to the base, it is said to be truncated as it is not allowed to grow uniformly around its axis. The inclination can be of any nature resulting in the removal of some portion from the base too. Figure 9.7 shows the pictorial views of the truncated prism, pyramid, cylinder, and the cone, respectively.

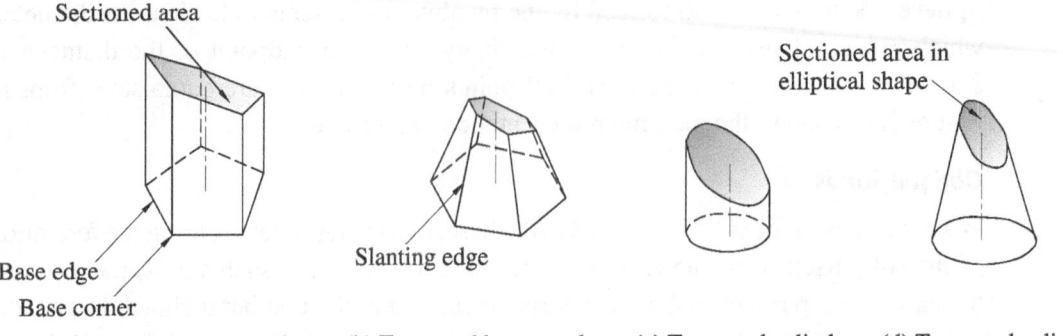

Sectioned area

Base edge

Base corner

Slanting edge

Sectioned area in elliptical shape

(a) Truncated pentagonal prism   (b) Truncated hexagonal pyramid   (c) Truncated cylinder   (d) Truncated cylinder

Fig. 9.7   Truncated solids

Frustum of Solids

The term frustum of a solid refers to the remaining portion of it, when the solid is divided by a plane parallel to the base. Figure 9.8 shows the pictorial views of the frustum of a pyramid and a cone, respectively.

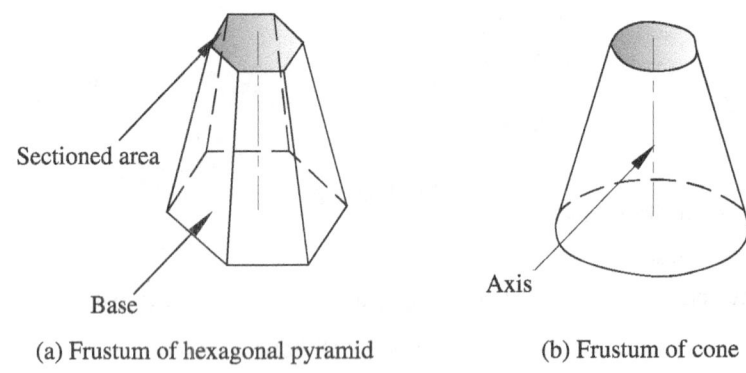

Sectioned area

Base

Axis

(a) Frustum of hexagonal pyramid   (b) Frustum of cone

Fig. 9.8   Frustum of solids

9.2.2 Nomenclature

In general, it is important to familiarize the user with the most commonly used terminologies or the nomenclature of the various terms of the solids, before discussing the projections of a solid with respect to the reference planes. These are presented in this section.

Axis of a Solid

An imaginary line connecting the centre of the base and that of the top (in the case of prisms and cylinders), or the apex/vertex (in the case of pyramids and cones) is known as axis of the solid, and it is always perpendicular to the base for all right and regular solids. (In this chapter, oblique solids are not discussed.) Figures 9.3(a) and 9.4(a) show the axis and its orientation with the base, in the case of a prism and pyramid, respectively.

Base of a Solid

It is the shape of the polygon representing the base on which the solid is built to form its required height. In case of prisms and pyramids, the base is a regular polygon such as triangle, square, rectangle, pentagon, and hexagon of the solid; and in case of cylinders and cones, it is the circle. When the solid is positioned vertically, the base and the top are the relevant terms, and when it is positioned horizontally, the ends are the appropriate terms to define the solid shape. As discussed earlier, the top or the other end is applicable only to prisms and cylinders. As discussed earlier, the base of the solid is always perpendicular to the axis. Figures 9.3(b) and 9.4(b) show the base of a prism and pyramid, respectively.

Base Edge or Side

It is defined as the side of the polygon representing the base, and it is the meeting edge of the base and the lateral face of the solid. When the solid is inclined to a reference plane, it will be tilted about a base edge (as one possibility), and in that case it becomes mandatory to keep this edge perpendicular to the reference line in the initial projections. (This was highlighted in Chapter 8 and will be explained in detail later). For cylinders and cones, this term is not applicable. Figures 9.3(b) and 9.4(c) show the base edges of a prism and pyramid, respectively.

Base Corner

It is the meeting point of the base edges of a polygon representing the base. It is also the point from which the lateral edge of the solid emanates, and often referred to as the base corner. When the solid is inclined to a reference plane, it will be tilted about a base corner (as another possibility), and in that case it becomes mandatory to keep the line joining this corner and the centre of the polygon (or sometimes the diagonal) parallel to the reference line in the initial projections. For cylinders and cones, this is always the diameter. (This was highlighted in Chapter 8 and will be explained in detail later). Figures 9.3(c) and 9.4(c) also represent the base corners of a prism and pyramid, respectively.

Lateral Face

This is a planar face representing the lateral surface of a solid which consists of the base edge and the lateral edges of the solid. In case of prisms, it is rectangular in shape and is perpendicular to the base of the solid. In pyramids, it is an isosceles triangle and is inclined to the base and is also referred as 'slant face'. When the solid is inclined to a reference plane and tilted about a base edge, the lateral face through that edge also gets inclined to the reference plane. For a cylinder and a cone, the lateral surface is curved. Figures 9.3(c) and 9.4(d) show the vertical rectangular lateral face of a prism and the slant lateral face of a pyramid, respectively.

Lateral Edge

This is a line or edge where two adjacent lateral faces of a solid meet and it emanates from the corresponding base corner. In prisms, it is perpendicular to the base of the solid. In pyramids, it is inclined to the base and is referred as 'slant edge'. When the solid is inclined to a reference plane and tilted about a base corner, the lateral edge through that corner also gets inclined to the reference plane. For a cylinder and a cone, this term is referred as the 'generator'. Figures 9.3(e) and 9.4(e) show the vertical lateral edge of a prism and the slant lateral edge of a pyramid, respectively.

9.3 PROJECTION OF SOLIDS IN SIMPLE POSITIONS

When a solid is kept with its axis perpendicular to one of the reference planes, it is said to be in a simple position. Since the axis being a straight line, when it is perpendicular to one of the reference planes, it is automatically parallel to the other reference planes. It will appear as a point in the plane to which it is perpendicular and its true length will be seen on the other reference planes.

In right and regular solid, the axis and its base are perpendicular to each other. Therefore, when the axis of the solid is perpendicular to one reference plane, its base is parallel to that reference plane, and hence, the true shape of the base is visualized on that reference plane.

For example, when the axis of a solid is perpendicular to horizontal plane (HP), its base is parallel to HP, and hence, the shape of the polygon corresponding to the base is visualized on HP or in the top view. The length or height of the axis of the solid will be visualized in the front view.

Similarly, when the axis of a solid is perpendicular to vertical plane (VP), its end is parallel to VP, and hence, the shape of the polygon corresponding to the end is visualized on VP or in the front view. The length or height of the axis of the solid will be visualized in the top view.

Similarly, when the axis of a solid is perpendicular to profile planes (PP), its end is parallel to PP, and hence, the shape of the polygon corresponding to the end is visualized on PP or in the side view. The length or height of the axis of the solid will be visualized in the front and top views.

The aforementioned observations are considered as hints for drawing the projections of the solid.

Figure 9.9(a) shows the pentagonal prism placed in the space between three reference planes system, that is, VP, HP, and PP.

Its front view, top view, and side views can be obtained by projecting it on the respective reference planes. Since the axis is perpendicular to HP, its base is parallel to HP, and hence, the pentagon appears in the top view. The top corners a, …, f are visible, and the bottom corners 1, …, 6 are hidden, and are shown in brackets in the top view. When the object is projected to VP, the front corners b', c', and d' at the top and 2', 3',

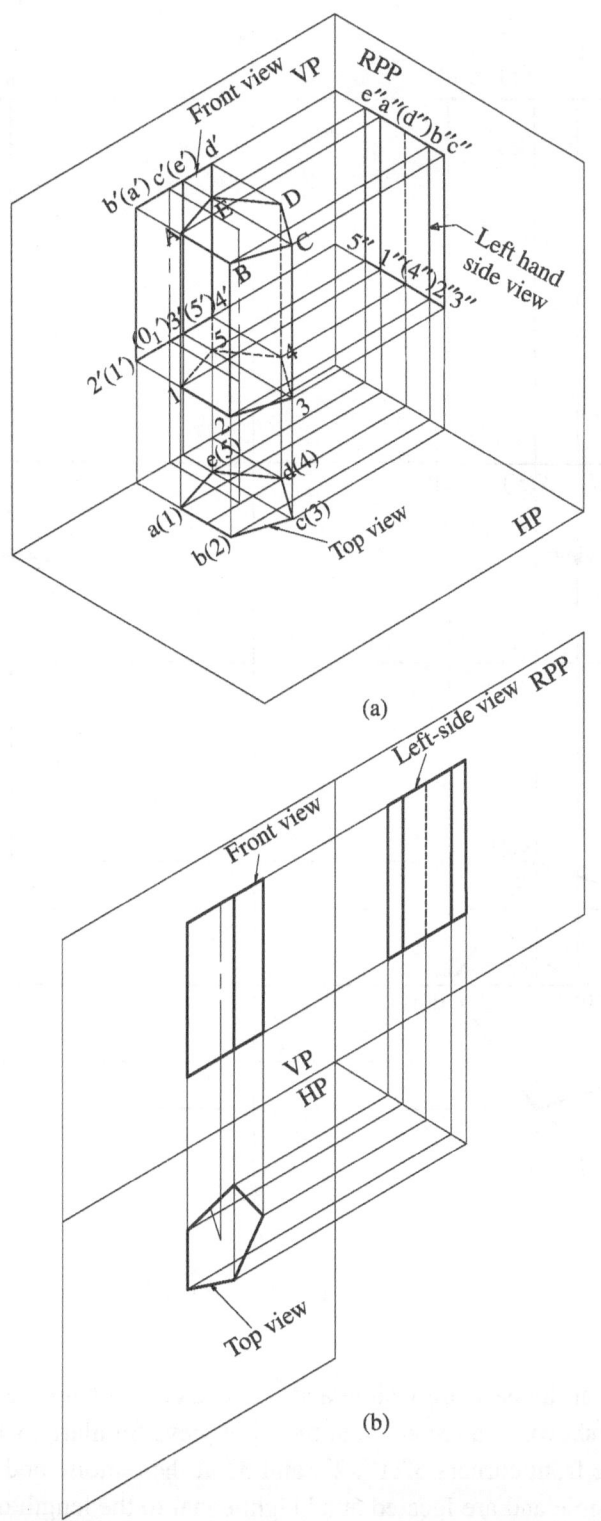

Fig. 9.9 Pentagonal prism with axis parallel to VP and perpendicular to HP (*Contd*)

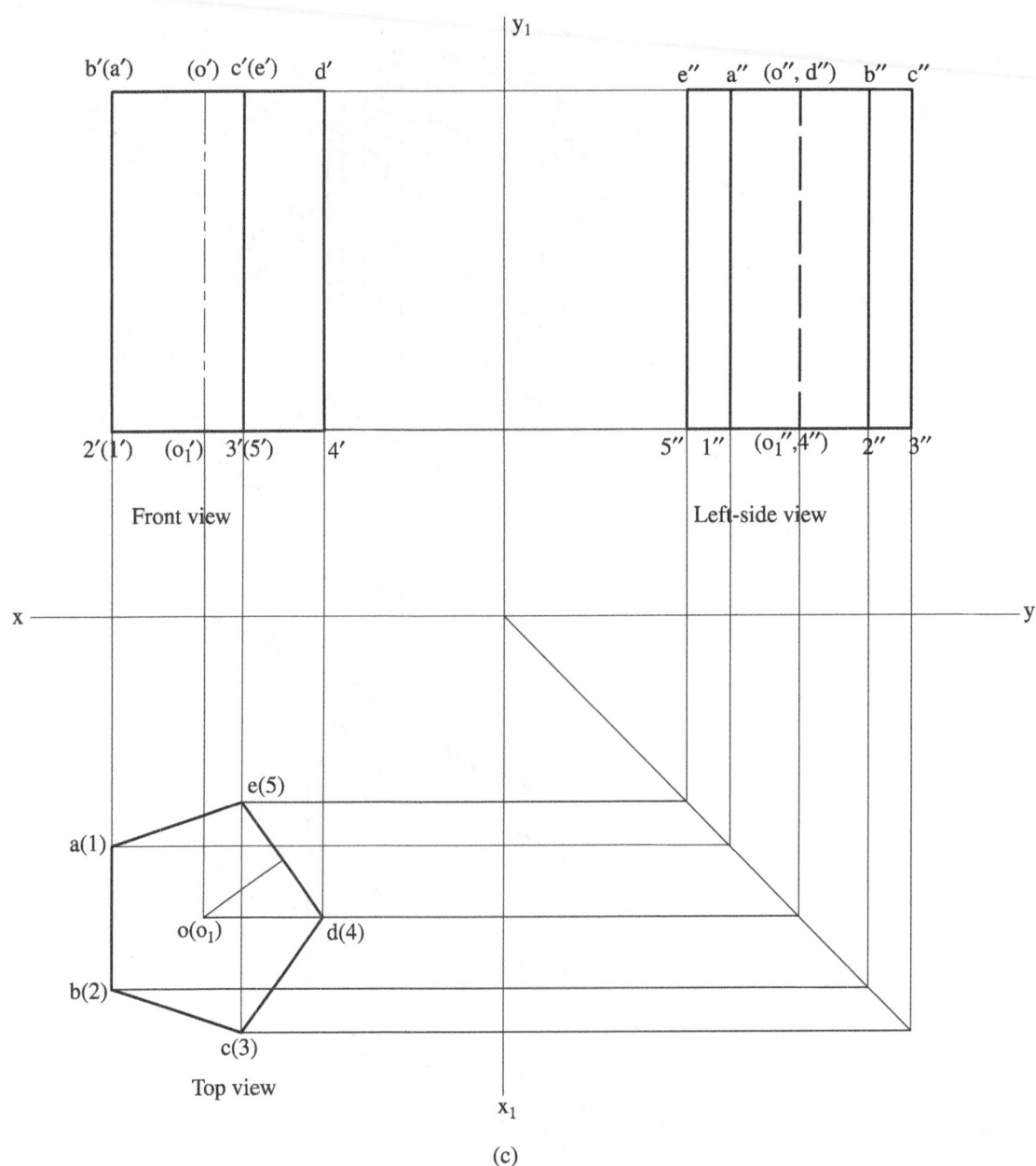

Fig. 9.9 *(Contd)*

and 4′ at the base are visible and are located at a height equal to the length of the axis and is shown as a rectangle in the front view. Similarly, when the object is projected to PP, the front corners 5″,1″, 2″, and 3″ at the bottom, and e″, a″, b″, and c″ at the top are visible and are located at a height equal to the length of the axis which is shown as a rectangle in the side view. Figure 9.9(b) shows the arrangement of these views on the respective planes that are brought in alignment with VP. Figure 9.9(c) shows the scheme of these views as per the orthographic projection layout.

Figure 9.10 shows the arrangement of the front and the top views, when the base of the solid is retained on the HP and is rotated such that the base occupies different orientations. It can be noted that the base polygon retains the same shape but gets rotated as per the following orientations.

1. Refers to the lateral face through the edge 'ab' is perpendicular to VP and HP, and hence, the base edge 'ab' is perpendicular to reference line. In the case of the pentagon shape, Fig. 9.10(a) also refers to the adjoining lateral faces equally inclined to VP and perpendicular to HP or the diagonal (line joining the centre and the corner d) parallel to the reference line.

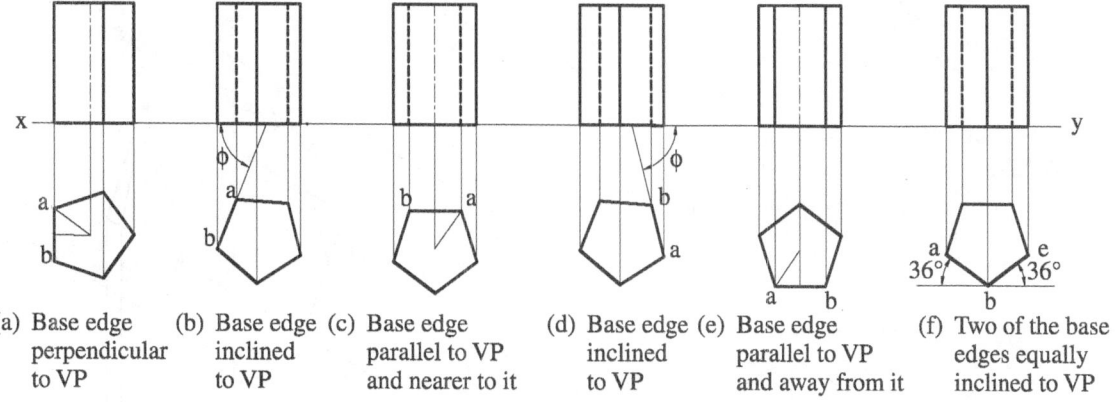

(a) Base edge perpendicular to VP

(b) Base edge inclined to VP

(c) Base edge parallel to VP and nearer to it

(d) Base edge inclined to VP

(e) Base edge parallel to VP and away from it

(f) Two of the base edges equally inclined to VP

Fig. 9.10 Various orientations of a pentagonal prism on HP

2. Refers to the lateral face through the edge 'ab' gets inclined to VP (ϕ) and perpendicular to HP, and hence, the base edge 'ab' is inclined at 'ϕ' to reference line.
3. Refers to the lateral face through the edge 'ab' parallel to VP and perpendicular to HP, and hence, the base edge 'ab' is parallel to reference line.
4. Refers to the lateral face through the edge 'ab' and perpendicular to HP.
5. Refers to the lateral face through the edge 'ab' is parallel to VP but far away from it.
6. Refers to the lateral faces through the edge 'ab', and 'bc' gets equally inclined to VP and perpendicular to the HP.

Figure 9.11(a) shows the projections of a pentagonal pyramid in a three plane system, while Fig. 9.11(b) describes the alignment of the projection planes with the VP. Figure 9.11(c) shows the orthographic views of the pyramid.

Figure 9.12 shows the arrangement of the front and top views, when the base of a hexagonal pyramid is retained on the HP and is rotated such that the lateral face (slant face) through the base edge 'ab' occupies different orientations. It can be noted that the base polygon retains the same shape but gets rotated as per the orientations similar to Fig. 9.10.

When the axis of solid is perpendicular to the VP, the polygon shape (pertaining to its ends) is visualized in the front view and similar orientations of the end polygons can be obtained as shown in Figs 9.10 – 9.12 with appropriate change of positions and front view notations.

Therefore, the reader can logically extend the projections and obtain the polygon views in the side view, when the axis of the solid is perpendicular to PP.

> **NOTE** *Among the various base orientations, the orientations of the base edge perpendicular to the reference line and a diagonal parallel to the reference line are very important and are used as the initial positions when the axis of the solid is inclined to one reference plane and parallel to the other. This will be explained in the next section in detail.*

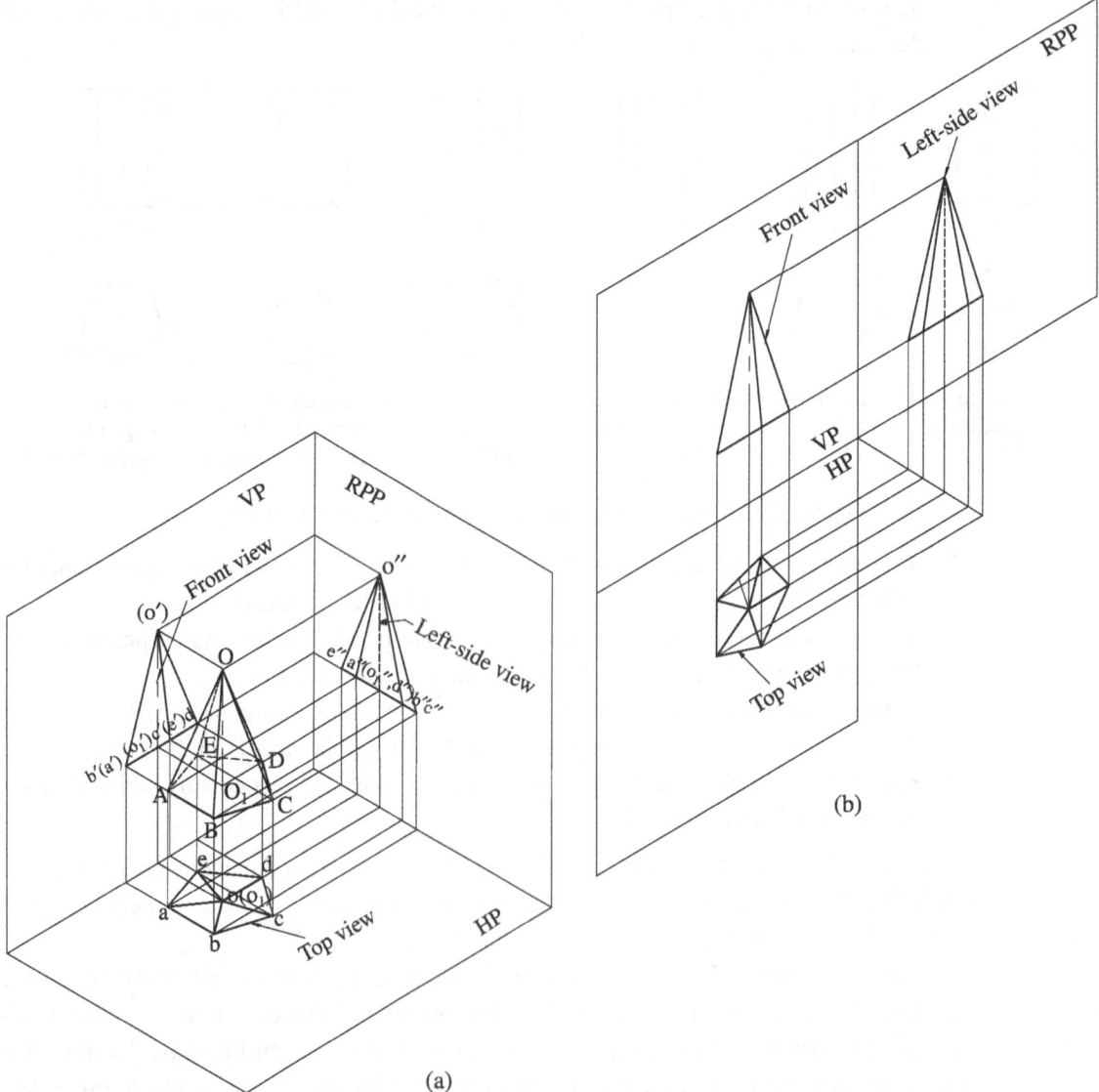

Fig. 9.11 Pentagonal pyramid axis parallel to VP and perpendicular to HP (*Contd*)

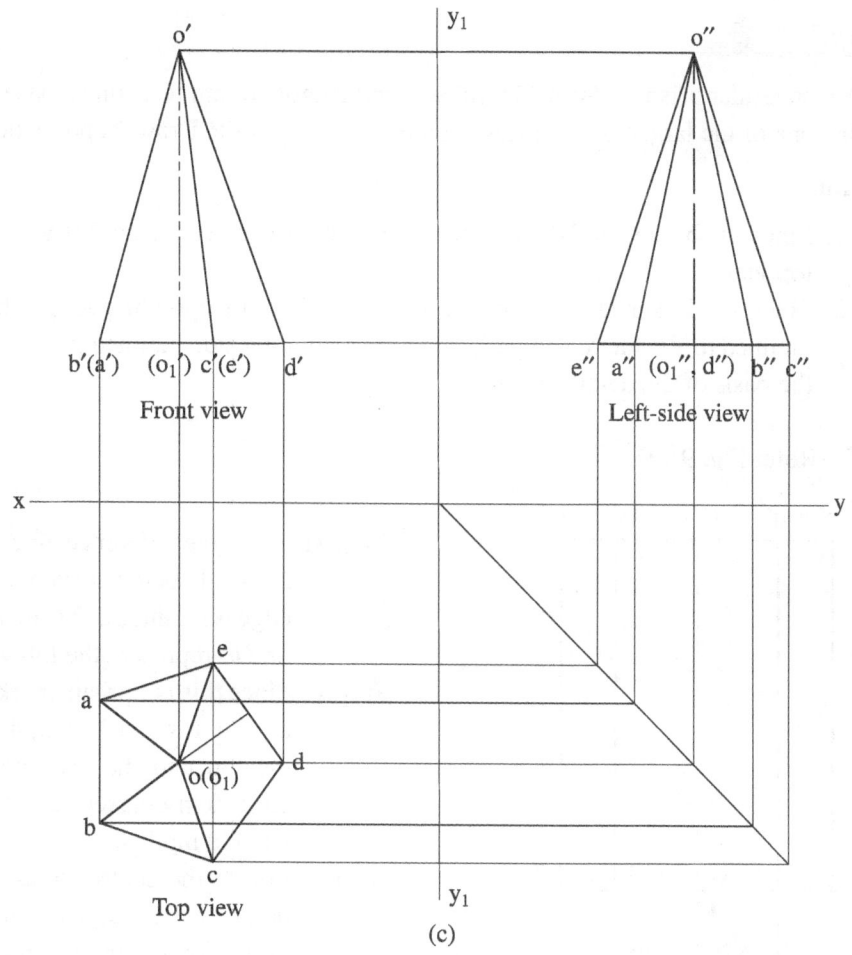

b'(a') (o₁') c'(e') d'
Front view

e" a" (o₁", d") b" c"
Left-side view

o(o₁)

Top view

(c)

Fig. 9.11 (*Contd*)

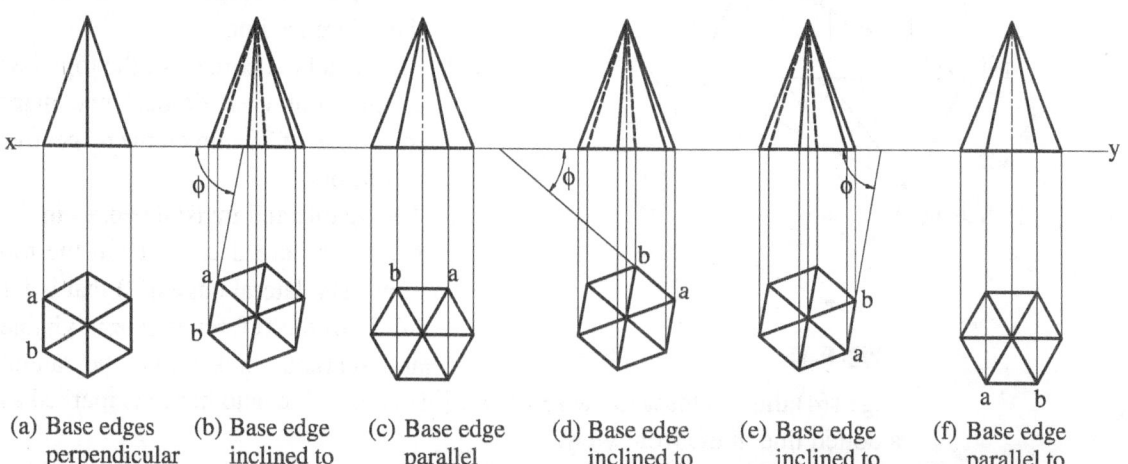

| (a) Base edges perpendicular to VP | (b) Base edge inclined to VP | (c) Base edge parallel to VP | (d) Base edge inclined to VP | (e) Base edge inclined to VP | (f) Base edge parallel to VP |

Fig. 9.12 Various orientations of a hexagonal pyramid on HP

Example 9.1

A rectangular prism of 50 × 25 mm base and length 70 mm rests on its base on HP such that one of the larger lateral faces is inclined at 27° to VP. Draw its projections.

Hint

1. Since the base is on HP, the corresponding polygon (rectangle) is visualized in the top view.
2. Since the larger face is inclined to VP or the corresponding longer base edge is inclined to VP, the rectangle is laid with the longer side inclined to reference line on the basis of the details given.

PROCEDURE (Refer Fig. 9.13)

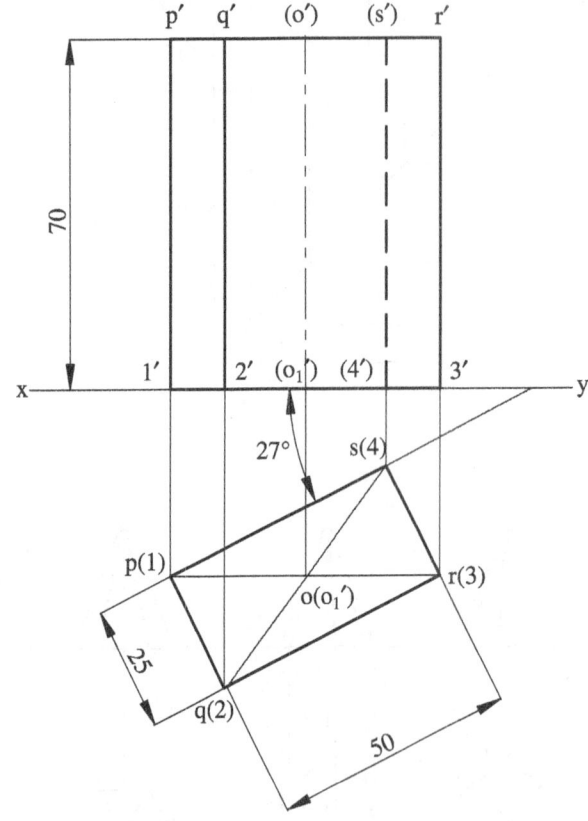

Fig. 9.13

Step 1: Mark the reference line and draw a rectangle below it such that its longer edge is inclined at 27° to the reference line to represent the top view.

Step 2: Since this is a prism, mark the corners as p, q, r, s and 1, 2, 3, 4 to represent the top and the base with the base corners in brackets as they are on the HP and hidden.

Step 3: Locate the centre 'o' as the meeting point of the diagonals and project it to the front view to indicate the axis and mark the length of 70 mm from the reference line.

Step 4: Project all the corners of the top view to the front view domain and mark the height of 70 mm on the respective projectors.

Step 5: The visible and invisible edges in the front view are decided from the top view. The lateral edge q(2) that is far off from the reference line is visible and marked thick while the lateral edge s(4) that is closer to the reference line is invisible, and hence is marked as a dotted line in the front view.

Step 6: The other edges that form the boundary in the front view are marked in thick lines as shown in Fig. 9.13.

Example 9.2

A pentagonal prism of 30 mm side of base and axis 70 mm long is resting on its base on HP in such a way that one of its faces is perpendicular to VP. Draw its projections.

Hint

1. Since the base is on HP, the corresponding polygon (pentagon) is visualized in the top view.
2. Since one of its face is perpendicular to VP, its corresponding base edge is perpendicular to VP, and hence, the pentagon is drawn with one side perpendicular to reference line.

PROCEDURE (Refer Fig. 9.14)

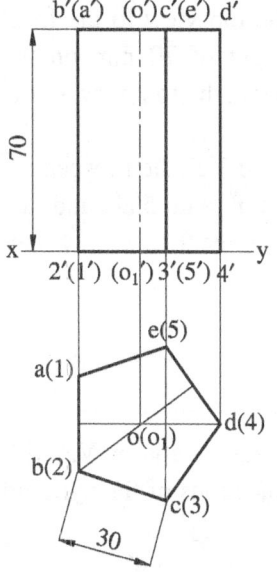

Fig. 9.14

Step 1: Mark the reference line and draw a pentagon of 30 mm sides below it with one side perpendicular to the reference line to represent the top view as shown by a, ..., e and 1, ..., 5 indicating the top and the base corners, respectively. The base corners are marked in brackets as they are on HP and hidden.

Step 2: Locate the centre 'o' as the meeting point of the diagonals, project it to the front view to indicate the axis, and mark the length of 70 mm from the reference line.

Step 3: Project all the corners of the top view to the front view domain, mark the height of 70 mm on the respective projectors, and obtain the font view of the prism.

Step 4: Mark the visible edge 3'c', its coincident invisible edge 5'e', and the boundary lines similar to Example 9.1.

Example 9.3

A hexagonal prism of 30 mm base edges and axis 70 mm long rests on one of its hexagonal ends on HP such that two of its lateral faces are parallel to VP. Draw its projections.

Hint

1. Since hexagonal end is on HP, it is visualized in the top view.
2. Since two of its lateral faces are parallel to VP, the hexagon is drawn with the two sides parallel to the reference line.

PROCEDURE (Refer Fig. 9.15)

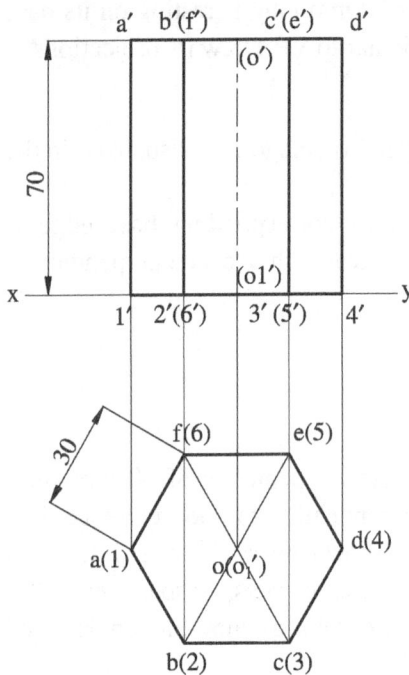

Fig. 9.15

Step 1: Mark the reference line and draw a hexagon of 30 mm sides below it with two sides parallel to the reference line to represent the top view as shown by a, ..., f and 1, ..., 6, indicating the top and base corners, respectively. The base corners are marked in brackets as they are hidden.

Step 2: Locate the centre 'o' as the meeting point of the diagonals, project it to the front view to indicate the axis, and mark the length of 70 mm from the reference line.

Step 3: Project all the corners of the top view to the front view domain, mark the height of 70 mm on the respective projectors, and obtain the font view of the prism.

Step 4: Mark the visible edges 2'b' and 3'c', their respective coincident invisible edges 6'f' and 5'e', and the boundary lines similar to Example 9.1. ▲

Example 9.4

A square pyramid of 30 mm sides and axis 70 mm long has its square end on VP with two of its adjoining edges equally inclined to HP. Draw the projections of the pyramid in this position.

Hint

1. Since square end is on VP, the corresponding polygon (square) is visualized in the front view.
2. Since adjoining edges are equally inclined to VP, two sides of square are equally inclined to the reference line.

PROCEDURE (Refer Fig. 9.16)

Step 1: Mark the reference line and draw a square of 30 mm sides with two sides equally inclined to the reference line to represent the front view as shown in Fig. 9.16.

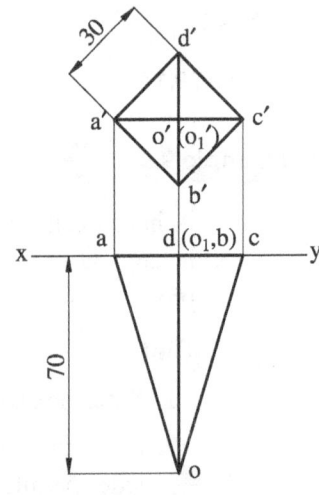

Fig. 9.16

Step 2: Locate the centre o′ as the meeting point of the diagonals, project it to the top view to indicate the axis, and mark the length of 70 mm from the reference line and obtain the apex 'o'.

Step 3: Project all the corners of the front view to the reference line, join these points with the apex, and obtain the top view of the pyramid.

Step 4: Mark the visible edge do, its respective coincident invisible edge bo, and the boundary lines similar to Example 9.1. ▲

Example 9.5 ◣

A pentagonal pyramid of 30 mm base edges and axis 70 mm long is lying on its base on HP such that one of its base edges is perpendicular to VP. Draw its projections.

Hint

1. Since the base is on HP, the corresponding polygon (pentagon) is visualized in the top view.
2. Since one of its base edges is perpendicular to VP, the pentagon is drawn with one side perpendicular to the reference line.

PROCEDURE (Refer Fig. 9.17)

Step 1: Mark the reference line and draw a pentagon of 30 mm sides below it with one side perpendicular to the reference line to represent the top view as shown in Fig. 9.17.

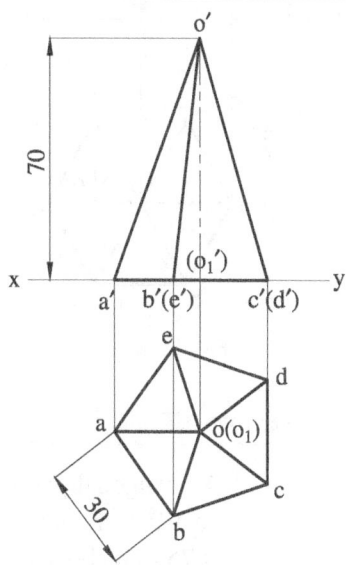

Fig. 9.17

Step 2: Locate the centre 'o' as the meeting point of the diagonals, project it to the front view to indicate the axis, mark the length of 70 mm from the reference line, and obtain the apex o′.

Step 3: Project all the corners of the top view to the reference line, join these points with the apex, and obtain the front view as in Fig. 9.17.

Step 4: Mark the visible edge b′o′, its respective coincident invisible edge e′o′, and the boundary lines similar to Example 9.1. ▲

Example 9.6 ◣

An inverted hexagonal pyramid of base edge 30 mm and axis 70 mm long has its apex on HP. Draw its projections, when its axis is perpendicular to HP and two of its hexagonal sides are parallel to VP.

Hint

1. Since the axis is perpendicular to the HP, the base is parallel to HP, and hence, the corresponding polygon (hexagon) is visualized in the top view, with two sides parallel to the reference line as noted in the problem.

PROCEDURE (Refer Fig. 9.18)

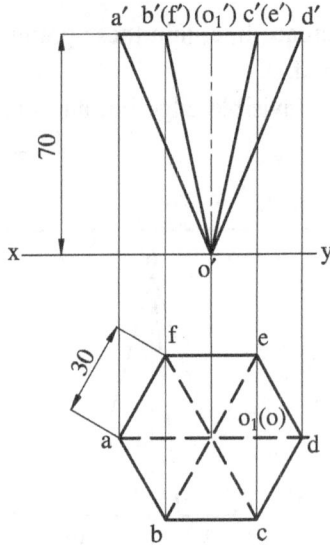

Fig. 9.18

Step 1: Mark the reference line and draw a hexagon of 30 mm sides below it with two sides parallel to the reference line to represent the top view as shown Fig. 9.18.

Step 2: Locate the centre 'o' as the meeting point of the diagonals, project it to the front view to indicate the axis, and mark the length of 70 mm from the reference line, and obtain the apex o′ on the reference line.

Step 3: Project all the corners of the top view to the front view domain to the length of 70 mm, join the corners with the apex, and obtain the front view as inverted triangle as in Fig. 9.18.

Step 4: Mark the visible edges b′o′ and c′o′ and their respective coincident invisible edges f′o′ and e′o′ and the boundary lines similar to Example 9.1. ▲

Example 9.7 ▲

A hexagonal pyramid of base sides 30 mm and axis 70 mm long has its hexagonal end on VP such that two of its sides are parallel to HP. Draw the projections when the axis is 75 mm above the HP.

Hint

1. Since hexagonal end is on VP, the corresponding polygon (hexagon) is visualized in the front view
2. Since two of its base sides are parallel to HP, hexagon is drawn with two sides parallel to reference line.

PROCEDURE (Refer Fig. 9.19)

Step 1: Mark the reference line and draw a hexagon of 30 mm sides above it with two sides parallel to the reference line to represent the front view as shown in Fig. 9.19.

Step 2: Locate the centre o′ as the meeting point of the diagonals, project it to the top view to indicate the axis, mark the length of 70 mm from the reference line, and obtain the apex o.

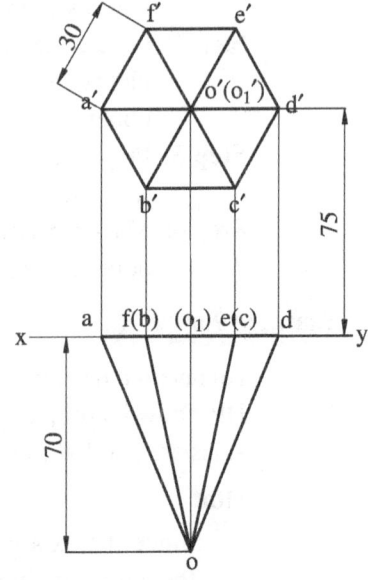

Fig. 9.19

Step 3: Project all the corners of the front view to the reference line, join these points with the apex, and obtain the top view as in Fig. 9.19.

Step 4: Mark the visible edges 'fo' and 'eo' and their respective coincident invisible edges 'bo' and 'co' and the boundary lines similar to Example 9.1. ▲

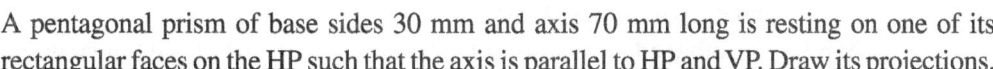

Example 9.8

A pentagonal prism of base sides 30 mm and axis 70 mm long is resting on one of its rectangular faces on the HP such that the axis is parallel to HP and VP. Draw its projections.

Hint

1. Since the axis is parallel to HP and VP, it is perpendicular to the profile plane, and hence, the cross section or the polygon (pentagon) is parallel to the profile plane and appears on it.
2. Since the prism rests on a rectangular face, the side of the base corresponding to it is on the reference line in the side view.

PROCEDURE   (Refer Fig. 9.20)

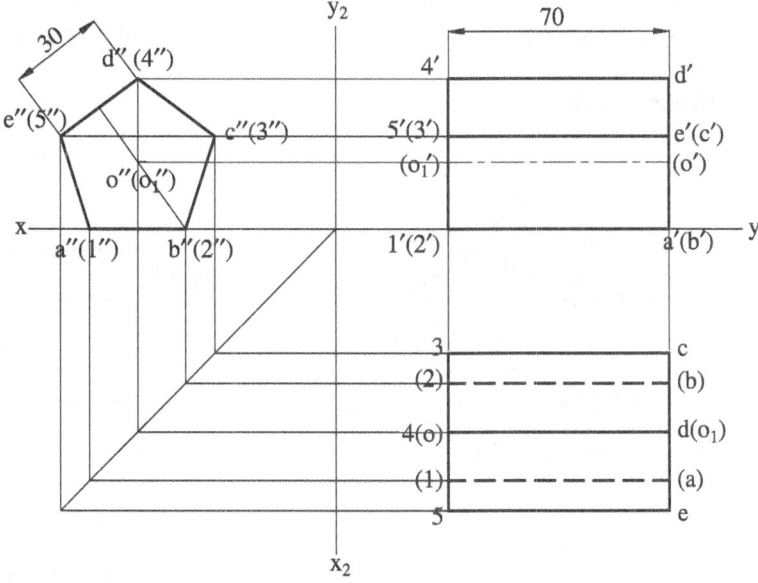

Fig. 9.20

Step 1: Mark the horizontal and vertical reference line and draw the pentagon of 30 mm sides, in the side view, such that one side is on the reference line

Step 2: Locate the centre, obtain the front view and top view from the side view, and mark the edges as shown in Fig. 9.20.

Step 3: The visible and invisible edges are marked in the front view with the side view as reference. The lateral edge e″5″ that is far off from the vertical reference line is visible, while the other lateral c″3″ that is closer to the vertical reference line is invisible, but this projection coincides with the former.

Step 4: Similarly, the visible and invisible edges are marked in the top view with the side view as reference. The lateral edge d″4″ that is far off from the horizontal reference line is visible, and hence marked in thick lines, while the lateral edges a″1″ and b″2″ that are closer and on the horizontal reference line are invisible and shown as dotted lines.

Step 5: The other edges that form the boundary in the front and top views are marked in thick lines as shown in Fig. 9.20.

Example 9.9

A hexagonal prism of base sides 30 mm and axis 70 mm long is resting on one of its lateral edges such that the axis is parallel to HP and VP. Draw its projections.

Hint

1. Since the axis is parallel to HP and VP, it is perpendicular to the profile plane, and hence the cross section or the polygon (hexagonal) is parallel to the profile plane and appears on it.
2. Since the prism rests on one of its lateral edges on HP, draw the hexagon with a corner on the reference line and with its corresponding two sides equally inclined to it.

PROCEDURE (Refer Fig. 9.21)

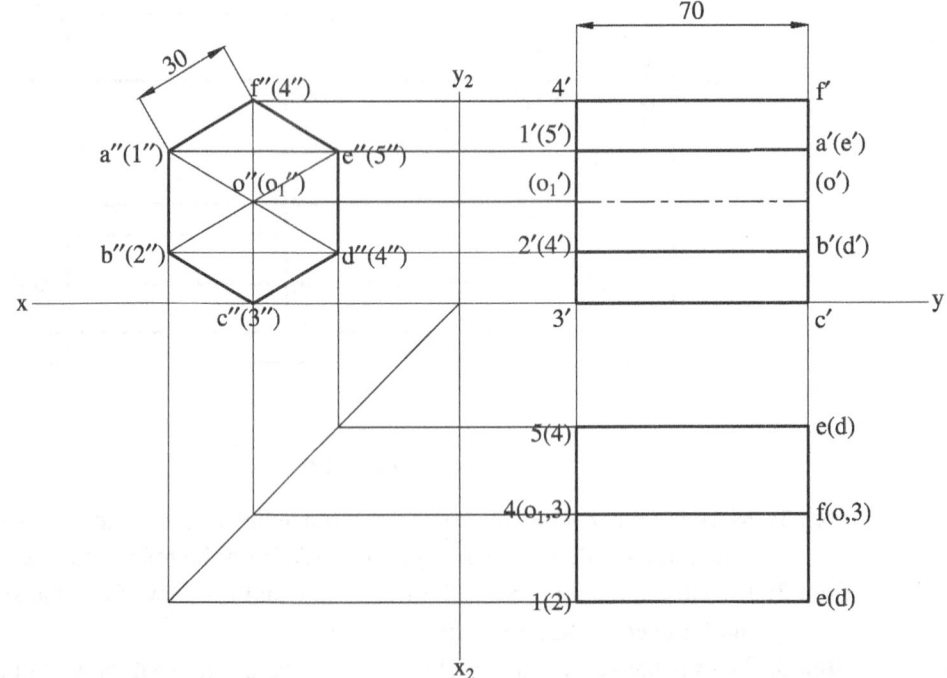

Fig. 9.21

Step 1: Mark the horizontal and vertical reference line and draw the hexagon of 30 mm sides, in the side view, such that one of its corners on the reference line and with corresponding two sides equal to it.

Step 2: Locate the centre to obtain the front view and top view and mark the edges as shown in Fig. 9.21.

Step 3: The visible and invisible edges are marked in the front view with the side view as reference. The lateral edges a″1″ and b″2″, which are far off from the vertical reference line, are visible and hence marked in thick lines, while the lateral edges e″5″ and d″4″, which are closer to the vertical reference lines, are invisible, but these projections coincide with the former.

Step 4: Similarly, the visible and invisible edges are marked in the top view with the side view as reference. The lateral edge f″4″ that is far off from the horizontal reference line is visible, while the other lateral edges c″3″ that is closer and on the horizontal reference line is invisible, but this projection coincides with the former.

Step 5: The other edges that form the boundary in the front and top views are marked in thick lines as shown in Fig. 9.21.   ▲

Example 9.10

A cylinder of base diameter 40 mm and axis 70 mm has one of its circular ends on VP such that the axis is at 40 mm from HP. Draw its projections.

Hint

1. Since the circular end is on VP the circle appears in front view.

PROCEDURE (Refer Fig. 9.22)

Step 1: Mark reference line, draw a circle from 60 mm diameter, with the centre located at 40 mm above it to obtain the centre o′, divide the circle into 12 parts, and mark the front and rear ends as a′, ..., l′ and 1′, ..., 12′, respectively.

Step 2: Project the point o′ and obtain the axis of length of 70 mm from the reference line.

Step 3: Project the other points in the front view to the top view domain and mark the length of 70 mm and complete the top view as shown in the Fig. 9.22.

Step 4: The visible and invisible generators in the top view are decided from the front view. The generators l12, k11, j10, i9, and h8, which are far off from the reference line, are visible and marked thick, while the respective generators coincident with them, and that are closer to the reference line, are invisible and marked in bracket.

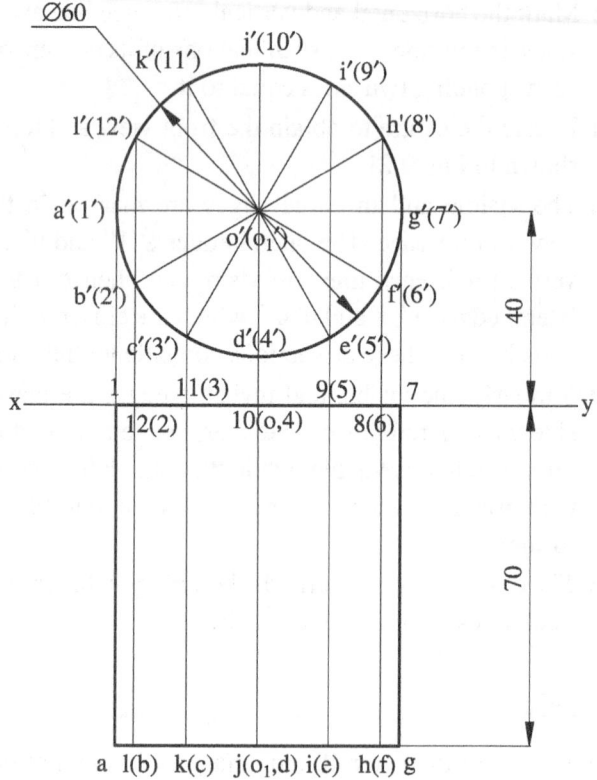

Fig. 9.22

Step 5: The extreme generators that form the boundary in the top view are marked in thick lines as shown in Fig. 9.22. ▲

Example 9.11

A cone of base diameter 60 mm and axis 70 mm has its circular end on the profile plane such that its axis is at 40 mm above HP and 50 mm in front of VP. Draw its projections.

Hint

1. As the circular end is on the profile plane, the circler appears in the side view.

PROCEDURE (Refer Fig. 9.23)

Step 1: Draw the horizontal and vertical reference lines, mark the centre of the circle in the side view at a distance of 40 mm above the horizontal reference line, 50 mm from the vertical reference line, and draw a circle of diameter 60 mm.

Step 2: Divide the circle into 12 equal parts and mark a'', ..., l'' in the side view.

Step 3: Project $o''(o_1'')$ to the front view and top view domains, mark the length of the cone, and get $o_1'o'$ and oo_1, respectively.

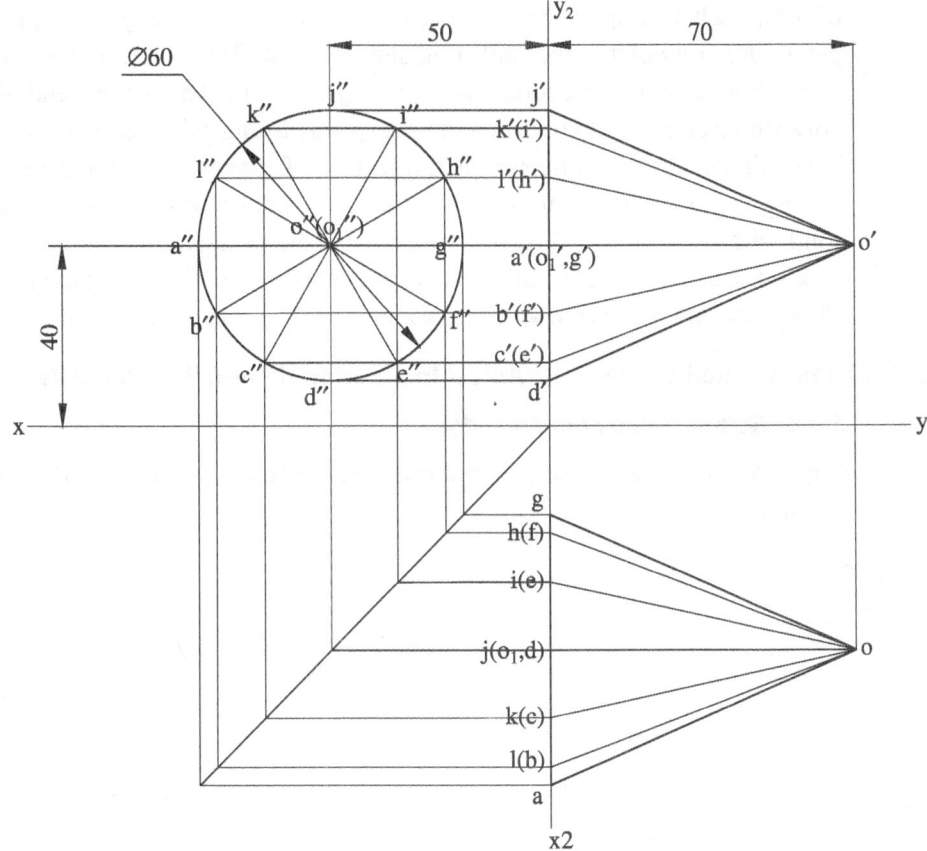

Fig. 9.23

Step 4: Project the other points in the side view to the vertical reference line, obtain a', ..., l' in the front view and a, ..., l in the top view, and join them with the apex in the respective views as shown in Fig. 9.23 to complete the projections.

Step 5: The visible and invisible generators are marked in the front view with the side view as reference. The generators $k''o''$, $l''o''$, $a''o''$, $b''o''$, and $c''o''$, which are far off from the vertical reference line, are visible, while the coincident generators closer to the vertical reference line are invisible and marked in brackets.

Step 6: The visible and the invisible generators are marked in the top view with the side view as reference in a similar way.

Step 7: The extreme generators that form the boundary in the front and top views are marked in thick lines as shown in Fig. 9.23. ▲

9.4 PROJECTION OF SOLIDS INCLINED TO ONE OF THE REFERENCE PLANES BY TILTING OBJECT METHOD

When the axis of solid is inclined to one of the reference planes and parallel to the other, the solid is in an inclined position. It means that the base or one end of the solid, instead of completely resting on that plane, gets lifted by retaining only one edge (in prisms and pyramids) or a point/corner (all solid shapes) on it. This is referred as the solid getting tilted about an edge or a corner, respectively. Since the base or one end of the solid is a polygon or circle (in cylinders and cones), this tilting procedure will follow the guide lines stipulated in the chapter on 'Projection of Planes'. The projections of the solid corresponding to the tilting or inclined position can be obtained initially by treating the solid with its base or end on the respective plane or in simple positions as done in the previous section and later tilting one of the views to the required inclination of the axis. This procedure is referred as tilting solid/object method or change of position method.

9.4.1 Projection of Solids when the Axis is Inclined to HP and Parallel to VP

Solids Getting Tilted about an Edge

Figure 9.24(a) shows a pentagonal prism standing on its base on HP or in the upright position.

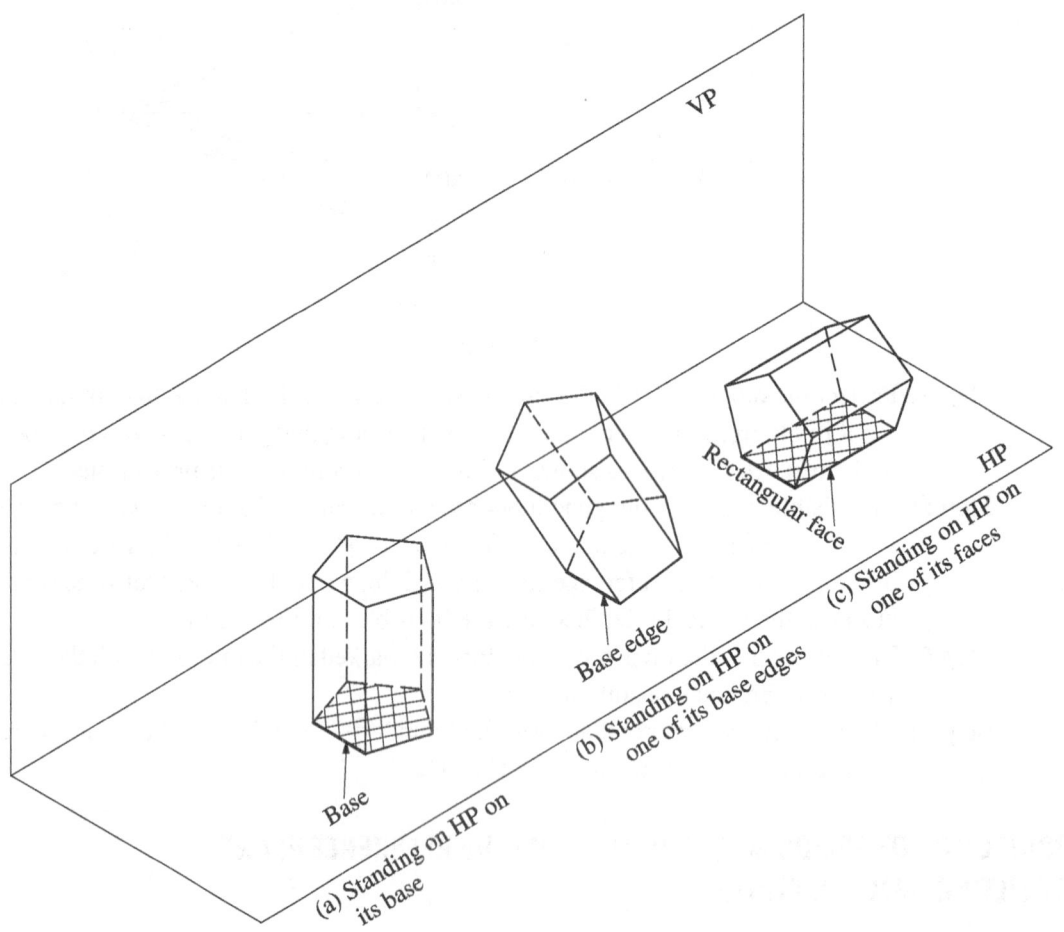

Fig. 9.24 Pentagonal prism placed on HP with different positions

Figure 9.25(a) shows the corresponding orthographic views.

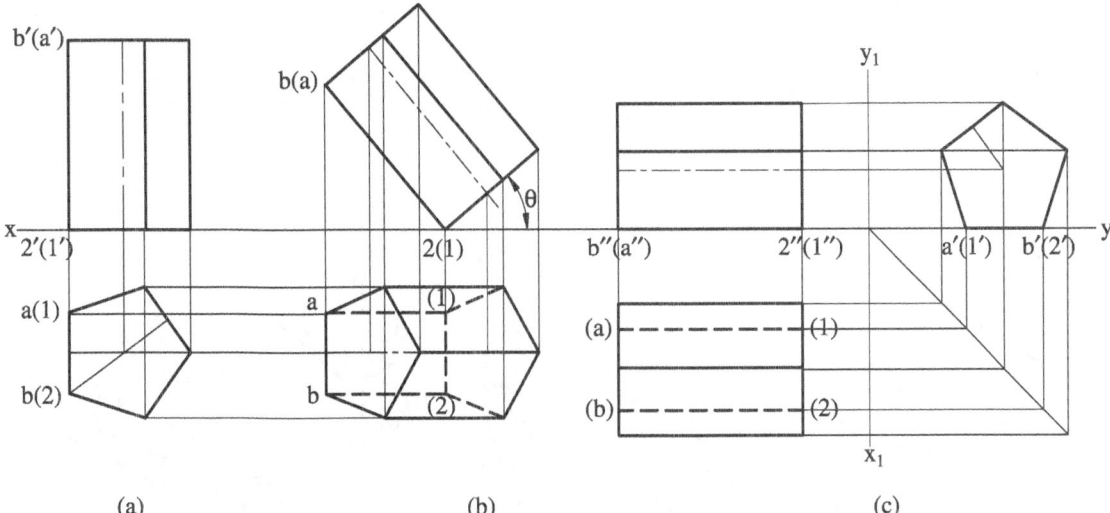

(a)   (b)   (c)

Fig. 9.25 Pentagonal prism resting on (a) its base on HP (b) one of its base edges and base inclined to HP (c) one of the rectangular faces on HP and axis parallel to both HP and VP

Figure 9.24(b) shows the base of the prism lifted from HP at an angle θ and resting on one of its edges on HP. When the solid gets tilted, the front view (rectangle) obtained in the upright position (Fig. 9.25a) also gets rotated and unchanged in its dimensions. This is reflected in Fig. 9.25(b). It can be noted that when the base of the solid gets lifted from HP, the rectangular face through that base edge also gets inclined to HP. The top view in this position is formed by two separate pentagons, one corresponding to the top and the other corresponding to the base and joined with lateral surfaces.

Logically, when the base of the solid gets lifted further, the angle made by the rectangular face with the HP gets reduced, and at one stage, the solid will have its rectangular face on HP itself, as shown in Figs 9.24(c) and 9.25(c). The front view noticed earlier gets rotated and falls on the reference line. In this case, top view also corresponds to the same length as that of the front view. Figure 9.26(a) to (d) shows similar cases of tilting in the case of pyramids (with hexagonal pyramid as an example). Figure 9.27(a) to (d) shows the orthographic views corresponding to various stages of tilting of the hexagonal pyramid. When the base is on HP, the slant face o′a′b′ is inclined to HP at its maximum angle and during the subsequent stages of tilting, this angle becomes 90°, < 90°, and finally zero degrees with the HP, when the slant face lies on HP. The projection of this face on HP, that is, 'oab' is visualized as a triangle, later as a line and then becomes fully hidden and dotted.

Solids Getting Tilted about a Corner

Figure 9.28(a) shows a pentagonal prism standing on its base on HP or in the upright position.

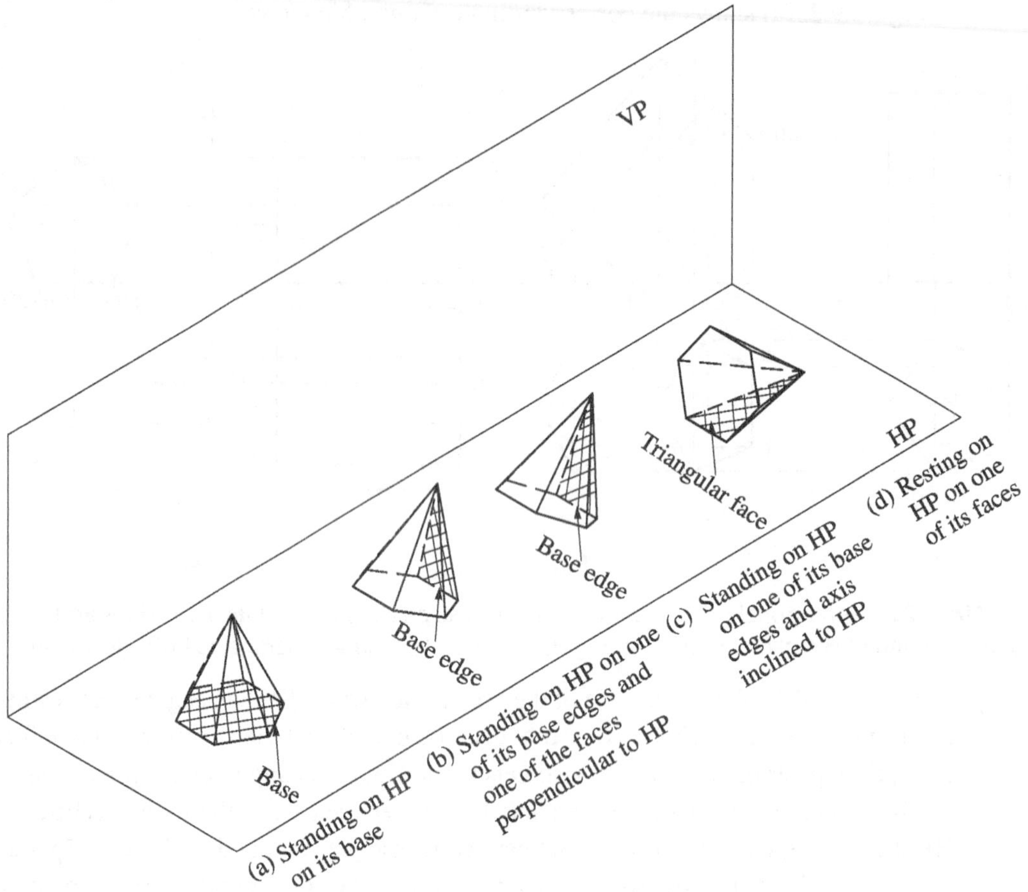

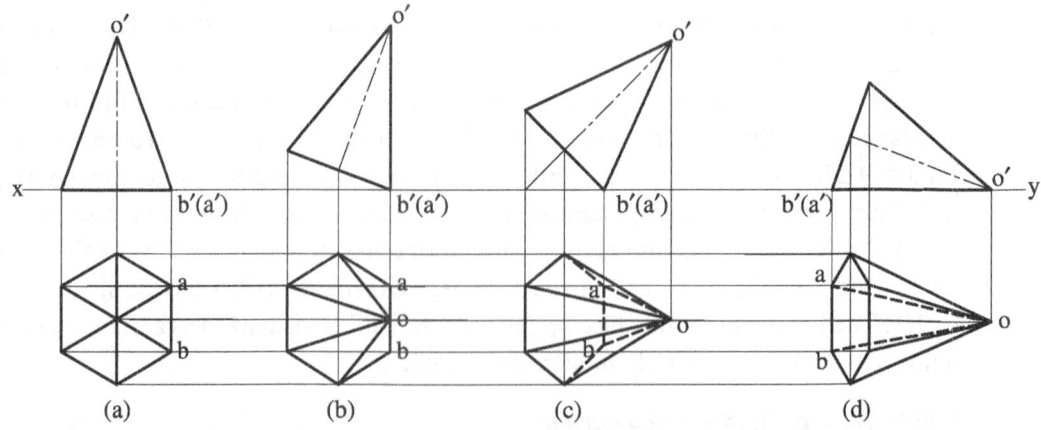

Fig. 9.26   Hexagonal pyramid placed on HP with different positions

Fig. 9.27   Hexagonal pyramid resting on (a) its base on HP (b) one of its base edges and the corresponding face perpendicular to HP (c) one of its base edges and the axis inclined to HP (d) one of the triangular faces on HP and axis parallel to VP

Figure 9.29(a) shows the corresponding orthographic views.

Figure 9.28(b) shows the base of the prism lifted from HP at an angle θ and resting on one of its corners on HP. When the solid gets tilted, the front view (rectangle) obtained in the upright position (Fig. 9.29a) also gets rotated, unchanged in its dimensions. This is reflected in Fig. 9.29(b). It can be noted that when the base of the solid gets lifted from HP, the lateral edge through that base corner also gets inclined to HP. The top view in this position is made up of two separate pentagons, one corresponding to the top, the other corresponding to the base, and joined with lateral surfaces.

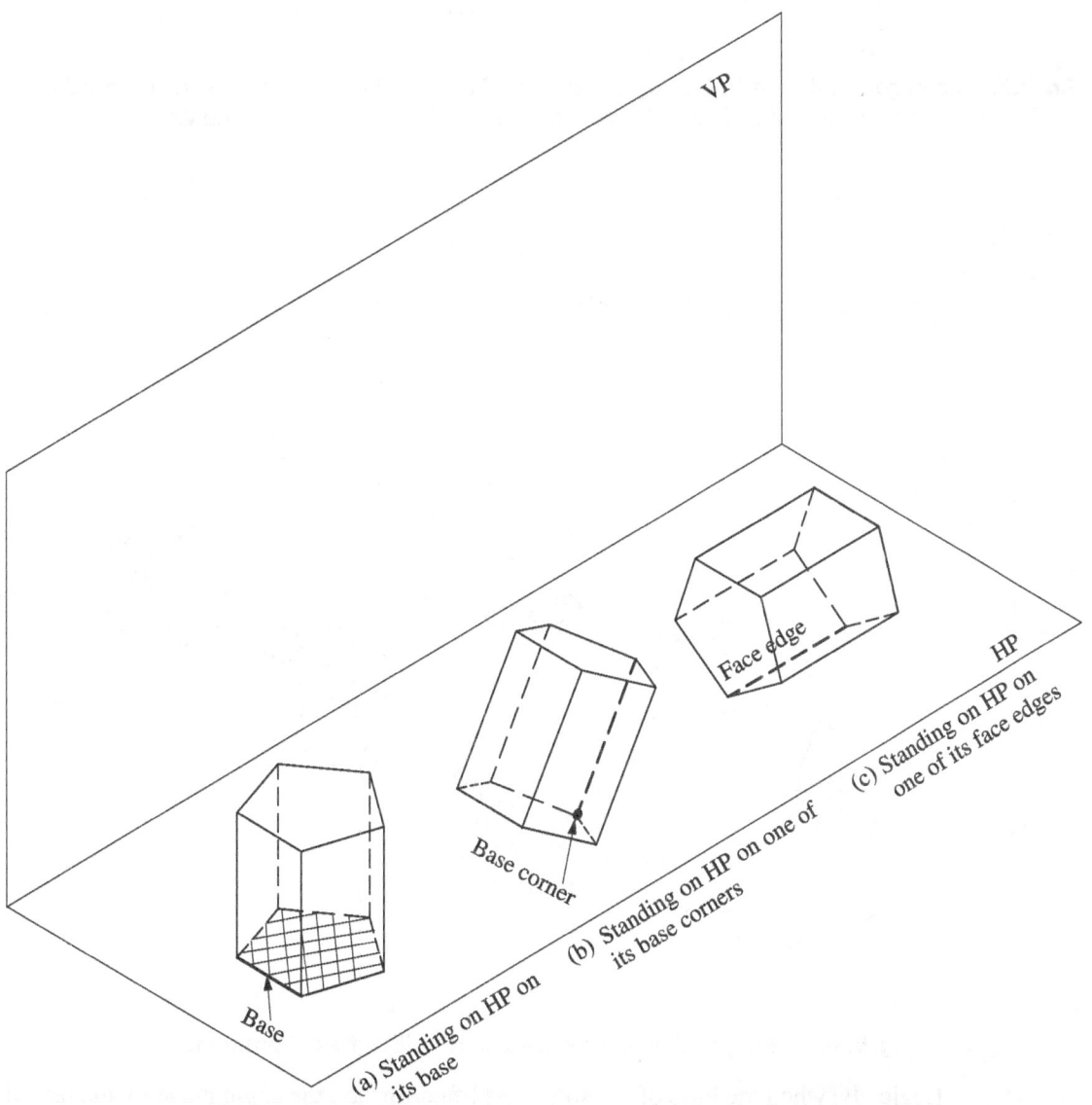

Fig. 9.28   Pentagonal prism placed on HP with different positions

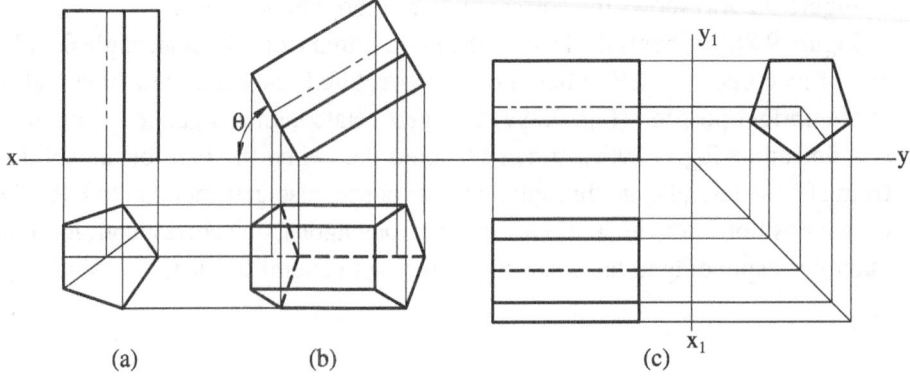

Fig. 9.29 Pentagonal prism resting on (a) its base on HP (b) one of its base corners and base inclined to HP (c) one of the face edges on HP and axis parallel to both HP and VP

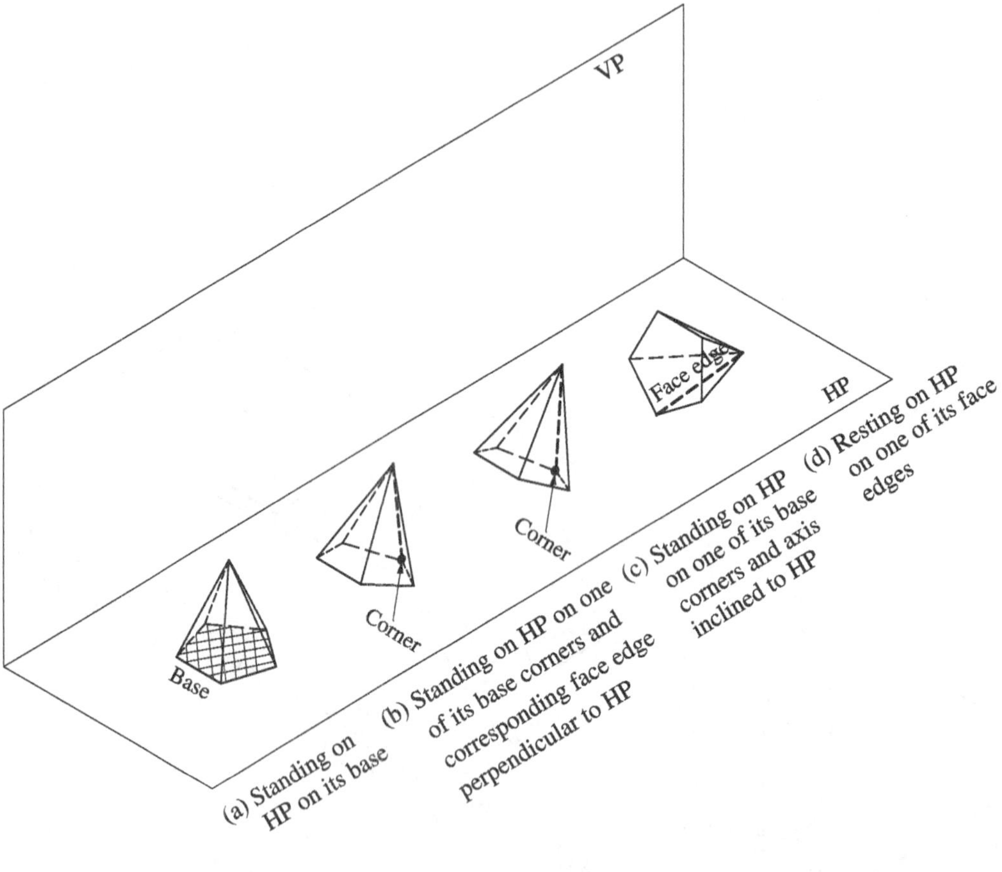

Fig. 9.30 Pentagonal pyramid placed on HP with different positions

Logically, when the base of the solid gets lifted further, the angle made by the lateral edge with the HP gets reduced, and at one stage, the solid will have its lateral edge on HP itself, as shown in Figs 9.28(c) and 9.29(c). The front view noticed earlier gets rotated and falls on the reference line. The top view in this case also corresponds to the same length as that of the front view. Figure 9.30(a) to (d) shows similar cases of tilting in the case of pyramids (with pentagonal pyramid as an example).

Figures 9.31(a)–(d) shows the orthographic views corresponding to various stages of tilting of the pentagonal pyramid. When the base is on HP, the slant edge o'a" is inclined to HP at its maximum angle, and during the subsequent stages of tilting, this angle becomes 90°, < 90°, and finally zero degrees with the HP, when the slant edge lies on HP. The projection of this edge on HP, that is, 'oa' is seen as a dotted line.

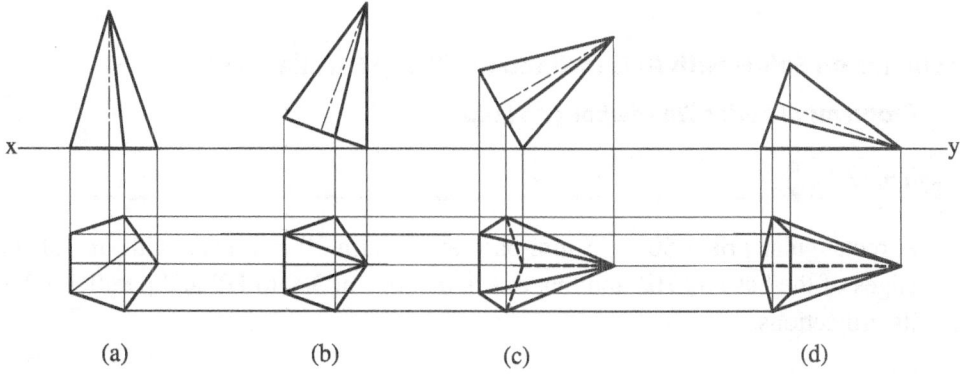

(a)   (b)   (c)   (d)

Fig. 9.31 Pentagonal pyramid resting on (a) Its base on HP (b) One of its base corners and corresponding face edge perpendicular to HP (c) One of its base corners and the axis inclined to HP (d) One of the face edges on HP and axis parallel to VP

9.4.2 Projection of Solids when the Axis is Inclined to VP and Parallel to HP

When the solid has an end representing the polygon on VP, its axis is perpendicular to VP and that forms the simple position or the initial position in this case. Subsequently, when the axis gets inclined to VP, the end of the solid moves itself from VP, and gets inclined to it. The tilting with respect to VP can occur either by a base edge on VP or by a corner on VP. In such cases, all the discussions mentioned in Section 9.4.1 hold good, mutually interchanging the shapes in the front and top views.

9.4.3 General Procedure for Projection of Solids Inclined to One Reference Plane and Parallel to the Other

1. Keep the axis perpendicular to the reference plane to which it is to be inclined later and draw the initial projections.
2. Draw the polygon representing the base or the end with one edge perpendicular to the reference line or one diagonal parallel to the reference line as per the tilting about an edge or corner, respectively, and draw the other non-polygonal view.
3. Tilt the non-polygonal view to the required inclination and draw vertical projectors from all its corners.
4. When the horizontal locus lines from the polygonal view in the first stage meet the vertical projectors, the final projections corresponding to the tilting of the solid are obtained.

5. The visible corners and edges in the final projections correspond to those located far off from the reference line in the non-polygonal view and those are closer to the reference line are invisible and marked as a dotted line.

6. Mark all the boundary edges of the final projection in thick lines as boundary existence is a reality while describing an object.

NOTE *The initial projections are drawn in all the problems by noting the various corners/points of the solid with alphabets/numerals using subscript 1, while their final views are shown with no subscripts.*

9.4.4 Problems on Solids with Axis Inclined to HP and Parallel to VP

Problems of Solids Tilted about an Edge

Example 9.12

A rectangular prism 50×25 mm base and length 70 mm rests with one of its longer edges of the base on HP, and the axis is inclined at 30° to HP and parallel to VP. Draw its projections.

Hint

Decide the initial projections based on the inclination of the axis.

1. Since the axis of the solid is inclined to HP, keep the axis perpendicular to HP. Therefore, its base is parallel to HP, and hence, the polygon (rectangle) is visualized in the top view.

2. Since the tilting of the solid happens with the longer edge of the base, lay the rectangle with the longer side perpendicular to the reference line.

PROCEDURE (Refer Fig. 9.32)

Step 1: Mark the reference line and draw the initial projections as mentioned earlier and as explained in Example 9.1.

Step 2: Since the solid is getting tilted about the longer edge $1_1'2_1'$, redraw the front view about this, such that the axis makes 30° with the reference line as shown in Fig. 9.32, and draw the vertical projectors from its various points.

Step 3: When the horizontal locus lines from the top view in the first stage meet these projector lines, the final top view is obtained.

Step 4: The visible and invisible edges are marked in the top view with the front view as reference. The rectangular end that is far off from the reference line (a', ..., d') is visible, while the other end (1', ..., 4') is marked as a dotted line. The other edges that form the boundary are marked in thick lines as shown in Fig. 9.32. ▲

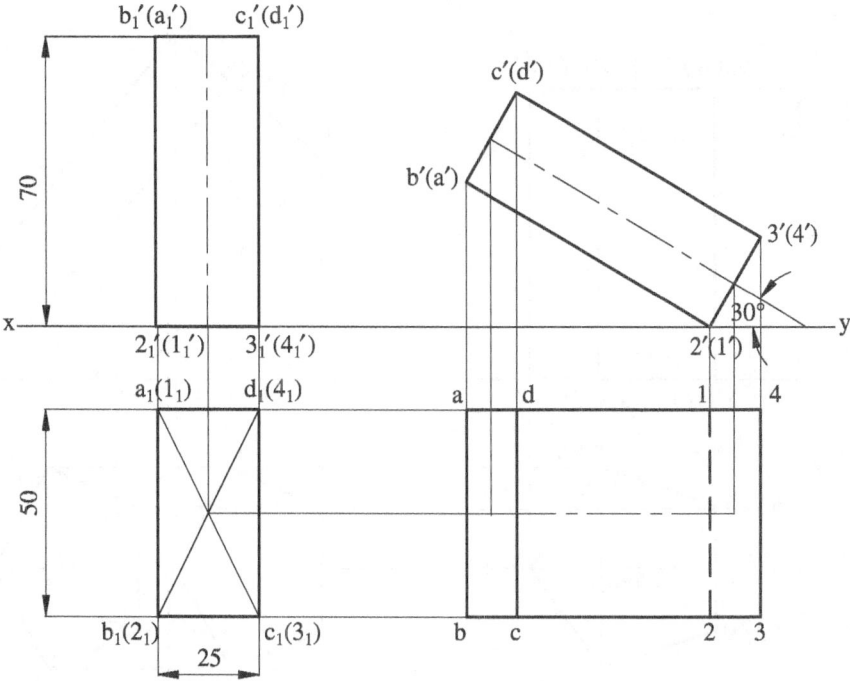

Fig. 9.32

Example 9.13

A pentagonal prism 30 mm side of base and axis 70 mm long is resting on one of its edges of the base, in such a way that the base makes an angle of 40° with HP. Draw the projections, if the axis is parallel to VP.

Hint

Decide the initial projections based on the inclination of the axis.

1. When the base of the solid is inclined to HP, the axis is also inclined to the HP. Therefore, keep the axis perpendicular to HP. Then its base becomes parallel to HP and hence the polygon of the base (pentagon) is visualized in the top view.
2. Since the tilting of the solid happens about the base edge, lay the pentagon with one edge perpendicular to the reference line.

PROCEDURE (Refer Fig. 9.33)

Step 1: Mark the reference line and draw the initial projections as mentioned earlier as explained in Example 9.2.

Step 2: Since the solid is getting tilted about the base edge $1_1'2_1'$, redraw the front view about this, such that the base makes 40° with the reference line as shown in Fig. 9.33, and draw the vertical projectors from its various points.

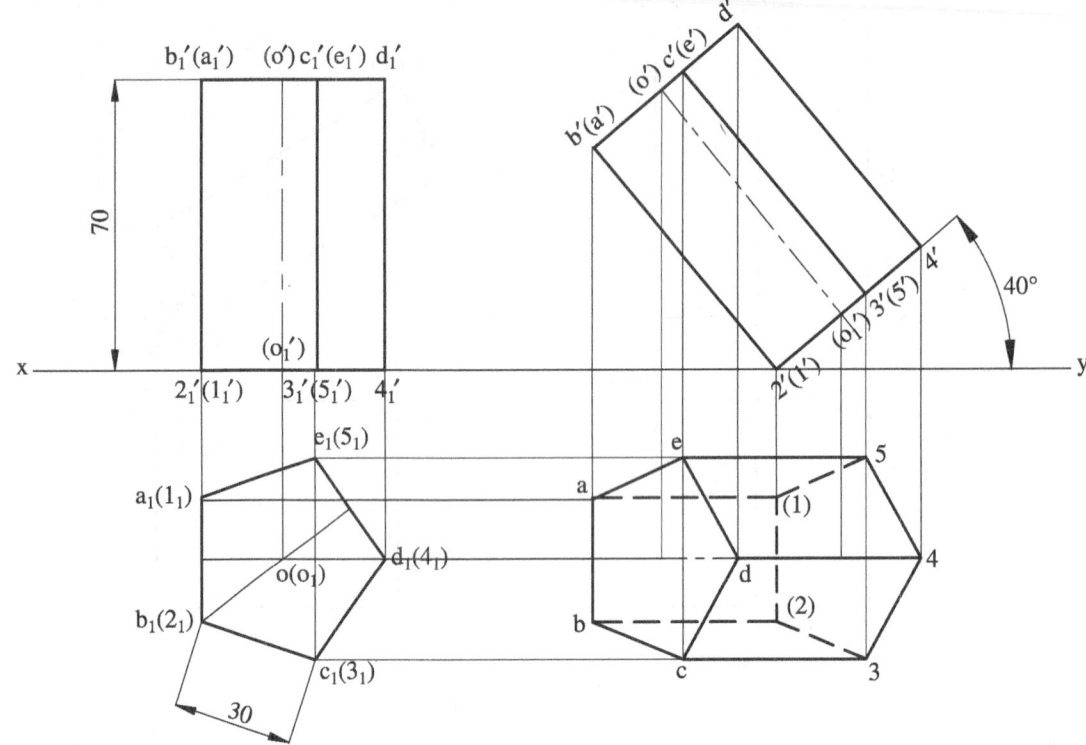

Fig. 9.33

Step 3: When the horizontal locus lines from the top view in the first stage meet these projector lines, the final top view is obtained.

Step 4: The visible and invisible edges are marked in the top view with the front view as reference. The pentagonal end that is far off from the reference line (a', ..., e') is visible, while the other end (1', ..., 5') is marked as a dotted line. The lateral edges corresponding to 1'a' and 2'b' also shown in dotted line as they are closer to the reference line. The other edges that form the boundary are marked in thick lines as shown in Fig. 9.33. ▲

Example 9.14

 A pentagonal pyramid of 30 mm base edges and axis 70 mm long is lying on one of its triangular faces on HP. Draw its projections when the base edge on this face is perpendicular to VP and the vertical plane containing the axis is parallel to VP.

Hint

Decide the initial projections based on the inclination of the axis.

1. When a pyramid is lying on one of its triangular faces on HP, its axis gets inclined to HP. Hence, keep the axis perpendicular to HP. Therefore, its base is parallel to HP, and hence, the polygon (pentagon) is visualized in the top view.

2. The tilting of the lateral face will happen only when the solid is tilted about the base edge. Hence lay the base shape, with one edge perpendicular to the reference line and tilt the concerned view about this edge.

PROCEDURE (Refer Fig. 9.34)

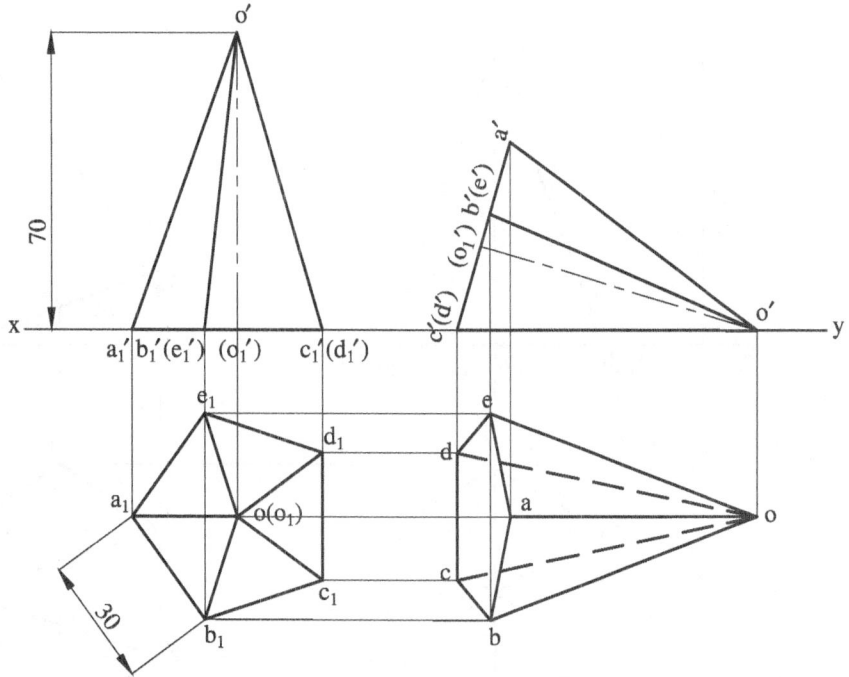

Fig. 9.34

Step 1: Mark the reference line and draw the initial projections as mentioned earlier as explained in Example 9.5.

Step 2: Since $o'c_1'\ d_1'$ constitutes a triangular face, the solid is getting tilted about the corresponding base edge $c_1'd_1'$; redraw the front view about this edge $(c_1'd_1')$ such that the triangular face $o'c_1'd_1'$ lies on the reference line as shown in Fig. 9.34, and draw the vertical projectors from its various points.

Step 3: When the horizontal locus lines from the top view in the first stage meet these projector lines, the final top view is obtained.

Step 4: The visible and invisible edges are marked in the top view with the front view as reference. The slant edge $o'a'$ that is far off from the reference line is visible, while the slant edges $o'c'$ and $o'd'$ are marked dotted. The other edges that form the boundary are marked in thick lines as shown in Fig. 9.34. ▲

Example 9.15 ▲

A hexagonal pyramid of 30 mm base edges and axis 70 mm long is tilted about one of its base edges such that the triangular face through that edge is vertical. Draw its projections.

Hint

Refer to the hint of Example 9.14.

PROCEDURE (Refer Fig. 9.35)

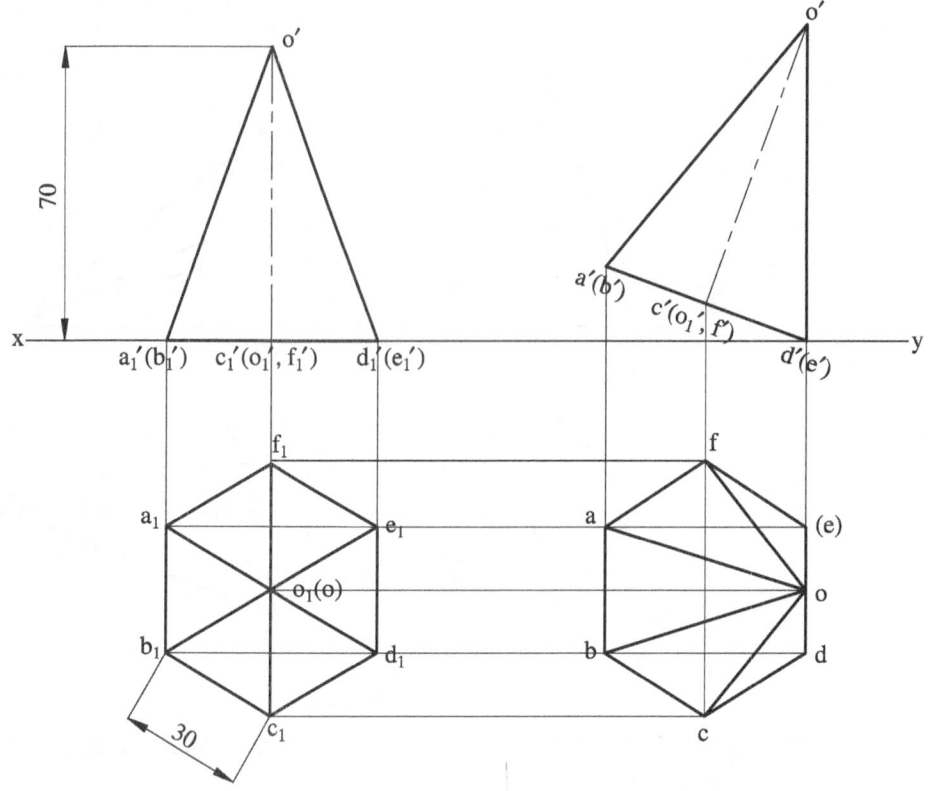

Fig. 9.35

Step 1: Mark the reference line and draw the initial projections as mentioned previously.

Step 2: Since $o'd_1'e_1'$ constitutes a triangular face in the front view, the solid is getting tilted about the corresponding base edge $d_1'e_1'$; redraw the front view about this edge such that the triangular face $o'd_1'e_1'$ is perpendicular to the reference line as shown in Fig. 9.35, and draw the vertical projectors from its various points.

Step 3: When the horizontal locus lines from the top view in the first stage meet these projector lines, the final top view is obtained.

Step 4: The visible and invisible edges are marked in the top view with the front view as reference. The slant edges $o'a'$, $o'b'$, oc', and $o'f'$, which are far off from the reference line, are visible and hence, mark them as dark lines. The other edges that form the boundary are marked in thick lines as shown in Fig. 9.35. ▲

Example 9.16

A rectangular prism 50×25 mm base and length 70 mm rests with one of its corners of the base on HP. The axis of the prism is inclined at 30° to HP and parallel to VP. Draw its projection.

Hint

Decide the initial projections based on the inclination of the axis.

1. Since the axis of the solid is inclined to HP, the axis is perpendicular to HP. Therefore, its base is parallel to HP, and hence, the polygon (rectangle) is visualized in the top view.
2. Since the tilting of the solid happens about a corner, lay the rectangle with the diagonal through that corner parallel to the reference line.

PROCEDURE (Refer Fig. 9.36)

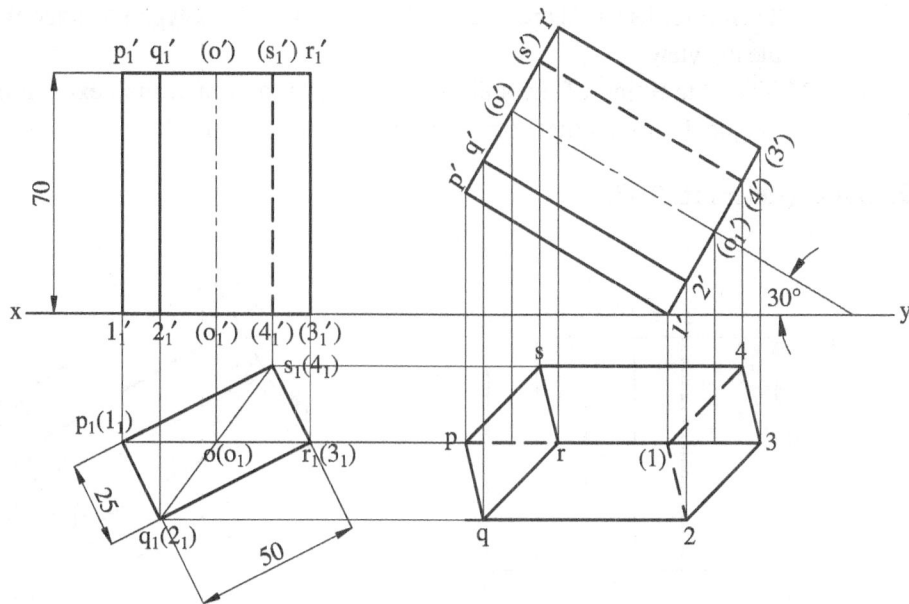

Fig. 9.36

Step 1: Mark the reference line and draw the initial projections as mentioned earlier.

Step 2: Since the solid is getting tilted about the base corner $1_1'$, redraw the front view about this, such that the axis makes 30° with the reference line as shown in Fig. 9.36, and draw the vertical projectors from its various points.

Step 3: When the horizontal locus lines from the top view in the first stage meet these projector lines, the final top view is obtained.

    The visible and invisible edges are marked in the top view with the front view as reference. The rectangular end, which is far off from the reference line (p′,

..., s'), is visible, while the other end (1', ..., 4') is marked dotted. The lateral edge far off from the reference line (3'r') is visible, while the lateral edge (1'p') appears dotted as it is closer to the reference line. The other edges that form the boundary are marked in thick lines as shown in Fig. 9.36. ▲

Problems of Solids Tilted about a Corner

Example 9.17 ▲

A hexagonal prism of 30 mm base edges and axis 70 mm long rests on one of its corners of base on HP. Draw its projections, when the lateral edge through that corner on HP is inclined at 30° to HP, and the vertical plane containing that lateral edge and the axis is parallel to VP.

Hint

Decide the initial projections based on the inclination of the axis.

1. Since the axis of the solid is inclined to HP, keep the axis perpendicular to HP. Therefore, its base is parallel to HP, and hence, the polygon (hexagon) is visualized in the top view.
2. Since the tilting of the solid happens about a corner, the hexagon is laid with the diagonal through that corner parallel to the reference line.

PROCEDURE (Refer Fig. 9.37)

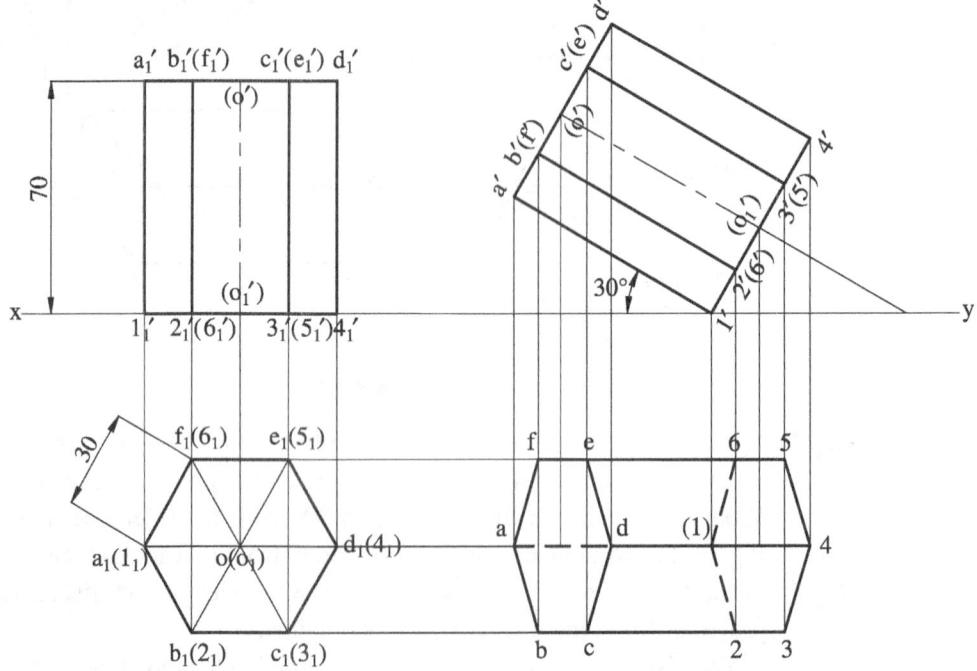

Fig. 9.37

Step 1: Mark the reference line and draw the initial projections as mentioned earlier.

Step 2: Since the solid is getting tilted about the base corner $1_1'$, redraw the front view about this, such that the lateral edge through $1_1'a_1'$ is inclined at 30° with the reference line as shown in Fig. 9.37, and draw the vertical projectors from its various points.

Step 3: When the horizontal locus lines from the top view in the first stage meet these projector lines, the final top view is obtained.

Step 4: The visible and invisible edges are marked in the top view with the front view as reference. The hexagonal end (a', ..., f'), which is far off from the reference line, is visible, while the other end (1', ..., 6') is marked dotted. The lateral edges far off from the reference line (4'd') and nearer (1'a') are, respectively, marked in thick and dotted lines. The other edges that form the boundary are marked in thick lines as shown in Fig. 9.37.   ▲

Example 9.18

Draw the top and front views of a right circular cylinder of base 45 mm diameter and axis 60 mm long resting on HP with its axis inclined at 30° to it.

Hint

Decide the initial projections based on the inclination of the axis.

1. Since the axis of the cylinder is inclined to HP, the axis should be perpendicular to HP. Therefore, its base is parallel to HP, and hence, the circle is visualized in the top view.

PROCEDURE (Refer Fig. 9.38)

Step 1: Mark the reference line and draw the initial projections by dividing the circle into eight equal parts as shown in Fig. 9.38.

Step 2: Since the solid is getting tilted about $5_1'$, redraw the front view about this, such that the axis is inclined at 30° with the reference line as shown in Fig. 9.38, and draw the vertical projectors from its various points.

Step 3: When the horizontal locus lines from the top view in the first stage meet these projector lines, the final top view is obtained with its ends appearing elliptical.

Step 4: The visible and invisible edges are marked in the top view with the front view as reference. The circular end (a', ..., h'), which is far off from the reference line is visible, while the other end (1', ..., 8') is marked dotted. The top generators are shown in thick lines along with their coincident hidden generators. The extreme generators that form the boundary are marked in thick lines as shown in Fig. 9.38.   ▲

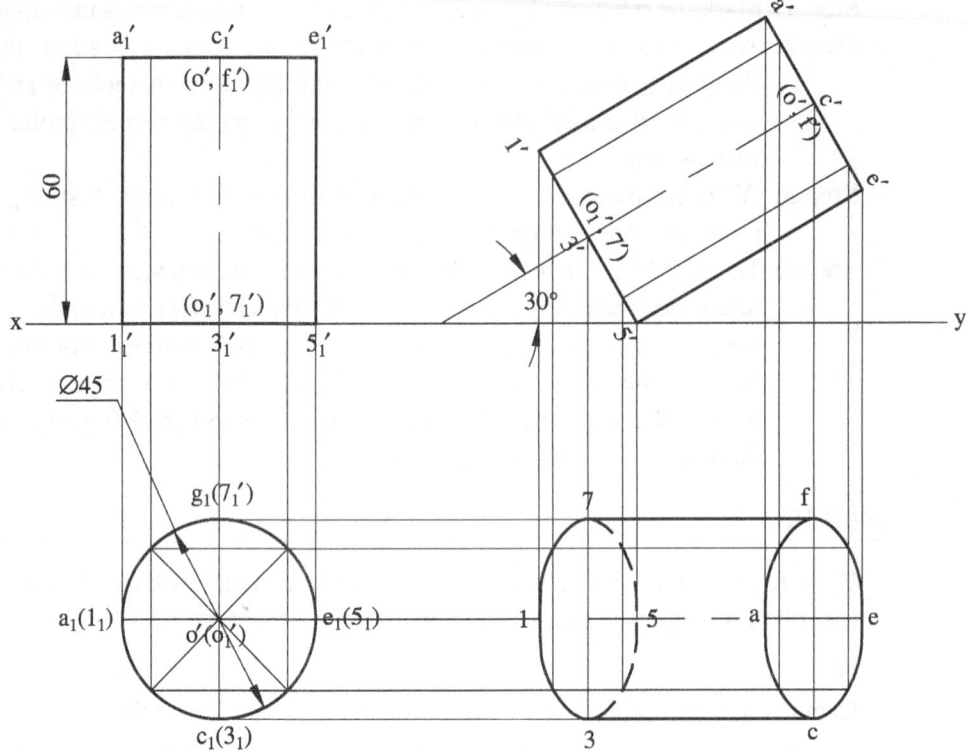

Fig. 9.38

Example 9.19

A cone of base 60 mm diameter and axis 80 mm long rests on HP with its axis inclined 45° to it. Draw its projections when the top view of the axis is parallel to VP.

Hint

Decide the initial projections based on the inclination of the axis.

1. Since the axis of the cone is inclined to HP, the axis should be perpendicular to HP. Therefore, its base is parallel to HP, and hence, the circle is visualized in the top view.

PROCEDURE   (Refer Fig. 9.39)

Step 1: Mark the reference line, draw the initial projections by dividing the circle into equal parts, and mark the generators as shown in Fig. 9.39.

Step 2: Since the solid is getting tilted about g_1', redraw the front view about this such that the axis is inclined at 45° with the reference line as shown in Fig. 9.39, and draw the vertical projectors from its various points.

Step 3: When the horizontal locus lines from the top view in the first stage meet these projector lines, the final top view is obtained with its bottom appearing elliptical.

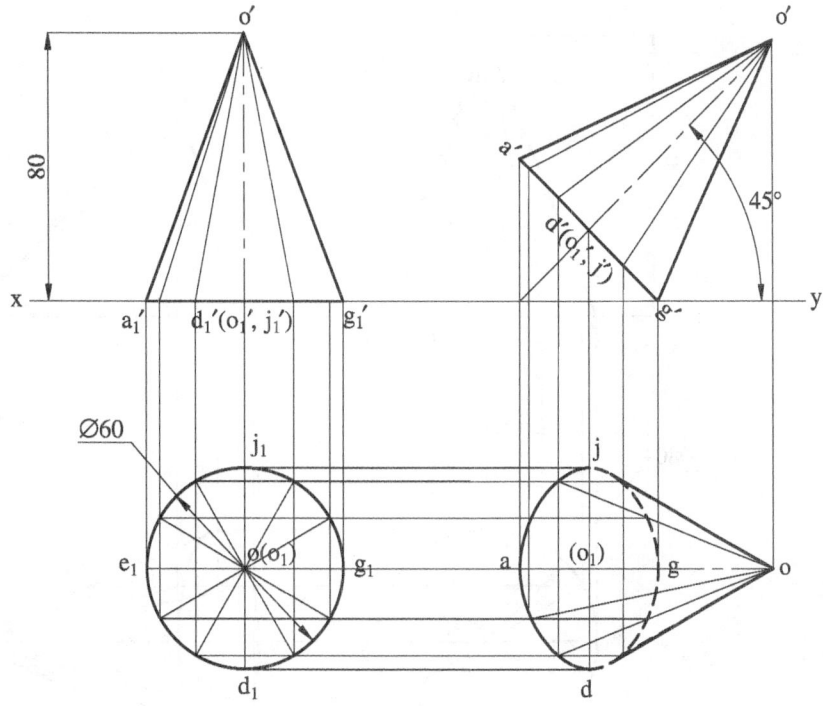

Fig. 9.39

Step 4: The visible and invisible edges are marked in the top view with the front view as reference. The dotted line represents circular end as it is nearer to the reference line. The extreme generators that form the boundary are marked in thick lines as shown in Fig. 9.39.

Example 9.20

A cone of base 80 mm diameter and height 100 mm is lying with one of its generators on HP and axis parallel to VP. Draw its projections.

Hint

Decide the initial projections based on the inclination of the axis.

1. Since the cone is lying on its generator on HP, the axis is inclined to HP, the axis is perpendicular to HP and its base is parallel to HP, and hence, the circle is visualized in the top view.

PROCEDURE (Refer Fig. 9.40)

Step 1: Mark the reference line, draw the initial projections by dividing the circle into 12 equal parts, and mark the generators as shown.

Step 2: Since the solid is lying on its generator on HP, redraw the front view such that $o'g_1'$ lies on the reference line as shown in Fig. 9.40, and draw the vertical projectors from its various points.

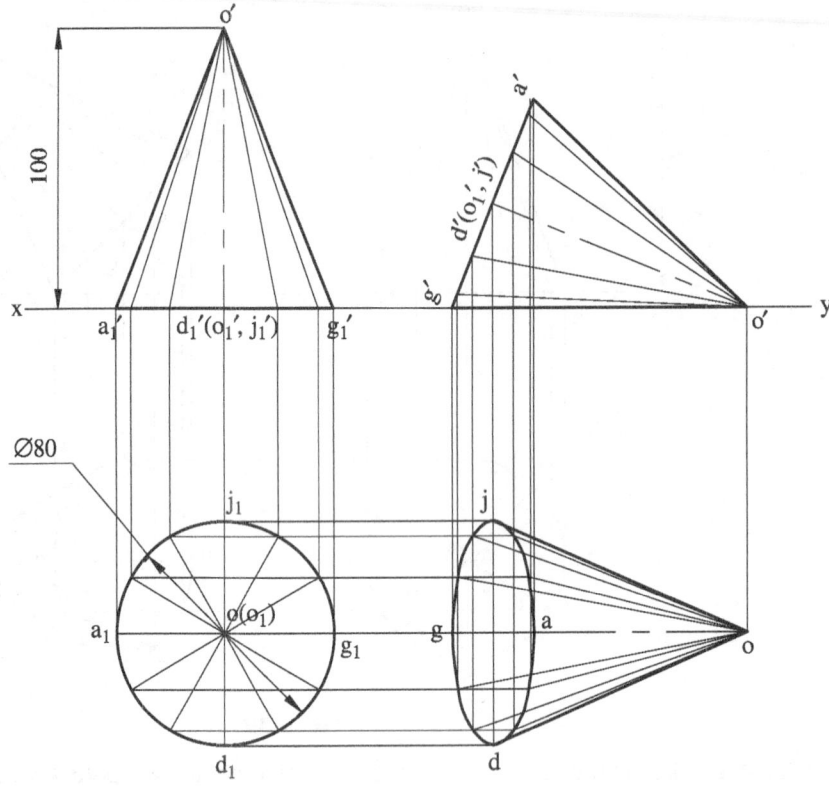

Fig. 9.40

Step 3: When the horizontal locus lines from the top view in the first stage meet these projector lines, the final top view is obtained with its circular end appearing elliptical.

Step 4: The extreme generators that form the boundary are marked in thick lines as shown in Fig. 9.40. ▲

Example 9.21

A pentagonal pyramid of base edges 30 mm and axis 70 mm long has a corner of base on HP. Draw its projections when the slant edge through corner lies on HP and is parallel to VP.

Hint

Decide the initial projections based on the inclination of the axis

1. When a pyramid is lying on one of its slant edges on HP, its axis gets inclined to HP. Hence, the axis is perpendicular to HP. Therefore, its base is parallel to HP, and hence, the polygon (pentagon) is visualized in the top view.

2. When the slant edge of a solid lies on a reference plane, it gets tilted about a corner of the base. Hence, the pentagon is laid with its diagonal parallel to the reference line.

PROCEDURE (Refer Fig. 9.41)

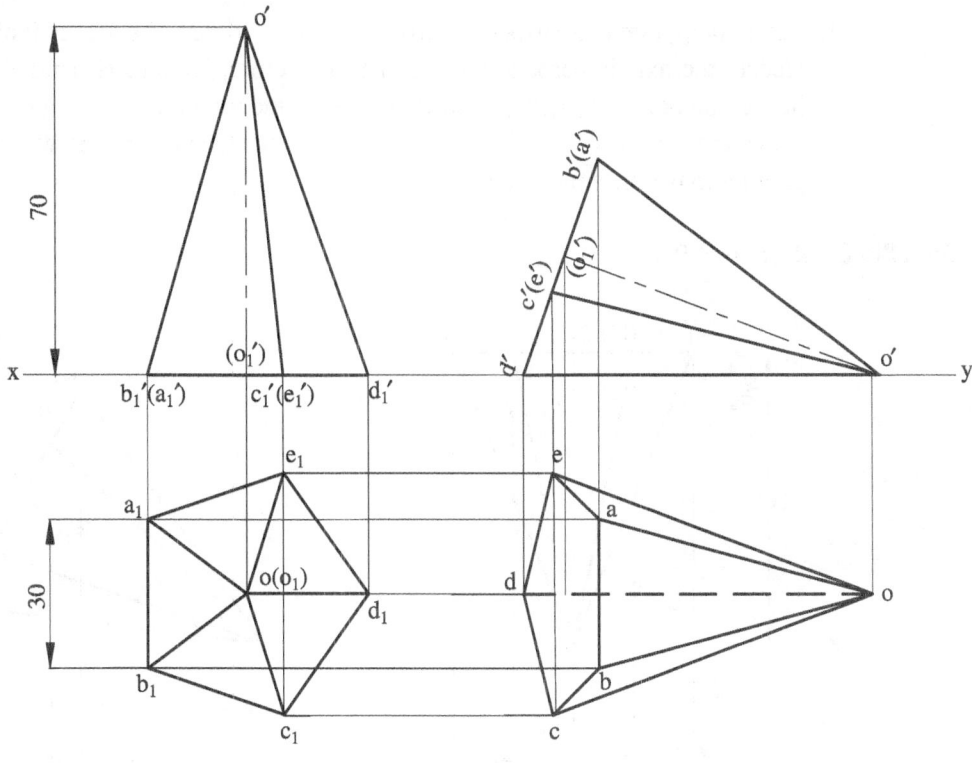

Fig. 9.41

Step 1: Mark the reference line and draw the initial projections as mentioned earlier.

Step 2: Since a slant edge has been lying on a reference line, tilt and redraw the front view such that $o'd_1'$ lies on the reference line, and draw the vertical projectors from its various points.

Step 3: When the horizontal locus lines from the top view in the first stage meet these projector lines, the final top view is obtained.

Step 4: The visible and invisible edges are marked in the top view with the front view as reference. The slant edges $o'a'$ and $o'b'$, which are far off from the reference line, are visible and marked in thick lines, while the slant edge $o'd'$, which lies on the reference line, is marked dotted. The other edges that form the boundary are marked in thick lines as shown in Fig. 9.41. ▲

Problems of Solids Suspended about a Corner

Example 9.22

A hexagonal pyramid of base edge 30 mm and axis 70 mm long is freely suspended from a corner of its base. Draw its projections, when the top view of the axis is parallel to the reference line.

Hint

Decide the initial projections based on the inclination of the axis.

1. Since the pyramid is suspended from one of its corners, its axis is inclined to HP. Hence, the axis is perpendicular to HP. Therefore, its base is parallel to HP, and hence, the polygon (hexagon) is visualized in the top view.
2. When the solid is getting tilted about a corner, the hexagon is laid with its diagonal parallel to the reference line.

PROCEDURE (Refer Fig. 9.42)

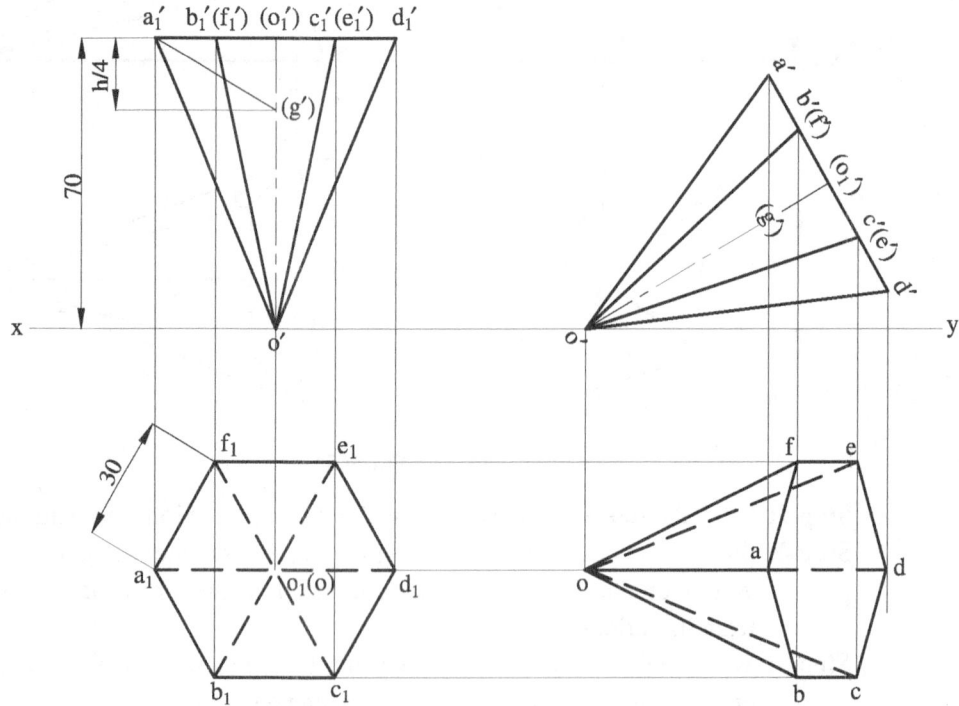

Fig. 9.42

Step 1: Mark the reference line and draw the initial projections keeping the hexagonal on the top and apex down to achieve the suspension about a corner.

Step 2: In the suspended position, the point of suspension (a') and the centre of gravity (G) of the body fall on a vertical line. The CG lies on the axis; for pyramids, it is at a distance of one-fourth of the length of the axis. This is marked as g' in the front view.

Step 3: The front view is redrawn such that the line a'g' is made vertical to the reference line as shown in Fig. 9.42. The vertical projectors are drawn from various points of the front view.

Step 4: When the horizontal locus lines from the top view in the first stage meet these projector lines, the final top view is obtained.

Step 5: The visible and invisible edges are marked in the top view with the front view as reference. The hexagonal end is visualized completely as it is farthest from the reference line. The farthest slant edge $o'a'$ are also completely visible. The closer slant edges $o'd'$, $o'c'$, and $o'e'$ are shown as dotted in the top view. The other edges that form the boundary are marked in thick lines as shown in Fig. 9.42.   ▲

9.4.5 Problems on Solids with Axis inclined to VP and Parallel to HP

Problems of Solids Tilted about an Edge

Example 9.23   ▲

A hexagonal prism of base edge 30 mm and axis 70 mm long has one of rectangular faces inclined at 45° to VP. Draw its projections, when the base edge on this face lies on VP and perpendicular to HP.

Hint

Decide the initial projections based on the inclination of the axis

1. When a rectangular face is inclined to VP, the axis of the solid is inclined to VP, keep the axis perpendicular to VP, and therefore, its base is parallel to VP. Hence, the polygon (hexagon) is visualized in the front view.
2. The tilting of the lateral face will happen only when the solid is tilted about the base edge. Hence, lay the base shape with one edge perpendicular to the reference line and tilt the concerned view about this edge.

PROCEDURE (Refer Fig. 9.43)

Step 1: Mark the reference line and draw the initial projections as mentioned earlier.

Step 2: Since the solid is getting tilted about the base edge $5_1 4_1$, redraw the top view about this, such that the rectangular face $5_1 e_1 d_1 4_1$ makes 45° with the reference line as shown in Fig. 9.43, and draw the vertical projectors from its various points.

Step 3: When the horizontal locus lines from the front view in the first stage meet these vertical projector lines, the final front view is obtained.

Step 4: The visible and invisible edges are marked in the front view with the top view as reference. The hexagonal end, which is far off from the reference line (a, ..., f) is visible, while the other end (1, ..., 6) is marked dotted. The lateral edges corresponding to 5e and 4d also appear dotted as they are closer to the reference line. The other edges that form the boundary are marked in thick lines as shown in Fig. 9.43.   ▲

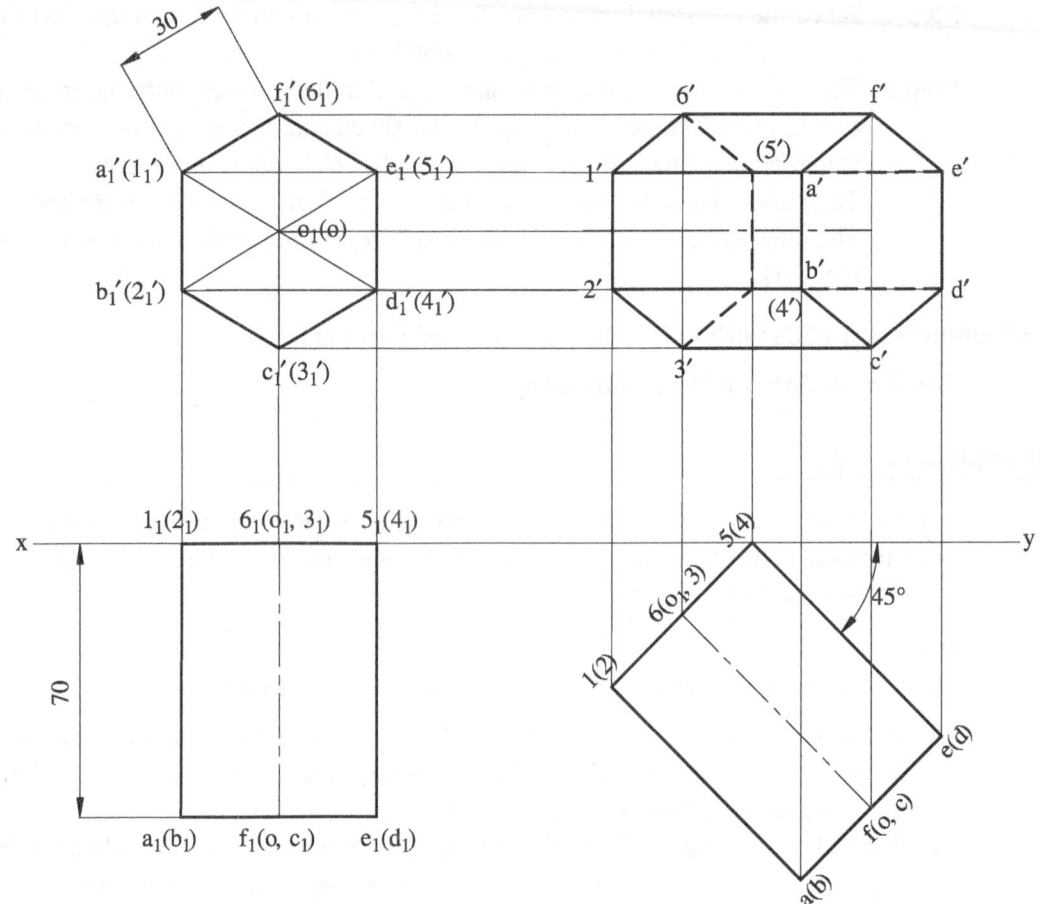

Fig. 9.43

Example 9.24

A pentagonal pyramid of base edges 30 mm and axis 70 mm long is tilted about one of its base edges lying on VP such that the triangular face passing through this edge is perpendicular to VP and the axis is parallel to HP. Draw its projections.

Hint

Decide the initial projections based on the inclination of the axis.

1. When the triangular face of the pyramid is perpendicular to VP, the axis of the pyramid is inclined to VP. Hence, keep the axis perpendicular to VP. Therefore, its base is parallel to VP, and hence, the polygon (pentagon) is seen in the front view.
2. Refer to the hint of Example 9.23.

PROCEDURE (Refer Fig. 9.44)

Step 1: Mark the reference line and draw the initial projections as mentioned previously.

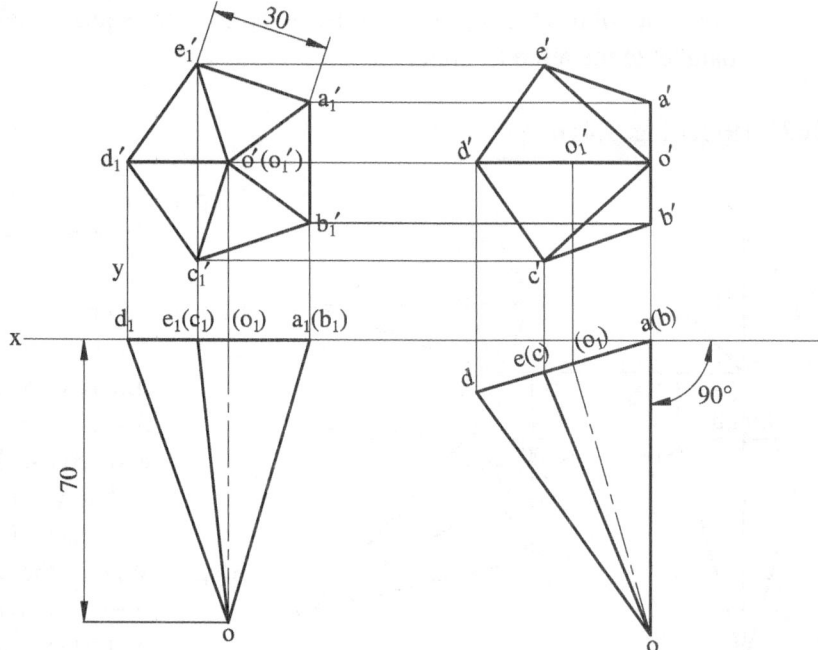

Fig. 9.44

Step 2: Since oa_1b_1 constitutes a triangular face in the top view, the solid is tilted about the corresponding base edge a_1b_1, redraw the top view about this edge such that the triangular face oa_1b_1 is perpendicular to the reference line as shown in Fig. 9.44, and draw the vertical projectors from its various points.

Step 3: When the horizontal locus lines from the front view in the first stage meet these vertical projector lines, the final front view is obtained.

Step 4: The visible and invisible edges are marked in the front view with the top view as reference. The slant edges do, co, and eo, which are far off from the reference lines are visible and hence, mark them as dark lines. The other edges that form the boundary are marked in thick lines as shown in Fig. 9.44.

Problems of Solids Tilted about a Corner

Example 9.25

A square pyramid of 30 mm base edges and axis 70 mm long has a corner of the base on VP. Draw its projections, when the axis is inclined at 30° to VP, and the plane containing that corner and the axis is horizontal.

Hint

Decide the initial projections based on the inclination of the axis.

1. When the axis is inclined to VP, keep the axis perpendicular to VP. Therefore, its base is parallel to VP, and hence, the polygon (square) is visualized in the front view.

2. Since the solid is tilted about a corner of the base, the square is laid with its diagonal parallel to the reference line.

PROCEDURE (Refer Fig. 9.45)

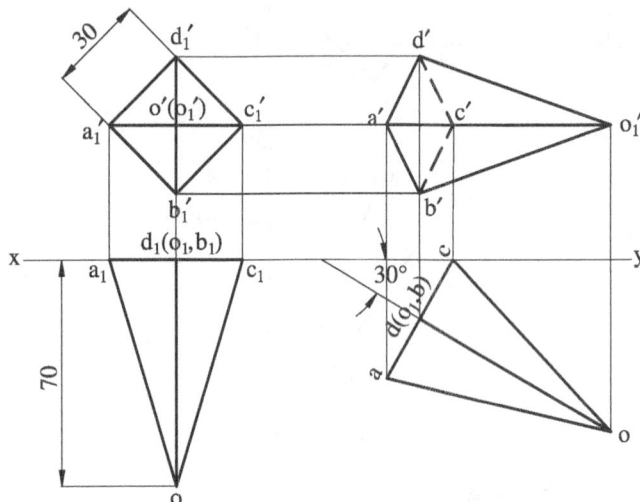

Step 1: Mark the reference line and draw the initial projections as mentioned earlier.

Step 2: Since the base corner of the solid lies on the reference line, tilt, redraw the top view such that c_1 lies on the reference line and axis inclined at 30° to it, and draw the vertical projectors from its various points.

Step 3: When the horizontal locus lines from the front view in the first stage meet these vertical projector lines, the final front view is obtained.

Fig. 9.45

Step 4: The visible and invisible edges are marked in the front view with the top view as reference. The slant edges ('oa', 'od', and 'ob') and base edges ('ab' and 'ad'), which are far off from the reference line are visible and marked in thick lines, while the slant edge 'oc' nearer to the reference is invisible and coincides with 'oa'. Similarly, the base edges 'cb' and 'cd' are nearer to the reference line which are invisible and marked dotted. The other edges that form the boundary are marked in thick lines as shown in Fig. 9.45. ▲

Example 9.26 ▲

A hexagonal pyramid of base sides 30 mm and axis 70 mm long is lying on VP on one of its slant edges. A plane containing this edge and the axis is perpendicular to VP and parallel to HP. Draw the projections of the pyramid.

Hint

Decide the initial projections based on the inclination of the axis.

1. When a pyramid is lying on one of its slant edges on VP, its axis gets inclined to VP. Hence, keep the axis perpendicular to VP. Therefore, its base is parallel to VP, and hence, the polygon (hexagon) is visualized in the front view.
2. Since the slant edge of a solid has been lying on a reference plane, it is tilted about a corner of the base. Hence, the hexagon is laid with its diagonal parallel to the reference line.

PROCEDURE (Refer Fig. 9.46)

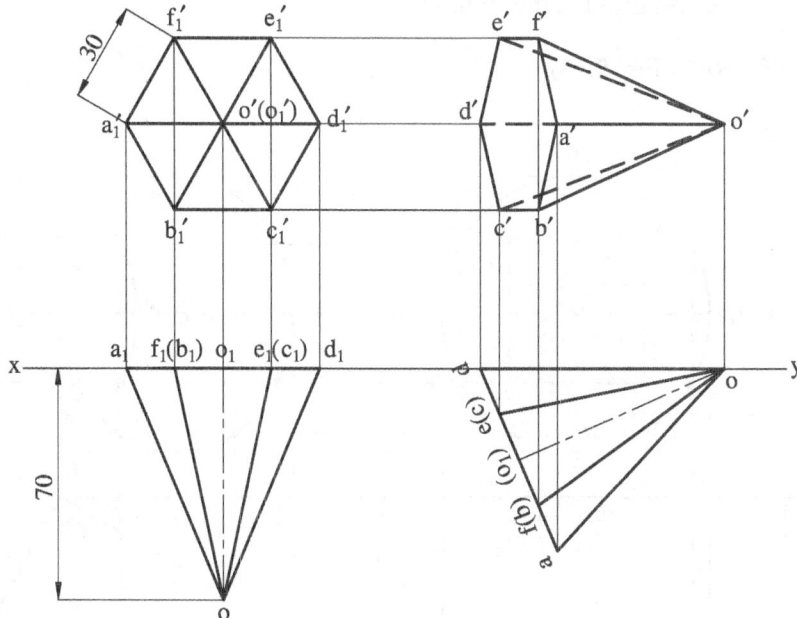

Fig. 9.46

Step 1: Mark the reference line and draw the initial projections as mentioned earlier.

Step 2: Since a slant edge has been lying on a reference line, the top view can be tilted and redrawn such that od_1 lies on the reference line and the vertical projectors from its various points can be drawn.

Step 3: When the horizontal locus lines from the front view in the first stage meet these vertical projector lines, the final front view is obtained.

Step 4: The visible and invisible edges are marked in the front view with the top view as reference. The slant edges 'oa', 'fo', and 'bo', which are far off from the reference line are visible and marked in thick lines, while the slant edge 'od', 'oe', and 'co', which are near to the reference line are invisible and hence marked dotted. The other edges that form the boundary are marked in thick lines as shown in Fig. 9.46.

Example 9.27

A cylinder of base diameter 30 mm and axis 70 mm long has its cylindrical end that is inclined at 30° to VP. Draw its projections, when the front view of the axis is parallel to the reference line.

Hint

Decide the initial projections based on the inclination of the axis.

1. Since the cylindrical end is inclined to VP, the axis of the cylinder is inclined to VP, keep the axis perpendicular to VP, and its base is parallel to VP, and hence, the circle is visualized in the front view.

PROCEDURE (Refer Fig. 9.47)

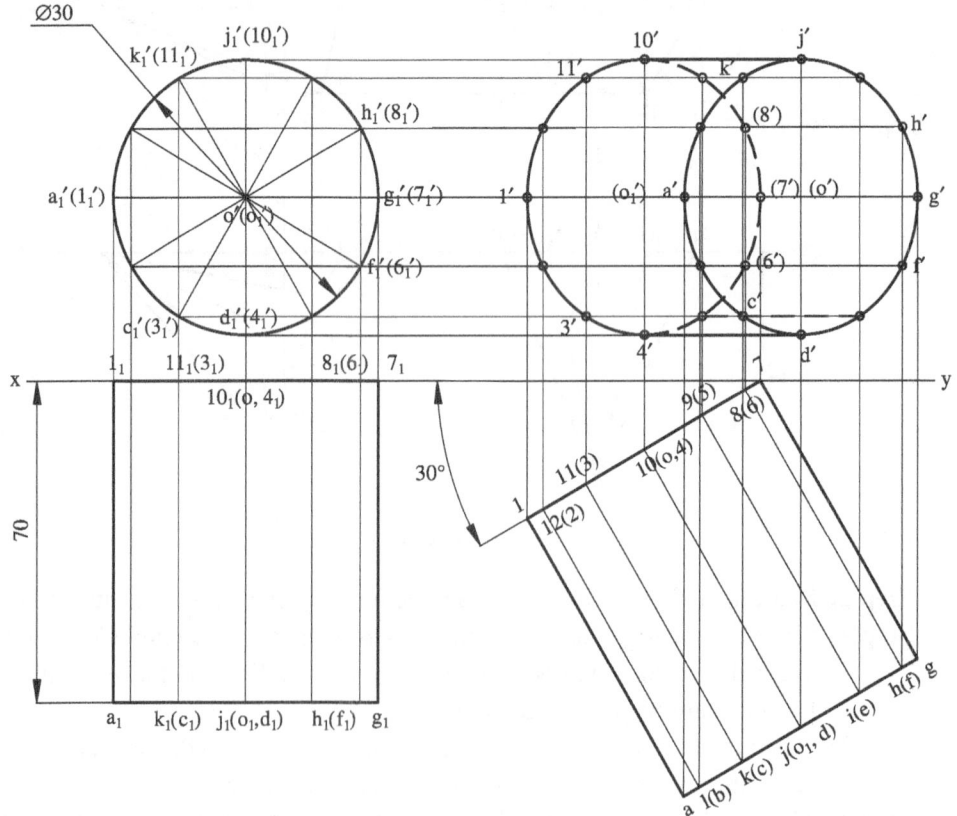

Fig. 9.47

Step 1: Mark the reference line and draw the initial projections by dividing the circle into 12 equal parts as shown in Fig. 9.47.

Step 2: Since the solid is getting tilted about 7_1, redraw the top view about this such that the base is inclined at 30° with the reference line as shown in Fig. 9.47, and draw the vertical projectors from its various points.

Step 3: When the horizontal locus lines from the front view in the first stage meet these projector lines, the final front view is obtained with its ends appearing elliptical.

Step 4: The circular end (a, ..., l), which is far off from the reference line is visible, while the other end (1, ..., 12) is marked dotted. The top generators are shown in thick lines along with their coincident hidden generators. The extreme generators that form the boundary are marked in thick lines as shown in Fig. 9.47.

Example 9.28

A cone of base diameter 30 mm and axis 70 mm long has one of its generators on VP. Draw its projections, when the axis is parallel to HP.

Hint

Decide the initial projections based on the inclination of the axis.

1. Since the cone is lying on its generator on VP, the axis is inclined to VP, keep the axis perpendicular to VP and its base is parallel to VP, and hence, the circle is visualized in the front view.

PROCEDURE (Refer Fig. 9.48)

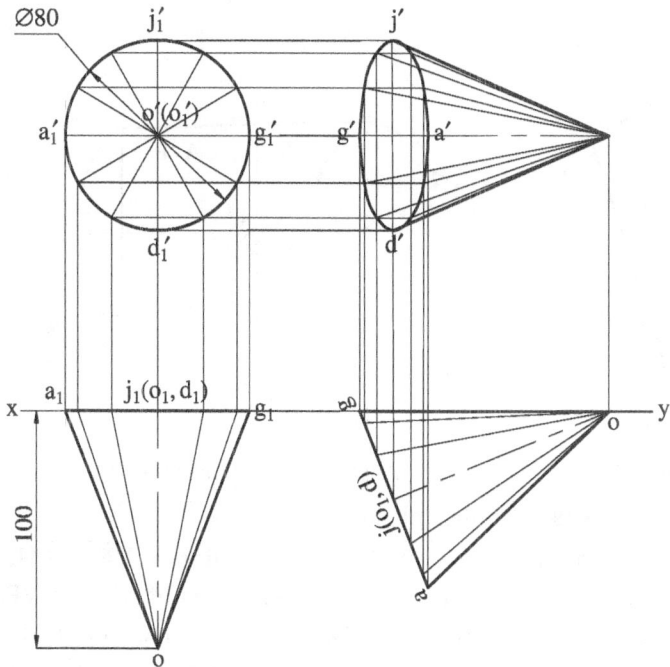

Step 1: Mark the reference line, draw the initial projections by dividing the circle into 12 equal parts, and mark the generators as shown.

Step 2: Since the solid is lying on its generator on VP, redraw the top view such that og_1 lies on the reference line as shown in Fig. 9.48, and draw the vertical projectors from its various points.

Step 3: When the horizontal locus lines from the front view in the first stage meet these projector lines, the final front view is obtained with its circular end visible and appearing elliptical.

Fig. 9.48

Step 4: The extreme generators that form the boundary are marked in thick lines as shown in Fig. 9.48.

9.5 PROJECTION OF SOLIDS INCLINED TO ONE OF THE REFERENCE PLANES BY AUXILIARY PLANE METHOD

9.5.1 Concept of Auxiliary Planes

Auxiliary planes are additional planes that are set up to obtain the true dimensions/ features of the object, otherwise not available in the three principal planes such as VP,

HP, and PP. The projections on auxiliary planes are called auxiliary/additional views. They may be used to get the true length of a straight line, the true size of a plane, and also to draw the projections of a solid when it is inclined to a reference plane without necessarily tilting the object.

9.5.2 Types of Auxiliary Planes and Auxiliary Views

The two types of auxiliary planes are as follows:

1. Auxiliary inclined plane (AIP)   2. Auxiliary vertical plane (AVP)

They are used to represent the projection of solids inclined to the reference planes.

Auxiliary Inclined Plane

This is an additional plane perpendicular to VP and inclined to HP (at an angle θ) as shown in Fig. 9.49(a).

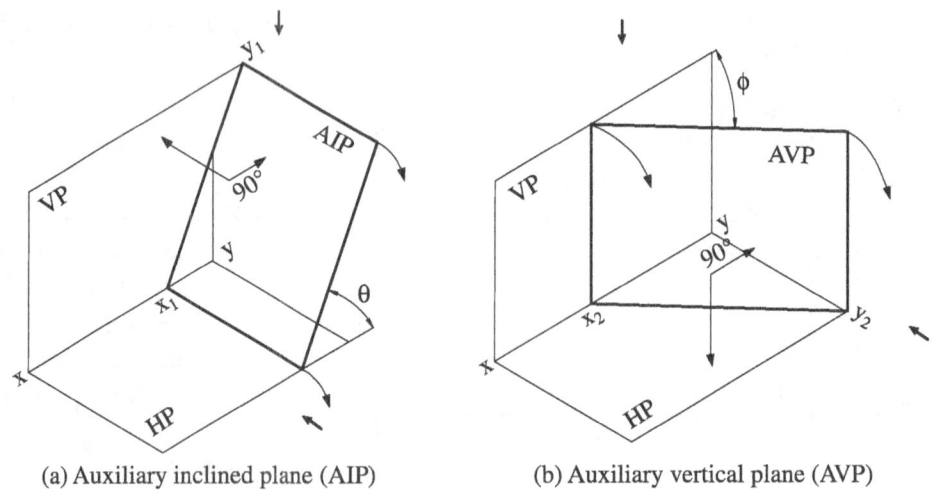

(a) Auxiliary inclined plane (AIP)   (b) Auxiliary vertical plane (AVP)

Fig. 9.49   Auxiliary planes

Similar to a reference line 'xy' that is used to mark the intersection of VP and HP earlier, an additional reference line x_1y_1 is used to mark the intersection of AIP and VP. When an object is held in the space between HP and VP and its axis is inclined to HP, instead of tilting the object and drawing the projections, the object can be held stationary and an auxiliary inclined plane can be created at the required angle. The object can then be projected to this AIP to get the new top view or the auxiliary top view, while the front view obtained earlier remains unchanged. The new or auxiliary reference line distinguishes the front view obtained earlier, and corresponding auxiliary top view is located on its other side when AIP is rotated and brought in alignment with the VP. It can be noted that when the AIP is rotated to make 90° with the HP, it becomes the PP, and the view obtained on it becomes the side view.

Auxiliary Vertical Plane

This is an additional plane perpendicular to HP and inclined to VP (at an angle ϕ) as shown in Fig. 9.49(b). As a reference line 'xy' is used to mark the intersection of VP and HP earlier, an additional reference line x_1y_1 is used to mark the intersection of AVP and HP. When an object is held in the space between HP and VP and its axis is inclined to VP, instead of tilting the object and drawing the projections, the object can be held stationary and an auxiliary vertical plane can be created at the required angle. The object can then be projected to this AVP to obtain the new front view or the auxiliary front view, while the top view obtained earlier remains unchanged. The new or auxiliary reference line distinguishes the top view obtained earlier and corresponding auxiliary front view is located on its other side when AVP is rotated and brought in alignment with the HP. It can be noted that when the AVP is rotated to make 90° with the VP, it becomes the profile plane and the view obtained on it becomes the side view.

9.5.3 Projection of a Solid on Auxiliary Planes

The problem of the projection of a solid inclined to both reference planes can be solved by the auxiliary plane method. In this method, the object is placed in the conventional space of VP and HP, and two auxiliary planes (AVP and AIP) are set up, and the solid will be projected on to these planes without tilting the solid with respect to the conventional reference planes. This is a very useful technique and helps to complete the projections at a very shorter time, as it avoids the repetitive labour involved in the redrawing of the views, which is usually done in the tilting object method. The auxiliary planes are incorporated by drawing additional or auxiliary reference lines and projections are drawn to these lines.

When the axis of the solid is inclined to one of the reference planes and parallel to the other, then either the AVP or AIP can be created as per the requirements and the object is projected on to the respective additional reference lines. The forthcoming section will explain the individual cases.

9.5.4 Projection of a Solid on AVP

NOTE *This is adopted when the axis of the solid is inclined to the VP and parallel to the HP.*

1. The initial projections of the solid are drawn as discussed in Section 9.3 and by drawing a conventional reference line 'xy'.
2. Retaining the non-polygonal top view obtained earlier as stationary, an AVP is obtained by drawing an auxiliary reference line x_1y_1 such that it is inclined to the axis in the top view and is drawn through the point representing the edge or the corner about which the solid is tilted.
3. From the various corners of the non-polygonal top view, draw the projector lines to the auxiliary reference line x_1y_1 to represent the projection of the solid on to AVP.
4. When the object is projected to AVP, its various corners maintain the same distance from HP (since the HP remains unchanged). To effect this, measure the distances of the various corners in the front view (polygonal view) from the xy line and mark them along the respective projectors drawn to the auxiliary reference line x_1y_1 and from it.

5. Join the meeting points thus obtained in a sequential way to get the auxiliary front view.
6. The visible and invisible edges/corners in the auxiliary front view are obtained by noting their locations from the auxiliary reference line to the top view. The points which are closer to x_1y_1 are invisible and marked hidden while that are far off are marked with visible representations.

9.5.5 Projection of a Solid on AIP

NOTE *This is adopted when the axis of the solid is inclined to the HP and parallel to the VP.*

1. The initial projections of the solid are drawn as discussed in Section 9.3 and by drawing a conventional reference line 'xy'.
2. Retaining the non-polygonal front view obtained earlier as stationary, an AIP is setup by drawing an auxiliary reference line x_1y_1 such that it is inclined to the axis in the front view and is drawn through the point representing the edge or the corner about which the solid is tilted.
3. From the various corners of the non-polygonal front view, draw the projector lines to the auxiliary reference line x_1y_1, to represent the projection of the solid on to AIP.
4. When the object is projected to AIP, its various corners maintain the same distance from VP (since the VP remains unchanged). To effect this, measure the distances of the various corners in the top view (polygonal view) from the 'xy' line and mark them along the respective projectors drawn to the auxiliary reference line x_1y_1 and from it.
5. Join the meeting points thus obtained in a sequential way to get the auxiliary top view.
6. The visible and invisible edges/corners in the auxiliary top view are obtained by noting their locations from the auxiliary reference line to the front view. The points that are closer to x_1y_1 are invisible and marked hidden while that are far off are marked with visible representations.

NOTE *The initial projections are drawn in all the problems by noting the various corners/points of the solid with alphabets/numerals using subscript 1, as in Section 9.4. However, in the final auxiliary views of the solids, no subscript reference is used, though they are projected from the views that have subscript notations. This is only to bring clarity to the auxiliary projection obtained and not to be construed as violation of notation norms of the orthographic projections.*

9.5.6 Problems on Solids with Axis Inclined to HP and Parallel to VP

Problems of Solids Tilted about an Edge

Example 9.29

(Example 9.12 by auxiliary plane method)

A rectangular prism 50×25 mm base and length 70 mm rests with one of its longer edges of the base on HP and the axis is inclined at 30° to HP and parallel to VP. Draw its projections.

Hint

1. Draw the initial projections based on inclination of the axis and as discussed in Example 9.12.
2. Draw a new reference line to the required inclination of the axis and passing through the edge about which it is tilted, in the non-polygonal view, and obtain its corresponding auxiliary view.

PROCEDURE (Refer Fig. 9.50)

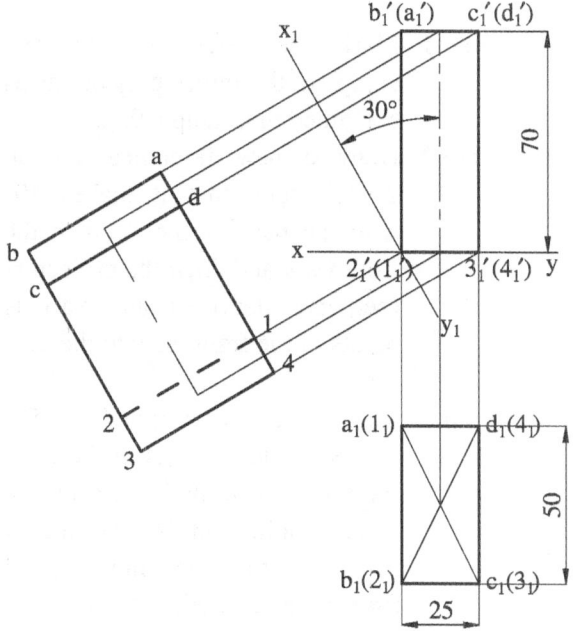

Fig. 9.50

Step 1: Mark the reference line and complete the initial projections as explained in Example 9.12.

Step 2: Draw a new reference $x_1 y_1$ at $2_1'(1_1')$ such that it makes 30° with the axis in the front view and draw the projectors (perpendiculars) from various points of the front view to the $x_1 y_1$ line.

Step 3: Measure the corresponding distances of these points in the first-stage top view and mark them on the corresponding projectors drawn to the new reference line $x_1 y_1$ and from it. For example, on the projector drawn at $1_1'$, mark the distance of its corresponding view, that is, 1_1 measured from the old reference line xy and locate 1. Similarly, the distances corresponding to the other points 2_1, ..., d_1 are measured from the xy line and are marked along the respective projectors and the points 2, ..., d are obtained in the new top view or generally known as the auxiliary top view.

Step 4: The visible and invisible edges are marked in the auxiliary top view with the front view and $x_1 y_1$ as reference. The rectangular end $(a_1', ..., d_1')$ that is far off from the new reference line $x_1 y_1$ is visible, while the other end $(1_1', ..., 4')$ is marked dotted. The other edges that form the boundary are marked in thick lines as shown in Fig. 9.50.   ▲

<hr>

Example 9.30 ◣

(Example 9.13 by auxiliary plane method)

A pentagonal prism 30 mm side of base and axis 70 mm long is resting on one of its edges of the base, in such a way that the base makes an angle of 40° with HP. Draw the projections, if the axis is parallel to VP.

Hint

1. Draw the initial projections based on inclination of the axis and as discussed in Example 9.13.
2. Draw a new reference line to the required inclination of the base and passing through the edge about which it is tilted, in the non-polygonal view and obtain its corresponding auxiliary view.

PROCEDURE (Refer Fig. 9.51)

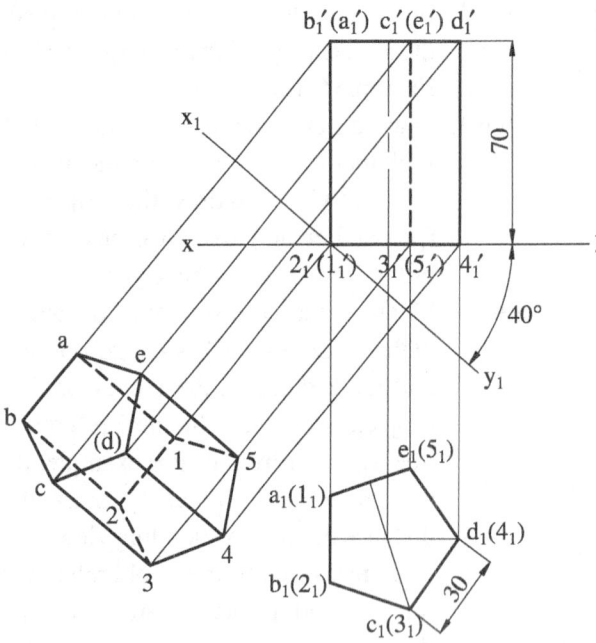

Fig. 9.51

Step 1: Mark the reference line and complete the initial projections as explained in Example 9.13.

Step 2: Draw a new reference x_1y_1 at $2_1'(1_1')$ such that it makes 40° with the base of the solid in the front view and draw the projectors (perpendiculars) from various points of the front view to the x_1y_1 line.

Step 3: Measure the corresponding distances of these points in the first-stage top view, mark them on the corresponding projectors drawn to the new reference line x_1y_1 and from it, and obtain the auxiliary top view in the similar lines of Example 9.29.

Step 4: The visible and invisible edges are marked in the auxiliary top view with the front view and x_1y_1 as reference. The pentagonal end $(a_1', ..., e_1')$ is visible, while the other end $(1', ..., 6')$ is marked dotted. The lateral edges corresponding to $1'a'$ and $2'b'$ also appear dotted as they are closer to the reference line. The other edges that form the boundary are marked in thick lines as shown in Fig. 9.51. ▲

Example 9.31(a) ◤

(Example 9.14 by auxiliary plane method)

A pentagonal pyramid of 30 mm base edges and axis 70 mm long is lying on one of its triangular faces on HP. Draw its projections, when the base edge on this face is perpendicular to VP and the vertical plane containing the axis is parallel to VP.

Hint

Decide the initial projections based on the inclination of the axis.

1. Since the triangular face is on HP, draw the new reference line to pass through that instead of the face on the conventional reference line and its corresponding auxiliary projection.

PROCEDURE (Refer Fig. 9.52)

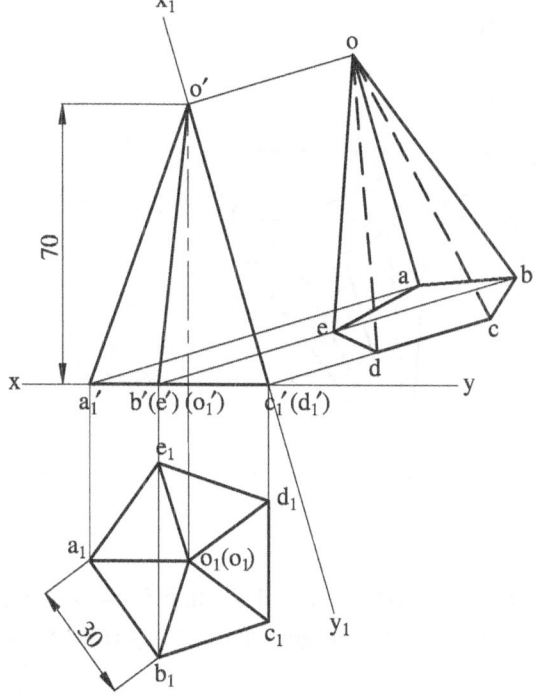

Fig. 9.52

Step 1: Mark the reference line and complete the initial projections as mentioned earlier and as explained in Example 9.14.

Step 2: Since $o'c_1'd_1'$ constitutes a triangular face and it has to lie on HP, draw the new reference line x_1y_1 to pass through that, and draw projectors from various points on to x_1y_1 and to the right side of it, since the top view has to be on the other side of the front view or x_1y_1 has to be in between them.

Step 3: Measure the corresponding distances of the various corner points in the first-stage top view, mark them on the corresponding projectors drawn to the new reference line x_1y_1 and from it, and obtain the auxiliary top view in the similar lines of Example 9.29.

Step 4: The visible and invisible edges are marked in the auxiliary top view with the front view and x_1y_1 as reference. The slant edge $o'a_1'$, which is far off from the reference line, is visible, while the slant edges $o'c_1'$ and $o'd_1'$ which are on the reference line are marked dotted. The other edges, which form the boundary are marked in thick lines as shown in Fig. 9.52. ▲

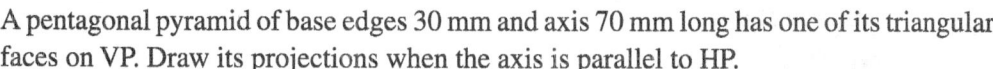

Example 9.31(b)

A pentagonal pyramid of base edges 30 mm and axis 70 mm long has one of its triangular faces on VP. Draw its projections when the axis is parallel to HP.

Hint

Decide the initial projections based on the inclination of the axis.

1. Since the triangular face is on VP, draw the new reference line to pass through that instead of the face on the conventional reference line, and obtain its corresponding auxiliary projection.

PROCEDURE (Refer Fig. 9.53)

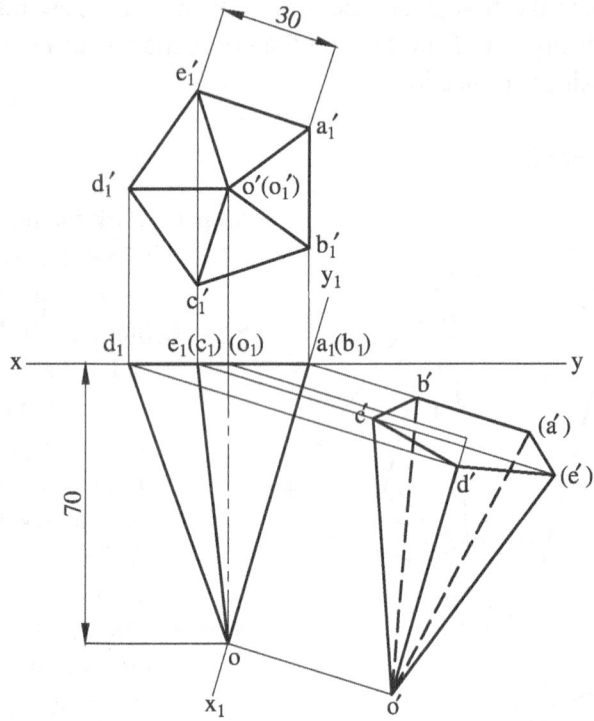

Fig. 9.53

Step 1: After drawing the initial projections as mentioned in Example 9.24, similar steps given in Example 9. 31(a) can be followed to get the auxiliary front view as shown in Fig. 9.53. ▲

Problems of Solids Tilted about a Corner

Example 9.32 ▲

(Example 9.16 by auxiliary plane method)

A rectangular prism 50×25 mm base and length 70 mm rests with one of its corners of the base on HP. The axis of the prism is inclined at 30° to HP and parallel to VP. Draw its projections.

Hint

1. Draw the initial projections based on inclination of the axis and discussed in Example 9.16
2. Draw a new reference line to the required inclination of the axis and passing through the corner about which it is tilted, in the non-polygonal view, and obtain its corresponding auxiliary view.

PROCEDURE (Refer Fig. 9.54)

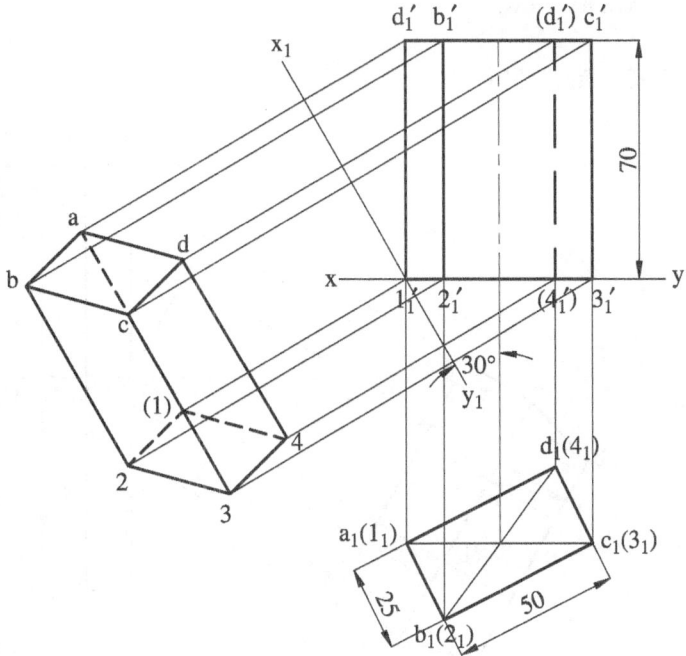

Fig. 9.54

Step 1: Mark the reference line and draw the initial projections as mentioned earlier.

Step 2: Since the solid is getting tilted about the base corner $1_1'$, draw the new reference line x_1y_1 through that point and inclined to the axis at 30° in the front view, and draw the projectors to the new reference line.

Step 3: Measure the corresponding distances of these points in the first-stage top view and mark them on the corresponding projectors drawn to the new reference line x_1y_1 and from it and obtain the auxiliary top view. The visible and invisible edges are marked in similar lines as explained in Example 9.29. ▲

Example 9.33 ◣

(Example 9.17 by auxiliary plane method)

A hexagonal prism of 30 mm base edges and axis 70 mm long rests on one of its corners of base on HP. Draw its projections, when the lateral edge through that corner on HP is inclined at 30° to HP and the vertical plane containing lateral edge and the axis is parallel to VP.

Hint

1. Draw the initial projections based on inclination of the axis as discussed in Example 9.17.

2. Draw a new reference line through the corner containing the lateral edge, to the required inclination in the non-polygonal view, and obtain its corresponding auxiliary view.

PROCEDURE (Refer Fig. 9.55)

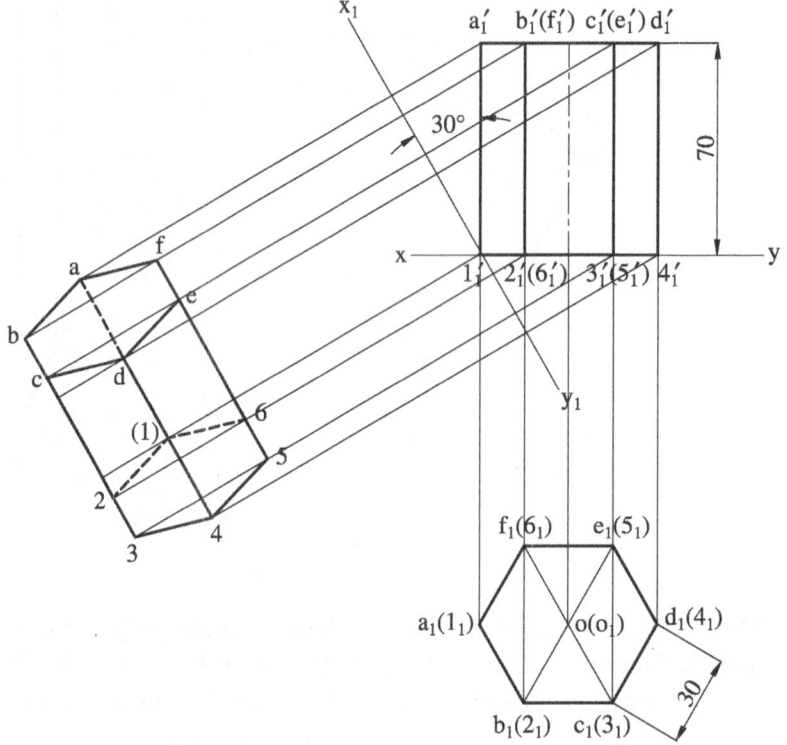

Fig. 9.55

Step 1: Mark the reference line and draw the initial projections as mentioned earlier.

Step 2: Since the solid is getting tilted about the base corner $1_1'$, draw the new reference line x_1y_1 through that point and inclined at $30°$ to the lateral edge $1_1'a_1'$ in the front view, and draw the projectors to the new reference line.

Step 3: Measure the corresponding distances of these points in the first-stage top view, mark them on the corresponding projectors drawn to the new reference line x_1y_1 and from it, and obtain the auxiliary top view. The visible and invisible edges are marked in similar lines as explained in Example 9.17. ▲

Example 9.34 ▲

(Example 9.18 by auxiliary plane method)

Draw the top and front views of a right circular cylinder of base 45 mm diameter and axis 60 mm long, when its axis is inclined at $30°$ to HP and parallel to the VP. Draw its projections.

Hint

1. Draw the initial projections based on inclination of the axis and as discussed in Example 9.18.
2. Draw a new reference line through a point containing the extreme generator to the required inclination of the axis in the non-polygonal view and obtain its corresponding auxiliary view.

PROCEDURE (Refer Fig. 9.56)

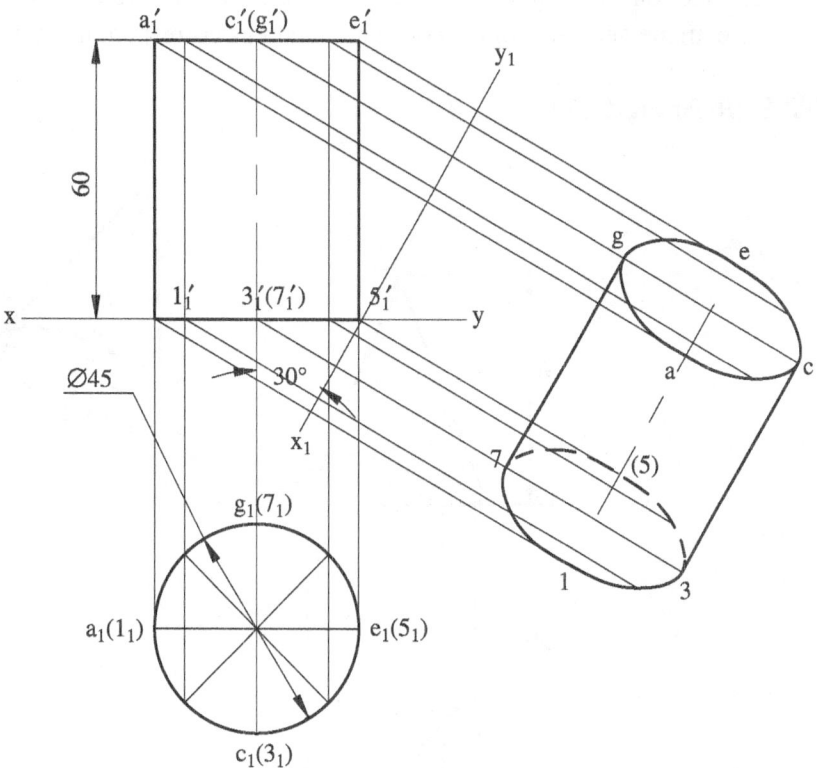

Fig. 9.56

Step 1: Mark the reference line and draw the initial projections as mentioned earlier.

Step 2: Since the solid is getting tilted about the base point $5_1'$, draw the new reference line x_1y_1 through that point and inclined at 30° to the axis in the front view, and draw the projectors to the new reference line.

Step 3: Measure the corresponding distances of these points in the first-stage top view, mark them on the corresponding projectors drawn to the new reference line x_1y_1 and from it, and obtain the auxiliary top view. The circular ends appear as ellipse, the farthest one is in thick line and nearest one in dotted lines. The extreme generators form the boundary as shown in Fig. 9.56. ▲

Example 9.35

(Example 9.20 by auxiliary plane method)

A cone of base 80 mm diameter and height 100 mm is lying with one of its generators on HP and axis parallel to VP. Draw its projections.

Hint

1. Draw the initial projections based on inclination of the axis and as discussed in Example 9.20
2. Since one of the generators lies on HP, draw a new reference line to pass through the extreme generator of the cone and obtain its corresponding auxiliary view.

PROCEDURE (Refer Fig. 9.57)

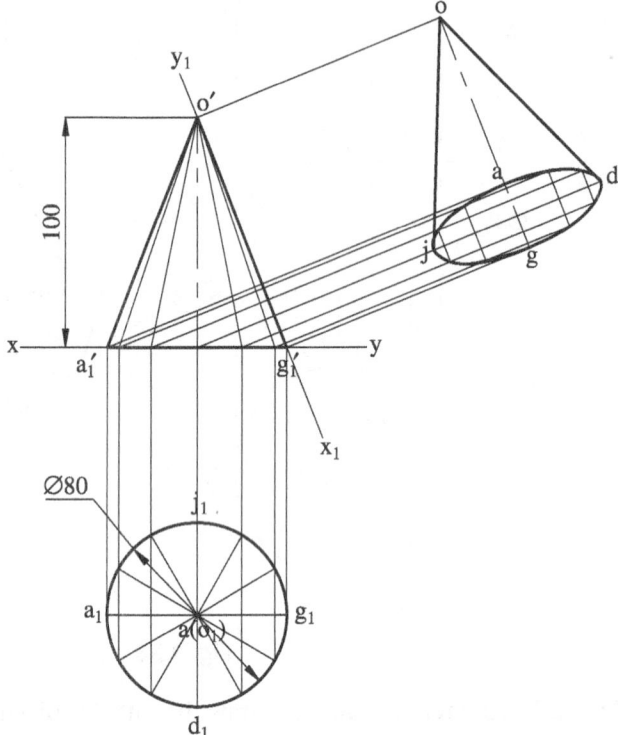

Fig. 9.57

Step 1: Mark the reference line and draw the initial projections as mentioned earlier.

Step 2: Since the generator is on HP, draw the new reference line $x_1 y_1$ through the extreme generator $o' g_1'$ and draw projectors from various points on to $x_1 y_1$ and to the right side of it, since the top view has to be on the other side of the front view.

Step 3: Measure the corresponding distances of the various points in the first-stage top view, mark them on the corresponding projectors drawn to the new reference line x_1y_1 and from it, and obtain the auxiliary top view in the similar lines discussed in previous problems.

Step 4: The circular end appears as an ellipse and marked in thick lines. The extreme generators form the boundary as shown in Fig. 9.57. ▲

Problems of Solids Suspended about a Corner

Example 9.36 ▲

(Example 9.22 by auxiliary plane method)

A hexagonal pyramid of base edge 30 mm and axis 70 mm long is freely suspended from a corner of its base. Draw its projections, when the top view of the axis is parallel to the reference line.

Hint

1. Decide the initial projections based on inclination of the axis.
2. In the suspended position, the point of suspension and the centre of gravity of the body fall on a vertical line, and hence, the new reference line is drawn perpendicular to this line, and auxiliary top view is obtained.

PROCEDURE (Refer Fig. 9.58)

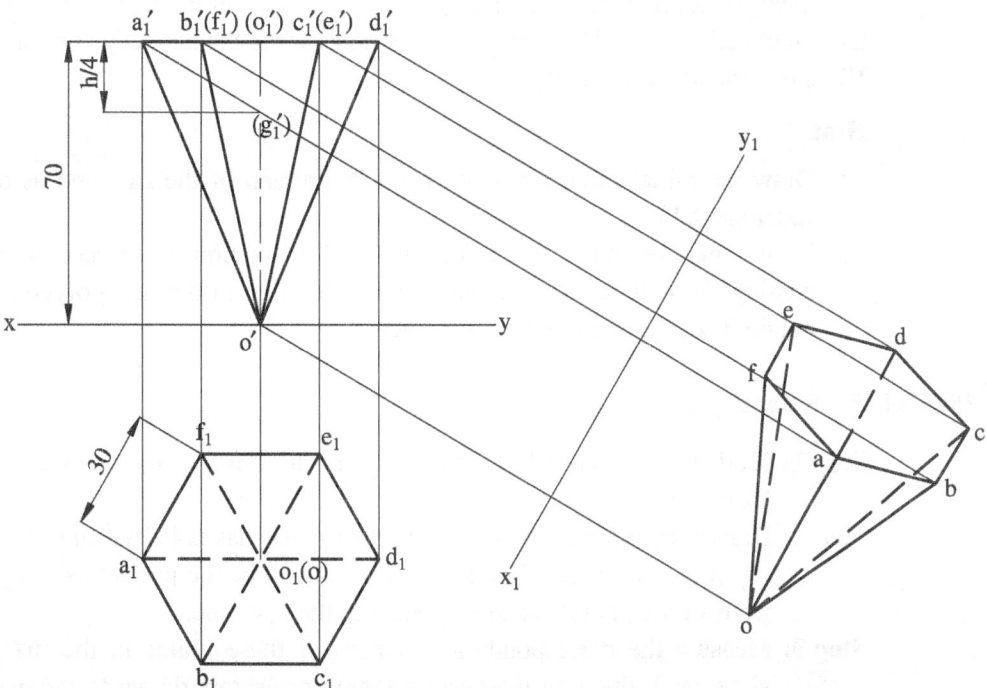

Fig. 9.58

Step 1: Mark the reference line and complete the initial projections as mentioned earlier and explained in Example 9.22.

Step 2: In the suspended position, the point of suspension (a_1') and the centre of gravity (G) of the body fall on a vertical line. The CG lies on the axis on the axis; for pyramids, it is at a distance of one-fourth of the length of the axis. This is marked as g_1' in the front view.

Step 3: Draw a new reference line x_1y_1 perpendicular to $a_1'g_1'$ and project the points to the new reference line.

Step 4: Measure the corresponding distances of the various points in the first-stage top view, mark them on the corresponding projectors drawn to the new reference line x_1y_1 and from it, and obtain the auxiliary top view in the similar lines discussed in previous problems.

Step 5: The visible and invisible edges are marked in similar lines as discussed in Example 9.22.   ▲

9.5.7 Problems on Solids with Axis Inclined to VP and Parallel to HP

Problems of Solids Tilted about an Edge

Example 9.37   ▲

(Example 9.23 by auxiliary plane method)

A hexagonal prism of base edge 30 mm and axis 70 mm long has one of rectangular faces inclined at 45° to VP. Draw its projections, when the base edge on this face lies on VP and perpendicular to HP.

Hint

1. Draw the initial projections based on inclination of the axis and as discussed in Example 9.23.
2. Draw a new reference line to the required inclination of the rectangular face and passing through the edge about which it is tilted in the non-polygonal view, and obtain its corresponding auxiliary view.

PROCEDURE   (Refer Fig. 9.59)

Step 1: Mark the reference line and complete the initial projections as explained in Example 9.23.

Step 2: Draw a new reference x_1y_1 at $5_1(4_1)$ such that it makes 45° with the rectangular face $(5_1e_1d_14_1)$ of the solid in the top view and draw the projectors (perpendiculars) from various points of the top view to the x_1y_1 line.

Step 3: Measure the corresponding distances of these points in the first-stage front view, mark them on the corresponding projectors drawn to the new reference line x_1y_1 and from it, and obtain the auxiliary front view similar to the previous problems.

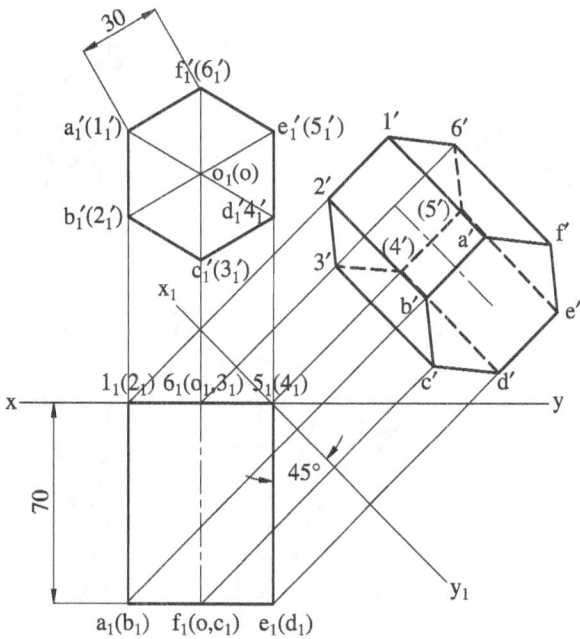

Fig. 9.59

Step 4: The visible and invisible edges are marked in the auxiliary front view with the top view and x_1y_1 as reference. The hexagonal end $(a_1, ..., f_1)$ is visible, while the other end $(1_1, ..., 6_1)$ is marked dotted. The lateral edges corresponding to 5_1e_1 and 4_1d_1 also appear dotted as they are closer to the reference line. The other edges that form the boundary are marked in thick lines as shown in Fig. 9.59. ▲

Problems of Solids Tilted about a Corner

Example 9.38 ▲

A pentagonal pyramid of base edges 30 mm and axis 70 mm long has one of its slant edges on VP. Draw its projections, when the axis is parallel to HP.

Hint

1. Decide the initial projections based on inclination of the axis.
2. Since the slant edge is on VP, draw the new reference line to pass through that instead of the slant edge on the conventional reference line and obtain its corresponding auxiliary projection.

PROCEDURE (Refer Fig. 9.60)

Step 1: Mark the reference line and complete the initial projections as mentioned earlier.

Step 2: Since the slant edge od_1 has to lie on VP, draw the new reference line x_1y_1 to pass through that and draw projectors from various points on to x_1y_1 and to the left side of it, since the front view has to be on the other side of the top view.

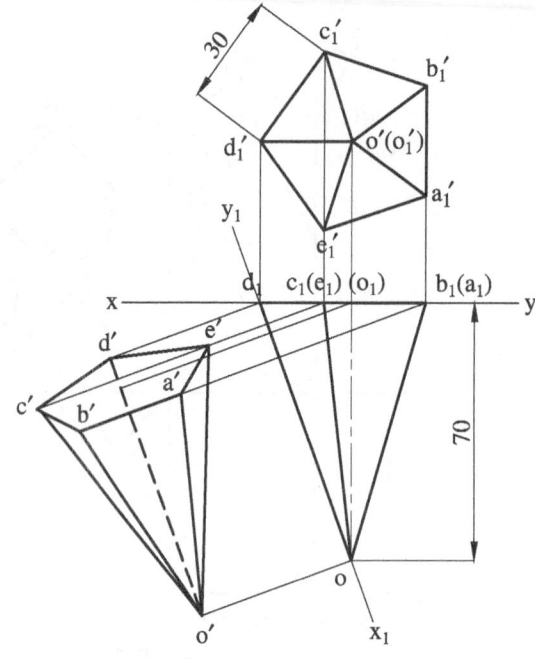

Fig. 9.60

Step 3: Measure the corresponding distances of the various corner points in the first-stage front view, mark them on the corresponding projectors drawn to the new reference line x_1y_1 and from it, and obtain the auxiliary front view similar to the previous problems.

Step 4: The visible and invisible edges are marked in the auxiliary front view with the top view and x_1y_1 as reference. The slant edges ob_1 and oa_1, which are far off from the reference line, are visible, while the slant edge od_1, which is on the reference line, is marked dotted. The other edges that form the boundary are marked in thick lines as shown in Fig. 9.60. ▲

Example 9.39 ◤

(Example 9.25 by auxiliary plane method)

A square pyramid of 30 mm base edges and axis 70 mm long has a corner of the base on VP. Draw its projections, when the axis is inclined at 30° to VP and the plane containing that corner and the axis is horizontal.

Hint

1. Draw the initial projections based on inclination of the axis and as discussed in Example 9.25

2. Draw a new reference line through the corner of the base such that it is inclined to the axis at the required inclination in the non-polygonal view, and obtain its corresponding auxiliary view.

PROCEDURE (Refer Fig. 9.61)

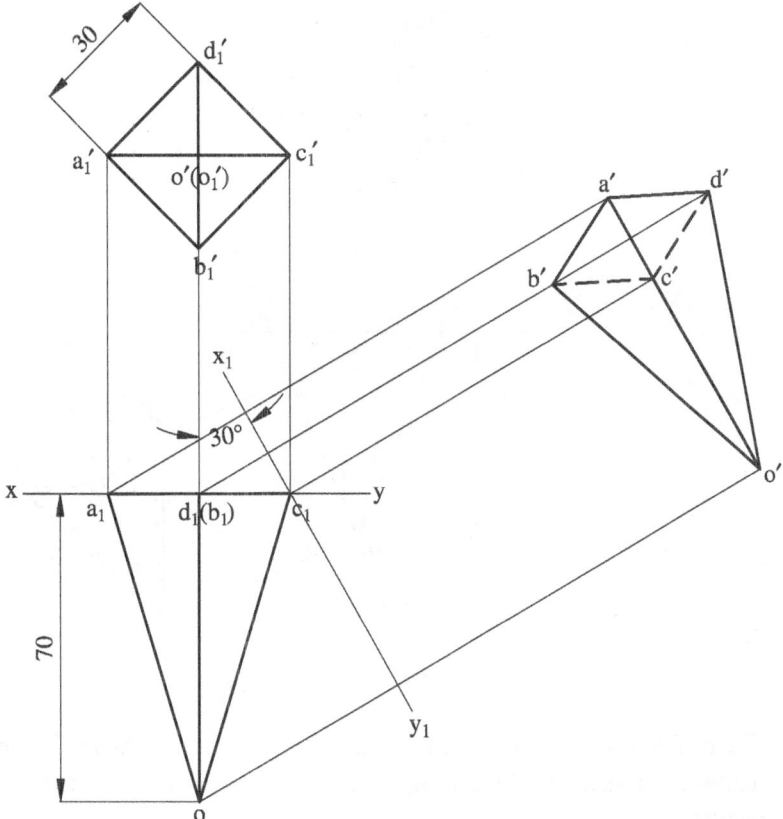

Fig. 9.61

Step 1: Mark the reference line and draw the initial projections as mentioned earlier.

Step 2: Since the solid is getting tilted about the base corner c_1, draw the new reference line x_1y_1 through that point and inclined at $30°$ to the axis in the top view and draw the projectors to the new reference line.

Step 3: Measure the corresponding distances of these points in the first-stage front view, mark them on the corresponding projectors drawn to the new reference line x_1y_1 and from it, and obtain the auxiliary front view. The visible and invisible edges are marked in similar lines as explained in Example 9.25.

Example 9.40

(Example 9.26 by auxiliary plane method)

A hexagonal pyramid of base sides 30 mm and axis 70 mm long is lying on VP on one of its slant edges. A plane containing this edge and the axis is perpendicular to VP and parallel to HP. Draw the projections of the pyramid.

PROCEDURE (Refer Fig. 9.62)

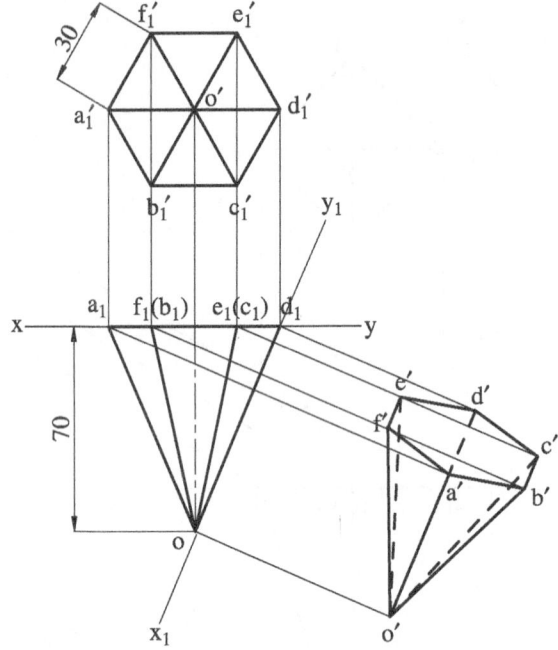

Fig. 9.62

The problem is similar to the Example 9.38 except the change in the polygon shape. The reader is advised to refer Example 9.38 for the complete procedure and Fig. 9.60 for the results. ▲

9.6 PROJECTION OF SOLIDS INCLINED TO BOTH REFERENCE PLANES BY TILTING SOLID METHOD AND AUXILIARY PLANE METHOD

Example 9.41

A pentagonal prism of 30 mm side of base and axis 70 mm long is resting on one of its base edges on HP. Draw its projections when the base makes an angle of 40° with HP and the top view of the axis is inclined at 45° to reference line.

Solid Tilting Method—Principle

1. Decide the initial projections based on the inclination of the axis and effect the second-stage projections as per the HP inclination and as discussed in Example 9.13.
2. Redraw the top view obtained in the second stage to the required VP inclination and obtain the final front view.

PROCEDURE (Refer Fig. 9.63a)

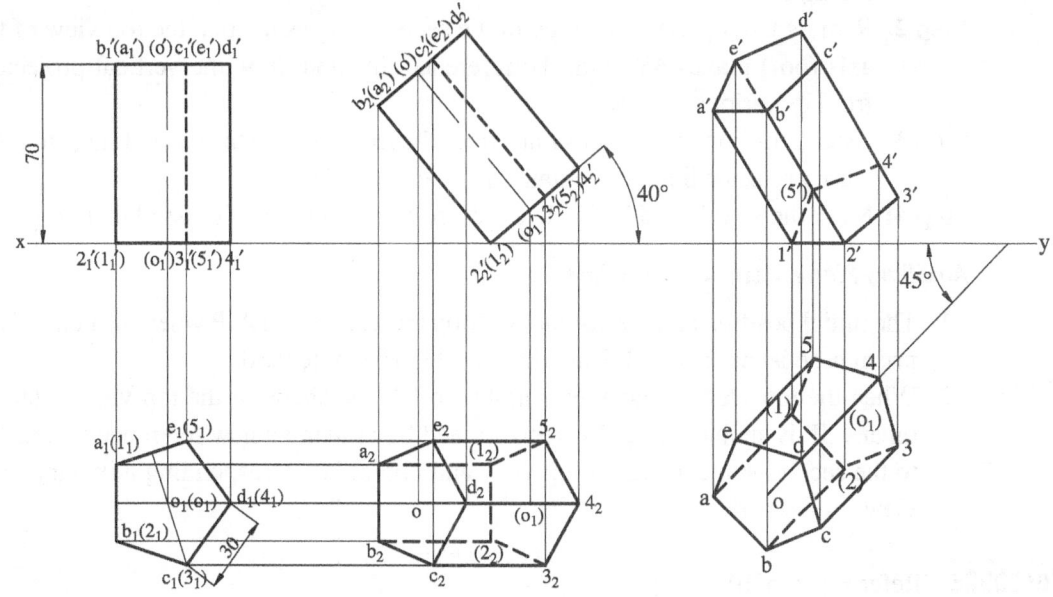

(a) Solution by solid tilting method

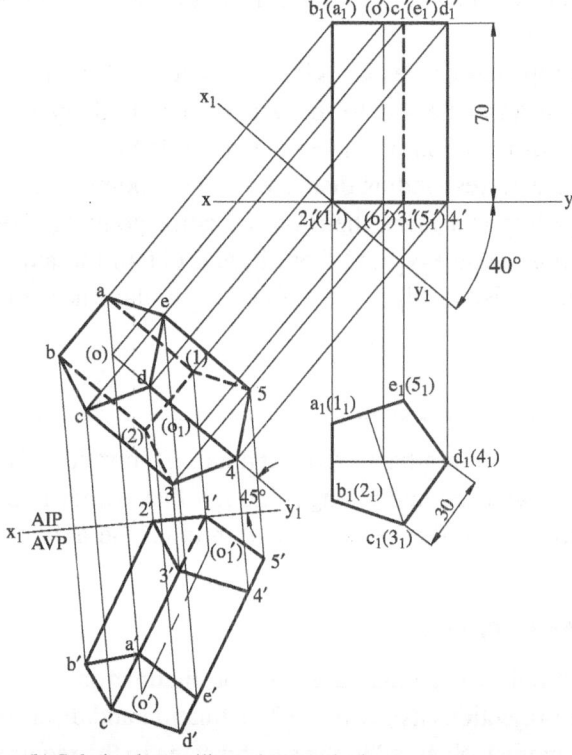

(b) Solution by auxiliary plane method

Fig. 9.63

Step 1: Draw the initial projections and the second-stage projections as explained in Example 9.13.

Step 2: Redraw the top view obtained in the second stage such that the top view of the axis (oo_1) makes 45° with the reference line and draw the vertical projectors from its various points.

Step 3: Extend the horizontal locus lines from the front view in the second stage to meet these projector lines and obtain the final front view.

Step 4: Mark the visible and invisible edges in similar lines as discussed earlier.

Auxiliary Plane Method—Principle

1. The initial position and the second position top view on an AIP when the pentagonal prism base is inclined to HP are discussed in Example 9.30.
2. When the top view of the axis is at 45° to VP, the shape of the top view obtained on an AIP is unchanged and is repositioned by drawing a new reference line at 45° to the second-stage auxiliary top view and obtain its corresponding auxiliary front view.

PROCEDURE (Refer Fig. 9.63b)

Step 1: The initial positions and the auxiliary top view on an AIP are explained in Example 9.30.

Step 2: Since the top view of the axis is inclined at an angle of 45°, draw a new reference line x_2y_2 at an angle 45° to the axis of the auxiliary top view, and draw the projectors from various points to the new reference line.

Step 3: Measure the corresponding distances of these points in the first-stage front view from x_1y_1 line and mark them on the corresponding projectors drawn to the new reference line x_2y_2 and from it, and obtain the auxiliary front view. The visible and invisible edges are marked in similar lines as explained in previous problems. ▲

Example 9.42 ▲

A hexagonal prism of 30 mm base edges and axis 70 mm long is resting on one of its base corners on HP and the lateral edge through that corner is inclined at 30° to HP. Draw its projections when the vertical plane through the axis makes an angle of 60° with the VP.

Solid Tilting Method—Principle

1. Decide the initial projections based on the inclination of the axis and effect the second-stage projections as per the HP inclination and discussed in Example 9.17.
2. Redraw the top view obtained in the second stage to the required VP inclination and obtain the final front view.

PROCEDURE (Refer Fig. 9.64a)

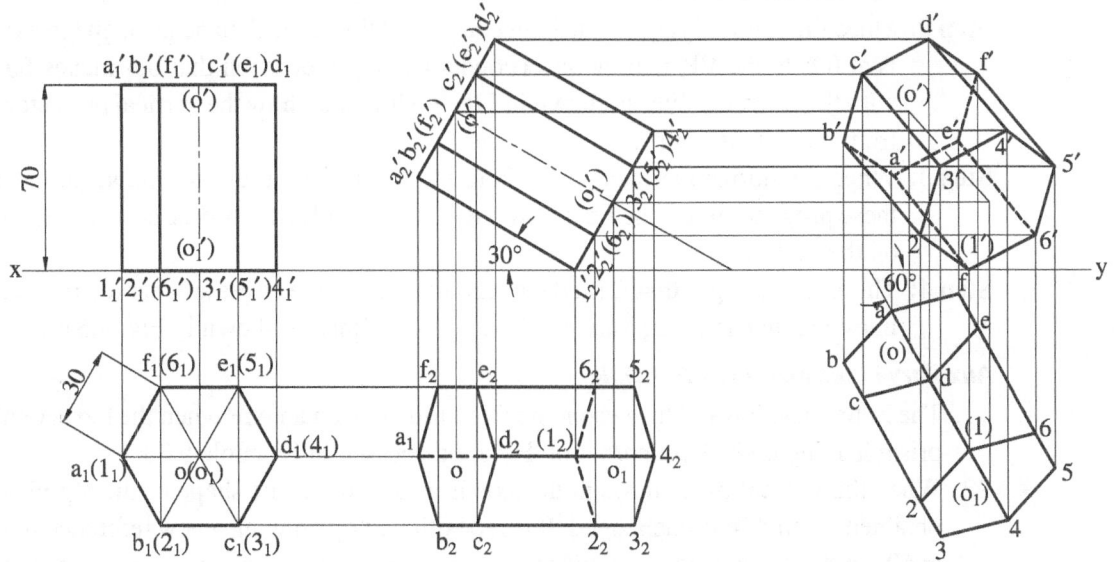

(a) Solution by solid tilting method

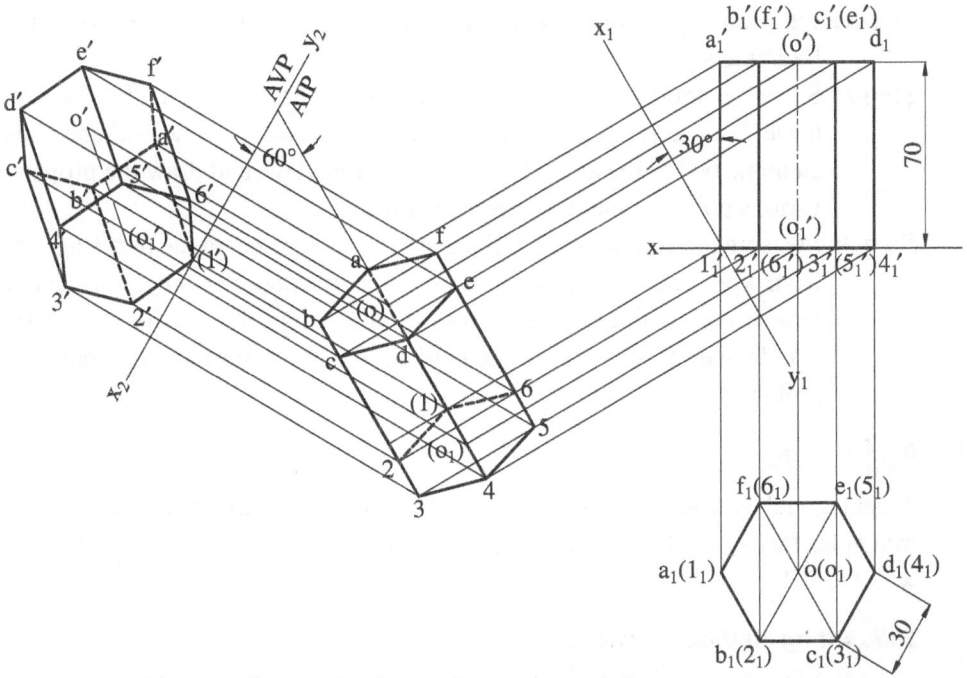

(b) Solution by auxiliary plane method

Fig. 9.64

Step 1: The first two stages of the solid positions and its respective projections are explained in Example 9.17.

Step 2: Since the solid is lying on the base corner and the vertical plane through the axis is at 60° to the VP, redraw the second-stage top view such that oo_1 makes 60° with the reference line as shown in Fig. 9.64(a), and draw the vertical projectors from its various points.

Step 3: When the horizontal locus lines from the front view in the second stage meet these projector lines, the final front view is obtained with its base corner 1′ lying on the 'xy' line.

Step 4: The extreme edges that form the boundary and visible edges are marked in thick lines and invisible edges are marked in dotted lines as shown in Fig. 9.64(a).

Auxiliary Plane Method—Principle

1. The initial position and the second position top view on an AIP when the hexagonal prism is lying on its base corner on HP are discussed in Example 9.33.

2. When the vertical plane through the axis is at 60° to VP, the shape of the top view obtained on an AIP is unchanged. It is repositioned by drawing a new reference line at 60° to the second-stage auxiliary top view, and its corresponding auxiliary front view.

PROCEDURE (Refer Fig. 9.64b)

Step 1: The initial positions and the auxiliary top view on an AIP are explained in Example 9.33.

Step 2: Since the solid is lying on its corners of the base on HP and the vertical plane through the axis is inclined at an angle of 60°, draw a new reference line x_2y_2 at an angle 60° to the axis of the auxiliary top view, and draw the projectors from various points to the new reference line.

Step 3: Measure the corresponding distances of these points in the first-stage front view from x_1y_1 line and mark them on the corresponding projectors drawn to the new reference line x_2y_2 and from it, and obtain the auxiliary front view. The visible and invisible edges are marked in similar lines as explained in previous problems. ▲

Example 9.43 ▲

A pentagonal pyramid of 30 mm base edges and axis 70 mm long is lying on one of its triangular faces on HP. Draw its projections, when the top view of the axis makes 19° with the VP.

Solid Tilting Method—Principle

1. Since the triangular face is lying on HP, the axis is inclined to HP, and hence, decide the initial projections with one base edge perpendicular to reference line, and tilt the solid about this edge as discussed in Example 9.14.

2. When the top view of the axis inclined to VP, the shape of the top view obtained in the second stage is unchanged and is repositioned to the required inclination from which the final front view can be obtained.

PROCEDURE (Refer Fig. 9.65a)

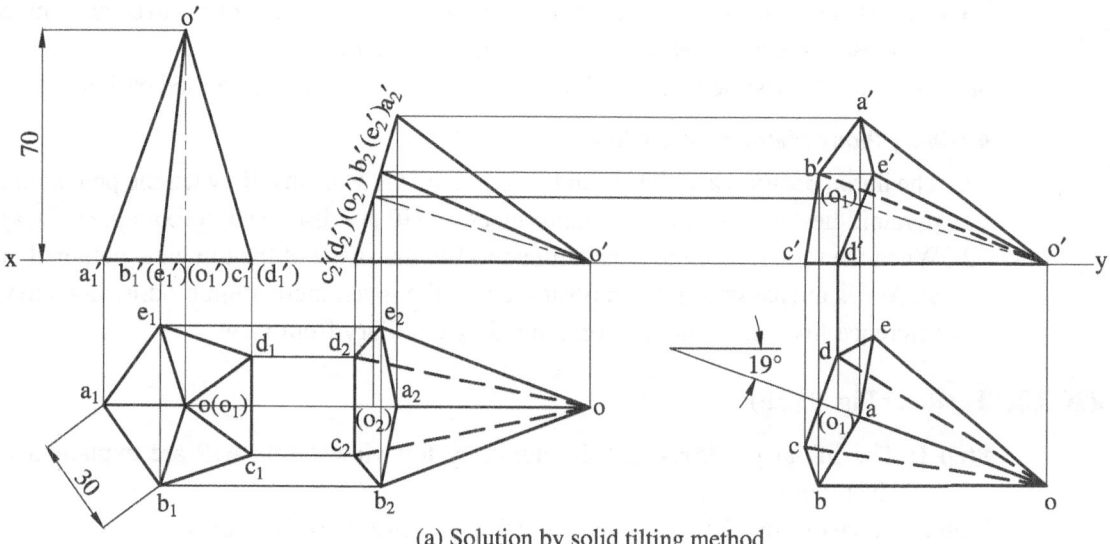

(a) Solution by solid tilting method

(b) Solution by auxiliary plane method

Fig. 9.65

Step 1: Draw the initial projections and the second-stage projections as explained in Example 9.14.

Step 2: Redraw the top view obtained in the second stage such that the top view of the axis (oo_1) makes 19° with the reference line and draw the vertical projectors from its various points.

Step 3: Extend the horizontal locus lines from the front view in the second stage to meet these projector lines and obtain the final front view.

Step 4: Mark the visible and invisible edges in similar lines as discussed earlier.

Auxiliary Plane Method—Principle

1. The initial position and the second position top view on an AIP when the pentagonal pyramid lies on one of its triangular faces on HP are discussed in Example 9.31(a).
2. When the top view of the axis inclined to VP, the shape of the top view obtained on an AIP is unchanged and is repositioned to the given inclination by drawing a new reference line and obtain its corresponding auxiliary front view.

PROCEDURE (Refer Fig. 9.65b)

Step 1: The initial positions and the auxiliary top view on an AIP are explained in Example 9.31(a).

Step 2: Since the top view of the axis is inclined at an angle of 19°, draw a new reference line x_2y_2 at an angle 19° to the axis of the auxiliary top view, and draw the projectors from various points to the new reference line.

Step 3: Measure the corresponding distances of these points in the first-stage front view and mark them on the corresponding projectors drawn to the new reference line x_2y_2 and from it, to obtain the auxiliary front view. The visible and invisible edges are marked in similar lines as explained in aforementioned problems. ▲

Example 9.44 ▲

A hexagonal pyramid of 30 mm base edges and axis 70 mm long is having on one of its slant edges on VP. Draw its projections, when (a) the front view of the axis makes 30° with the reference line and (b) the axis is inclined to the HP at an angle of 30°.

Solid Tilting Method (Case A and Case B)—Principle

1. When a pyramid is lying on one of its slant edges on VP, its axis gets inclined to VP. Hence, the axis is kept perpendicular to VP, and the initial projections are drawn by laying the hexagon with its diagonal parallel to the reference line in the front view.
2. When the front view of the axis is at 30° to the reference line, the shape of the front view obtained in the second stage is unchanged and is repositioned such that the axis of the front view is 30° to the reference line from which the final top view can be obtained for the 'Case A'.
3. Similarly, for 'Case B', when the axis makes 30° to HP, find the apparent angle of inclination as discussed in Chapter 7, reposition the front view to this angle, and obtain the final front view.

PROCEDURE (Refer Fig. 9.66a)

The first two stages of the projections are explained in Example 9.26.

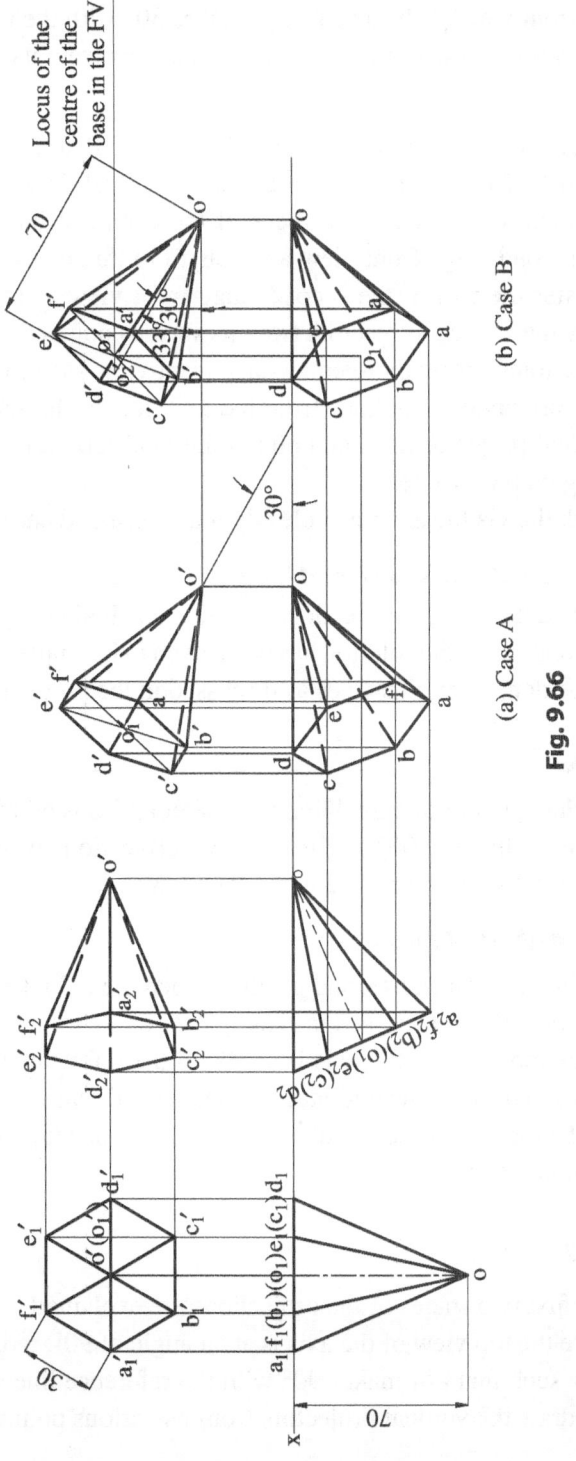

(a) Case A

(b) Case B

Fig. 9.66

Case A

Step 1: Since the solid is lying on one of its slant edges and the axis of the front view is at an angle of 30° to the HP, redraw the second stage front view such that the front view of the axis $(o'o_1')$ makes 30° with the reference line as shown in Fig. 9.66(a), and draw the vertical projectors from its various points.

Case B

Step 1: Since the axis is at an angle of 30° to the HP, draw a 30° line $o'o_2'$ of length equal to 70 mm (true length of the axis), and draw a horizontal line through the point o_2' which is the locus of the centre of the base in the FV. Transfer the second-stage front view axis length to this locus line and obtain o_1'. Then transfer the rest of the second-stage front view as shown in Fig. 9.66(b) and draw the vertical projectors from its various points.

Step 2: After transferring the second-stage front view as per Case A and Case B, draw the horizontal locus lines from the top view in the second stage to meet these vertical projector lines and obtain the final top view with its slanting edge do lying on the xy line.

Step 3: Mark the visible and invisible edges as discussed earlier.

Auxiliary Plane Method (Case A and Case B)

The second and third stage projections can also be obtained by auxiliary plane method. Up to the second stage, the solution has been given in Example 9.40. The final projections on AIP can be done in similar lines as discussed in the previous examples.   ▲

Example 9.45   ▲

A right circular cylinder of base 45 mm diameter and axis 60 mm long rests on HP, such that its axis is inclined at 30° to HP. Draw its projections when the axis appears to be perpendicular to the reference line in the top view.

Solid Tilting Method—Principle

1. The initial position and second position when the cylinder is lying on the corner of the base on HP are discussed in Example 9.18
2. When the top view of the axis is at 90° to the reference line, the shape of the top view obtained in the second stage is unchanged and is repositioned such that the axis of the top view is perpendicular to the reference line from which the final front view can be obtained.

PROCEDURE   (Refer Fig. 9.67a)

Step 1: The first two stages of the projections are explained in Example 9.18.

Step 2: Since the top view of the axis is at an angle of 90°, redraw the second-stage top view such that oo_1 makes 90° with the reference line as shown in Fig. 9.67(a), and draw the vertical projectors from its various points.

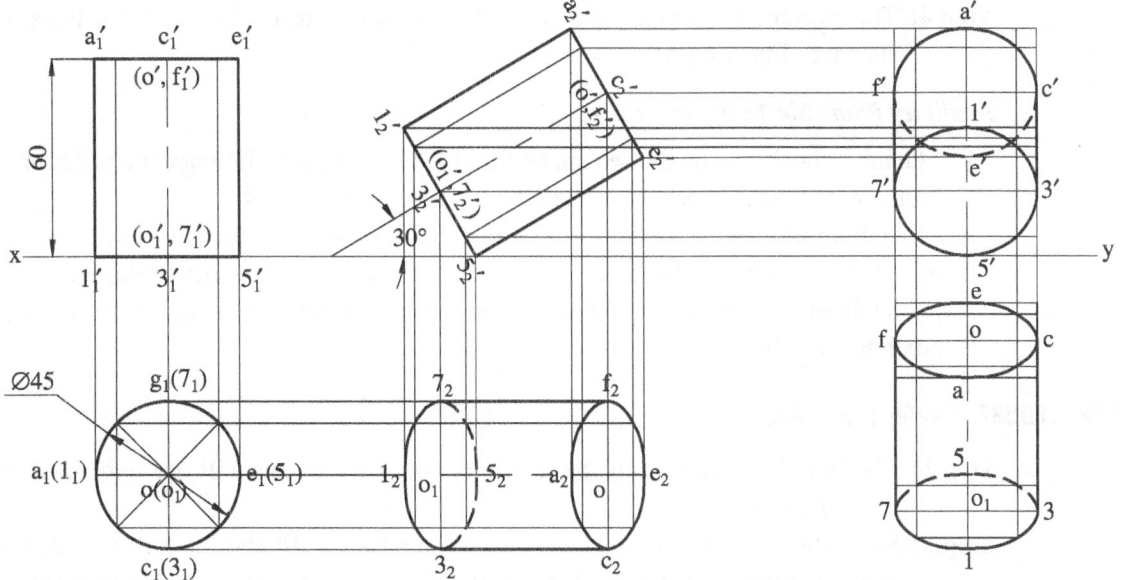

(a) Solution by solid tilting method

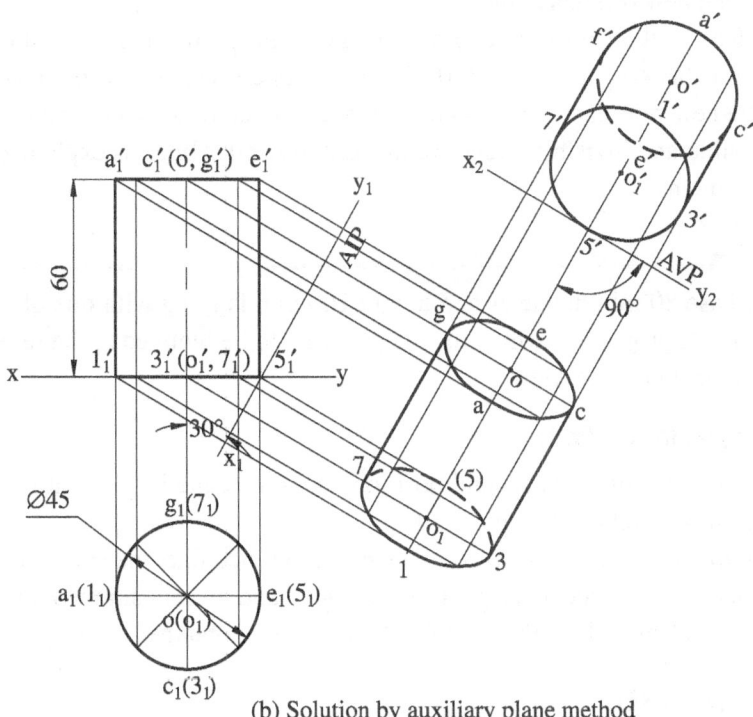

(b) Solution by auxiliary plane method

Fig. 9.67

Step 3: When the horizontal locus lines from the front view in the second stage meet these projector lines, the final front view is obtained with its circular end appearing elliptical and the corner $5'$ lying on the xy line.

Step 4: The extreme generators that form the boundary are marked in thick lines as shown in Fig. 9.67(a).

Auxiliary Plane Method—Principle

1. The initial position and the second position top view on an AIP when the cylinder is lying on its base corner on HP are discussed in Example 9.34.
2. When the top view of the axis is at 90° to VP, the shape of the top view obtained on an AIP is unchanged and is repositioned by drawing a new reference line perpendicular to the second stage auxiliary top view and obtain its corresponding auxiliary front view.

PROCEDURE (Refer Fig. 9.67b)

Step 1: The initial positions and the auxiliary top view on an AIP are explained in Example 9.34.

Step 2: Since the solid is lying on its point of the base on HP and the top view of the axis is inclined at an angle of 90°, draw a new reference line x_2y_2 at an angle 90° to the axis of the auxiliary top view and draw the projectors from various points to the new reference line.

Step 3: Measure the corresponding distances of these points in the first stage front view with the x_1y_1 line and mark them on the corresponding projectors drawn to the new reference line x_2y_2 and from it, to obtain the auxiliary front view. The visible and invisible edges are marked in similar lines as explained in previous problems. ▲

Example 9.46 ▲

A cone of base 80 mm diameter and height 100 mm is lying with one of its generators on HP. Draw its projections when the axis appears to be inclined to the reference line at an angle of 40° in the top view.

Solid Tilting Method—Principle

1. The initial position and second position when the cone is lying on its generator on HP are discussed in Example 9.20.
2. When the top view of the axis inclined to reference line, the shape of the top view obtained in the second stage is unchanged and is repositioned to the required inclination from which the final front view can be obtained.

PROCEDURE (Refer Fig. 9.68a)

Step 1: The first two stages of the projections are explained in Example 9.20.

Step 2: Since the top view of the axis is inclined at an angle of 40°, redraw the second-stage top view such that oo_1 makes 40° with the reference line as shown in Fig. 9.68(a), and draw the vertical projectors from its various points.

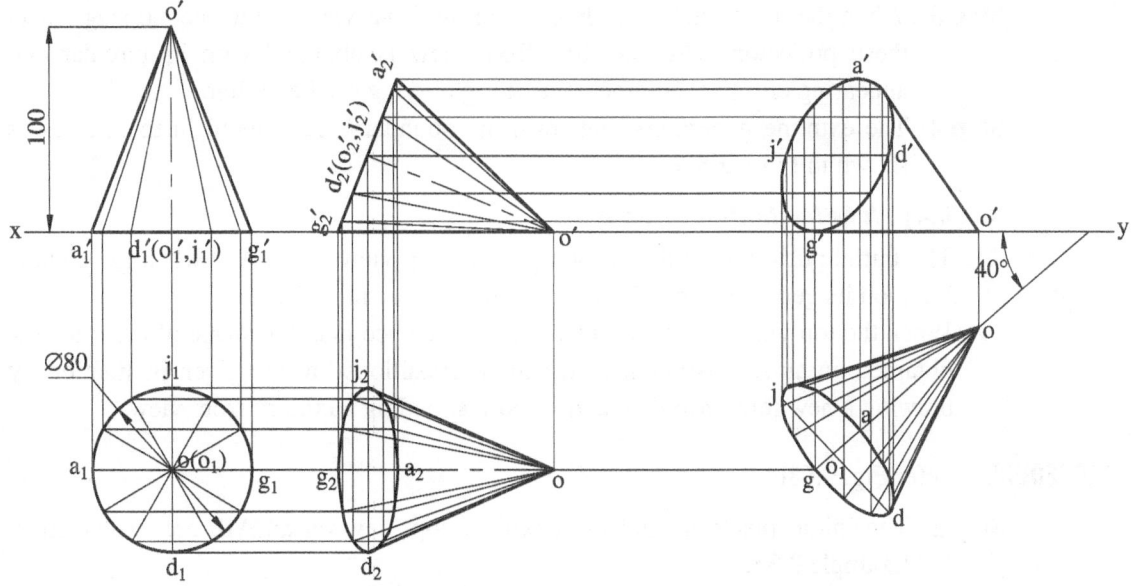

(a) Solution by solid tilting method

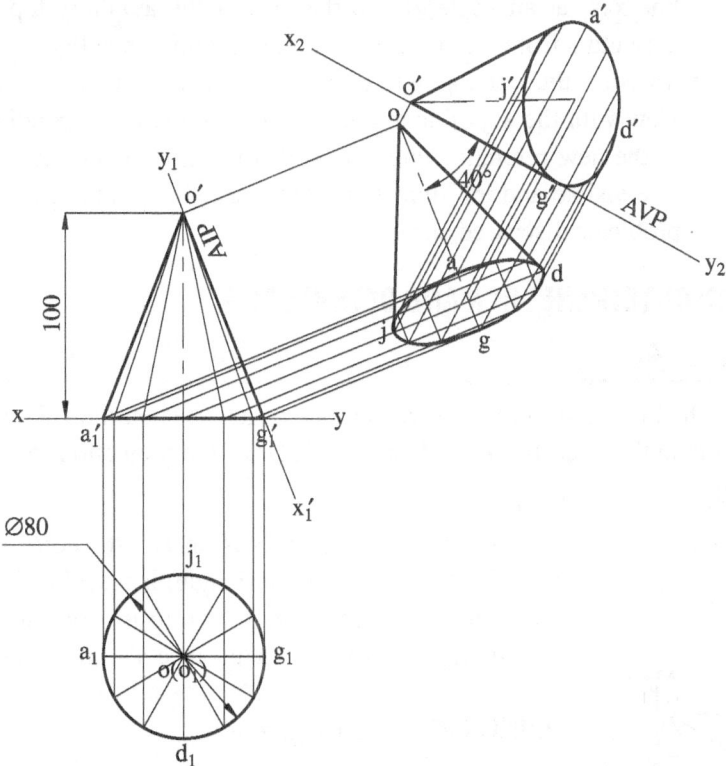

(b) Solution by auxiliary plane method

Fig. 9.68

Step 3: When the horizontal locus lines from the front view in the second stage meet these projector lines, the final front view is obtained with its circular end appearing elliptical and the generator g'o' lying on the xy line.

Step 4: The extreme generators that form the boundary are marked in thick lines as shown in Fig. 9.68(a).

Auxiliary Plane Method—Principle

1. The initial position and the second position top view on an AIP when the cone is lying on its generator on HP are discussed in Example 9.35
2. When the top view of the axis inclined to reference line, the shape of the top view obtained on an AIP is unchanged and is repositioned to the given inclination by drawing a new reference line, and its corresponding auxiliary front view.

PROCEDURE (Refer Fig. 9.68b)

Step 1: The initial positions and the auxiliary top view on an AIP are explained in Example 9.35.

Step 2: Since the top view of the axis is inclined at an angle of 40°, draw a new reference line x_2y_2 at an angle 40° to the axis of the auxiliary top view and draw the projectors from various points to the new reference line.

Step 3: Measure the corresponding distances of these points in the first-stage front view with the x_1y_1 line and mark them on the corresponding projectors drawn to the new reference line x_2y_2 and from it, to obtain the auxiliary front view. The visible and invisible edges are marked in similar lines as explained in the problems discussed earlier.   ▲

9.7 PROJECTION OF TETRAHEDRON AND OCTAHEDRON

Example 9.47

A tetrahedron of 40 mm sides lies on one of its faces on the HP, with the nearest edge containing that face 10 mm in front of VP. Draw the projections of the solid.

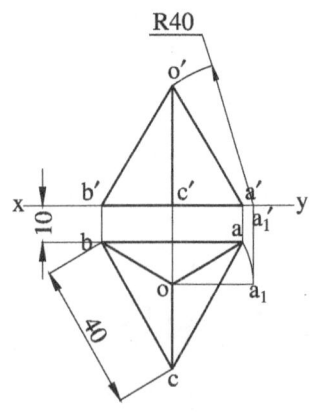

Fig. 9.69

Hint

1. Since one of its faces is on HP, the corresponding polygon of that face (equilateral triangle) is visualized in the top view.
2. Since the face edge is 10 mm in front of VP, the edge of the triangle is marked at a distance of 10mm from the reference line.

PROCEDURE (Refer Fig. 9.69)

Step 1: Mark the reference line and draw an equilateral triangle abc of side 40mm, with one side, say ab, parallel to the reference line and 10 mm from it. Bisect the internal angles and locate the centre o. Join oa, ob, and oc to complete the top view.

Step2: Project o and locate a′, ..., c′ on the reference line. To locate o′ (the height of the solid), make the line oa parallel to the reference line and obtain a_1' on the reference line. With a_1' as centre and radius equal to the length of one side, draw an arc to cut the projector through o and obtain o′. Join o′ with all the corners and complete the front view.

Example 9.48

A tetrahedron of 40mm sides rests on one of its edges on the HP and perpendicular to the VP. Draw its projections when the triangular face containing that edge makes 45° with the HP.

Hint

Decide the initial projections based on the inclination of the axis.

1. Since the triangular face (base) is inclined to HP, its axis is inclined to HP. Therefore, keep the axis perpendicular to HP, then the base is parallel to HP and the shape of the base (equilateral triangle) is visualized in the top view.
2. Since tilting happens with the edge of the base, the triangle is drawn with one edge perpendicular to the reference line and the front view is tilted about this edge in the appropriate view.

PROCEDURE (Refer Fig. 9.70)

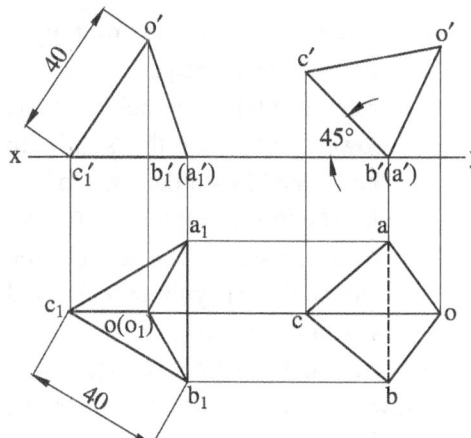

Fig. 9.70

Step 1: Mark the reference line and draw an equilateral triangle $a_1b_1c_1$ of side 40mm, with one side, say a_1b_1, perpendicular to the reference line. Bisect the internal angles and locate the centre o (o_1). Join oa_1, ob_1, and oc_1 to complete the top view.

Step2: Project $o(o_1)$ and other corners and locate a_1', ..., c_1' on the reference line. To locate o′ (the height of the solid), draw an arc with c_1' as centre and radius equal to the length of one side, to cut the projector through o and obtain o′ (since the line oc is parallel to the reference line). Join o′ with all the corners and complete the front view.

Step3: Tilt or redraw the front view about $b_1'(a_1')$ such that the line $b_1'(a_1') - c_1'$ (representing the face) makes 45° with the reference line and draw vertical projectors to meet the horizontal locus lines from the first stage top view and obtain the final top view as shown in Fig. 9.70. The edge ab in the top view is marked dotted as it is invisible.

Example 9.49

An octahedron of 34 mm sides is resting on the HP on one of its triangular faces with an edge of that face perpendicular to the VP. Draw its projections.

Hint

Decide the initial projections based on the inclination of the axis.

1. Since the triangular face is on the HP, its axis is inclined to HP. Therefore, keep the axis perpendicular to HP, then the square connecting the two portions of the octahedron is parallel to HP and is visualized in the top view.
2. In order to make the face lie on the HP, the solid should be tilted about its edge. Hence lay the square with one edge perpendicular to the reference line and the front view is tilted about this edge in the appropriate view.

PROCEDURE (Refer Fig. 9.71)

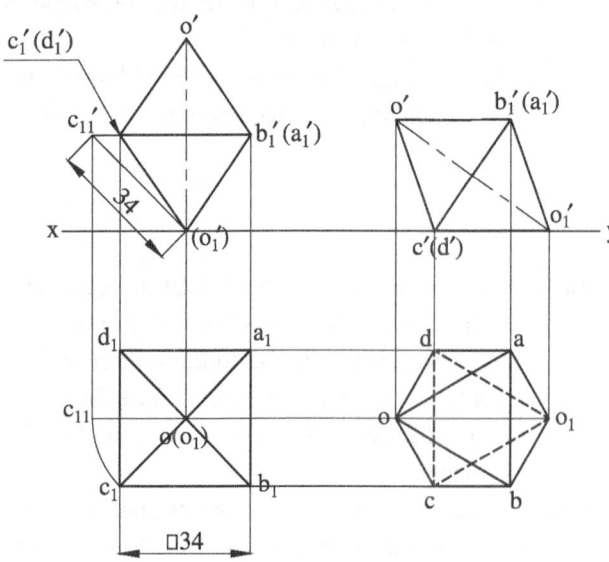

Fig. 9.71

Step 1: Mark the reference line and draw the square of $a_1 b_1 c_1 d_1$ of side 34 mm, with one side, say $c_1 d_1$, perpendicular to the reference line. Locate the centre o (o_1) and join the corners to complete the top view.

Step 2: Project $o(o_1)$ to the front view domain and locate o_1' on the reference line. The height of the lower portion of the solid can be located in similar lines of the tetrahedron (Example 9.16), but by inverting it. The top portion can be drawn symmetrically and o' can be located.

Step 3: Tilt or redraw the front view about $c_1'(d_1')$ such that the line o_1'–$c_1'(d_1')$ (the triangular face) is on the reference line, draw vertical projectors to meet the horizontal locus lines from the first stage top view, and obtain the final top view as shown in Fig. 9.71.

Step 4: As the corners a′, b′, and o′ are far away from the reference line, these points and their connecting lines are visible in the final top view. Similarly the corners o_1', c′, and d′ are on the reference line, these points and their connecting lines in the top view are invisible and are marked dotted. The other lines that lie on the boundary are marked in thick lines as the boundary is always real and existent for any object.

Classification of Solids

- The solids are three-dimensional objects formed by the arrangement of plane surfaces and are known as polyhedra. When similar polygon shapes are used for the bottom and top and bounded with rectangular faces laterally, they are known as prisms. When the polyhedral surface representing the face is bounded with triangular lateral surfaces, all converging to a single point or vertex at the top, it is called a pyramid. The axis of a solid is an imaginary line connecting the centre of the bottom and the top or the bottom and the vertex. When the sides of the top and the base are equal and the axis is perpendicular to the base, they are known as right and regular solids.
- Solids obtained by the revolution of a plane surface (rectangle, right-angled triangle) about one of its sides result in a cylinder or a cone. The lateral surface of such solids is curved and the axis is perpendicular to the base.

Position of Solids

- The orthographic projection or drawing the orthographic views of these solids are discussed as per three orientation schemes, namely,
  - The axis of the solid perpendicular to one reference plane and parallel to the other.
  - The axis of the solid inclined to one reference plane and parallel to the other.
  - The axis of the solid inclined to both reference planes.

Projection of Solids in Simple Positions

- When a solid is kept with its axis perpendicular to one reference plane, it is set to be in a simple position.
- For right and regular solids, the base and the axis are perpendicular to each other, and hence, when the axis is perpendicular to one reference plane, the base is automatically parallel to that reference plane, and hence, its true shape is visualized on that reference plane.

- The length of the axis or the height of the solid is seen in the other view.
- The projections of the solid are drawn accordingly.

Projection of Solids Inclined to One Reference Plane

- When the axis of a solid is inclined to one reference plane and parallel to the other, its base or one end of the solid gets lifted and the solid occupies a tilted position. Its projections can be drawn by two methods, namely, tilting object method and auxiliary plane method.

Projections of Solids by Tilting Object Method or Change of Position Method

- The projections of the solid are drawn in two stages. In the first stage, the axis is kept perpendicular to the reference plane to which it is to be inclined later, and the initial projections are drawn similar to that mentioned in simple positions. If the solid is tilted about an edge or corner, the true shape of the polygon is laid as discussed in planes and the other view is obtained.
- In the second stage, the non-polygonal view is tilted to the required inclination and its projections are completed from that obtained in the first stage.
- Since the solid and its views are tilted to achieve the projections, this is called tilting object method.

Projections of Solids by Auxiliary Plane Method or Change of Reference Line Method

- Auxiliary planes are the additional planes set up apart from HP and VP to get the views of the solid without necessarily tilting it.
- Auxiliary vertical plane (AVP) is an additional plane perpendicular to HP and inclined to VP (at an angle φ) used to get a new front view or auxiliary front view, when the object is inclined to VP, without tilting the solid with respect to VP.
- Auxiliary inclined plane (AIP) is an additional plane perpendicular to VP and inclined to HP (at an angle θ) used to get a new top view or

auxiliary top view, when the object is inclined to HP, without tilting the solid with respect to HP.

- These auxiliary planes are distinguished by additional or new reference lines x_1y_1 apart from the conventional reference line xy.
- After drawing the initial projections of the solid as mentioned in the previous method, a new reference line x_1y_1 can be drawn with the required inclination at appropriate view and passing through the feature of interest (edge/corner).
- The non-polygonal view when projected to this new reference line and linked with the distances from the conventional reference lines yields the required projections in the inclined positions.

Projection of Solids Inclined to Both Principal Planes by Tilting Solid Method and Auxiliary Plane Method

- When the axis of the solid is inclined to both reference planes, it involves the physical tilting of the axis with respect to one reference plane and the solid getting further tilted or rotated about the edge or the corner with which it has turned in the first stage. This results in the repositioning of the axis obtained in the final projections of the earlier section.
- The second-stage projections are repositioned to suit the inclination of the edge/diagonal to get the final views of the object.

Marking Visible and Invisible Edges

- The visible corners and edges in the final projections correspond to those located far off from the reference line or from the new reference line (auxiliary projection method) in the non-polygonal view and are marked in thick lines. The other points/edges are invisible and are marked using dotted lines. However, the boundary lines in the final drawings always appear in thick lines to ensure the solid out line as a reality.

MULTIPLE-CHOICE QUESTIONS

9.1 Name the regular polyhedron that is formed with four equilateral triangles.
    (a) Tetrahedron   (c) Dodecahedron
    (b) Triangular pyramid (d) Icosahedron

9.2 A hexahedron is also known as
    (a) a cube   (c) a rectangular prism
    (b) a square prism   (d) a hexagonal prism

9.3 Dodecahedron is a regular polyhedron formed with
    (a) five equal pentagonal planes
    (b) 10 equal pentagonal planes
    (c) 15 equal pentagonal planes
    (d) 12 equal pentagonal planes

9.4 Name the solids that are categorized under polyhedron formed with dissimilar plane surfaces.
    (a) Cone and cylinder
    (b) Prisms and pyramids
    (c) Icosahedron and dodecahedron
    (d) Tetrahedron and octahedron

9.5 A cylinder rests on HP on one of its points of the base which is inclined to HP and perpendicular to VP. The ends of the cylinder in the top view will appear as
    (a) a circle   (c) a straight line
    (b) an ellipse   (d) a point

9.6 The frustum of a solids is obtained when it is cut by
    (a) a plane inclined to the base
    (b) a plane parallel to the base
    (c) a plane passing through the base
    (d) a plane perpendicular to the base

9.7 The axis of a hexagonal prism is parallel to both HP and VP. The true shape of the hexagon will appear in the
    (a) top view   (c) side view
    (b) front view   (d) auxiliary view

9.8 An auxiliary inclined plane is a plane
    (a) perpendicular to VP and inclined to HP
    (b) perpendicular to HP and inclined to VP
    (c) inclined to both HP and VP
    (d) perpendicular to VP and parallel to HP

9.9 An auxiliary vertical plane is a plane
    (a) perpendicular to VP and inclined to HP

(b) perpendicular to HP and inclined to VP
(c) inclined to both HP and VP
(d) perpendicular to VP and parallel to HP

9.10 A pentagonal pyramid rests on HP on one of its triangular faces such that its axis is parallel to VP. This position can be achieved from the following options of the initial position.
(a) The base is on HP with the edge representing

the triangular face perpendicular to VP.
(b) The base is on HP with the edge representing the triangular face parallel to VP.
(c) The base is on HP with the edge representing the triangular face inclined to VP.
(d) The polygonal end on VP with the edge representing the triangular face perpendicular to HP.

WORK PRACTICE LEVEL – 1

Solid Resting on its Base on HP/Parallel to HP

9.1 A square prism of 35 mm base sides and 75 mm long rests on its base on HP such that two of its lateral faces are equally inclined to VP. Draw its projections.

9.2 A pentagonal prism of 35 mm sides of base and axis 75 mm long is resting on its base on HP in such a way that one of its faces is perpendicular to VP. Draw its projections.

9.3 A hexagonal prism of 35 mm base edges and axis 75 mm long rests on HP on its base and one of its lateral faces is inclined at 30° to VP. Draw its projections.

9.4 A triangular pyramid of 35 mm base sides and axis 75 mm long rests on HP on its base and one of its base edges inclined at 45° to VP. Draw the projections of the pyramid in this position.

9.5 A pentagonal pyramid of 35 mm base edges and axis 75 mm long is resting on HP on its base such that one of its slant edges is parallel to VP. Draw its projections.

9.6 A cylinder of base diameter of 60 mm and axis 75 mm long has one of its circular ends on HP such that the axis is 50 mm from VP. Draw its projections.

9.7 A cone of base diameter 60 mm and axis 75 mm has its circular end on HP and the axis is 40 mm in front of VP. Draw its projection.

Solid having One of its Polygonal Ends on VP/parallel to VP

9.8 A rectangular prism 60 mm × 30 mm base and 75 mm long is placed such that its

larger rectangular face is parallel to HP and perpendicular VP and its one end is 20 mm in front of VP. Draw its projections.

9.9 A hexagonal prism of 35 mm base edges and axis 75 mm long is placed such that one of its lateral edges is on HP. Draw its projections, when two of its adjacent lateral faces are equally inclined to HP and perpendicular to VP.

9.10 A pentagonal pyramid of 35 mm base edges and axis 75 mm long is placed such that its pentagonal end is on the VP. Draw its projections, when and the axis is parallel to HP and 60 mm above it and one of the sides of the pentagonal end is parallel to HP.

9.11 A cone of base diameter 50 mm and axis 75 mm long has its circular end on VP and axis is 50 mm above HP. Draw its projections.

Solid having its Axis Parallel to Both Reference Planes

9.12 A pentagonal prism is resting on HP on one of its longer face edges such that rectangular face opposite to this edge is parallel to HP and the axis is parallel to both HP and VP. Draw its projections.

9.13 The polygonal end of a hexagonal pyramid is resting on one of the profile planes such that one of its corners is on HP and the side containing that corner is inclined at 20° to the HP. Draw its projections.

9.14 A cylinder of base diameter 50 mm and axis 75 mm has one of its circular ends 40 mm in front of PP. Draw its projections, when one of its generators lie on HP. Draw its projections.

Problems of Solids whose Axes are Inclined to HP and Parallel to VP

Problems of Solids Tilted about an Edge

(The reader is advised to draw the projections by the tilting object method and by the auxiliary projection method)

9.1   A square prism of 35 mm base sides and length 75 mm rests on one of its base edges on HP, and the axis is inclined at 30° to HP and parallel VP. Draw its projections.

9.2   A pentagonal prism of 30 mm sides of base and axis 80 mm long is resting on one of its base edges on HP, and the axis is inclined at 45° to HP. Draw its projections, when the base edge on HP is perpendicular to VP.

9.3   A hexagonal prism of 30 mm base edges and axis 75 mm long has one of its base edges on HP. Draw its projections, when the rectangular face containing this base edge makes 45° with HP and perpendicular to to VP.

9.4   A square pyramid of 35 mm base sides and axis 75 mm long rests on one of its base edges on HP, and the axis is inclined at 30° to HP. Draw its projections, when the base edge on HP is perpendicular to VP.

9.5   A hexagonal pyramid of 35 mm base edges and axis 75 mm long has one of its triangular faces on HP. Draw its projections, when the vertical plane containing the axis is parallel to VP.

9.6   A pentagonal prism of 35 mm sides of base and axis 75 mm long is resting on one of its rectangular faces on VP. Draw its projections, when the top end of the prism makes 45° to HP.

Problems of Solids Tilted about a Corner

9.7   A pentagonal prism of 35 mm sides of base and axis 75 mm long is resting on one of its base corners on HP, and the axis is inclined at 30° to HP and parallel to VP. Draw its projections.

9.8   A hexagonal prism of 35 mm base edges and axis 75 mm long rests on one of its base corners on HP. Draw its projections, when the lateral edge through this corner is inclined at 30° to HP and adjacent rectangular faces are equally inclined to VP.

9.9   A square pyramid of 40 mm base edges and axis 75 mm long is resting on a corner of its base on HP. Draw its projections, when the base makes 30° with HP and slant faces adjacent to the corner on HP make equal inclinations with VP.

9.10 A pentagonal pyramid of 30 mm base edges and axis 75 mm long rests on one of its slant edges on HP. Draw its projections, when the vertical plane containing the axis and this slant edge is parallel to VP and 60 mm in front of it.

9.11 A cylinder of base diameter of 60 mm and axis 75 mm long rests on a point of its base on HP. Draw its projections, when the generator through that point on HP is inclined at 45° to HP parallel to VP.

9.12 A cone of base diameter of 60 mm and axis 75 mm long has a point on its base resting on HP, and the axis is inclined at 30° to HP parallel to VP. Draw its projections.

9.13 A cone of base diameter of 60 mm and axis 75 mm long rests on one of its generators on HP. Draw its projections, when the vertical plane through this generator is parallel VP.

9.14 A hexagonal prism of 35 mm sides of base and axis 75 mm long is having one of its lateral edges on VP and inclined at 30° to HP. Draw its projections, when its adjacent lateral faces are equally inclined to VP.

9.15 A frustum of a right circular cone of base 60 mm diameter and top 40 mm diameter and axis 55 mm long is resting on one of its generators such that the vertical plane containing the axis is parallel to VP. Draw its projections.

Problems of Solids Suspended from a Corner

9.16 A hexagonal prism of 35 mm base edges and axis 75 mm long is suspended from one of its top corners. Draw its projections, when the vertical plane containing this corner and the axis is parallel to VP and 60 mm in front of it.

9.17 A pentagonal pyramid of 35 mm base edges and axis 75 mm long is suspended from one of its base corners such that the top view of the slant edge through this corner is parallel to VP. Draw its projections.

9.18 A cone of base diameter of 60 mm and axis 75 mm long is freely suspended from a point on its circumference of the base such that the axis is parallel to VP. Draw its projections.

Problems of Solids whose Axes are inclined to VP and Parallel to HP

Problems of Soilds Tilted about an Edge

9.19 A pentagonal prism of 35 mm sides of base and axis 75 mm long is resting on one of edges of its end on VP, and the axis is inclined at 30° to VP and parallel to HP. Draw its projections.

9.20 A hexagonal prism of 35 mm base edges and axis 75 mm long rests on one of the edges of its end on VP and the lateral face through this edge makes 45° to VP. Draw its projections, when the edge on VP is perpendicular to HP.

9.21 A pentagonal pyramid of 35 mm base edges and axis 75 mm long has one of its slant faces on VP. Draw its projections, when the axis is parallel to HP.

9.22 A hexagonal prism of 35 mm sides of base and axis 75 mm long is resting on one of its rectangular faces on HP, and the axis is inclined at 30° to VP and parallel to HP. Draw its projections.

Problems of Solids Tilted about a Corner

9.23 A pentagonal prism of 30 mm sides of base and axis 75 mm long is having one of its end corners on VP. Draw its projections, when that end is inclined at 30° to VP and the plane containing the axis and the lateral edge passing through that corner is horizontal and 60 mm above HP.

9.24 A hexagonal prism of 35 mm base edges and axis 75 mm long has one of the corners of the hexagonal end on VP. Draw its projections, when the lateral edge through that corner is inclined at 45° to VP and parallel to HP.

9.25 A pentagonal pyramid of 35 mm base edges and axis 75 mm long has one of its slant edges on VP and parallel to HP. Draw its projections.

9.26 A cylinder of base diameter of 60 mm and axis 75 mm long has a point of its circular end on VP. Draw its projections, when the circular end is inclined at 45° to VP and perpendicular to HP.

9.27 A cone of base diameter of 60 mm and axis 75 mm long has on one of its generators on VP. Draw its projections, when the plane containing this generator and axis is horizontal and 60 mm above HP.

9.28 A pentagonal prism of 30 mm sides of base and axis 75 mm long is having on one of its lateral edges on HP and inclined at 30° to VP. Draw its projections, when the adjacent lateral faces of this edge are equally inclined to HP.

WORK PRACTICE LEVEL – 3

Engineering Applications

9.1 A table weight is in the form of a cylindrical slab with a centrally cut coaxial hole. The cylindrical slab stands on HP with its axis making 60° to HP. One of the faces of the square hole is parallel to VP. Draw the projections of the solid, if its thickness is 20 mm. The diameter of the outer cylinder is 60 mm, and the sides of the square hole are 30 mm.

9.2 A bucket made of thin sheet metal is in the form of an inverted frustum of a cone. Its top is 300 mm diameter, bottom 180 mm diameter, and height 325 mm. A circular ring of 180 mm diameter and 60 mm high is attached to the frustum at the bottom. Draw the projections of the bucket, when it is resting on the ground on the rim of the ring with its axis inclined at 30° to the ground. The vertical plane containing the axis of the bucket and the generator passing through the point touching the ground is parallel to the observer.

9.3 A hexagonal-headed bolt without chamfer, is 22 mm in diameter, and the cylindrical portion is 50 mm long. The side of the hexagonal head is 22 mm having a thickness of 18 mm. The bolt is placed such that one of the lateral faces of the hexagonal head is on HP and perpendicular to VP. Draw its projections, when the axis of the bolt is inclined at 30° to VP. The cylindrical portion of the bolt (shank) may be assumed to be horizontal. The effect of the threads may be neglected in the drawing.

9.4 A brass spherical flower vase with flat top and spherical bottom is placed over a hollow square prism of outer base edge 50 cm, inner base edge 36 cm, and height 20 cm. The greatest diameter of the flower vase is 56 cm, and the flat top face is of diameter 40 cm. Draw its projections, when the top face is horizontal. Draw an auxiliary top view of the combination of solids on a plane inclined at 40° to HP.

9.5 A knob of machine handle consists of 12 mm diameter × 70 mm long cylindrical portion and 38 mm diameter spherical portion. The centre of the sphere lies on the axis of the cylindrical portion. Draw the projections, when its axis is inclined at 30° to HP and parallel to VP. The overall length of the handle is 100 mm.

WORK PRACTICE LEVEL – 4

Projection of Solids Inclined to Both Principal Planes

(The reader is advised to draw the projections by the tilting object method and by the auxiliary projection method)

9.1 A pentagonal prism 35 mm side of base and axis 75 mm long is resting on one of its base corners on HP such that the two of the base edges passing through this corner make equal inclination with the VP. Draw its projections, when the lateral edge through this corner is inclined at 45° to HP and the top view of the axis makes 30° with the reference line.

9.2 A hexagonal prism 35 mm side of base and axis 75 mm long is resting on one of its base edges on HP such that the base is inclined at 45° to HP and the axis is inclined at 30° to VP. Draw its projections.

9.3 A pentagonal pyramid 30 mm side of base and axis 75 mm long is resting on one of its base corners on HP. Draw its projections, when its axis is inclined at 45° to HP and 30° to VP.

9.4 A hexagonal pyramid 35 mm side of base and axis 75 mm long is resting on one of its triangular faces on HP. Draw its projections, (i) when the axis appears to be inclined at an angle of 40° to the VP and (ii) the axis is inclined an angle of 40° to the VP.

9.5 A cone of base 60 mm diameter and height 75 mm is having with one of its generators on VP. Draw its projections, (i) when the axis appears to be inclined at an angle of 30° to the HP and (ii) the axis is inclined at an angle of 30° to the HP.

9.6 A cylinder of base 60 mm diameter and height 75 mm is resting on a point on the circumference of the base on HP. The base solid makes 40° with the HP. Draw its projections (i) when the axis appears to be inclined at an angle of 30° to the VP and (ii) the axis is inclined at an angle of 30° to the VP.

9.7 A hollow hexagonal prism of height 80 mm has outside hexagon of 40 mm sides and inside hexagon of 20 mm sides. It rests on a corner of its base on HP such that two of its faces are parallel to VP, and the axis is inclined at 45° to HP. Draw its projections, when the parallel faces make 30° with the VP.

9.8 A pentagonal pyramid of 40 mm base edges and axis 75 mm long is suspended from one of its base corners. Draw its projections, when the top view of the slant edge through this corner makes 30° with VP.

Projection of Sectioned Solids and Obtaining their True Shape of Sections

10

OBJECTIVES

This chapter will help the reader to understand the following:

- Need for sectioning an object and the methods of sectioning
- Types of section/cutting planes and their representation by means of drawings
- Changes to be made in the orthographic views of the object caused by the sectioning and preparation of the sectional views at the preferred locations
- Shape and size of the cut surfaces and their true shapes based on the orientation of section planes
- Section interface to combine an another solid, thereby extending the knowledge to assemble various components as encountered in many machine components and other engineering applications
- Representation of the views that explain the uncut and the cut portions together

10.1 INTRODUCTION

The main objective of projecting a solid to the reference planes is to enable the appearance of its outer surfaces. These views are sufficient if the solid does not possess any other feature inside it. In the case of a solid and a hollow cylinder, there is no change in the appearance of the cylindrical surface in regular orthographic views, as details of the inner feature are not visible and hence not reflected. Although the invisible features are represented by dotted lines, it is difficult to show while inner features are many.

To distinguish the details of inner features of an object, it is customary to imagine an object as being cut by imaginary section planes. To visualize the inside feature, the portion of an object that lies between the observer and the cutting plane has to be removed. The visualization of the remaining portion of the solid together with the cut surface is known as 'sectional view'.

This concept can be observed while cutting vegetables or fruits into smaller parts or pieces by knifes or blades to visualize the defects inside. The knives or blades are real-life cutting planes. When an object is projected to visualize the inner details, it is assumed to be cut by section planes that are imaginary in nature.

10.2 SECTION PLANES AND THEIR REPRESENTATIONS

10.2.1 Types of Section/Cutting Planes

The section/cutting planes are classified by their locations with respect to the principal planes as follows:

1. Section/Cutting planes parallel to one reference plane and hence perpendicular to the other reference plane.
2. Section/Cutting planes inclined to one reference plane and perpendicular to the other reference plane.

Since the section plane is basically a two-dimensional plane surface, the facts that were discussed in Chapter 8 hold good here too. From the earlier discussions, it can be observed that a plane surface perpendicular to a reference plane will appear as a straight line on that reference plane, the section plane will also be visualized as straight line depending on its perpendicularity with a reference plane. If the section plane is perpendicular to horizontal plane (HP), it will be visualized as a line on HP or in the top view, and if it is perpendicular to vertical plane (VP), it will be realized as a line in the front view. This line will be drawn parallel to the reference line or inclined, depending on its orientation as being parallel or inclined to the other reference plane. Therefore, for the cases of section planes classified earlier, the following views will appear.

1. If the section plane is parallel to VP (and hence perpendicular to HP), it is visualized as a straight line in the top view and is drawn parallel to the reference line, since it is parallel to VP.
2. If the section plane is inclined to VP and perpendicular to HP, it is visualized as a straight line in the top view, but this line will appear inclined to the reference line, according to its inclination with VP.

Similarly, if the section plane is parallel to HP, it is visualized as a straight line parallel to reference line in the front view and if it is inclined to HP, it is drawn inclined to the reference line at the given orientation. It can be remembered that the trace of a plane is a straight line on the respective reference plane, when the plane is extended to meet the reference plane. Hence, the view, in which the section plane is visualized as a line, can be referred as its trace. If the section plane is visualized as a line in the top view, it is known as horizontal trace (HT), and if observed in the front view, it is known as vertical trace (VT). The HT and VT of a section plane can be parallel or inclined to the reference line as discussed earlier.

It can be noted that the sectioning is carried out after successful completion of projections of the solid, and hence, their representations will be made after drawing the orthographic projections.

10.2.2 Representation of Section Planes

Since the section plane will appear as a straight line in the view corresponding to its perpendicularity with a reference plane, the following representation schemes are to be noted based on the BIS convention (SP46-2003). It is represented by a long dashed dotted narrow line or chain narrow line, the ends being marked thick (see Line Type 4.15 in Table 1.5 and Fig. 1.14a in Chapter 1). In Fig. 10.1(a), a parallel cutting plane represented, while Fig. 10.1(b) is for an inclined cutting plane. Figure 10.1(c) shows the representation when the position of the cutting plane passes through two locations.

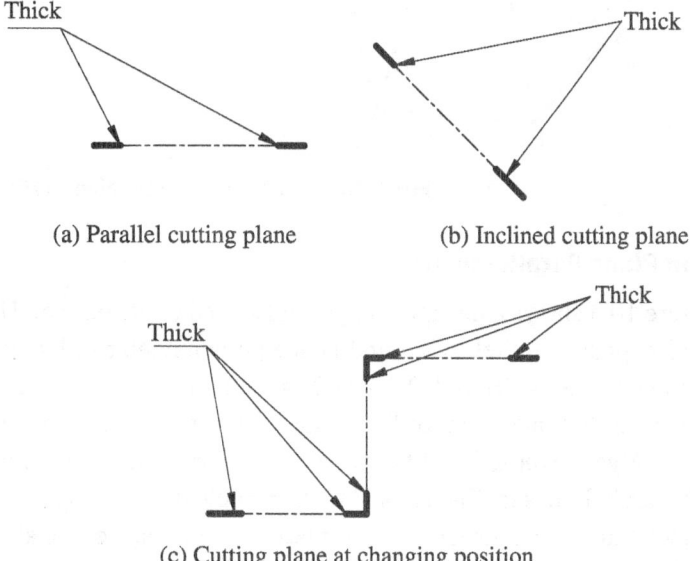

(a) Parallel cutting plane   (b) Inclined cutting plane

(c) Cutting plane at changing position

Fig. 10.1   Representation of cutting planes

10.3 PROJECTION OF SECTIONED SOLIDS

Based on the orientations of the section plane discussed in Section 10.2.1, the projections of the solids when sectioned are discussed in the following sections.

10.3.1 Section Plane Parallel to HP

Figure 10.2 shows a pentagonal pyramid seated on its base on HP and is cut by a section plane parallel to the base or parallel to the HP and hence perpendicular to VP. This results in cutting all the slant edges OA, ..., OF of the pyramid at points 1, ..., 5, respectively. When the top portion between the observer and the section plane is removed, according to the first-angle projection method, the cut surface 1, ..., 5 gets exposed, becomes visible from the top and is shown by hatching the area with a series of thin parallel lines inclined at 45° to the principal outer edges. The retained portion of the object and the cut surface representation forms the sectional top view. Since the cut surface is parallel to HP (to the base), it reveals the true shape of the cut portion.

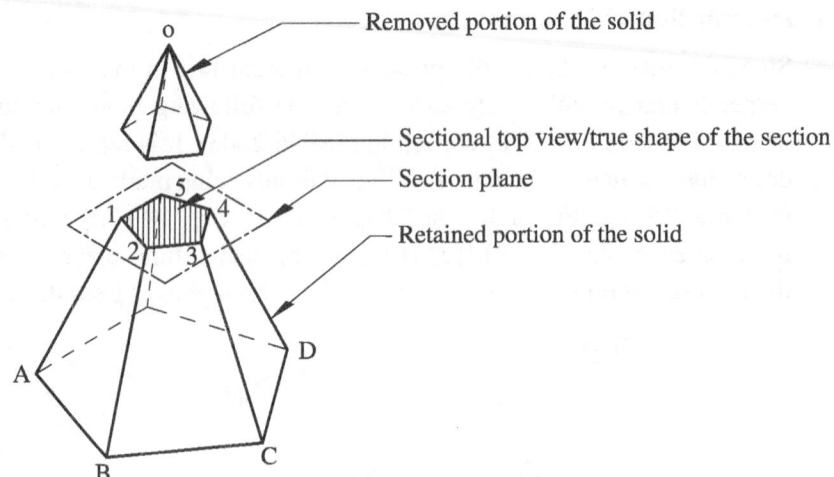

Fig. 10.2 Cutting plane parallel to HP

10.3.2 Section Plane Parallel to VP

Figure 10.3 shows a hexagonal pyramid seated on its base on HP and is cut by a vertical section plane parallel to VP and hence perpendicular to HP. It results in cutting of the front slant edges OB and OC at points 4 and 3 and extending to the base edges AB and CD at points 1 and 2. According to the first-angle projection method, the vertical surface 1, ..., 4 gets exposed and becomes visible from the front and is shown by a series of thin parallel lines inclined at 45° to the principal outer edges. The retained portion of the object and the cut surface representation forms the sectional front view. Since the cut surface is parallel to VP, it reveals the true shape of the cut portion.

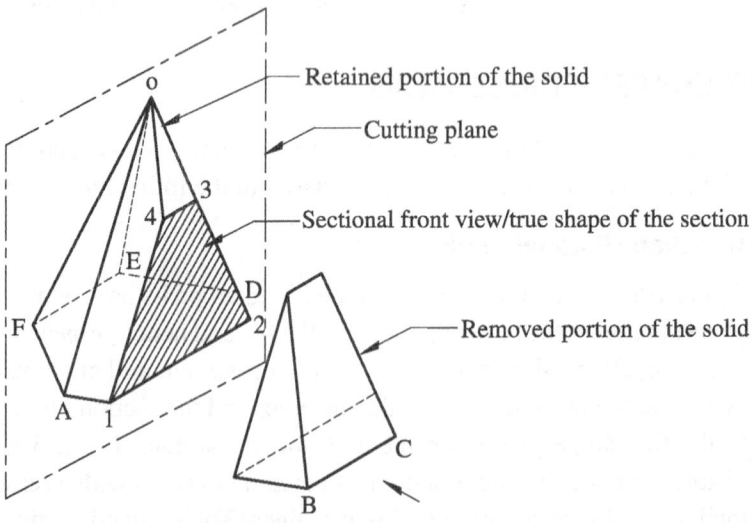

Fig. 10.3 Cutting plane parallel to VP

10.3.3 Section Plane Inclined to VP and Perpendicular to HP

Figure 10.4 shows a cube with the bottom surface ABCD and the top surface PQRS seated on its base on HP. A vertical section plane inclined to VP at ϕ degrees and perpendicular to HP, cuts the solid at points 1, ..., 4, located on the base edges AD, BC, and the top edges PS, QR. Based on the first-angle projection method, the vertical surface 1, ..., 4 gets exposed and becomes visible from the front and is shown by a series of thin parallel lines inclined at 45° to the principal outer edges. The retained portion of the object and the cut surface representation forms the sectional front view. Since the cut surface is not parallel to VP, it does not reflect its true shape. The true shape can be obtained if the cut surface is projected to an auxiliary vertical plane, which is parallel to the section plane.

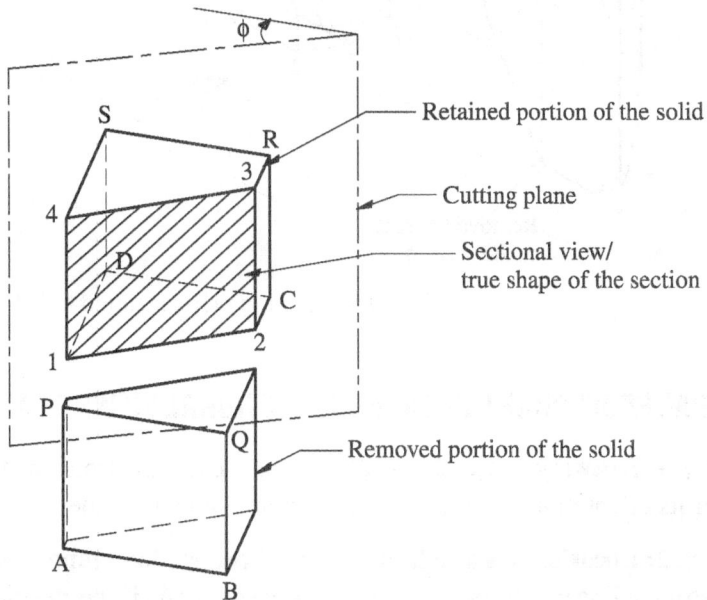

Fig. 10.4   Cutting plane inclined to VP

10.3.4 Section Plane Inclined to HP and Perpendicular to VP

Figure 10.5 shows a hexagonal prism seated on its base on HP. A section plane inclined to HP at θ degrees and perpendicular to VP cuts the solid on its base, the top, and the lateral edges through B, C, E, and F at points 1, 8 and 4, 5 and at 2, 3, 6, and 7, respectively. When the cut portion between the observer and section plane is removed and as per the first-angle projection method, the retained portion of the object and the cut surface representation can form the sectional top view drawn below the reference line. The cut surface is shown by a series of thin parallel lines inclined at 45° to the principal outer edges. Since the cut surface is not parallel to HP, it does not reflect its true shape. The true shape can be obtained if the cut surface is projected to an auxiliary inclined plane, which is parallel to the section plane.

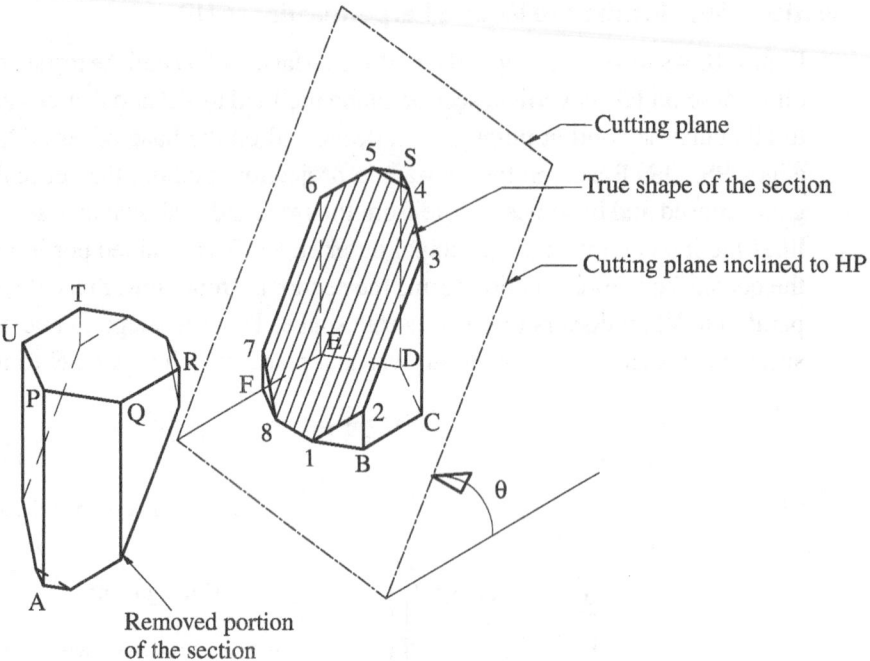

Fig. 10.5   Cutting plane inclined to HP

10.4 GENERAL PROCEDURE FOR DRAWING SECTIONAL VIEWS OF AN OBJECT

Step 1: Drawing projections of the solid   Draw the front and top views of the solid, as per its orientations with the reference planes and draw the side view if required.

Step 2: Location of straight line view of the section plane   Mark the section plane as a straight line in the appropriate view, depending on its perpendicularity with a reference plane. This can be drawn parallel or inclined to the reference line depending on the orientation of the section plane, being parallel or inclined to the other reference plane. The portion of the view of the solid in between this line and the reference line can be made thicker, as it is retained.

Step 3: Identification of section or cutting points   Identify the cutting points of the straight line view representing the section plane with concerned view of the solid and project them to meet their corresponding locations in the other view.

Step 4: Obtaining the cut surface   Join the points obtained in the other view of the solid in the sequential order and obtain a closed bounded area that represents the cut surface. Shade the bounded area by means of regular uniformly spaced thin hatching lines, inclined at 45° to the reference line or the main outline of the drawing.

Step 5: Obtaining the sectional view   The shaded area along with the portion of the object retained gives the sectional view of the object.

Step 6: Obtaining the true shape of the section   If the section plane is parallel to reference plane, the shaded area obtained gives the true shape of the cut surface or the section. If the section plane is inclined to the reference plane, the shaded area obtained now does not reveal true shape of the cut surface or the section. The true shape can be obtained by drawing an auxiliary line or new reference line 'x_1y_1' parallel to the straight line view of the section plane at any convenient distance. Project the cutting points to the auxiliary line and measure the distance of the cutting points from the old reference line, that is, 'xy' and mark them along the respective projector line drawn from 'x_1y_1' and from it. Join these points in sequential order and resulting shape gives the true shape of the section. Shade the area as carried out in Step 4.

> **NOTE**   *The auxiliary line can be drawn to the opposite side, where the object is removed. This will ensure that the sectional views obtained in the problem and the true shapes of section are drawn similar to the removal of portion of the object and observing the remaining in the same sequence. There is also a practice of drawing the auxiliary line on the same side of the removal of the object, because it brings more clarity. This option can be used only if the true shape of section alone is required. If the sectional view of the remaining portion of the object is also required, then the procedure suggested in this paragraph is recommended.*

10.5 SOLIDS IN SIMPLE POSITION AND CUTTING PLANE PARALLEL TO ONE OF THE PRINCIPAL PLANES

10.5.1 Cutting Plane Parallel to HP

Example 10.1

A pentagonal pyramid of side with base 30 mm and axis 70 mm long is standing on its base on HP with one of its base edges inclined at 6° to VP. It is sectioned by a horizontal section plane passing through a point on the axis and 25 mm from the base. Draw the sectional top view.

Hint

1. Since the section plane is horizontal, it is perpendicular to VP, and hence, it is visualized as a line on VP and therefore, it represents the section plane as a line in the front view.
2. Since SP is parallel to HP, draw this line parallel to the reference line.

PROCEDURE   (Refer Fig. 10.6)

Step 1: Complete the projections of the solid according to the details given and as discussed in Chapter 9.

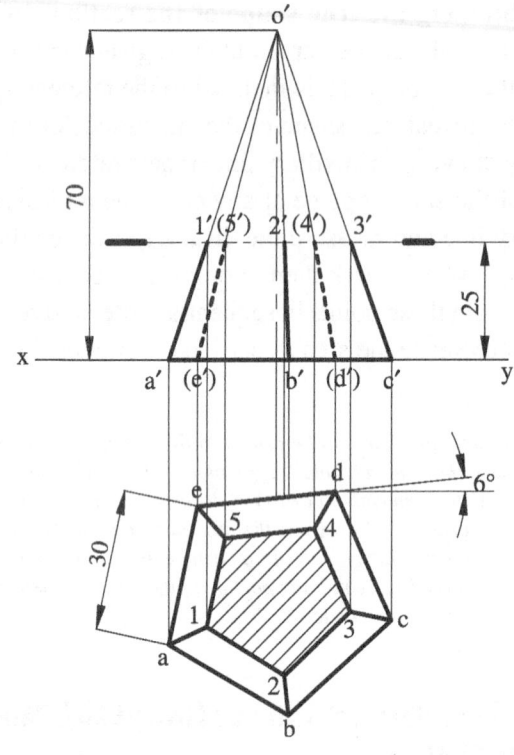

Fig.10.6

Step 2: Mark the section plane as a straight line, parallel to the reference line, 25 mm above it in the front view, and obtain the cutting points 1′, ..., 5′ in the sequential order of the edges.

Step 3: Project the cutting points to meet the corresponding edges in the top view and join them by straight lines which gives a closed bounded area 1, ..., 5.

Step 4: Shade the bounded area by means of regular and uniformly spaced hatching lines, inclined at 45° to the reference line, as per the standard conventions, after removing the visible edges within, if any. This is known as sectional top view.

Step 5: Since the section plane is parallel to HP, this also gives the true shape of the section. ▲

Example 10.2 ▲

A cone of base diameter 50 mm and altitude 55 mm rests on its base on the HP. It is sectioned by a horizontal section plane 30 mm above the base. Draw the sectional top view.

Hint The hint given in Example 10.1 may be referred.

PROCEDURE (Refer Fig. 10.7)

Step 1: Complete the projections of the solid according to the details given and as discussed in Chapter 9.

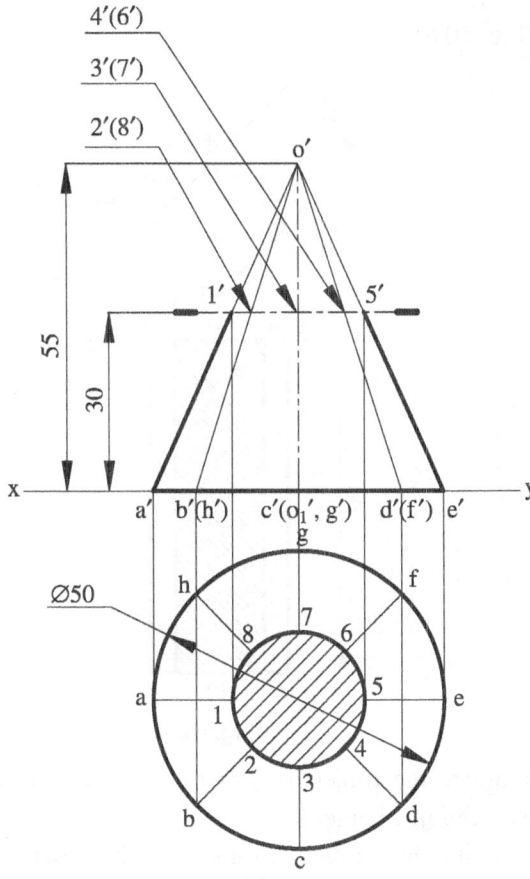

Fig. 10.7

Step 2: Mark the section plane as a straight line, parallel to the reference line, 30 mm above it in the front view, and obtain the cutting points in sequential order of the generators.

Step 3: Project the cutting points to meet the corresponding edges in the top view and join them to form a circle.

Step 4: Shading of the bounded area can be performed as mentioned in Example 10.1.

Step 5: Since the section plane is parallel to HP, this also gives the true shape of the section. ▲

Example 10.3

A right square prism of 40 mm side square base and axis 100 mm long has its square end on VP such that its faces are equally inclined to HP. A horizontal section plane cuts the prism at a distance of 7.5 mm above the axis. Draw the sectional top view indicating the cut surface.

Hint   The hint given in Example 10.1 may be referred.

PROCEDURE (Refer Fig. 10.8)

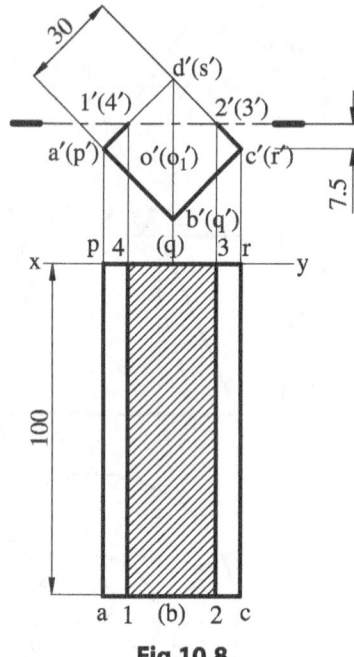

Fig.10.8

Step 1: Complete the projections of the solid according to the details given and as discussed in Chapter 9.

Step 2: Mark the section plane as a straight line, parallel to the reference line and 7.5 mm above the axis in the front view, and obtain the cutting points 1', ..., 4' in the sequential order.

Step 3: Project the cutting points to meet the corresponding edges in the top view, and join them by straight lines, and obtain a closed bounded area 1, ..., 4.

Step 4: Shading of the bounded area can be performed as mentioned in Example 10.1.

Step 5: Since the section plane is parallel to HP, it also results in the true shape of the section.

Example 10.4

A cylinder of 60 mm diameter and axis 70 mm long has one of its circular ends on VP. It is sectioned by a horizontal section plane 20 mm above the axis. Draw the sectional top view of the remaining solid.

Hint The hint given in Example 10.1 may be referred.

PROCEDURE (Refer Fig. 10.9)

Step 1: Complete the projections of the solid according to the details given and as discussed in Chapter 9.

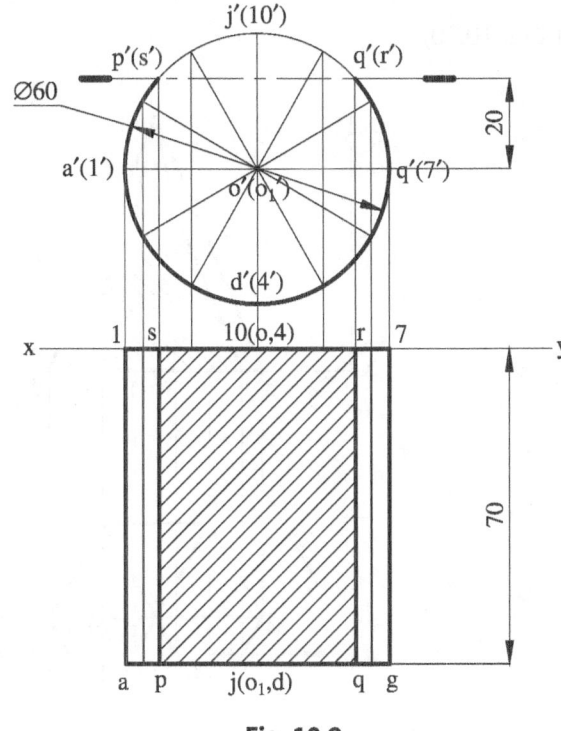

Fig. 10.9

Step 2: Mark the section plane as a straight line, parallel to the reference line and 20 mm above the axis in the front view, and obtain the cutting points p′ ,..., s′ in the sequential order.

Step 3: Project the cutting points to meet the corresponding points of the circular ends in the top view, join them by straight lines, and obtain a closed bounded area p, ..., s.

Step 4: Shading of the bounded area can be carried out as mentioned in Example 10.1.

Step 5: Since the section plane is parallel to HP, it results in the true shape of the section. ▲

10.5.2 Cutting Plane Parallel to VP

 Example 10.5

A hexagonal pyramid of 30 mm base edges is placed on HP with two of its base ends parallel to VP. A vertical section plane parallel to VP passes through a point 10 mm in front of the axis. Draw the sectional front view of the remaining solid.

Hint   Since the section plane is vertical and parallel to VP, it is visualized as a straight line, parallel to reference line in the top view.

PROCEDURE (Refer Fig. 10.10)

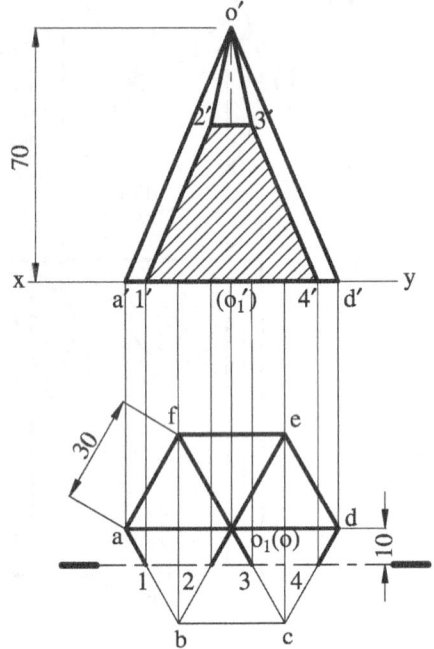

Fig. 10.10

Step 1: Complete the projections of the solid according to the details given and as discussed in Chapter 9.

Step 2: Mark the SP as a straight line, parallel to the reference line, 10 mm below the axis in the top view, and obtain the cutting points 1, ..., 4, in the base edges and on the slant edges, as shown.

Step 3: Project the cutting points to meet the corresponding edges in the front view, join them by means of a straight line, and obtain a closed bounded area $1'$, ..., $4'$.

Step 4: Shade the bounded area by means of regular and uniformly spaced hatching lines, inclined at 45° to the reference line, as per the standard conventions after removing the visible edges within, if any. This is known as sectional front view.

Step 5: Since the section plane is parallel to VP, it results in the true shape of the section. ▲

<hr>

Example 10.6 ◢

A cone of base diameter 60 mm and altitude 80 mm rests on its base on the HP. It is sectioned by a vertical section plane parallel to VP passing through a point 10 mm in front of the axis. Draw the sectional front view.

Hint Since the section plane is vertical and parallel to VP, it is visualized as a straight line, parallel to reference line in the top view.

PROCEDURE (Refer Fig. 10.11)

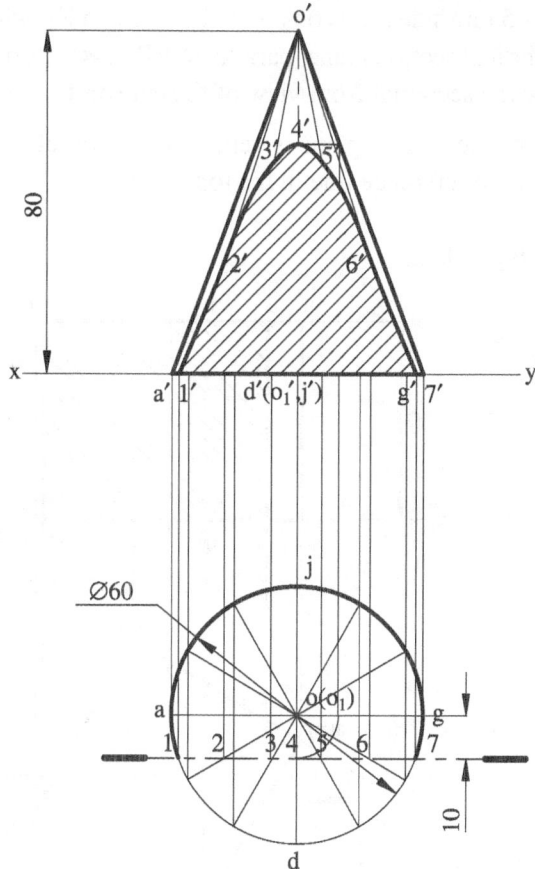

Fig. 10.11

Step 1: Complete the projections of the solid according to the details given and as discussed in Chapter 9.

Step 2: Mark the SP as a straight line, parallel to the reference line, 10 mm below the centre of the circle in the top view, and obtain the cutting points 1, ..., 7, as indicated.

Step 3: Project the cutting points to meet the corresponding edges in the front view, join all points by means of a curve, and obtain a closed bounded area.

Step 4: Shade the bounded area by means of regular and uniformly spaced hatching lines, inclined at 45° to the reference line, as per the standard conventions after removing any visible edges within, if any. This is referred as the sectional front view.

Step 5: Since the section plane is parallel to VP, it results in the true shape of the section. This shape is also known as 'hyperbola' as discussed in Chapter 2 on Conics.

Example 10.7

A cube of 45 mm sides rests on one of its faces on HP with a vertical face inclined at 30° to VP. A vertical section plane, parallel to VP passes through a point 5 mm in front of the axis. Draw the sectional front view of the cut solid.

Hint   Since the section plane is vertical and parallel to VP, it is visualized as a straight line, parallel to reference line in the top view.

PROCEDURE   (Refer Fig. 10.12)

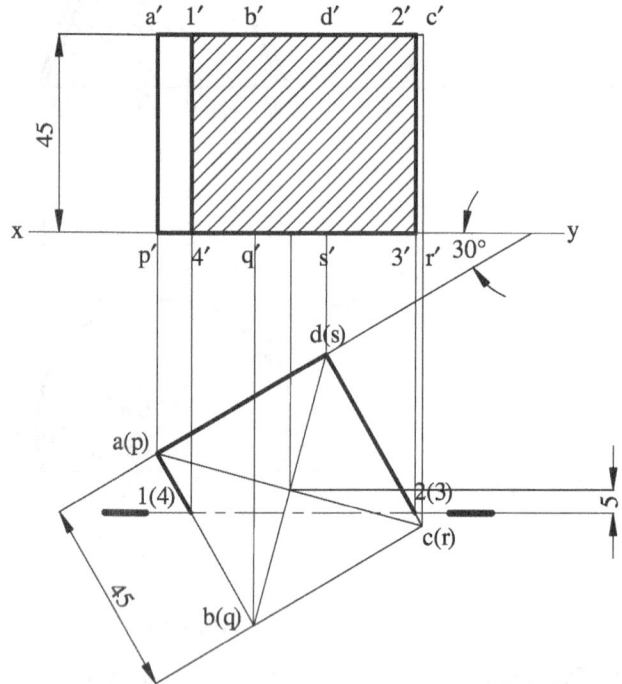

Fig. 10.12

Step 1: Complete the projections of the solid as per the details given and as discussed in Chapter 9.

Step 2: Mark the SP as a straight line, parallel to the reference line, 5 mm below the axis of the solid in the top view, and obtain the cutting points 1, ..., 4, on the edges of the cube as indicated.

Step 3: Project the cutting points to meet the corresponding edges in the front view, join them by means of a straight line, and obtain a closed bounded area 1', ..., 4'.

Step 4: Shading of the bounded area can be done as mentioned in Example 10.6. This is known as sectional front view.

Step 5: Since the section plane is parallel to VP, it results in the true shape of the section. Since the edges through b and c are removed, they are not visualized and not marked.

Example 10.8

A pentagonal prism of 30 mm base edges and axis 70 mm long lies on its base on HP with one of the lateral faces inclined at 30° VP. A vertical section plane, parallel VP passes through the axis and cuts the prism into two halves. Draw the sectional front view of the remaining solid.

Hint   Since the section plane is vertical and parallel to VP, it is visualized as a straight line, parallel to reference line in the top view.

PROCEDURE   (Refer Fig. 10.13)

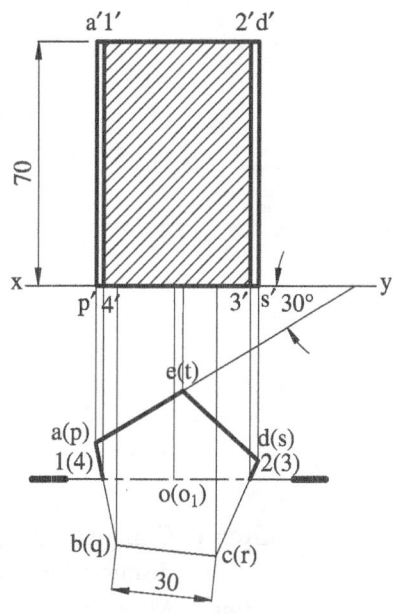

Fig. 10.13

Step 1: Complete the projections of the solid according to the details given and as discussed in Chapter 9.

Step 2: Mark the SP as a straight line, parallel to the reference line, passing through the centre of the pentagon in the top view and obtain the cutting points 1, ..., 4, on the edges of the two ends of the solid, as shown in Fig. 10.13.

Step 3: Project the cutting points to meet the corresponding edges in the front view, and join them by means of straight lines and obtain a closed bounded area 1',..., 4'.

Step 4: Shading of the bounded area can be done as mentioned in Example 10.6. This is known as sectional front view.

Step 5: Since the section plane is parallel to VP, it results in the true shape of the section.

10.6 SOLIDS IN SIMPLE POSITION AND CUTTING PLANE INCLINED TO ONE OF THE PRINCIPAL PLANES

10.6.1 Cutting Plane Inclined to HP

Example 10.9

A cube of 45 mm sides rests on one of its faces on HP with a vertical face inclined at 30° to VP. A section plane perpendicular to VP and inclined at 27° to HP cuts the solid into two equal halves. Draw the sectional top view and the true shape of the section.

Hint   Since the section plane is perpendicular to VP, it is visualized as a straight line in the front view and drawn as an inclined line to the reference line due to its inclination with HP.

PROCEDURE (Refer Fig. 10.14)

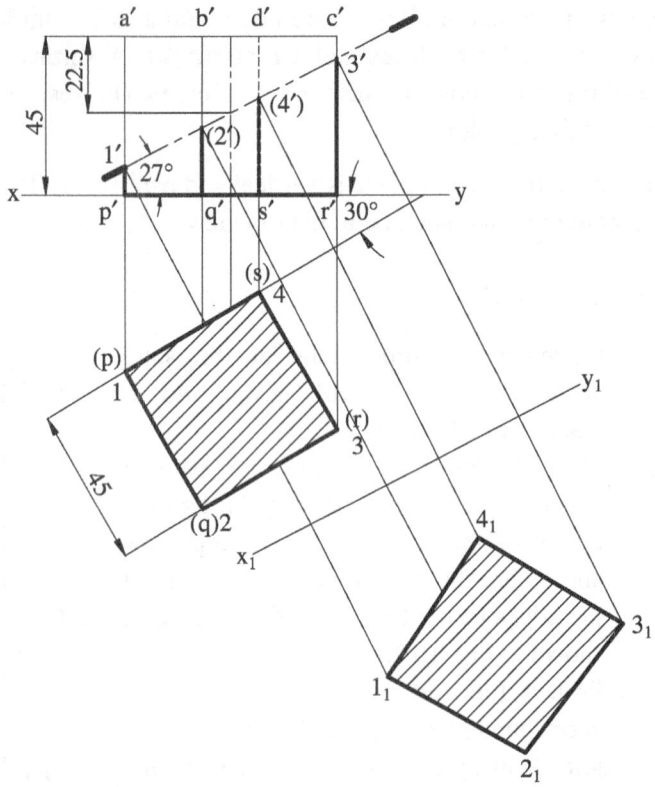

Fig. 10.14

Step 1: Complete the projections of the solid according to the details given and as discussed in the Chapter 9.

Step 2: Mark the SP as a straight line, inclined at 27° to the reference line, and passing through the midpoint of the axis in the front view, and obtain the cutting points 1′, ..., 4′, in the sequential order of the edges.

Step 3: Project the cutting points to meet the corresponding edges in the top view and join them and obtain a closed bounded area that coincides with the top view of the solid itself, as the section plane cuts only the lateral edges.

Step 4: Shade the bounded area by means of regular and uniformly spaced hatching lines, inclined at 45° to the reference line, as per the standard conventions after removing the visible edges within, if any. This is known as sectional top view.

Step 5: Since the SP is inclined to HP, the sectional top view does not give the true shape of the cut portion. Therefore, draw a new reference line 'x₁y₁' parallel to the line representing the SP, draw it opposite to the part removed in the front view (as mentioned in Section 10.4), and draw projections to 'x₁y₁' as shown in Fig. 10.14. As done in the auxiliary projections of the solids, measure the distances of the projections of these cutting points, that is, 1, 2, ..., 4 from 'xy' line and mark along the new projectors from 'x_1y_1' line as 1_1, ..., 4_1. The resulting shape obtained by joining these points gives the true shape of the section. Shade it as done in Step 4. ▲

> **NOTE** *Since the top portion is removed, the auxiliary line is drawn on the opposite side well below, so that it does not interfere with the top view already drawn and some of the projection lines from the cutting points which cross the top view are to be broken at the crossing region of the view.*

Example 10.10 ▲

A hexagonal prism of 30 mm sides and axis 80 mm long is resting on HP on its base with two of its lateral faces parallel to VP. The prism is sectioned by an inclined section plane that passes through the midpoint of the axis and makes 60° with the HP. Draw the sectional top view and the true shape of the section if the section plane is perpendicular to VP.

Hint Since the section plane is perpendicular to VP, it is visualized as a straight line in the front view and drawn as an inclined line to the reference line due to its inclination with HP.

PROCEDURE (Refer Fig. 10.15)

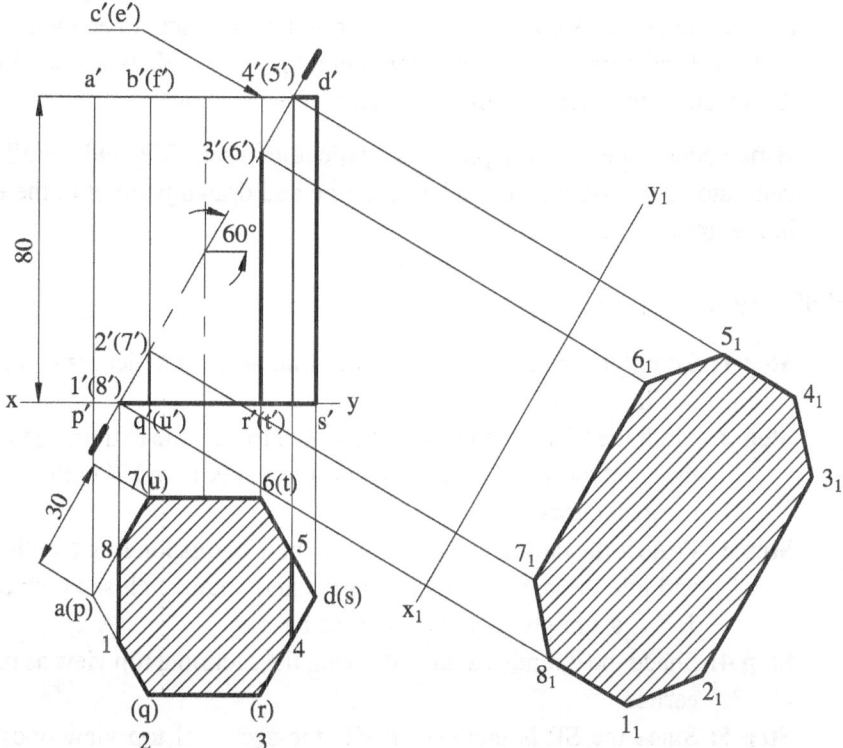

Fig. 10.15

Step 1: Complete the projections of the solid according to the details given and as discussed in Chapter 9.

Step 2: Mark the SP as a straight line, inclined at 60° to the reference line and passing through the midpoint of the axis, in the front view and obtain the cutting points 1', ..., 8', on the base, the top and the lateral edges in the sequential order of the edges.

Step 3: Project the cutting points to meet the corresponding edges in the top view and join them and obtain a closed bounded area 1, ..., 8. Since the lateral edge a'p' and a part of the base are removed, these details do not find a place in the top view.

Step 4: Shading of the bounded area can be done as mentioned in Example 10.9 in the sectional top view.

Step 5: Since the SP is inclined to HP, the sectional top view does not give the true shape of the cut portion. Therefore, draw a new reference line 'x₁y₁' as described in Example 10.9 and obtain the cutting points 1_1, ..., 8_1. The resulting shape obtained by joining these points gives the true shape of the section. Shade it as done in Step 4.   ▲

Example 10.11 ▲

A right circular cone of base diameter 60 mm and vertical height 65 mm placed with its axis parallel to the VP and perpendicular to the HP is sectioned by a plane perpendicular to the VP and parallel to the contour generator and at a distance of 12 mm from it. Draw the sectional top view and the true shape of the section.

Hint   Since the section plane is perpendicular to VP and parallel to the contour generator, it is visualized as a straight line and drawn parallel to the extreme generator in the front view.

PROCEDURE   (Refer Fig. 10.16)

Step 1: Complete the projections of the solid as per the details given and as discussed in Chapter 9.

Step 2: Mark the SP as a straight line parallel to any one extreme generator (i.e., o'e') in the front view and obtain the cutting points 1' and 7' in the base and 2', ..., 6' in the generators.

Step 3: Project the cutting points to meet their corresponding locations in the top view and join them by means of a curve and obtain a closed bounded area. Note that 1–7 is a straight line as it cuts the base.

Step 4: Shade the bounded area and obtain the sectional top view as problems discussed earlier.

Step 5: Since the SP is inclined to HP, the sectional top view does not give the true shape of the cut portion. Therefore, draw a new reference line 'x₁y₁' parallel to

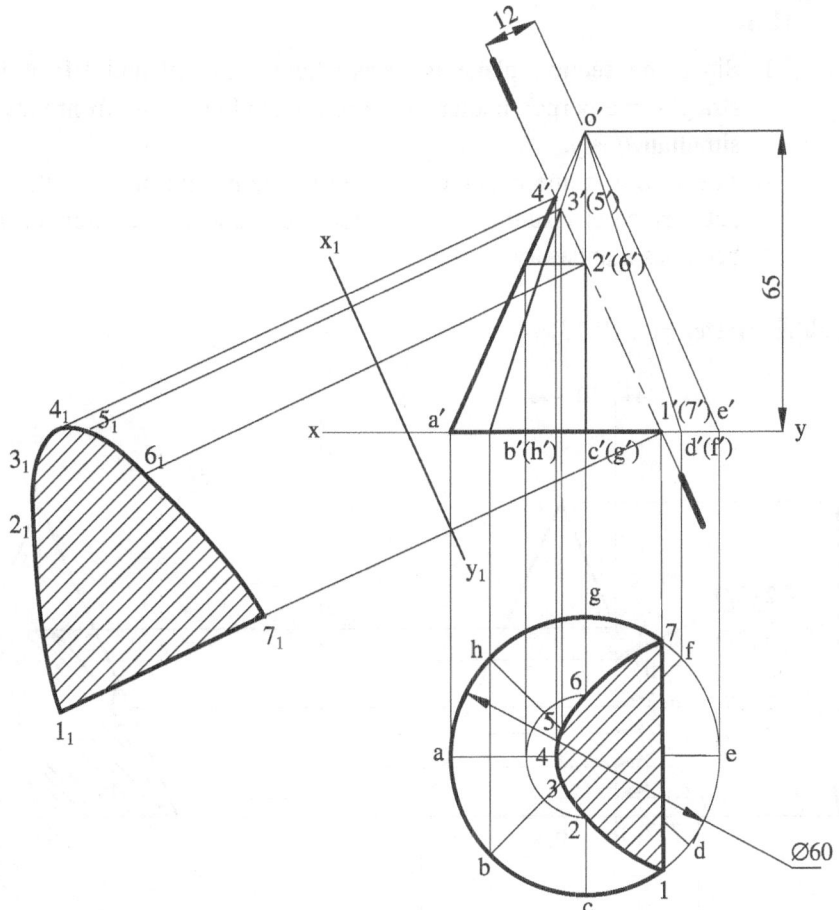

Fig. 10.16

the SP in the front view on the opposite side of the removal of the cut portion as shown in Fig. 10.16. As performed in the auxiliary projections of the solids, measure the distances of the projections of these cutting points, that is, 1, 2, ..., 7 from 'xy' line and mark along the new projectors from 'x_1y_1' line as 1_1, ..., 7_1. The resulting shape obtained is by joining these points by means of a curve with the base portion corresponding to the sector length in the top view. Shade it as done in Step 4. As discussed in Chapter 2, the true shape obtained is a parabola, whose base is equal to the chord of the base circle and the height or altitude is equal to the length of the section plane (1'4') contained within the cone. ▲

Example 10.12 ◢

A cone of base diameter 50 mm and altitude 55 mm rests on its base on the HP. It is sectioned by a section plane perpendicular to both the HP and the VP and is placed at 10 mm to the left of the axis. Draw the top view, front view, and sectional end view.

Hint

1. Since the section plane is perpendicular to HP and VP, it is visualized as a straight line perpendicular to the reference line in the front view and the top view simultaneously.
2. The section plane becomes parallel to the profile plane or the side plane and the cut surface is visualized in the side view, and hence, drawing the side view also becomes mandatory.

PROCEDURE (Refer Fig. 10.17)

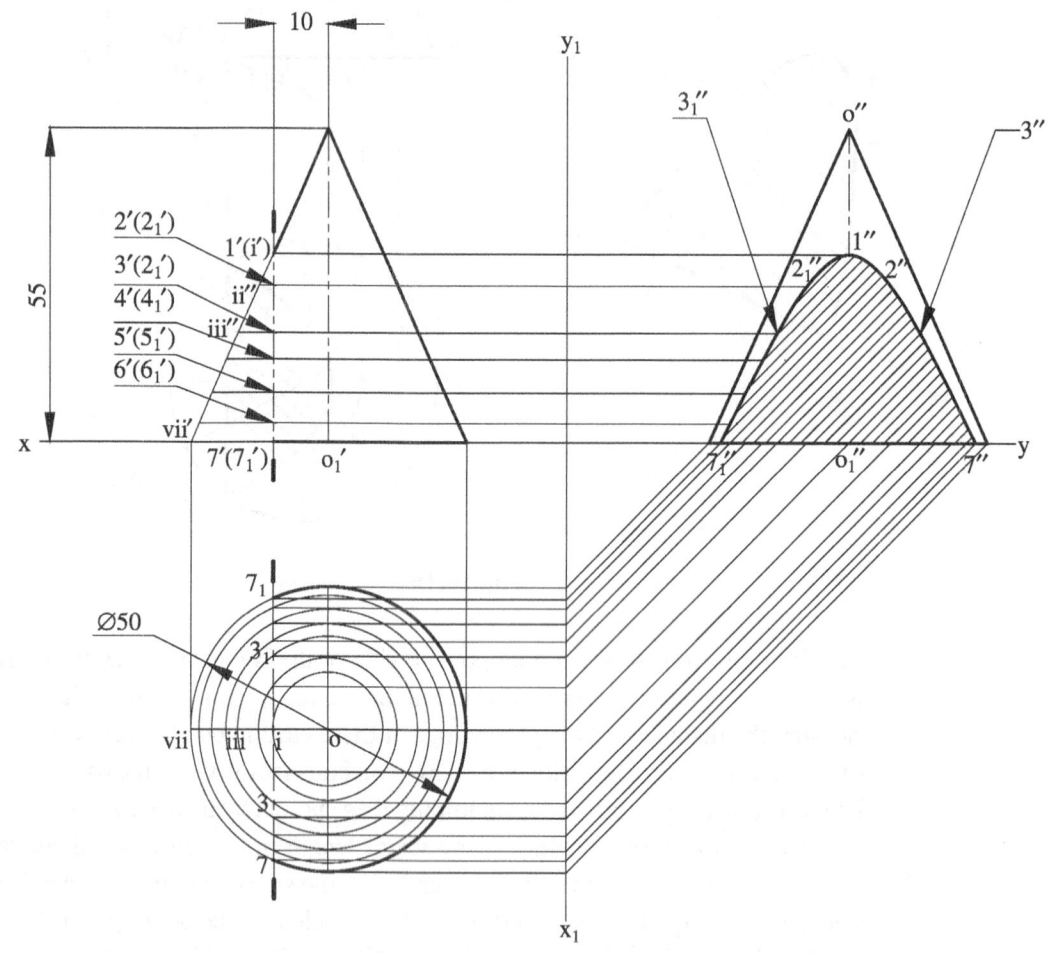

Fig. 10.17

Step 1: Complete the projections of the solid including the right-side view according to the details given and as discussed in Chapter 9.

Step 2: Mark the SP as a straight line parallel to the axis (i.e. $o'o_1'$) and 10 mm to the left of it in the front view and also in the top view from the centre, which represents the axis. Select the points $1'$, ..., $7'$ arbitrarily on the cutting plane in the front view. If these cutting points are projected to obtain their top views, they do not meet, as all of them fall on the same line. Every cutting point in the front view such as $1'$, $2'$ yield two points in the top view corresponding to the length of the horizontal line passing through them and contained within the cone. For example, the length of the horizontal line through $3'$ is used to get two points 3_1, 3 in the top view section line, when a corresponding arc is drawn. Similarly, all the cutting points yield their respective twin points, when concentric arcs are drawn with 'o' as centre.

Step 3: The sectional view is drawn in the side view. Project the cutting points from the front view horizontally to the side view and mark the distance of the respective points in the top view on them. For example, the horizontal projector line through $3'$ is drawn and the distance of the top view points 3_1, 3 is measured from 'xy' line and transferred to the side view reference line 'x_1y_1' to obtain $3_1''$, $3''$, as carried out in any auxiliary projection method. In fact, the side view itself is an auxiliary projection method, where the auxiliary reference line is at $90°$ to the conventional reference line. Join the points $1_1''$, ..., $7_1''$ and $1''$, ..., $7''$ by means of a curve and shade the area based on the sectional guidelines. This gives the true shape of the section as section is parallel to the side or profile plane. The curve obtained is known as a hyperbola, as discussed in Chapter 2. ▲

Example 10.13 ▲

A right regular hexagonal pyramid side of base 30 mm and height 80 mm is resting on its base on the HP with two of its adjacent lateral faces equally inclined to VP. It is sectioned by a horizontal section plane and an inclined section plane thereafter. The two section planes meet at the midpoint of the axis in the front view. The inclined section plane makes $70°$ with the HP and perpendicular to the VP. Draw the projections indicating the cut surfaces and also represent the true shape of the cut portion corresponding to the inclined section plane.

Hint   Since section plane is perpendicular to the VP and is composed of a horizontal and an inclined portion, it is represented as a straight line similarly in the front view, with the horizontal and the inclined portions both meeting at the midpoint of the axis.

PROCEDURE   (Refer Fig. 10.18)

Step 1: Complete the projections of the solid according to the details given and as discussed in Chapter 9.

Step 2: Mark the SP as a straight line parallel to the reference line in the front view and up to the midpoint of the axis and then as an inclined line at $70°$ as shown in Fig. 10.18. Obtain the cutting points for both the portions of the SP as $1'$, ..., $6'$ in the sequential order of the lateral edges and the base.

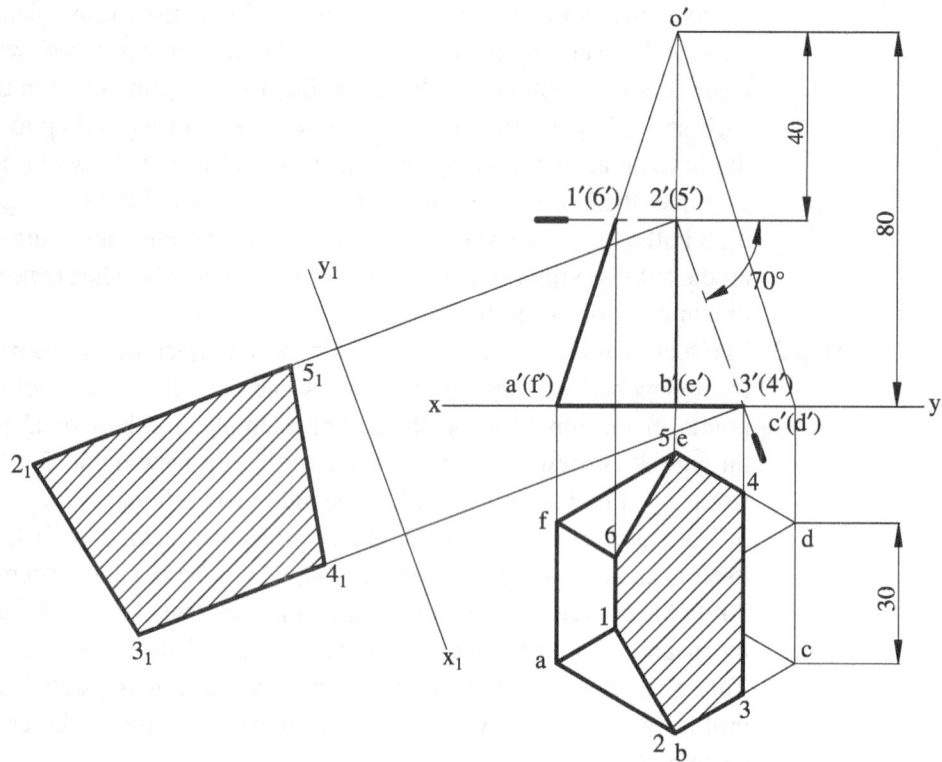

Fig. 10.18

Step 3: Project the cutting points to meet the corresponding edges in the top view and join them to obtain a closed bounded area and shade it as done in the problems discussed earlier to get the sectional top view. The portion of the section in the sectional top view marked as 1–2–5–6–1 belongs to the horizontal section plane.

Step 4: To obtain the true shape of section for the inclined cut portion, draw a new reference line 'x_1y_1' parallel to the inclined portion of the SP in the front view and draw projections from those points situated in this portion (2', ..., 5') to 'x_1y_1'. Measure the distances of these cutting points, that is, 2, 3, 4, and 5 from 'xy' line in the top view and mark along the new projectors from 'x_1y_1' line as 2_1–3_1–4_1–5_1–2_1. The resulting shape obtained by joining these points gives the true shape of the portion of the section that is inclined to HP and perpendicular to VP. Shade it as per the conventions. ▲

Example 10.14 ▲

A right square pyramid of 30 mm side base and axis 70 mm long has its square end on VP such that its sides are equally inclined to HP. An inclined section plane, perpendicular to VP and inclined at 40° to HP cuts the pyramid at a distance of 7.5 mm from the axis. Draw the sectional top view and the true shape of the section if the larger portion of the solid is retained.

Hint   Since the section plane is perpendicular to VP, it is visualized as a straight line in the front view and drawn as an inclined line to the reference line due to its inclination with HP.

PROCEDURE (Refer Fig. 10.19)

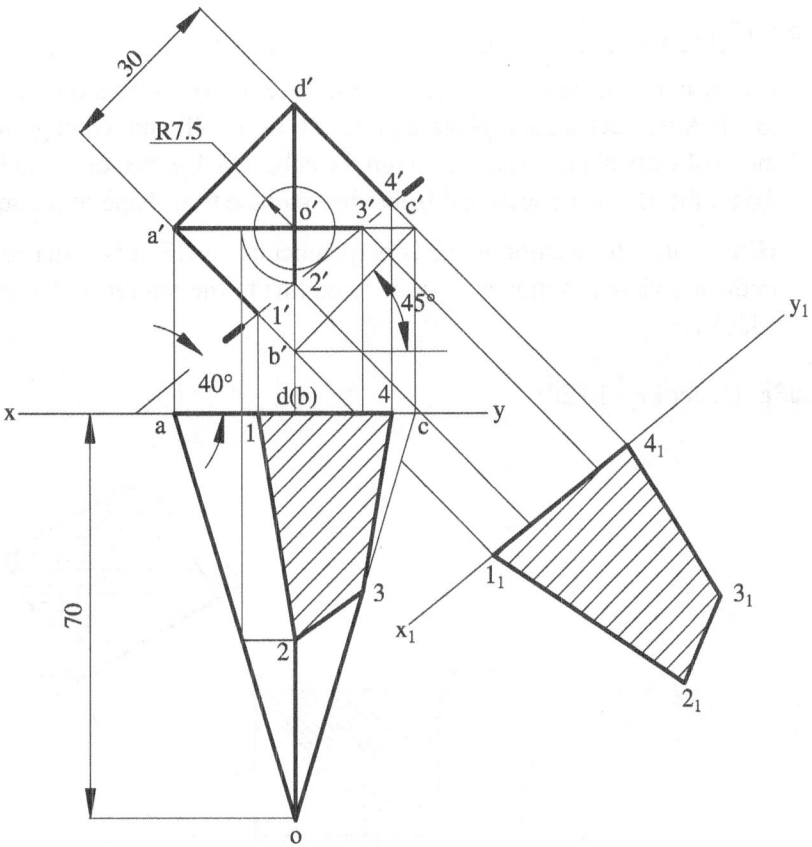

Fig. 10.19

Step 1: Complete the projections of the solid on the basis of the details given and as discussed in Chapter 9.

Step 2: Mark the SP as a straight line inclined at 40° to the reference line in the front view. Since it is at a distance of 7.5 mm from the axis, it can be oriented by drawing a circle at the centre with a radius of 7.5 mm and by drawing a tangent to the circle to the left or to the right. Since the larger portion is retained, the SP is indicated as shown in the Fig. 10.19. Obtain the cutting points of the SP as 1′, ..., 4′ in the sequential order of the edges.

Step 3: Project the cutting points to meet the corresponding edges in the top view and join them to obtain a closed bounded area. The sectional top view 1–2–3–4–1 is the apparent section.

Step 4: To find the true shape of the section, draw a new reference line 'x₁y₁' parallel to the SP in the front view, draw projections to 'x₁y₁', and mark the points 1_1, ..., 4_1 as discussed earlier. The resulting shape obtained by joining these points gives the true shape of the section. Shade it as in earlier problems.   ▲

10.6.2 Cutting Plane Inclined to VP

Example 10.15 ◢

A cube of 45 mm sides rests on one of its faces on HP, with a vertical face inclined at 30° to VP. A vertical section plane inclined at 30° to VP and leaning opposite to the above inclined vertical face passes through the cube at a distance of 5 mm from the axis and in front of it. Draw the sectional front view and the true shape of the section.

Hint   Since the section plane is perpendicular to HP, it is visualized as a straight line in the top view and drawn as an inclined line to the reference line due to its inclination with VP.

PROCEDURE   (Refer Fig. 10.20)

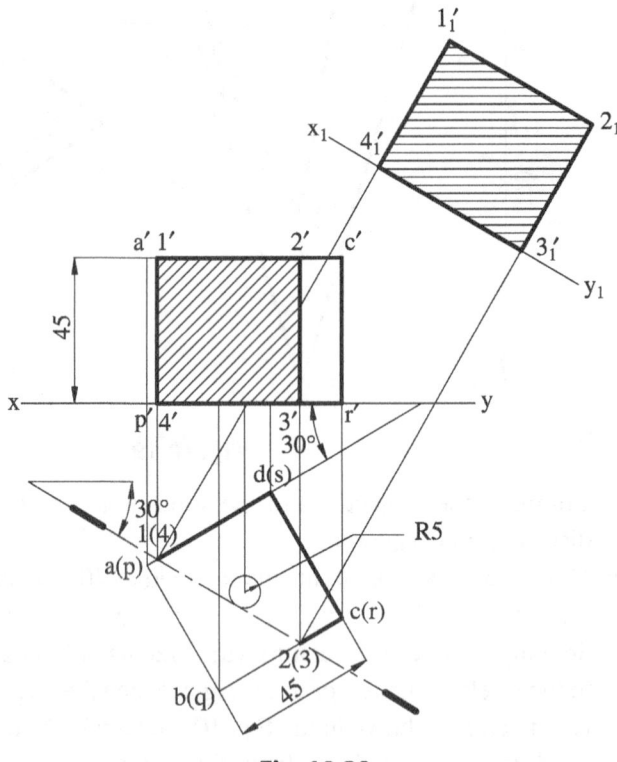

Fig. 10.20

Step 1: Complete the projections of the solid with one face inclined according to the details given and as discussed in Chapter 9.

Step 2: Draw a circle with a radius of 5 mm from the axis in the top view and mark the SP as a straight line tangential to the circle and inclined at 30° to the reference line in the opposite direction to the inclined face as shown in Fig. 10.20 Obtain the cutting points of the SP as 1, ..., 4 in the sequential order of the edges.

Step 3: Project the cutting points to meet the corresponding edges in the front view and join them to get the sectional front view $1'-2'-3'-4'-1'$ that is apparent. Since the outer edge 'ap' is removed in the sectioning, it is not shown in the front view.

Step 4: Draw a new reference line 'x_1y_1', project the cutting points, and transfer their distances from the front view. The resulting shaded area $1_1'-2_1'-3_1'-4_1'-1_1$ represents the true shape of the cut section. ▲

Example 10.16 ▲

An equilateral triangular pyramid of 45 mm side of base and axis 100 mm long is resting on HP on its base with one of the base edges perpendicular to VP. A vertical section plane inclined at 40° to the VP passes through the pyramid at a distance of 10 mm from the axis and in front of it. Draw the sectional front view and the true shape of the section.

Hint   Since the section plane is perpendicular to HP, it is visualized as a straight line in the top view and drawn as an inclined line to the reference line due to its inclination with VP.

PROCEDURE (Refer Fig. 10.21)

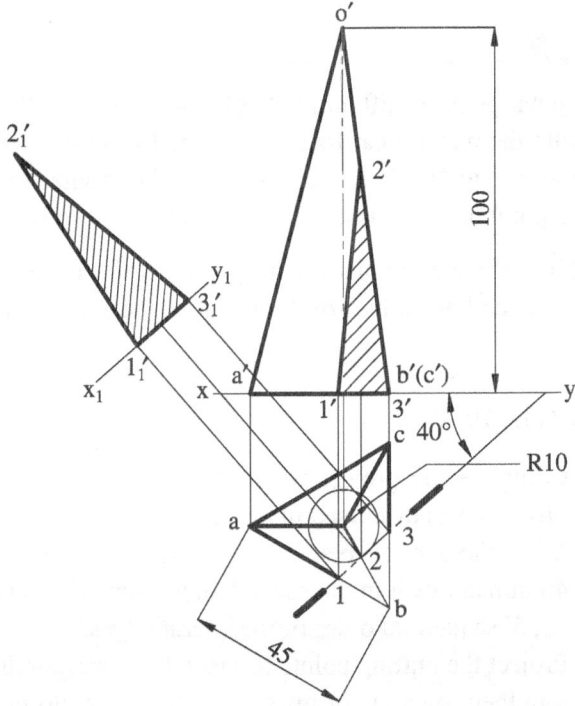

Fig. 10.21

This example is similar to Example 10.15 with the exception of a triangular base in place of the square base, and hence, the reader is advised to refer the working procedure mentioned therein for solving this example. ▲

10.7 SOLIDS IN INCLINED POSITIONS AND CUTTING PLANE PARALLEL TO ONE OF THE PRINCIPAL PLANES

In the foregoing section, the solid will be discussed in the tilted positions and the cutting plane will be parallel to one of the reference planes. As discussed in Chapter 9, the projection of the solid can be obtained in two stages. The initial stage projections are obtained by keeping the axis perpendicular to the reference plane to which it is to be inclined later. The hints of keeping the edge perpendicular or the diagonal parallel to the reference line are applicable, depending on the tilting about the edge or about a corner, respectively. The projections can be drawn by the tilting object method or by the auxiliary plane method. After completing the projections of the solid, the section plane is represented in the appropriate view as a straight line as discussed in the current chapter. Examples 10.17–10.26 are drawn by tilting object method. To save time and labour, the reader is advised to use auxiliary projection method for drawing the projection of the solid.

10.7.1 Cutting Plane Parallel to HP

Example 10.17 ◢

A pentagonal prism of 30 mm side of base and axis 70 mm long rests on a base edge on HP with the rectangular face containing this base edge perpendicular to the VP and inclined at 50° to HP. It is sectioned by a horizontal section plane whose VT passes through a point on the axis and located 40 mm above HP. Draw the sectional top view.

Hint Since the section plane is horizontal, VT is represented as a straight line parallel to the reference line in the front view after the final projections of the uncut solid are drawn.

PROCEDURE (Refer Fig. 10.22)

Step 1: Complete the projections of the solid in two stages according to the details given and as discussed in Chapter 9.

Step 2: Mark the section plane as a straight line, parallel to the reference line and 40 mm above it in the second-stage front view and obtain the cutting points 1′, ..., 5′ sequential order of the lateral edges.

Step 3: Project the cutting points to meet the corresponding edges in the top view, and join them by straight lines, and obtain the closed bounded area formed by the points 1, ..., 5 after removing the edges present within. The sectioned area can

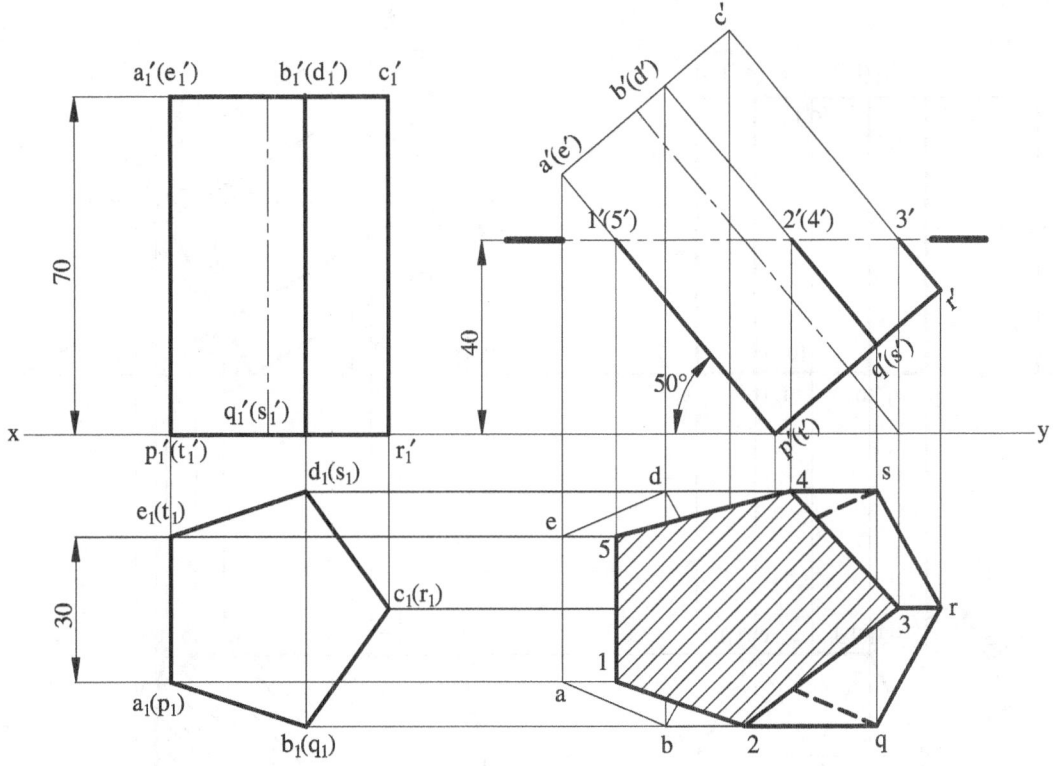

Fig. 10.22

be shaded as discussed earlier. It can be noted that the top pentagonal end a, ..., e is removed and hence is not shown in the top view. ▲

NOTE *The retained portion along with the corners and the edges are to be drawn in visible thick lines and all the edges or corners that are removed are to be strictly erased. However, in all the examples, they will be shown very light, only to indicate to the reader the procedures that are followed.*

Example 10.18 ▲

A hexagonal prism of 30 mm sides and axis 70 mm long is resting on HP on one of its corners of the base and the axis is inclined at 30° to HP and the vertical plane containing the axis and the lateral slant edge passing through the base corner on HP is parallel to VP. It is cut by a horizontal section plane passing through the midpoint of the axis. Draw the sectional top view of the cut arrangement.

Hint The hint given in Example 10.17 may be referred.

PROCEDURE (Refer Fig. 10.23)

Step 1: This example is similar to Example 10.17 with the exception that this is a hexagonal prism resting on a corner, and hence, the reader is advised to refer the working procedure mentioned therein for solving this example.

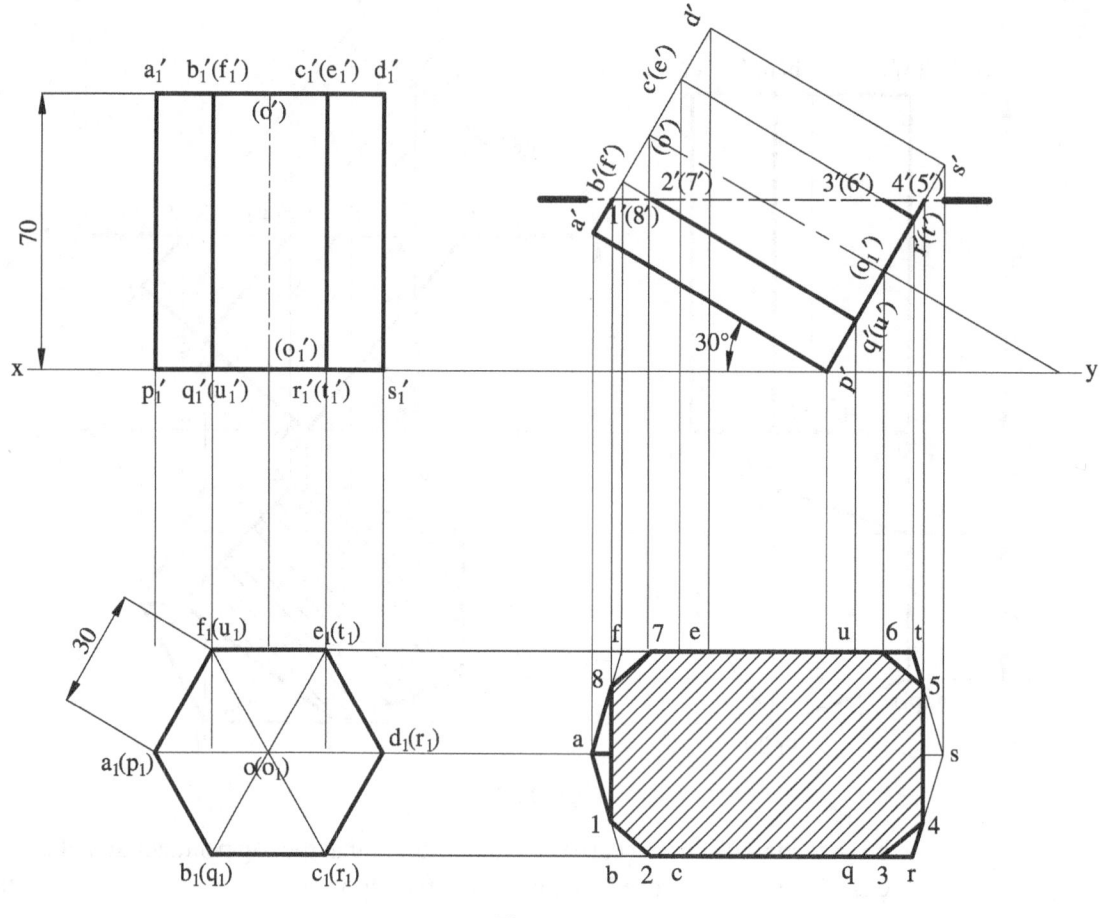

Fig. 10.23

Step 2: In the sectional top view, the corner S corresponding to the base hexagon has been removed, and hence, that point and the connecting edges with that point are all removed. ▲

10.7.2 Cutting Plane Parallel to VP

Example 10.19 ▲

A hexagonal prism of 30 mm side of base and axis 70 mm long has an edge of one hexagonal end on VP with a rectangular face containing this edge perpendicular to HP and inclined at 45° to VP. The solid is sectioned in this position by a vertical section plane whose HT is parallel to the reference line and passes through a point on the axis and bisecting it. Draw the sectional front view of the retained solid.

Hint Since the section plane is vertical, its HT is visualized as a line in the top view and this is parallel to the reference line as cited in this example.

PROCEDURE (Refer Fig. 10.24)

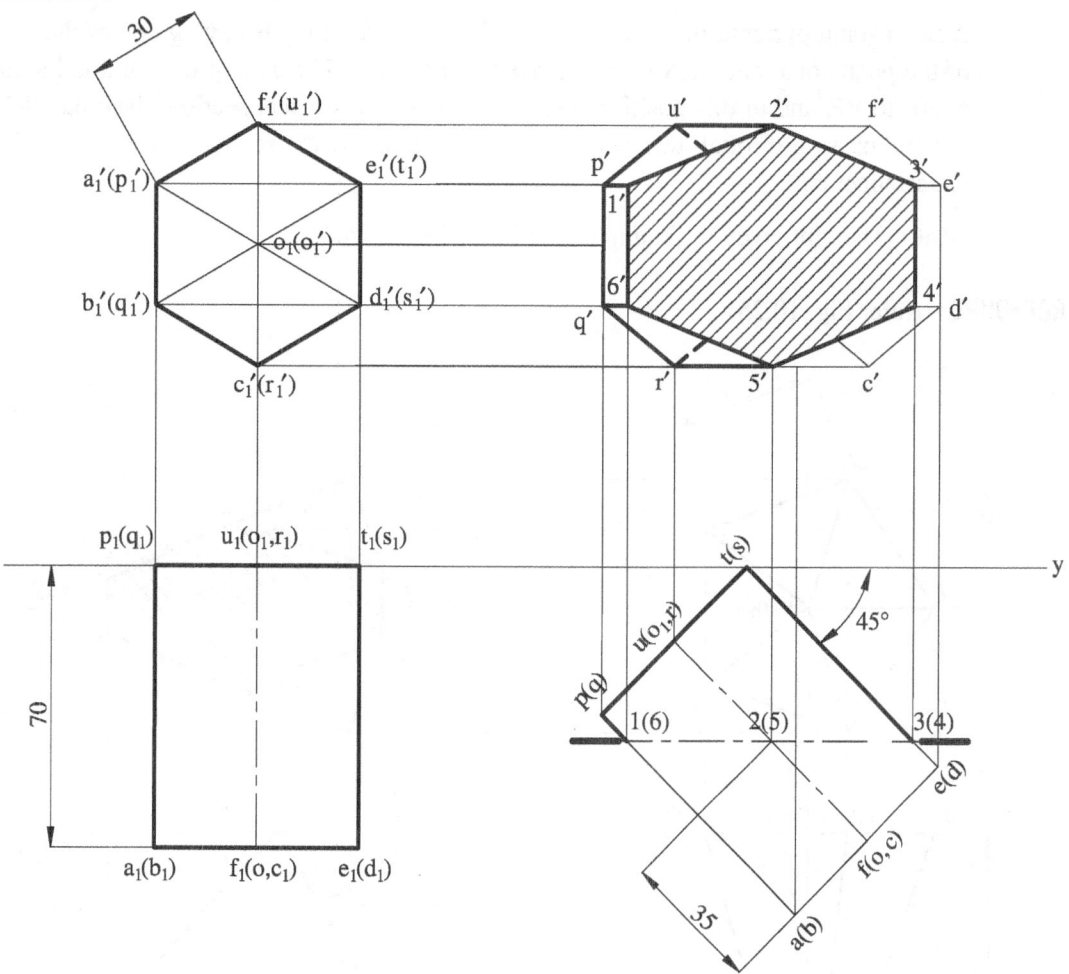

Fig. 10.24

Step 1: Since the axis of the solid is inclined to the VP, the hexagonal end is drawn in the front view. To facilitate the tilting of the solid about an edge, the hexagon is drawn with one edge perpendicular to the reference line in the initial projections. The views corresponding to the tilted position of the solid are completed in the second stage as shown in Fig. 10.24.

Step 2: Mark the section plane as a straight line, parallel to the reference line and below it, passing through the midpoint of the axis in the second-stage top view, and obtain the cutting points 1, ..., 6 in the sequential order of the lateral edges.

Step 3: Project the cutting points to meet the corresponding edges in the front view and obtain the closed bounded area, which is shown as per sectional conventions. Since the front hexagonal end is removed, the corners and the edges pertaining to that are not shown in the front view.

Example 10.20

A pentagonal pyramid of 30 mm sides and axis 70 mm long is having one of the edges of the pentagonal end on VP and perpendicular to HP. The axis of the solid is inclined at 30° to VP, and in this position, the solid is cut by a vertical section plane parallel to VP and passing through the midpoint of the axis. Draw the sectional front view of the cut arrangement.

Hint The hint given in Example 10.19 may be referred.

PROCEDURE (Refer Fig. 10.25)

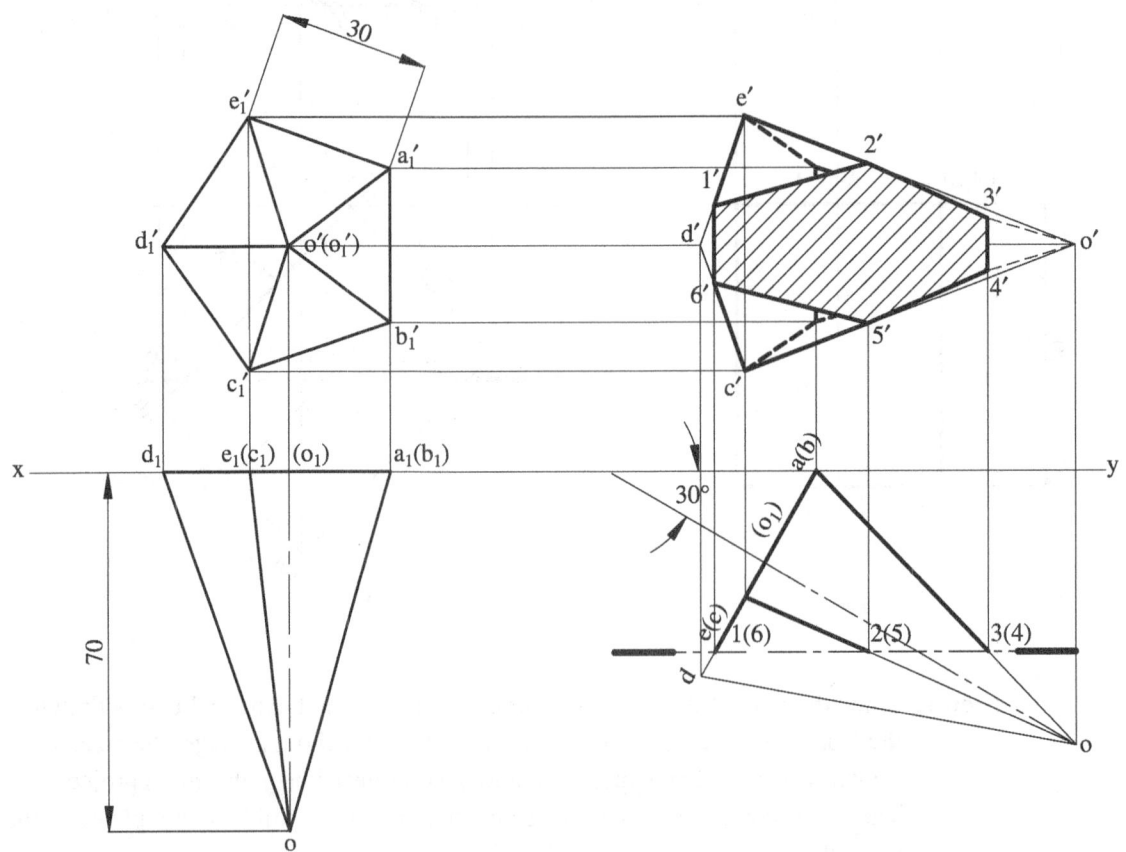

Fig. 10.25

This problem is similar to that discussed in Example 10.19 with the exception being the solid as a pentagonal pyramid. The complete procedure mentioned therein can be referred. Since the apex 'o' is removed, the edges connected with that up to the section plane are also shown removed in the final sectioned front view.

10.8 SOLIDS IN INCLINED POSITIONS AND CUTTING PLANE ALSO INCLINED TO ONE OF THE PRINCIPAL PLANES

In continuation of Section 10.7, both the solids and the cutting or section planes are discussed in inclined orientations in this section. As before, after drawing the projections of the solids in two or the required number of stages, the cutting or the section plane is finally represented as a straight line inclined to the reference line, in the appropriate view (on the plane to which the cutting plane is perpendicular). The other procedures such as the identification of the cutting points by the section plane and obtaining their projections with the other view remain the same as discussed in the earlier section.

10.8.1 Cutting Plane Inclined to HP

Example 10.21

A pentagonal pyramid base 30 mm side and axis 70 mm long is resting on HP on one of its triangular faces with its axis parallel to VP. It is sectioned by a section plane perpendicular to VP and inclined at 45° to HP and passing through the midpoint of the axis. Draw the sectional top view, front view, and the true shape of the section if the apex portion is retained.

Hint   Since the section plane is perpendicular to VP, it is visualized as a straight line in the second-stage front view and is drawn as an inclined line to the reference line due to its inclination with HP.

PROCEDURE   (Refer Fig. 10.26)

Step 1: Complete the projections of the solid in two stages according to the details given and as discussed in Chapter 9.

Step 2: Mark the SP as a straight line, inclined at 45° to the reference line and passing through the midpoint of the axis, in the second-stage front view, and obtain the cutting points $1'$, ..., $5'$, on the lateral edges.

Step 3: Project the cutting points to meet the corresponding edges in the top view, and join them to obtain the closed bounded area by joining the points 1, 2, ..., 5, and shade it as per the conventions. Since the apex portion is retained, the sectioned area connecting the apex is shown in thick lines. The other portion can be erased or to be shown very light, removing all the corresponding corners.

Step 4: To find the true shape of the section, draw a new reference line 'x_1y_1' parallel to the SP in the front view and draw projections to 'x_1y_1' and mark the points 1_1, ..., 5_1 as discussed earlier. The resulting shape obtained by joining these points gives the true shape of the section. Shade it as in earlier problems.   ▲

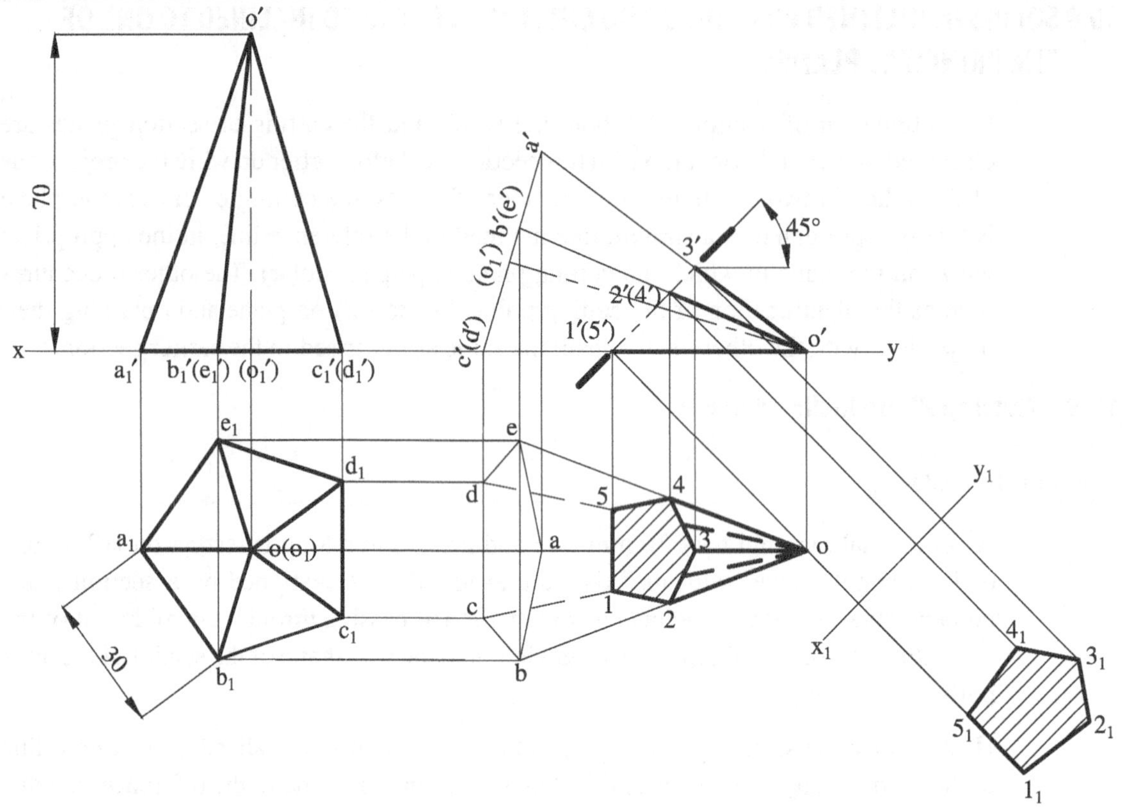

Fig. 10.26

Example 10.22

@ A right circular cone of 80 mm diameter base and 100 mm height has one of its generators on HP and axis parallel to VP. An auxiliary plane normal to VP cuts the cone such that its VT is making an angle of 45° with the reference line and intersecting the axis of the cone at a distance of 40 mm from the vertex. Draw the sectional top view and the true shape of the section.

Hint   The hint given in Example 10.21 may be referred.

PROCEDURE (Refer Fig. 10.27)

Step 1: This problem is similar to that discussed in Example 10.21 with the exception being the solid referred now is a cone. The complete procedure mentioned therein can be referred. Since the apex portion is retained, the sectioned area connecting the apex is shown in thick lines. The other portion can be erased or to be shown very light, removing all the corresponding generators and the ellipse representing the base.

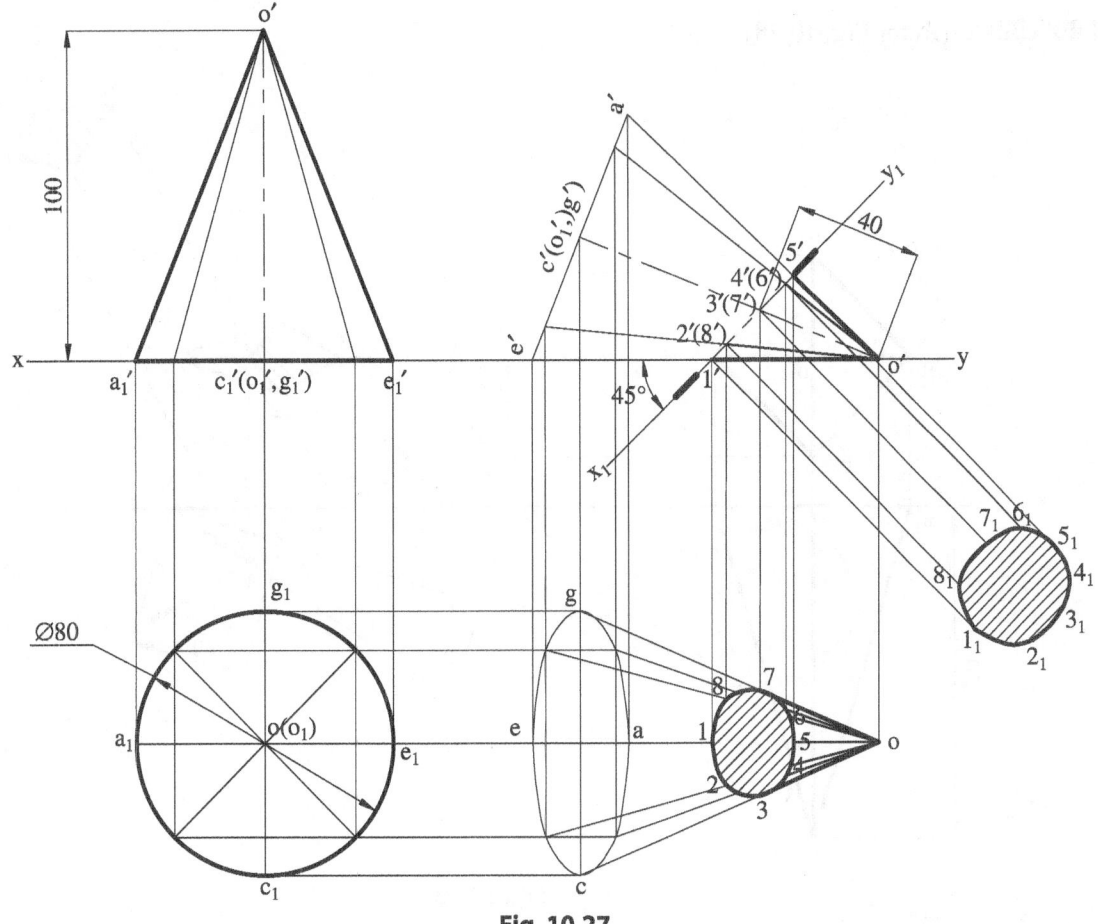

Fig. 10.27

Step 2: To find the true shape of the section, draw a new reference line 'x_1y_1' parallel to the SP in the front view, draw projections to 'x_1y_1', and mark the points 1_1, ..., 8_1, as discussed earlier. The resulting shape obtained by joining these points gives the true shape of the section. Shade it as in earlier problems. ▲

10.8.2 Cutting Plane Inclined to VP

Example 10.23 ▲

A right square pyramid of 30 mm side of base and axis 70 mm long has one of its lateral edges on VP such that two of its slant faces adjacent to this lateral edge are equally inclined to HP. A vertical section plane, inclined at 30° to VP passes through a point on the top view of the axis at a distance of 25 mm from the vertex. Draw the sectional front view and the true shape of the section.

Hint Since the section plane is perpendicular to HP, it is visualized as a straight line in the second-stage top view and drawn as an inclined line to the reference line due to its inclination with VP.

PROCEDURE (Refer Fig. 10.28)

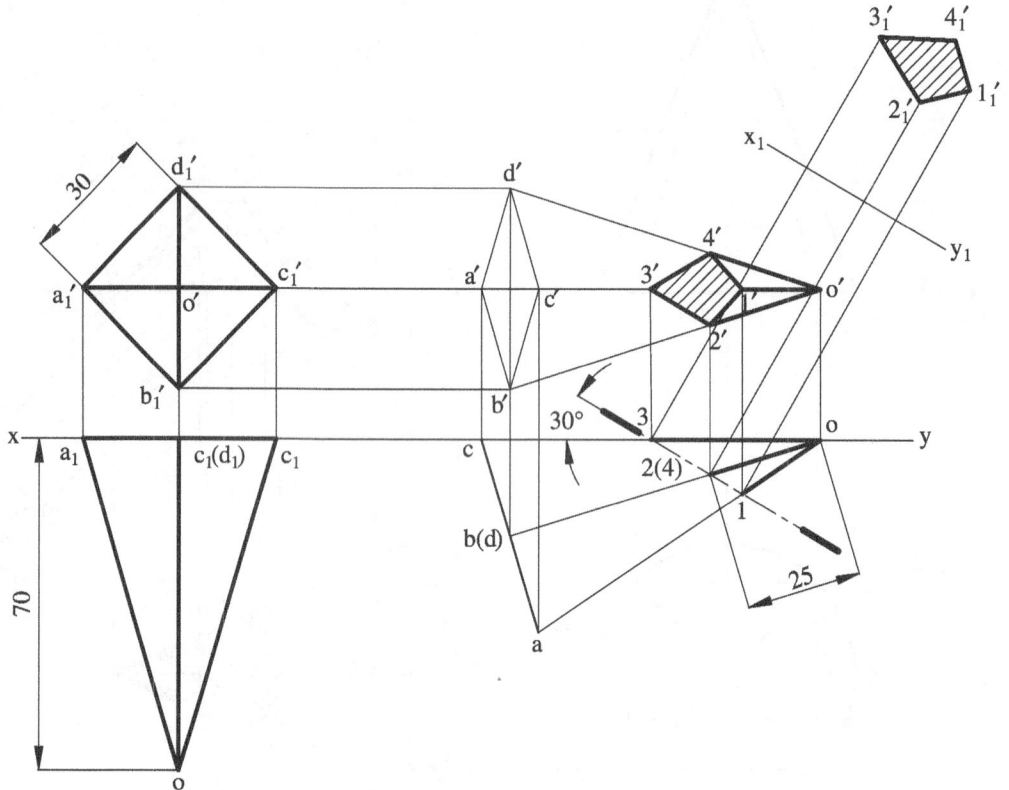

Fig. 10.28

Step 1: Since the slant edge has to lie on VP, the initial projections are drawn with the square end in the front view with a diagonal parallel to the reference line. The slant edge through this corner in the top view is tilted to lie on the reference line and the corresponding front view is drawn. This completes the second-stage projections.

Step 2: Mark the SP as a straight line inclined at 30° to the reference line and passing through a point on the axis at 25 mm from the apex in the second-stage top view, and obtain the cutting points 1, ..., 4, along the lateral edges.

Step 3: Project the cutting points to meet the corresponding edges in the front view and join them to obtain a closed bounded area through the points $1'$, $2'$, ..., $4'$.

Step 4: Since the SP is inclined to VP, the sectional front view does not give the true shape of the cut portion. Therefore, draw a new reference line 'x_1y_1' parallel to the SP in the top view and draw projections to 'x_1y_1'. As done in the auxiliary projections of the solids, measure the distances of the projections of these cutting points, that is, $1'$, $2'$, ..., $4'$ from 'xy' line and mark along the new projectors from 'x_1y_1' line as $1_1'$, ..., $4_1'$. The resulting shape obtained by joining these points gives the true shape of the section.

Example 10.24

A right regular hexagonal pyramid side of base 30 mm and height 80 mm is resting on one of its slant faces on HP. It is sectioned by a vertical section plane, inclined at 30° to VP and perpendicular to HP, bisecting the axis. Draw the projections and the true shape of the section.

Hint   Since the section plane is perpendicular to HP, it is visualized as a straight line in the second-stage top view and drawn as an inclined line to the reference line due to its inclination with VP.

PROCEDURE   (Refer Fig. 10.29)

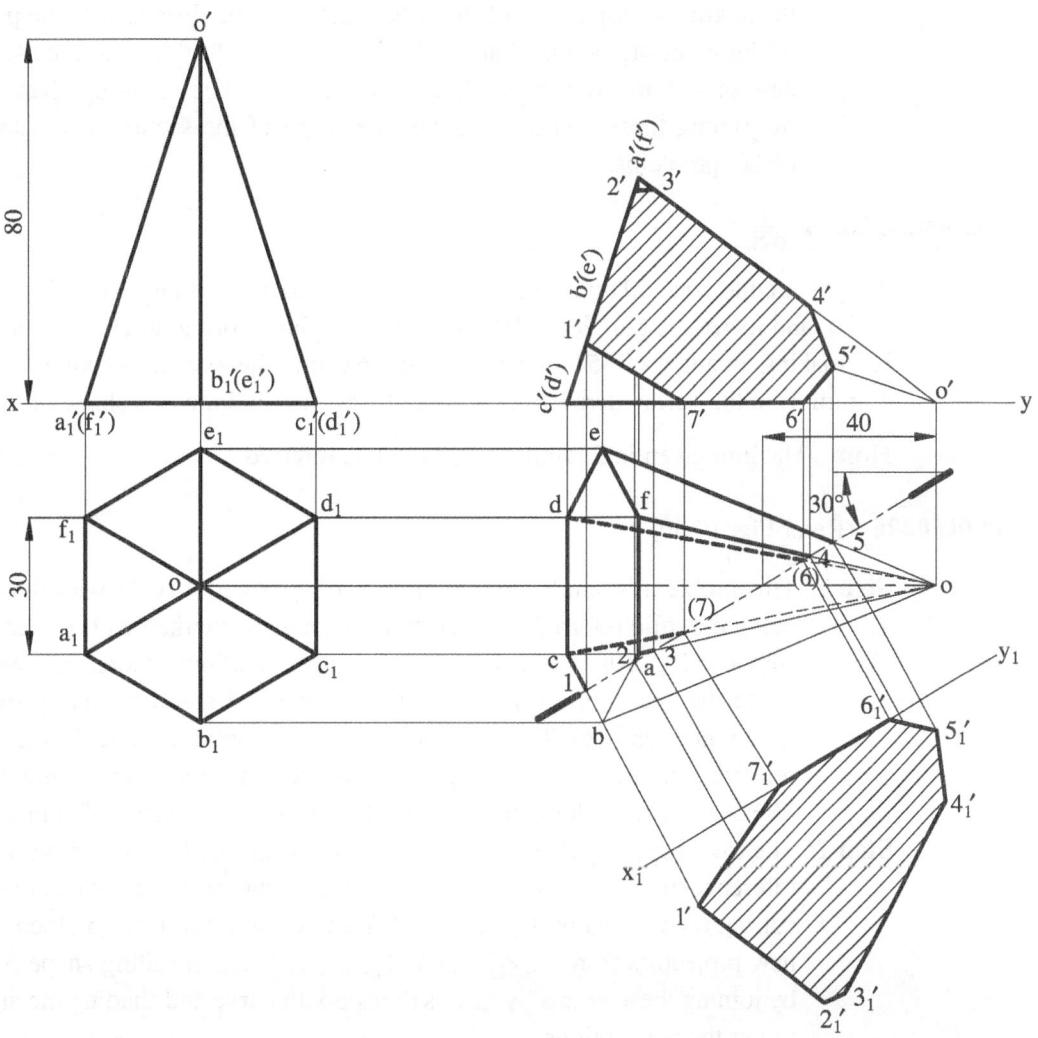

Fig. 10.29

Step 1: The projections of the solid can be drawn as per the two stages indicated in Example 10.21, replacing the solid with a hexagonal pyramid.

Step 2: Mark the SP as a straight line inclined at 30° to the reference line and passing through the midpoint of the axis in the second-stage top view, and obtain the cutting points $1'$, ..., $7'$, in the hexagonal end and along the slant edges.

Step 3: Project the cutting points to meet the corresponding edges in the front view and join them to obtain a closed bounded area, and shade as per the conventions. Since the apex is removed, the edges connected with that will not appear in the front view.

Step 4: Since the SP is inclined to VP, the sectional front view does not give the true shape of the cut portion. Therefore, draw a new reference line 'x_1y_1' parallel to the SP in the top view and draw projections to 'x_1y_1'. As done in the auxiliary projections of the solids, measure the distances of the projections of these cutting points, that is, $1'$, $2'$, ..., $7'$ from 'xy' line and mark along the new projectors from 'x_1y_1' line as $1_1'$, ..., $7_1'$. The resulting shape obtained by joining these points gives the true shape of the section and shade it as in earlier problems. ▲

Example 10.25 ▲

A right circular cone 80 mm diameter of base and 100 mm high has one of its generators on VP and its axis is parallel to HP. An auxiliary plane normal to HP cuts the cone, its HT making an angle of 45° with VP and intersecting the axis of the cone at a distance 43 mm from the vertex. Draw the sectional front view and the true shape of section.

Hint The hint given in Example 10.23 may be referred.

PROCEDURE (Refer Fig. 10.30)

Step 1: This problem is similar to that discussed in Example 10.23 with the exception being the solid referred now is a cone and may be marked with 12 generators as shown in Fig. 10.30. The complete procedure mentioned therein can be referred. Since the apex portion is retained, the sectioned area connecting the apex is shown in thick lines. The other portion can be erased or to be shown very light, removing all the corresponding generators and the ellipse representing the base. The true shape of the section is also drawn in similar lines of Example 10.23.

Step 2: The true shape of the section can be drawn in similar lines by drawing auxiliary reference line 'x_1y_1' and by transferring the distances of the projections of these cutting points, that is, $1'$, $2'$, ..., $12'$ from 'xy' line and marking them along the new projectors from 'x_1y_1' line as $1_1'$, ..., $12_1'$. The resulting shape is obtained by joining these points by means of a smooth curve and shading the area within as per the conventions. ▲

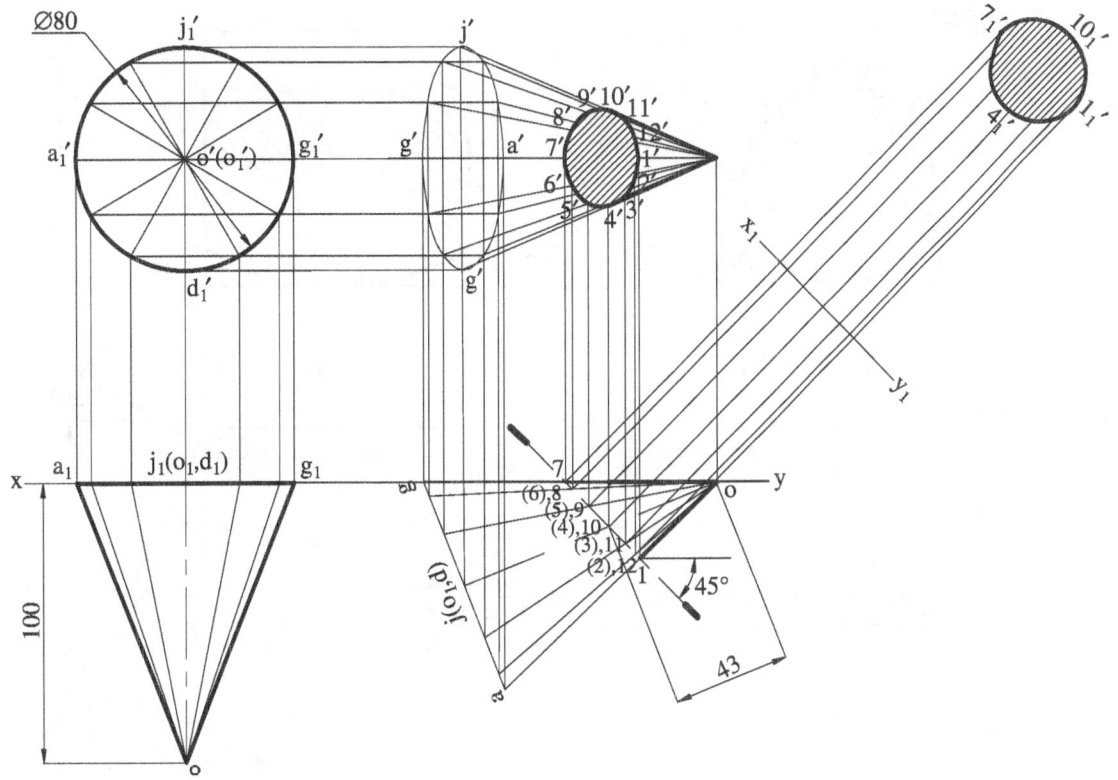

Fig. 10.30

Example 10.26

A cylinder of base diameter 50 mm and axis 70 mm is lying on one of its generators on HP such that its axis is parallel to VP. A vertical section plane inclined at 30° to VP passes through the midpoint of the axis. Draw the sectional front view, the end view, and the true shape of the section.

Hint   Since the section plane is perpendicular to HP, it is visualized as a straight line in the top view and is drawn as an inclined line to the reference line due to its inclination with VP.

PROCEDURE (Refer Fig. 10.31)

Step 1: Since the axis of the solid is parallel to both HP and VP, the views can be drawn by drawing the circle in the side view and projecting the front and the top view from it. Draw the circle inside view and divide into eight equal parts, which represent the generators. Draw the front and the top views and mark the generators in the respective views.

Step 2: Mark the SP as a straight line inclined at 30° to the reference line and passing through the midpoint on the axis, in the top view, and obtain the cutting points 1, …, 10, in the sequential order of the generators and the circular ends.

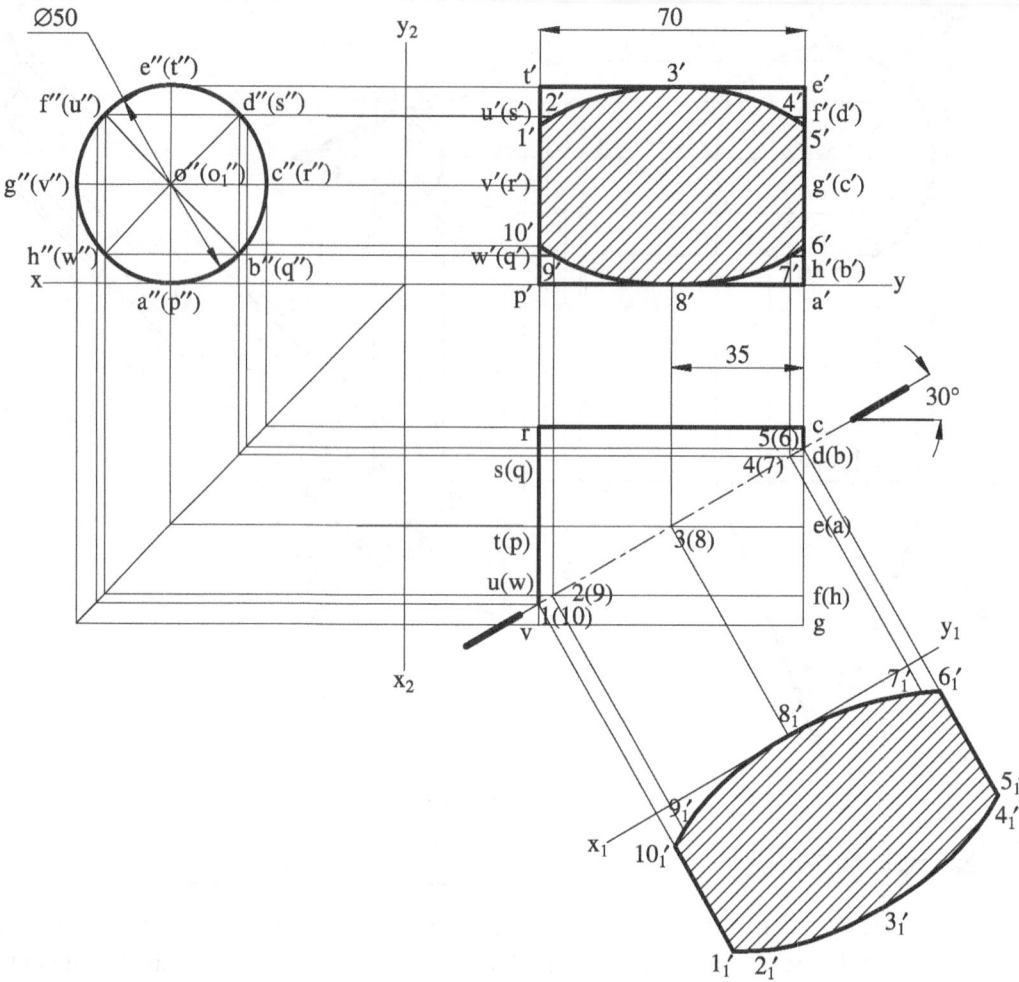

Fig. 10.31

Step 3: Project the cutting points to meet the corresponding locations in the front view and obtain the sectional front view as discussed in similar examples.

Step 4: Since the SP is inclined to VP, the sectional front view does not give the true shape of the cut portion. Therefore, draw a new reference line 'x_1y_1' parallel to the SP in the top view and draw projections to 'x_1y_1'. As done in the auxiliary projections of the solids, the distances of the projections of these cutting points can be measured, that is, $1'$, $2'$, ..., $10'$ from 'xy' line and mark along the new projectors from 'x_1y_1' line as $1_1'$, ..., $10_1'$. The resulting shape obtained by joining these points gives the true shape of the section. ▲

10.9 SECTION PLANE DETAILS FROM THE TRUE SHAPE OF SECTION

This section will deal with the examples where the true shape of section will be given, and it is required to find the location and the inclination of the section plane which produces that true shape. When the true shape of the section is given, the section plane orientation is decided such that the number of cutting points is equal to the number of corners that make the true shape, if the true shape is a polygon. For example, if the true shape section of a solid is a hexagon, the position of the line representing the section plane is decided such that it passes through six points in the concerned view of a solid. This will be illustrated in the following examples.

Example 10.27

A vertical square prism when cut by a section plane yields the true shape of section of size 60 mm × 40 mm. The section plane passes through one of the lateral faces of the solid at a height of 20 mm from the base. Determine the minimum size of the solid and inclination of the section plane.

Hint

1. Since the true shape of section is a rectangle having four corners, the line representing the section plane has to pass through four points in the solid.

2. Since the section plane has to pass through one of the lateral faces, two cutting points have to emerge from this face.

3. Two more remaining corners that make the true shape have to necessarily lie on the opposite parallel lateral face. The larger length of the true shape, that is, 60 mm has to be equal to the length of the section plane contained between the lateral faces.

4. Therefore, the other dimension of the true shape, that is, 40 mm has to correspond to the side of the square base of the prism.

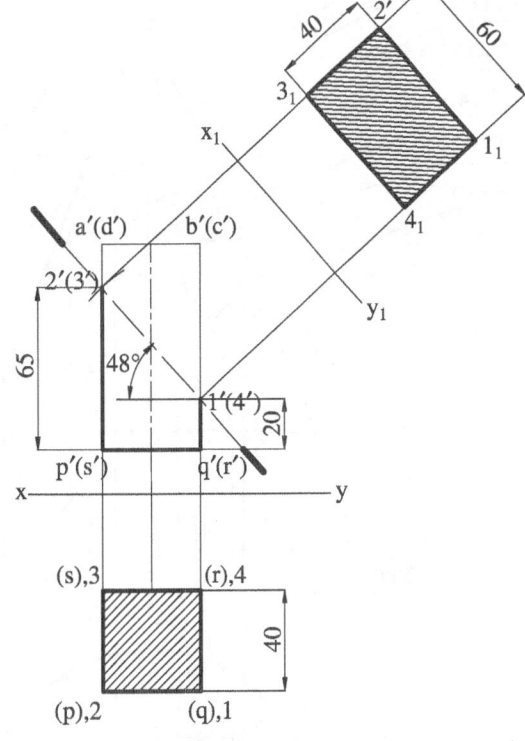

Fig. 10.32

PROCEDURE (Refer Fig. 10.32)

Example 10.28

A cone of base diameter 80 mm and axis 100 mm long is sectioned by a section plane whose vertical trace cuts one of the extreme generators at 35 mm below the apex. If one of the axes of the ellipse representing the true shape of the section measures 55 mm, determine the length of the other axis of the ellipse. Draw the sectional top view and true shape of the section obtained.

Hint

1. Since the true shape of section is an ellipse, the line representing the section plane has to pass through the extreme generators of the cone and its length corresponds to the length of one of the axes of the ellipse representing the true shape of the section.
2. The other axis lies at the midlength of this axis but perpendicular to it.

PROCEDURE (Refer Fig. 10.33)

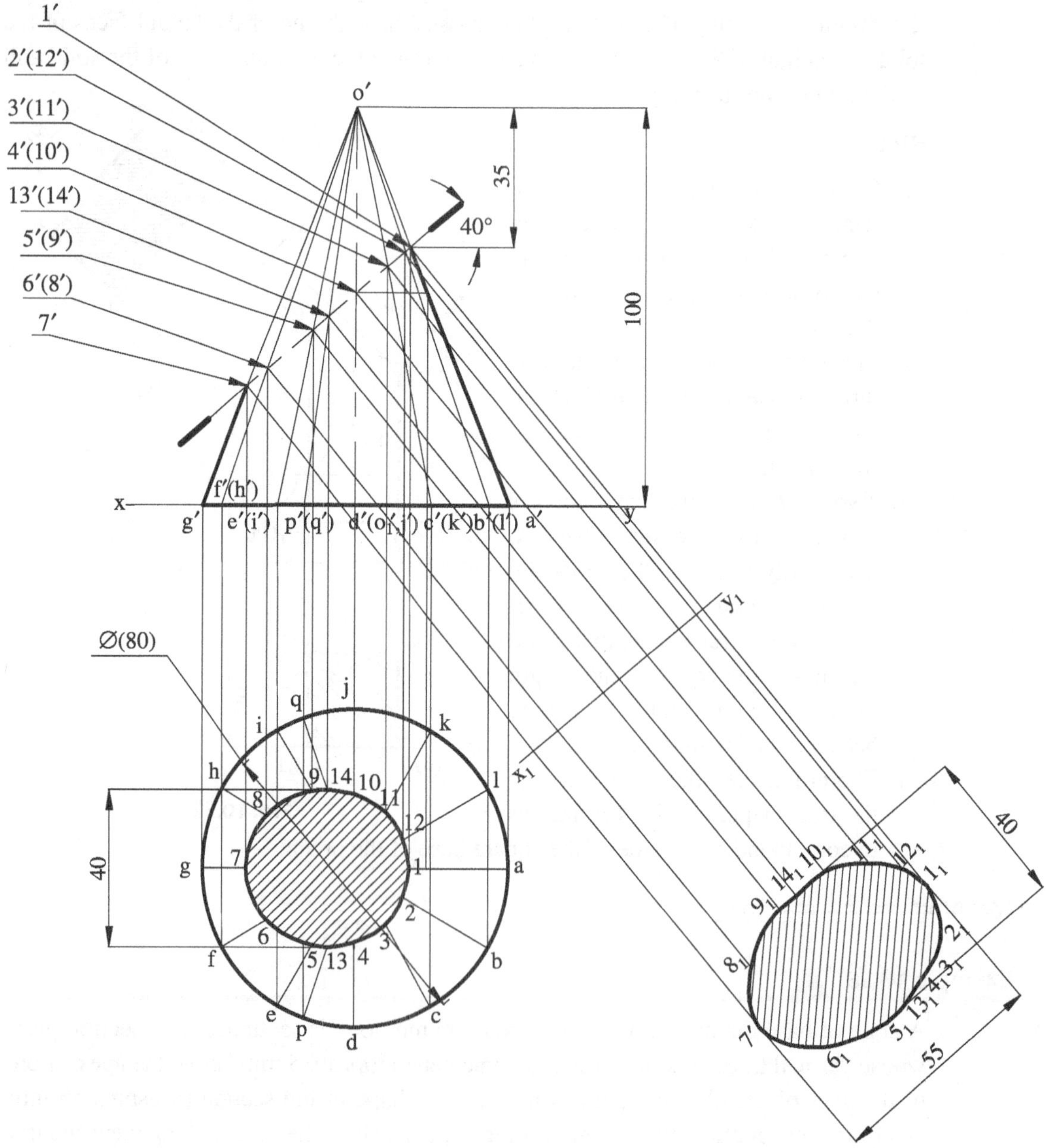

Fig. 10.33

Step 1: Draw the front and top views of the cone and mark the various generators a, ..., l.

Step 2: In concurrence with the hints mentioned earlier, locate 1′ on one of the extreme generators of the cone at a distance of 35 mm from the apex and draw an arc equal to the given length of the axis of the true shape, that is, 55 mm to cut the other extreme generator at 7′.

Step 3: Identify the cutting points 1′, ..., 12′ on the respective generators. Locate the centre of the section line contained within the cone and mark its cutting points 13′ and 14′. Obtain their corresponding base location points by joining them with the apex and extending them to cut the base in the front view. Name these points as p′ and q′ and obtain their top view points as p and q.

Step 4: Project all these cutting points including 13′ and 14′ and obtain the sectional top view as discussed in earlier problems.

Step 5: The distance between 13 and 14 is measured as 40 mm which gives the length of the other axis of the true shape in the top view.

Step 6: Complete the true shape of the section as discussed before. ▲

10.10 ENGINEERING APPLICATIONS

The concept of sectioning has been extensively used in domestic applications such as cutting of vegetables and fruits with knife or blades, which represents the section plane, to visualize defects inside if any. The cutting of solid in a particular orientation and joining with another at the cut interface is often found in the assembly of various components or parts. Engineering drawings that represent the orthographic views of buildings or machine components are often explained to show the complete sectioned assemblies of various inner parts. Sectional views are often used in machine drawings to give the complete cut sectional views. For symmetric objects, half sectional views are adopted to give the external appearance in one half and the internal cut details in the other. Castings, machine frames and beds, gear boxes, building plans with the presence of windows, and their supports are the engineering examples of sectional views.

RECAPITULATION

- The concept of sectioning an object enables to distinguish the details of the features located inside it.
- By assuming imaginary planes called section planes to pass through the required location of the inside details, the object is cut open to visualize them.

- The portion of the object that lies in between the observer and the section plane is assumed to be removed, and the remaining portion of the object along with the cut surface is projected to get the sectional views, as per the first-angle projection method.
- The section planes can be classified into two categories, namely,

➤ section plane parallel to one reference plane and perpendicular to the other

➤ section plane inclined to one reference plane and perpendicular to the other

• A section plane is visualized as a straight line on the reference plane to which it is perpendicular, and this line will appear parallel or inclined to the reference line, depending on its orientation as parallel or inclined to the other reference plane.

• The section plane is represented as a thin chain line with single dots, the ends being marked thick.

This also represents the trace of the section plane in that view.

• The projection of the sectioned solid is attempted only after completing the projection of the solids in the uncut stage.

• The cutting points of the section plane are identified and projected to the other view and a closed bounded area that represents the cut surface is identified and presented with a series of thin parallel lines inclined at 45° to the outline.

• If the section plane is not parallel to the reference plane, then this area does not represent the true cut surface. The true cut surface is identified by drawing a new reference line 'x_1y_1' parallel to the section plane and transferring the cutting points to this line as in auxiliary projection method. The area thus obtained is the true shape of the section and can be shown as per the conventions.

MULTIPLE-CHOICE QUESTIONS

10.1 The concept of sectioning an object enables
   (a) to identify the features located inside
   (b) to identify the features located outside
   (c) to erase the features located inside
   (d) to erase the features located outside

10.2 The trace of a section plane is
   (a) a curve   (c) a circle
   (b) a straight line   (d) a rectangle

10.3 The section plane is represented in drawing as
   (a) a continuous line
   (b) a continuous dashed line
   (c) a thick line with dots
   (d) a long, dashed, dotted, narrow line, the ends being marked thick

10.4 A section plane perpendicular to HP and parallel to VP will appear as
   (a) a straight line parallel to the reference line in the top view
   (b) a straight line parallel to the reference line in the front view
   (c) a straight line perpendicular to the reference line in the top view

   (d) a straight line perpendicular to the reference line in the front view

10.5 A section plane inclined at 30° to HP and perpendicular to VP will appear as
   (a) a straight line 30° to the reference line in the top view
   (b) a straight line 30° to the reference line in the front view
   (d) a straight line 60° to the reference line in the front view
   (e) a straight line 30° to the reference line in the side view

10.6 As per the first-angle projection method during sectioning the portion of the object in between the observer and the section plane is assumed to be
   (a) included for the sectional view
   (b) removed and the rest looked upon for the sectional view
   (c) unaffected for the sectional view
   (d) tilted at 90° to the observer

10.7 The representation of the cut surface is marked by drawing
    (a) a series of dotted lines inclined at 45°
    (b) a series of continuous lines inclined at 90°
    (c) a series of uniformly spaced thin lines inclined at 30°
    (d) a series of uniformly spaced thin lines inclined at 45°

10.8 When the section plane is inclined to a reference plane and perpendicular to the other, the projected cut surface represents
    (a) the true shape   (c) a reduced shape
    (b) an enlarged shape (d) enlarged twice

10.9 The sectioning of regular pyramid parallel to its base yields a cut surface
    (a) as equal to that of the base

    (b) as a reduced polygon
    (c) as a reduced circle
    (d) as an enlarged polygon

10.10 When the section plane is inclined to a reference plane is and perpendicular to the other, the true shape of the section is obtained by
    (a) projecting the cut surface perpendicular to the section plane
    (b) projecting the cut surface on an auxiliary reference line perpendicular to the section plane
    (c) projecting the cut surface on an auxiliary reference line parallel to the section plane
    (d) projecting the cut surface from the side view

WORK PRACTICE LEVEL – 1

Solids in Simple Positions and Cutting Plane Parallel or Inclined to the Reference Plane

10.1 A cube of 50 mm long edge rests with one of its faces on HP such that one of its vertical faces is inclined at 40° to VP. A vertical section plane inclined at 50° to VP passes through the axis of the cube. Draw sectional front view and true shape of the section.

10.2 A hollow square prism of base 60 mm sides at the outside, axis 80 mm, and thickness 15 mm is resting on its base on HP with its one of the vertical faces inclined at 40° to VP. A vertical section plane perpendicular to VP and inclined at 40° to HP passes through the axis at a point 20 mm from its top end and cuts the prism. Draw the sectional top view and true shape of the section.

10.3 A pentagonal pyramid of base 30 mm and axis 65 mm has its base on HP with one of the edges of the base perpendicular to VP. It is cut by a section plane making 40° with HP perpendicular to VP and meeting the axis 35 mm below the vertex. Draw the sectional top view and true shape of the section.

10.4 A square pyramid of base 30 mm and axis 50 mm rests on its base on HP and perpendicular to

VP with one of the base edges of the pyramid inclined to VP by 30°. A sectional plane inclined at 50° to HP passes through a point on the axis and 20 mm from the base. Draw the sectional top view and true shape of the section.

10.5 A right hexagonal pyramid of base 40 mm and axis 75 mm rests with its base on HP and one of base edges inclined at 30° to VP. An auxiliary vertical plane inclined at 50° to VP cuts the pyramid and contains its axis. Draw the sectional front view and true shape of the section.

10.6 A hexagonal prism of base 35 mm side and axis 70 mm is resting on HP on one of its polygonal ends with two vertical faces being parallel to VP. It is cut by a vertical sectional plane inclined at 45° to VP and is 10 mm away from the axis and in front of it. Draw the sectional front view and true shape of the section.

10.7 A cylinder of base diameter 60 mm and axis 75 mm is resting on HP on its base. It is cut by a section plane perpendicular to VP and inclined at 60° to HP and passes through the axis at a distance 50 mm above the base of the cylinder. Draw the sectional top view and true shape of the section.

10.8 A right circular cone diameter of base 60 mm and axis 80 mm is resting on HP on its base. It is cut by a sectional plane perpendicular to VP and inclined at 50° to HP and passes through the axis at a distance 30 mm below the apex. Draw the sectional top view and true shape of the section.

10.9 A hexagonal pyramid of base 30 mm sides and axis 70 mm long has its hexagonal end on VP with the two sides of the hexagon parallel to HP. It is cut by a vertical section plane inclined at 30° to VP and passing through the midpoint of

the axis. Draw the sectional front view and true shape of the section.

10.10 A cone of 60 mm base diameter and axis 80 mm long has its circular end on the right profile plane. A vertical section plane inclined at 30° to VP passes through a point on the axis, 20 mm from the apex. Draw the sectional front view and true shape of the section if the apex portion is removed.

WORK PRACTICE LEVEL – 2

Solids in Tilted Positions and Cutting Plane Parallel or Inclined to the Reference Plane

10.1 A cube of 35 mm sides rests with an edge on HP. This edge is perpendicular to VP and one of the square faces containing this edge is inclined at 40° to HP. A section plane perpendicular to VP and inclined at 50° to HP bisects the axis of the cube. Draw the front view, sectional top view, and the side view showing the section.

10.2 A cube of 35 mm sides rests with one of its corners on HP such that the vertical edge through this corner makes 40° with the HP. The vertical plane containing this lateral edge and the axis is parallel to the VP. A horizontal section plane passing through the midpoint of the axis cuts the solid into two halves. Draw the front view and the sectional top view of the remaining portion of the cube.

10.3 A square prism of base sides 55 mm and axis 90 mm rests with one of its base edges on HP and is tilted about the base edge such that the axis makes 50° to HP and parallel to VP. A section plane perpendicular to HP and inclined at 40° to VP cuts the top view of the axis of the prism at a distance of 30 mm from its top end. Draw the sectional front view, the top view, and the true shape of the section.

10.4 A pentagonal prism of 25 mm side of base and axis 70 mm long rests on a base edge on HP with the rectangular face containing this base edge perpendicular to VP and inclined at 60° to HP. It is cut by a horizontal section plane whose VT passes through a point on the axis and located

40 mm above HP. Draw the sectional top view and the true shape of the section.

10.5 A hexagonal prism of 30 mm sides and axis 80 mm long has a coaxial circular hole of 35 mm diameter. The prism is resting on HP on one of its rectangular faces and axis inclined at 30° to VP and is cut by a vertical section plane, parallel to VP, passing through the midpoint of the axis. Draw the sectional front view and the true shape of the section.

10.6 A right square pyramid of 40 mm side square base and axis 100 mm long has one of its longer edges on VP such that the sides of the base are equally inclined to HP. A vertical section plane, inclined at 30° to VP passes through a point on the top view of the axis at a distance of 20 mm from the vertex. Draw the sectional front view and the true shape of the section.

10.7 A right regular hexagonal pyramid side of base 30 mm and height 80 mm is resting on one of its slant faces on HP with the vertical plane containing the axis inclined at 60° to VP and the apex nearer to the observer. It is cut by a vertical section plane parallel to VP, bisecting the axis. Draw the sectional front view and the true shape of the cut surface.

10.8 A pentagonal pyramid of 30 mm base sides and axis 70 mm long is having one of its slant faces on VP and the axis is 60 mm above HP. A vertical section plane inclined at 45° to VP bisects the axis. Draw the sectional front view and top view. Find the true shape of the section.

10.9 A right circular cone of base diameter 65 mm and the axis 80 mm lies with its axis inclined at 40° to HP and parallel to VP. The cone is cut by an auxiliary vertical plane 40° to VP and passes through the midpoint of the axis of the cone. Draw the sectional front view, the top view, and the true shape of the section.

10.10 A cylinder of base 50 mm diameter and axis 85 mm long has a square hole of 25 mm sides cut through it. The axes of the two shapes coincide with each other. The solid is tilted such that the cylindrical end makes 30° with VP and axis parallel to HP. The adjacent faces of the square hole are equally inclined to HP in this position. A vertical section plane also inclined at 30° to VP but sloping opposite to the cylindrical ends cuts the solid into two halves. Draw the sectional front view and true shape of the section.

10.11 A frustum of a cone, top diameter 70 mm, base diameter 90 mm, and height 90 mm has a coaxial through hole of 45 mm diameter and it rests on HP on its base. It is cut by a section plane which is inclined at 50° to HP and passes through the midpoint of the axis of the frustum. Draw the sectional top view and true shape of section.

10.12 A frustum of a pentagonal pyramid of base 40 mm sides and height 65 mm rests on HP on one of its trapezoidal faces such that its axis is parallel to VP. It is cut by a section plane perpendicular to VP and inclined at 45° to the ground, and passing through the mid-point of the axis. Draw the sectional top view and true shape of the section. The imaginary position of the apex is at 80 mm from the base in the upright position.

Problems on Typical Examples

10.1 A cube of 40 mm sides is cut by a section plane perpendicular to VP so that the true shape of the section is found to be an equilateral triangle of sides of maximum length. Draw the sectional top view and the true shape of section. Find the inclination of the section plane with HP and find the length the sides of the equilateral triangle.

10.2 A rectangular prism of height 90 mm and cross section 55 mm × 25 mm is standing with its base on HP. It is cut by a plane in such a way that

the true shape of section is a square of sides of maximum dimension. Find the inclination of the cutting plane with the HP. Draw the sectional top view and the true shape of the section.

10.3 A cone diameter of base 70 mm and axis 80 mm long is resting on its base on HP. It is cut by an AIP such that the true shape of section is an isosceles triangle having 45 mm base. Draw the front view, sectional top view, and sectional side view. In addition, find the true length of the sides of isosceles triangle in the true shape of section.

10.4 A cone diameter of base 60 mm and axis 70 mm long resting on its base on HP. It is cut by a section plane perpendicular to both the reference planes in such a way that the true shape of section is rectangular hyperbola having 45 mm base. Draw the front view, top view, and sectional side view.

10.5 A vertical cylinder 60 mm diameter is cut by an AVP making 40° to VP in such a way that the true shape of the section is a rectangle of 45 mm and 90 mm sides. Draw the projections and true shape of the section.

Problems on Engineering Applications

10.1 A plastic bucket of height 450 mm, top and bottom diameters of 350 mm and 250 mm, respectively, is filled with water to a certain height. The water from the bucket is about to come out when the bucket gets tilted to an angle of 40° with the HP, about a point on the circumference of the base. Draw the front view, top view with the section showing the water surface, and true shape of the free surface of the water which is about to come out from the bucket.

10.2 A wooden waste paper basket is in the form of the frustum a hexagonal pyramid of base sides 250 mm and top sides 500 mm. The height of the basket is 750 mm. It has a uniform wall of thickness 20 mm. It is kept leaning to the floor such that one of its slant faces is inclined at 30° to the ground and parallel to the wall. It is cut by a horizontal section plane passing through the lower edge of the top end. Draw the sectional top view and front view of the basket.

10.3   A crucible is in the form of truncated cone of top diameter 60 mm and base diameter 40 mm and height of 50 mm. It is cut by a section plane when the solid lies inverted on HP such that the VT is inclined at 30° to the reference line and passes through the midpoint of the axis of the crucible. Draw the sectional top view and the true shape of the section.

Answers for Multiple-choice Questions

10.1 (a)	10.2 (b)	10.3 (d)	10.4 (a)	10.5 (b)	10.6 (b)
10.7 (d)	10.8 (c)	10.9 (b)	10.10 (c)		

Intersection of Surfaces

11

OBJECTIVES

This chapter will help the reader to understand the following:

- Basic principles of intersection of two solids of same and different natures and their importance in engineering applications
- Different orientations of the intersecting solids and the influence of their views at intersecting junctions
- Building up the intricate dimensional details that occur in the interface of two solids
- Evolving the development of surfaces of the individual solids of intersection

11.1 INTRODUCTION

In Chapter 9, the procedure for obtaining different views of a solid was discussed. It was also shown how these views get altered when such a solid was fully or partially cut by a section plane located in different orientations as discussed in Chapter 10. Sectioning of a solid has helped us to understand the changes that are occurring at the surface of the solid, when it is cut by a planar surface. In practice, such a requirement happens only to connect or assemble another solid at the cut surface. The assembly of one object or solid with another is vital to achieve certain functional requirements. The common examples can be found in domestic plumbing and sewerage practice, where the pipe lines are connected with transition pieces, such as the couplings, elbows, tees, and reducers, to divert or alter the flow of water. In industrial applications, the water lines in the boilers use such joints. The gas pipe lines, ducts, smoking chutes, exhaust, and ventilating devices are the engineering applications that use the assembly of two solid shapes at their interfaces. The connections between a cylindrical container and the tap, the arrangement of cylindrical and conical shapes in conventional bottles are some of the common examples. In certain applications, the interconnections are purposely done with a view of enhancing the appearance, look, appeal or to improve the aesthetics.

These inventions would have become possible by observing the nature, where the trees with a trunk portion branch out in different directions with uniform or non-uniform-shaped surfaces. Even the human body depicts such knowledge in its assembly of different shapes such as the neck connecting the trunk and the head, the interfaces of the legs and arms and other organs. The human anatomy will reveal the intricate interfaces that are involved in joining of various organs as observed in the junctions of the heart and the arteries and other vital veins.

All the aforementioned examples clearly highlight the practical importance of joining different solid shapes to achieve a purpose or a function out of it. The intersecting surfaces are to be realized with exactness and accuracy to make the intended function effective. This chapter deals with the geometry of two interesting surfaces and the methods of achieving the intersection curves at their interfaces.

11.2 INTERSECTION POINT

An intersection point is a point at which an edge or an imaginary line parallel to an edge or a generator (in the case of cylindrical/conical surfaces) lying on one solid surface meets that of the other solid. A series of intersection points will emerge depending on the edges/generators present or assumed in the surfaces of the two solids.

11.3 LINE OR CURVE OF INTERSECTION

When the surface of one solid meets that of the other, the lines or the curves along which the two surfaces meet each other is known as the 'curve(s) of intersection'. This is obtained by joining the intersection points in proper sequence. The curve of intersection is common to both the surfaces and is always in a closed form. When the two intersecting surfaces are plane surfaces, the intersection curve will be a set of lines. When one or both the intersecting surfaces are curved, the intersection will be a set of curves. When one solid penetrates into the other completely and comes out, the intersection curves will appear at the entry and at the exit surfaces.

11.4 TYPES OF INTERSECTION

The following classifications can be made depending on the position of the axes of the two intersecting solids:

1. Solids whose axes are perpendicular to each other and intersecting or non-intersecting
2. Solids whose axes are inclined to each other and intersecting or non-intersecting
3. Solids whose axes are parallel to each other and coinciding or non-coinciding

The nature of the solids that are intersecting with each other can be prismatic, pyramidal, cylindrical, conical, or spherical or their combinations.

The types of surfaces that are intersecting can be composed of plane surfaces as in the cases of prisms, pyramids, or single- or double-curved surfaces in the cases of cylinders, cones, spheres or their combinations too.

11.5 BASIC PRINCIPLES OF OBTAINING CURVES OF INTERSECTION

The curves of intersection can be procedurally obtained by retaining one solid usually with its axis vertical, and the axis of the other solid can be oriented in different positions to ensure the possibilities mentioned in Section 11.4. For convenience, the vertical solid is identified as the first solid, and the solid that intersects with the vertical solid can be named as the intersecting solid or the second solid.

11.5.1 Intersection Cases where Vertical Solids are Prismatic or Cylindrical

It can be noted that when the prismatic or the cylindrical shaped solids are retained with their axes vertical, the polygon or the circle representing their cross section will appear in the top view and does not change with respect to the height of the solid. The intersecting solid meets the vertical solid at the same points in the top view irrespective of its height. This has an inherent advantage to locate the intersecting points in the front view by simply projecting the intersecting points identified in the top view. Based on the nature of the intersecting solid, the procedures are outlined as follows:

Case 1   If the intersecting solid is also of prismatic or cylindrical nature and its axis is horizontal, then the side view or the profile view represents its cross section and that contained within that of the vertical solid represents the curve of intersection in that view. Suitable points can be chosen at the boundary of the intersecting solid contained within the vertical solid in the side view and can be transferred to the top view and the corresponding intersection points of the respective edges/generators can be obtained in the front view. However, if the axis of the intersecting solid is inclined, an auxiliary top view (in the direction of the axis) can be drawn for the intersecting solid and the intersecting points can be obtained similar to the horizontal case. The intersection curves will appear in the front view distinctly on its right and left halves. If the axis of the interesting solid is offset and the solid bulges out, the intersection curves present in two halves will try to join in the front view.

Case 2   If the intersecting solid is pyramidal or conical nature and its axis is horizontal, then the intersection points can be identified as discussed in Case 1 or by noting the edges/surface lines/generators of the intersecting solid in the top view. The intersection curves can be obtained in the front view on the corresponding lateral edges or the generator lines. However, the left and the right half curves of the front view will vary since the surface lines of the horizontal solid are not parallel but converge to the vertex. The intersection curves in the inclined and offset cases can be obtained in similar lines of the previous case.

11.5.2 Intersection Cases where Vertical Solids are Pyramidal or Conical

It can be noted that when the pyramidal or conical shaped solids are retained with their axes vertical, the top view of the intersecting points depend on the height at which each edge/generator/surface line meets the vertical solid. These lead to locating the intersecting points in the top view and obtain an intersection curve first, and hence, locate the corresponding intersection points in the front view to obtain the front view intersection curve. On the basis of the nature of the intersecting solid, the procedures are outlined as follows:

Case 1 If the intersecting solid is of prismatic or cylindrical nature and its axis is horizontal, suitable points can be chosen at the boundary of the intersecting solid contained within the vertical solid in the side view or by noting the meeting points of the edges/surface lines/generators of the vertical solid with the intersecting solid in the side view. The intersection points of the horizontal section planes passing through these points and the vertical solid are obtained in the top view and then projected to yield the intersection curves in the front view. When the axis of the intersecting solid is inclined, an auxiliary top view is drawn to locate the intersecting points and the corresponding curves of intersection.

Case 2 If the intersecting solid is also pyramidal or conical as that of the vertical solid, the procedures outlined in Sections 11.5.1 and 11.5.2 (Case 1) can be combined suitably to get the curves of intersection.

11.6 METHODS OF OBTAINING INTERSECTION CURVES

11.6.1 Line/Edge or Generator Method

Identification of the meeting points of the various edges of both the solids with the lateral surfaces and joining them in proper sequence to obtain the curves of intersection in the three views is done based on this method. For solids that do not have specific edges, number of lines parallel to the generators can be drawn in the region of intersection and their meeting points on the surface of the other solid can be obtained to get the curves of intersection. This method will be convenient for the problems on interpenetration of prisms, cylinders, etc.

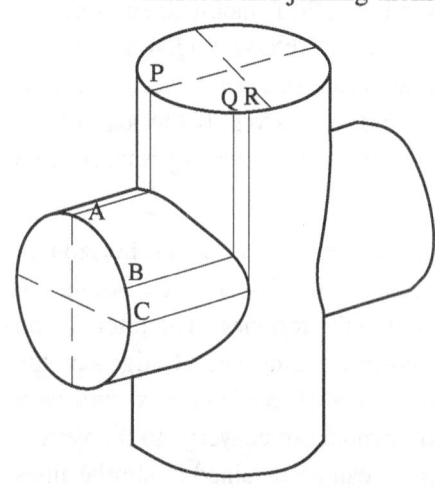

Fig. 11.1 Axes of two cylinders perpendicular to each other

Figure 11.1 shows two intersecting cylinders. Since cylinders do not have specific lateral edges on their surfaces, generators are assumed. The horizontal cylinder is marked with the generators A, B, and C, whereas generators from P, Q, and R mark their corresponding ones on the vertical cylinder. Their common intersecting points are identified as 1, 2, 3 and the sequential curve through them represents the intersection curve.

11.6.2 Cutting Plane Method

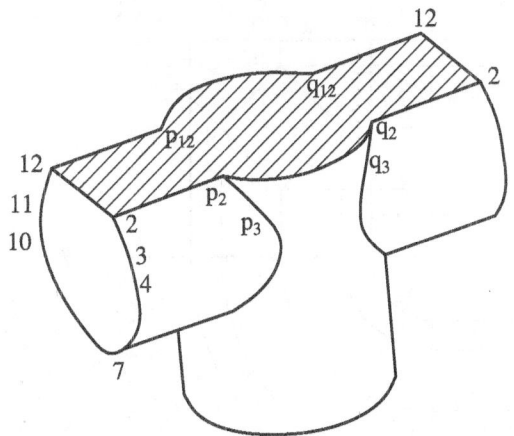

Fig. 11.2   Axes of two cylinders perpendicular to each other and sectioned by horizontal plane

In this method, both the intersecting solids are cut by a series of imaginary cutting planes and the meeting points of the corresponding cut surfaces obtained with each section plane are identified in different views and are joined in proper sequence to get the curves of intersection. The imaginary cutting planes or the section planes can be horizontal, vertical, or inclined at any angle. This method will be convenient for the problems on interpenetration of pyramids, cones, cylinders, spheres, etc.

Figure 11.2 shows two intersecting cylinders. The horizontal cylinder end surface is divided at angular spacing as shown by points 1 to 12. A horizontal section plane passing through one set of such points (2 and 12) yields a rectangular surface as shown by the points 2–12–12–2. The intersection of this surface with the vertical solid yields the corresponding points p_2, p_{12}, q_{12}, and p_2. Similarly, the common points p_3 to p_{12} and q_3 to q_{12} are obtained and joined sequentially to yield the curves of intersection.

11.7 ILLUSTRATIVE EXAMPLES

11.7.1 Intersection of Prismatic Solids

Example 11.1

Intersection of Two Square Prisms with Axes Perpendicular to Each Other

A square prism of base edges 60 mm sides and axis 100 mm long is standing on its base with its faces equally inclined to the VP. It is completely penetrated by a horizontal square prism whose axis is perpendicular and intersecting that of the vertical solid at a height of 60 mm. Draw the curves of intersection of the solids at their interfaces, if the faces of the horizontal prism are equally inclined to the HP. The ends of the horizontal solid are of 40 mm square and axis is 100 mm long.

PROCEDURE   (Refer Fig. 11.3)

Draw the views of the solids

Step 1: Draw the three views of the vertical solid by placing the square in the top view with its sides equally inclined at 45° to the reference line.

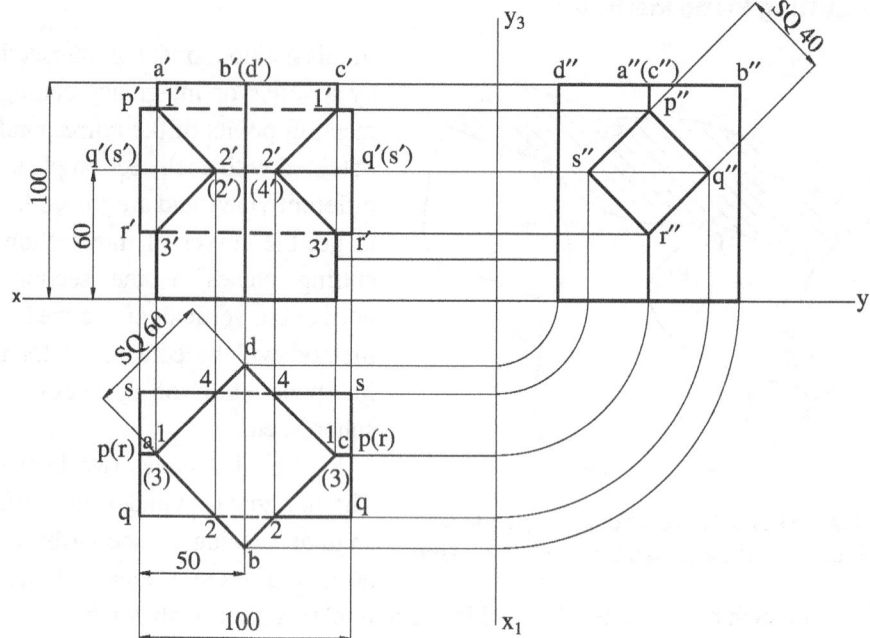

Fig. 11.3

Step 2: Draw the three views of the intersecting (horizontal) solid by placing the square in the end view with its sides equally inclined (at 45°) to the reference line such that its centre is located on axis of the vertical solid at the required height.

Obtain the intersection points

Step 1: The intersection points of the horizontal solid in the side view correspond to that of the lateral edges through p'', ..., s'', while in the top view they are obtained as 1 to 4 as the meeting points of the various edges such as pp and qq of the horizontal solid with the vertical prism.

Step 2: The edges of the vertical solid through a'' and c'' also meet the horizontal solid at the points that coincide with 1 and 3 and hence are not marked separately.

Step 3: All the intersection points in the top view are projected to meet the corresponding edges in the front view.

Draw the intersection curves

Step 1: The curves of intersection are realized by joining the intersection points obtained in the front view in a sequential manner.

Step 2: The line $1'2'3'$ indicates the intersection curve in the front face, while the line $3'4'1'$ indicates the intersection curve in the rear face.

Step 3: As the points $2'$ and $4'$ coincide, these curves also coincide with each other in the front view.

These can be obtained on both the left and the right halves as the solid is penetrating fully. ▲

Example 11.2

Intersection of Two Square Prisms with Unequal Face Inclinations and
Axes Perpendicular

A square prism of base edges 50 mm sides and axis 100 mm long is standing on its base with one of its faces inclined at 30° to the VP. It is completely penetrated by a horizontal square prism whose axis is perpendicular and intersecting that of the vertical solid at its mid-height. Draw the curves of intersection of the solids at their interfaces, if the faces of the horizontal prism are equally inclined to the HP. The ends of the horizontal solid are of 40 mm square and axis is 150 mm long.

PROCEDURE (Refer Fig. 11.4)

Draw the views of the solids

Step 1: Draw the three views of the vertical solid by placing the square in the top view with one side inclined at 30° to the reference line.

Step 2: Draw the three views of the intersecting (horizontal) solid by placing the square in the end view with its sides equally inclined to the reference line as in Example 11.1.

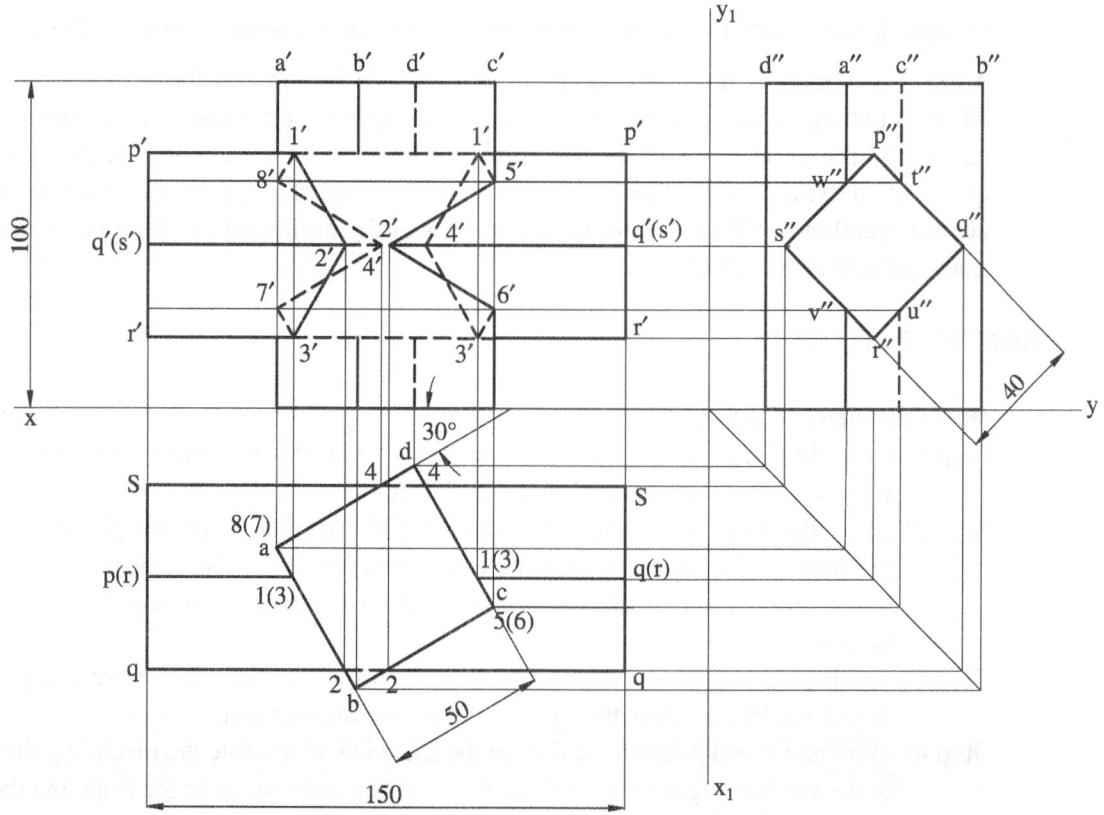

Fig. 11.4

Obtain the intersection points

Step 1: The intersection points 1 to 4 are obtained as explained in Example 11.1. In addition, the edges of the vertical solid through a″ and c″ also meet the horizontal solid at the points 5, 6, 7, and 8 in the top view.

Step 2: All the intersection points in the top view are projected to meet the corresponding edges in the front view.

Draw the intersection curves

Step 1: The curves of intersection are realized by joining the intersection points obtained in the front view in a sequential manner.

Step 2: The line 5′2′6′ indicates the intersection curve in the front face, while the line 6′3′4′1′5′ indicates the intersection curve in the rear face on the left half.

Step 3: Similarly, the line 1′2′3′ indicates the intersection curve in the front face, while the line 3′7′4′8′1′ indicates the intersection curve in the rear face on the left half.

Step 4: As the points 2′ and 4′ do not coincide, the front and the rear face curves appear separately in the front view. The curves in the front face are visible, while that at the rear are invisible and hence marked dotted. ▲

Example 11.3 ▲

Intersection of Two Different Shaped Prisms with Axes Perpendicular to Each Other

A hexagonal prism of base edges 30 mm sides and axis 75 mm long has its base on the HP with two of its faces parallel to the VP. It is completely penetrated by a horizontal pentagonal prism. The axes of the two solids are intersecting at right angles. Draw the curves of intersection of the solids at their interfaces, if one of the faces of the horizontal prism is parallel to VP and nearer to it. The ends of the horizontal solid are of 25 mm sides and axis is 90 mm long.

PROCEDURE (Refer Fig. 11.5)

Draw the views of the solids

Step 1: Draw the three views of the vertical solid by placing the hexagon in the top view with two of its sides parallel to the reference line.

Step 2: Draw the three views of the intersecting (horizontal) solid by placing the pentagon in the end view with one of its sides parallel to the vertical reference line and the centre located on axis of the vertical solid at any convenient height.

Step 3: The intersection points of the horizontal solid in the side view correspond to that of the lateral edges through the corners of the polygon.

Step 4: Mark additional points u″ and v″ in the side view to indicate the meeting points of the vertical edges through a″ and d″. Project these points in the front and the top view to realize the surface lines through them.

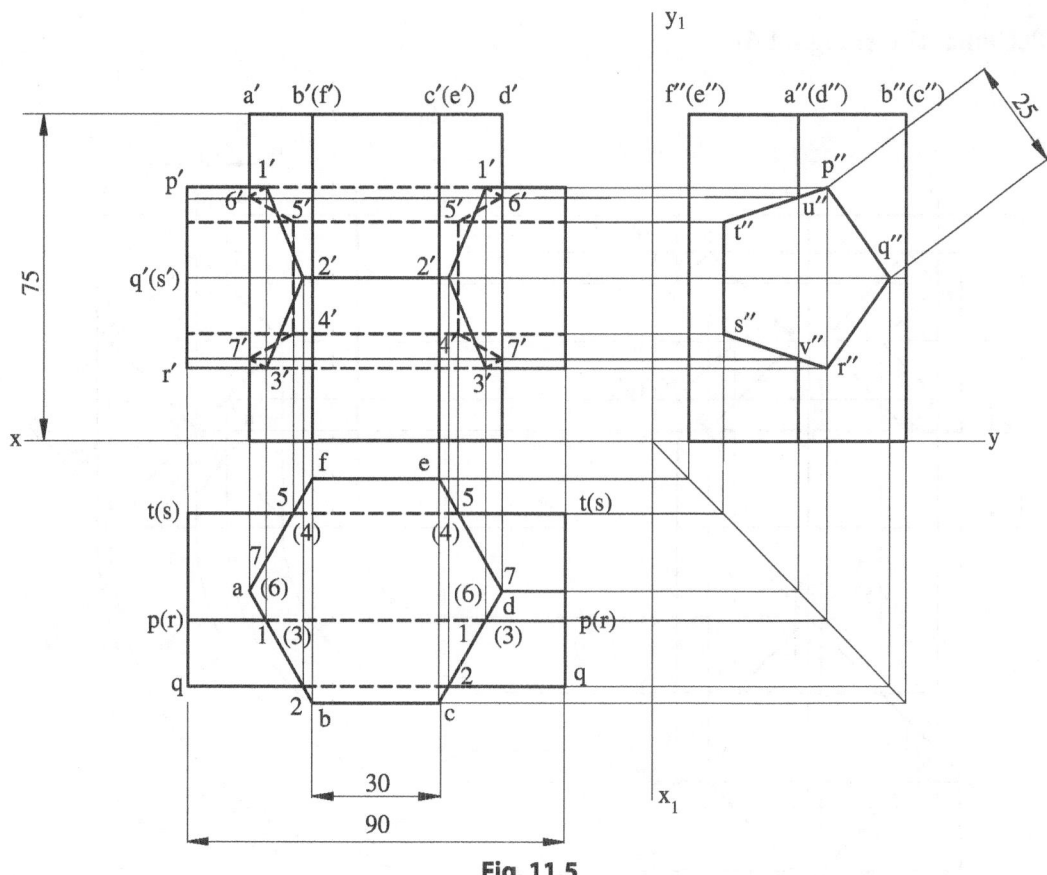

Fig. 11.5

Obtain the intersection points

Step 1: The intersection points 1 to 5 corresponding to the edges in the top view are obtained as explained in Example 11.1.

Step 2: Obtain additional intersection points 6 and 7, pertaining to the surface lines through u and v.

Step 3: All these points are projected to meet the corresponding edges/surface lines in the front view.

Draw the intersection curves

Step 1: Join the points 1'2'3' to indicate the intersection curve in the front face.

Step 2: Similarly, join the points 3'7'4'5'6'1' by a dotted line to indicate the intersection curve in the rear face.

Example 11.4

Intersection of Two Square Prisms with Axes Perpendicular to Each Other but Offset

The dimensions and the position of the solids are the same as in Example 11.1, but the axis of horizontal solid is 10 mm in front of that of the vertical solid. Draw the curves of intersection.

PROCEDURE (Refer Fig. 11.6)

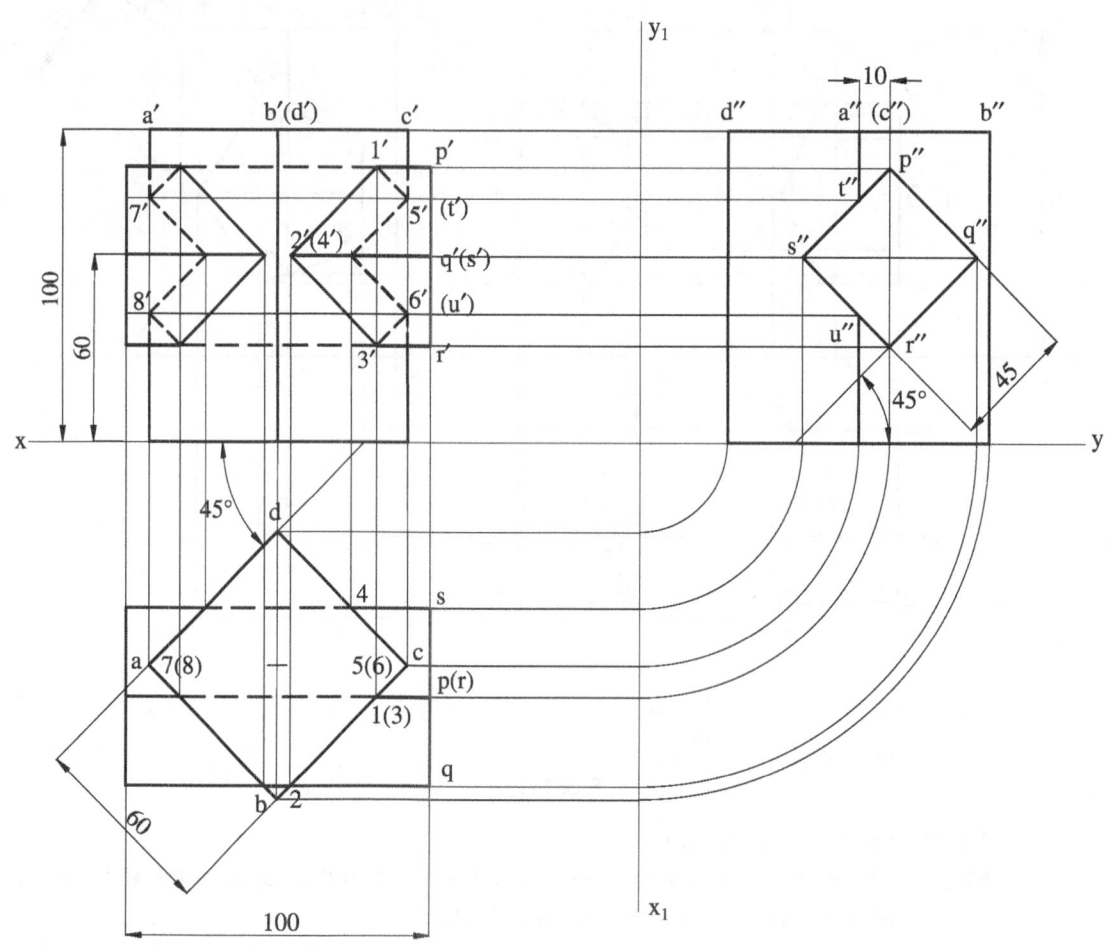

Fig. 11.6

Draw the views of the solids

Step 1: Draw the three views of the vertical solid as in Example 11.1.

Step 2: Draw the three views of the intersecting (horizontal) solid as in Example 11.1, by placing its centre 10 mm from the axis of the vertical solid in the side view. Mark additional points t″ and u″ in the side view to indicate the meeting points of the vertical edges through a″ and c″. Project these points in the front and the top view to realize the surface lines through them.

Obtain the intersection points

Step 1: The intersection points 1 to 4 are obtained as explained in Example 11.1.

Step 2: In addition, the edges of the vertical solid through a″ and c″ also meet the horizontal solid at the points 5 and 6, and 7 and 8 in the top view.

Step 3: All these points are projected to meet the corresponding edges/surface lines in the front view.

Draw the intersection curves

Step 1: The curves of intersection are realized by joining the intersection points obtained in the front view in a sequential manner.

Step 2: The line 1'2'3' indicates the intersection curve in the front face, while the line 3'6'4'5'1' indicates the intersection curve in the rear face. As the points 2' and 4' do not coincide, the front and the rear face curves appear separately in the front view.

Step 3: It can be noted that the points 5' and 6', and 7' and 8' located on the outer edges of the vertical solid indicate that intersection curve turns at these locations to obtain a closed loop. ▲

11.7.2 Intersection of Cylindrical Solids

> **NOTE** *The procedure of obtaining intersection curves of cylindrical solids is similar to that of the prisms. While prisms intersect at their definite edges or surface lines, the intersection of cylinders is obtained by the mutual intersection of various generators created on their surfaces. The intersection points are joined by a curve.*

Example 11.5 ▲

 Intersection of Two Cylinders with Axes Perpendicular to Each Other

A cylinder of base diameter 50 mm and axis 75 mm long is standing on its base on the HP. It is completely penetrated by a horizontal cylinder of 45 mm diameter and axis 80 mm long, such that their axes intersect at right angles and at 40 mm above the base. Draw the curves of intersection of the solids at their interfaces.

PROCEDURE (Refer Fig. 11.7)

Draw the views of the solids

Step 1: Draw the three views of the vertical solid by placing the circle in the top view.

Step 2: Draw the three views of the intersecting (horizontal) solid by placing the circle in the end view, such that its centre is located on axis of the vertical solid as the required height. Divide the circle into a set of equal parts (at p″, q″, …, w″) and obtain generators in the top and front view.

Obtain the intersection points

Step 1: The intersection points are obtained as the meeting points of the various generators such as pp and qq of the horizontal solid with the vertical solid in the top view. These points are respectively marked as 1 to 8 and are realized on both

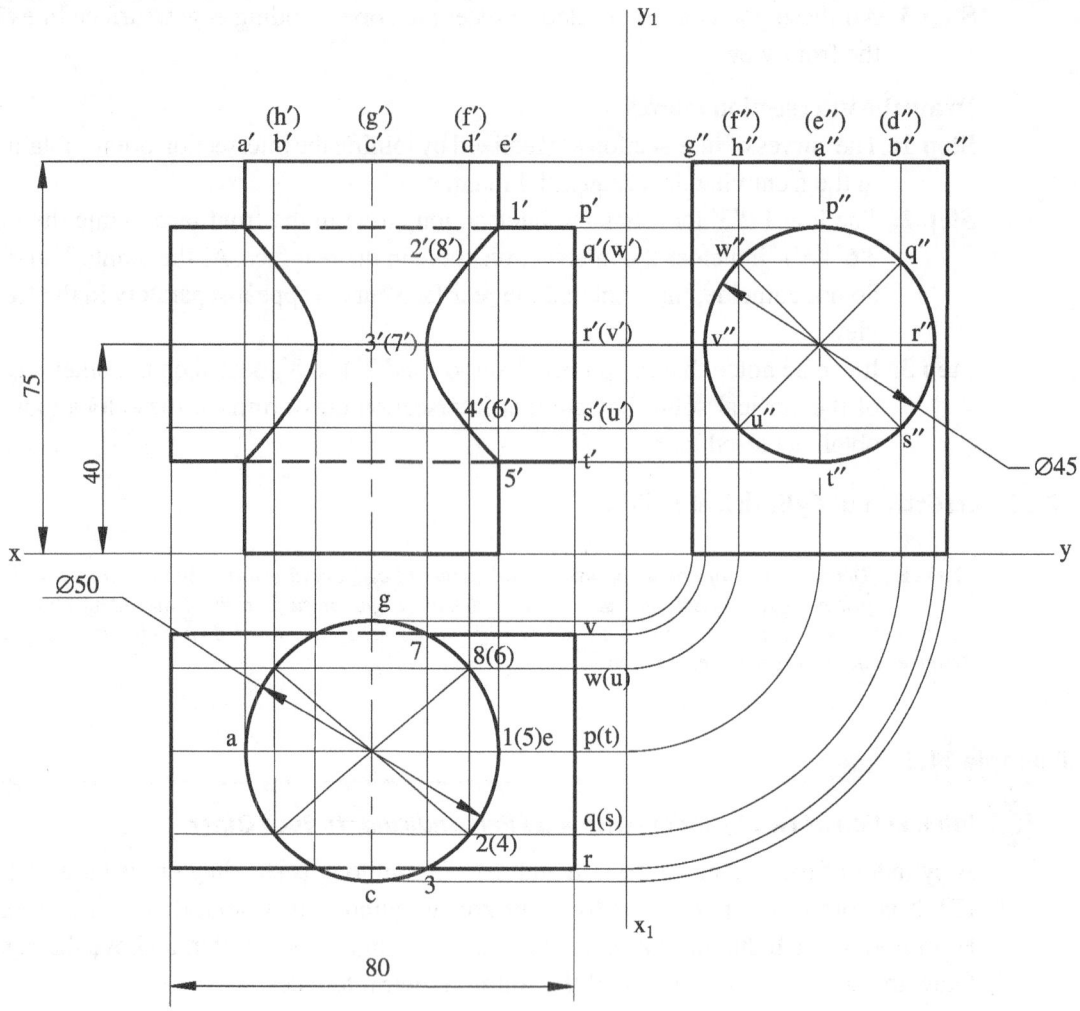

Fig. 11.7

the left and the right halves in the top view as the horizontal solid is completely penetrating the vertical solid.

Step 2: Obtain the corresponding points in the front view.

Draw the intersection curves

Step 1: The curves of intersection are realized by joining the intersection points obtained in the front view by means of a curve in a sequential manner.

Step 2: Join 1'2'3'4'5' by means of a curve. Note that the curve joining the points 5'6'7'8'1' denotes the intersection curve at the rear face and does not appear separately in the front view, as the respective points coincide.

Step 3: Draw the curves on both the left and the right halves. ▲

11.7.3 Intersection of Prisms and Cylinders

Example 11.6

Intersection of a Vertical Pentagonal Prism and a Cylinder when their Axes are Intersecting and Perpendicular to Each Other

A vertical pentagonal prism of 60 mm base edges and axis 180 mm long is completely penetrated by a horizontal cylinder of 60 mm diameter and 225 mm long. One of the rectangular faces of the prism is parallel to VP and nearer to it. Draw the curves of intersection of the two solids.

PROCEDURE (Refer Fig. 11.8)

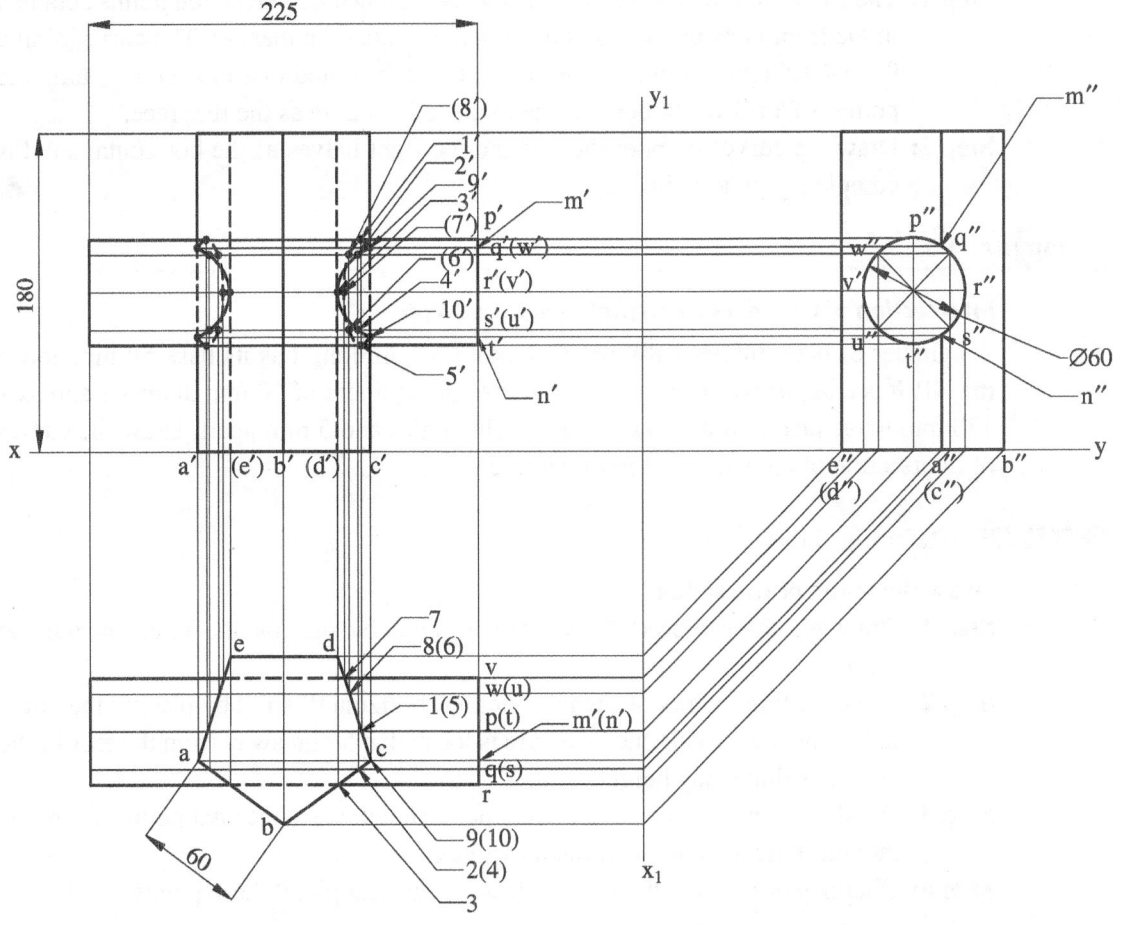

Fig. 11.8

Draw the views of the solids

Step 1: Draw the three views of the vertical pentagonal prism with one face (passing through ed) parallel to VP as shown in the top view.

Step 2: Draw the three views of the intersecting (horizontal) solid by placing the circle in the end view, such that its centre is located on axis of the vertical solid. Divide the circle into a set of equal parts (at p'', q'', ..., w'') and obtain generators in the top and front view.

Obtain the intersection points

Step 1: The intersection points (1 to 8) are obtained as the meeting points of the various generators such as pp and qq of the horizontal solid with the vertical solid in the top view. Additional points 9 and 10 are realized due to the surface lines at m and n.

Step 2: Obtain the corresponding points in the front view.

Draw the intersection curves

Step 1: The curves of intersection are realized by joining the intersection points obtained in the front view by means of a curve in a sequential manner. The curve joining $9'2'3'4'10'$ appears in the front face.. It can be noted that the curve joining the points $10'5'6'7'8'1'9'$ denotes the intersection curve as the rear face.

Step 2: Draw the curves on both the left and the right halves as the horizontal solid is completely penetrating. ▲

Example 11.7 ▲

Intersection of Cylinders when their Axes are Offset

A cylinder of base diameter 80 mm and axis 100 mm long has its base 50 mm above the HP. It is completely penetrated by a horizontal cylinder of 50 mm diameter and axis 120 mm long, such that their axes are at right angles and 5 mm apart. Draw the curves of intersection of the solids at their interfaces.

PROCEDURE (Refer Fig. 11.9)

Draw the views of the solids

Step 1: Draw the three views of the vertical solid by placing the circle in the top view.

Step 2: Draw the three views of the intersecting (horizontal) solid by placing the circle in the end view, such that its centre is located at 5 mm away from the axis of the vertical solid at any height.

Step 3: Divide the circle into a set of equal parts and create additional points a'' and b'' that meet the axis of the vertical cylinder.

Step 4: Obtain generators in the top and front view through all these points.

Obtain the intersection points

Step 1: Obtain the intersection points 1 to 12 as the meeting points of the various generators such as pp and qq in the top view.

Step 2: Obtain the additional intersection points 13 and 14 corresponding to the generators through a and b.

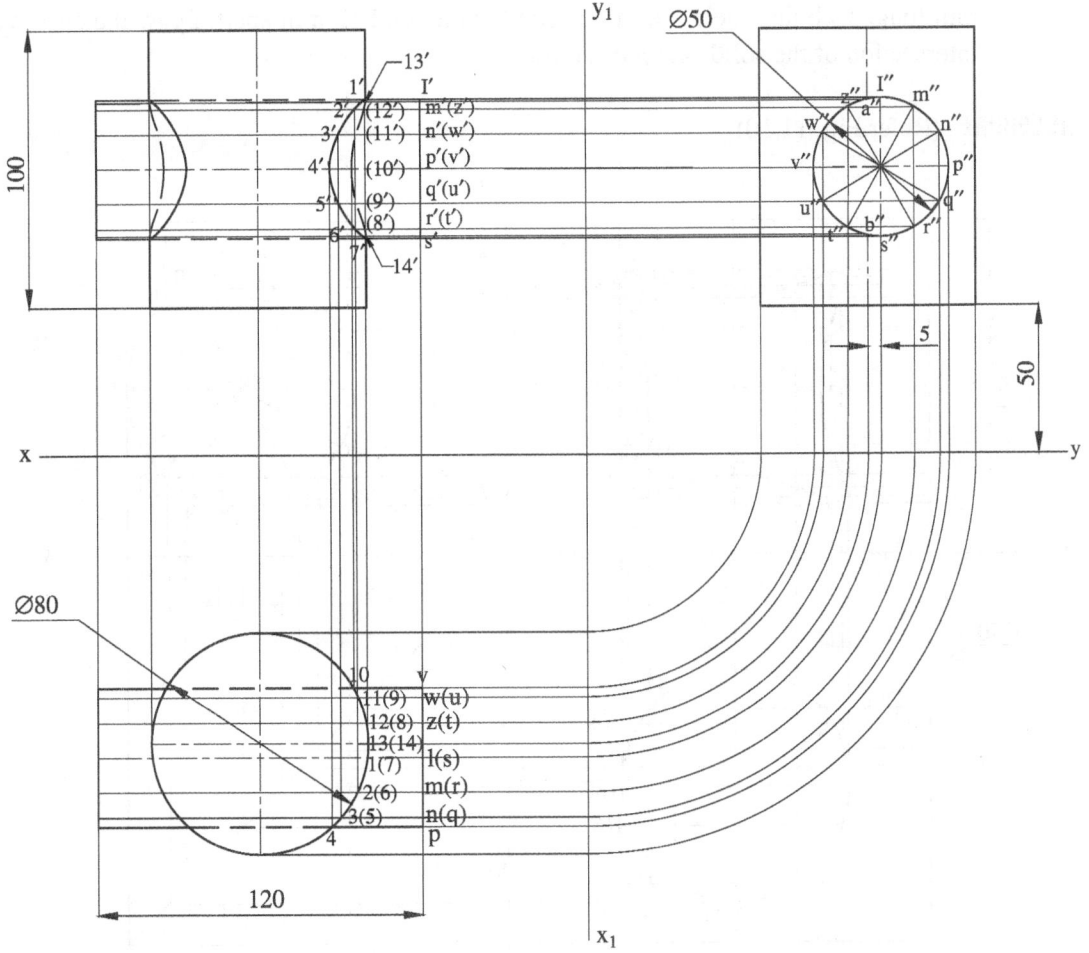

Fig. 11.9

Step 3: Obtain the front view corresponding to all these points.

Draw the intersection curves

Step 1: The curves of intersection are realized by joining the intersection points obtained in the front view by means of a curve in a sequential manner.

Step 2: The curve joining 1′2′ … 7′ is drawn thick, while the curve 7′14′8′ … 13′1′ is drawn dotted as it represents the intersection curve at the rear face. The points 13′ and 14′ are located at the outer edges and cause the trend change.

Step 3: The curves are drawn on both the left and the right halves as the solid is completely penetrated. ▲

Example 11.8 ◣

Intersection of Cylinders when their Axes are Offset and Bulging

A cylinder of base diameter 70 mm and axis 100 mm long is standing on its base on the HP. It is completely penetrated by a horizontal cylinder of 70 mm diameter and axis 120 mm long, such that their axes are at right angles and 12 mm apart. Draw the curves of intersection of the solids at their interfaces.

PROCEDURE (Refer Fig. 11.10)

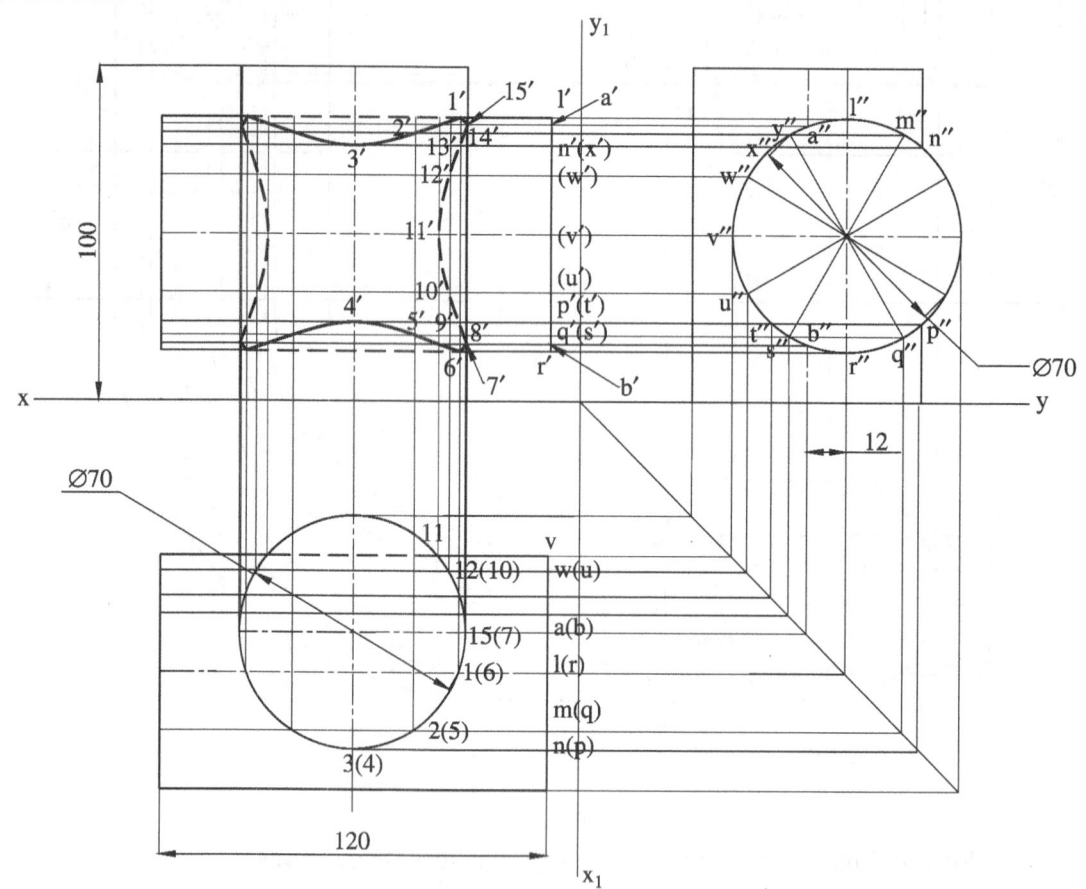

Fig. 11.10

Draw the views of the solids

Step 1: Draw the three views of the vertical solid by placing the circle in the top view.

Step 2: Draw the three views of the intersecting (horizontal) solid by placing the circle in the end view. Divide the circle contained in the vertical cylinder into a set of parts and create additional points a″ and b″ as in Example 11.7.

Step 3: As the intersecting cylinder bulges out, create special points n″ and p″ corresponding to that location.

Step 4: Obtain generators in the top and front view through all these points.

Obtain the intersection points

Step 1: Obtain the intersection points 1 to 15 as the meeting points of the various generators including the ones created through the additional and special points.

Step 2: Join these points in a sequential manner.

Draw the intersection curves

Step 1: The curves connecting 1'2'3' in the front face and lying on the left and the right halves join together and so do the curves 4'5'6' since the horizontal cylinder bulges out in the front.

Step 2: Project these points to obtain the front view and join them in a sequential manner. ▲

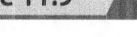 Example 11.9

Intersection of Two Cylinders with their Axes Intersecting and Inclined to Each Other

A vertical cylinder of base diameter 50 mm and axis 120 mm long is intersected by another cylinder of 35 mm diameter and 150 mm long such that the axes of the two solids intersect and make 60° with each other. Draw the intersection curves at the meeting surfaces.

PROCEDURE (Refer Fig. 11.11)

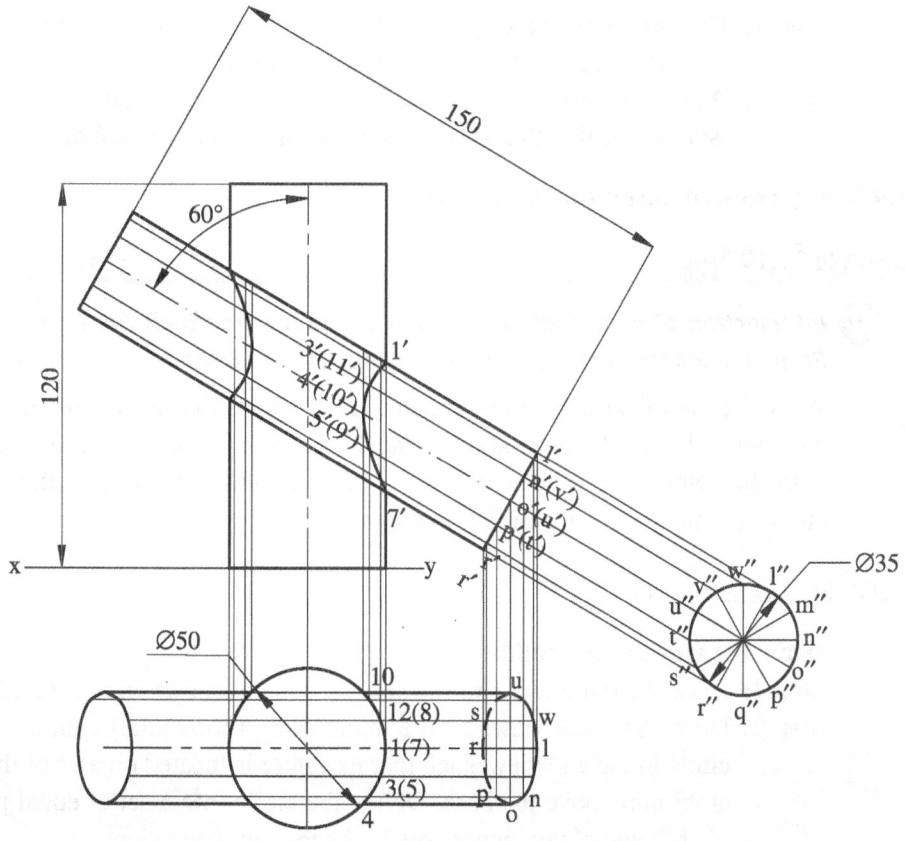

Fig. 11.11

Draw the views of the solids

Step 1: Draw the two views of the vertical solid by placing the circle in the top view as shown.

Step 2: Draw the intersecting cylinder outline at an angle of 60° to the vertical cylinder in the front view and add an end view circle at a convenient location. The centre of the circle corresponds to the position of the axis of the vertical cylinder too.

Step 3: Divide the circle into a set of equal parts (at l″, m″, ..., w″) and obtain the generators corresponding to that in the top and front view.

Obtain the intersection points

Step 1: The intersection points are obtained in the top view in similar lines of Example 11.5.

Step 2: These points are respectively marked as 1 to 12 and are realized on both the left and the right halves in the top view as the horizontal solid are completely penetrating the vertical solid.

Step 3: Obtain the corresponding points in the front view.

Draw the intersection curves

Step 1: The curves connecting 1′, 2′, ..., 7′ represent that at the front face, while that joining 7′, ..., 12′ represent that at the rear face.

Step 2: Draw the curves on both the left and the right halves. These will appear asymmetrical with respect to the axis of the inclined solid. ▲

11.7.4 Intersection of Cones and Cylinders

Example 11.10 ▲

 Intersection of a Vertical Cone and a Cylinder when their Axes are Intersecting and Perpendicular to Each Other Using Cutting Plane Method

A solid cone of 80 mm base diameter and 90 mm high is penetrated by a horizontal cylinder of 40 mm diameter and 150 mm long. The axes of the two solids are intersecting with the centre of the horizontal solid 25 mm above the base of the cone. Draw the curves of intersection by cutting plane method.

PROCEDURE (Refer Fig. 11.12)

Draw the views of the solids

Step 1: Draw the three views of the vertical cone as shown in Fig. 11.12.

Step 2: Draw the three views of the intersecting (horizontal) cylinder by placing the circle in the end view, such that its centre is located on axis of the vertical solid at 25 mm above the base. Divide the circle into a set of equal parts (at a″, b″, ..., h″) and obtain generators in the top and front view.

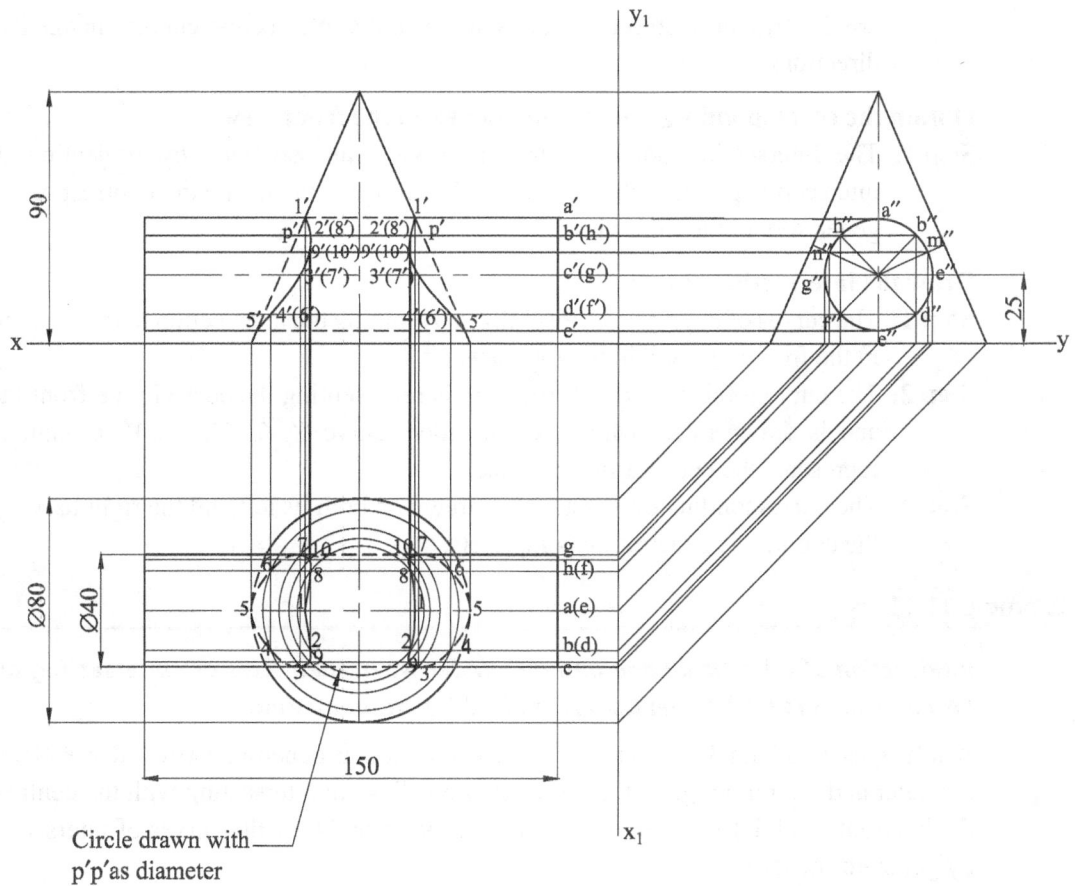

Fig. 11.12

Obtain the intersection points in the top view

Step 1: Assume a horizontal section plane passing through the division points b″ and h″. The cylinder when cut by this plane yields a rectangle of width b″h″ and length equal to the length of the intersecting solid in the top view.

Step 2: The vertical cone when cut by this horizontal plane gives a circle of diameter p′p′ (contained within the generator b′b′) in the front view. The intersection points of the cutting plane through b″h″ are obtained when these two sections, that is, rectangle and the circle meet in the top view. Therefore, draw a circle in the top view with p′p′ as diameter to cut the generators through b and h at the points 2 and 8 simultaneously in the left and the right halves.

Step 3: Similarly, the other points 1 to 7 are obtained by marking the appropriate diameters on the respective generators.

Step 4: Also obtain the additional intersection points 9 and 10 corresponding to the generators through m′ and n′ in a similar manner described previously. These are the critical intersection points at which the intersection curves change their directions.

Obtain the corresponding intersection points in the front view

Step 1: The intersection points in the front view are obtained by projecting the intersection points of the top view and identifying them with their corresponding generators in the front view.

Draw the intersection curves

Step 1: The curves of intersection are realized by joining the intersection points obtained in the front view and in the top view.

Step 2: The curve joining 1' to 5' through 9' is representing the curve in the front face and is drawn thick, while the coincident curve 5', 6', 7', ..., 1' through 10' represents the curve at the rear face.

Step 3: The corresponding curves are also drawn on both its left and the right halves as the cylinder is penetrating completely. ▲

Example 11.11 ▲

Intersection of a Vertical Cone and a Cylinder when their Axes are Intersecting and Perpendicular to Each Other Using the Line/Generator Method

A solid cone of 80 mm base diameter and 100 mm high is penetrated a cylinder of 40 mm diameter and 90 mm long. The axes of the two solids are intersecting with the centre of the horizontal solid 30 mm above the base of the cone. Draw the curves of intersection by generator method.

PROCEDURE (Refer Fig. 11.13)

Draw the views of the solids

Step 1: Draw the three views of the vertical cone and mark its generators in all the three views as shown in Fig. 11.13.

Step 2: Draw the three views of the intersecting (horizontal) cylinder by placing the circle in the end view, such that its centre is located on axis of the vertical solid at 30 mm above the base. Mark the points 1″ to 12″ as cutting points of the circle and the generators of the vertical cone and project them to the top and front views.

Obtain the intersection points

Step 1: The intersection points of the cylinder and the cone are obtained by locating the meeting points of the circle and the various generators in the side view. For example, points 2″ and 3″ are located on the circle, when the generator o″b″ meets it. Mark also the respective coincident points 8″ and 9″ on the generator o″f″. Similarly, the other intersection points are obtained.

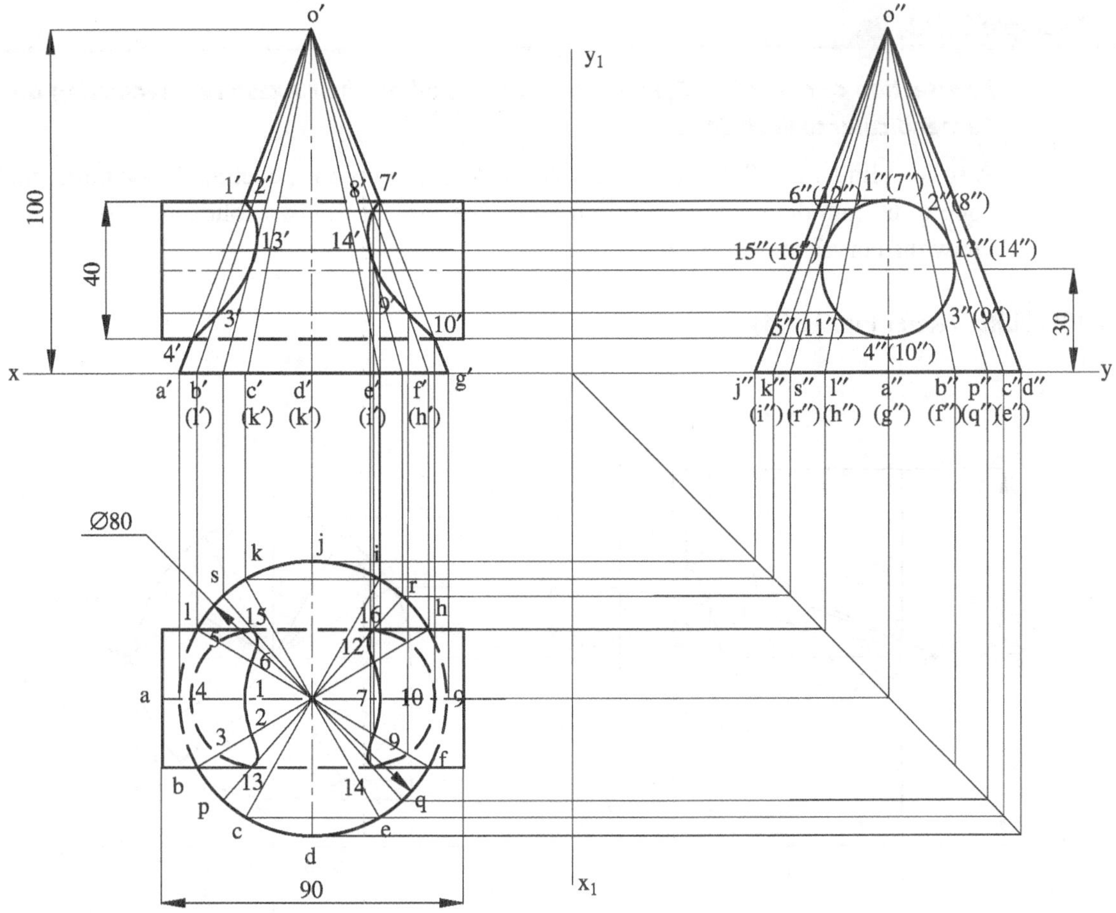

Fig. 11.13

Step 2: Mark also two sets of critical points 13″ and 15″ along with the coincident points 14″ and 16″ by drawing additional generators tangential to the circle.

Step 3: Project all the intersection points on the respective generators in the front and the top view.

Draw the intersection curves

Step 1: The intersection points obtained in the front view and in the top views are joined in a sequential manner to obtain the curves of intersection.

Step 2: The curves joining 1′2′3′4′ through the critical point 13′and 4′5′6′1′ through the critical point 15′ represent the front and the rear face curves in the left half of the front view. These curves coincide as the axes of both the solids are intersecting.

Step 3: Similarly, 7′8′14′9′10′ and 10′11′16′12′7′ show the right half curves in the front view.

Step 4: The corresponding curves are also drawn in the top view.

Example 11.12

Intersection of a Vertical Cylinder and a Cone when their Axes are Intersecting and Perpendicular to Each Other

A solid cylinder of 50 mm base diameter and 80 mm high is penetrated by a horizontal cone of 60 mm diameter and 70 mm long. The axes of the two solids are intersecting. Draw the curves of intersection.

PROCEDURE   (Refer Fig. 11.14)

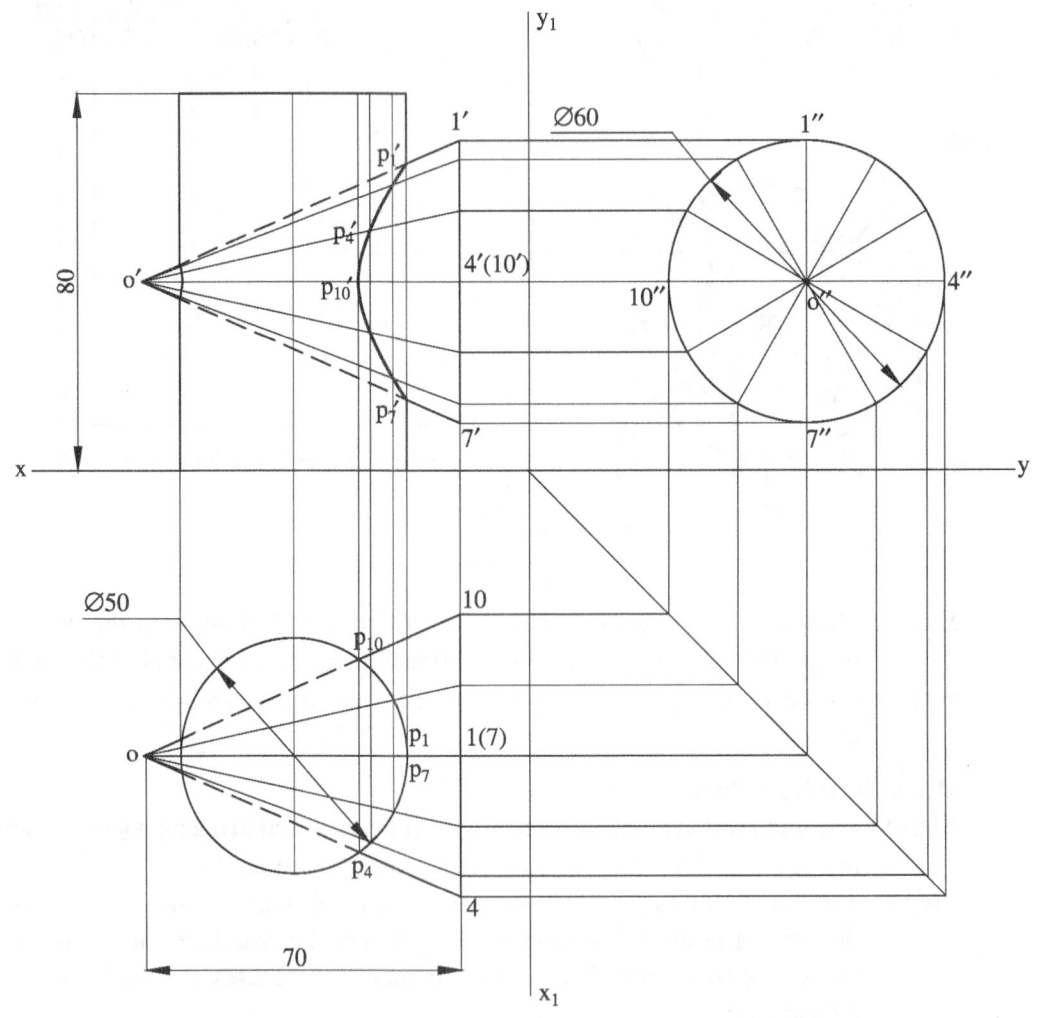

Fig. 11.14

Draw the views of the solids

Step 1: Draw the front and top views of the vertical cylinder as shown in Fig. 11.14.

Step 2: Draw a circle of 50 mm diameter corresponding to the horizontal cone by locating the centre o″ along the axis line of the vertical solid as shown in Fig. 11.14

Step 3: Draw the front and top view of the cone for a length of 70 mm and mark the generators are 1″ to 12″.

Step 4: Mark the points p_1 to p_{12} in the top view corresponding to the intersection of the generators of the horizontal cone with the circle pertaining to the cylinder.

Step 5: Project these points to meet the corresponding generators in the front view.

Draw the intersection curves

Step 1: The curves of intersection are realized by joining the intersection points obtained in the front view by means of a curve in a sequential manner.

Step 2: The curve joining p_1' to p_7' is drawn thick, while the curve p_7', p_8', ..., p_1' is hidden as it represents the intersection curve at the rear face.

Step 3: Draw the curves on both the left and the right halves. It can be noted that the left intersection curves are smaller due to the tapering nature of the cone. ▲

11.7.5 Intersection of Solids when their Axes are Parallel

Example 11.13

Intersection of a Vertical Cone and a Vertical Cylinder

A solid cone of 60 mm base diameter and 70 mm high is penetrated by a vertical cylinder of 40 mm diameter such that their axes are parallel to each other and 7 mm apart. If the plane containing the axes is inclined at 45° to VP, draw the curves of intersection.

PROCEDURE (Refer Fig. 11.15)

Draw the views of the solids

Step 1: Draw the front and the top views of the cone of base circle diameter 60 mm and height 70 mm.

Step 2: Divide the circle into many parts (say, twelve equal parts) and mark the radial lines 'oa' to 'ol' to represent the generators in the top view.

Step 3: Obtain the views of the generators in the front view by joining the base points with the apex.

Step 4: Draw a circle of 40 mm diameter with its centre located at a distance of 7 mm, along a line inclined at 45° to the reference line to represent the top view of the cylinder.

Step 5: Draw its outline in the front view as shown in Fig. 11.15.

Obtain the intersection points

Step 1: Mark the points of intersection of the circle and the radial lines in the top view as o–1 to o–12.

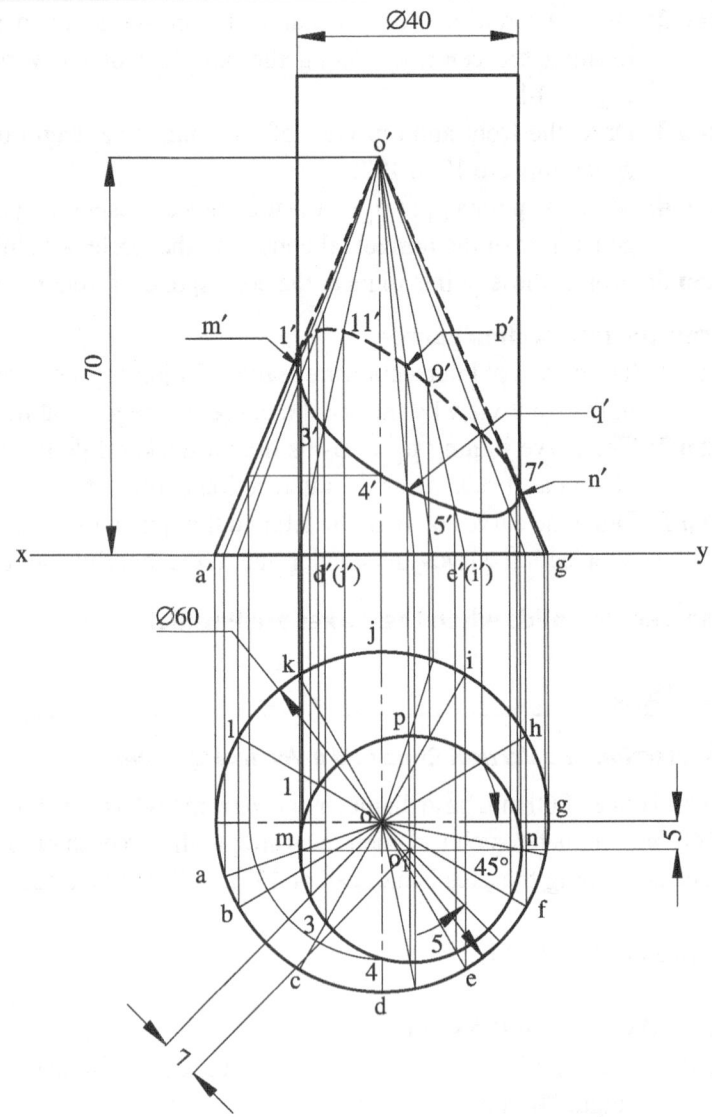

Fig. 11.15

Step 2: Project these intersection points to the respective generators in the front view as 1′, 2′, ..., 12′.

Step 3: The points 4 and 10 situated on the vertical radial lines will not meet their corresponding projectors in the front view. Such points are transferred to any other radial line, projected to the corresponding generator in the front view, and then transferred by a parallel line to the vertical generators to yield 4′ and 10′, respectively.

Draw the intersection curves

Step 1: The curves of intersection are realized by joining the intersection points obtained in the front view.

Step 2: The curve joining 1′ to 7′ represents the intersection curve in the front portion, while the dotted curve 7′8′ … 1′ represents that at the rear portion. ▲

Example 11.14 ▲

Intersection of a Vertical Cone and a Vertical Prism

A solid cone of 60 mm base diameter and 60 mm high is penetrated by a vertical square prism of 20 mm sides such that their axes are parallel to each other and 5 mm apart and lie in a plane parallel to VP. Draw the curves of intersection.

PROCEDURE (Refer Fig. 11.16)

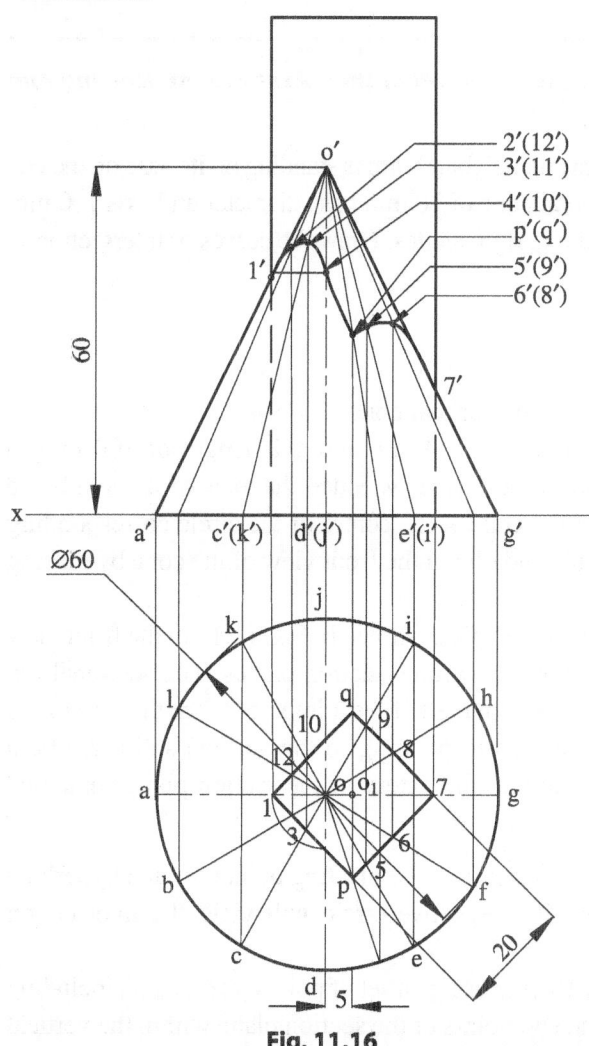

Fig. 11.16

Draw the views of the solids

Step 1: Divide the circle into many parts (12 equal parts) and mark the radial lines 'oa' to 'ol' to represent the generators in the top view.

Step 2: Obtain the views of the generators in the front view by joining the base points with the apex.

Step 3: Draw a square of 20 mm sides with its centre located at a distance of 5 mm on the diameter 'ag' as shown in Fig. 11.16 and draw its outline in the front view.

Obtain the intersection points

Step 1: Obtain the intersection points in the top view and get them projected in the front view.

Step 2: Mark the points of intersection of the square and the radial lines in the top view as o–1 to o–12.

Step 3: Project these intersection points to the respective generators in the front view as 1′, 2′, …, 12′.

Step 4: The points 4 and 10 situated on the vertical radial lines will not meet their corresponding projectors in the front view. Such points are transferred to any other radial line, projected to the corresponding generator in the front view, and then transferred by a parallel line to the vertical generators to yield p′ and q′, respectively. To locate the corners p and q in the front view, realize them in radial lines in the top view and extend the similar procedure adopted for any other intersection point.

Draw the intersection curves

Step 1: The curves of intersection are realized by joining the intersection points obtained in the front view.

Step 2: The curve joining 1' to 7' (including p') represents the intersection curve in the front portion, while the curve 7'8' ... 1' (including q') corresponding to the rear portion is coincident. ▲

11.7.6 Intersection of Vertical Cone by Horizontal Cone

Example 11.15 ▲

Intersection of a Vertical and Horizontal Cone when their Axes are Intersecting and Perpendicular to Each Other

A vertical cone of base diameter 80 mm and height 80 mm is standing on its base on the HP. It is completely penetrated by a horizontal cone of 70 mm base diameter and axis 100 mm long, such that their axes are intersecting at right angles. Draw the curves of intersection.

PROCEDURE (Refer Fig. 11.17)

Draw the views of the solids

Step 1: Draw the top and front views of the vertical cone.

Step 2: Draw a triangle with a base length of 70 mm and axis length of 100 mm to represent the horizontal cone in the top view. Since the axes of the solids are intersecting, place the axis of the cone at the centre of the circle corresponding to the vertical solid. Project this and obtain the front view of the cone by placing it at the required height.

Step 3: Mark a series of horizontal lines I to VIII as shown in Fig. 11.17 in the front view of the intersecting solid to represent various section planes. The sectional top view corresponding to each section plane will be a hyperbola for the horizontal solid in the top view, while that corresponding to the vertical solid will be a circle of diameter of length equal to the extent of the section plane contained within the cone.

Obtain the intersection points in the top view The meeting points of each hyperbola corresponding to a section plane and its respective circle will yield the intersection points in the top view as follows:

Step1: Select one section plane , say III, and assign a set of points p', q', ..., u', including v' and w' corresponding to the end points of the section plane within the vertical cone. Draw a set of semicircles with the diameter of the cone at these points and let 1', 2', ..., 5' be the points where the corresponding semicircles meet the section plane III.

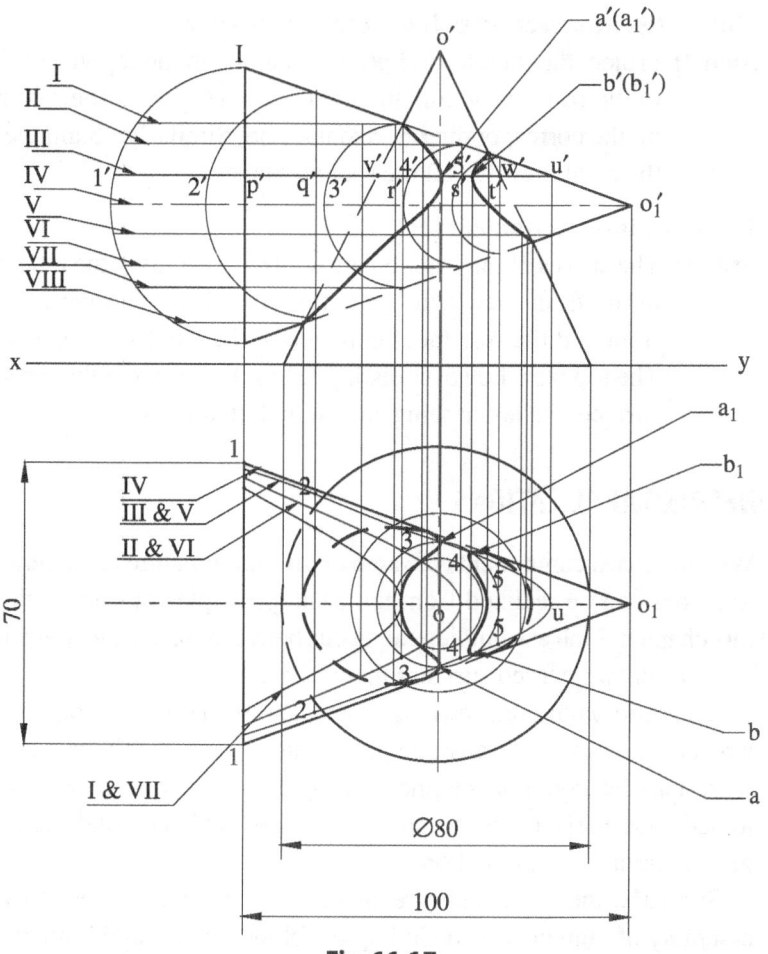

Fig. 11.17

Step 2: Mark points 1–1 on the projector through p′ and symmetrical on both sides of the horizontal axis in the top view so that 1–1 = twice p′1′. Similarly obtain 2–2, ..., 5–5 on the respective projectors through q′, ..., t′. Draw a hyperbola through the points thus obtained.

Step 3: Since the section plane III cuts the vertical cone at v′ and w′, its corresponding top view will be a circle of diameter v′w′. Therefore draw a circle with centre o and diameter v′w′, cutting the hyperbola obtained for III at the points a, a_1 and b, b_1.

Step 4: Placing a section plane symmetrically from the axis and on its other side in the front view will enable the usage of the same hyperbola obtained earlier. For example, section planes III and V yield the same hyperbola in the top view (by virtue of being at the same distance from the axis) but help to get different cutting points, as the diameters of these two planes within the cone are different. Obtain the cutting points in the top view for all the section planes in similar lines as described here.

Obtain the intersection points in the front view

Step 1: Project the intersection points obtained in the top view for the section plane III, to the front view domain and obtain $a'(a_1')$ and their coincident points $b'(b_1')$ on the corresponding horizontal line. Similarly obtain the intersection points in the front view for all the section planes.

Draw the intersection curves

Step 1: The curves of intersection are realized by joining the intersection points obtained in the front view and in the top view. Note that the intersection curves of the front and the rear faces coincide in the front view as the axes are intersecting. The top view however distinguishes the curves in the top surface and the bottom surface, the latter being shown in dotted lines.

11.8 ENGINEERING APPLICATIONS

Whenever two components of same or dissimilar nature are assembled, their intersecting junctions can be realized from the basic principles of intersection explained earlier in this chapter. Such assemblies can occur between that of two flat surfaces, flat and curved surfaces or two curved surfaces themselves.

Domestic plumbing, sewerage lines, and water and gas pipe lines used in the boilers when connected with transition pieces such as the couplings, elbows, tees, and reducers represent the common engineering applications where two cylindrical surfaces are joined. Assembly of ducts, smoking chutes, exhaust, and ventilating devices are the general industrial applications.

Similarly, these concepts are also used in medical science that involves the study of assembly of human parts including the blood vessels and implants.

RECAPITULATION

- Engineering applications involve assembly of two or more solid shapes or surfaces, as the existence of one solid shape is seldom useful practically. The knowledge of understanding the interface of such solids or surfaces is very vital in the assembly procedure.
- Accurate geometrical prediction of the intersecting surfaces of the solids not only helps the assembly, but also plays a very important role in the manufacture of the individual objects.
- Two major methods, that is, line or edge or generator method and the cutting plane method are commonly used to obtain the intersection curves that occur at the interfaces.

- The curves of intersection can be procedurally obtained by retaining one solid in the vertical position and orienting the other solid as per the requirements. The vertical solid is termed as the main solid, while the other one is known as the intersecting solid.
- In the line or edge or generator method, the cutting points of the various edges or the generators (in the case of cylindrical or conical solids) or additional surface lines of the intersecting solid, with the main solid are identified and joined to obtain the curves of intersection.
- In the cutting plane method, the imaginary cutting planes passing through the generators

or surface lines or edges of the intersecting solid are identified in one view (usually the front view) and the intersection points corresponding to these lines and the cutting planes are obtained in the other view (top view). Joining of these intersection points results in a curve in the top view. By suitably projecting these points to the front view results in the intersection curve in that view also. This procedure is more appropriate when the main solid is pyramidal or conical in nature.

- When the vertical solids are prismatic or cylindrical, the curves of intersection appear only in the front view.

- When the vertical solid is pyramidal or conical, the intersection curves are obtained in the top and in the front views, as the assessment of intersection points vary depending on the height at which they occur.

- The accurate prediction of the intersection of curved surfaces or contours not only becomes useful in engineering applications but has great appeal in medical science, particularly in grafting, facial surgeries, and fine surface and contour mapping.

MULTIPLE-CHOICE QUESTIONS

11.1 The junction of two interpenetrating plane surfaces perpendicular to each other produces
(a) a straight line   (c) a curve
(b) a point   (d) a circle

11.2 The base of a circular shaped light house is built on a hexagonal raft foundation. The junction is
(a) a straight line   (c) a curve
(b) a point   (d) a circle

11.3 The junction of two interpenetrating prisms with their axes perpendicular to each other produces
(a) a closed polygon   (c) a curve
(b) a point   (d) a circle

11.4 The smaller diameter of the frustum of a hollow cone is joined to the same diameter tube to form a funnel. The junction is
(a) a closed polygon   (c) a curve
(b) a point   (d) a circle

11.5 The ends of the horizontal and vertical circular pipes are cut at an angle of 45° and joined together. The interface junction forms
(a) a closed polygon   (c) a curve
(b) a point   (d) an ellipse

11.6 A horizontal cylinder penetrates through one of the rectangular faces of a vertical square prism. The junction of the two solids appears as
(a) a closed polygon   (c) a curve
(b) a point   (d) a circle

11.7 A tap with a circular cross section is to be fitted with a cylindrical boiler. The minimum length

of the intersection curve can be obtained from which of the following cases
(a) A horizontal cylinder penetrating into a vertical cylinder with their axes intersecting each other
(b) A horizontal cylinder penetrating into a vertical cylinder with their axes offset to the required distance
(c) An inclined cylinder penetrating into a vertical cylinder with their axes intersecting each other
(d) An inclined cylinder penetrating into a vertical cylinder with their axes offset to the required distance

11.8 A hollow cylindrical chimney is placed on the top of a hemi spherical shaped dome of a building. The interface of the two solids appears as
(a) a closed curve   (c) a closed polygon
(b) a point   (d) an ellipse

11.9 A horizontal cylinder of smaller diameter penetrates into a vertical cylinder with their axes intersecting at each other. The junction of the two solids appears as
(a) a closed polygon   (c) a curve
(b) a point   (d) two identical curves in the front and the rear views

11.10 An inverted conical funnel has a cylindrical branch at its bottom. The cylindrical branch and the conical portion envelop each other such that their axes intersect at right angles. The interpenetration curve appears as

(a) a closed polygon   (c) an open curve
(b) an ellipse   (d) a closed curve

WORK PRACTICE LEVEL – 1

Intersection of Two Square Prisms

11.1 A square prism of base edges 50 mm sides and axis 100 mm long is standing on its base with its faces equally inclined to the VP. It is completely penetrated by a horizontal square prism such that the axes of two prisms bisect each other at right angles. Draw the curves of intersection of the solids at their interfaces, if the faces of the horizontal prism are equally inclined to the HP. The ends of the horizontal solid are of 50 mm square and axis is 100 mm long.

Intersection of Cylindrical Solids

11.2 A vertical cylinder of base diameter 55 mm and axis 80 mm long is completely penetrated by a horizontal cylinder of 40 mm diameter and axis 80 mm long, such that their axes bisect each other at right angles. Draw the curves of intersection of the solids at their interfaces.

Intersection of Prisms and Cylinders

11.3 A vertical cylinder of 50 mm diameter and 75 mm long is completely penetrated by a horizontal square prism of 30 mm edge of the square and axis 75 mm long such that the axis of the two solids intersect each other at right angles. The lower front rectangular face of the horizontal prism is inclined at 50° to HP. Draw the curves of intersection of the two solids.

WORK PRACTICE LEVEL – 2

Intersection of Cylindrical Solids

11.1 A vertical cylinder of base diameter 60 mm and axis 80 mm long is completely penetrated by a horizontal cylinder of 45 mm diameter and axis 80 mm long, such that their axes intersect each other at right angles and at 45 mm above the base. Draw the curves of intersection of the solids at their interfaces.

11.2 A vertical cylinder of base diameter 70 mm and axis 90 mm long is standing on HP on its base. A circular hole of 50 mm diameter is drilled through it such that the axis of the hole is parallel to both HP and VP and perpendicular to the axis of the vertical cylinder. Draw the curves of intersection of the cylinder and hole at their interfaces.

Intersection of Cones and Cylinders

11.3 A solid cone of 75 mm base diameter and 100 mm high is standing on HP on its base and horizontal cylinder of 40 mm diameter and 130 mm long is completely penetrated into it. The axes of the two solids are intersecting at right angles with the centre of the horizontal solid 35 mm above the base of the cone. Draw the curves of intersection by cutting plane method.

11.4 A solid cone of 70 mm base diameter and 90 mm high is standing on HP on its base and horizontal cylinder of 35 mm diameter and 130 mm long is completely penetrated into it. The axes of the two solids are intersecting with the centre of the horizontal solid 30 mm above the base of the cone. Draw the curves of intersection by generator method.

11.5 Solid cylinder of 55 mm base diameter and 90 mm high is standing on HP and on its base and a horizontal cone of 50 mm diameter and 80 mm long is completely penetrated into it. The axes of the two solids are intersecting at right angles. Draw the curves of intersection.

Intersection of Two Square Prisms when their Axes are Offset

11.6 A vertical square prism of side of base 55 mm is penetrated by a horizontal square prism of end faces of 50 mm side. The axis of the horizontal prism is 7 mm in front of the axis of the vertical prism and lower front rectangular face of the horizontal prism is inclined at 40° to HP. Draw the curves of intersection of the solids at their interfaces, if the faces of the horizontal prism are equally inclined to the HP. The axis is 100 mm long for both the prism.

Intersection of Cylinders when their Axes are Offset

11.7 A cylinder of base diameter 60 mm and axis 90 mm long is standing on its base on the HP. It is completely penetrated by a horizontal cylinder of 50 mm diameter and axis 110 mm long, such that their axes are at right angles and 7 mm apart. Draw the curves of intersection of the solids at their interfaces.

11.8 A vertical cylinder of base diameter 65 mm and axis 90 mm long is standing on its base on the HP. It is completely penetrated by a cylinder of 35 mm diameter which is parallel to HP and inclined to 30° to VP and axis 110 mm long such that their axes 10 mm apart. Draw the curves of intersection of the solids at their interfaces.

11.9 Two sheet metal pipes of diameter meet at right angles to each other. The diameter of the vertical pipe is 55 mm. The axis of the 30 mm diameter pipe is horizontal and is 7 mm in front of the axis of the other pipe. Draw the curves of intersection of the solids at their interfaces.

11.10 A vertical pipe of 70 mm diameter has a branch of 35 mm diameter. The axis of the branch is 40° to the axis of the vertical pipe and 10 mm in front of it and parallel to VP. Draw the curves of intersection of the solids at their interfaces.

Intersection of Cone and Cylinder when their Axes are Parallel and Offset

11.11 A cone of base diameter 80 mm and 70 mm high rests with its base on HP. A vertical cylinder of 35 mm diameter has its axis 5 mm in front of cone and both the axes are contained in a plane making an angle of 40° with VP. Draw the curves of intersection of the solids at their interfaces.

Answers for Multiple-choice Questions

11.1 (a)	11.2 (d)	11.3 (a)	11.4 (d)	11.5 (d)	11.6 (d)
11.7 (a)	11.8 (a)	11.9 (d)	11.10 (b)		

Development of Surfaces of Solids

12

OBJECTIVES

This chapter will help the reader to understand the following:

- Importance of development of lateral surfaces of a solid in engineering
- Major applications of the sheet metal works in the field of fabrication, construction, storage, packing and transportation, and the need for pattern making as a prerequisite
- Different methodologies to be used for obtaining the two-dimensional layout of the lateral surfaces of uncut solids
- Changes the development sketches when the solids are cut by section planes in the form of straight lines, arcs, cut-outs, slots, holes, etc.
- Development of lateral surfaces of the assembly of different solids and their interfaces in engineering applications

12.1 INTRODUCTION

It is well known that a three-dimensional solid, such as a prism or a pyramid, consists of lateral surfaces in the form of rectangles or triangles, respectively. In the case of solids of revolution, such as a cylinder or a cone, the lateral surfaces are curved in nature. In engineering applications, these lateral surfaces are to be accurately obtained in order to define the shape of the solid properly. When such lateral surfaces are very thin and are made of metallic sheets, it is known as 'sheet metal work'. Products manufactured with sheet metal are extensively used in the form of hollow containers, vessels, buckets, trays, computer and machine storage cabinets and thin casings, thin tubes, air-conditioning ducts, automobile, air craft and ship body panels, storage tanks, hoppers, bins, funnels, chimneys, boilers, etc. Thin card board and plastic sheets are abundantly used in packaging industries to pack and transport materials, medicines, etc.

These lateral surfaces can be obtained and laid out accurately in the form of a plain sheet and then folded, bent or rolled appropriately. Then, free edges of the sheet can be joined by adhesives, riveting, soldering, or welding to maintain the outer shape. The layout of the lateral surface of a solid on a plain sheet is known as its 'development', and when it is developed with suitable allowances to achieve the required folding and jointing, it is known as the 'pattern'. Therefore, the development of a lateral surface of an object and its pattern making are considered to be important in the fabrication and production of the objects that are manufactured with sheet metal. The importance of this topic can be well understood and appreciated by observing the practices of a tailor during the preparation of garments. Clothes not only aim for the right fit, but are also ruled by fashion to a large extent. In this chapter, different methodologies used for the development of lateral surfaces of simple solid shapes such as prisms, cylinders, pyramids, and cones are discussed. When these solids are cut or sectioned by plane surfaces or by curved surfaces in the form of holes and cut-outs, the appropriate changes that can be made in their uncut development profiles are also discussed.

12.2 PRINCIPLE OF DEVELOPMENT

The principle followed in the development of the lateral surface of a solid lies in wrapping a thin sheet or paper around its surface such that their free edges touch each other and then unwrapping it to lay on a flat plane surface. Therefore, the area of the plain sheet obtained is equal to the surface area of the lateral surface, as it corresponds to the perimeter of the base or the top of the solid as one linear dimension and the length of the lateral edge or surface line as the other linear dimension. Since the presence of one-to-one correspondence between development sketch and lateral surface of the solid has to exist, true dimensions of the solid should only be reflected in the development sketch. Instead of wrapping a sheet on the lateral surface and unwrapping it, the development sketch can also be obtained by rolling the solid on a flat plane sheet or surface such that its lateral surfaces come in contact with the sheet for one complete revolution of the solid and cutting the two-dimensional sheet accordingly. The development of lateral surface of the cut solid will correspondingly get reduced, when the cutting points are transferred to the uncut development sketch and trimmed accordingly.

12.3 METHODS OF DEVELOPMENT

To obtain the development sketch for the lateral surface of a solid, the following methods are used.

12.3.1 Parallel Line Method for Prisms and Cylinders

This method is used for the development of prisms and cylinders or solids that have parallel top and bottom surfaces. When a prism or a cylinder rolls over a plane surface for one complete revolution as shown in Fig. 12.1 or Fig. 12.2, the total area of the impressions obtained by it corresponds to the development of its lateral surface. Therefore, one dimension of the layout sheet is equal to the perimeter of the polygon as shown in Fig. 12.1 or the circumference of the cylinder as shown in Fig. 12.2. The other dimension corresponds to the length or the height of the solid. The rolling can be accounted from any lateral edge or generator, but it has to be ended with one complete revolution.

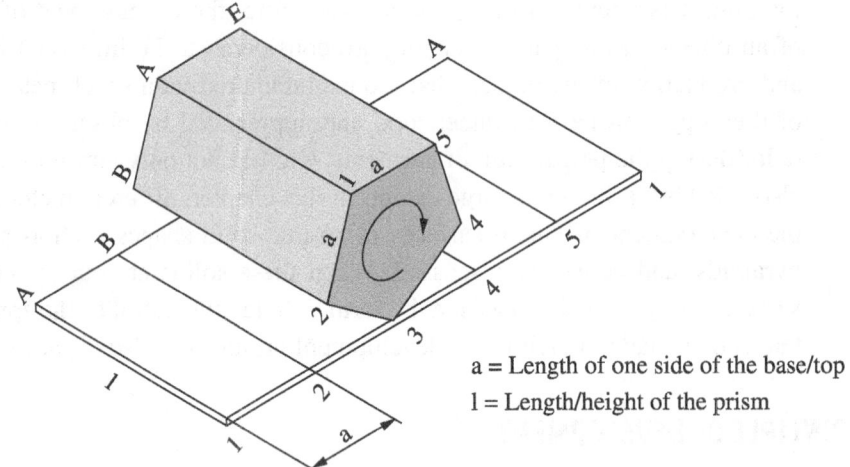

a = Length of one side of the base/top
l = Length/height of the prism

Fig. 12.1   Development of a lateral surface of a pentagonal prism

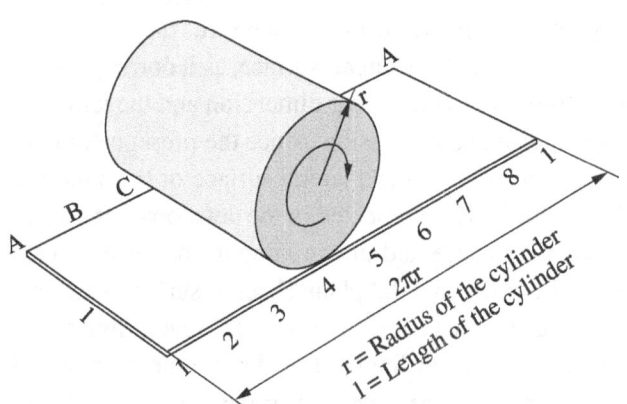

Fig. 12.2   Development of a lateral surface of a cylinder

In Fig. 12.1, the rolling of a prism has been accounted from the first lateral edge 1-A and has ended with that. As it rolls the other lateral edges in between also make their impressions, which are shown as 2-B, 3-C, ..., 5-E. Therefore, the development sketch for the lateral surface of the pentagonal prism is made up five rectangles of size $a \times l$ corresponding to the length of one side of the polygon (a) and the length or height (l) of the prism. In general, the development sketch for the lateral surface of a prism of base or top of n sides is composed of n rectangles of size equal to the side of the base of the prism and its length or height laid adjacent to each other.

When a cylinder starts rolling on the sheet from one of its lateral generators (1-A), its development sketch is completed when the same generator touches the sheet after one complete revolution. Therefore, the development sketch for the lateral surface of a cylinder corresponds to the perimeter of the circular end, given by $2\pi r$, where 'r' is the radius of the cylinder and the other dimension is equal to the length or height 'l' of the cylinder as shown in Fig. 12.2. Since the cylinder does not have any specific lateral edges, many surface lines or generators can be created conveniently. In Fig. 12.2, when the circle is divided into eight equal parts, the corresponding generators will get laid in the development sketch as 1–A, 2–B, ..., 8–H, thus making the length 1–1 as the circumference.

In the developmental sketches shown in Figs 12.1 and 12.2, any number of surface lines parallel to the lateral edges or the generators can be created for better accuracy and use, and their corresponding impressions can be accounted in the solid or vice versa. This will help locate any specific point or critical points on the lateral faces, when the solid is sectioned in particular.

Since the solids mentioned previously roll on a straight path, the lines 1–1 and A–A correspond to the path generated by the ends of the solid in the development sketch are parallel to each other and are known as stretch out lines. Any point on the lateral surface of the solid can also be located on the development sketch by drawing a line parallel to the stretch out line depending on its height, and hence, this method is known as parallel line method of development.

12.3.2 Radial Line Method for Pyramids and Cones

This method is adopted for the development of pyramids and cones or solids that radially converge to the apex or vertex. When a pyramid or a cone rolls over a plane surface for one complete revolution as shown in Fig. 12.3 or Fig. 12.4, the total area of the impressions made by it corresponds to the development of its lateral surface and this area is in the form of a sector, converging to a point, which denotes the apex, functioning as a hinge. Therefore, one dimension of the layout sheet is equal to the perimeter of the base of the solid and the other dimension being the length of the slant edge or the slant generator as the case may be. As discussed in Section 12.3.1, the rolling can be started

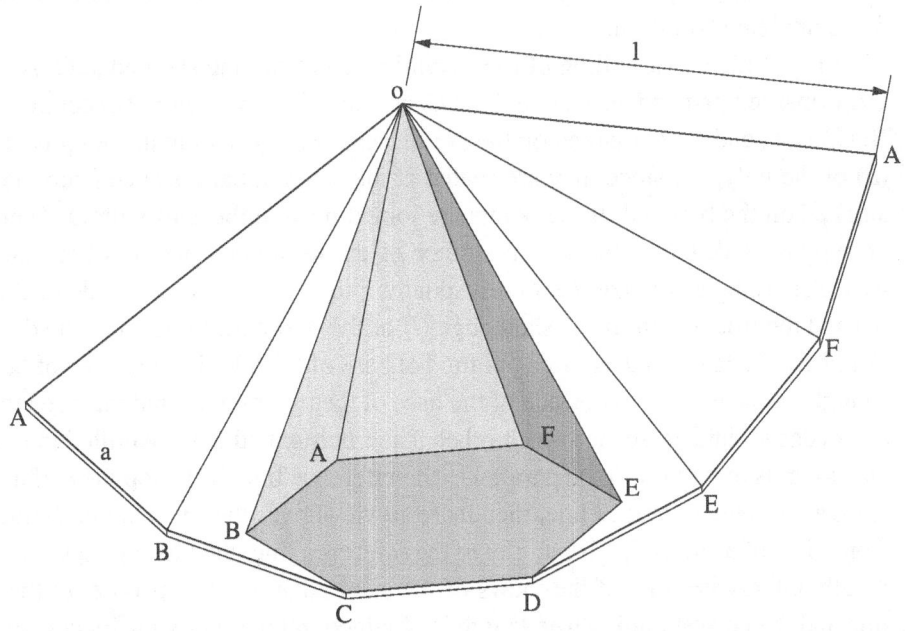

a = Length of one side of the base pyramid
l = Length of the slant edge of the pyramid

Fig. 12.3   Development of lateral surface of a hexagonal pyramid

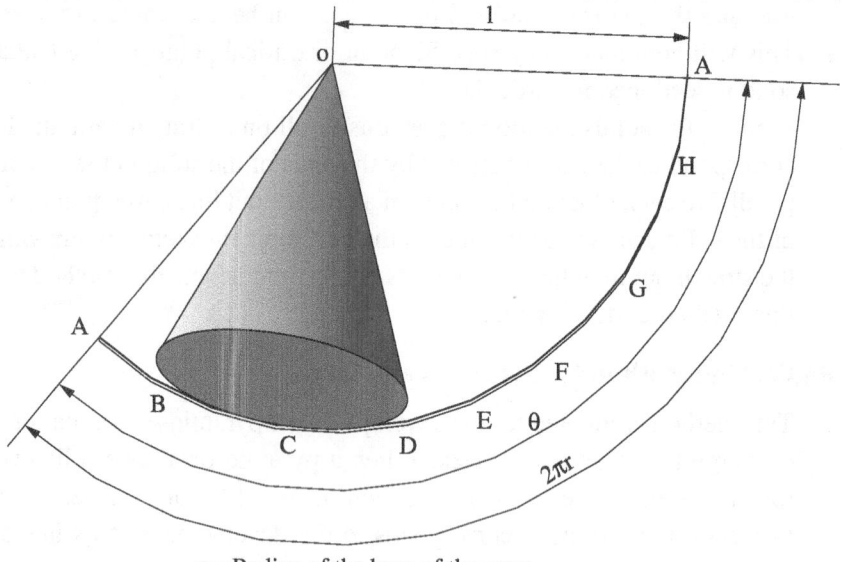

r = Radius of the base of the cone
l = Length of the slant edge of the cone
θ = Solid angle of the cone = (r/l) × 360°

Fig. 12.4   Development of a lateral surface of a cone

from any lateral edge or generator but has to be ended with the same, as the solid makes one complete revolution.

In Fig. 12.3, as the rolling of the pyramid starts from the slant edge O–A, the length OA is marked first and an arc equal to OA as radius is drawn with O as centre. The points B, C, ..., F, and A are marked on the arc successively by cutting the lengths of each side (*a*) of the polygon, since they are seated at the same radial distance from the apex and situated on the base. All these points are joined by straight lines with O. Therefore, the development sketch of the lateral surface of the hexagonal pyramid is made up of six isosceles triangles of size *a* × *l* corresponding to the length of one side of the polygon (*a*) and the true length of the slant edge (*l*) of the pyramid. In general, the development sketch for the lateral surface of a pyramid of base of '*n*' sides is composed of '*n*' isosceles triangles of size equal to the side of the base of the pyramid (*a*) and the true length of the slant edge (*l*) laid adjacent to each other. The true slant edge is identified in any view by checking its other view to be parallel to the reference line. If the top view of a slant edge is parallel to the reference line, then its front view gives the true length. Similarly, if the front view of a slant edge is parallel to the reference line, then its top view gives the true length. Otherwise, one of the views of the slant edge is made parallel to the reference line and the corresponding true length is obtained in the other view by construction.

Similarly, if a cone starts rolling on a plane surface with one of its slant generators, say OA, as in Fig. 12.4, its development sketch is completed when the arc length becomes equal to the circumference of the base and the sector O–A–A results in. Though the circumference of the base can be calculated by a formula, it is not possible to lay it along the arc. Hence, the required arc length is set by using the angle subtended by the arc at its centre. The subtended angle (θ) can be calculated using the mathematical relation $\theta = (r/l) \times 360°$, where '$r$' is the radius of the base of the cone and 'l' is the true length of the generator of the cone. Physically, 'θ' is the angle subtended by the solid cone at its apex, and hence, it is known as solid angle. It can be obtained from the logical principle that when the base radius 'r' of the cone completes one revolution (360°), the slant generator 'l' through that subtends a solid angle θ at its apex, and hence, the mathematical relation $r \times 360° = \theta \times 1$ is established. It can be noted that in the case of the cone, its extreme generators indicate the true length, as their projections are always parallel to the reference line.

Similar to cylinder, the cone can also be assumed with the desired number of generators to improve the accuracy of the development profile when it is sectioned in particular. For example, in Fig. 12.4, when the base circle of the cone is divided into eight equal parts, the corresponding generators will get laid in the development sketch as OA, OB, ..., OH, and OA, thus making up the arc length, A–A. Instead of dividing the arc into certain number of parts, the subtended angle can be dividing easily. The reader is advised to refer Chapter 2 on the division of an angle into the desired number of equal parts.

In Figs 12.3 and 12.4, the base points A, B, etc., are all located radially equidistant from the point O. Similarly, any other point on the lateral surface of the solid can be located by drawing a surface line through that and marking in the appropriate segment of the face or the curved surface. The points at the same level can be located from the apex O by drawing an arc with the same radius, and hence, this method is known as the radial line method for development.

12.3.3 Triangulation Method

This method is adopted for the development of transition pieces that connect two different sizes and shapes. For example, when the base of a solid is a square and its top is a circle, the lateral surface connecting such ends cannot conform to the ones discussed in the previous sections. The development of such surface is attempted by splitting the relevant curved portions as consisting of triangles. The transition pieces are very common in air-conditioning ducts, pipe joints, fancy containers, etc. However, in this book, the development done by this method is out of scope.

12.3.4 Approximate Method

This method is used to draw the development of double curved or warped surfaces such as hemisphere, sphere, ellipsoid, paraboloid, hyperboloid, helicoid, etc. These applications are found in storage tanks, antennas, transmitters and receivers' satellite dishes, tubes and tyres for wheels, elbows, springs, cables, and ropes. There is no appropriate method to reproduce these physical surfaces. The solid is sliced into many portions, and the development of each portion is drawn independently and then combined to represent the whole solid. For example, a sphere that resembles an orange can be developed by creating lunes, a shape with which the orange pieces are present segment wise naturally. From the description of this method, the development is not accurate but totally approximate, and hence, a judicial choice of attributing shapes will help a great extent to minimize the errors in the development of such surfaces. This topic is out of scope in this book.

12.4 DEVELOPMENT OF LATERAL SURFACES OF SIMPLE SOLIDS

The development of lateral surfaces of simple solids such as prisms and cylinders involve rectangles of size equal to the perimeter and the height or length. The prisms involve as many subrectangles as the number of sides of the base or the top. Generally, in the case of prisms and cylinders, there is no need to draw their orthographic views, get the dimensions, and then draw the development, as the length and the sides of the prism are self-explanatory from the geometrical details given in the problem.

But while drawing the development sketches for the pyramid and the cone, the side of the base is available from the geometry, while the true length of the slant edge or slant generator is to be obtained only by drawing the orthographic views of these solids in the given position. The true length of the slant edge is obtained by making its corresponding view parallel to the reference line, if not available in the given position.

Example 12.1

A pentagonal prism of 30 mm side of base and axis 70 mm long is resting on its base on horizontal plane (HP), in such a way that one of its faces is perpendicular to vertical plane (VP). Draw the development of the lateral surface of the solid.

PROCEDURE (Refer Fig. 12.5)

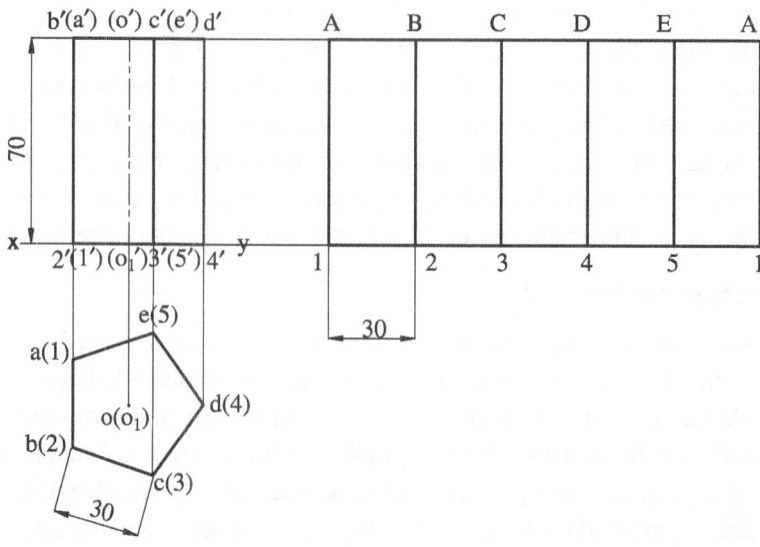

Fig. 12.5

Step 1: Draw the orthographic views of the prism with one of its base edges perpendicular to the reference line.

NOTE   *The orthographic views are not mandatory in this problem, as the length and height of the side are available from the problem itself for drawing the development.*

Step 2: Draw the bottom stretch out line 1–1 and mark the points 1, 2, ..., 5, and 1 successively at lengths equal to the side of the prism.

Step 3: Draw the top stretch out line A–A at a distance equal to the height of the solid and mark points A, B, ..., E, and A corresponding to the points located in the bottom stretch out line and join them to mark the lateral edges.

Step 4: The area 1–A–A–1 gives the development of the whole solid.   ▲

<hr>

Example 12.2   ▲

<hr>

A hexagonal pyramid of base sides 30 mm and axis 70 mm long has its hexagonal end on the HP such that two of its sides are parallel to VP. Draw the development of lateral surface of the solid.

PROCEDURE   (Refer Fig. 12.6)

Step 1: Draw the top view with two sides of the hexagon parallel to the reference line and the corresponding front view.

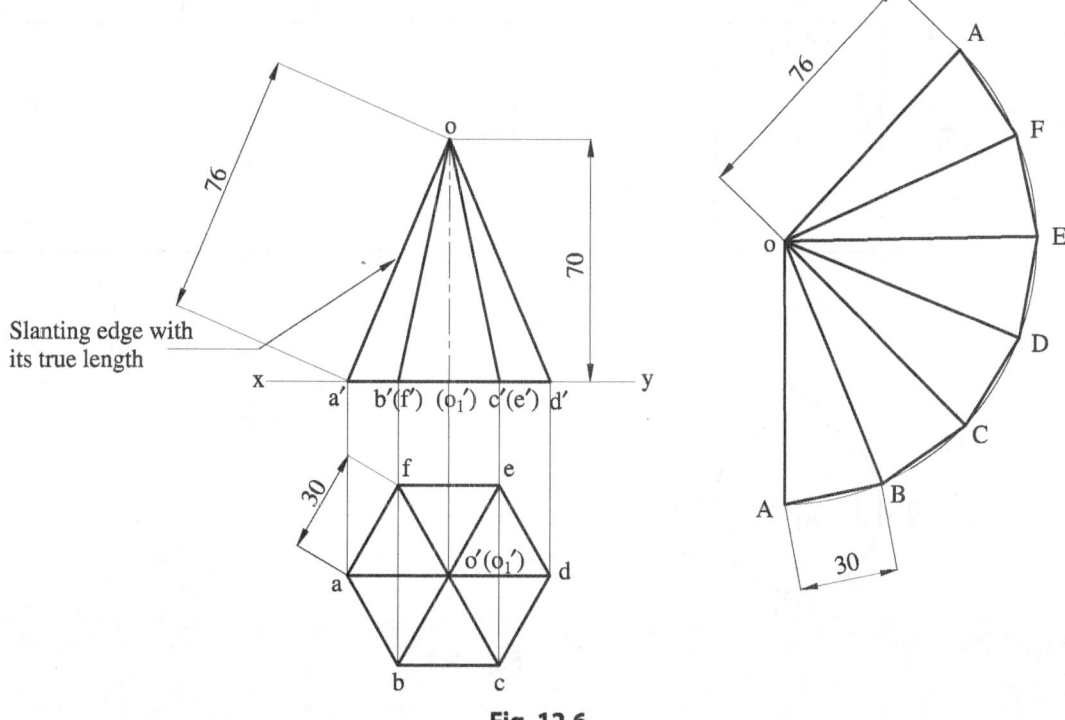

Fig. 12.6

Step 2: Since the top view of the slant edges oa and od are parallel to the reference line, their corresponding front views o'a' and o'd' give the true length of the slant edge.

Step 3: With O as centre and o'a' or o'd' as radius, draw an arc and mark the starting point A. Set the points B, C, ..., F, and A at lengths equal to the side of the prism and join them by straight lines.

Step 4: Draw the radial lines OA, OB, ..., OF, and OA to mark the lateral edges of the pyramid.

Step 5: The area O–A–A–O gives the development of the whole solid. ▲

Example 12.3

A right circular cylinder of base 45 mm diameter and axis 60 mm long is resting on HP on its base with its axis parallel to VP. Draw the development of lateral surface of the cylinder.

PROCEDURE (Refer Fig. 12.7)

Step 1: Draw the orthographic views of the cylinder. Divide the circle into eight equal parts and obtain the generators a'1', b'2', ..., h'8'.

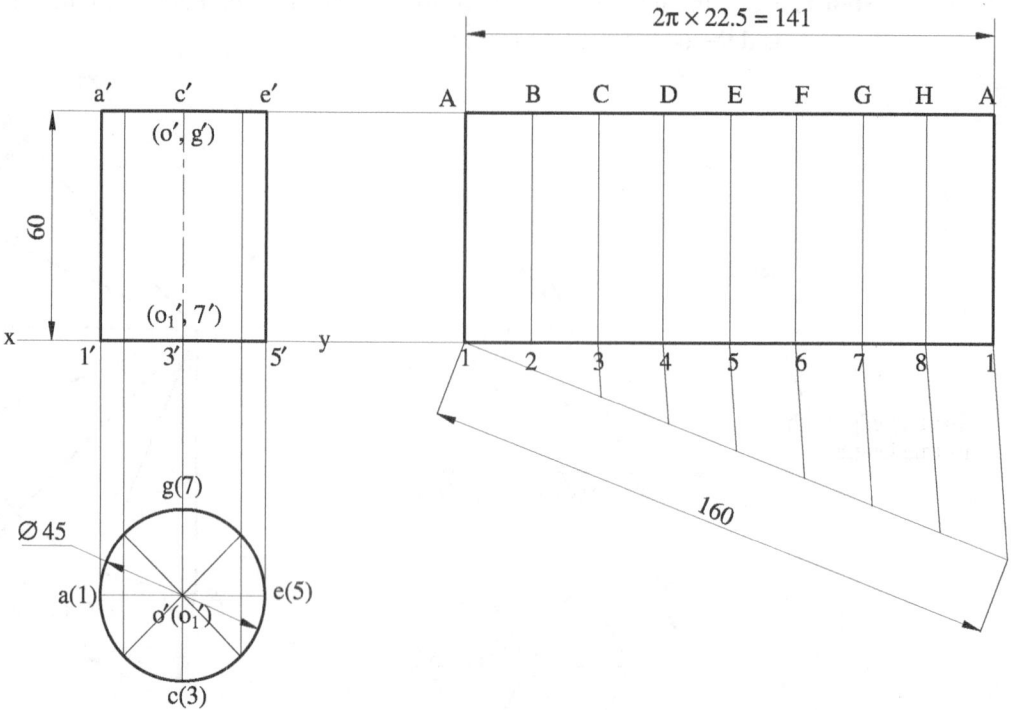

Fig. 12.7

NOTE   *The orthographic views are not mandatory in this problem, as the length and height of the side are available from the problem itself for drawing the development.*

Step 2: Calculate the circumference of the circle using $\pi \times 45$ mm, mark it on the bottom stretch out line 1–1, and divide it into eight equal parts by choosing a known length divisible by eight. Mark the points 1, 2, ..., 8, and 1.

Step 3: Draw the top stretch out line A–A at a distance equal to the height of the solid and mark points A, B, ..., H, and A corresponding to the points located in the bottom stretch out line and join them to mark the lateral generators.

Step 4: The area 1–A–A–1 gives the development of the whole solid.   ▲

Example 12.4

A cone of base 50 mm diameter and height 65 mm rests with its base on HP and axis parallel to VP. Draw the development of lateral surface.

PROCEDURE   (Refer Fig. 12.8)

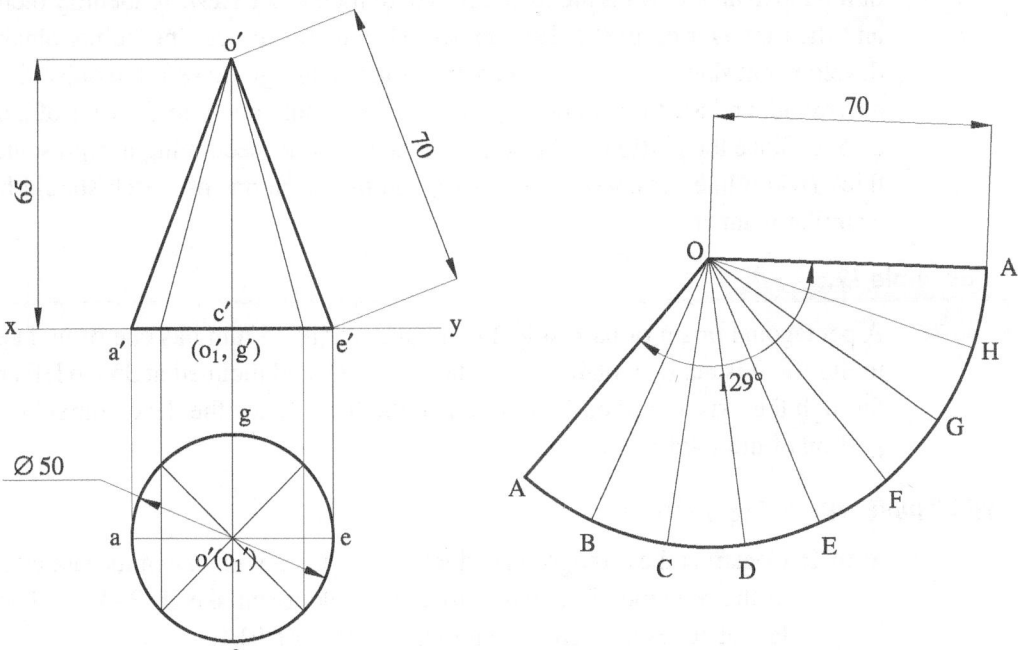

Fig. 12.8

Step 1: Draw the top and front views of the cone. Divide the circle into eight equal parts and obtain the generators $o'a'$, $o'b'$, ..., $o'h'$.

Step 2: Since the top view of the extreme generators oa and oe are parallel to the reference line, their corresponding front views $o'a'$ and $o'e'$ give the true length of the slant edge.

Step 3: With O as centre and o'a' or o'e' as radius, draw an arc, and mark the starting point A.

Step 4: Calculate the solid angle $\theta = (r/l) \times 360°$ or $\theta = (25/70) \times 360° = 129°$, set it from the line OA, and divide it into eight equal parts as per the procedure in Chapter 2. Mark the division points B, C, ..., H, and A on the arc through A.

Step 5: Draw the radial lines OA, OB, ..., OF, and OA to mark the lateral generators of the cone.

Step 6: The sector area O–A–A–O gives the development of the whole solid. ▲

12.5 DEVELOPMENT OF LATERAL SURFACES OF SECTIONED OR TRUNCATED SOLIDS

When a solid is sectioned, the cutting plane is marked by a straight line on the plane to which it is perpendicular. All the cutting points should be transferred to the development sketch of the uncut solid and joined sequentially. If the cutting points are situated on the lateral edges or generators, they can be straight away transferred in the case of prisms and cylinders, while in the case of pyramids and cones, they are transferred to the true slant edge or generator, and then, their radial distances are to be marked in the development sketch at their appropriate locations. The cutting points situated in the base or the top in all the solids are to be brought to their other views to identify their locations and then transferred to the development sketch. All the cutting points obtained in the development sketch are to be joined by means of straight lines if the object is a prism or a pyramid, and are to be joined by means of a smooth curve in the case of a cylinder or a cone. Since the portion of the solid retained after the sectioning is represented by bold thick (wide) lines or curves, the corresponding development sketch should be made in a similar manner.

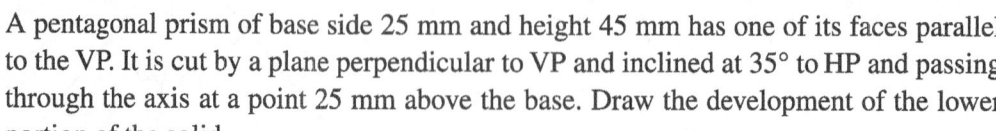

Example 12.5

A pentagonal prism of base side 25 mm and height 45 mm has one of its faces parallel to the VP. It is cut by a plane perpendicular to VP and inclined at 35° to HP and passing through the axis at a point 25 mm above the base. Draw the development of the lower portion of the solid.

PROCEDURE (Refer Fig. 12.9)

Step 1: Construct the orthographic views of the prism with one of its base edges parallel to the reference line and draw the development sketch P–A–A–P with all the lateral edges marked in it similar to Example 12.1.

Step 2: Mark the section plane as a straight line in the front view, as it is perpendicular to the VP and orient it at 35° to the reference line and passing through a point on the axis at 25 mm above the base.

Step 3: Mark the cutting points 1', 2', ..., 5' as shown. As all of them are situated in the lateral edges through p', q', ..., t', transfer them straight to the corresponding lateral edges in the development sketch, obtain the points 1, 2, ..., 5, and join them by means of straight lines.

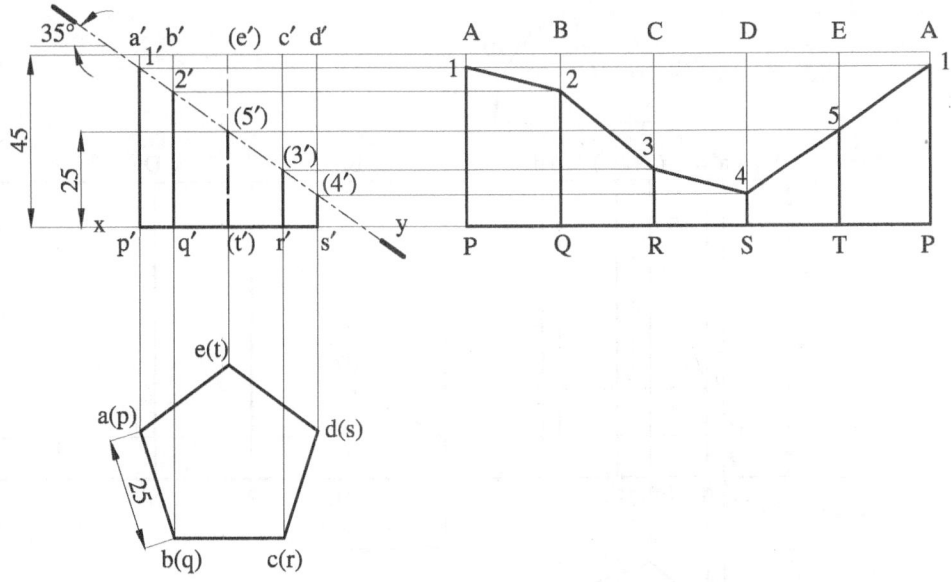

Fig. 12.9

Step 4: The area P–1–2–3–4–5–1–P is the development sketch of the required lower portion of the solid.

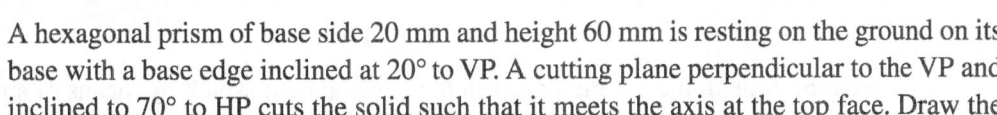

Example 12.6

A hexagonal prism of base side 20 mm and height 60 mm is resting on the ground on its base with a base edge inclined at 20° to VP. A cutting plane perpendicular to the VP and inclined to 70° to HP cuts the solid such that it meets the axis at the top face. Draw the development of the larger portion of the solid.

PROCEDURE (Refer Fig. 12.10)

Step 1: Construct the orthographic views of the prism with one of its base edges inclined at 20° to the reference line and draw the development sketch P–A–A–P with all the lateral edges marked in it similar to Example 12.1.

Step 2: Mark the section plane as a straight line in the front view, as it is perpendicular to the VP and orient it at 70° to the reference line and passing through the top point on the axis.

Step 3: Mark the cutting points 1′, 2′, ..., 5′ as shown. As the points 1′, 2′, and 5′ are situated in the lateral edges through p′, q′, and u′, transfer them straight to the corresponding lateral edges in the development sketch, and obtain the points 1, 2, and 5.

Step 4: Since the cutting points 3′ and 4′ are in the top end, project them to the top view and obtain their locations as 3 and 4 on the top edges bc and ef. Measure the distances of b3 and e4 and mark them from B and E in the development sketch.

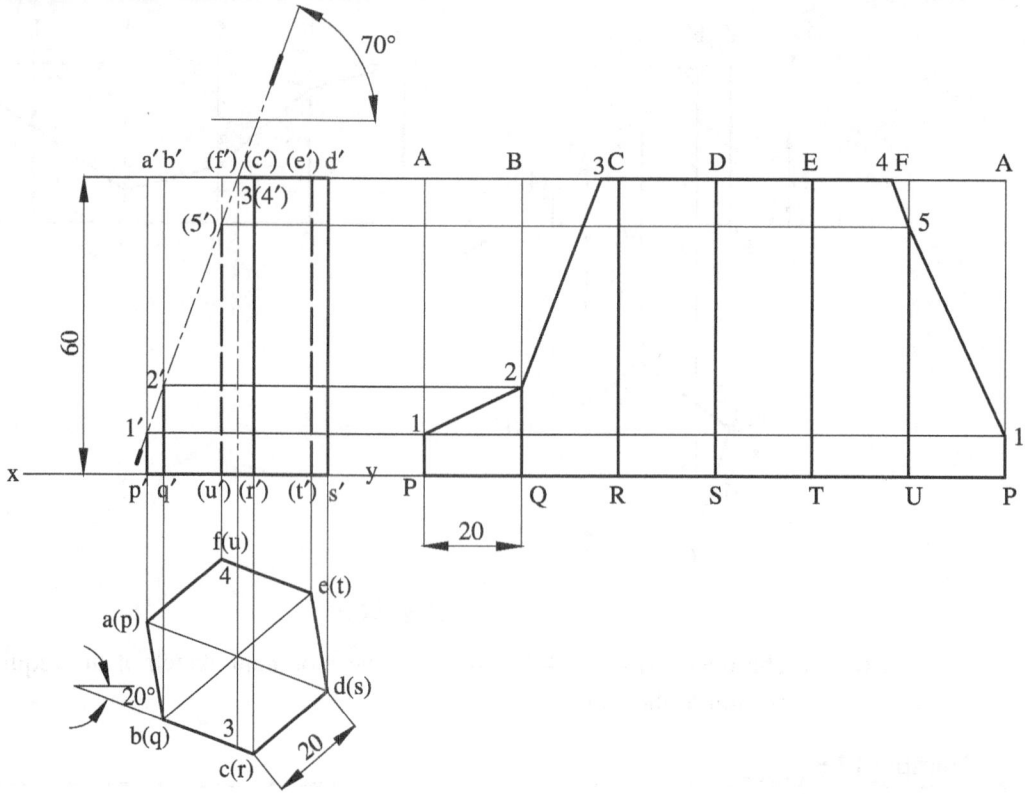

Fig. 12.10

Step 5: Join all the cutting points in the development sketch by means of straight lines and the area P–1–2–3–4–5–1–P is the development sketch of the required larger portion of the solid. ▲

Example 12.7 ▲

A vertical chimney of circular section of 400 mm diameter is located on the rooftop sloping at 35° to the horizontal. If the shortest portion of the chimney is 300 mm high, then determine the shape of the sheet metal area from which the chimney can be made. Use 1:10 scale.

PROCEDURE (Refer Fig. 12.11)

Step 1: Adopt a scale of 1:10, construct the orthographic views of the cylinder of 400 mm diameter by dividing the circle into eight equal number of parts and mark the generators through a′, …, h′. The height of the cylinder can be kept arbitrary.

Step 2: Mark the sloping roof at a distance of 300 mm below the top end and at an inclination of 70° to the reference line to meet the other extreme generator at 1′.

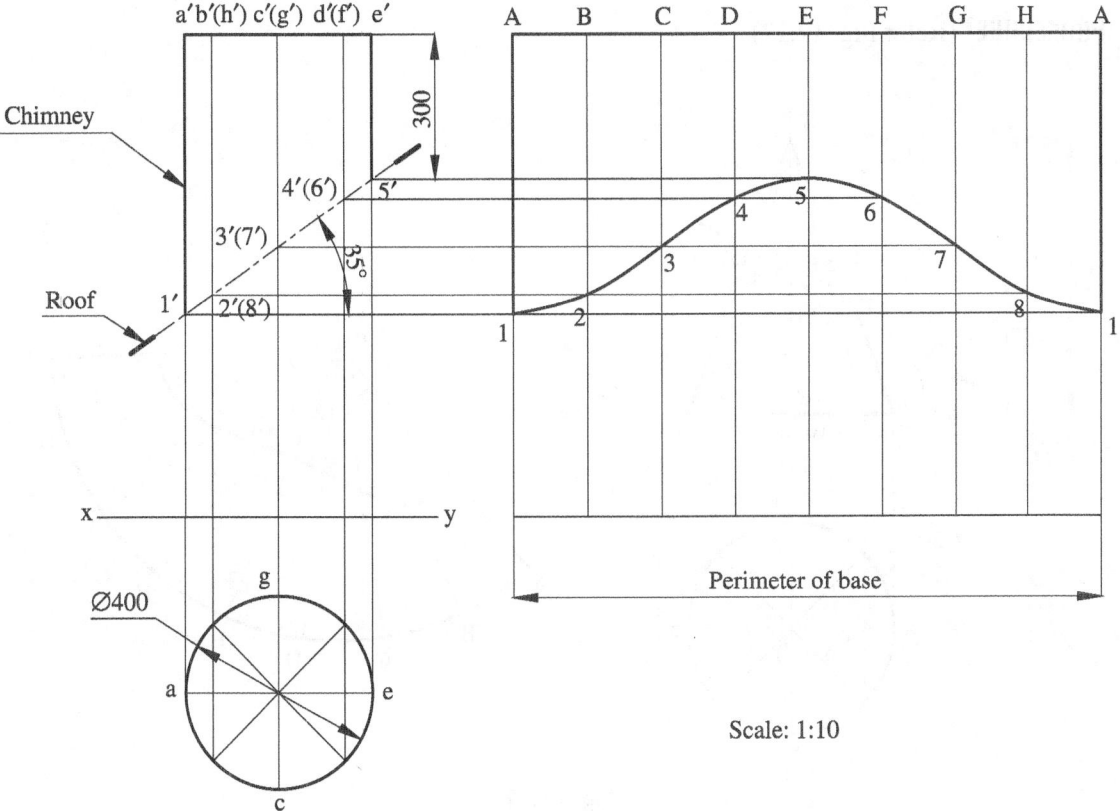

Fig. 12.11

The cylinder length can be limited to this point, and the horizontal stretch out line can be drawn from this point. The portion of the cylinder above the sloping line represents the chimney above the roof. Draw the development sketch of the uncut cylinder 1–A–A–1 with all the lateral generators marked in it similar to Example 12.3.

Step 3: The sloping roof represents the section plane in the front view. Hence, the cutting points are identified as 1', 2', ..., 8'.

Step 4: Since all the cutting points are situated in the lateral generators through a', b', and h', transfer them straight to the corresponding lateral lines in the development sketch, and obtain the points 1, 2, ..., 8, and 1.

Step 5: Join the above points by means of a smooth curve, and the portion above this curve represents the development profile of the chimney. ▲

Example 12.8 ◢

A cone of base 50 mm diameter and height 65 mm rests with its base on the HP. A section plane perpendicular to VP and inclined at 30° to the HP bisects the axis of the cone. Draw the development of the lateral surface of the truncated cone.

PROCEDURE (Refer Fig. 12.12)

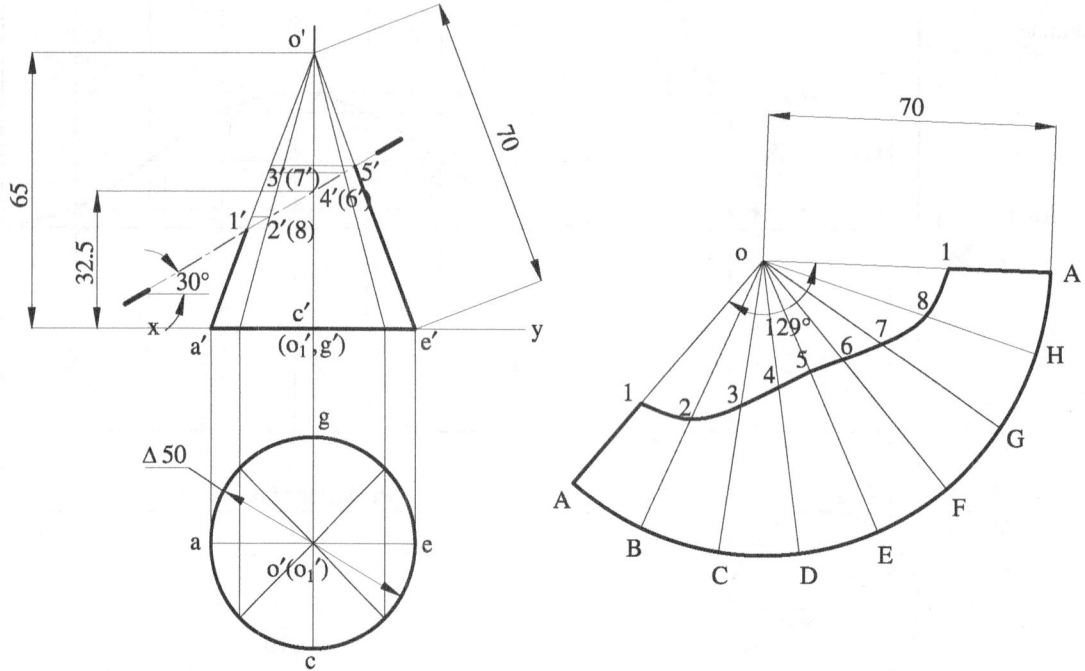

Fig. 12.12

Step 1: Draw the top and front views of the cone and obtain the generators o'a', o'b', …, o'h'. With O as centre and o'a' or o'e' as radius, draw an arc at a solid angle θ = $(r/l) \times 360°$ or $\theta = (25/70) \times 360° = 129°$, and obtain the development sketch O–A–A–O with all the generators marked in it similar to Example 12.4.

Step 2: Mark the section plane as a straight line in the front view, as it is perpendicular to the VP and orient it at 30° to the reference line and passing through the midpoint of the axis.

Step 3: Mark the cutting points 1', 2', …, 8' as shown. As all of them are situated in the lateral edges through a', b', …, h', transfer them to the true slant edge o'a' (or o'e'), set the corresponding radial distances on the respective generators in the development sketch, obtain the points 1, 2, …, 8, and 1, and join them by means of a smooth curve. The area below the curve gives the development sketch. ▲

Example 12.9 ▲

@ A lamp shade is formed by cutting a cone of base 144 mm diameter and 174 mm height by a horizontal plane at a distance of 72 mm from the apex and by another plane inclined at 30° to HP and passing through one extremity of the base. Draw the development of the lamp shade.

PROCEDURE (Refer Fig. 12.13)

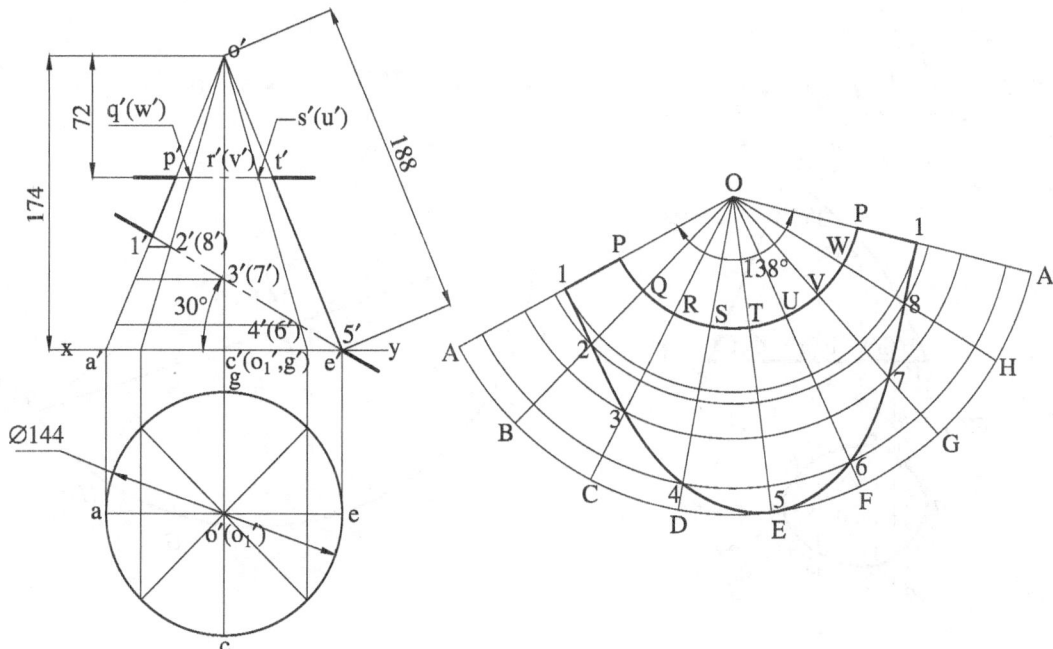

Fig. 12.13

Step 1: Draw the top and front views of the cone and obtain its development sketch at a solid angle $\theta = (r/l) \times 360°$ or $\theta = (72/188) \times 360° = 138°$. Mark all the generators OA, OB, ..., OH, and OA similar to Example 12.8.

Step 2: Draw the horizontal section plane at a distance of 72 mm from the apex and mark its cutting points p', q', ..., v' in the front view. Similarly, draw the 30° inclined section plane passing through the base point e' in the front view and mark the cutting points $1'$, $2'$, ..., $8'$. The shape of the cone in between the two section planes is of our interest.

Step 3: Since all the cutting points are situated in the lateral generators, transfer them to the true slant edge $o'a'$ (or $o'e'$), set the corresponding radial distances on the respective generators in the development sketch, obtain the top curve P–Q–, ..., W, and P, and the bottom curve 1–2–,..., 8–1.

Step 4: The portion in between the two curves is the development sketch for the lamp shade. ▲

<hr>

Example 12.10 ▲

<hr>

A right circular cone of base 60 mm diameter and axis 70 mm long is resting on the ground on its base on the HP. Calculate the shortest length of a string required to be wound over the lateral surface of the solid, starting from one extreme point of the base and ending at the same point. Obtain the path of the string in the front view and top view.

PROCEDURE (Refer Fig. 12.14)

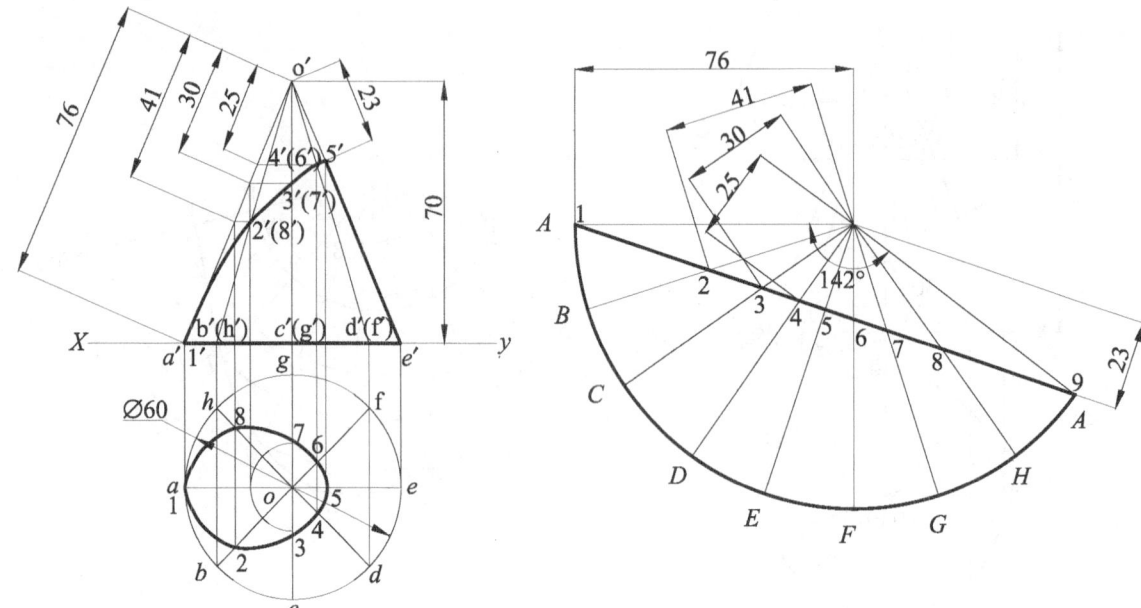

Fig. 12.14

Step 1: Draw the top and front views of the cone and obtain its development sketch with a solid angle $\theta = (r/l) \times 360°$ or $\theta = (30/76) \times 360° = 142°$. Mark all the generators, OA, OB, …, OH, and OA, similar to Example 12.8.

Step 2: The shortest length of the string wound around the lateral surface of the cone starting from the base point A and back to it can be identified by the shortest line connecting these points in the development sketch. Therefore, the straight line joining the points A and A gives the shortest length.

Step 3: Tracing back the points cut by this line with the lateral generators and transferring them to their respective locations in the front view gives the path of the string in that view. Projecting such points to the other view yields its path in the top view.

Step 4: The points cut by the line A–A are identified as 1, 2, …, 8, and 1 on the lateral generators OA, OB, …, OH, and OA in the development sketch. Measure the distances O1, O2, …, O8 and mark them in the true length generator o'a', and then transfer these points to the respective generators in the front view as 1', 2', …, 8'. Join the points 1'–2', …, 8', and 1' by means of a smooth curve and obtain the path of the string in the front view.

Step 5: Project the points 1', 2', …, 8' to their corresponding locations in the top view and join them to obtain the path of the string in that view. ▲

12.6 DEVELOPMENT OF LATERAL SURFACES OF SOLIDS WITH CUT-OUTS AND HOLES

In engineering applications, the sectioning of the objects are carried out with a purpose to connect another object or to facilitate the flow of liquid or gas from one direction to the other when the objects are hollow. Pipe lines branching in different directions, vessels, water jugs and kettles attached with handles and tee connections of taps and valves are some of the examples where the lateral surface of one solid has to be provided with a hole or opening to attach another solid. The reader can note many examples in Chapter 11 'for this purpose. Sometimes, a solid may be provided with certain openings or cut-outs in the form of arcs to enhance the appearance or fulfil a specific purpose. In such cases, the openings or holes or arcs are also considered as section planes and the same principle of identifying the cutting points and their locations in the development sketch are followed. More number of points on the arcs or holes can be assumed to improve the accuracy of the developed surface. The examples described in the following section explain these features and their method of development.

Example 12.11

A pentagonal prism of side of base 25 mm and height 65 mm is resting on the HP on its base with a face parallel to the VP. It is cut by two planes: one perpendicular to the VP, inclined at 50° to the HP, passing through the right extreme corner of the top face, and the other also perpendicular to the VP but in the form of a circular arc of radius 25 mm with the bottom right corner as its centre. Draw the development of the lateral surface of the portion of the solid entrapped between these two cutting planes.

PROCEDURE (Refer Fig. 12.15)

Step 1: Draw the top and front views of the prism with one of its base edges parallel to the reference line and draw the development sketch with all the lateral edges marked in it similar to Example 12.5.

Step 2: Draw the inclined section plane in the front view at an angle of 50° to the reference line and mark the cutting points 1′, 2′, ..., 5′ on the lateral edges through a′, b′, ..., e′.

Step 3: Similarly, draw an arc of radius 25 mm with its centre located at the bottom right corner d′ to indicate the other cutting plane. The cutting points of this arc with the solid are identified as p′ and t′ in the base and as q′, r′, and s′ in the lateral edges through c′, d′, and e′.

Step 4: The cutting points located in the lateral edges can be directly transferred to the development sketch and joined by straight lines or by a smooth curve, depending on the nature of the section plane and the solid surface.

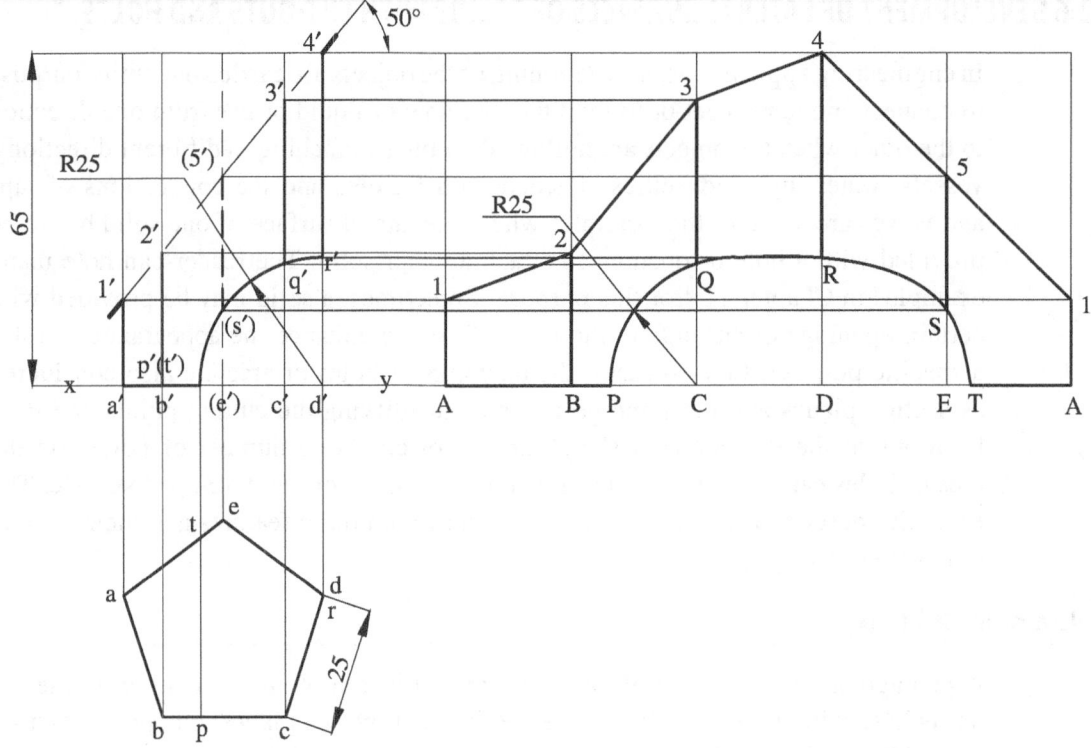

Fig. 12.15

Step 5: The cutting points p′ and t′ located in the base are projected to the top view and their locations on the edges bc and ea are obtained. The distances of p and t are measured from b and e and are marked in the development sketch from B and E, respectively. The points P, Q, ..., T are joined by means of a smooth curve.

Step 6: The portion in between the two curves is the development sketch of the lateral surface desired.

In order to get a more accurate development curve, some more points can be assumed in the arc and can be located in the top view and transferred to the development sketch at their respective base positions and the lateral heights marked. ▲

Example 12.12 ▲

Draw the development of the lateral surface of the part 'A' of the cylinder shown in Fig. 12.16.

PROCEDURE (Refer Fig. 12.16)

Step 1: Draw the top and front views of the cylinder and obtain the development sketch of the uncut solid with its generators marked.

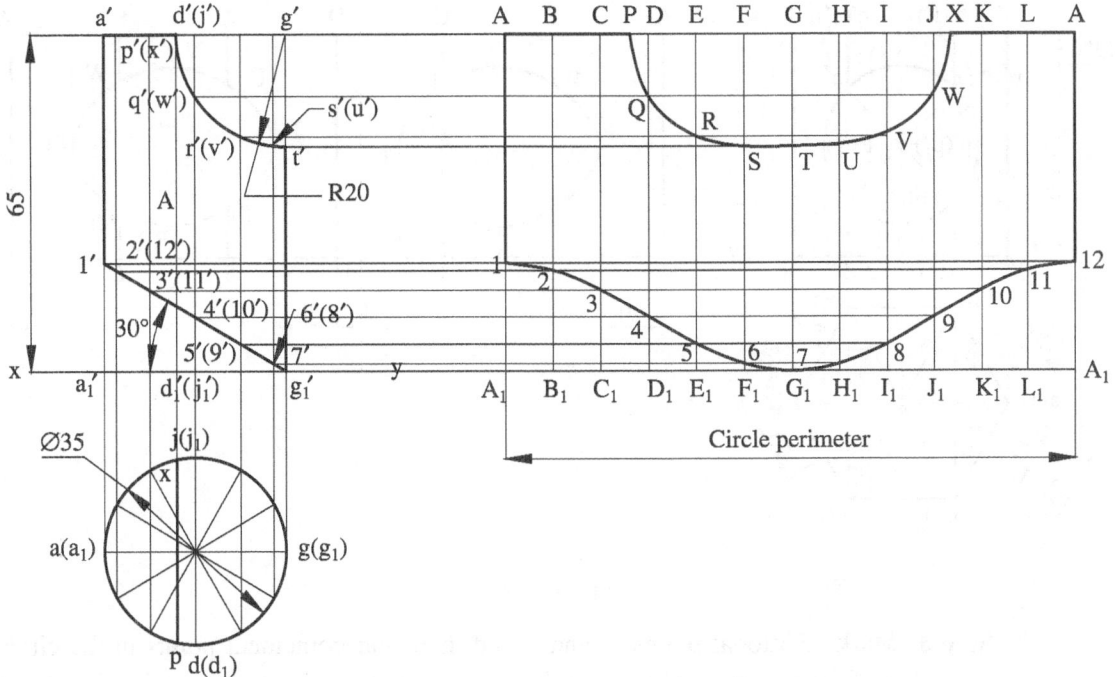

Fig. 12.16

Step 2: The procedure of locating the inclined section plane and the arc and obtaining the cutting points are similar to Example 12.11. The cutting points in the lateral generators are transferred directly.

Step 3: The cutting points at the top end p′ and x′ are projected to the top view and then transferred in between their respective positions CD and JK in the development sketch. The portion in between the two curves is the development sketch of the lateral portion A of the solid. ▲

Example 12.13 ◢

A solid hexagonal prism of 35 mm base sides and axis 70 mm long has a circular hole of 50 mm diameter, drilled at its mid-height. Draw the development of lateral surface of the solid with the hole.

PROCEDURE (Refer Fig. 12.17)

Step 1: Draw the top and front views of the hexagonal prism and obtain the development sketch of the uncut solid with its lateral edges marked.

Step 2: Draw the circle of diameter 50 mm to mark the hole in the front view. Mark the cutting points of the circle with the front and the rear lateral edges and transfer them directly to the development sketch.

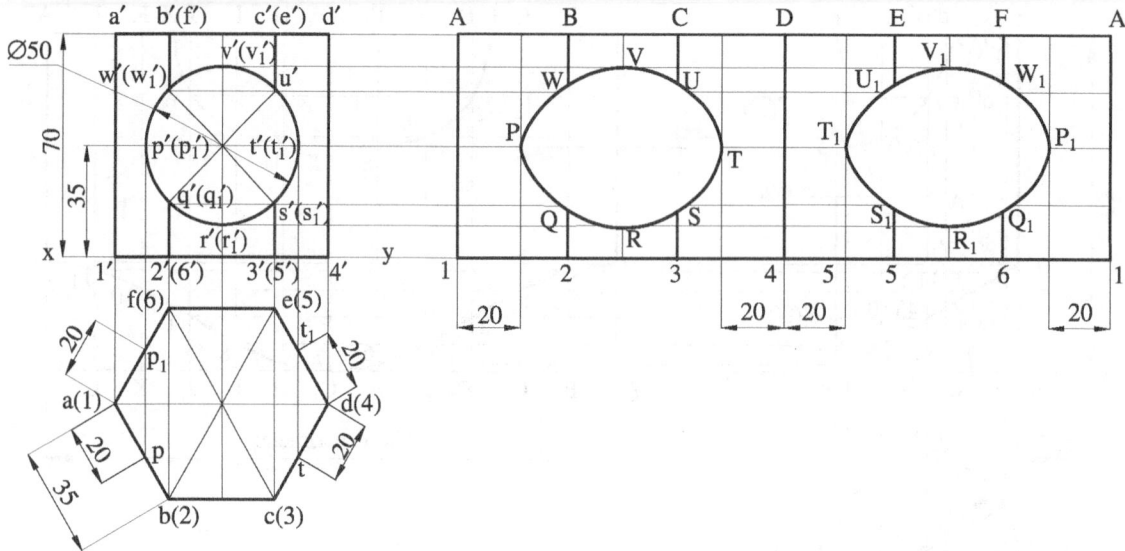

Fig. 12.17

Step 3: Mark additional points p′ and t′ and their rear coincident points in the circle by drawing tangent lines parallel to the lateral edges. As these points lie on the lateral faces, they are projected to the top view to obtain their locations.

Step 4: The distances of p, t, t_1, and p_1 are measured from 1, 3, 4, and 6, respectively, and are transferred to their corresponding locations in the bottom stretch out line. When lateral surface lines are realized through these points and are associated with the horizontal lines from p′, the points P, T, T_1, and P_1 are obtained in the development sketch.

Step 5: When the points pertaining to the holes are joined by means of smooth curves, two elliptical curves belonging to the front and rear portions of the hole emerge in the development sketch. As already mentioned, the accuracy of the elliptical curves can be improved by choosing more points in the circle in the front view. ▲

Example 12.14 ◢

A solid hexagonal pyramid of 35 mm base edges and axis 70 mm long has a circular hole of 30 mm diameter drilled along the axis at a distance of 43 mm from the apex. Draw the development of lateral surface of the solid with the hole.

PROCEDURE (Refer Fig. 12.18)

Step 1: Draw the top and front views of the hexagonal pyramid and obtain the development sketch of the uncut solid with its lateral edges marked. The lateral lines o′a′ and o′d′ give the true length of the slant edge.

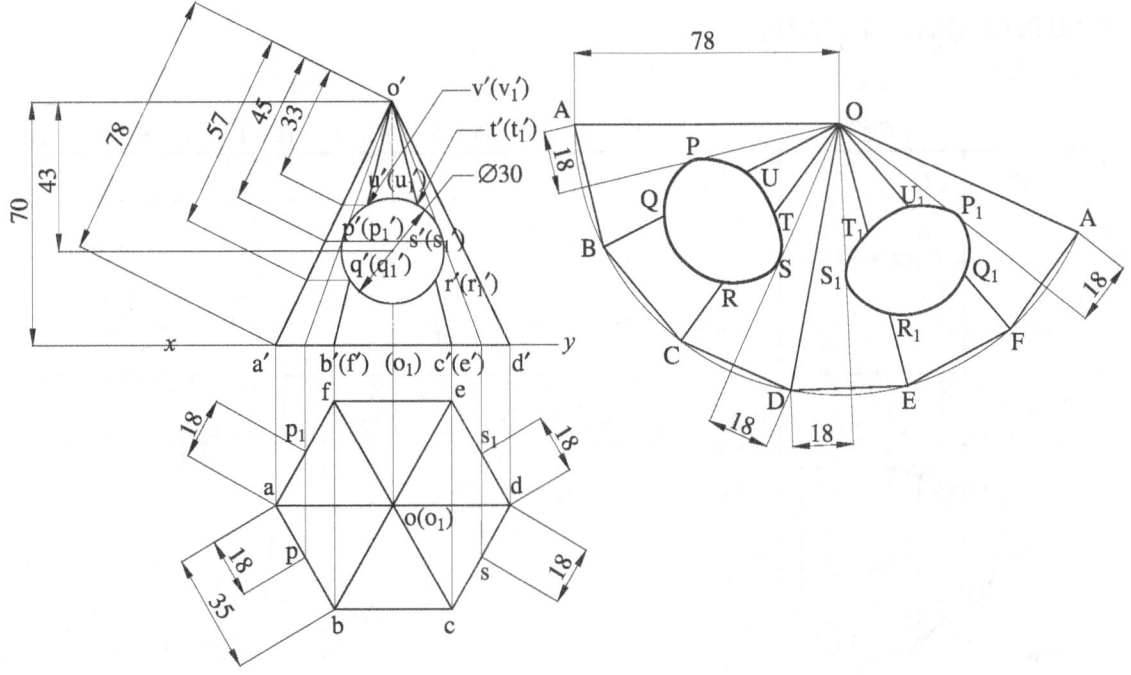

Fig. 12.18

Step 2: Draw the circle of diameter 30 mm to mark the hole in the front view. Mark the cutting points of the circle with the front and the rear lateral slant edges, transfer them to the true slant edge, and then mark radially to their respective locations in the development sketch.

Step 3: Mark additional points p′ and s′ and their rear coincident points in the circle by drawing tangent lateral lines. As these points lie on the lateral faces, they are projected to the top view to obtain their locations.

Step 4: The distances of p, s, s_1, and p_1 are measured from a, c, d, and f, respectively, and are transferred to their corresponding locations in the base line in the development sketch. When lateral surface lines are realized through these points and are associated with the radial distances of p′, etc., the points P, S, S_1, and P_1 are obtained in the development sketch.

Step 5: When the points pertaining to the holes are joined by means of smooth curves, two elliptical curves belonging to the front and rear portions of the hole emerge in the development sketch. As already mentioned, the accuracy of the elliptical curves can be improved by choosing more points in the circle in the front view. ▲

Example 12.15 ▲

A solid cylinder of 60 mm diameter and 70 mm high is drilled with a hexagonal slot of 23 mm sides at its mid-height. Draw the development of lateral surface of the cylinder with its hexagonal slot.

PROCEDURE (Refer Fig. 12.19)

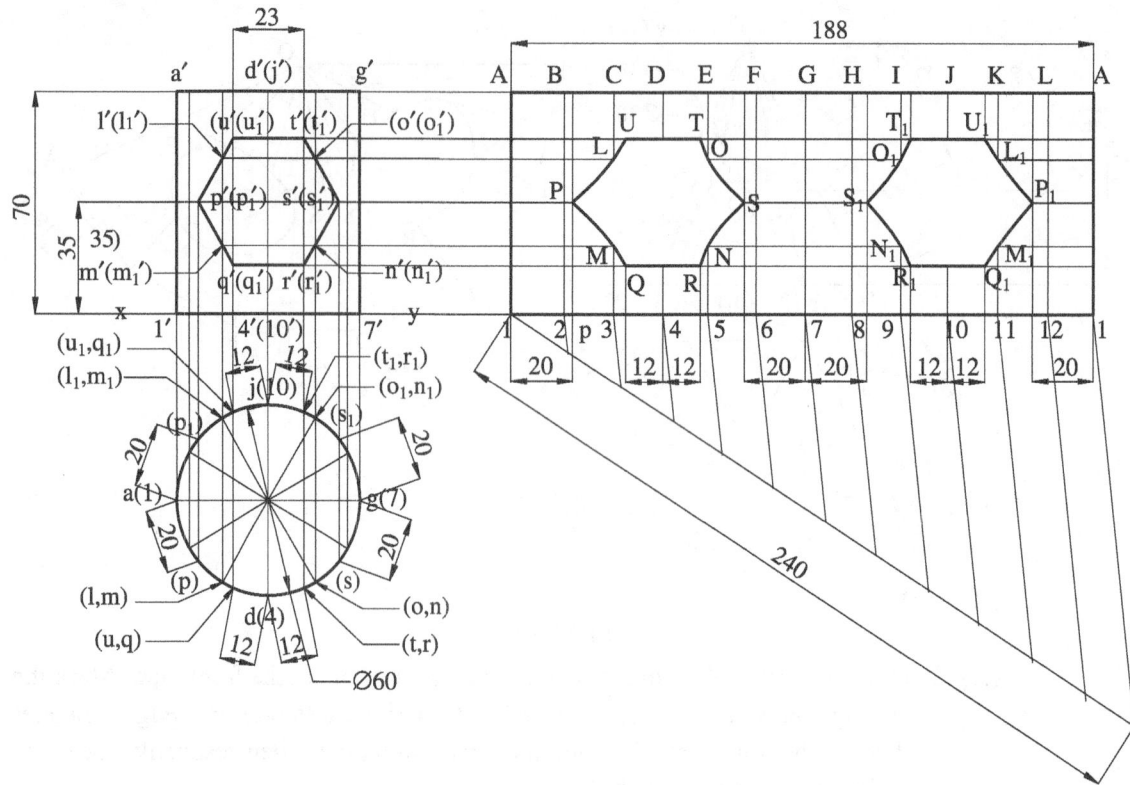

Fig. 12.19

Step 1: This problem is similar to Example 12.13 with the replacement of the vertical solid as a cylinder and the circular hole as a hexagonal slot.

Step 2: The development of the uncut cylinder is drawn first and the points pertaining to the hexagon are located as discussed earlier. The additional corner points p', s', s_1', and p_1' located in the lateral faces are brought to the development sketch following the same procedure indicated in Example 12.13.

Step 3: The points pertaining to the hexagonal slot are joined by means of straight lines and smooth curves belonging to its front and rear portions. As already mentioned, the accuracy of the curves can be improved by choosing more points on the inclined sides of the hexagon in the front view. ▲

Example 12.16 ▲

A sheet metal cone assembly with a circular base of 60 mm diameter and height 70 mm long has a circular opening of 30 mm diameter on its lateral curved surface. The centre of the hole lies on the axis of the cone and 23 mm from its base. Draw the development of lateral surface of the sheet metal assembly with the hole.

PROCEDURE (Refer Fig. 12.20)

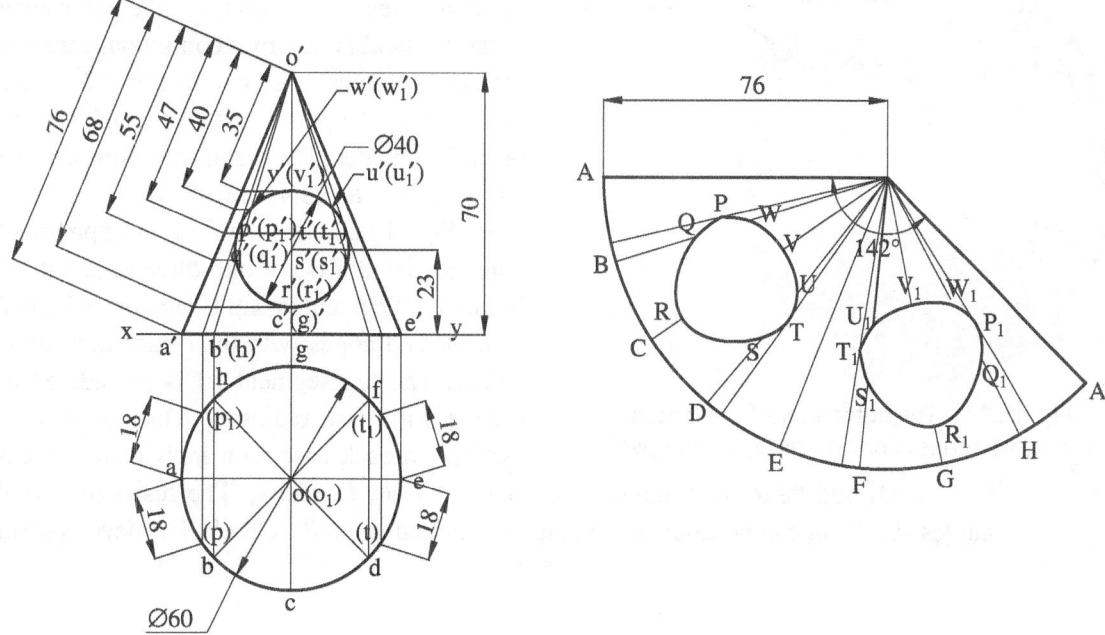

Fig. 12.20

Step 1: This problem is similar to Example 12.14 with the replacement of the vertical solid as a cone.

Step 2: The development of the uncut cone is drawn first after calculating the solid angle and the points pertaining to the hole are located as discussed earlier. The additional points p', t', t_1', and p_1' located in the curved face are brought to the development sketch following the same procedure indicated in Example 12.14.

Step 3: When the points pertaining to the circular hole are joined by means of smooth curves, two elliptical curves belonging to its front and rear portions emerge in the development sketch. As already mentioned, the accuracy of the elliptical curves can be improved by choosing more points in the circle in the front view.   ▲

12.7 DEVELOPMENT OF SPHERES

As mentioned earlier, a sphere is a double curved surface and cannot be developed by a single layout sketch. It can be approximately developed by slicing it into a set of horizontal segments or considering it as an assembly of lune-shaped pieces. Slicing by a set of horizontal planes makes each portion resemble a truncated cone, while the lune-shaped strips when bent form a semicylinder. Hence, these are known as polyconic or zone method and polycylindric or lune method, respectively.

12.7.1 Zone, Polyconic, or Cutting Plane Method

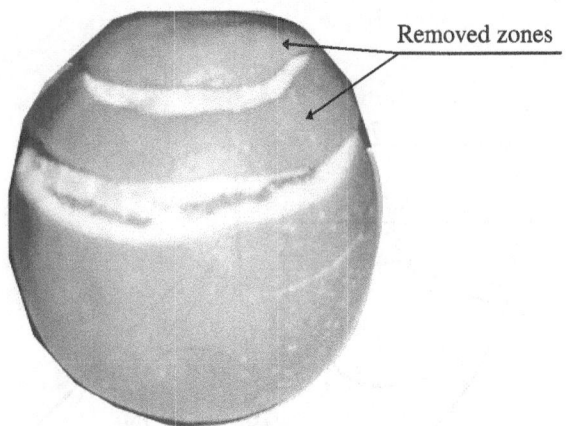

Fig. 12.21 The outer layer of the Mosambi is cut by a series of horizontal layers and peeled

In this method, the sphere is cut by a series of horizontal section planes (e.g., the outer layer of the Mosambi is cut by a continuous series of horizontal planes as shown in Fig. 12.21 and peeled off). The portion between each section plane is considered as a truncated cone and the top portion as an approximate cone.

In Fig. 12.22, one half of the sphere of diameter 70 mm is cut into three segments, I, II, and III. Segments I and II are approximated as truncated cones with their base radii as r_1 and r_2. The top segment III is considered as a cone with base radius r_3. The full cone is realized in each case with individual vertices $O_1, \ldots, O_3$ and their slant lengths are obtained as $l_1, l_2,$ and l_3. The respective solid angles $\theta_1, \ldots, \theta_3$ can be calculated using the relation $\theta = (r/l) \times 360°$. The development

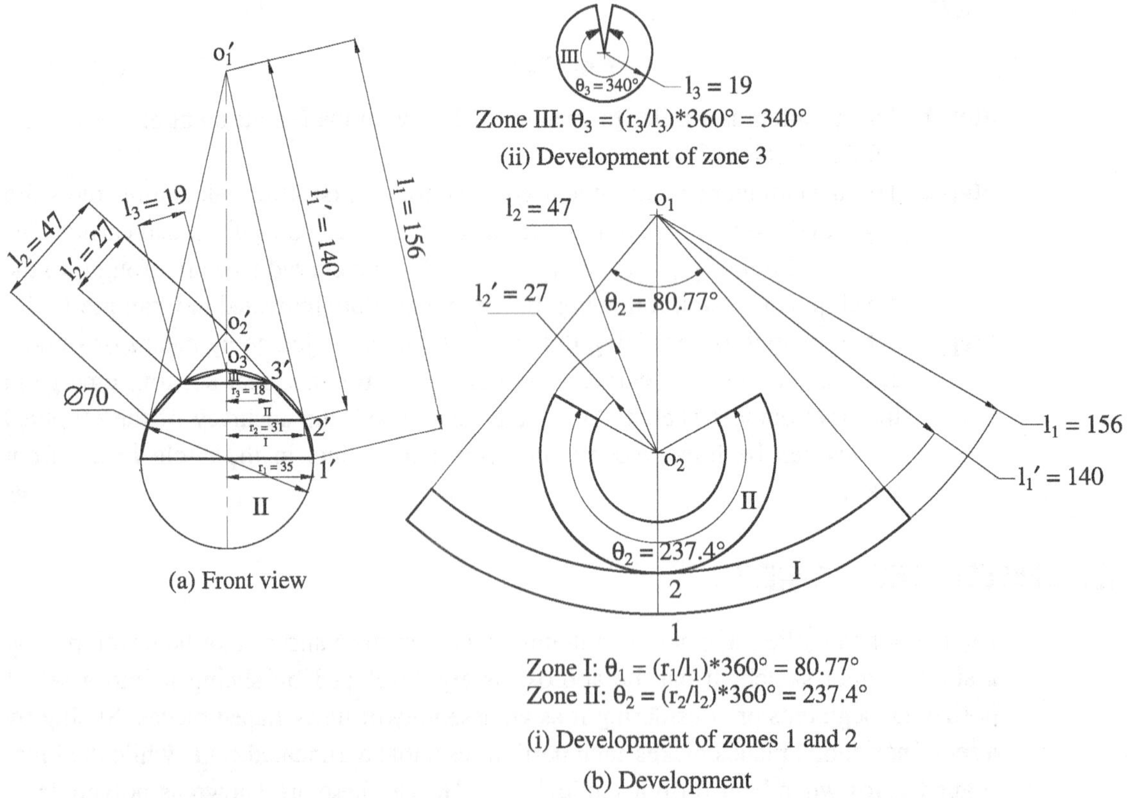

Zone III: $\theta_3 = (r_3/l_3)*360° = 340°$

(ii) Development of zone 3

(a) Front view

Zone I: $\theta_1 = (r_1/l_1)*360° = 80.77°$
Zone II: $\theta_2 = (r_2/l_2)*360° = 237.4°$

(i) Development of zones 1 and 2

(b) Development

Fig. 12.22 Surface development of the sphere by polyconic or zone method

of each portion is shown as a sector with the respective solid angles and arc lengths corresponding to the circumference of their base radii. Parallel arcs are shown in each segment to denote the top portion of that segment. The development of the portion III is shown as a complete sector with the solid angle θ_3 and the slant length l_3.

12.7.2 Lune Method

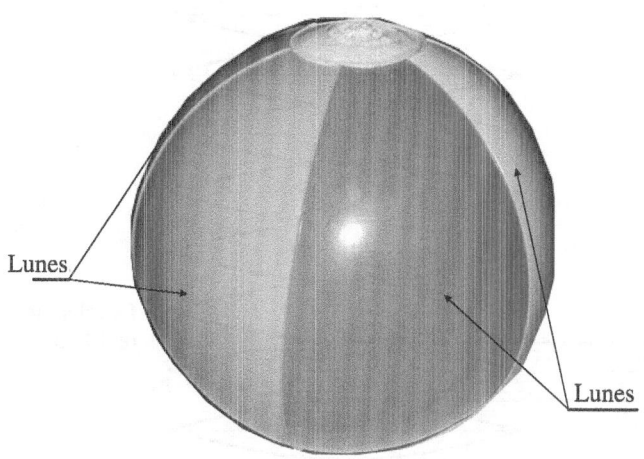

Lunes

Lunes

Fig. 12.23 Play ball made with the assembly of lunes

In this method, the spherical surface is considered to be an assembly of lunes or strips that converges to the centre point of the top and the bottom of the sphere (e.g., the children play ball that is made by assembling of number of lunes as shown in Fig. 12.23). These lunes are made up of a number of surface meridian lines. Such lines can be visualized in a globe model in natural spherical-shaped fruits like oranges. When pumpkin is cut, it is made in the form of lunes or segments such that all of them join at the centre. Each piece is also known as 'Gore'. These lunes or gores, when bent or rolled into semicylindrical shapes, it joins with top and bottom of the sphere. Since this lune arrangement is semicylindrical, this method is also known as polycylindric method. Therefore, the development of each lune stretches out for a length equal to half the circumference of the diameter that makes the sphere in one dimension and the maximum height of the lune as the other dimension.

PROCEDURE (Refer Fig. 12.24)

Step 1: Draw the two views of the sphere

Step 2: Divide one of the views (the top view) into equal number of parts (12) and name the angular sectors as I, II, ..., XII. Each sector is a lune or gore. Let us consider the half of the lune III which is represented as 0–4_{11}–4_{22}.

Step 3: Since one lune stretches to half the circumference, the corresponding semi circle in the front view gives its projection. Divide that semicircle into eight equal parts and name them as $0'$, a', ..., h'.

Step 4: Set at any convenient distance a length equal to half the circumference, adjacent to the lune and mark it as 0_1–h_1, divide into eight equal parts, and erect perpendiculars at each division points.

Step 5: Project the points $0'$, a', ..., h' from the semicircle to the lune 0–4_{11}–4_{22} and mark them as 0, a, ..., d.

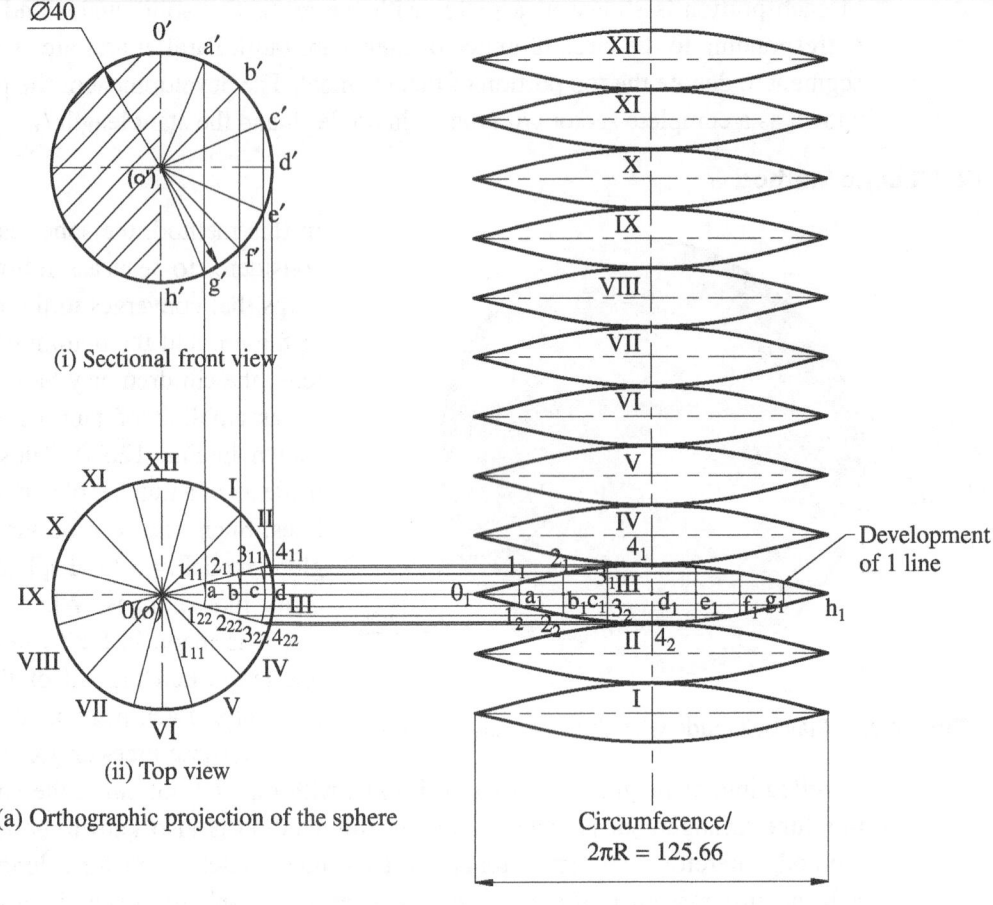

(i) Sectional front view

(ii) Top view

(a) Orthographic projection of the sphere

(b) Development of lunes of the sphere

Circumference/
$2\pi R = 125.66$

Fig. 12.24   Surface development of the sphere by polycylindric or gore method

Step 6: With 0 as centre and 0a, 0b, etc., as radius, draw concentric arcs to cut the lune at 1_{11}–1_{22}, 2_{11}–2_{22}, 3_{11}–3_{22}, and 4_{11}–4_{22}.

Step 7: Project the points 1_{11}–1_{22}, 2_{11}–2_{22}, etc., horizontally to meet the perpendiculars erected at a_1, ..., h_1.

Step 8: Join the points, thus, obtained by a smooth curve. This gives the development of one lune arrangement.

Step 9: Since the circle is divided into 12 equal parts (12 lunes), 12 such development profiles are to be cut and bent and to be joined to get the development of the sphere.

12.8 ENGINEERING APPLICATIONS

Engineering applications of the development of solid surfaces involve the basic solid shapes discussed earlier getting assembled with each other to facilitate a purpose.

Sheet metal assemblies are often found in funnels, metallic hoppers to facilitate the flow of granular particles, flow of liquids, etc. Air-conditioning ducts with conduits of curved shapes are used to make smooth flow of air and to avoid the turbulence effects. The examples shown below discuss the assembly two or three solid shapes. The development sketches of the individual parts are prepared by considering the interfaces as the section or cutting planes and are laid by the side of their lateral surfaces for easy identification.

Example 12.17

A sheet metal funnel is made up of a truncated conical portion of 60 mm top diameter and 17 mm base diameter and joins with a cylindrical portion at its lower end. If the conical and the cylindrical portions are each of length 50 mm, draw the layout of the sheet metal required to fabricate the funnel.

PROCEDURE (Refer Fig. 12.25)

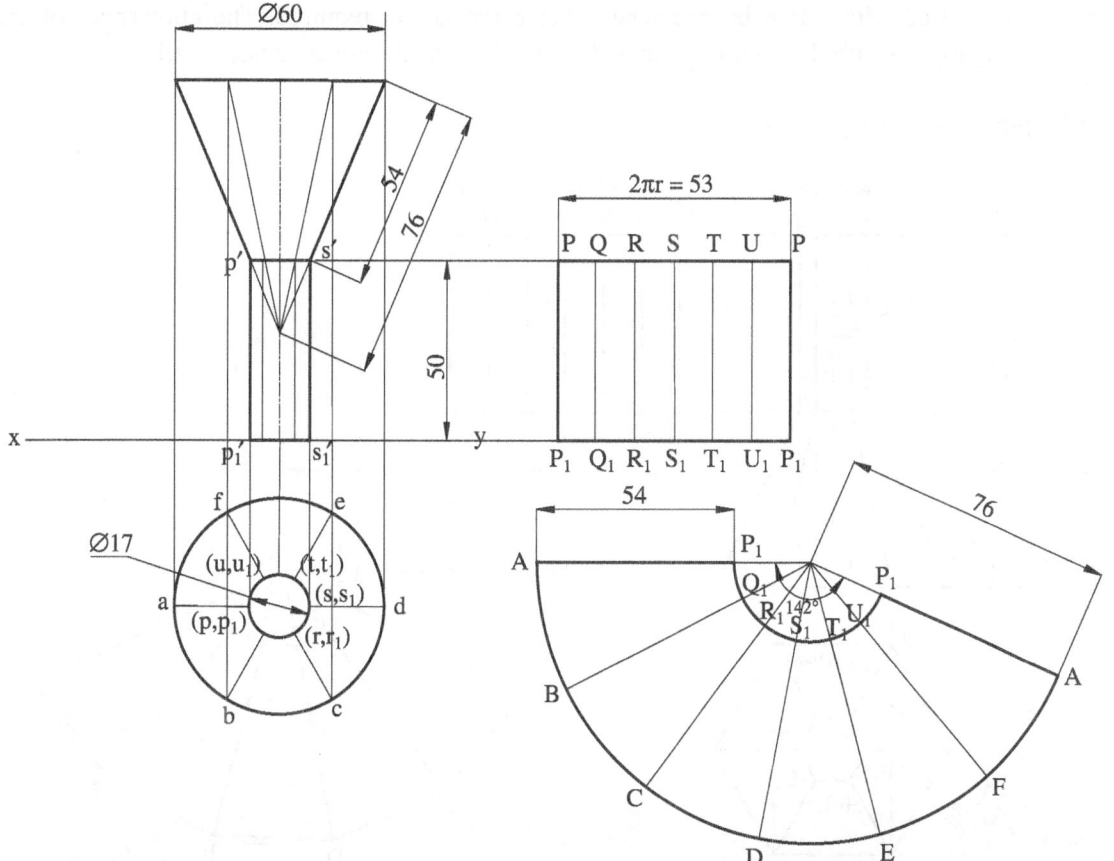

Fig. 12.25

Step 1: Draw the circle of 60 mm diameter in the top view and obtain the front view of the full cone, by keeping the larger end above. Draw the development sketch of the uncut cone by setting the true length of the slant generator.

Step 2: Represent the truncated cone by a horizontal section plane at a distance of 50 mm from the top end and mark it in the development as an arc parallel to the top end and at a distance of 54 mm.

Step 3: Draw the front view of the lower cylindrical portion in the front view and mark the development of its lateral surface with its generator lines as shown in Fig. 12.25.

Step 4: The two individual developmental sketches are the sheet metal areas meant for the funnel. ▲

Example 12.18 ◢

A wooden table weight has an upper portion made of a hexagonal prism arrangement, followed by a truncated hexagonal pyramid as its lower portion. The hexagonal upper portion has 10 mm base sides, while the base of the lower portion is of 35 mm sides. The height of both portions is equal to 50 mm. If the wooden table weight is to be covered with a thin bronze sheet, sketch the layout required. The allowances of the margin required for joining the ends of the bronze sheet can be neglected.

PROCEDURE (Refer Fig. 12.26)

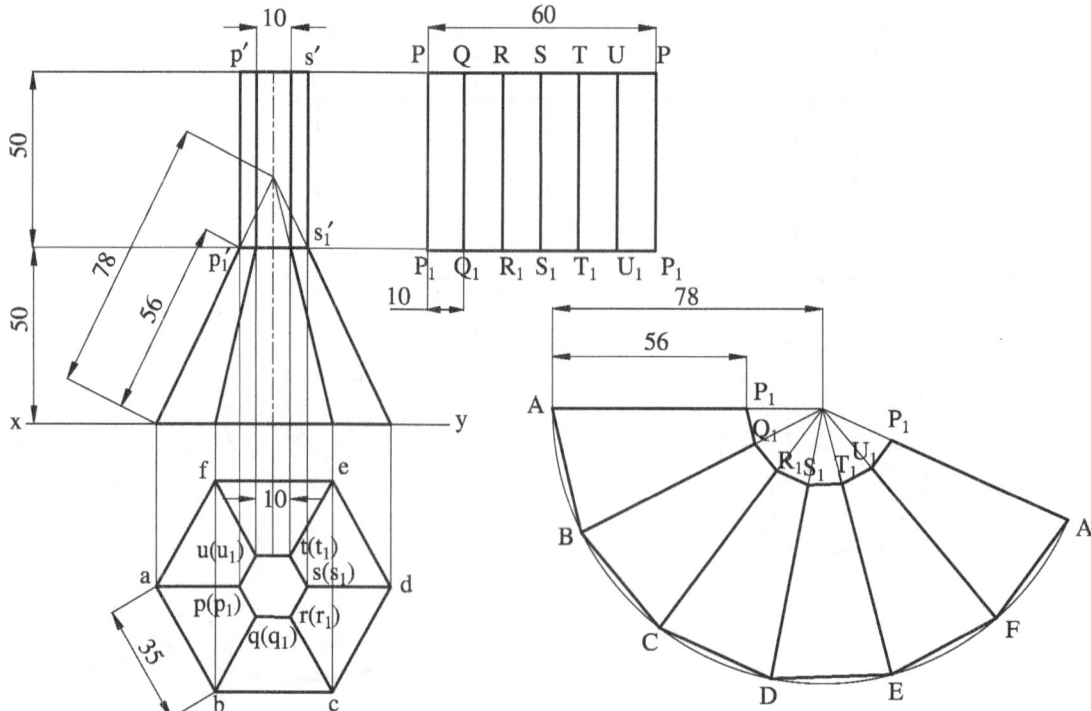

Fig. 12.26

Step 1: Draw the hexagon of 35 mm sides in the top view and obtain the front view of the full pyramid. Draw the development sketch of the uncut solid after obtaining the true length of the slant edge.

Step 2: Represent the truncated arrangement by a horizontal section plane at a distance of 50 mm above the base and mark it in the development sketch after setting the true length of the cut slant edge.

Step 3: Add front view of the prism portion for a height of 50 mm, after drawing a concentric hexagon of 10 mm sides in the top view. Complete the development of its lateral surface with its lateral edges as shown in Fig. 12.26.

Step 4: The two individual developmental sketches are the sheet metal areas meant for the table weight. ▲

Example 12.19 ◢

A sheet metal assembly as shown in Fig. 12.27(a) is to be fabricated. Sketch the layout of the sheet metal required, neglecting the allowances for the joints.

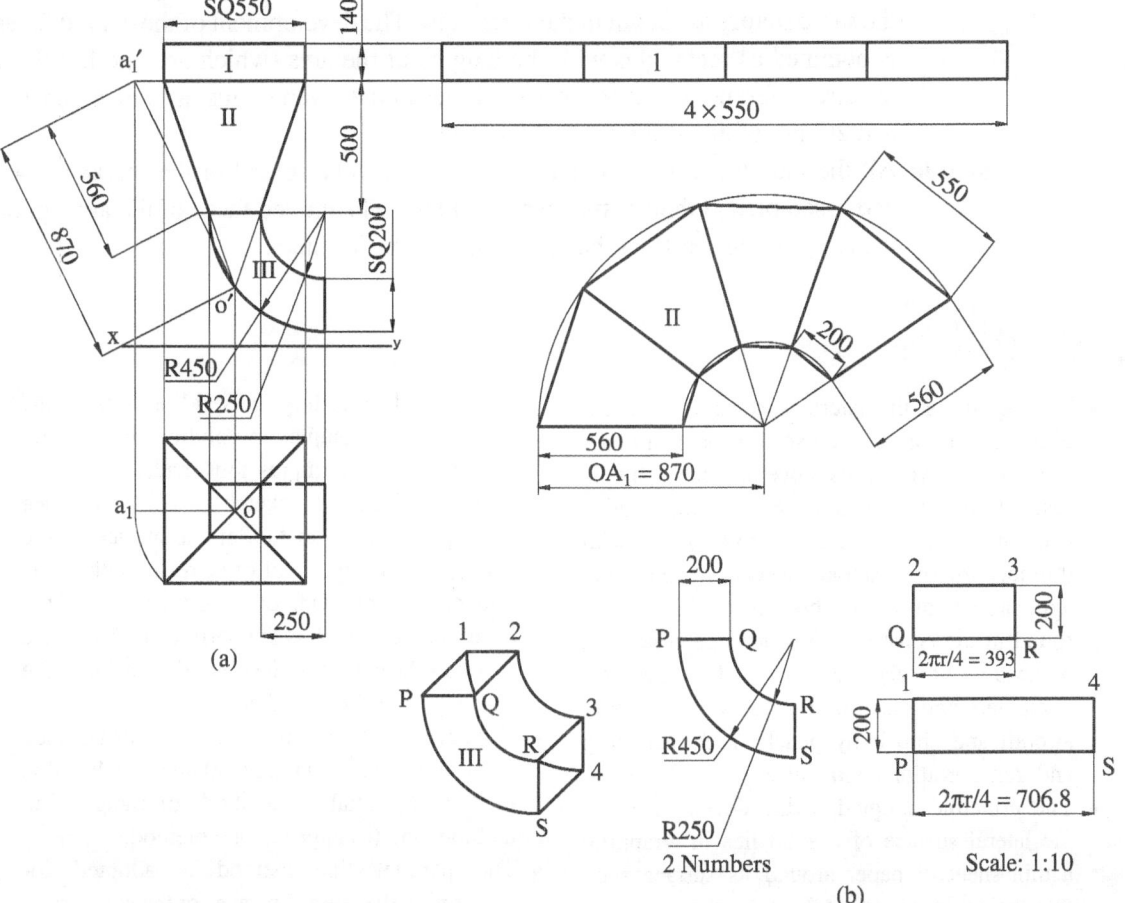

(a)

(b)

Fig. 12.27

PROCEDURE   (Refer Fig. 12.27b)

Step 1: Draw the front and top views of the given assembly which involves three parts I, II, and III.

Step 2: Draw the development of the first portion as four rectangles of 550 mm sides for a height of 140 mm.

Step 3: The II portion is a truncated inverted square pyramid. Construct the full pyramid in the front view and obtain the top view. The slant edge in the front view does not give the true length as its top view is not parallel to the reference line. Therefore, make the top view parallel to the reference line and obtain its true length. Set it in the development sketch and mark the lateral edges.

Step 4: Extend the bottom end of the pyramid to the true slant edge, obtain its length and set it in the development sketch.

Step 5: The pictorial view of the IIIrd portion is shown in figure for reference. It involves two curved surfaces, namely 2–Q–R–3 and 1–P–S–4 and two flat surfaces for the front and the rear. The development of the front and the rear portions are of the same shapes as shown in the front view. The development of the curved faces is obtained as rectangles with the lengths of the arcs (which are one fourth of the circumference of the respective circles of the given radius) as one dimension and 20 mm width as the other dimension.

Step 6: All the individual development sketches are to be folded in the marks shown and assembled at their respective interfaces. The curved faces of the arc can be made by suitably rolling the rectangles meant for them.   ▲

RECAPITULATION

- The layout of the lateral surface of a three-dimensional object on a two-dimensional plane sheet is known as its development. This is essential for the development of hollow solids and whose outer surfaces are to be made with thin metallic or cardboard sheets. The common applications of these sheet metal works are found in the construction of storage container, domestic vessels, casings of machines, computers and electronic gadgets, automobile, aircraft and ship body panels, and packaging and transportation containers.

- The principle adopted in the development of the lateral surface of a solid lies in wrapping a thin sheet or paper around its surface such that their free edges touch each other and then unwrapping it to lay on a flat plane surface. Alternatively, the development sketch can also be obtained by rolling the solid on a flat plane sheet for one complete revolution of the solid and obtaining the impressions made.

- The development sketch has one-to-one correspondence with the lateral surface of the solid, and hence, it always reflects the true dimensions only. One of the dimensions of the development sketch is the perimeter of the base or top of the solid and the other dimension is the true length of the lateral edge or generator.

- The development profile of the solid can be made by four methods, namely (a) the parallel line method, (b) radial line method, (c) triangulation method, and (d) approximate method.

- The parallel line method is adopted for developing the lateral surfaces of the solids such as prisms and cylinders that involve the top and the base ends parallel to each other. Any point on the lateral surface can be located parallel to the locus of these ends at its respective height. The development of the prism of *n* sides consists of *n* rectangles of size equal to the side of the base of the prism and length or height as the other dimension.

- The radial line development method is adopted for solids such as pyramids and cones that have their base points located radially and equidistant from the apex. Therefore, the development sketch consists of a sector that represents the perimeter and the true length of the slant edge or generator. The development of the lateral surface of a pyramid of base *n* sides consists of *n* isosceles triangles of size equal to the base of the pyramid and the true length of the slant edge laid adjacent to each other.

- The development of the lateral surface of a cone is obtained as a sector with an arc length equal to the true length of the slant generator (l) and an arc angle, $\theta = (r/l) \times 360°$, where '$r$' is the radius of the base of the cone.

- The triangulation method is used for the development of the lateral surfaces of transition pieces that connect different sizes and shapes. The curved surface is attempted by splitting them into triangular portions. Many air-conditioning ducts involve such shapes.

- The development of the surfaces of the objects that involve double curvature or warping like sphere, ellipsoid, paraboloid, helicoids, etc., are attempted by approximation process. The object is sliced into different portions, and each portion is approximated with the nearest solid shape, their development profiles are drawn and combined judiciously.

- When the solids are sectioned or cut with conventional two-dimensional section planes, curved sections such as arcs, cut-outs, and holes, the cutting points of such planes are identified as discussed in the Chapter 10 on 'Section of Solids'. The cutting points are appropriately transferred to the uncut development sketch of the object and the resultant shapes are identified for the cut solids.

MULTIPLE-CHOICE QUESTIONS

12.1 The development sketch of the lateral surface of a solid reveals
   (a) the largest dimension of the solid
   (b) all three dimensions of the solid
   (c) any two dimensions of the solid
   (d) the two dimensions of the lateral surface only

12.2 The area of the development profile of the lateral surface of a solid involves
   (a) one side of the base and the height of the solid
   (b) the perimeter of the base and the true length of the lateral edge or generator
   (c) the perimeter of the base and half the height of the solid
   (d) half the perimeter of the base and the height of the solid

12.3 The parallel line method of development is adopted for

   (a) all the solids
   (b) the prisms only
   (c) the prism and cylinders
   (d) the prisms and the pyramids

12.4 The development sketch for the lateral surface of a cone involves
   (a) an arc of length equal to the radius of the base
   (b) an arc of length equal to the circumference of the base
   (c) a cord of length equal to the circumference of the base
   (d) an arc of length equal to the true length of the slant generator

12.5 The development of the lateral surface of a right and regular pyramid consists of
   (a) a set of isosceles triangles adjacent to each other and converging with the apex

(b) a set of isosceles triangles opposite to each other

(c) a set of equilateral triangles adjacent to each other and converging with the apex

(d) a set of oblique triangles adjacent to each other and converging with the apex

12.6 The solid angle (θ) subtended at the apex of a cone is related to the base radius (r) and the true length of the slant generator (l) and is given by
(a) $\theta = (r/l) \times 90°$ (c) $\theta = (r/l) \times 360°$
(b) $\theta = (l/r) \times 360°$ (d) $\theta = (r/l) \times 180°$

12.7 The triangulation method is used for the development of
(a) the triangular pyramids
(a) the triangular prisms
(c) the sphere
(d) the transition pieces

12.8 The cutting points on the lateral edges or generators of a pyramid or cone are transferred
(a) to the true slant edge or generator and then transferred to the development sketch

(b) directly to the development sketch as per their distances from the base

(c) directly to the base and then transferred to the development sketch as per their heights

(d) always to the axis line and then transferred to the development sketch

12.9 The true length of the slant edge or generator is identified in one view if its corresponding other view is
(a) perpendicular to the reference line
(b) inclined at 45° to the reference line
(c) parallel to the reference line
(d) inclined at 30° to the reference line and passing through the apex

12.10 The development of solids with double curved and warped surfaces are obtained by
(a) parallel line method
(b) triangulation method
(c) approximate method
(d) radial line method

WORK PRACTICE LEVEL – 1

12.1 Draw the development of the lateral surface of a pentagonal prism when it has one of its faces perpendicular to HP and VP and has its base on HP. The base edges measure 30 mm and its axis is 80 mm long.

12.2 A hexagonal pyramid of base sides 30 mm and axis 70 mm long is standing on its base on HP with two triangular faces equally inclined to VP. It is cut by a section plane whose VT is at 30° reference line and passes through a point on the axis and 25 mm from the base. Draw the development of the lateral surface of the upper portion of the pyramid.

12.3 A square pyramid of side of base 40 mm and axis 80 mm long has its base on HP with two base edges parallel to VP. A section plane perpendicular to VP and inclined at 45° to HP bisects the axis. Draw the development of the lower portion of the pyramid on the HP.

12.4 A thin lamp shade in the frustum of a cone has its larger end of 150 mm diameter and smaller

end of 50 mm diameter height 150 mm. Draw the development of its lateral surface.

12.5 The top and bottom surfaces of a hopper are regular pentagons with sides of 400 mm and 300 mm, respectively. If the height of the hopper is 400 mm, draw the shape of the layout of the sheet metal required to fabricate it.

12.6 Draw the development of the lateral surface of the cylinder as shown in Fig. 12.28.

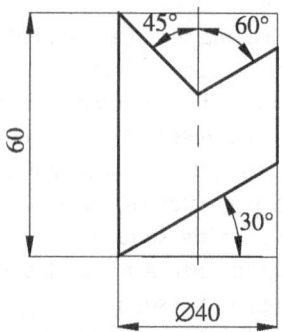

Fig. 12.28

12.7 Draw the development of the lateral surface of the cylinder as shown in Fig. 12.29.

12.8 Draw the development of the lateral surface of the cone as shown in Fig. 12.30.

12.9 A cylinder of 40 mm diameter and 60 mm high has a transverse square hole of 20 mm size, whose centre is at the midpoint of the axis. Draw the development of the lateral surface of the cylinder with the hole if its edges are equally inclined to the HP.

12.10 Draw the development of the lateral surface of a hexagonal pyramid when it lies on one of its triangular faces on HP and axis parallel to VP and when it rolls on HP for one complete revolution. The base edges measure 30 mm and its axis is 70 mm long.

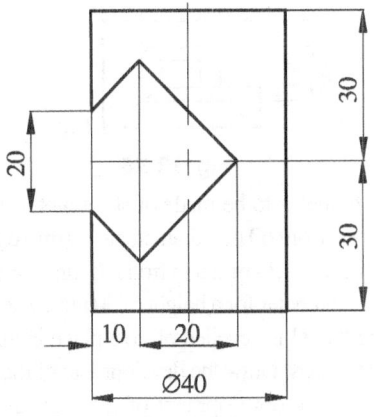

Fig. 12.29

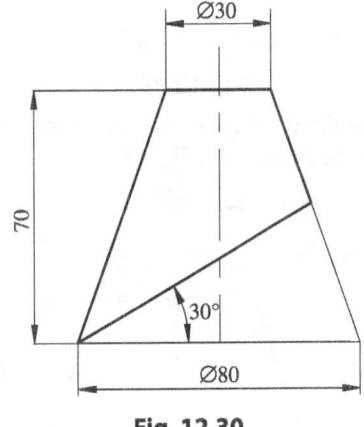

Fig. 12.30

WORK PRACTICE LEVEL – 2

12.1 A hexagonal solid block of 20 mm sides and axis 80 mm long is drilled with a circular hole of 30 mm diameter passing through two of its parallel flat faces centrally. Draw the development of the solid surface with the hole.

12.2 Figure 12.31 shows the front view of a cut hexagonal pyramid. Draw the development of the lateral surface of the portion that is retained.

12.3 Draw the development of the lateral surface of the cylinder as shown in Fig. 12.32.

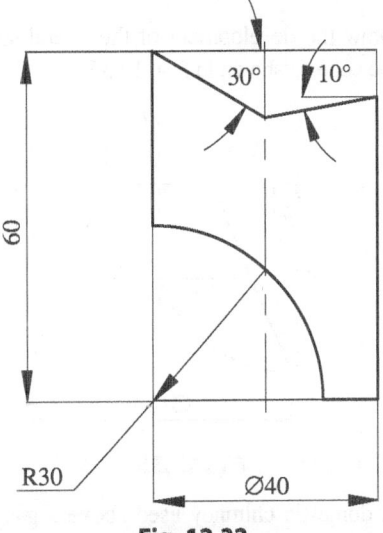

Fig. 12.32

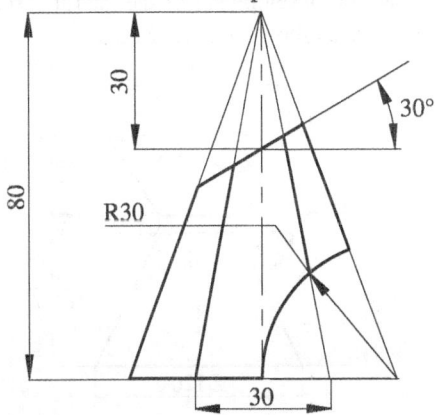

Fig. 12.31

12.4 Draw the development of the lateral surface of the cylinder as shown in Fig. 12.33.

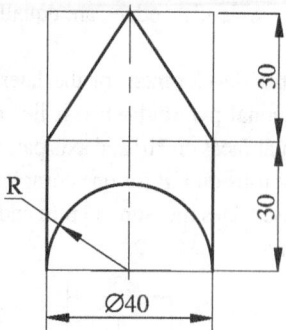

Fig. 12.33

12.5 Draw the development of the lateral surface of the cone as shown in Fig. 12.34.

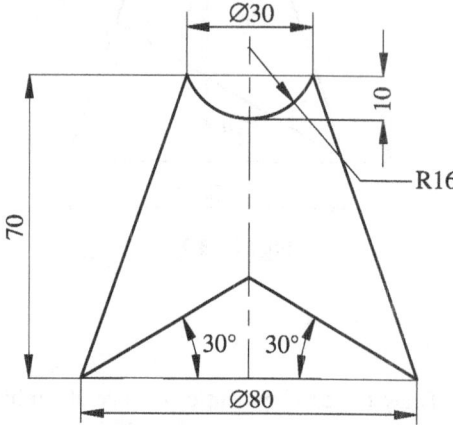

Fig. 12.34

12.6 Draw the development of the lateral surface of the cone as shown in Fig. 12.35.

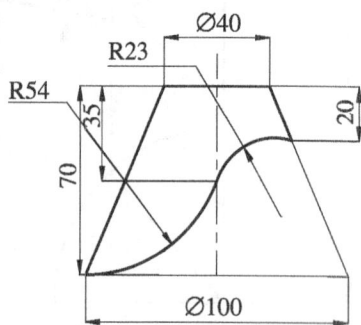

Fig. 12.35

12.7 A domestic chimney used above a gas stove is shown in Fig. 12.36. Draw the development of sheet metal required to fabricate it. The margins for joining the edges can be neglected.

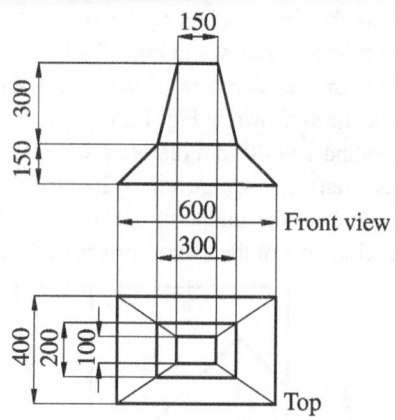

Fig. 12.36

12.8 A funnel is to be made of sheet metal. The funnel tapers from 60 mm diameter to 30 mm diameter for a height of 25 mm and from 30 mm diameter to 20 mm diameter for a height of 50 mm. The bottom of the funnel is bevelled off to a plane inclined at 45° to the axis. Draw the development of the funnel.

12.9 Draw the development of the joint as shown in Fig. 12.37.

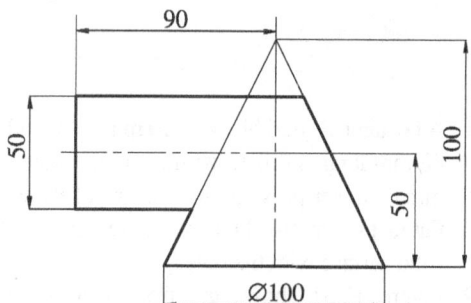

Fig. 12.37

12.10 Draw the development of the lateral surface of the cone as shown in Fig. 12.38.

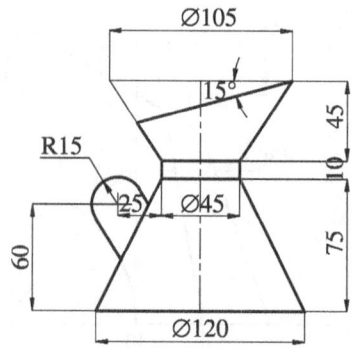

Fig. 12.38

12.11 Two pieces A and C are connected by the piece B as shown in Fig. 12.39. Draw the development of the three pieces.

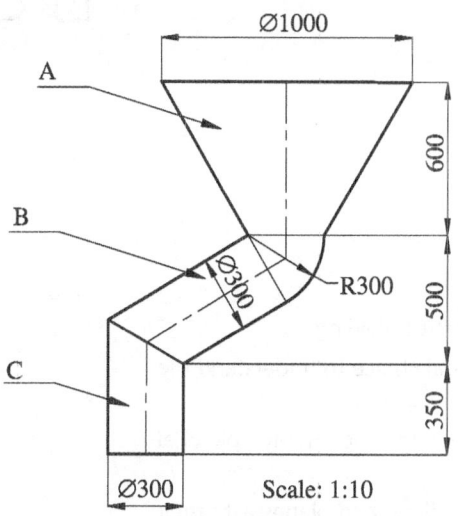

Fig. 12.39

Scale: 1:10

12.12 Draw the development of the duct assembly as shown in Fig. 12.40.

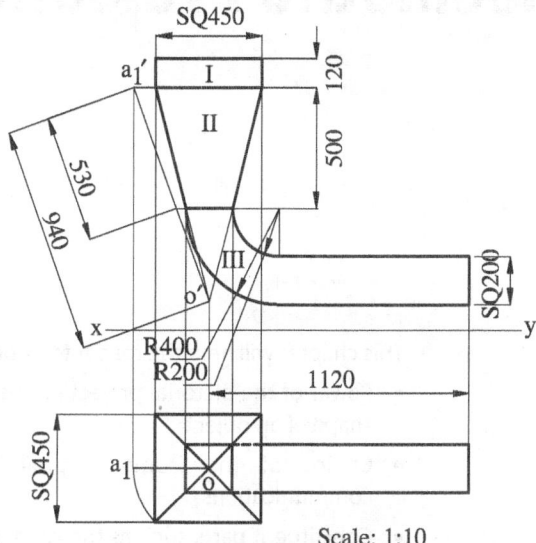

Scale: 1:10

Fig. 12.40

Answers for Multiple-choice Questions

12.1 (d)	12.2 (b)	12.3 (c)	12.4 (b)	12.5 (a)	12.6 (c)
12.7 (d)	12.8 (a)	12.9 (c)	12.10 (c)		

Isometric Projections

13

OBJECTIVES

This chapter will help the reader to understand the following:

- Power of the pictorial projections as a general choice to understand the shape of an object
- Choice of the isometric projection among the many pictorial communications
- Constituent parts such as the isometric axes, lines, and planes with their orientations and the fitment scheme of the relevant orthographic views in them
- Distinguish the isometric projections and the isometric views
- Method of construction of the isometric projections of the planar objects
- Method of construction of the isometric projections of three-dimensional objects in simple positions and cut orientations
- Isometric projections or views of combined solids
- Method of drawing the isometric views of solids from their orthographic projections
- Engineering applications related to machine components, castings, and bearings

13.1 INTRODUCTION

In this book, from Chapters 5 to 12, the orthographic projections have been elaborately discussed in order to enable the reader to obtain complete details of different features involved in the projection of a three-dimensional (3D) object in its true form. This has necessitated the preparation of multiple views such as front view, top view, and side views for an object. In some cases, additional views in specific direction may also be required as in obtaining the true shape of a cut surface. Such a thorough knowledge is essential as engineers are required not only to understand the drawings of an object but also have the responsibility to make or manufacture them. Usually, the first-level understanding and communication of the shape of the object with a larger group (technical and non-technical) becomes easier and faster, when the shape is presented in a single picture or in a pictorial form, fairly bringing all the three dimensions of the object. The pictorial representation of sketches or drawings in 3D form enhance the visualization and thinking capabilities, since the holistic idea of all the three surfaces and their dimensions are fairly available in one view. As discussed in Chapter 4, there are three major types of pictorial projections, namely axonometric, oblique, and perspective projections, which are used to convey the appearance of an object. In this chapter, axonometric projections are discussed in general, and isometric projections, a subgroup of axonometric projections, are explained in particular, since it is used widely.

13.2 TYPES OF AXONOMETRIC PROJECTIONS

The axonometric projection of an object presents its appearance on a vertical plane (VP), when the object is tilted about a corner of its base on the horizontal plane and is closer to the observer such that all the three front mutually perpendicular edges are inclined to the vertical plane or the plane of projection. Since the front edges of the object are inclined to the plane of projection, the axonometric projection is always foreshortened.

If all the three mutually perpendicular edges of 3D object are inclined at different angles with the plane of projection, then in the corresponding axonometric projection, the perpendicular edges also appear at different inclinations and their lengths though foreshortened are in different proportions. Such an axonometric projection is known as a *trimetric projection.*

If two of the three mutually perpendicular edges of the 3D object make equal inclinations with the plane of projection, then in the corresponding axonometric projection, these two perpendicular edges alone appear equally inclined and are foreshortened in the same proportion, though the third edge appears at a different angle and is in a different proportion. Such an axonometric projection is known as *dimetric projection.*

If all the three mutually perpendicular edges of the 3D object are making equal inclinations with the plane of projection, then in the corresponding axonometric projection, these edges appear equally inclined and are foreshortened in the same and equal proportion. Such an axonometric projection is known as *isometric projection.*

In the dimetric and trimetric projections, since the edges are not equally shortened, representations of symmetric features, such as circles, are not possible, and hence, these projections are not in use often. Since all the three sides of an object are equally inclined and are equally foreshortened, isometric projections are often preferred. Figure 13.1 shows the relative comparison of the different types of the axonometric projections of a cube, when it is resting of one of its corners Y and towards the observer and the three perpendicular edges OX, OY, and OZ through that are inclined to the vertical plane of projection with the axonometric angles α, β, and v between them.

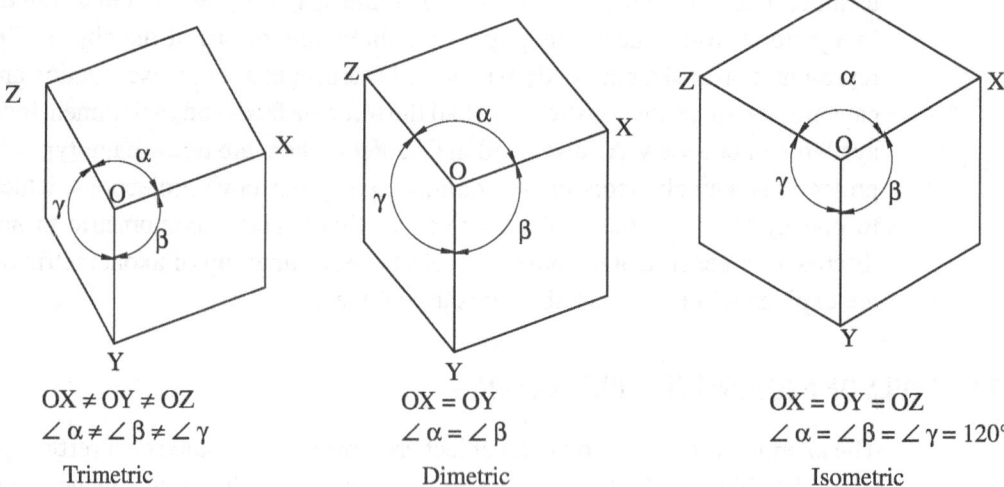

$$OX \neq OY \neq OZ \qquad OX = OY \qquad OX = OY = OZ$$
$$\angle\alpha \neq \angle\beta \neq \angle\gamma \qquad \angle\alpha = \angle\beta \qquad \angle\alpha = \angle\beta = \angle\gamma = 120°$$

Trimetric   Dimetric   Isometric

Fig. 13.1   Types of axonometric projections

13.3 ISOMETRIC PROJECTIONS AND RELATED FACTS

13.3.1 Basic Properties

In isometric projection, the three mutually perpendicular edges of an object are equally inclined to each other at an angle of 120°, and also, the lengths of these edges are also equally foreshortened by the same proportion. For this reason, isometric projection is one of the most preferred pictorial projections, as it enables the dimensions to be measured from the drawing itself. It has the convenience of observing the dimensions without any ambiguity and projecting the shape with all its three faces and hence has an inherent visual appeal.

13.3.2 Verification of Basic Properties

The basic property mentioned in Section 13.3.1 can be verified by considering the cube as an example.

In Fig. 13.2(a), the various stages of the projection of a cube resting on a base corner (1_1) on horizontal plane (HP) and getting tilted towards the observer are shown. In the first stage, the cube is placed such that two of its faces are equally inclined to the VP with one of the other faces resting on the HP. In the second stage, the cube is tilted towards the left such that it rests with one of the bottom corners $(1_2')$ on HP and the solid diagonal opposite to it $(a_2'3_2')$ becomes parallel to HP and VP. In the third stage, the solid is rotated about the base corner (1_1) lying on HP such that the edge 1'-a' is moved towards the observer, retaining the other inclinations of the solid. This will result in the solid diagonal (a'-3') becoming perpendicular to the VP, as it appears as a point in the front view. Now, the perpendicular edges through the bottom corner or its corresponding top corner are equally inclined to the VP or the plane of projection. It can be easily proved

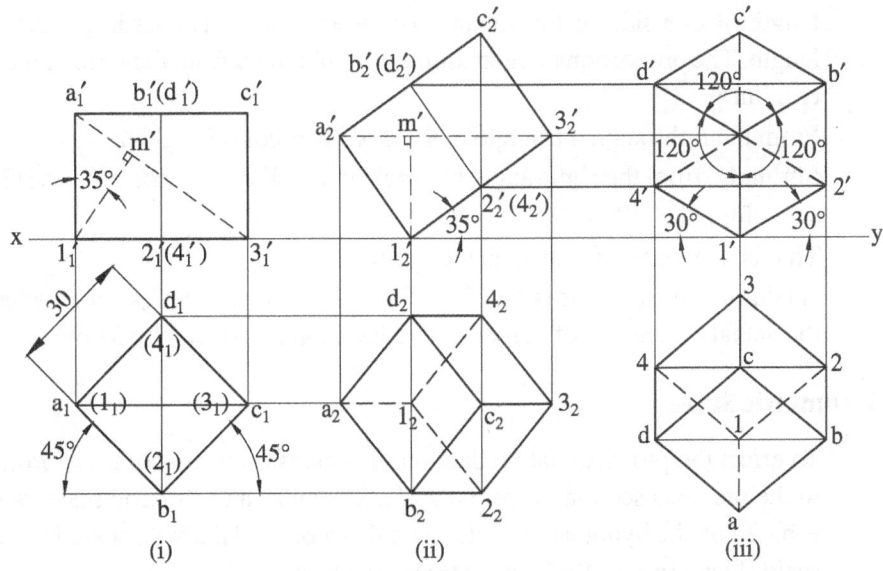

(a) Different stages of orthographic projections

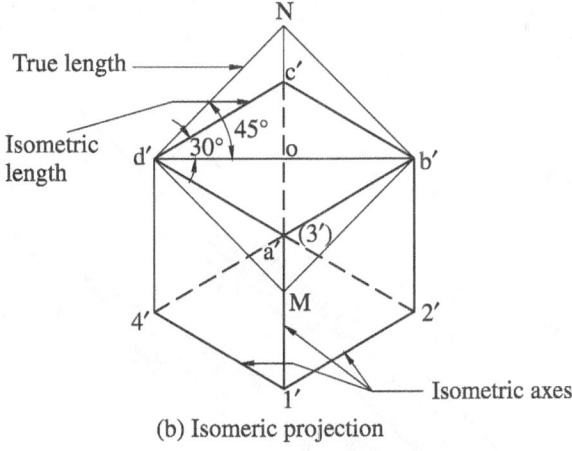

(b) Isomeric projection

Fig. 13.2   Evolution of isometric projection of a cube

that the inclination of these physical edges with the VP is 35°. In accordance with their projections on VP, these edges make equal angles of 120° among themselves and also get foreshortened in the same proportion. Hence, the front view obtained for the cube satisfies the requirement of isometric projection and therefore can be claimed as its isometric projection.

For clarity, the front view obtained in the third stage (i.e., the isometric projection) is once again discussed in Fig. 13.2(b) to quantify the proportionate reduction in the lengths of the edges. For this purpose, construct a square d′Mb′N with sides equal to the side of the cube and with d′b′ as the diagonal. While the line d′N is equal to the true length of one side of the square, d'c' represents its isometric projection of a reduced length. The proportionate reduction can be obtained from the ratio of the lengths of these two lines.

From the right-angled triangle d'oN, d'o/d'N = cos 45° = 1/√2

Similarly, from the right-angled triangle d'oc', d'o/d'c' = cos 30° = √3/2

Therefore, d'c'/d'N = (2/√3) × (1/√2) = √2/√3 = 0.816

That is, isometric length/actual length = 0.816

As discussed earlier, it is concluded that the isometric length of an edge is 0.816 times the actual dimension of that edge and hence gets reduced by 81.6%.

13.3.3 Isometric Scale

To effect the proportionate reduction of dimension in an isometric projection discussed in the previous section, a special scale, known as an isometric scale, is used. This scale is based on the hypotenuse of the triangles d'oc' and d'oN discussed in Fig. 13.2(b). This scale shown in Fig. 13.3 is constructed as follows:

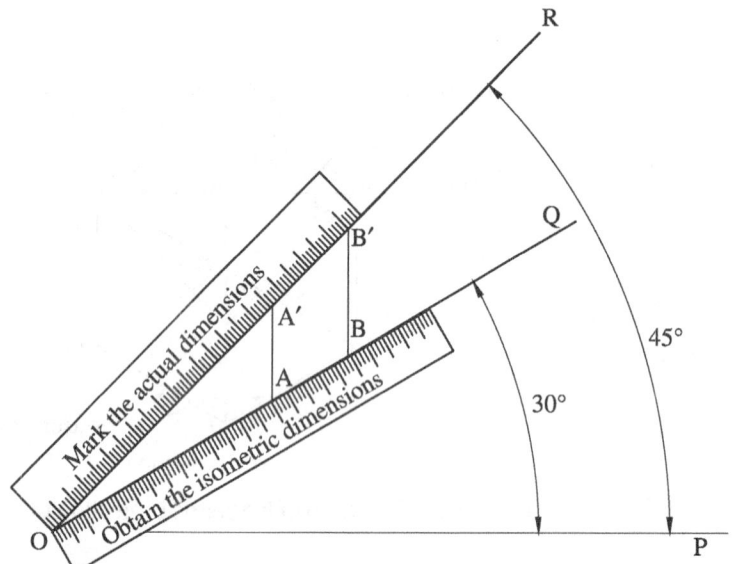

Fig. 13.3   Isometric scale

1. Draw a horizontal line OP for any convenient length.
2. Set angles of 30° and 45° at O as shown by the lines OQ and OR.
3. Mark the actual dimensions of the solid whose isometric projection is required on the 45° line as indicated by OA' and OB'.
4. Project the points A' and B' by erecting perpendiculars to the horizontal line OP.
5. When the perpendiculars meet the 30° line, at the points A, B, etc., the distances OA, OB, etc., give the isometric lengths to be used in the isometric projection.

13.3.4 Isometric Axes, Lines, and Planes

The mutually perpendicular edges of the cube a'1', a'b', and a'd' that meet at the corner a' at an angle of 120° to each other in an isometric projection (Fig. 13.2b) are known as 'isometric axes'. The other edges are lines in the object that are parallel to these isometric axes and are known as isometric lines. Similar to isometric axes, the isometric lines are also foreshortened in the same proportion. Any other lines or edges in the object that are not parallel to the isometric axes, neither follow the same inclination nor the same proportion and hence are marked in the drawing only by locating their ends on the appropriate isometric axes or isometric lines. The faces adjoining the isometric axes or any planes parallel to them are known as isometric planes, and they follow the same equal inclination as the edges.

13.3.5 Isometric Projection and Isometric View

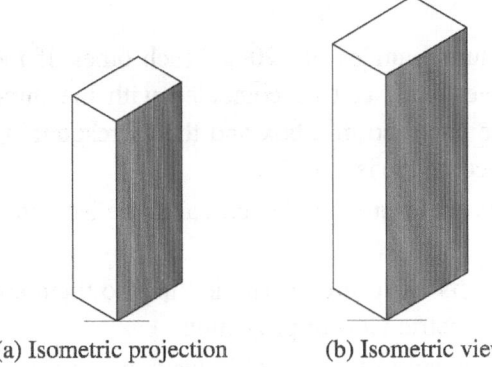

(a) Isometric projection   (b) Isometric view

Fig. 13.4   Isometric projection vs isometric view

In isometric projection, the dimensions of an object are foreshortened as per the isometric scale, when represented along the isometric axes or isometric lines. If the actual dimensions of the object are used along the isometric axes and the isometric lines, then the resultant drawing is known as isometric drawing or isometric view. Both isometric projection and isometric view depict the same shape of the object, but the later one appearing larger in proportion compared with the isometric projection as shown in Fig. 13.4. Furthermore, in the isometric view, the distances measured between any two points along the isometric axes or lines give directly that occurs in the original object. Hence, the isometric views serve the purpose of the pictorial appearance of the object along with the facility to get the actual dimensions of the object and is generally preferred than the isometric projection. The reader should note that unless and otherwise indicated specifically, only the isometric projections are to be drawn for academic classroom practice.

13.3.6 Preferred Isometric Positions of an Object

An object can be shown in the isometric projection in different positions. In general, the isometric projection of an object is drawn such that the longer edge of the object is kept along either the vertical or the 30° isometric axes to make the larger face or an important detail visible more prominently. For example, the isometric projection of a rectangular prism is shown in Fig. 13.5(a) to (d) in four different positions. In Fig. 13.5(a) and (b), the object is shown to lie on its base on the ground or on HP with the longer edge 1A vertical and the larger face 1–2–B–A visible on the right side or on the left side, respectively. In Fig. 13.5(c) and (d), the object is shown with its longer edge 1A along the 30° line such that the larger face 1–2–B–A is visible more prominently.

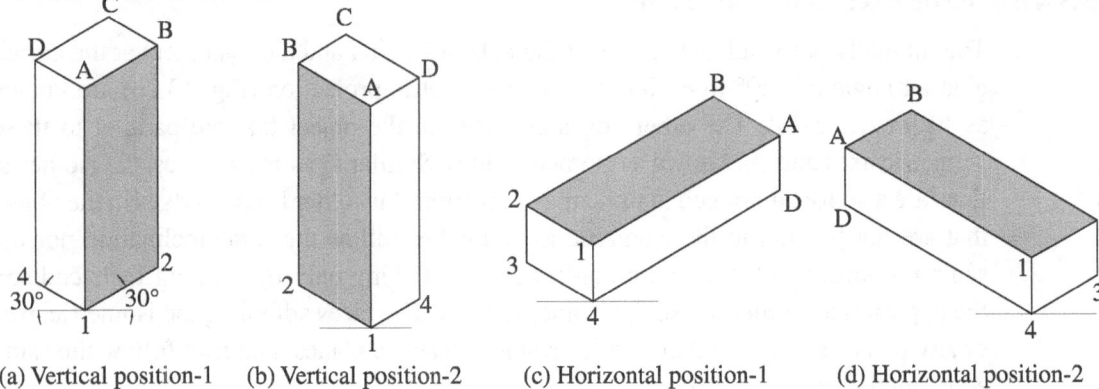

(a) Vertical position-1   (b) Vertical position-2   (c) Horizontal position-1   (d) Horizontal position-2

Fig. 13.5 Preferred isometric positions of an object

13.3.7 General Guidelines for Isometric Drawings

1. The isometric axes can be drawn at mutual angles of 120° to each other. If the object has got three perpendicular edges, they can be coincided with the three isometric axes. If not the object can be enclosed in a box and the corresponding three perpendicular edges of the box become the isometric axes.

2. All the vertical edges of the object or its enclosure remain vertical in the isometric view or projection.

3. All the horizontal edges of the object or its enclosure or parallel lines to them are marked at 30° to the horizontal in the isometric view or projection.

4. The inclined edges of the object are obtained in the isometric view or projection only after marking their end points suitably on the relevant isometric axes or lines.

5. The front view of the object can be marked in the right- or left-side face of the isometric, depending on the visibility desired. Then, the other side face will reveal the side view of the object.

6. If both the front and the side views are given for an object, then the corresponding side face of the isometric is chosen for the side view, and in the other side face, the front view is accommodated. For example, if the left-side view of the object is given, the left-side face of the isometric is chosen to accommodate that.

7. The top view of the object is always accommodated along the 30° lines.

8. All construction lines are retained very thin in the isometric view or projection and are not to be erased. The visible lines alone are made to be thicker. No dotted lines are to be drawn to mark the hidden edges but can be shown very thin.

9. Dimensioning is not done usually in the isometric view or projection.

10. Drawing of orthographic views of an object is not mandatory for the isometric sketch. However, to locate some points or edges in isometric, the orthographic views can be drawn.

11. The isometric projection is drawn only after converting the given dimensions through the isometric scale. Isometric views or drawings can be drawn as per the given dimensions.

In the following sections, the constructional procedures for isometric projections or views are explained for planar straight edged objects, planar objects with curved boundaries, 3D objects with straight edge and curved boundaries, objects with holes and cut-outs, and combination of solids. Finally, the illustrations are prepared in order to explain how to convert a given set of orthographic views of an object and make their corresponding isometric projections or views.

13.4 PLANAR OBJECTS

In this section, the constructional procedure for thin planar-shaped objects with straight edges and curved boundaries are explained in detail by considering them in different positions. The planar surfaces can be made up with two perpendicular edges as in square or rectangle or can be made up with many inclined edges as in any polygon. Depending on their positions, these shapes may appear in the front view, the top view, or in the side views. Since the object has got only two dimensions, two isometric axes are constructed for drawing their isometric projections or views. The two isometric axes may be inclined at 120° among themselves to accommodate the top view or one isometric axis perpendicular and the other inclined at 30° to the horizontal to accommodate the front view or the side views. The isometric projection of squares or rectangles can be drawn by locating their perpendicular edges along the two isometric axes. The views of other shapes that do not have perpendicular edges have to be enclosed in a square or rectangle as the case may be. The isometric projection of the enclosed figure is drawn first, and the various corners of the shape are identified along the enclosed edges or on suitable lines parallel to them and then transferred to the isometric sketch in similar manner. For shapes that have circular holes or curved features in the form of semicircles or arcs, a similar procedure of enclosing their views in a square or rectangle is adopted and the various points on the circles or arcs are realized by creating a set of lines parallel to the outer edges and then transferring them in the isometric sketch. Instead of creating many points on a circle and realizing them in the isometric sketch, a set of suitable arcs can be constructed to yield the isometric sketch. This procedure is explained in detail in the examples. Table 13.1 explains the isometric sketches for different plane shapes with straight edges. The basic shape of each planar object is shown in the first column and the isometric sketches corresponding to that are shown, depending on the basic shape appearing as a front view, a top view, or side views. The square and the rectangle are considered as the basic examples, and other polygons are explained by enclosing them in square or rectangle as the case may be and their corners obtained suitably.

Table 13.1 Planar straight edge objects and their isometric shapes

Shape of the object	Isometric shapes when viewed from			
	Front	Top	Left side	Right side
Rectangle				
Triangle				
Square				
Pentagon				
Hexagon				

Regular Polygons

Example 13.1

Draw the isometric projections of a regular pentagonal lamina of 22 mm side, when it is placed with its surface (i) vertical (ii) horizontal.

PROCEDURE (Refer Fig. 13.6)

Step 1: Draw a regular pentagon abcde of 22 mm sides as shown in Fig. 13.6(a) and enclose it in a rectangle $1_12_13_14_1$. This can be considered as the front view if the plane surface is vertical or a top view if the plane surface is horizontal.

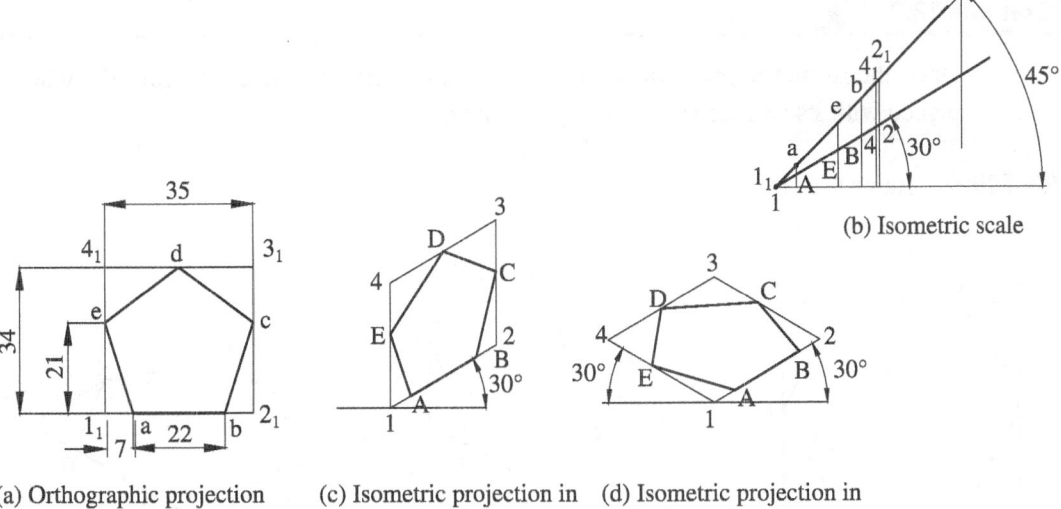

(a) Orthographic projection

(b) Isometric scale

(c) Isometric projection in vertical position

(d) Isometric projection in horizontal position

Fig. 13.6

Step 2: Construct the isometric scale with 30° and 45° lines as shown in Fig. 13.6(b). Using the divider, note the distances of the points a, b, e, 2_1, and 4_1 from the point 1_1 in Fig. 13.6(a), and set them along the 45° line. Draw perpendiculars from these points and obtain A, B, E, 2, and 4 along the 30° line.

NOTE *The distances need not be measured using scale but can be noted using the divider and can be transferred. The dimensions are written only for reference.*

Step 3: Construct one isometric edge 1–4 vertical and the other edge 1–2 inclined at 30° to the horizontal as discussed in section 13.3.7 and complete the parallelogram 1–2–3–4 as shown in Fig. 13.6(c).

Step 4: Note that the distances of the corner points A, B, and E on the isometric scale (30° line) using a divider and set them along their corresponding isometric edges or lines. For example, measure 1_1A and 1_1B using the divider, transfer them on the isometric line 12, and obtain A and B. Similarly, obtain the other corner points and join them by straight lines.

Step 5: ABCDE gives the isometric projection of the pentagon when it is vertical (Fig. 13.6 c).

Step 6: Construct the isometric edges 1–4 and 1–2 inclined at 30° to the horizontal as discussed in section 13.3.7 and complete the parallelogram 1–2–3–4 as shown in Fig. 13.6(d).

Step 7: Mark the points A, ..., E in the parallelogram in similar lines as in Step 4.

Step 8: ABCDE gives the isometric projection of the pentagon, when its surface is horizontal (Fig. 13.6d).

Example 13.2

Draw the isometric projections of a regular hexagonal lamina of 18 mm side, when it is placed with its surface (i) vertical (ii) horizontal.

PROCEDURE (Refer Fig. 13.7)

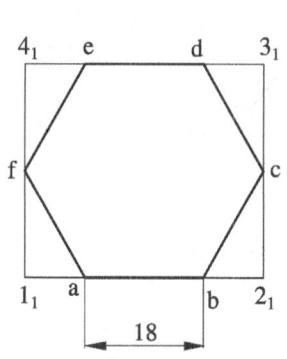

(a) Orthographic projection

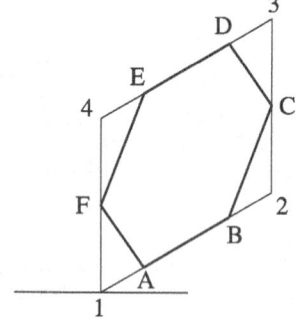

(b) Isometric projection in vertical position

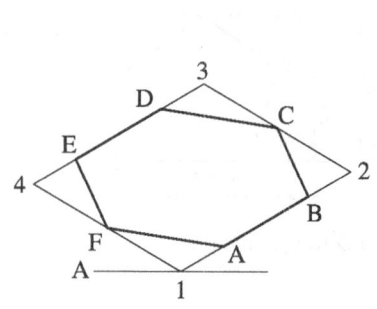

(c) Isometric projection in horizontal position

Fig. 13.7

Step 1: Construct a regular hexagon abcdef of 18 mm sides as shown in Fig. 13.7(a) and enclose it in a rectangle $1_1 2_1 3_1 4_1$. Construct the isometric scale and obtain the reduced dimensions in similar lines of Example 13.1.

NOTE *The isometric scale is not shown in Fig. 13.7, but the reader is advised to draw it. The drawing shown, however, adopts the reduced dimensions only.*

Step 2: Figures 13.7(b) and (c) show the isometric projections of the hexagonal plane when its surface is vertical and horizontal, respectively.

Irregular Polygons

Example 13.3

Draw the isometric projections of an irregular plane surface as shown in Fig. 13.8(a) when it is placed with its surface (i) vertical (ii) horizontal.

PROCEDURE (Refer Fig. 13.8)

Step 1: Construct the plane surface abcde with its given dimensions and enclose it in a rectangle $1_1 2_1 3_1 4_1$ as shown in Fig. 13.8(a).

Step 2: Construct the isometric scale and obtain the reduced dimensions in similar lines of Example 13.1.

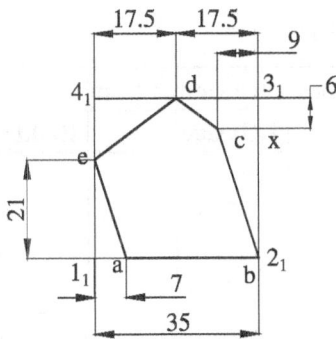

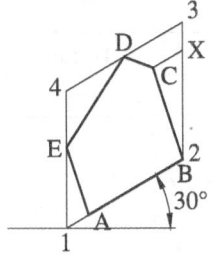

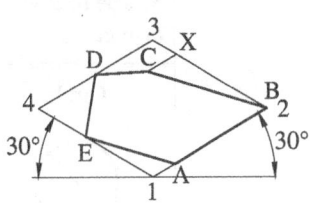

| (a) Orthographic projection | (b) Isometric projection in vertical position | (c) Isometric projection in horizontal position |

Fig. 13.8

Step 3: Fig. 13.8(b) and (c) show the isometric projections of the plane when its surface is vertical and horizontal, respectively.

Circular, Semicircular, Quarter circular, or any Arc Section

The circle in isometric projection appears as an ellipse. The following are two methods of construction:

1. Four centres method
2. Method of points

The general principle of the four centres method has been explained in Chapter 2 and in Example 2.16 and is again explained here. The longer diagonal of the rhombus constructed between the isometric axes for accommodating the ellipse is taken as a reference and the two corners opposite to that on either side, are the two centres for drawing two larger arcs. The radii of the larger arcs are decided by the length of the lines joining that centre and the midpoints of the sides opposite to them. The meeting point of these lines with the longer diagonal becomes the other two centres for the smaller arcs. The radii of smaller arcs are decided by the lengths of the lines joining these centres and the midpoints of the sides opposite to them.

Table 13.2 shows the various plane surfaces such as the circle, semicircle, and quarter circle, and the isometric sketches corresponding to their front view, top view, and side views. These curves are enclosed in a square meant for the full circle and the ellipse or the required portions of the ellipse are constructed following the general procedure of the four centres method explained earlier.

The method of points works on the principle of choosing many points on the circle or arc, whose isometric sketch is required. These points are identified by the set of parallel lines passing through them and locating them correspondingly in the isometric sketch.

However, the procedure of the four centre method has been explained to construct an ellipse for the isometric projection of a circle.

Table 13.2 Planar curved objects and their isometric shapes

Shape of the object	Isometric shapes when viewed from			
	Front	Top	Left side	Right side
Circle				
Semicircle				
Quarter circle				

Example 13.4

Draw the isometric projections of a circular lamina of 26 mm diameter when it is placed with its surface as vertical and horizontal using four centre method.

PROCEDURE (Refer Fig. 13.9)

Step 1: Draw a circle of diameter 26 mm as shown in Fig. 13.9(a), enclose it in a square $1_1 2_1 3_1 4_1$, and mark the midpoint of the sides $1_1 2_1$, $2_1 3_1$, $3_1 4_1$, and $4_1 1_1$ as a, b, c, and d, respectively.

Step 2: Draw the isometric projection of the square surface 1234 in the required position after constructing the isometric scale. Figures 13.9(b) and (d) show the isometric projections of the circular surface when it is vertical. Figure 13.9(b) shows the isometric projection accommodated in the left face of the isometric, while Fig. 13.9(d) shows its location on the right face.

Step 3: Figure 13.9(c) shows the isometric projection when its surface is horizontal and is accommodated between the isometric axes inclined at 30° to the horizontal.

Step 4: Join the longest diagonal 2 and 4 in the respective rhombuses and identify the opposite corners 1 and 3. Join 1 and 3, respectively, with the midpoints B and D of the opposite sides, and get P and Q as their meeting points with the longest diagonal.

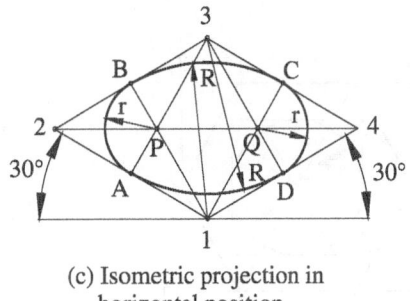

(c) Isometric projection in
horizontal position

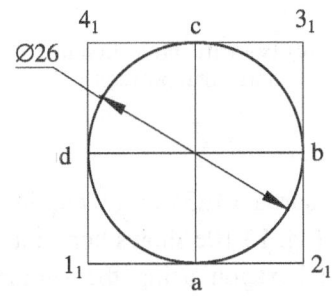

(a) Orthographic projection

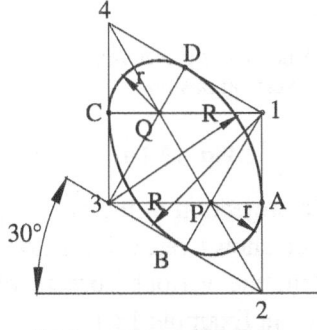

(b) Isometric projection in
vertical position-1

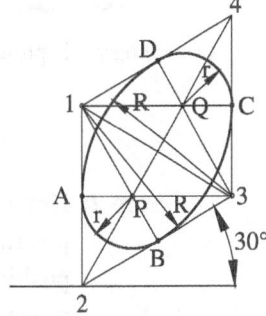

(d) Isometric projection in
vertical position-2

Fig. 13.9

Step 5: With the points 1 and 3 as centres and the lengths 1–B and 3–D as radii, draw the larger arcs. With the points P and Q as centres and the lengths P–B and Q–D as radii, draw the smaller arcs.

Step 6: When the four arcs are joined, it gives the ellipse ABCD which is the isometric projection of the circle. ▲

Combination of Regular Polygon and Circle

In this section, the constructional procedure for the combination of regular polygons with circle is explained.

The following example illustrates about the construction procedure for isometric projection of a hexagonal plate with a central hole.

Example 13.5

Draw the isometric projections of a thin hexagonal plate of 18 mm sides with a central hole of 20 mm diameter when it is placed with its surface as vertical and horizontal.

PROCEDURE (Refer Fig. 13.10)

Step 1: Construct a regular hexagon of sides 18 mm as shown in Fig. 13.10(a), enclose it in an outer square $1_1 2_1 3_1 4_1$, mark the centre, and draw a circle with diameter equal to 20 mm.

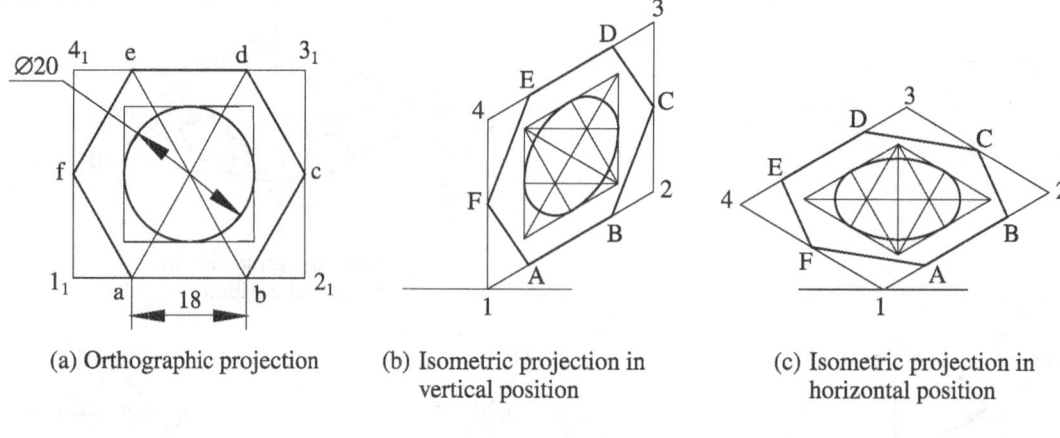

(a) Orthographic projection

(b) Isometric projection in
vertical position

(c) Isometric projection in
horizontal position

Fig. 13.10

Step 2: Draw the isometric projection of the outer square surface 1234 in the required position (Fig. 13.10b shows vertical position and Fig. 13.10c shows horizontal position) and obtain the various corners of the hexagon using the similar procedure explained in Example 13.1.

Step 3: Enclose the circle in a square and obtain its isometric projection as a rhombus. Construct the ellipse within the rhombus as explained in the four centres method. ▲

13.5 SIMPLE AND TRUNCATED SOLIDS

Since the solids are 3D objects, three isometric axes equally inclined at angles of 120° to each other to be used to draw their isometric projections or views. The isometric sketches of the simple solids, such as square, rectangular prisms, and cubes, that have their three mutually perpendicular principal edges can be drawn directly as these edges can be located on the three isometric axes. In solids where the base or the top edges are not parallel to the isometric axes but the face edges are vertical and parallel to each other as in the case of pentagonal, hexagonal prisms, and cylinders, the solid is to be enclosed in an equivalent rectangular scaffold box and the isometric sketch is to be constructed such that all the principal edges of the rectangular box coincide with the isometric axes. This method is known as the *box method*. Before drawing the isometric sketches, the orthographic views of the solid are to be drawn and the polygonal views are to be enclosed in a rectangle or a square appropriately. The isometric projections of the polygonal ends are drawn as per the procedure outlined for polygons in Section 13.4, and then, the solid shape is built over that. The various corners, edges, and faces of the solid, which lie on the different faces or within the enclosed box, are to be located from that of the measurements made from the outer edges of the box.

For drawing the isometric sketches of the solids in which the base or the top edges are not parallel to the isometric axes and the face edges are slanting as in the case of pyramids and cones, the base is to be enclosed in a square or a rectangle and its isometric projections are to be drawn. The coordinates of the various points that lie on the solid surface are obtained from their orthographic views, by measuring their location in the base and by noting the corresponding height at which they are located from the base. These are correspondingly realized in the isometric sketch from the respective isometric axes or the relevant isometric lines. The coordinates of any point on the solid referred in the isometric sketch are known as offsets, and hence, this method is known as the *offset method*.

It can be noted that for both methods, the isometric projection of the polygon representing the base, the top, or the ends of the solid is drawn first. If the polygon appears in the top view, then its isometric sketch is drawn in between the two isometric axes inclined at 30° to the horizontal and the third isometric axis is drawn vertical to the required height of the solid and the solid shape is built along that. In contrast, if the polygon appears in the front view, then its isometric sketch is drawn in the left- or the right-side face of the isometric plane arrangement keeping one isometric axis vertical and the other inclined at 30° to the horizontal. The third isometric axis is also drawn inclined at 30° to the horizontal but on the other side to the required length of the solid and the solid shape is built along that. The construction of the box or the setting of the coordinates of the relevant points on the surface of the solid can be made depending on the requirement of the isometric sketches of prismatic, cylindrical, or the pyramidal or conical solids, respectively.

A sphere will appear as a circle equal to the radius of the sphere in all the three orthographic views. Hence, the isometric projection of a sphere will also be a circle of same radius as the sphere. When the sphere rests on a flat surface, the vertical height of the centre of the sphere will, however, get reduced by the isometric scale. It is to be noted that the isometric scale has to be used whenever objects in conjunction with spheres or spherical parts are involved and isometric projections are only to be drawn for such objects.

Isometric Projection of Prisms

Example 13.6

Draw the isometric projection of a pentagonal prism of 22 mm base edges and axis 50 mm long when it rests on its base on HP with a rectangular face parallel to VP.

PROCEDURE (Refer Fig. 13.10)

Step 1: Draw the top and the front views of the regular pentagonal prism of side 22 mm and height 50 mm as shown in Fig. 13.11(a). Enclose the polygonal view in a rectangle $p_1q_1r_1s_1$/pqrs. Since the polygonal view appears in the top view, its isometric projection is to be drawn first along the 30° inclined isometric axes, and the solid is to be built over that by constructing the vertical isometric axis.

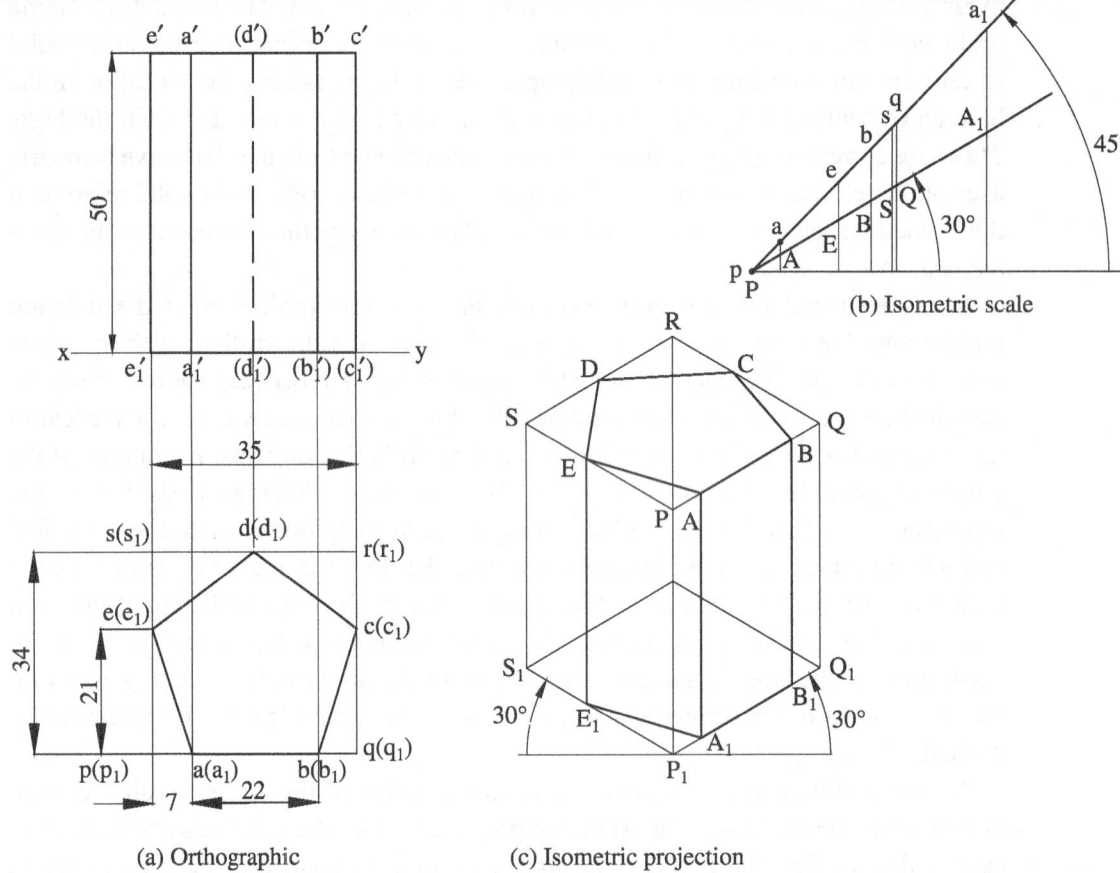

(a) Orthographic

(b) Isometric scale

(c) Isometric projection

Fig. 13.11

Step 2: Draw the three isometric axes P_1Q_1, P_1S_1, and P_1P through the corner P_1 at the mutual angles of 120° to each other. The dimensions of the isometric axes or lines are obtained after reducing them as per the isometric scale as shown in Fig. 13.11(b).

Step 3: Mark the vertical and the inclined isometric lines and obtain the rectangular box $P_1Q_1R_1S_1$–PQRS as shown in Fig. 13.11(c) for the isometric height of the solid.

Step 4: Obtain the isometric projection of the bottom surface (pentagon) in between the inclined isometric axes as shown by A_1, ..., E_1 and as discussed in Example 13.1.

Step 5: Similarly, obtain the isometric projection of the top surface (pentagon) in between the inclined isometric axes as shown by A, ..., E and as discussed in Example 13.1.

Step 6: Join the vertical edges A_1A, ..., E_1E and retain the front visible edges in thick lines to complete the isometric projection of the pentagonal prism. ▲

Example 13.7 ◤

Draw the isometric projection of a pentagonal prism of 22 mm base edges and axis 50 mm long when it has one of its rectangular faces on HP and axis perpendicular to the VP.

PROCEDURE (Refer Fig. 13.12)

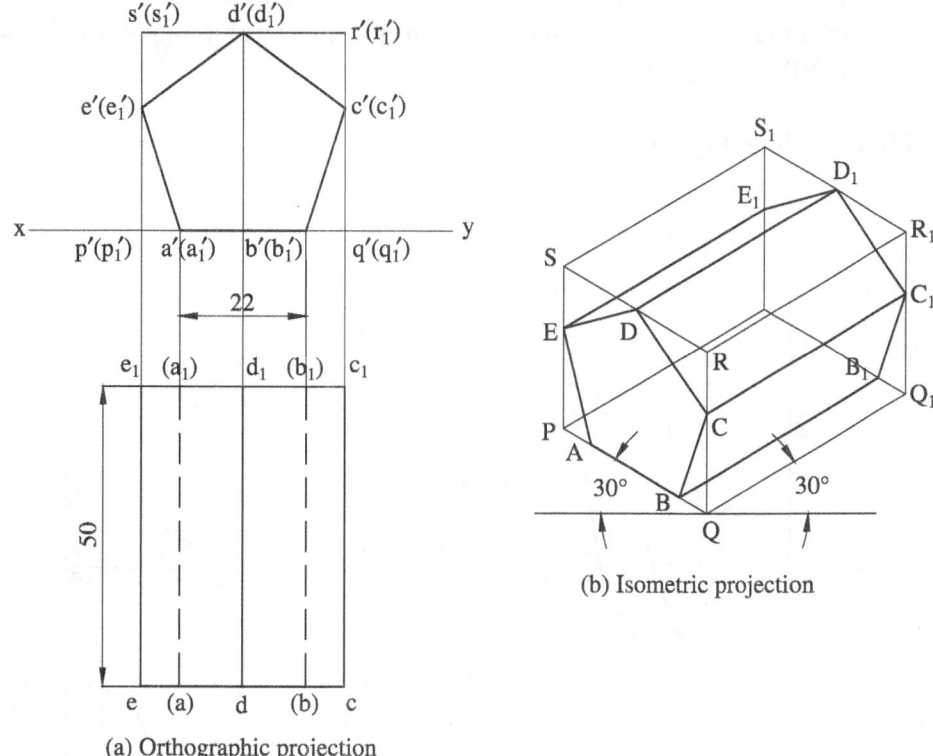

(a) Orthographic projection

(b) Isometric projection

Fig.13.12

Step 1: Draw the front and the top views of the regular pentagonal prism of side 22 mm and length 50 mm as shown in Fig. 13.12(a). Enclose the polygonal view in a rectangle $p_1'q_1'r_1's_1'/p'q'r's'$. Since the polygonal view appears in the front view, its isometric projection can be drawn in one of the side isometric planes, and the solid can be built along its length on the third isometric axis inclined at 30° to the horizontal.

Step 2: Draw the three isometric axes QR, QP, and QQ_1, respectively, vertical and inclined at 30° to the horizontal line, after reducing the dimensions as per the isometric scale. Mark the relevant isometric lines and obtain the rectangular box $PQRS-P_1Q_1R_1S_1$ as shown in Fig. 13.12(b).

Step 3: Obtain the isometric projection of the front end (pentagon) in the left-side isometric plane as shown by A, …, E and as discussed in Example 13.1.

Step 4: Similarly, obtain the isometric projection of the rear end (pentagon) as shown by A_1, …, E_1 and as discussed in Example 13.1.

Step 5: Join the longitudinal edges AA_1, BB_1, …, EE_1 and retain the visible edges in thick lines to complete the isometric projection of the pentagonal prism. ▲

Example 13.8

Draw the isometric projection of a hexagonal pyramid of 20 mm base edges and height 60 mm long when it rests on its base on HP with two of its opposite sides of base parallel to VP.

PROCEDURE (Refer Fig. 13.13)

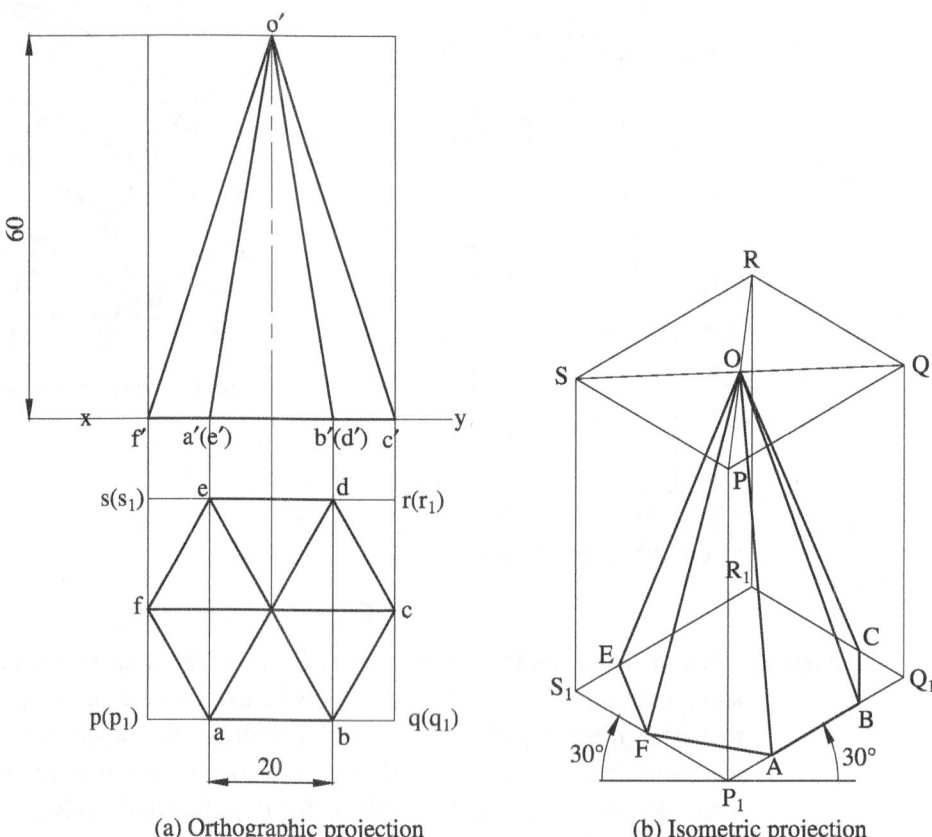

(a) Orthographic projection   (b) Isometric projection

Fig. 13.13

Step 1: Draw the top and front views of the hexagonal pyramid of side 20 mm and height 60 mm as shown in Fig. 13.13(a). Enclose the polygonal view in a rectangle $p_1q_1r_1s_1$/pqrs. Since the polygonal view appears in the top view, its isometric projection is to be drawn first along the 30° inclined isometric axes, and the solid is to be built over that by constructing the vertical isometric axis.

Step 2: Draw the three isometric axes, P_1S_1, P_1P, and P_1Q_1 through the corner P_1 after reducing the dimensions as per the isometric scale. Mark the vertical and the inclined isometric lines and obtain the rectangular box $P_1Q_1R_1S_1$–PQRS as shown in Fig. 13.13(b).

Step 3: Obtain the isometric projection of the bottom surface (hexagon) in between the inclined isometric axes as shown by A, ..., F and as discussed in Example 13.1.

Step 4: Mark the apex O in the isometric projection on the top parallelogram PQRS by intersecting its two diagonals. And join base edges A, B, C, ... to the apex O which completes the isometric projection of the pyramid. It is to be noted that only the visible edges are retained by thick lines in the isometric projection. ▲

Example 13.9 ◢

Draw the isometric projection of a vertical cylinder of 26 mm base diameter and height 50 mm when it has its base on HP.

PROCEDURE (Refer Fig. 13.14)

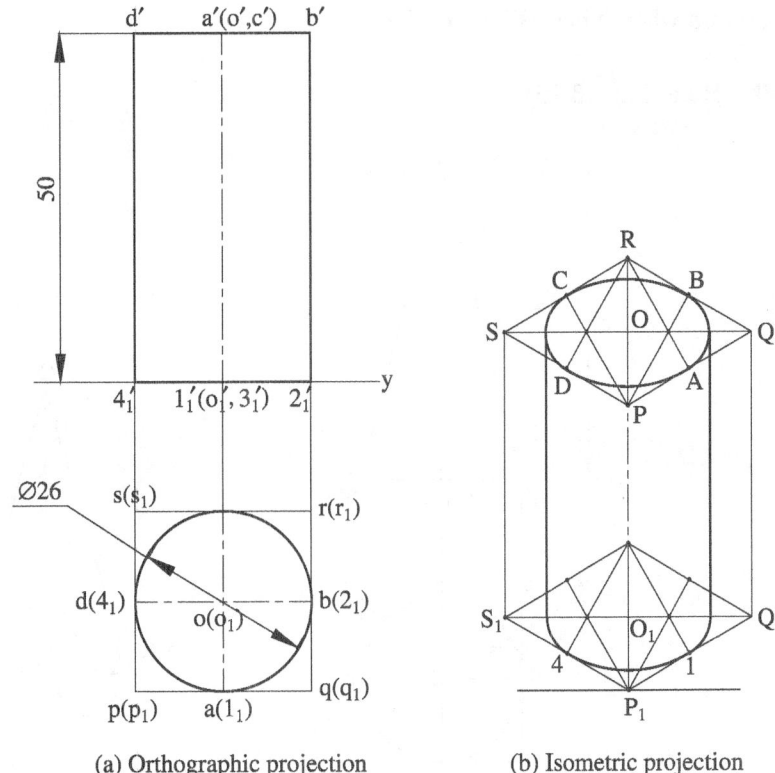

(a) Orthographic projection   (b) Isometric projection

Fig.13.14

Step 1: Draw the top and the front views of the cylinder as shown in Fig. 13.14(a). Enclose the circle in a square $p_1q_1r_1s_1/pqrs$. Since the circle appears in the top view, its isometric projection is to be drawn first along the 30° inclined isometric axes and the solid is to be built over that by constructing the vertical isometric axis.

Step 2: Draw the three isometric axes P_1S_1, P_1P, and P_1Q_1 through the corner P_1 after reducing the dimensions as per the isometric scale. Mark the vertical and the inclined isometric lines and obtain the rectangular box $P_1Q_1R_1S_1$–PQRS as shown in Fig. 13.14(b).

Step 3: Obtain the isometric projection of the bottom surface (circle) in between the inclined isometric axes as an ellipse by using the four centres method discussed in Example 13.4.

Step 4: Similarly, obtain the ellipse ABCD in the top surface and join it with the bottom ellipse by means of two vertical tangents on either side as shown in Fig. 13.14(b). ▲

Example 13.10 ▲

Draw the isometric projection of a vertical cone of 26 mm base diameter and height 50 mm when it has its base on HP.

PROCEDURE (Refer Fig. 13.15)

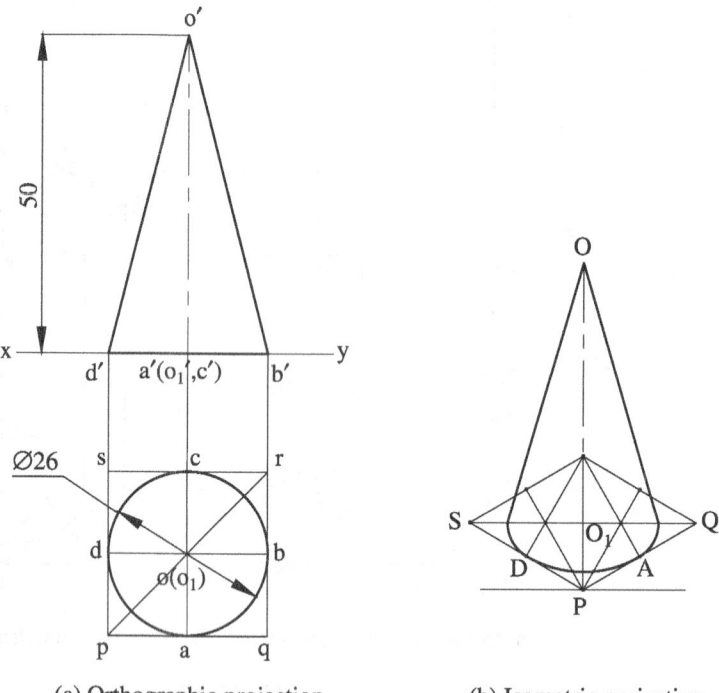

(a) Orthographic projection   (b) Isometric projection

Fig. 13.15

Step 1: Draw the top and the front views of the cylinder as shown in Fig. 13.15(a). Enclose the circle in a square pqrs. Since the circle appears in the top view, its isometric projection is to be drawn first along the 30° inclined isometric axes, and the solid is to be built over that by constructing the vertical isometric axis.

Step 2: Draw the isometric axes PS and PQ at an angle of 30° to the horizontal and construct the rhombus PQRS being the isometric projection of the enclosed square. Adopt the reduced dimensions as per the isometric scale.

Step 3: Obtain the isometric projection of the bottom surface (circle) in between the inclined isometric axes as an ellipse by using the four centres method as shown in Fig. 13.15(b) and as discussed in Example 13.4.

Step 4: Mark the centre O_1 as the intersection of the diagonals of the rhombus and from this point, draw a vertical line of length equal to the isometric length of the cone and mark the apex as O.

Step 5: Draw the tangents to the ellipse on either side from the apex O and complete the isometric projection of the cone.

> **NOTE**   *The hexagonal pyramid shown in Example 13.8 can also be drawn as mentioned in Example 13.10.*

Example 13.11

Draw the isometric projection of the frustum of a hexagonal pyramid of base edges 20 mm sides and top edges 8 mm sides and axis 55 mm long when its base is on HP. Two of the base edges are parallel to the VP.

PROCEDURE   (Refer Fig. 13.16)

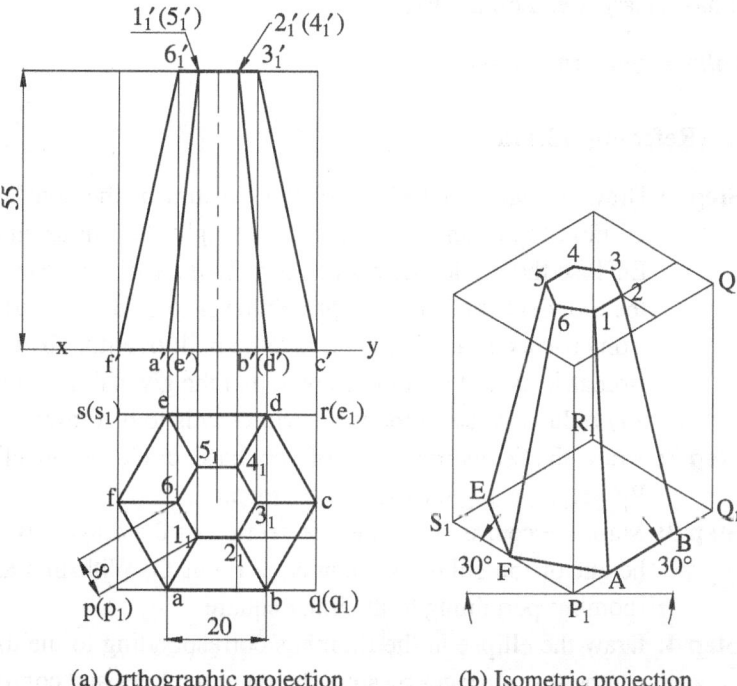

(a) Orthographic projection   (b) Isometric projection

Fig. 13.16

Step 1: Draw the top and front views of the frustum of a regular hexagonal pyramid of base sides 20 mm, top sides 8 mm, and height 55 mm as shown in Fig. 13.16(a). Enclose the polygonal view in a rectangle $p_1q_1r_1s_1$/pqrs. Since the polygonal view appears in the top view, its isometric projection is to be drawn first along the 30° inclined isometric axes and the solid is to be built over that by constructing the vertical isometric axis.

Step 2: Draw the three isometric axes, P_1S_1, P_1P, and P_1Q_1 through the corner P_1 after reducing the dimensions as per the isometric scale. Mark the vertical and the inclined isometric lines and obtain the rectangular box $P_1Q_1R_1S_1$–PQRS as shown in Fig. 13.16(b).

Step 3: Obtain the isometric projection of the bottom surface (hexagon) in between the inclined isometric axes as shown by A, ..., F and as discussed in Example 13.1.

Step 4: Obtain the isometric projection of the top hexagon in the parallelogram PQRS by locating its corners 1, 2, ..., 6. For example, the corner 2 is located by setting of its distances from the top view and by drawing isometric lines.

Step 5: Join the corresponding top and the bottom corners and complete the isometric projection. ▲

Example 13.12 ◢

Draw the isometric views of the frustum of the conical solid of base diameter 26 mm and top diameter 13 mm and 25 mm height, (a) when it rests on its base on the HP (b) when it has its larger end on the VP.

Solid Resting on its Base on HP

PROCEDURE (Refer Fig. 13.17a)

Step 1: Draw top and front views of the frustum of the cone with the base diameter 26 mm, top diameter 13 mm, and height 25 mm as shown in Fig. 13.17(a-i). Enclose the circles with squares $p_1q_1r_1s_1$ and pqrs. Since the circle appears in the top view, the isometric projection is to be drawn first along the 30° inclined isometric axes, and the solid is to be built over that by constructing the vertical isometric axis. It should be noted that the given dimensions can be used without any reduction, since the isometric view is to be drawn.

Step 2: Draw the isometric view of the base circle as an ellipse, in the rhombus $P_1Q_1R_1S_1$, using the four centre method.

Step 3: Mark the centre O_1 as the intersection of the diagonals of the rhombus, set the height of the solid, and mark the point O. With this as centre, construct the rhombus pertaining to the inner square.

Step 4: Draw the ellipse in the rhombus corresponding to the top end and join the top and bottom surfaces by smooth tangential lines and complete the isometric view of the frustum as shown in Fig. 13.17(a-ii).

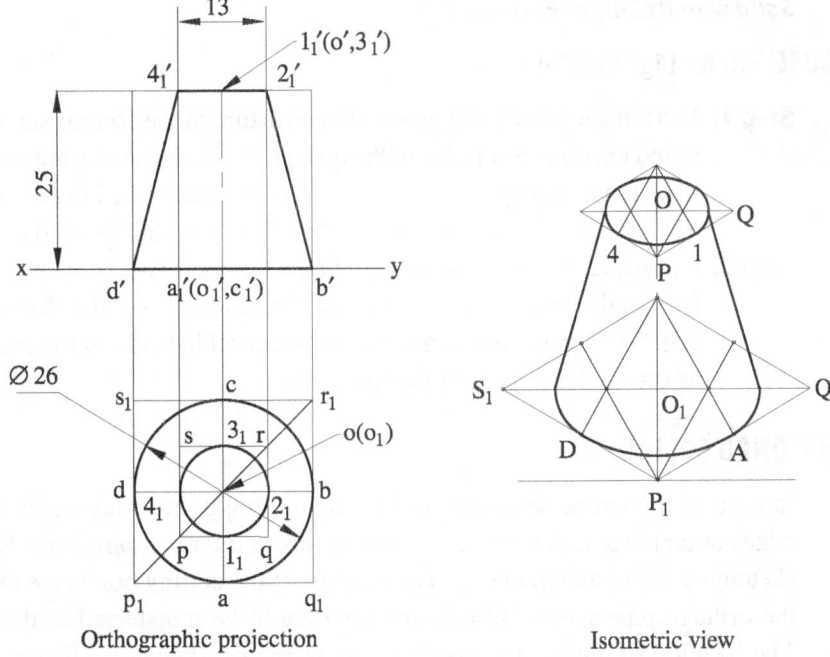

Orthographic projection   Isometric view

(a) When it rests on its base on HP

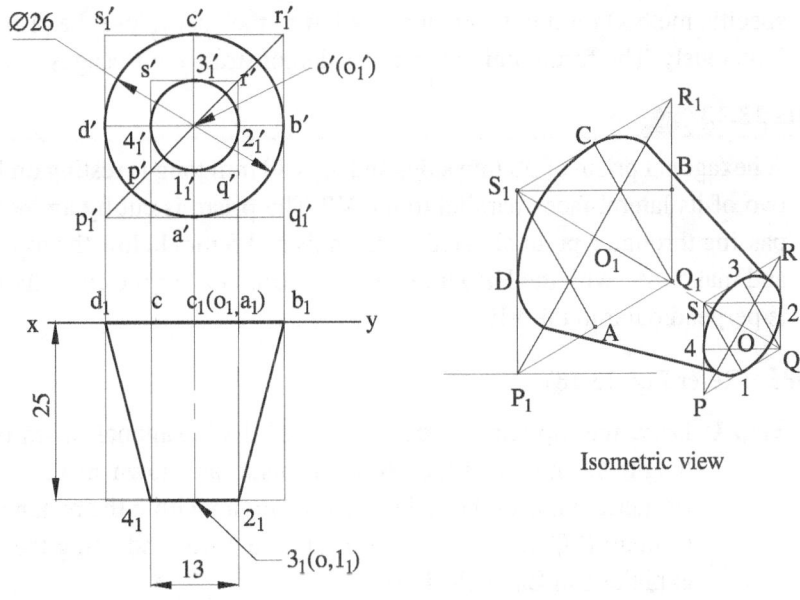

Orthographic projection

Isometric view

(b) When it has its larger end on VP

Fig. 13.17

Solid with its Larger End on VP

PROCEDURE   (Refer Fig. 13.17b)

Step 1: Draw the top and front views of the frustum of the cone as shown in Fig. 13.17(b-i) and enclose the circles with squares. Since the circles appear in the front view, the isometric projection of the rear end is constructed in the rhombus $P_1Q_1R_1S_1$. The centre O_1 is located and the length of the solid is marked to get O.

Step 2: Construct a rhombus around O and obtain the front end ellipse. Join the two ends and complete the isometric view of the frustum as shown in Fig. 13.17(b-ii). Note that the construction is made in the side isometric plane, as the circles appear in the front view.   ▲

13.6 SECTIONED SOLIDS

In case of the cut or sectioned solids, the cutting points may be located on the lateral edges or on the lateral surfaces or on the ends or their combinations. Since the isometric sketch is a 3D arrangement, the coordinates of the cutting points are to be obtained from the orthographic views initially and then are to be transferred to the isometric sketch. The location of the cutting points in the isometric sketch can be made by drawing the necessary isometric lines on the base or the end and another one in the longitudinal direction. Either the box method or the offset method can be followed. There is no specific method for any solid, and it is left to the user to use the principles behind them judiciously. The forthcoming examples demonstrate their usages.

Example 13.13   ◣

A hexagonal prism of 30 mm sides and axis 80 mm long is resting on HP on its base with two of its lateral faces parallel to the VP. The prism is cut by an inclined section plane passing through a point situated on the axis and 5 mm below the top surface of the solid and makes 45° with the HP. Draw the isometric view of the cut solid if the section plane is perpendicular to the VP.

PROCEDURE   (Refer Fig. 13.18)

Step 1: Draw the top and the front views of the hexagonal prism of side 30 mm and height 80 mm with the sectional plane as shown in Fig. 13.18(a). Draw the isometric view of the full hexagonal prism within the rectangular enclosure box through $P_2Q_2R_2S_2$ as shown in Fig. 13.18(b) and using the same procedure as explained in Example 13.6.

Step 2: The cutting points $1_1'$, $2_1'$, and $5_1'$ situated on the lateral edges can be marked in the isometric view on the respective longer edges through P, Q, and U by measuring their distances from the bottom end in the front view. The coordinates for the cutting points 3_1 and 4_1 can be found from the top view and transferred to the top end in the isometric view to get the points 3 and 4.

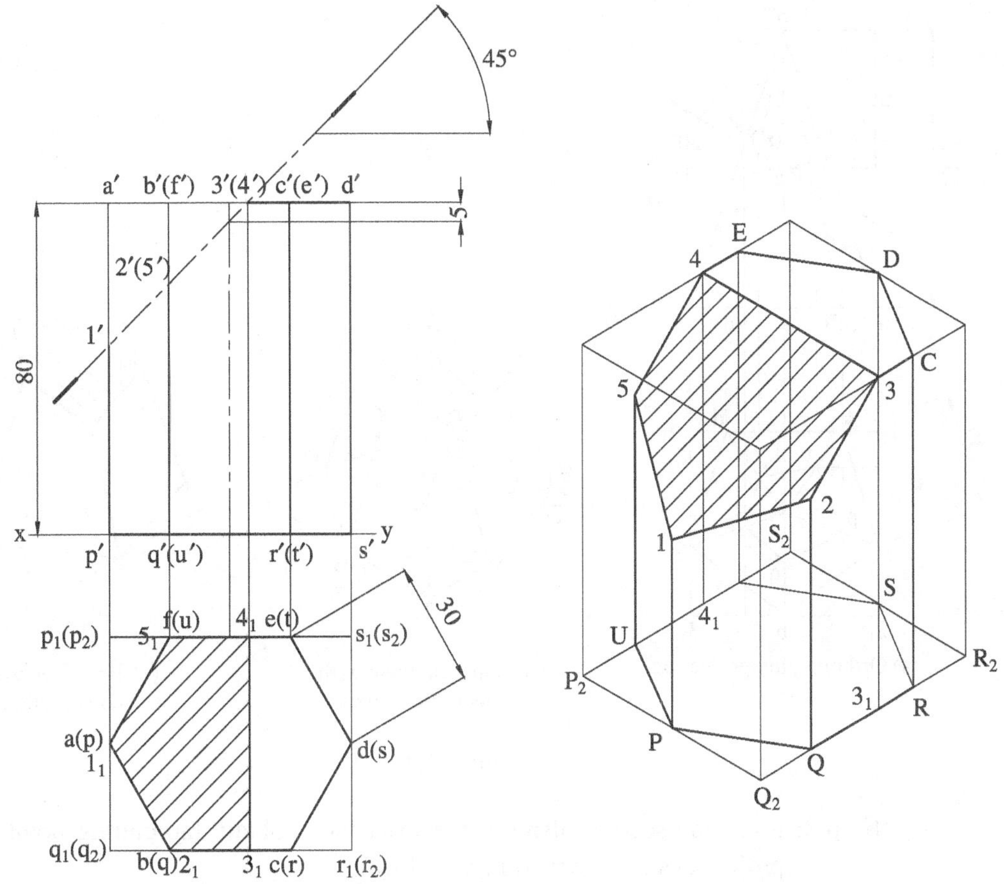

45°

a'   b'(f')   3'(4') c'(e') d'

2'(5')

1'

80

5

E

4

D

C

3

5

2

x

p'   q'(u')   r'(t')   s'   y

S₂

1

S

p₁(p₂)   f(u)   4₁ e(t)   s₁(s₂)

U

4₁

30

P₂

R₂

a(p)   d(s)   3₁

1₁

R

P

Q

q₁(q₂)   b(q)2₁   3₁ c(r)   r₁(r₂)

Q₂

(a) Orthographic projection

(b) Isometric view

Fig. 13.18

Step 3: Join all the cutting points 1–2–3–4–5–1 in the solid and mark the retained portion of the cut solid with thick lines. Hatch the sectioned portion of the solid as shown in Fig. 13.18(c). ▲

Example 13.14 ◣

 A pentagonal pyramid of base sides 30 mm and axis 70 mm long is resting on its base on HP with one of its base edges parallel to VP and nearer to it. It is cut by a section plane perpendicular to VP, inclined at 30° to HP and passing through the midpoint of the axis. Draw the isometric view of the cut solid.

PROCEDURE (Refer Fig. 13.19)

Step 1: Draw the top and the front views of the pyramid and obtain the isometric view of the base of the pyramid using the same procedure as explained in Example 13.8.

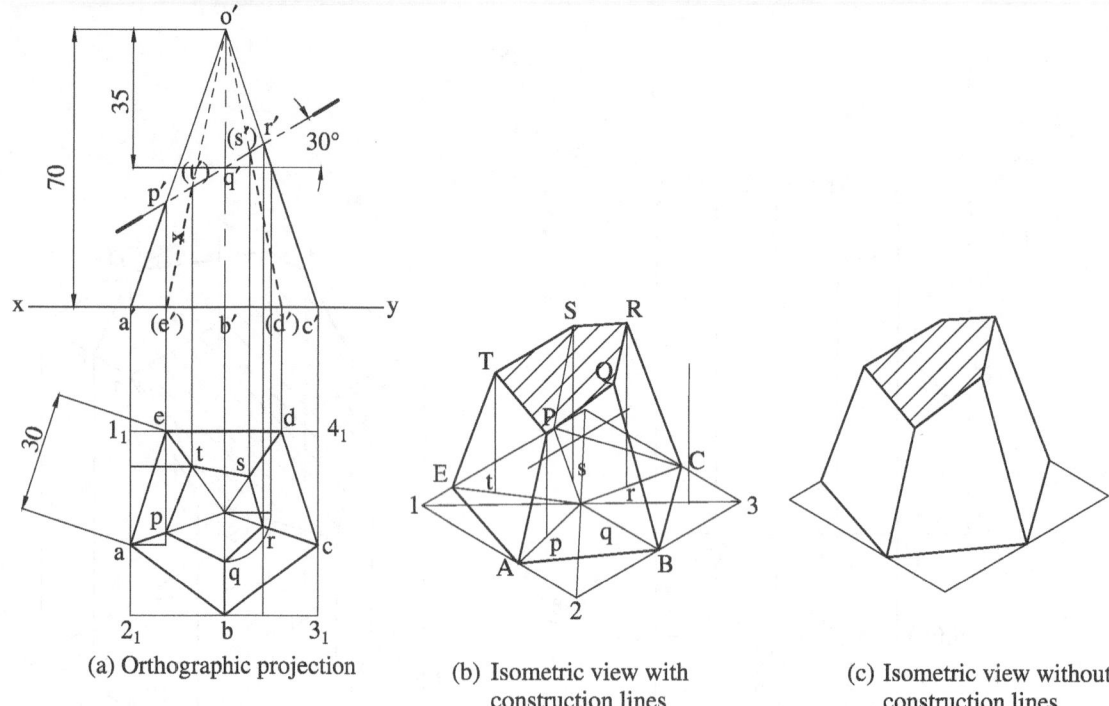

(a) Orthographic projection

(b) Isometric view with construction lines

(c) Isometric view without construction lines

Fig. 13.19

Step 2: Mark the section plane in the front view, obtain the cutting points and their projections as shown in Fig. 13.19(a).

Step 3: The positions of the cutting points on the base can be marked in the isometric view by transferring their respective distances from the top view. The vertical distances of the cutting points from the base can be measured from their respective front view and transferred on to the perpendicular isometric lines erected from their respective base positions in the isometric view.

Step 4: Therefore, the cutting points P, Q, ..., T can be located, joined, and shaded to indicate the cut surface. Joining these points with the corresponding base corners A, B, ..., E yields the isometric view of the solid portion that is retained. Figures 13.19(b) and (c) show the cut solid with and without the construction lines.

Example 13.15 ▲

@ A cylinder of base diameter 80 mm and axis 80 mm long is lying on its base on HP. A section plane perpendicular to VP and inclined at 45° to HP passes through a point situated on the axis and at 20 mm below the top surface of the cylinder. Draw the isometric view of the cut solid.

PROCEDURE (Refer Fig. 13.20)

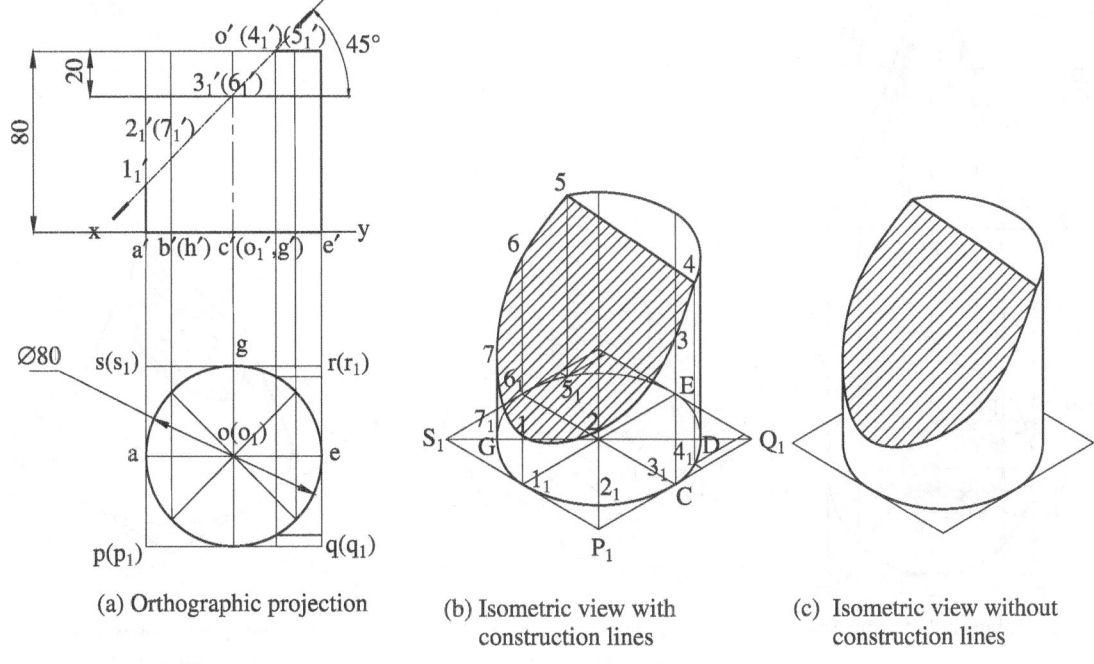

(a) Orthographic projection

(b) Isometric view with construction lines

(c) Isometric view without construction lines

Fig. 13.20

Step 1: Draw orthographic projection of the cylinder by dividing the circle into eight equal parts. Mark the section plane and identify the cutting points as shown in Fig. 13.20(a).

Step 2: The procedure of obtaining the isometric view is similar to Example 13.13, except the lateral edges are replaced by the lateral generators and the cut surface is joined by a smooth curve.

Step 3: Figures 13.20(b) and (c) show the isometric view of the cut solid with and without the construction lines. ▲

Example 13.16 ▲

A right circular cone of 80 mm diameter of base and 100 mm high is standing on its base on the HP. The vertical trace of a section plane perpendicular to VP makes an angle of 45° with the reference line and intersects the front view of the axis of the cone at a distance of 40 mm from the vertex. Draw isometric view of the cut solid.

PROCEDURE (Refer Fig. 13.21)

Step 1: Draw the top and front views of the cone by dividing the circle into eight equal parts and obtain the isometric view of the base of the cone using the four centre method.

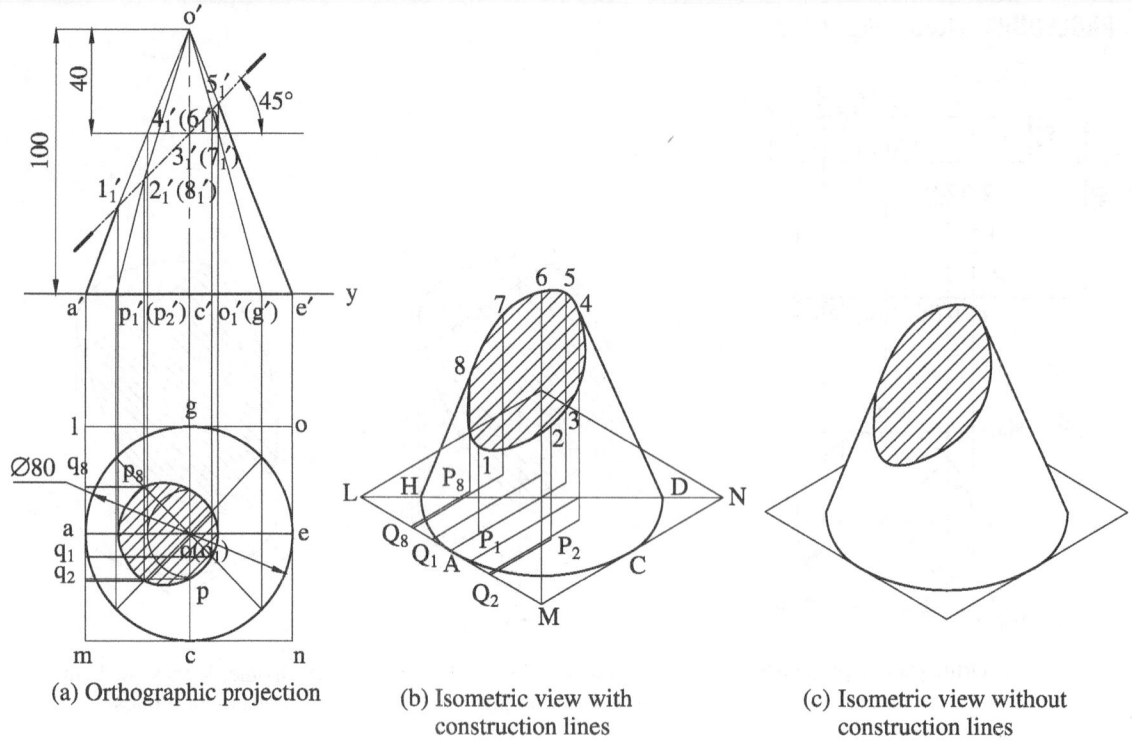

(a) Orthographic projection
(b) Isometric view with construction lines
(c) Isometric view without construction lines

Fig. 13.21

Step 2: Mark the section plane in the front view and obtain the cutting points and their projections as shown in Fig. 13.21(a).

Step 3: The projection of the cutting points on the base can be marked in the isometric view by transferring their respective distances from the top view. These distances can be noted by the set of perpendicular lines to these points from the outer edge l–m of the enclosed square in the top view as shown by the lines q_1p_1, q_2p_2, ..., q_8p_8. Therefore, the cutting points in the base P_1, P_2, ..., P_8 are located in the isometric view.

Step 4: The vertical distances of the cutting points from the base can be measured from their respective front view, transferred on to the perpendicular isometric lines erected from their respective base positions P_1, P_2, ..., P_8, and the cutting points 1, 2, ..., 8 of the solid are obtained in the isometric view.

Step 5: Join all the cutting points 1, 2, ..., 8 in the solid and mark the retained portion of the cut solid with thick lines. Hatch the sectioned portion of the solid as shown in Figs. 13.21(b) and (c). ▲

Example 13.17 ◢

A hexagonal prism of 20 mm base edges and axis 45 mm long has a through square hole of 16 mm sides. The axis of the hole coincides with axis of the prism. Two lateral faces of the hole are parallel to that of the prism. Draw the isometric view of the solid when it is standing upright.

PROCEDURE (Refer Fig. 13.22)

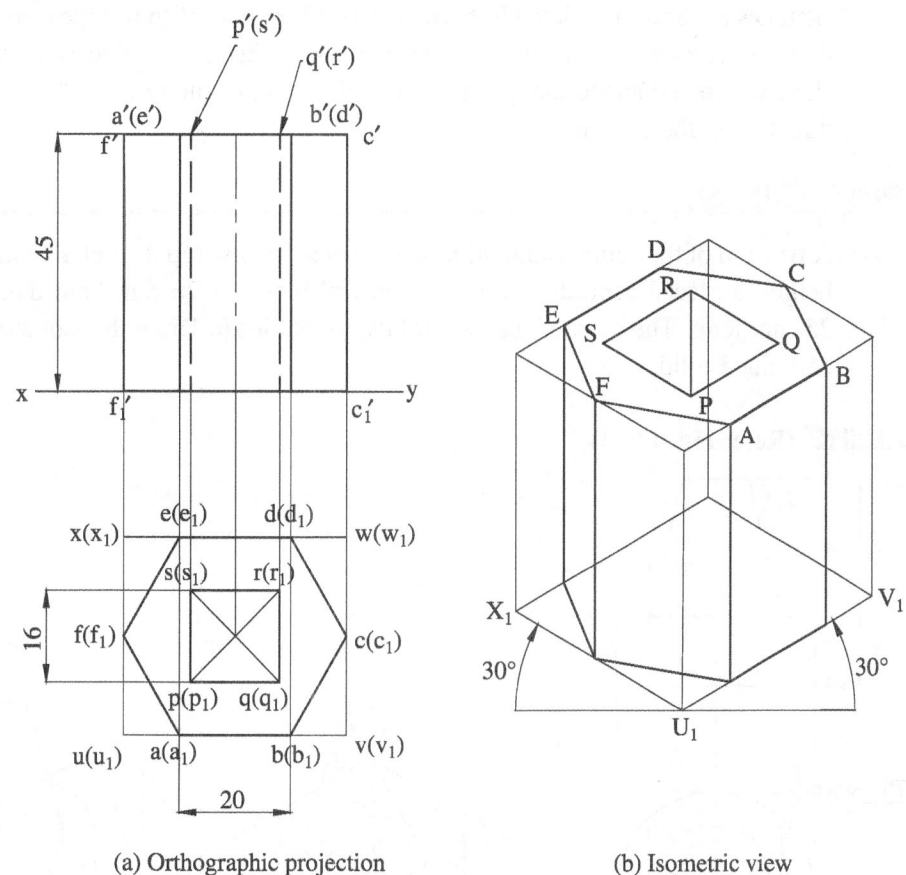

(a) Orthographic projection   (b) Isometric view

Fig. 13.22

Step 1: Draw the orthographic views of the solid as shown in Fig 13.22(a).

Step 2: Obtain the isometric view of the outer solid as discussed in Example 13.6.

Step 3: Obtain the isometric view of the top surface of the square hole (PQRS) by locating its corners on the corresponding isometric lines. Draw a perpendicular line through R to indicate the visible vertical edge of the hole as shown in Fig. 13.20(b). The edges through other points P, Q, and S are not visible and hence, are not marked.

13.7 TWO AND THREE SOLID OBJECTS

The isometric projections or the views of two or three solid objects are attempted by taking one object at a time and drawing their isometric sketches and superimposing them, as present in the assembly. In general, the bottom most solid or the base solid is considered first and other solids are built one over the other. The orthographic views of these solids are drawn and the respective polygonal views are enclosed in squares or rectangles as applicable. If the polygonal views appear in the top view, the isometric sketches are accommodated in between the 30° axes or if they appear in the front view, then the isometric sketches are drawn in the right- or left-side isometric plane. The sketches are expanded along the vertical direction or along the 30° inclined isometric direction as the case may be.

Example 13.18

A frustum of the conical solid of base diameter 50 mm, top diameter 26 mm, and 50 mm height is placed centrally over a cylindrical block of 76 mm base diameter and axis 25 mm long. The axes of the two solids are collinear. Draw the isometric view of the combined solid.

PROCEDURE (Refer Fig. 13.23)

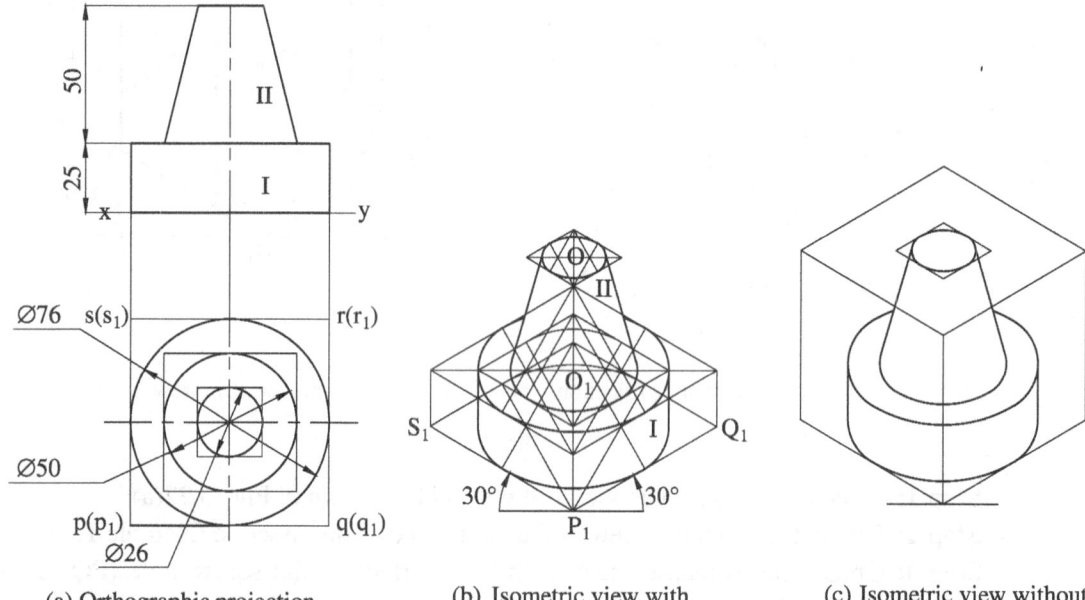

(a) Orthographic projection

(b) Isometric view with construction lines

(c) Isometric view without construction lines

Fig. 13.23

Step 1: Draw the top view and front view of the combined solid and enclose the circles in the top view in squares of respective sizes as shown in Fig. 13.23(a).

Step 2: Draw the isometric view of the base cylindrical block using the four centre method in similar lines as explained in Example 13.9.

Step 3: Locate the centre O_1 of the top end of the base block and draw the rhombus intended for the base of the frustum. Erect a perpendicular at O_1 and mark the height of the frustum and locate O. Construct ellipses at the top and bottom ends, connect them with tangential lines, and complete the isometric projection of the combination of the solid as shown in Fig. 13.23(b) and (c).   ▲

Example 13.19 ◢

Draw the isometric projection of a sphere of a 40 mm diameter resting centrally on the top of a square prism of base 50 mm sides and height 70 mm.

PROCEDURE (Refer Fig. 13.24)

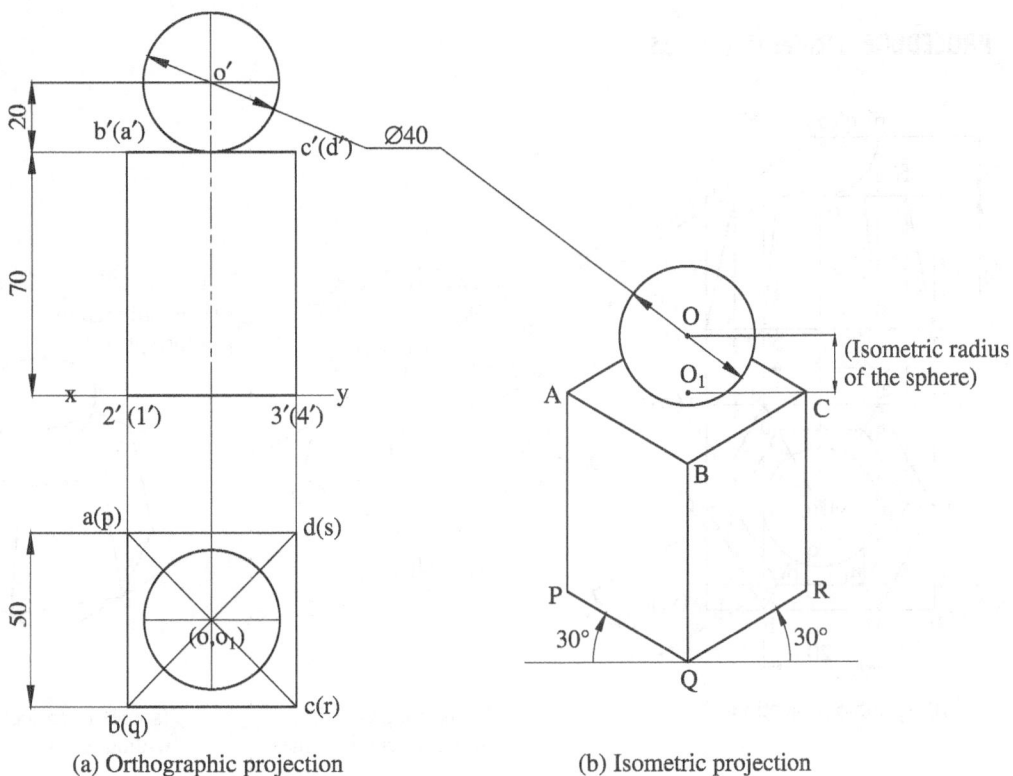

(a) Orthographic projection   (b) Isometric projection

Fig. 13.24

Step 1: Draw the top view and front view of the combination of the solid as shown in Fig. 13.24(a).

Step 2: Draw the isometric projection of the square prism in similar lines as explained in Example 13.6. The construction of the isometric scale has to be done though not shown here. Locate the centre O_1 of the top surface.

Step 3: As the sphere is placed on the top of the prism, the centre of the sphere lies at a distance equal to the isometric length of the radius of the sphere from the from point O_1. Hence, erect a perpendicular at O_1 for a length equal to the isometric radius of the sphere and mark the point O.

Step 4: Draw a circle with O as centre and radius equal 20 mm (the actual radius of the sphere) and complete the isometric projection of the combination of the solid as shown in Fig. 13.24(b). ▲

Example 13.20 ▲

A brass hollow hemispherical bowl of 20 mm diameter is placed centrally on the top of the frustum of a hexagonal solid pyramid of base dimensions of 20 mm sides and top dimensions of 15 mm sides. The height of the frustum is 22 mm. Draw the isometric projection of the above arrangement.

PROCEDURE (Refer Fig. 13.25)

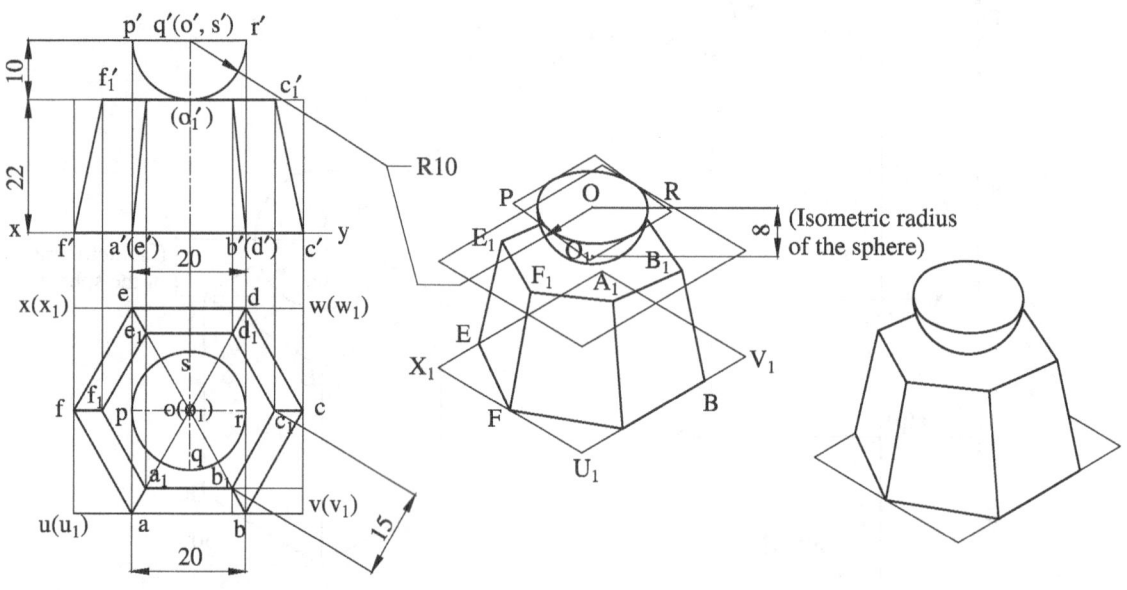

(a) Orthographic projection

(b) Isometric projection with construction lines

(c) Isometric projection without construction lines

Fig. 13.25

Step 1: Draw the top view and front view of the combination of the solid as shown in Fig. 13.25(a).

Step 2: Draw the isometric projection of the frustum of the pyramid in similar lines as explained in Example 13.11, after adopting the isometric scale. Locate the centre O_1 of the top surface.

Step 3: As the hemisphere is placed on the top of the prism, the centre of the sphere lies at a distance equal to the isometric length of the radius of the hemisphere from the from point O_1. Hence, erect a perpendicular at O_1 for a length equal to the isometric radius of the hemisphere and mark the point O.

Step 4: Draw a semicircle with O as centre and radius equal 10 mm (the actual radius of the hemisphere) and complete the isometric projection of the combination of the solid as shown in Figs 13.25(b) and (c).   ▲

Example 13.21 ▲

A square pyramid of height 60 mm and edge of base 26 mm rests symmetrically on the top of a cube of edges 52 mm. If the cube is mounted on a cylinder of 100 mm diameter and 40 mm thickness, draw the isometric view of the above arrangement. The axes of all the three solids lie in one line.

PROCEDURE (Refer Fig. 13.26)

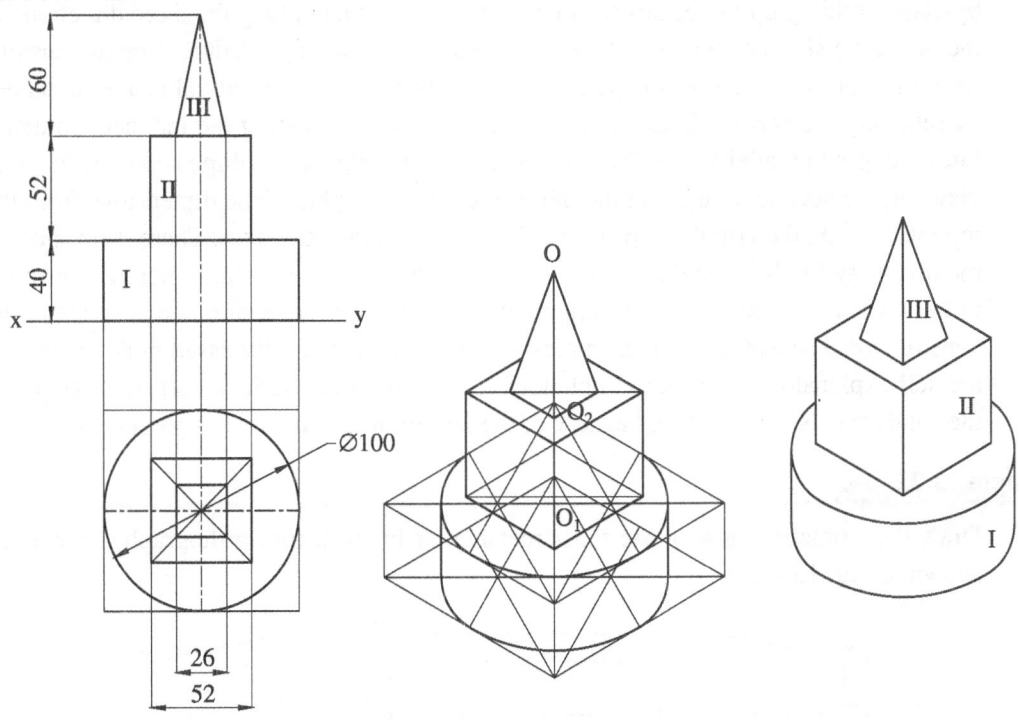

| (a) Orthographic projection | (b) Isometric view with construction lines | (c) Isometric view without construction lines |

Fig. 13.26

Step 1: Draw the top view and front view of the combination of the solids as shown in Fig. 13.26(a).

Step 2: Draw the isometric view of the base cylindrical block using the four centre method and locate the centre O_1 of the top surface of the cylindrical block.

Step 3: Erect a perpendicular at this point O_1 to a distance equal to the height of the cube and locate the centre of the top surface as O_2 and construct the cube with O_1O_2 as the axis.

Step 4: Draw a perpendicular from O_2 to a length equal to the height of the square pyramid and locate the apex as O and construct the square pyramid and complete the isometric of the combination of the solid as shown in Fig. 13.26(b) and (c). ▲

13.8 CONVERSION OF ORTHOGRAPHIC VIEWS OF OBJECTS INTO ISOMETRIC VIEWS

It was shown that the isometric projection or the isometric view of an object or any assembly can be developed from the available orthographic projections of the object. Simple object shapes were discussed in earlier sections. In the current section, the objects with many features such as holes, slots, chamfers, and slopes are discussed. The base block arrangement of the given object can be identified, and the various features can be considered to be built over the base block at their appropriate locations. In fact, stage-by-stage building up of various features and realizing them along the three directions in the isometric sketches will enable quick grasp of the whole procedure. The dimensions from the front view can be transferred on to the one of the side vertical isometric planes by retaining the vertical lines as vertical and by accommodating the inclined isometric lines along or parallel to the 30° isometric axis. Similarly, the dimensions of the side view will be accommodated on the other side isometric plane. The dimensions from the top view are marked on the appropriate 30° inclined isometric axes. There is no specific methodology for drawing the isometric sketches, because each practitioner may have his or her own way of doing this, depending on the level of imagination one possesses. The constructional schemes of the examples shown below are not discussed in detail as they are self-explanatory. The formal notations used to refer the various corners or edges of the solids are also not adopted because of the familiarity the reader has possessed by now.

Example 13.22 ◢

Draw the isometric view of the taper rectangular block whose orthographic views are shown in Fig. 13.27.

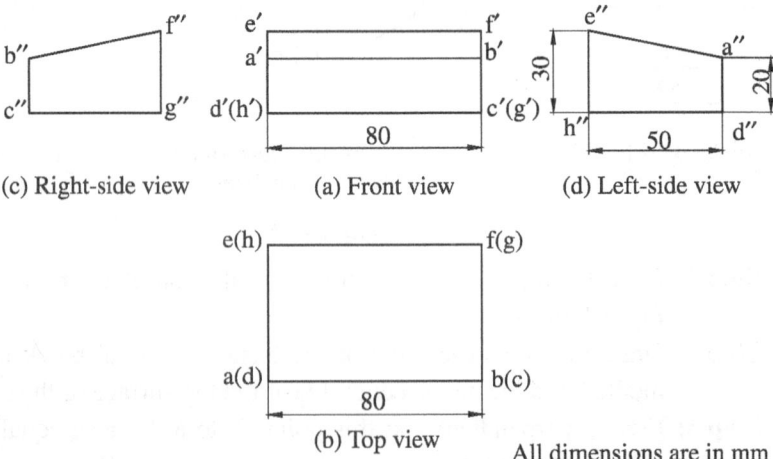

(c) Right-side view   (a) Front view   (d) Left-side view

(b) Top view   All dimensions are in mm

Fig. 13.27   Orthographic projections of a tapered rectangular block

PROCEDURE (Refer Fig. 13.28)

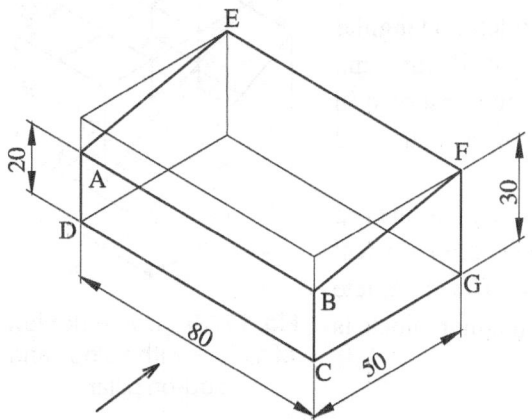

Fig. 13.28 Isometric view of a tapered rectangular block

Step 1: Draw the isometric view of the rectangular box with the base dimensions equal to that observed in the top view (80 mm × 50 mm) and height equal to the maximum height of 30 mm as observed in the front view.

Step 2: The side view of the object can be drawn on the right vertical isometric plane and with the sloping line to indicate the slanting surface and the right isometric plane can be trimmed to the height of 20 mm. The complete isometric view of the taper rectangular block is shown in Fig. 13.28. ▲

Example 13.23 ◣

Draw the isometric view of the object whose orthographic views are shown in Fig. 13.29.

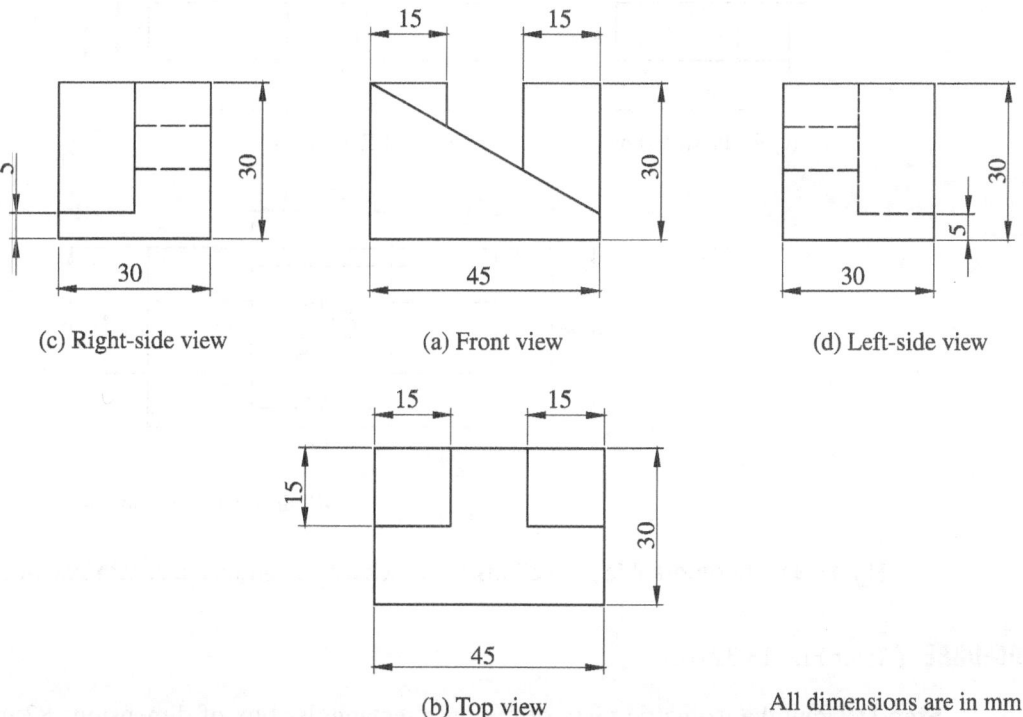

(c) Right-side view   (a) Front view   (d) Left-side view

(b) Top view   All dimensions are in mm

Fig. 13.29 Orthographic projections of a block with a slope and add-on pillars

PROCEDURE   (Refer Fig. 13.30)

Step 1: Draw the isometric view of the complete rectangular box of base dimensions 45 mm × 45 mm and height equal 30 mm. Mark the slope observed in the front view in the left-side isometric plane and complete the sloping block.

Step 2: Set the two squares of 15 mm × 15 mm at the top surface to represent the pillars and drop vertical lines to meet the sloping block. The complete isometric view of the taper rectangular block is shown in Fig. 13.30.

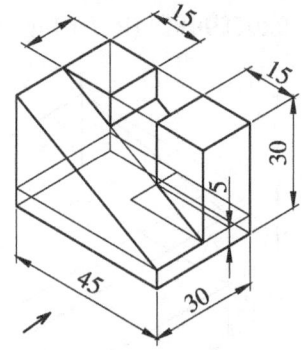

Fig. 13.30   Isometric view of a block with a slope and add-on pillars

Example 13.24

Draw the isometric view of the block whose orthographic views are shown in Fig. 13.31.

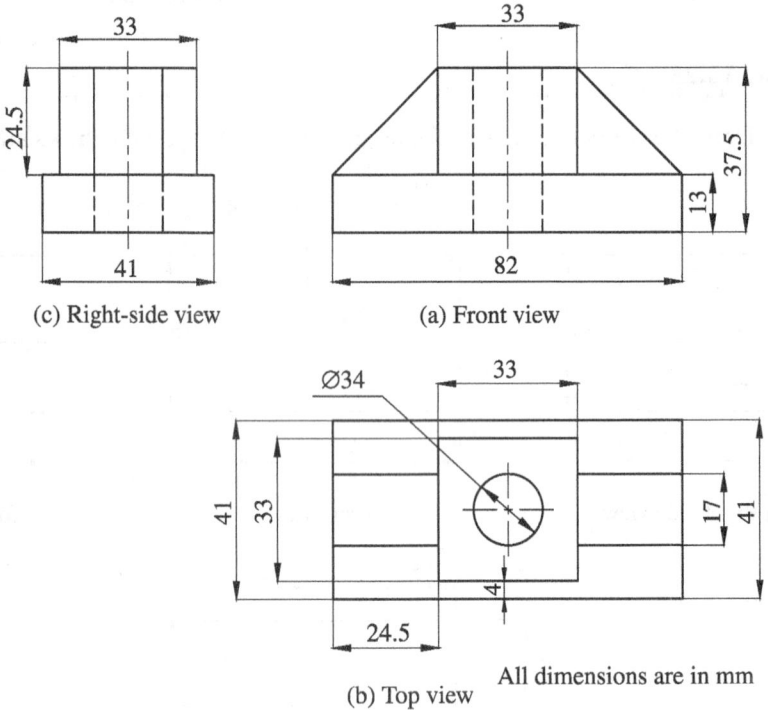

(c) Right-side view   (a) Front view

(b) Top view

All dimensions are in mm

Fig. 13.31   Orthographic projections of a block with a central block with side ribs

PROCEDURE   (Refer Fig. 13.32)

Step 1: Draw the isometric view of the base rectangular box of dimensions 82 mm × 41 mm and height 13 mm, with the left-side isometric plane for the front view.

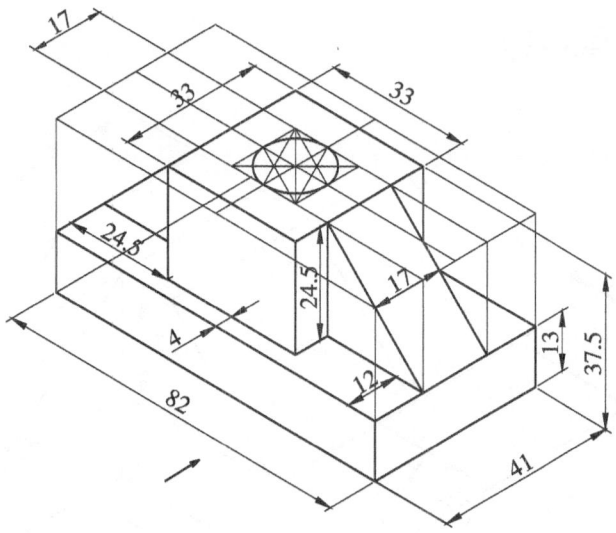

Fig. 13.32 Isometric view of a block with a central block with side ribs

Step 2: Set the square of 33 mm centrally on the top surface and raise it by 24.5 mm and obtain the isometric view of the centre block. Construct the ellipse on its top surface as discussed earlier. Mark the side ribs and complete the isometric view as shown in Fig. 13.32.   ▲

Example 13.25 ◤

Draw the isometric view of the block with V-cuts and as shown in Fig. 13.33.

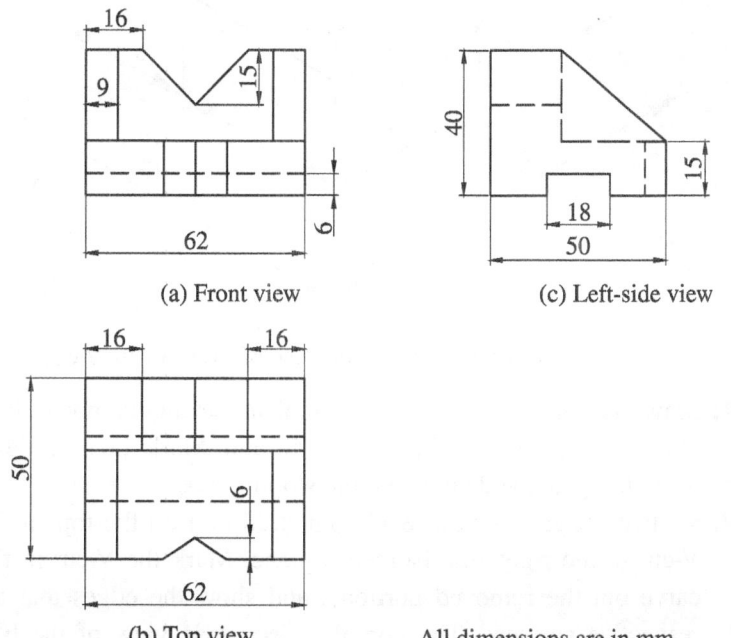

(a) Front view   (c) Left-side view

(b) Top view   All dimensions are in mm

Fig. 13.33 Orthographic projections of a block with V-cuts

PROCEDURE (Refer Fig. 13.34)

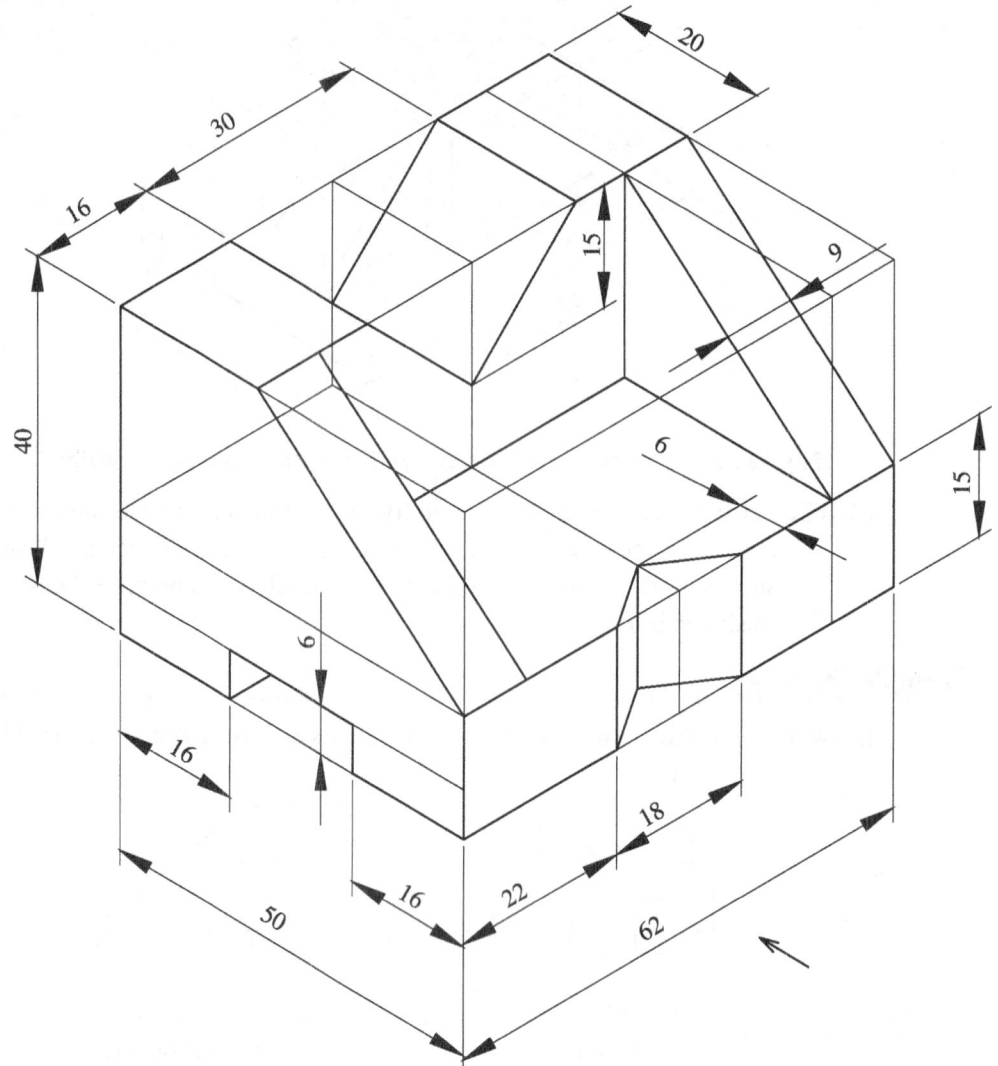

Fig. 13.34 Isometric view of a block with V-cuts

Step 1: Draw the isometric view of the complete rectangular box of base dimensions 50 mm × 62 mm and height equal to 40 mm. Mark the side view on the left-side isometric plane and complete the sloping block.

Step 2: Set two rectangles of size 16 mm × 20 mm on the top surface and mark the V-cut in the right-side isometric plane. Mark the V-cut in the bottom block, carve out the removed portions, and show the edges that are visible in the respective surfaces. The complete isometric view of the block is shown in Fig. 13.34. ▲

Example 13.26

Draw the isometric view of the bearing block with the supporting rib as shown in Fig. 13.35.

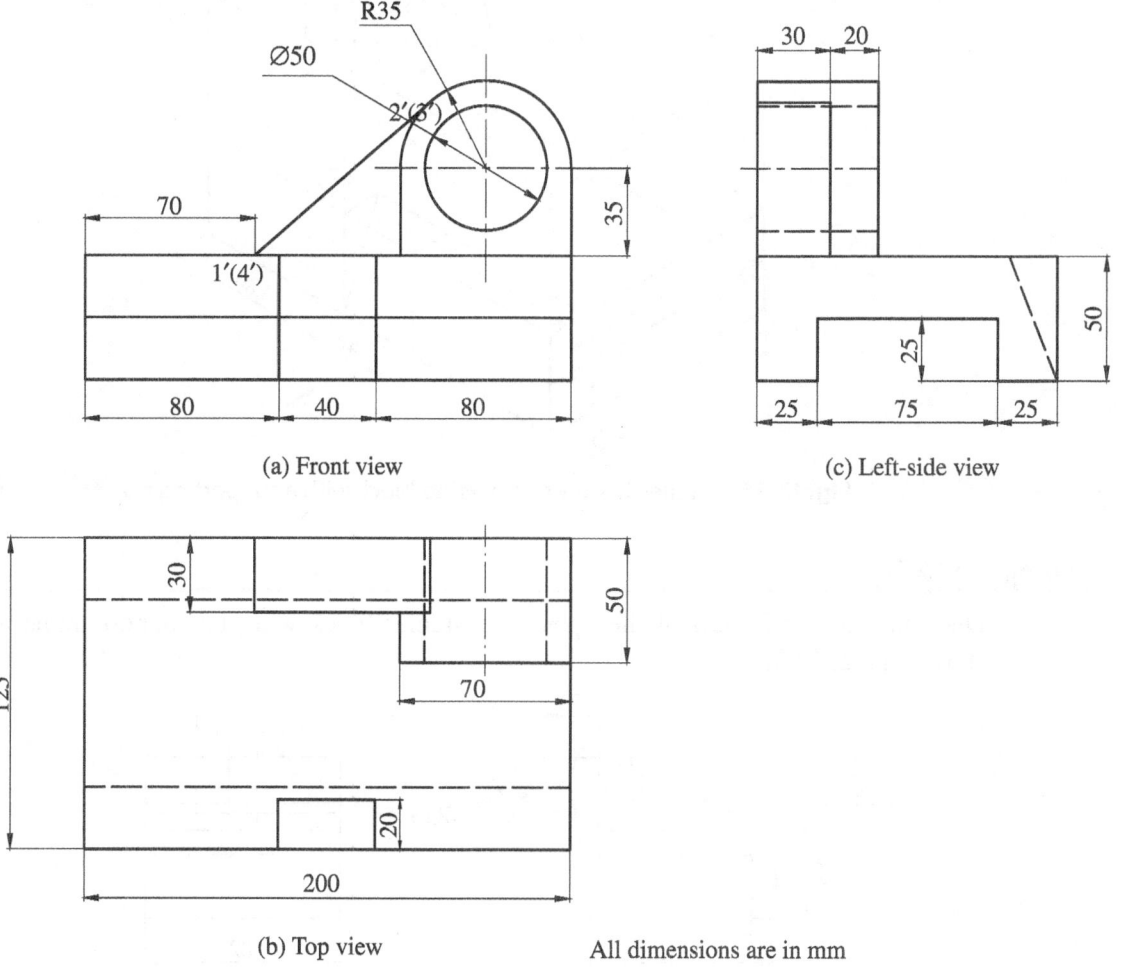

(a) Front view

(c) Left-side view

(b) Top view   All dimensions are in mm

Fig. 13.35   Orthographic projections of a bearing block with a supporting rib

PROCEDURE (Refer Fig. 13.36)

Step 1: Draw the isometric view of the base rectangular box of dimensions 125 mm × 200 mm and height 50 mm, with the right-side isometric plane for the front view.

Step 2: Set on the right-side isometric plane the semicircular block as discussed earlier and mark the rib. Mark the circular hole as an ellipse using the four centre method. The complete isometric view of the block is shown in Fig. 13.36.

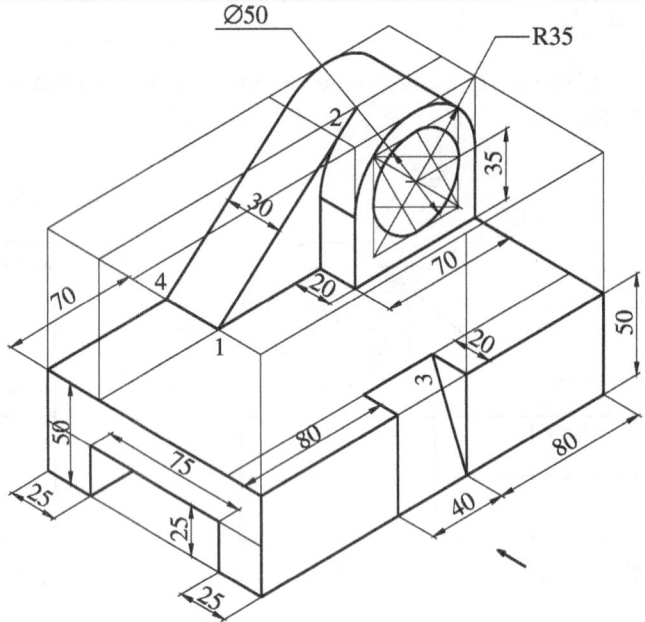

Fig. 13.36 Isometric view of a bearing block with a supporting rib ▲

Example 13.27 ◣

Draw the isometric view of the open semicircular block with the support frame as shown in Fig. 13.37.

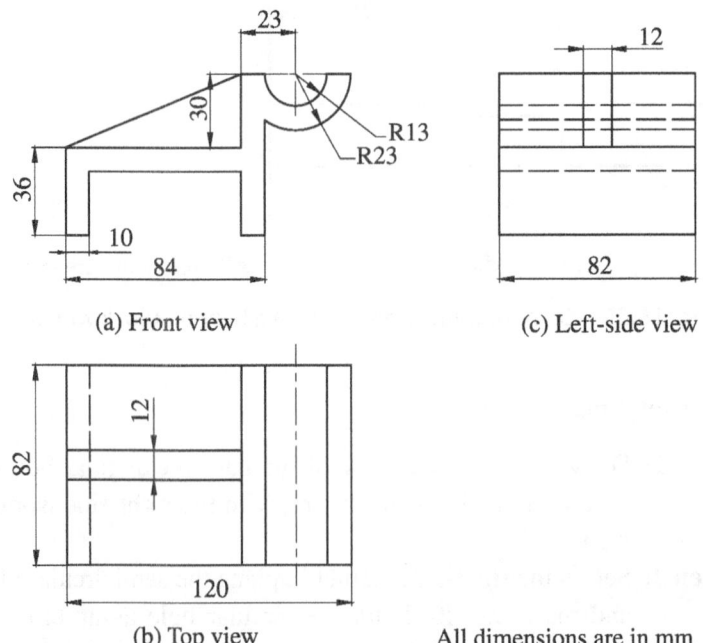

(a) Front view

(c) Left-side view

(b) Top view

All dimensions are in mm

Fig. 13.37 Orthographic projections of an open semicircular block with support frame

PROCEDURE (Refer Fig. 13.38)

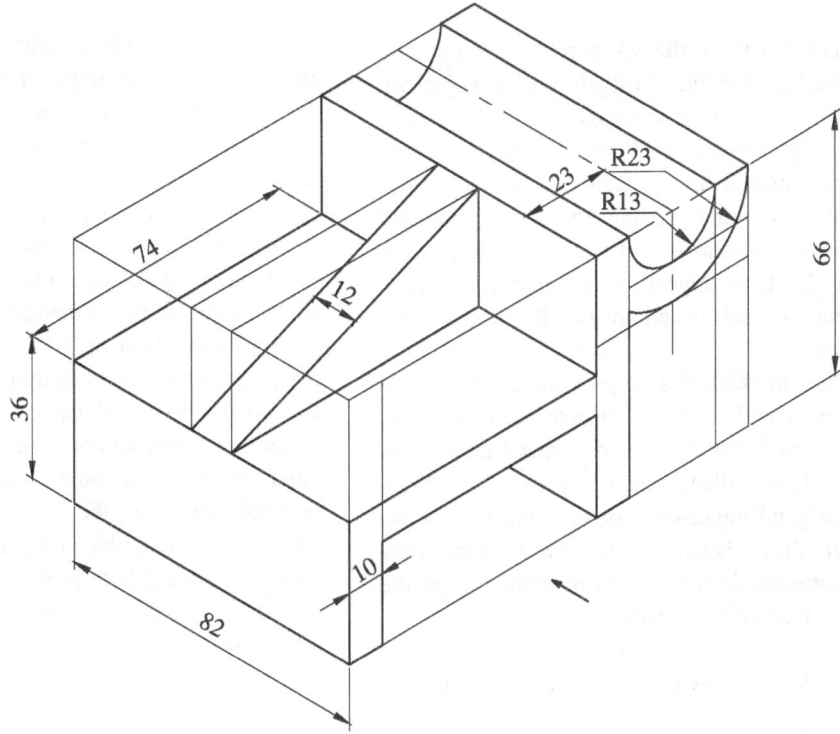

Fig. 13.38   Isometric view of an open semicircular block with support frame

Step 1: Draw the isometric view of the complete rectangular box of base dimensions 82 mm × 120 mm and height equal to 66 mm.

Step 2: Set the front view in the right-side isometric plane and extend the various features along the other isometric axis.

Step 3: Set the top view of the rib on the top surface and mark its height and the slope in the right-side isometric plane. The complete isometric view of the block is shown in Fig. 13.38.

13.9 ENGINEERING APPLICATIONS

Since the isometric projection or the isometric view is the pictorial projection of a 3D object with fair representation of all its dimensions, it is used as one of the powerful means of communication of the shape of an object. Though it cannot technically bring all the internal features of an object, it serves fairly well as for as external appearance is concerned. Simple machine components, castings, brackets and holding devices, furniture, vessels, and containers are usually represented in isometric sketches for easy appeal and understanding. Simple assemblies involving two or three components can also be shown in isometric sketches.

- Isometric sketch is the type of pictorial projection in which all the three dimensions of an object are not only available in one single picture but appear in equal proportion, enabling the reader to observe the dimensions directly from that sketch.
- In the isometric sketch, the three mutually perpendicular faces, and hence, the three axes of a 3D object appear equally inclined at 120° among themselves and are equally shortened by 81.6%.
- In order to effect this proportionate reduction, a special scale known as isometric scale is drawn. When the isometric sketch is made as per this scale, it is called isometric projection. When the original dimensions of the object are used, the resulting sketch is called the isometric view or isometric drawing. Both of them convey the same shape of the object.
- The isometric projection or the isometric view of an object can be made from its details that are generally available in its orthographic views. The above sketch can be prepared by accommodating the three views in a three adjacent plane assembly held along three isometric axes inclined at 120° to each other.
- The top view of the object can be accommodated in between the two isometric axes inclined at 30° to the horizontal, while the front view is accommodated among the vertical isometric axis and the inclined one at 30° to the horizontal. The front view can be in the left-side isometric plane or on the right-side plane.
- To locate the various points or edges of the object other than those situated on the three mutually perpendicular axes, the object is enclosed in an imaginary rectangular or square box and these points are identified from the common axes and are then transferred to the isometric sketch appropriately.

MULTIPLE-CHOICE QUESTIONS

13.1 In an isometric view or projection, the principal edges of the object are drawn along
   (a) three mutually perpendicular axes
   (b) isometric axes
   (c) 45° with +ve x-axis, –45° with –ve x-axis, and 90°
   (d) 30° with +ve x-axis, –45° with –ve x-axis, and 90°

13.2 In an isometric view, the dimensions of the object are equal to
   (a) the actual dimensions of the object
   (b) 81.6% of actual dimension
   (c) 70% of actual dimension
   (d) 60% of actual dimension

13.3 The isometric projection of an object is one of the axonometric projections of the object in which its principal coordinate axes are
   (a) equally inclined at 30° to the plane of projection

   (b) equally inclined at 45° to the plane of projection
   (c) equally inclined at 60° to the plane of projection
   (d) equally inclined at 45° to the HP

13.4 In an isometric projection, the dimensions of the object along the isometric axes or isometric lines are reduced by
   (a) 70%   (c) 81.6%
   (b) 50%   (d) 0%

13.5 The isometric projection of a circle is
   (a) the same circle itself
   (b) an ellipse
   (c) a circle of diameter reduced to 70% of actual diameter
   (d) a circle of diameter reduced to 81.6% of actual diameter

13.6 The four centre method is used to draw the isometric projection of

(a) an ellipse   (c) a parabola

(b) a circle   (d) a hyperbola

13.7 The isometric projection of a curve can be drawn by using

(a) four centre method

(b) offset method

(c) parallelogram method

(d) rectangular method

13.8 The isometric projection of a sphere placed on a flat surface will appear as a circle of diameter _____ the diameter of the sphere.

(a) greater than   (c) equal to

(b) less than   (d) 81.6% of

13.9 The isometric projection of a prismatic solid can be done by

(a) Box method

(b) Parallelogram method

(c) Rectangular method

(d) Visual ray method

13.10 The top view of a square when projected in the isometric horizontal plane takes the shape of

(a) rhombus   (c) parabola

(b) square   (d) rectangle

WORK PRACTICE LEVEL – 1

13.1 Draw the isometric projection of a square lamina of 40 mm sides, when it is 20 mm above the HP with one of its sides inclined at 30° to the VP.

13.2 A thin pentagonal plate of 30 mm sides has one of its edges on the HP. Draw its isometric projection, when its surface is vertical and parallel to the VP and 25 mm in front of it.

13.3 A thin metallic washer is of the hexagonal shape with a circular hole punched in it centrally. Draw its isometric projection when its surface is 20 mm above the HP. The sides of the hexagon are 30 mm long and the central hole is 30 mm diameter.

13.4 A thin brass decorative plate is made up of two semicircular ends with a square portion in between. Draw its projection, when its surface is vertical and is 15 mm in front of the VP. The square side of the plate is 300 mm and is 25 mm above the HP.

13.5 Draw the isometric projection of a vertical hexagonal prism of 30 mm sides and 100 mm high, when it rests on its base on the HP with two of its faces perpendicular to the VP.

13.6 Draw the isometric projections of a pentagonal pyramid of 30 mm base sides and axis 70 mm long, when it rests on its base on the HP and axis 30 mm in front of VP with two of its adjacent slant faces equally inclined to the VP.

13.7 A waste paper basket is in the form of the frustum of an inverted square pyramid with base 100 mm sides and top 150 mm sides. Draw the isometric view of the basket if its height is 250 mm. The wall thickness can be neglected.

13.8 A cylindrical block is 80 mm diameter and 50 mm high. It has a central hexagonal shaped hole of 20 mm sides. Draw the isometric view of the solid if its surface is parallel to the VP. Two of the lateral surfaces of the hexagon are parallel to the HP.

13.9 A sheet metal funnel is in the form of a frustum of a cone. The larger end is of 80 mm diameter and is on the VP. If the shorter end is of 50 mm diameter and its length is 80 mm, draw the isometric projection of the funnel.

13.10 A brass cylindrical bush has an external diameter of 60 mm and length of 120 mm. If the annular wall thickness of the bush is 10 mm, draw the isometric view when it rests with its axis parallel to the HP and the VP.

WORK PRACTICE LEVEL – 2

13.1 A pentagonal prism of 30 mm base sides and axis 80 mm long is resting on its base on the HP with one of its base edges parallel to the VP and closer to it. It is cut by a section plane, perpendicular to the VP and inclined at 60° to the HP and passing through a point on the axis and 20 mm from the top end. Draw the isometric view of the cut solid, when its larger portion is retained.

13.2 A hexagonal pyramid of 25 mm base edges and axis 60 mm long has its hexagonal end on the VP, with two of its base sides parallel to the HP. A vertical section plane inclined at 30° to the VP bisects the axis. Draw the isometric view of the portion of the solid on the VP.

13.3 A cylinder of 50 mm diameter and axis 80 mm long has its axis parallel to the HP and the VP. It is cut by a section plane whose horizontal trace is inclined at 30° to the reference line and passes through a point 15 mm from one of its ends and situated on the axis. Draw isometric view of the cut solid with its cut section.

13.4 A cone of base diameter 60 mm and height 60 mm rests on the HP on its base. A vertical section plane cuts the cone such that it passes through the cone at a distance of 10 mm in front of the axis. If the section is parallel to the VP, draw isometric view of the portion of the solid that is closer to the VP.

13.5 A thin cylindrical plastic bucket of 300 mm base diameter and axis 500 mm long is seated on its base on the ground and filled with water to its full height. When the bucket is tilted such that its base makes an angle of 30° with the ground, represent its position with the water in the isometric view. Adopt a convenient scale.

13.6 A square pyramid of base edges 30 mm and axis 50 mm long is placed on the top of a cube whose sides are 45 mm long. The edges of the cube are parallel to the base edges of the pyramid. Draw the isometric projection of the two solids when their axes are collinear.

13.7 A hexagonal pyramid of base edges 25 mm and axis 60 mm long rests symmetrically on a cylindrical slab of 80 mm diameter on one of its flat ends. Draw the isometric view of the solids if the thickness of the slab is 20 mm.

13.8 A compound solid consists of a base solid in the shape of a pentagonal prism and a cone placed

on its top. Draw the isometric view of their arrangement, when their axes are in the same line. The base solid has 30 mm sides and is of 30 mm high. The diameter of the base of the cone is 40 mm and its axis is 50 mm long.

13.9 Draw the isometric projection of a sphere of 50 mm diameter resting centrally on the frustum of a hexagonal pyramid with a top face of 30 mm sides, a base of 40 mm sides, and axis 80 mm long.

13.10 A monument consists of three solid shapes, namely a bottom cylindrical slab, a top square pyramid, and a frustum of a hexagonal pyramid in between. The base diameter of the slab is 1000 mm and its height is 300 mm. The intermittent solid has a base of 400 mm sides, a top of 300 mm sides, and a height of 500 mm. The base sides of the top solid are 400 mm long and its height is 600 mm. The solids are symmetrically placed with their base edges parallel to each other. Draw the isometric view of the monument. Adopt a suitable scale.

13.11 Draw the isometric view of the object whose projections are shown in Fig. 13.39.

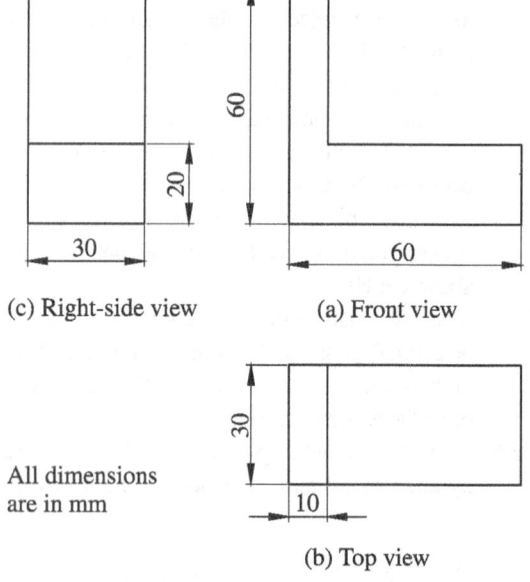

(c) Right-side view   (a) Front view

All dimensions
are in mm

(b) Top view

Fig. 13.39   L-shaped object

13.12 Draw the isometric view of the object whose projections are shown in Fig. 13.40.

13.13 Draw the isometric view of the object whose projections are shown in Fig. 13.41.

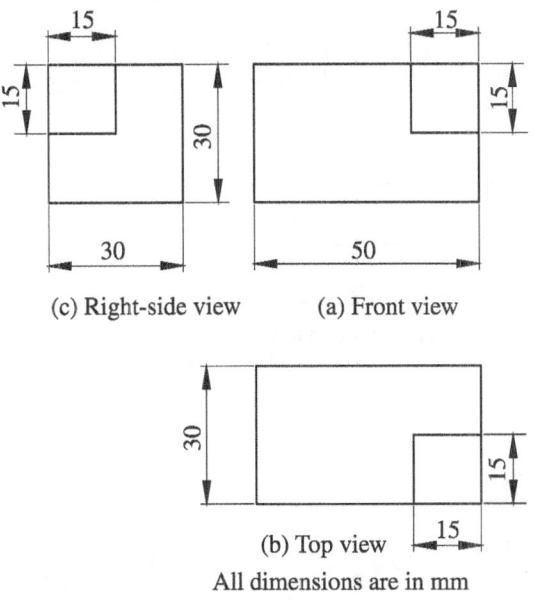

(c) Right-side view   (a) Front view

(b) Top view

All dimensions are in mm

Fig. 13.40   A cube with cut-out

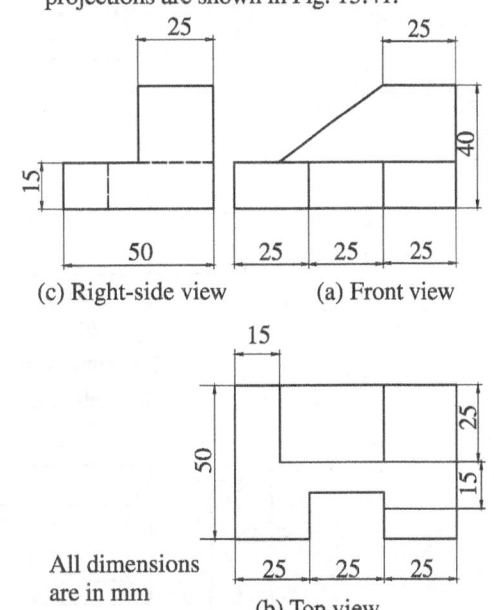

(c) Right-side view   (a) Front view

All dimensions are in mm

(b) Top view

Fig. 13.41   Block with cut-out and built up slope

13.14 Draw the isometric view of the object whose projections are shown in Fig. 13.42.

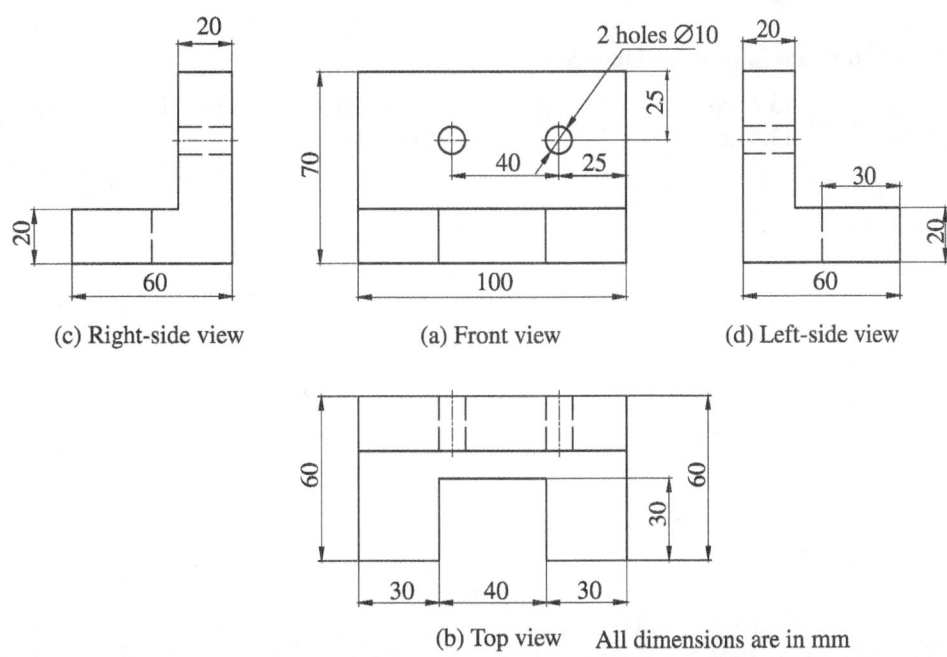

(c) Right-side view   (a) Front view   (d) Left-side view

2 holes Ø10

(b) Top view   All dimensions are in mm

Fig. 13.42   L-shaped block with rectangular cut-out

13.15 Draw the isometric view of the object whose projections are shown in Fig. 13.43.

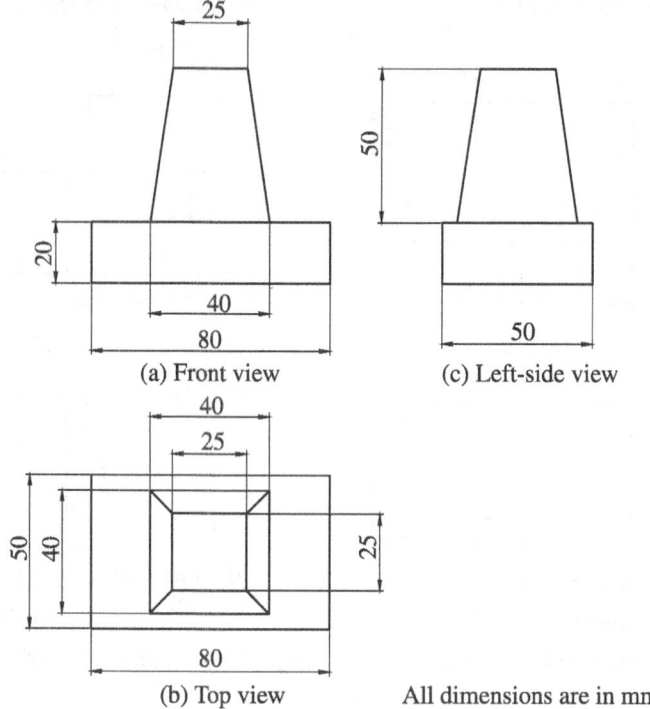

(a) Front view   (c) Left-side view

(b) Top view   All dimensions are in mm

Fig. 13.43   Rectangular block with a centre pillar with sloping surfaces

Answers for Multiple-choice Questions

13.1 (b)   13.2 (a)   13.3 (a)   13.4 (c)   13.5 (b)   13.6 (b)

13.7 (b)   13.8 (c)   13.9 (a)   13.10 (a)

Transformation of Projections into Orthographic Views

14

OBJECTIVES

This chapter will help the reader to understand the following:

- Identifying different surfaces and their views of an object with respect to the direction of vision
- Decomposing a three-dimensional object by understanding their sub-shapes and obtaining the views of element shapes
- Organizing the views of element shapes and obtaining the resultant orthographic views of overall object

14.1 INTRODUCTION

In Chapter 4, the basic visualization procedure was explained to identify different surfaces of a three-dimensional (3D) object and make the freehand sketching arrangement of their views. In order to obtain true shape of the different surfaces, the principle of orthographic projection was underlined in Chapter 6 and a combination of procedures to visualize an object orthogonally and draw the views of the surfaces was discussed elaborately. This chapter deals with the transformation techniques to convert the pictorial sketches of complicated shapes and features that are used in many engineering applications into their detailed orthographic views with proper instruments and constructional procedures.

14.2 BASIC PRINCIPLES OF TRANSFORMATION

When an object is viewed along the direction of one of the three perpendicular axes, the dimensions that are parallel to the direction of vision will not be visualized, but only other two dimensions will be visible in that view, which forms a plane surface.

The plane surfaces that are normal to the direction of vision, appear in true sizes, whereas the plane surfaces that are inclined to the direction of visualization will appear smaller than the original sizes.

The curved surfaces of the object, when viewed normal to the direction of vision, appear as plane surfaces in that view and the nature or details of the curve will be observed in one of the other views.

14.3 LAYOUT OF VIEWS

In order to convert a pictorial view into the orthographic view, the direction from which the object to be viewed is indicated by an arrow, and in general, the view corresponding to this arrow is known as the front view. The top view is realized when the direction of view is rotated by 90° and the object is viewed from its top surface. The side view is recognized by the surface that is visualized adjacent to the arrow in the pictorial view. This can be predicted on the right or left side of the picture, and accordingly, the side view is drawn.

The front view reveals the two dimensions, namely the length and the height, of the object. The top view gives the length and the other dimension (i.e., width of the object). The side views reveal the height/depth and the width of the object (i.e., the length of the object). The layout of views is arranged based on first-angle projection method.

14.4 GUIDELINES FOR TRANSFORMATION

The basic guideline to transform a pictorial view of an object into orthographic views is to understand the number of element shapes with which the object is realized. By drawing the three views (front, top, and side views) of the element shapes and by suitably combining them, one can get the orthographic views of the whole object. Whenever it involves the addition of element shapes, their overall dimensions increase in two views, and in the third view, the cross-sectional shape of that element appears. When the removal of an element shape takes place within an object, the boundaries of the element are not visible, and hence, they are marked as 'hidden lines or curves'. The examples discussed in the following sections will explain the methodology of the transformation of a pictorial sketch into the orthographic views.

14.5 ILLUSTRATIVE EXAMPLES

14.5.1 L-shaped Block

Example 14.1

Draw the orthographic projections (front, top, and side views) of the L-shaped block as shown in Fig. 14.1(a).

Observation

The object consists of an L-shaped block with a groove at the bottom, and a small rectangular block is added at the rear edge on the top left corner. The front view appears

L-shaped along with a small rectangle added at its top surface. The other views show the outlines with the transition edges.

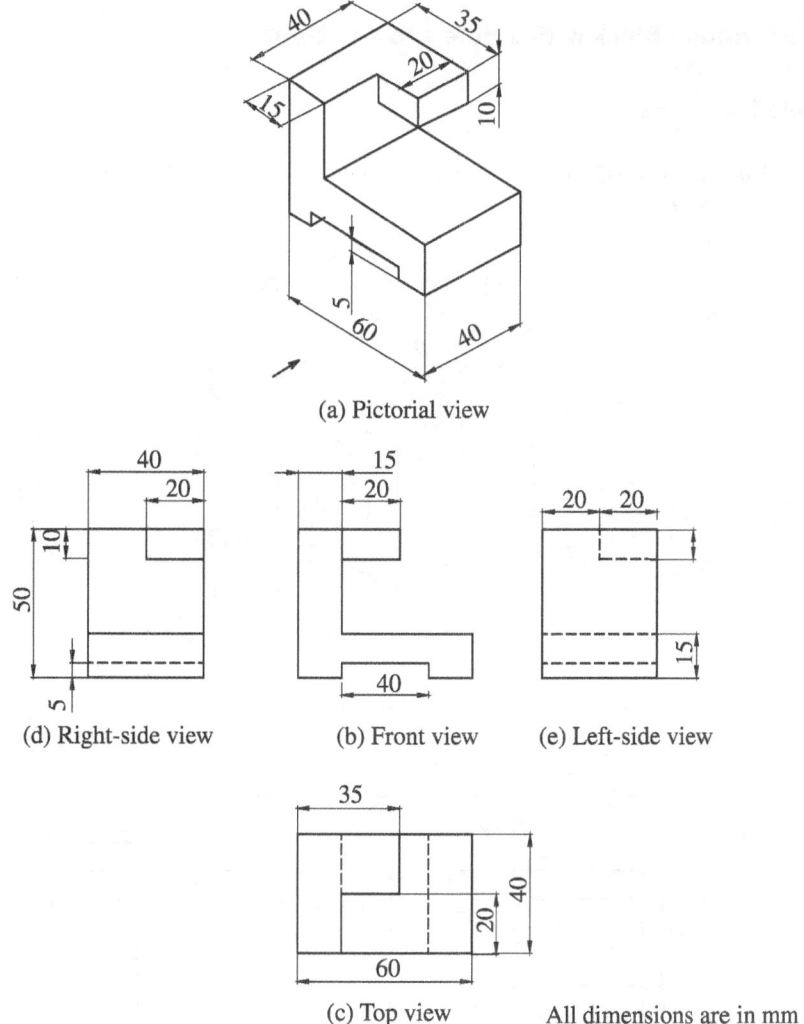

(a) Pictorial view

(d) Right-side view (b) Front view (e) Left-side view

(c) Top view All dimensions are in mm

Fig. 14.1 Orthographic projections of L-shaped block

PROCEDURE

Step 1: Draw the front view as shown in Fig 14.1(b) with an L-shaped appearance and mark the bottom groove. Add a rectangle at the vertical leg for a length of 20 mm and height of 10 mm to represent the upper block.

Step 2: Draw a rectangle of 60 mm length and 40 mm width below the front view and project the top surface as shown in Fig 14.1(c). Mark the bottom groove in dotted lines.

Step 3: Draw a rectangle of height 50 mm and width 40 mm to represent the side views and mark the transition edge of the L-shaped block. Add a rectangle corresponding to the top block for a height of 10 mm. The right- and left-side views are shown in Fig. 14.1(d) and (e). ▲

14.5.2 Rectangular Block with a Hole and a Cut-out

Example 14.2 ▲

Draw the orthographic projections (front, top, and side views) of the block as shown in Fig. 14.2(a).

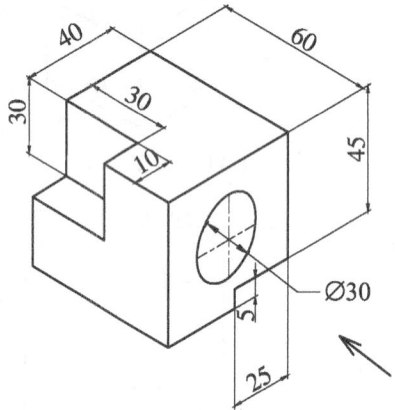

(a) Pictorial view

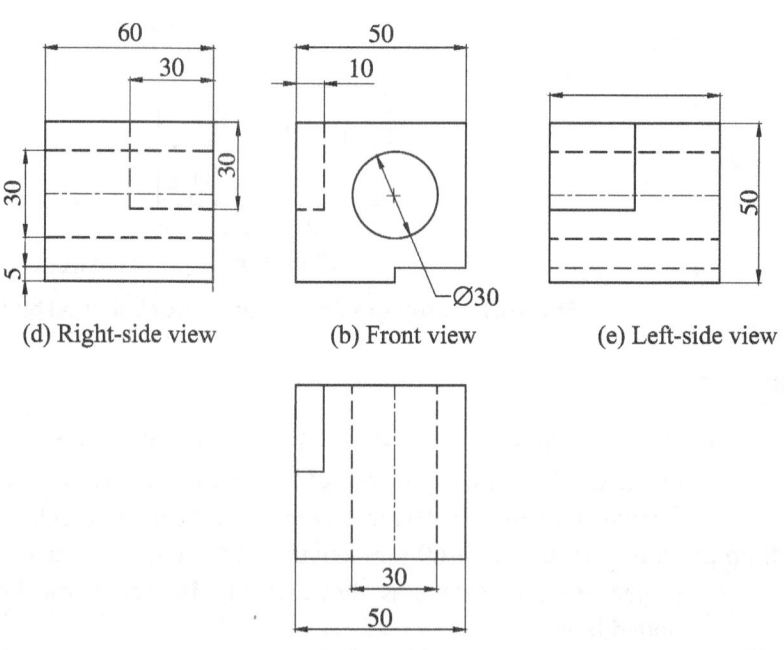

(d) Right-side view   (b) Front view   (e) Left-side view

(c) Top view   All dimension are mm

Fig. 14.2 Orthographic projection of a block with a hole and cut-out

Observation

The object represents a rectangular block with a cylindrical hole. A cut-out at the rear edge and a groove at the bottom surface are provided. The front view reveals a square with a circle placed centrally. The bottom right-hand edge is removed due to the presence of the groove and the top left corner is removed at the rear face due to the cut-out.

PROCEDURE

Step 1: Draw a square of size 50 mm with an opening for the bottom groove as shown in the front view in Fig 14.2(b). Mark the circular hole centrally and draw a rectangle of length 10 mm and height 30 mm at the top left corner in dotted lines to represent the cut-out at the rear face.

Step 2: Draw a rectangle of 50 mm length and 60 mm width and represent the cut-out as shown in the top view in Fig 14.2(c). Mark the axis line of the cylindrical hole and its edges in dotted lines. It may be noted that the bottom groove is merged with the axis line.

Step 3: Draw a rectangle to represent the overall dimensions and mark the edges of the slot, the bottom groove, and the cylindrical hole as shown in the side views in Fig. 14.2(d) and (e). ▲

14.5.3 Block with Add-on Cylindrical Surfaces

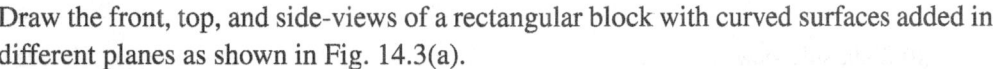

Example 14.3

Draw the front, top, and side-views of a rectangular block with curved surfaces added in different planes as shown in Fig. 14.3(a).

Observation

Figure 14.3(a) shows a rectangular block with three cylindrical surfaces, namely a cylinder, a semicircular groove, and two semi-cylinders introduced in the front, top, and right side, respectively. Both cylinder and groove yield their cross sections in the front view, while the others yield in the top view.

PROCEDURE

Step 1: Draw a rectangle of length 70 mm and height 60 mm with a rectangular cut-out at the top left corner and a semicircular opening at the top surface as shown in the front view in Fig. 14.3(b). The semi-cylindrical surfaces are added as rectangles at the right edge. It can be noted that the transition edge of the lower cylindrical block does not appear as it merges smoothly.

Step 2: Project the front view directly below and obtain the top view as shown in Fig 14.3(c). The edges corresponding to the cut-outs are marked in thick lines.

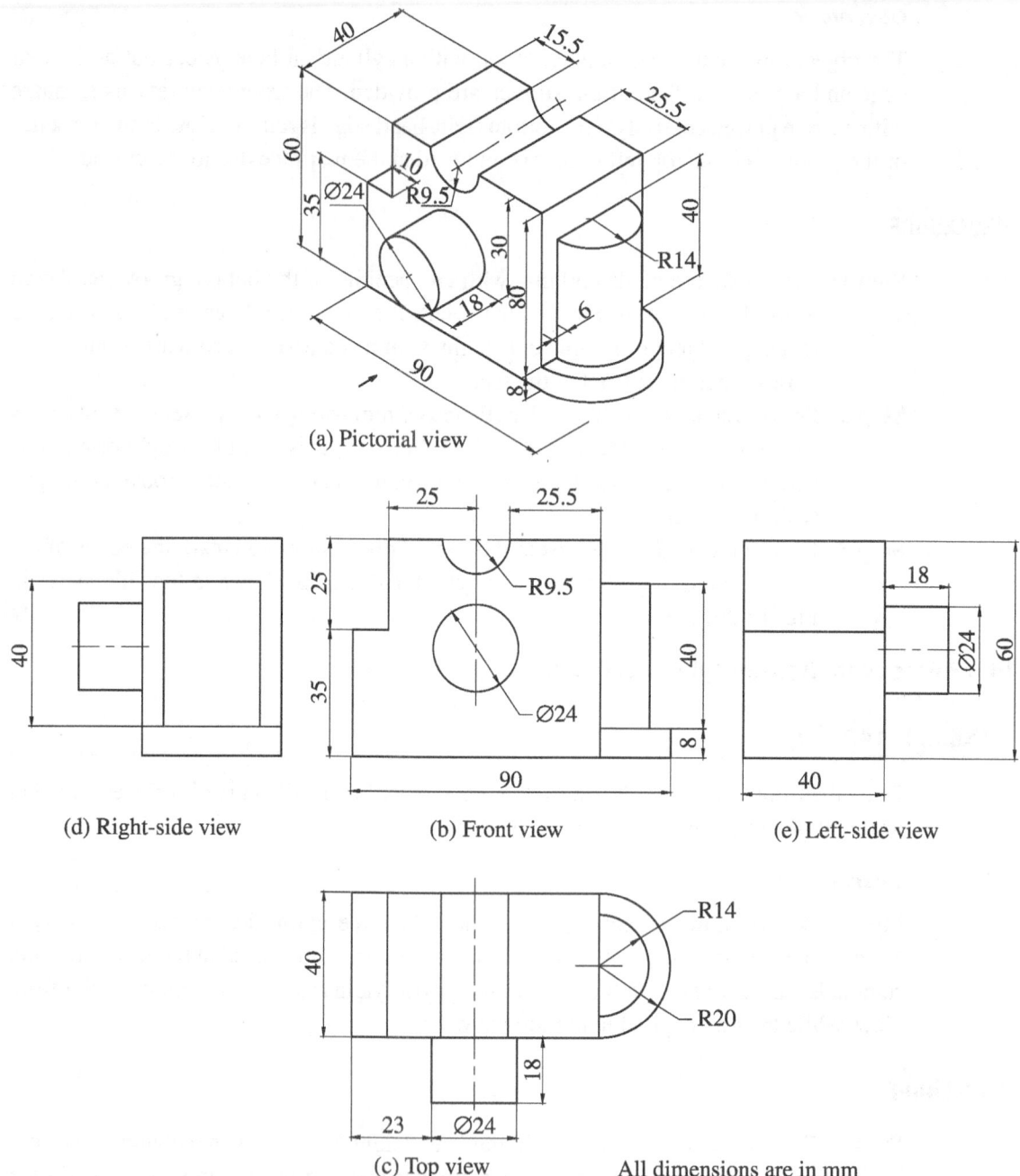

(a) Pictorial view

(d) Right-side view   (b) Front view   (e) Left-side view

(c) Top view   All dimensions are in mm

Fig. 14.3 Orthographic projections of rectangular block with curved surfaces

Step 3: The side views (shown in Fig. 14.3d and e) are drawn as rectangles of 40 mm width and 60 mm height. The semi-cylindrical blocks are addcd as rectangles within the right-side view, while the cut-out is drawn visible in left-side view.

14.5.4 Block with Convex- and Concave-shaped Surfaces

Example 14.4

Draw the front view, top view, and right-side views of the block that has convex- and concave-shaped surfaces as shown in Fig. 14.4(a).

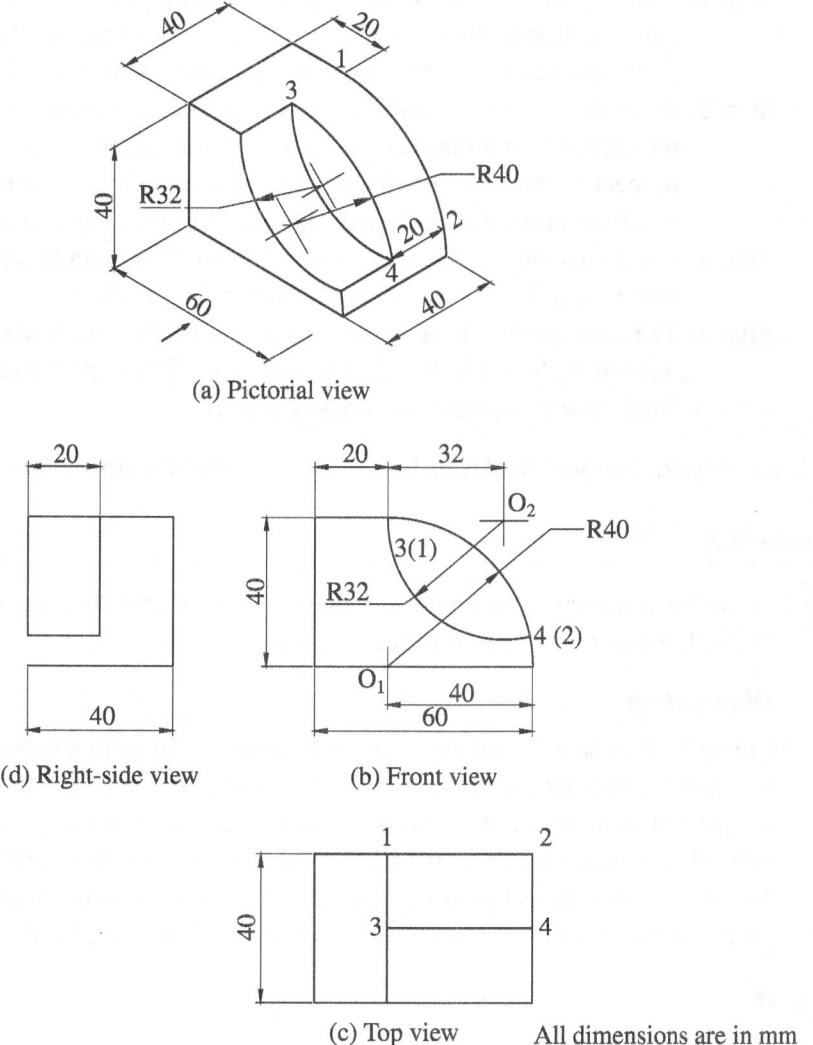

(a) Pictorial view

(d) Right-side view

(b) Front view

(c) Top view   All dimensions are in mm

Fig. 14.4   Orthographic projections of the block with convex- and concave-shaped surfaces

Observation

Figure 14.4(a) shows a rectangular block with a curved convex surface of radius 40 mm along its width. A portion of the curved surface is removed or carved out in the front at a radius of 32 mm and the nature of this surface is concave. Both the curved surfaces appear in the front view, with curves opposite fashioned, while the other views show the outline of the rectangular block.

PROCEDURE

Step 1: Draw a rectangle of 60 mm length and 40 mm height with the top right corner removed. Locate the centre O_1 at a distance of 40 mm on the bottom edge and draw an arc to meet the top edge at point 1 as shown in the front view in Fig. 14.4(b).

Step 2: Since the convex surface also meets the top edge at a coincident point 3, mark its centre O_2 at a distance of 32 mm, and draw an arc in the opposite fashion to meet the curve 1–2 at the point 4. The curve 1–2 represents the rear surface, while the curve 3–4 represents the front surface as shown in Fig. 14.4(b).

Step 3: Draw a rectangle of length 60 mm and width 40 mm to represent the top view and project the points 1 to 4 as shown in Fig. 14.4(c).

Step 4: The right-hand side view is shown in Fig. 14.4(d) as a square of 40 mm size and a rectangle for a width of 20 mm and to a height as projected from point 4 in the front view to represent the convex surface.

14.5.5 Bracket with Cut-outs and Arcs in Horizontal, Vertical, and Side Surfaces

Example 14.5

 Draw the front, top, and right–side views of the bracket with cut-outs and arcs in the vertical, horizontal, and side surfaces as shown in Fig. 14.5(a).

Observation

Figure 14.5(a) shows a bracket with three portions, namely a horizontal, a central, and a vertical surface parallel to the viewer. The horizontal portion is made up of an arc and a taper towards the rear and with a circular hole, and these details are visible in the top view. The central vertical portion has a curved surface whose details are visible in the side views. The vertical portion, parallel to the viewer, has an arc at the top right corner and a rectangular cut-out, and these details are visible in the front view.

PROCEDURE

Step 1: The front views of the three portions are shown in Fig. 14.5(b) with an arc of 26 mm at the top right corner.

Step 2: The corresponding top views of the three portions can be obtained as shown in Fig. 14.5(c). The bottom left-hand corner has been rounded off with an arc of 22 mm radius whose centre is located equidistant from the front and left edges. The taper is obtained by locating the point 1 and drawing a tangent to the arc at the point 2.

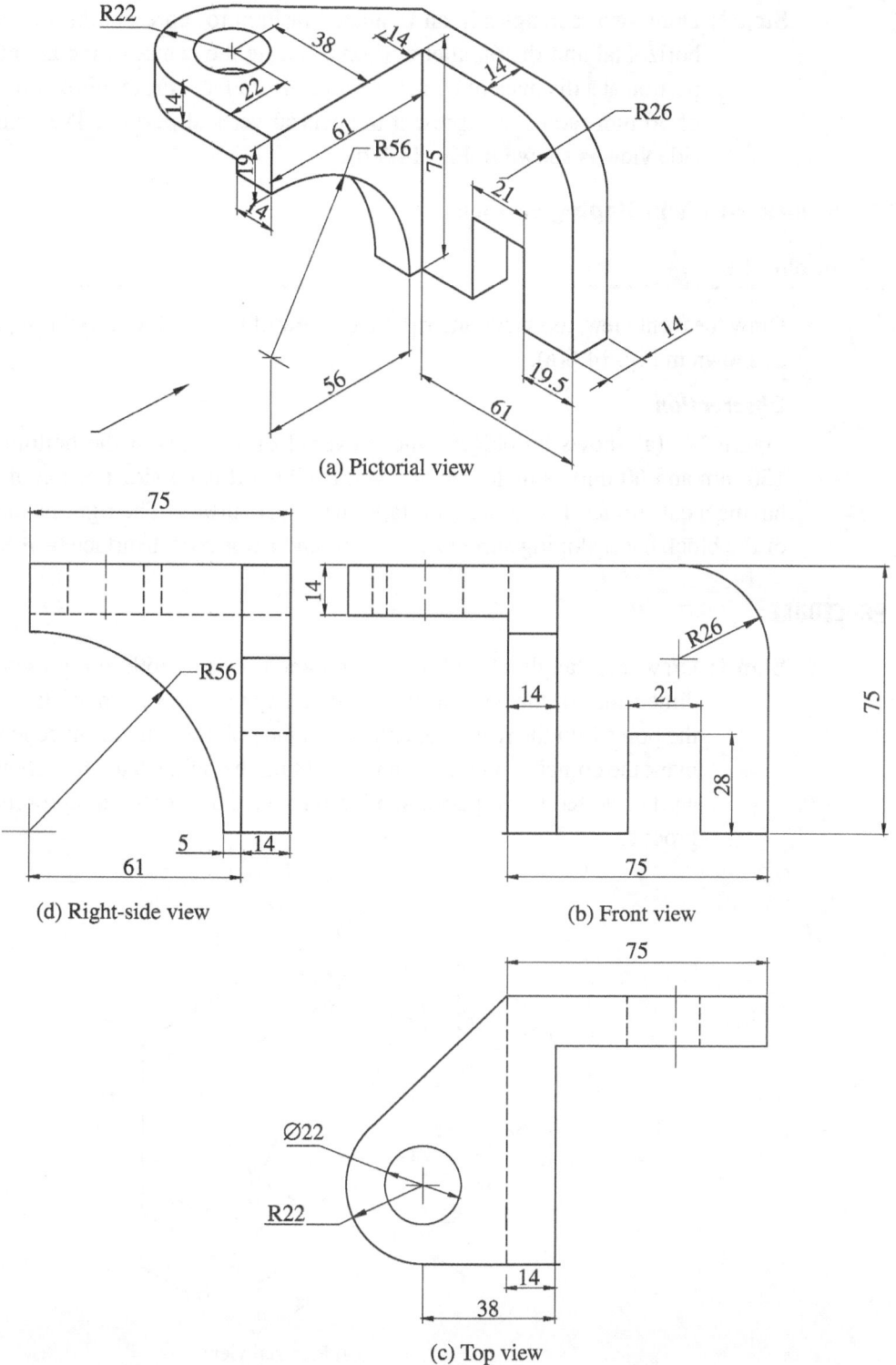

(a) Pictorial view

(d) Right-side view

(b) Front view

(c) Top view

Fig. 14.5 Orthographic projections of bracket with cut-outs and arcs in horizontal, vertical, and side surfaces

Step 3: Draw two rectangles in an L-shaped fashion to represent the side views of the horizontal and the parallel portions. Locate the centre of the arc of the middle portion at a distance of 61 mm from the front face. From this centre, draw an arc of 56 mm radius to represent the central vertical portion. The complete right-side view is shown in Fig. 14.5(d). ▲

14.5.6 Block with Twin Sloping Surfaces

Example 14.6 ▲

Draw the front view, top view, and right-side view of the block with twin sloping surfaces as shown in Fig. 14.6(a).

Observation

Figure 14.6(a) shows an object, whose overall dimensions at the bottom surface are 150 mm and 60 mm as its length and width. The left hand-side portion of the block is having a cut surface 1–2–3–4 at 60° taper with rear surface. The right-hand side portion of the block has a sloping surface 5–6–9–10 and a horizontal surface 6–7–8–9.

PROCEDURE

Step 1: Draw a rectangle of 150 mm length and 60 mm width to represent the overall dimensions of the bottom surface in the top view as shown in Fig. 14.6(c). Mark the point 1 on the rear edge and obtain the point 2 in the front edge at 60°. Also, mark the coincident points 3 and 4. Obtain the points 5 to 10 as shown and draw a set of dotted lines spaced at 50 mm apart in the centre to represent the bottom groove.

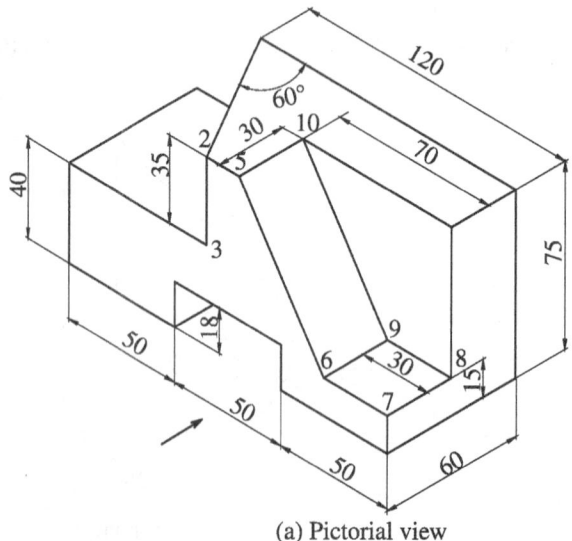

(a) Pictorial view

Fig. 14.6 Orthographic projections of block with twin sloping surfaces (*Contd*)

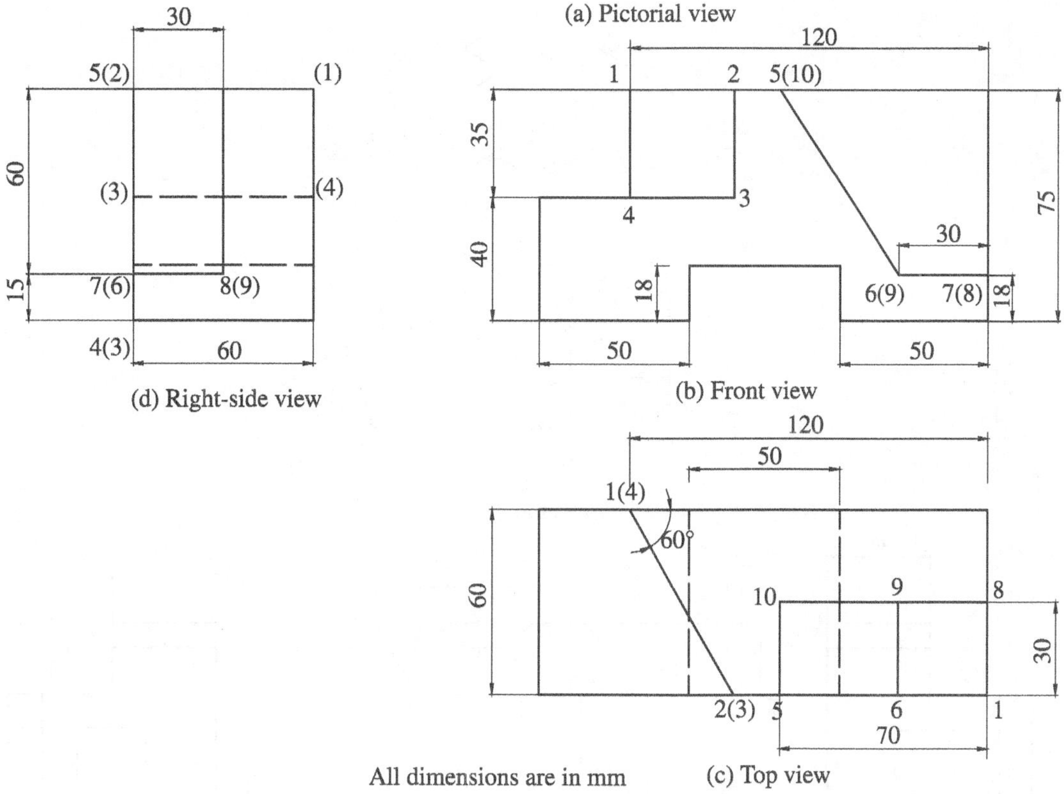

(a) Pictorial view

(d) Right-side view

(b) Front view

All dimensions are in mm   (c) Top view

Fig. 14.6   (*Contd*)

Step 2: The contour of the front view is drawn above the top view by projecting the points 1 to 10 at their respective heights as shown in Fig. 14.6(b). The groove at the bottom surface is marked at a height of 18 mm.

Step 3: The right-side view is drawn for an overall height of 75 mm and width 60 mm. The corners 1 to 10 located at the cut surfaces are projected and the sloping surface is identified by 2–6–9–10. Refer Fig. 14.6(d).

14.5.7 Block with Sloping Surfaces and Cut-outs

Example 14.7

Draw the front view, top view, and side views of the block with sloping surfaces and cut-outs as shown in Fig. 14.7(a).

Observation

The object shown in Fig. 14.7(a) has got a horizontal portion with a vertical cut in its front face and a vertical portion having a vertical cut. Since the side surfaces are inclined, their inclinations or the slope is visible in the front view.

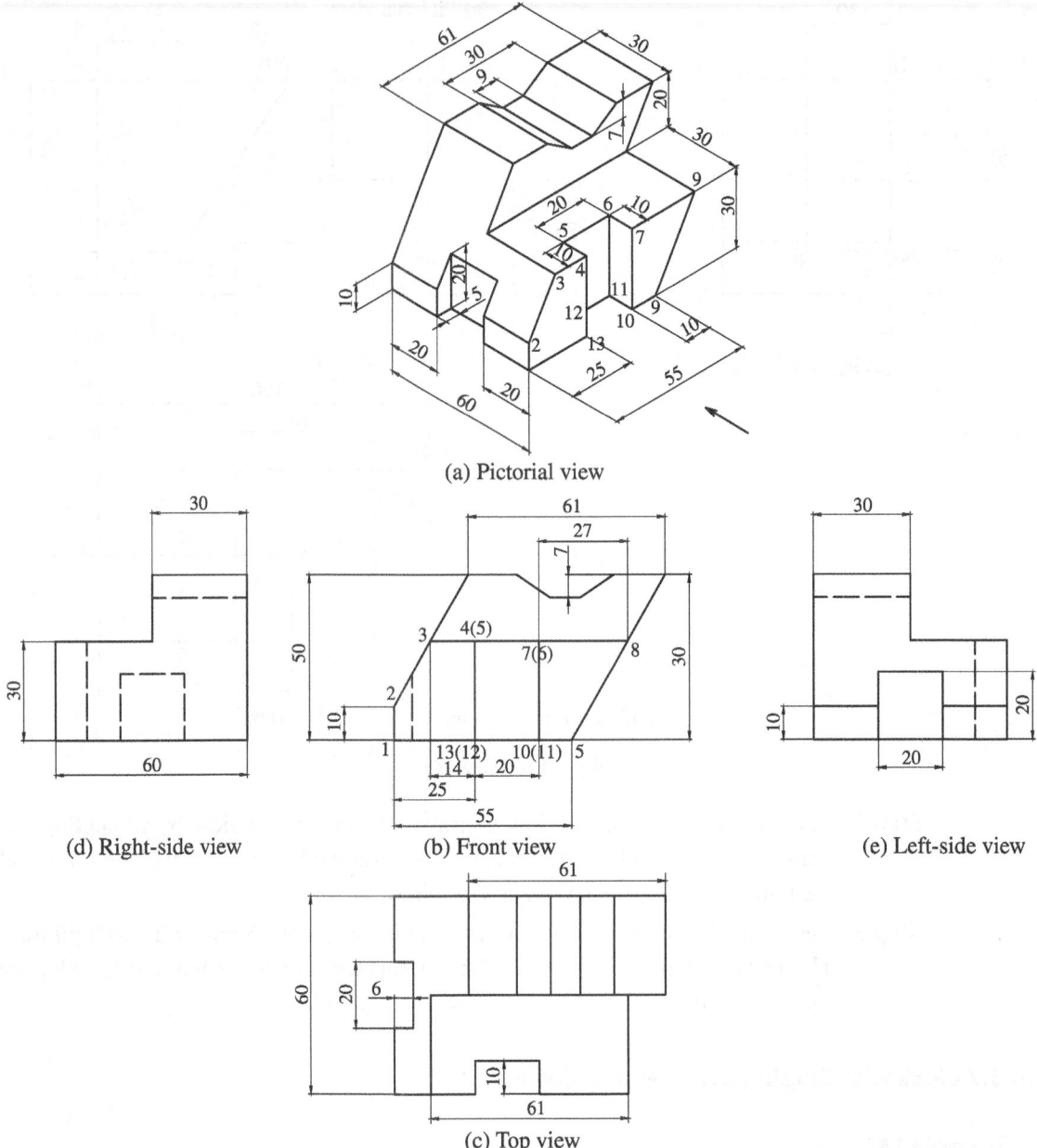

(a) Pictorial view

(d) Right-side view   (b) Front view   (e) Left-side view

(c) Top view

Fig. 14.7 Orthographic projections of the block with sloping surfaces and cut-outs

PROCEDURE

Step 1: The front view of the horizontal block is obtained by the contour lines joining the points 1 to 13 as shown in Fig. 14.7(b). The edges of the central cut-out are shown as thick lines.

Step 2: The front view of the vertical block is obtained by extending the former to a height of 50 mm with a central opening corresponding to the cut portion.

Step 3: The top view of the respective blocks is obtained at a width of 30 mm as shown in Fig. 14.7(c).

Step 4: The side views are obtained as L-shaped portions. While the right-side view (Fig. 14.7d) gives the overall boundary, the left-side view (Fig. 14.7e) includes the side cut-out. The change of slope in the side surface is indicated by a straight line at a height of 10 mm.   ▲

14.5.8 Semicircular Arch with Truncated Centre Pillar

Example 14.8 ▲

Draw the front view, top view, and right-side view of the block with a semicircular arch and truncated centre pillar as shown in Fig. 14.8(a).

Observation

The object has a semicircular arch with two horizontal wings like arrangements at the base unit. Since the top unit is uniformly tapering on all the four sides, the slopes of these surfaces appear in the front view and in the side view, while its top view gives the cross sections of the top and bottom surfaces, which are rectangles centrally arranged.

PROCEDURE   (Refer Fig. 14.8)

Step 1: The front view of the base block is drawn as a semicircle with horizontal portions at its ends. The top and bottom edges of the centre block are marked at their respective heights symmetrically above the base block as shown in Fig. 14.8(b).

Step 2: The overall top view is drawn with the larger dimensions of the base the unit as shown in Fig. 14.8(c). The top and the bottom surfaces of the centre block are represented as rectangles corresponding to their widths and by joining their corners.

Step 3: Draw the side view of the base unit corresponding to the height of the semicircular arch as shown in Fig. 14.8(d). Project the top and the bottom surfaces of the centre block from its front view.

Step 4: Mark the cross sections of the cylindrical hole in the base unit and in the central block as circles in the top view and as rectangles in dotted lines in the other views.

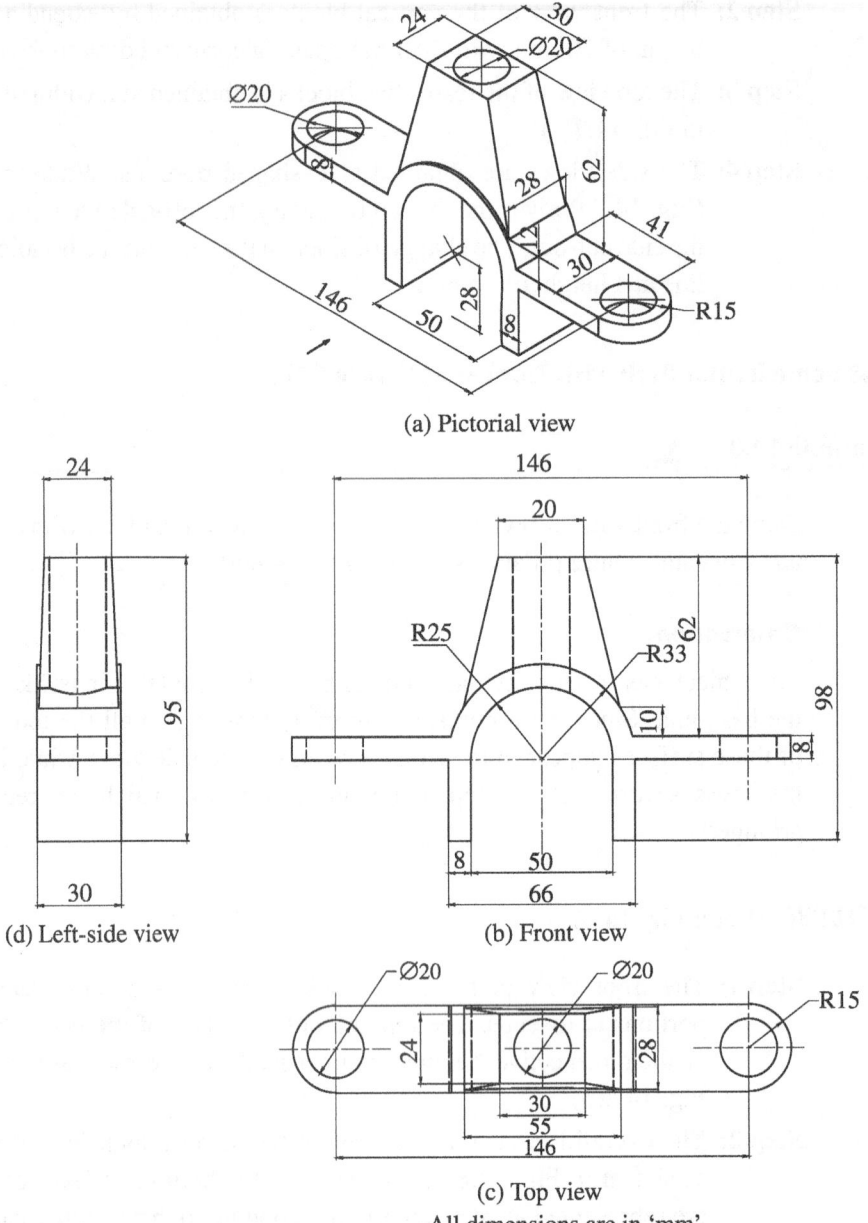

(a) Pictorial view

(d) Left-side view

(b) Front view

(c) Top view

All dimensions are in 'mm'

Fig. 14.8 Orthographic projections of semicircular arch with truncated centre pillar ▲

14.5.9 Horizontal and Vertical Semicircular Blocks with an Intermittent Connecting Piece

Example 14.9 ◢

Draw the front view, top view, and left-side views of the arrangement of the horizontal and vertical semicircular blocks with an intermittent connecting piece as shown in Fig. 14.9(a).

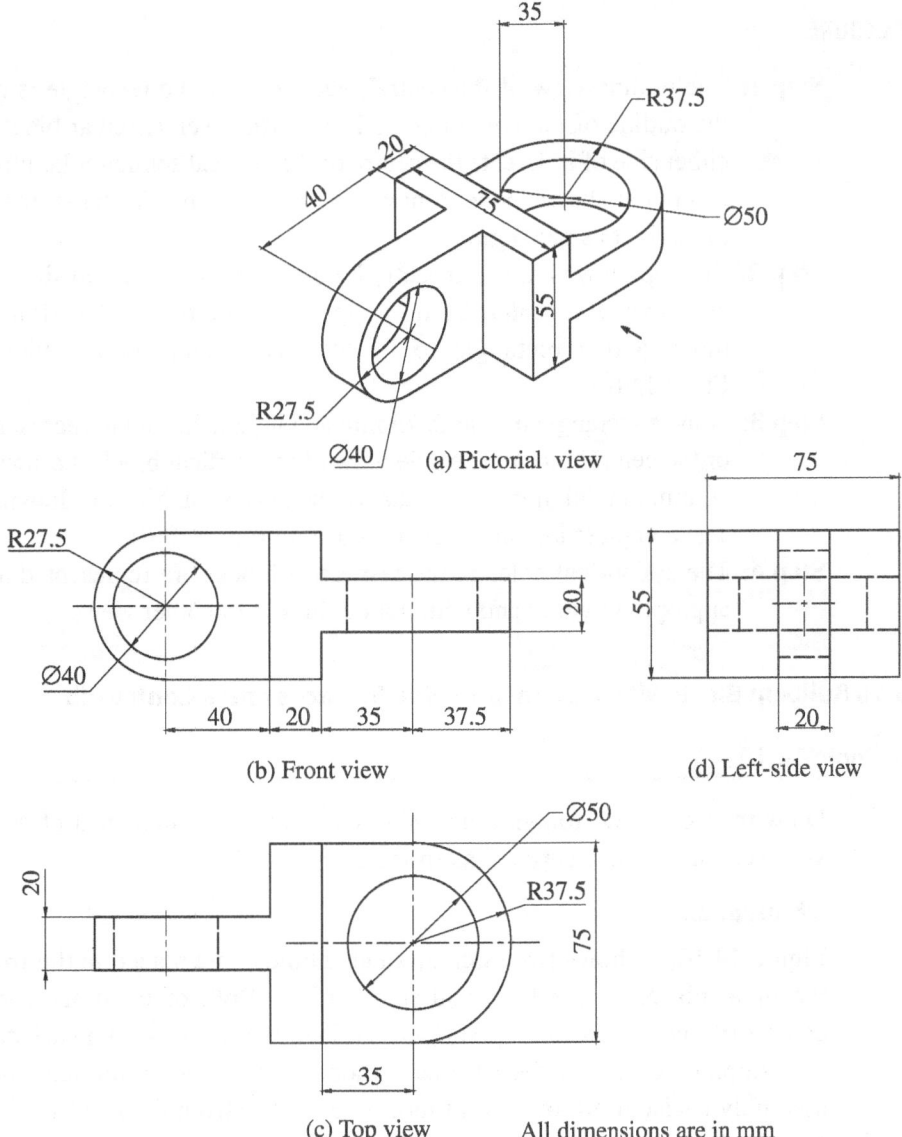

Fig. 14.9 Orthographic projections of horizontal and vertical semicircular blocks with an intermittent connecting piece

Observation

Figure 14.9(a) shows a horizontal and a vertical semicircular block connected with a rectangular-shaped intermittent block in between. The semicircular profile of the vertical block is seen in the front view, while that of the horizontal block is visualized in the top view. Such an assembly is generally used to connect devices to transmit motion from the vertical axis to the horizontal axis or vice versa.

PROCEDURE

Step 1: The outline view of the central piece, that is, the rectangle is drawn first, and the outline of the horizontal and the vertical semicircular blocks are drawn on either side of it. Locate the centre of the vertical semicircular block and draw its contour as shown in the front view in Fig. 14.9(b). The horizontal block is added to its right as rectangle.

Step 2: The top view is drawn directly below the front view and similar construction as above is repeated by drawing the semicircle to the right of the central block and a rectangle to its left. The complete top view is shown in Fig. 14.9(c).

Step 3: Draw a rectangle of length 75 mm and height 55 mm to represent the side view of the central block. The side view of the vertical block is drawn for a width of 20 mm in thick lines, while that of the horizontal block is drawn in dotted lines. The complete left-side view is shown in Fig. 14.9(d).

Step 4: The cylindrical holes in the respective blocks are represented as circles in the appropriate views and with dotted lines in the other views.   ▲

14.5.10 Built-up Block with Vertical and Side Surfaces and a Centre Rib

Example 14.10 ◢

Draw the front view, top view, and side views of the arrangement of the built up block with vertical and side surfaces as shown in Fig. 14.10(a).

Observation

Figure 14.10(a) shows two circular-shaped holes, one placed in the front surface and the other placed in the left hand-side surface. Both of them are connected with a central rib, which has its top surface inclined while its bottom surface is horizontal. The sloping surface of the rib joins smoothly with the cylindrical surface. Such an assembly is also used to transmit motion or power from the front vertical plane to the side vertical plane.

PROCEDURE

Step 1: Two concentric circles of 55 mm and 80 mm diameter are drawn to represent the front view of the vertical cylindrical block. Locate the axis of the side block at a height of 22.5 mm from the centre and draw its outline rectangle at the left as shown in the front view in Fig. 14.10(b).

Step 2: The outlines of the central rib are drawn as tangential lines to the circle as shown in the front view. The inclined surface of the rib is identified as 1–2–3–4.

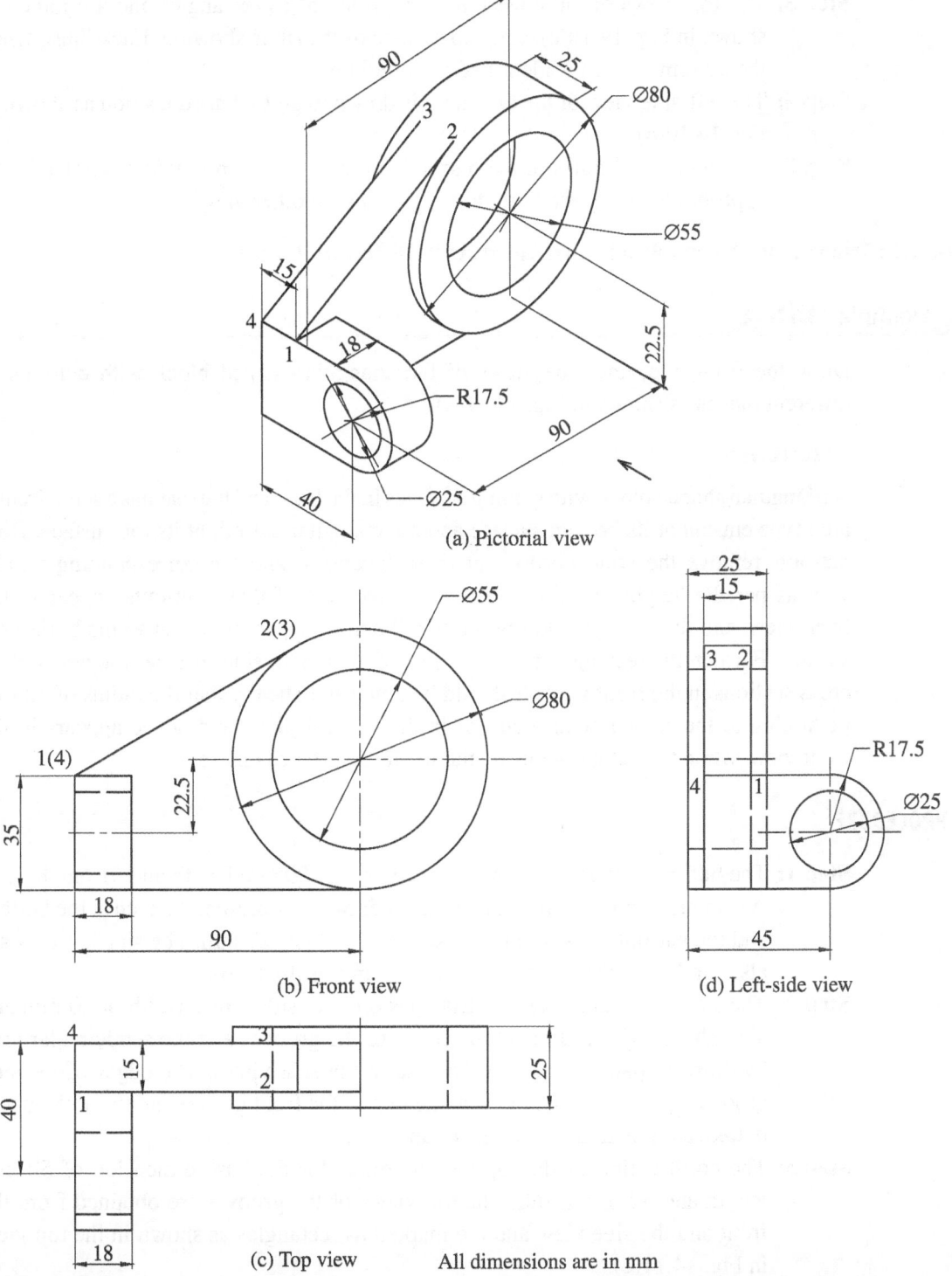

(a) Pictorial view

(b) Front view

(c) Top view   All dimensions are in mm

(d) Left-side view

Fig. 14.10   Orthographic projections of built up block with vertical and side surfaces and a centre rib

Step 3: The top views of the three units are projected as rectangles and are joined as shown in Fig. 14.10(c). The top surface of the rib is shown in thick lines, while the bottom surface is identified in dotted lines.

Step 4: The left-side view of the assembly is drawn in an L-shaped fashion as shown in Fig. 14.10(d).

Step 5: The cylindrical holes in the respective blocks are represented as circles in the appropriate views and with dotted lines in the other views. ▲

14.5.11 Triangular-shaped Block with Cut-outs in Different Planes

Example 14.11 ▲

Draw the front, top, and side-views of the triangular-shaped block with cut-outs in different planes as shown in Fig. 14.11(a).

Observation

A triangular-shaped block with sloping surfaces in the front and the rear has a semicircular arch type cut-out at its bottom surface and a rectangular cut-out at its top surface. Both cut-outs remove the solid portions at these locations, and the corresponding widths vary as per the height prevailing. The cross sections of these cut-outs appear in the front view, and their heights depend on the dimensions perpendicular to the horizontal surface. Further, the rectangular cut-outs provided on both side surfaces too reveal their cross sections in the front view. It should be noted specifically that the radius of 18 mm pertaining to the semicircular arch lies in the vertical plane and hence appears in the front view, while the actual radius on the inclined surface is more.

PROCEDURE

Step 1: The outline of the front view is a rectangle of 50 mm length and 51 mm height. A semicircle with a radius of 18 mm is drawn to represent the arch at the bottom and the cut-out at the top is marked at a depth of 22 mm. The side cut-outs are also marked in the front view as shown in Fig. 14.11(b).

Step 2: The side views are drawn as triangles on both sides for a width of 60 mm and a height of 51 mm. Project the edges of the grooves and the semicircular arch by drawing parallel lines to the base in the side views. The edges of the side grooves appear as thick lines, while that of the front groove and the arch appear dotted, as shown in Fig. 14.11(d) and (e).

Step 3: The construction of the top view is made for the base dimension of 50 mm length and 60 mm width. The top views of the grooves are obtained from the front and the side view and are marked as rectangles as shown in the top view in Fig. 14.11(c).

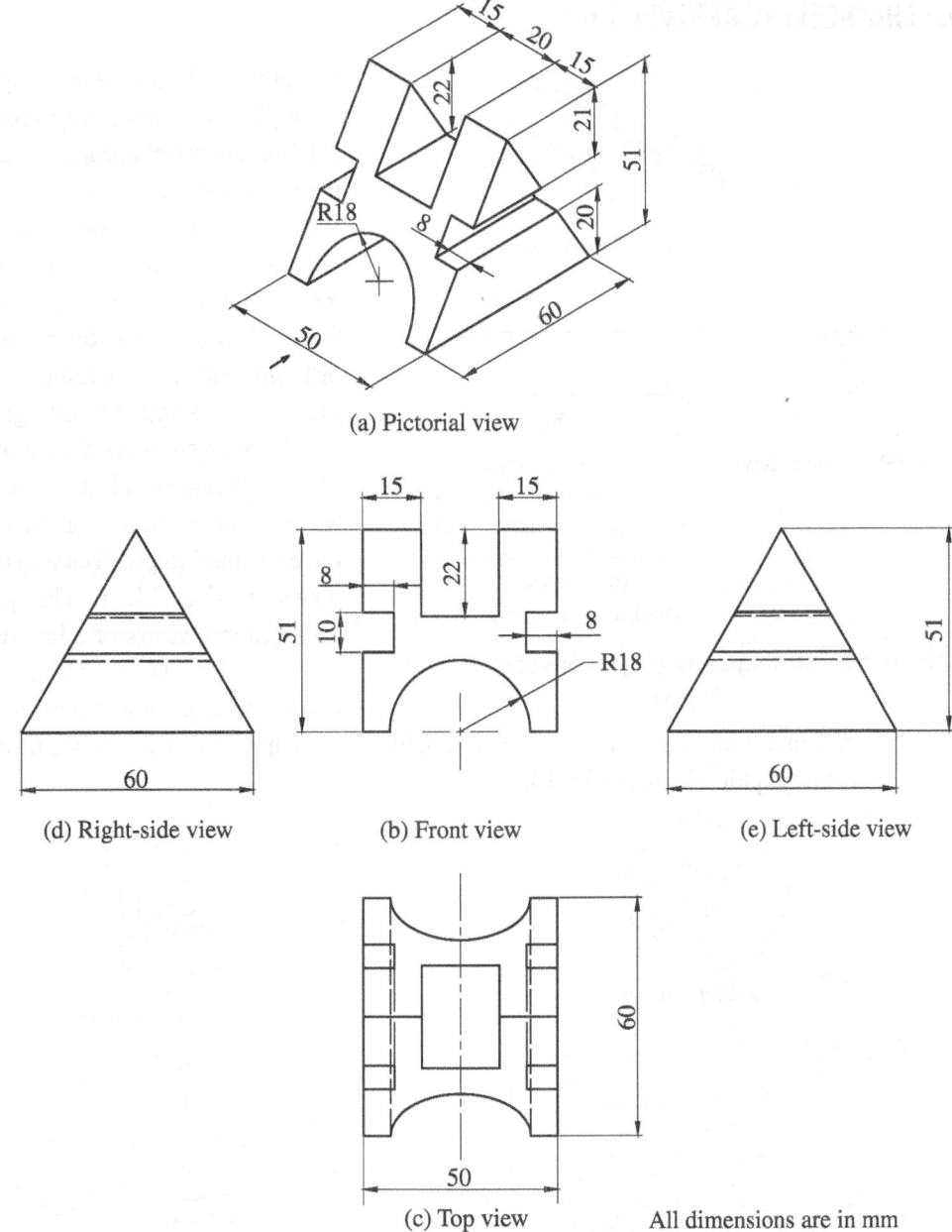

(a) Pictorial view

(d) Right-side view   (b) Front view   (e) Left-side view

(c) Top view   All dimensions are in mm

Fig. 14.11   Orthographic projections of triangular-shaped block with cut-outs in different planes

Step 4: The top view pertaining to the semicircular arch can be obtained by selecting many points from the front view and projecting them to the inclined lines in the side view. When such points are transferred from the front and the side views to the top view location, the top view of the arcs are obtained as shown in the Fig. 14.11(c).

14.6 ENGINEERING APPLICATIONS

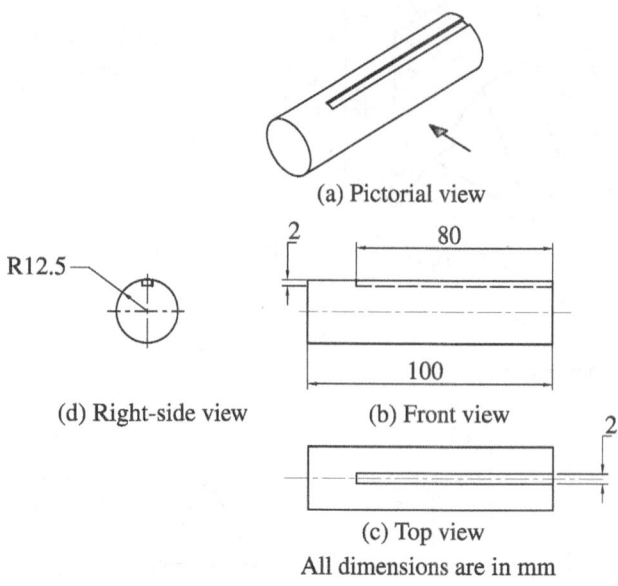

(a) Pictorial view

R12.5

(d) Right-side view   (b) Front view

80

100

(c) Top view

All dimensions are in mm

Fig. 14.12 Orthographic projections of shaft with key way

Certain simple machine components with their pictorial views are presented in this section. Their orthographic views are made in similar lines as discussed in the previous section, with representative dimensions, only to bring clarity of the elements and their interfaces. Figure 14.12 shows pictorial view and their corresponding orthographic views of a shaft with key way. The pictorial and the orthographic views of a 6-spline shaft are shown in Fig. 14.13. The applications of an overhung crank adopted in internal combustion engines are explained in pictorial and orthographic views in Fig. 14.14. The pictorial and orthographic views of a bearing block are shown in Fig. 14.15, while that of a half journal bearing are shown in Fig. 14.16. A general casting bracket used in machine tool applications is shown in pictorial and orthographic views in Fig. 14.17.

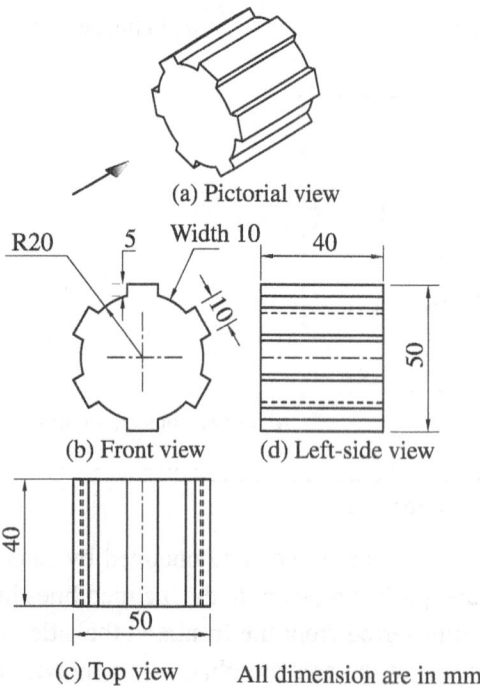

(a) Pictorial view

R20   5   Width 10   40

50

(b) Front view   (d) Left-side view

40

50

(c) Top view   All dimension are in mm

Fig. 14.13 Orthographic projections of a 6-spline shaft

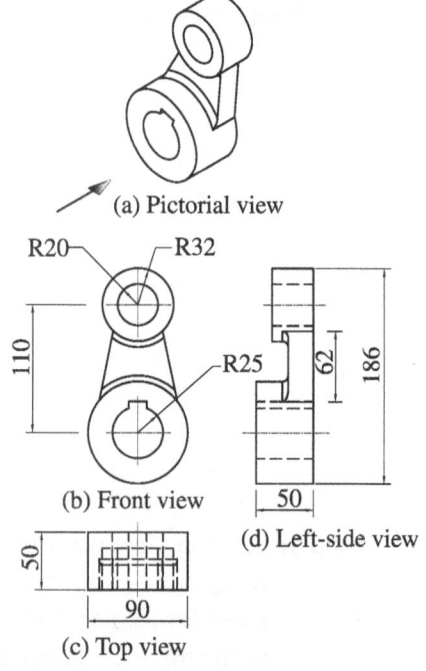

(a) Pictorial view

R20   R32

110   R25

62   186

(b) Front view   50

(d) Left-side view

50

90

(c) Top view

Fig. 14.14 Orthographic projections of overhung crank

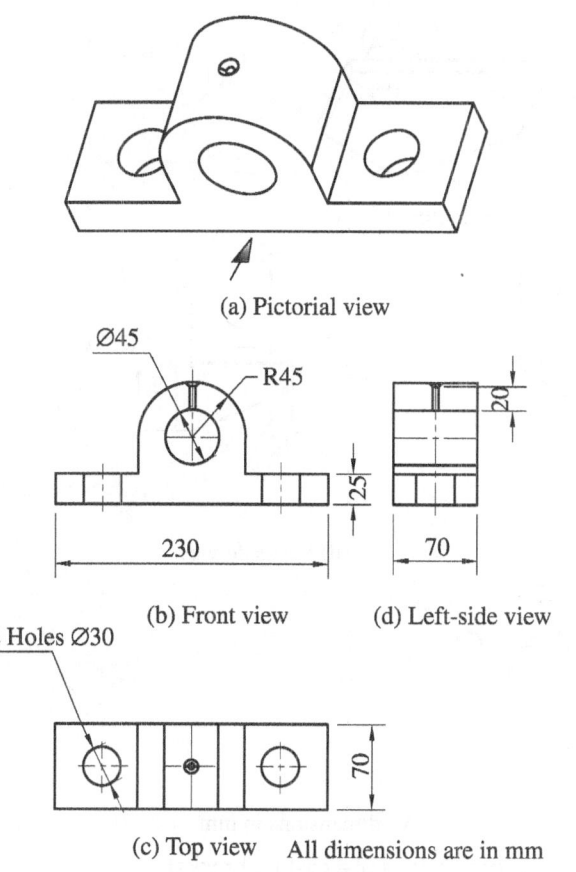

(a) Pictorial view

(b) Front view   (d) Left-side view

(c) Top view   All dimensions are in mm

Fig. 14.15   Orthographic
projections of the bearing block

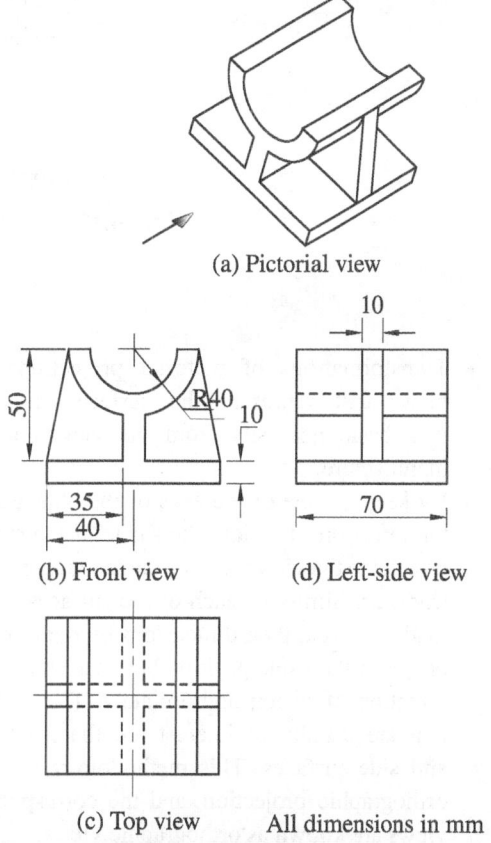

(a) Pictorial view

(b) Front view   (d) Left-side view

(c) Top view   All dimensions in mm

Fig. 14.16   Orthographic
projections of half journal bearing

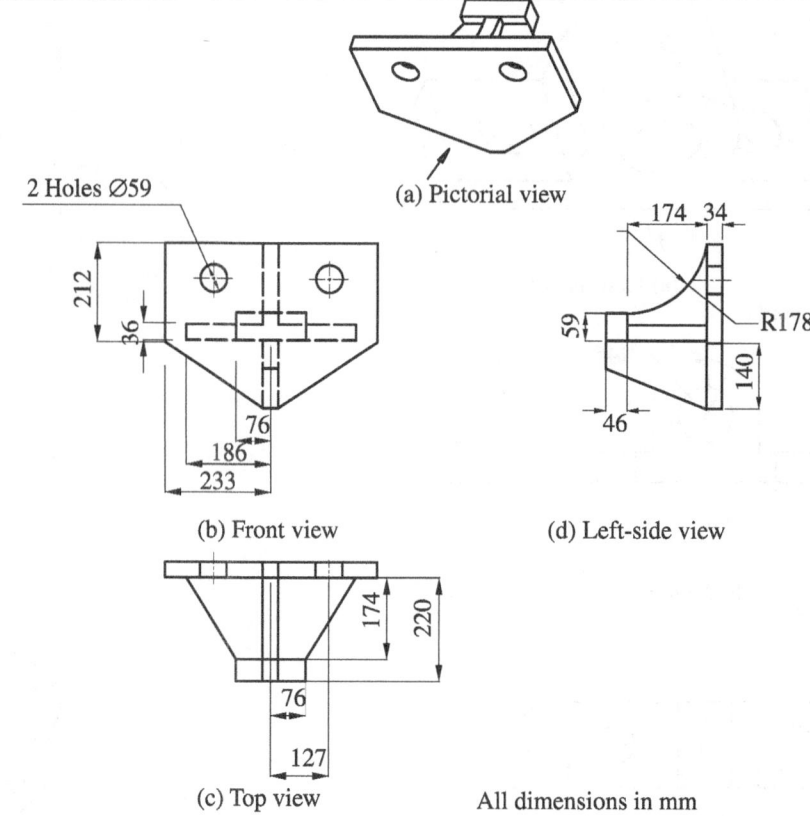

2 Holes Ø59

212

36

76
186
233

(b) Front view

174 34

R178

59

46

140

(d) Left-side view

174

220

76

127

(c) Top view

All dimensions in mm

Fig. 14.17 Orthographic projections of a casting bracket

- Transformations of pictorial projections into views that represent the surfaces in detail is a basic necessity from the view point of manufacture.
- By keeping one of the axes of the object along the direction of vision, the shape of the surface parallel to the viewer can be made to represent its true size. Similarly, each of the surfaces can be made to reveal their dimensions by retaining the object in the same position but by changing the direction of vision appropriately. The surfaces that are usually of interest are the front, top, and side surfaces. This methodology is called orthographic projection, and the corresponding views are known as orthographic views.
- Usually, the front view is decided by the direction of the arrow shown in the pictorial view and the

top and side views are decided by rotating the arrow upwards and on to the available side of the pictorial view, respectively. If the arrow is not available, the lager surface that is visible in the picture is considered to represent the front view.
- The views are laid out as per the first-angle projection principle, as a practice in this book.
- The plane surfaces that are normal to the direction of vision reveal the true sizes and the plane surfaces that are inclined to the direction of vision will appear in sizes smaller than the true sizes.
- When the curved surfaces of the object are normal to the direction of vision, they will appear as plane surfaces in that view and the nature or details of the curve will be seen in one of the other views.

14.1 Study the pictorial view of the objects shown in the following exercises A1 to A6 and choose the correct views—(a), (b), or (c)

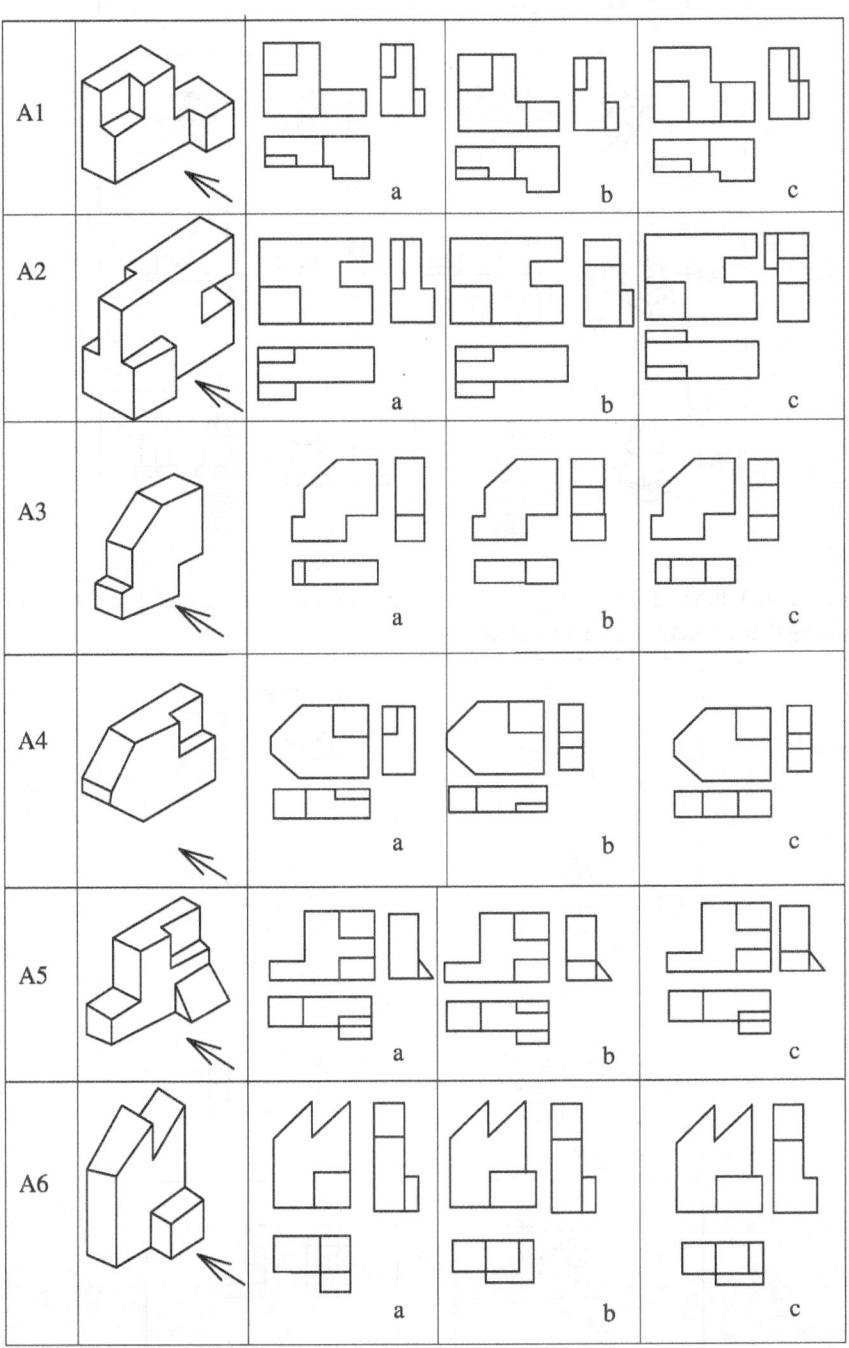

14.2 Study the pictorial drawings of the objects shown in the following exercises B1 to B4 and decide which multiview drawing (a), (b), or (c) describes the object in the best way, with the correct minimum number of views.

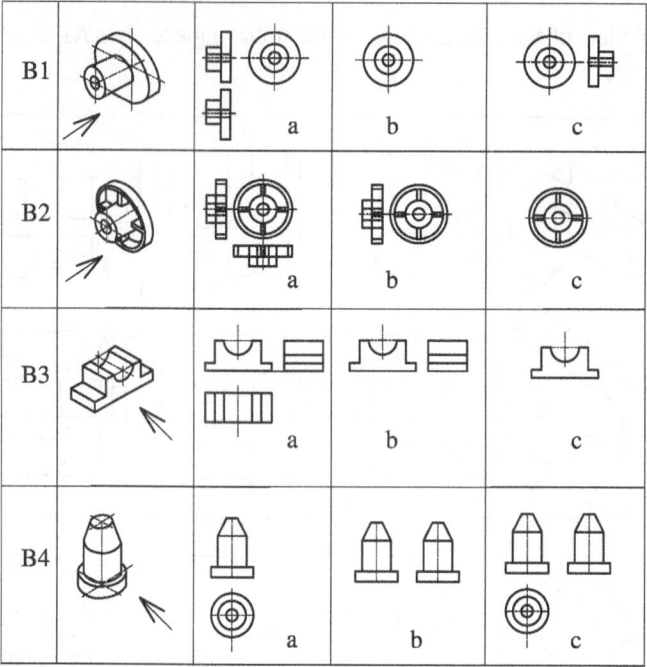

14.3 Study the pictorial drawings of the objects shown in the following exercises C1 to C4 and match the multiview drawing that correctly suits each object.

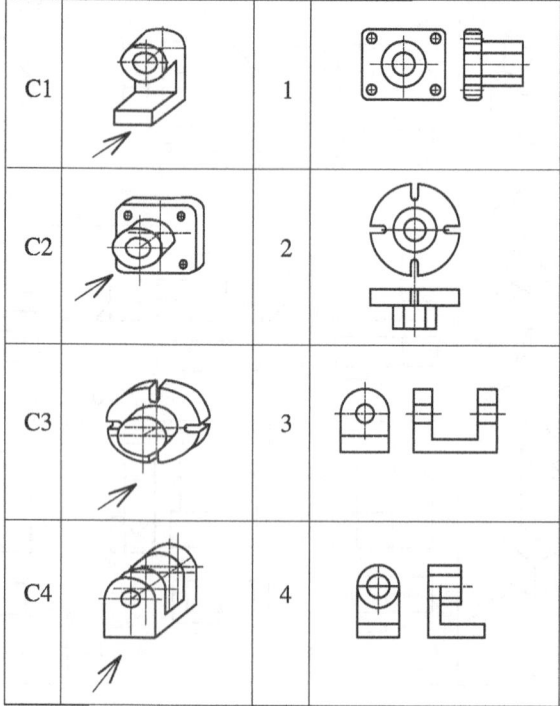

14.1 Draw the orthographic views for the following
      problem given with their pictorial views.

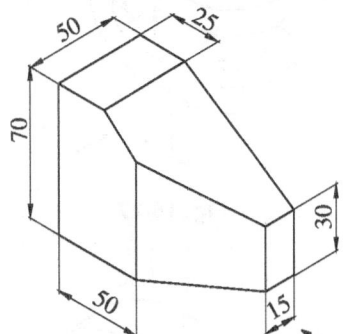

Fig. 14.18

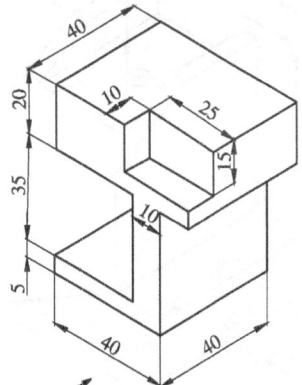

Fig. 14.20

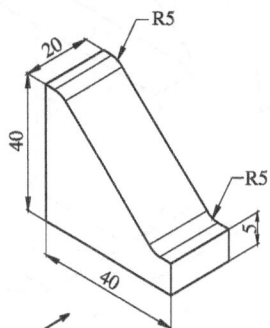

Fig. 14.19

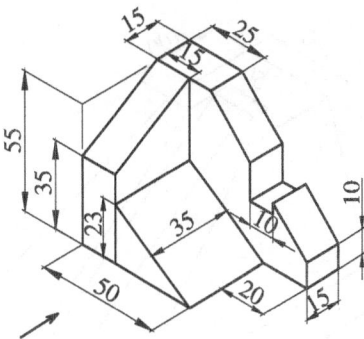

Fig. 14.21

14.1 Draw the orthographic views for the following
      objects given with their pictorial views.

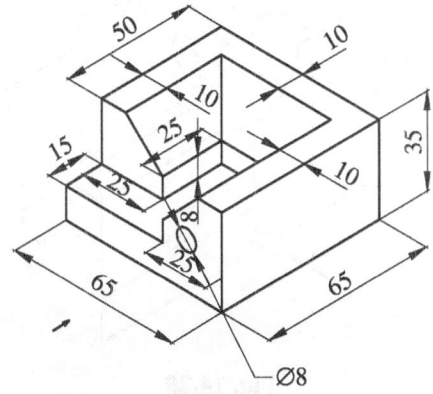

Fig. 14.22

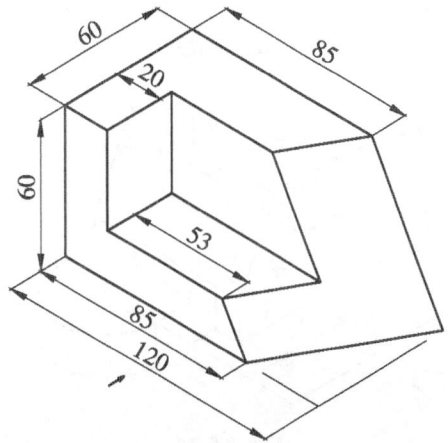

Fig. 14.23

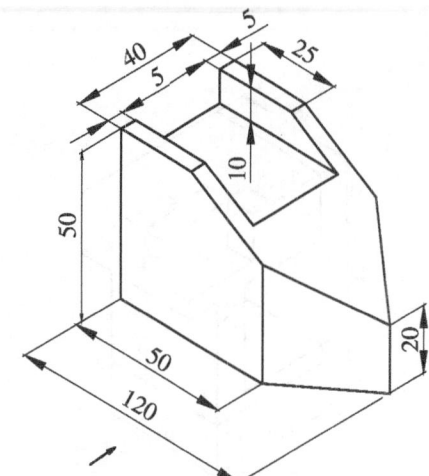

Fig. 14.24

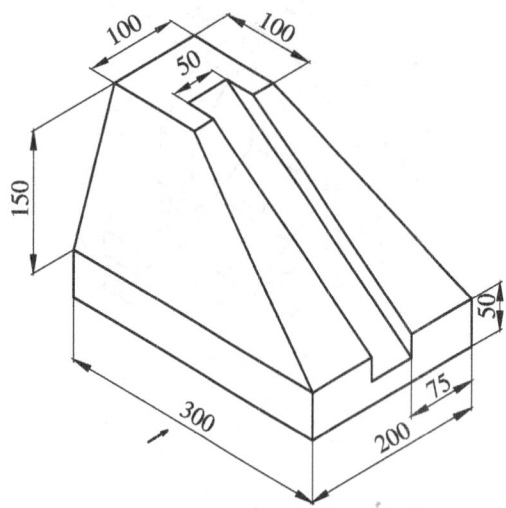

Fig. 14.25

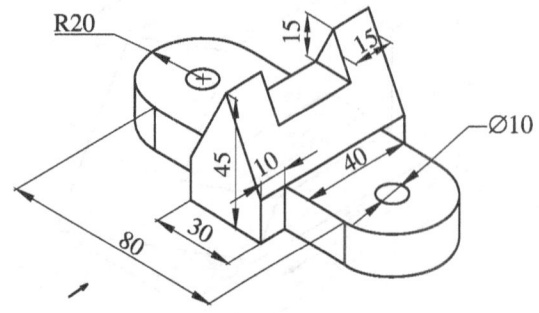

Fig. 14.26

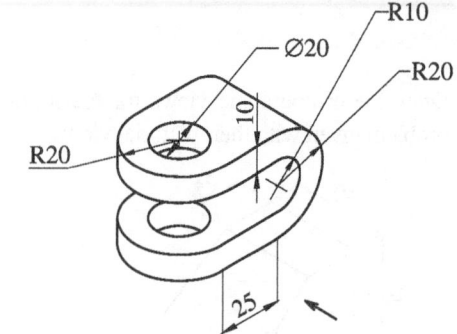

Fig. 14.27

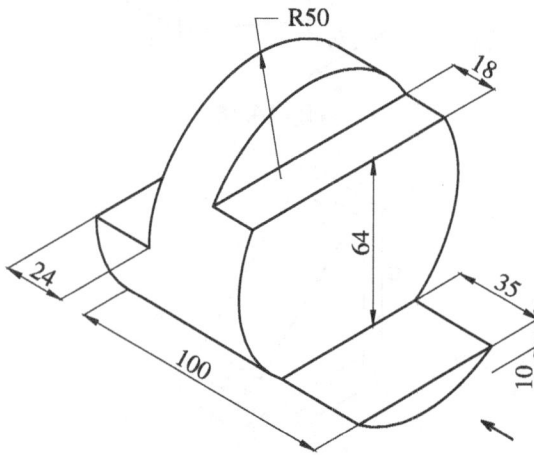

Fig. 14.28

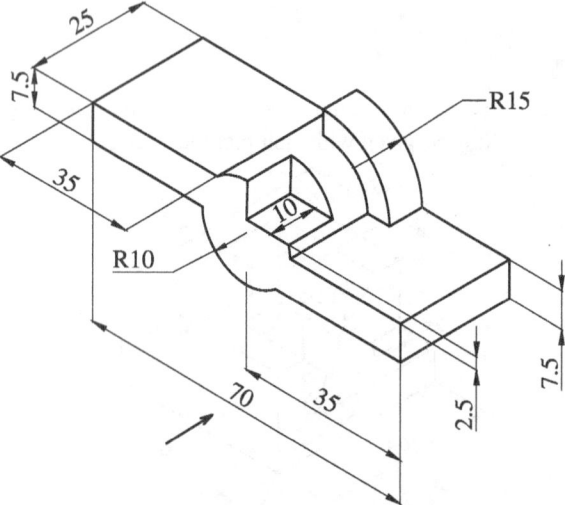

Fig. 14.29

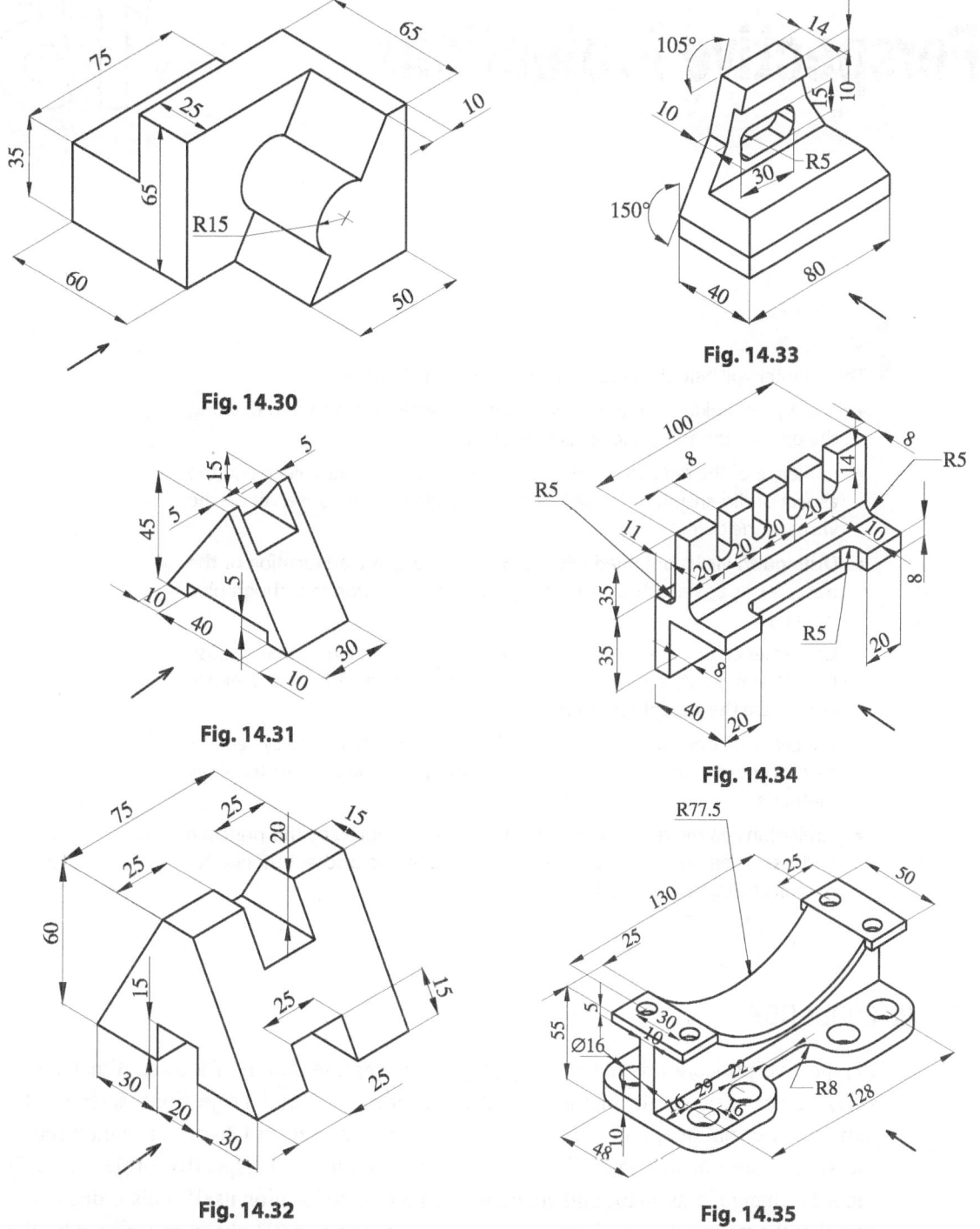

Fig. 14.30

Fig. 14.33

Fig. 14.31

Fig. 14.34

Fig. 14.32

Fig. 14.35

Answers for Multiple-choice Questions

14.1 A1 (b)   A2 (a)   A3 (c)   A4 (b)   A5 (c)   A6 (a)
14.2 B1 (c)   B2 (a)   B3 (a)   B4 (c)
14.3 C1 matches with 4; C2 matches with 1; C3 matches with 2; C4 matches with 3

Perspective Projections

OBJECTIVES

This chapter will help the reader to understand the following:

- Concept of linking the observer and object while viewing and drawing the projections as per the realism involved in
- Influence of the relative position of the observer and the object; and the aspect of dimension of the object and the height of the observer on the views of the object
- Distinguish views obtained with respect to the specified location of the observer, as against that observed from infinite location in orthographic projection
- Convenience of representing the object in a single picture that is close to that viewed by a human eye or eye lens as in a camera and hence appreciate the appeal caused in
- Concepts of optical illusion and make use of them in the systematic development of the views of an object using vanishing ray and visual ray methods
- Limitations on the facts conveyed in the pictures caused by this principle and use them only for admiration and appeal purpose but not for manufacturing requirements

15.1 INTRODUCTION

Perspective projection is defined as the pictorial representation of a three-dimensional (3D) object in a single picture. Unlike the other pictorial projections such as the axonometric, oblique, and isometric projections discussed in Chapter 4, which retain the same shape of the object in a specific orientation, the perspective projection will record a change in its shape and appearance in the same location itself. This is due to the fact that the perspective projection is the visualization of a 3D object as realized by the observer from a specified location. The position and size or the height of the observer and the position and size of the object together decide the perspective view. The most important difference in the appearance of perspective view from other pictorial drawings is caused by a common phenomenon known as *optical illusion*.

In day-to-day practice, two types of optical illusions are noticed. First illusion is due to the fallacy in the register of heights of objects held in a row and behind each other. The heights of the objects appear shorter than their true heights and the extent of shortening increases as the distance of the object from the observer increases. This kind of illusion can be noticed by the observer when series of lamp posts or telephone masts of equal height on a road side are viewed simultaneously (Fig. 15.1). The far off objects in such cases appear shorter than the nearer ones.

The other type of optical illusion occurs due to the fallacy in the vision of parallel edges or lines of the objects in a horizontal plane (HP) such as rail tracks, appearing to converge to a point (Fig. 15.2). These two effects are simultaneously observed by a human eye when we visualize the objects and are also reflected when viewed through a lens or camera (Fig. 15.3). Hence, the photograph of an object also gives its perspective view.

Fig. 15.1 Optical illusion on heights

Courtesy: www.shutterstock.com

Fig. 15.2 Optical illusion on widths (Railway track)

Courtesy: www.shutterstock.com

Fig. 15.3 Simultaneous optical illusions on heights and widths

Courtesy: www.shutterstock.com

The appearance of the object as observed through a photograph can be made more appealing by appropriately selecting the location of the camera lens or the proper position of the viewer. Hence, the perspective view of an object is usually drawn at a position that makes it more visually impressive.

This concept is widely used in civil engineering, architectural drawings, and advertising, where the commercialization of the product appearance plays a vital role. As much as the appeal matters in perspective drawings, the unrealism connected with that makes it inoperative for technical evaluation. Therefore, perspective drawings are used mostly for appeal and hence are not used in manufacturing or for any other requirements of technical importance.

15.2 DEFINITION OF PERSPECTIVE PROJECTION

Perspective projection of an object is a two-dimensional (2D) drawing obtained on an imaginary vertical plane (VP) known as picture plane (PP), as viewed by an observer/eye, located in front of it. The observer/eye is assumed to be situated at a finite distance and the diverging visual rays from the observer's eye reach the various corners of the object and meet the perspective plane to form an image known as perspective view. Depending on the position of PP and object, the size of perspective view will be determined. If the object is behind the PP, the image formed on PP will be reduced in size, and while the object is in front of PP, the perspective view will appear larger in shape. Figure 15.4 shows the perspective view of an object formed on PP.

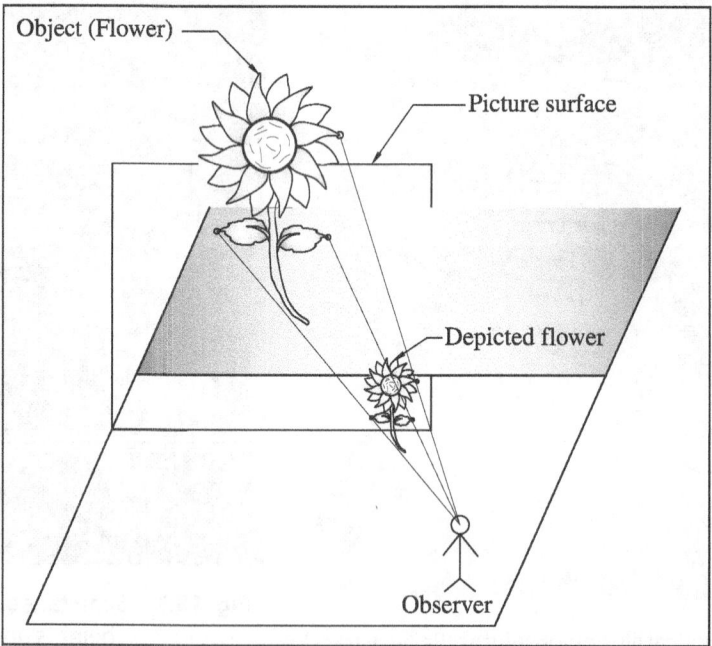

Fig. 15.4   Visualization of an object through picture plane

15.3 ELEMENTS OF PERSPECTIVE PROJECTION

As the perspective projection depends on the relative positions, sizes of the object, observer, and reference planes, the following terminologies of the various elements used are as shown in Fig. 15.5.

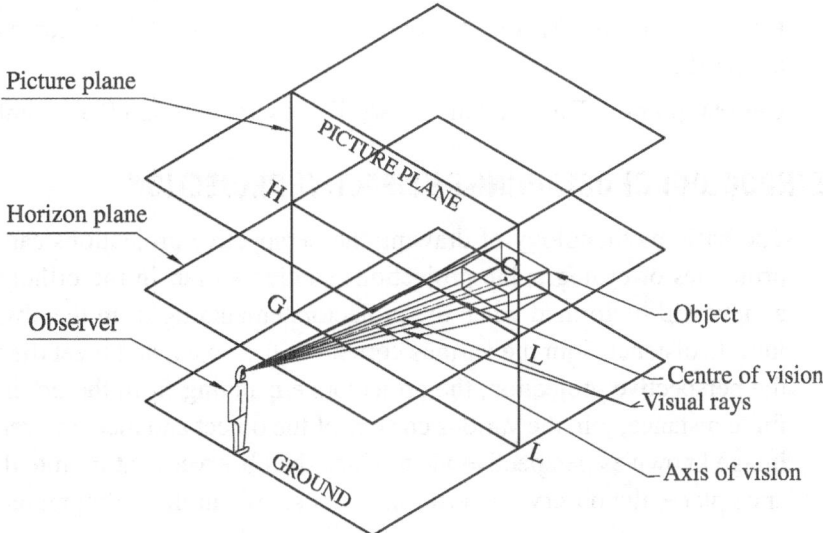

Fig. 15.5   Elements of perspective projection

Picture plane (PP)   It is an imaginary vertical transparent plane on which the object is projected and the perspective view is formed. It will be placed between the observer and the object.

Ground plane and ground line (GL)   It is a horizontal plane on which the object is placed. The line of intersection of the ground plane with the PP is known as ground line (GL).

Station point (S)   The observer is characterized as the station point, and the location of the observer is marked as the station point in particular, usually with the capital letter 'S'.

Horizon plane and horizon line (HL)   The imaginary horizontal plane passing through the eye level of the observer is known as the 'horizon plane', and it is above the ground plane. The line of intersection of the horizon plane with the PP is known as the 'horizon line' (HL).

Auxiliary ground plane (AGP)   It is an imaginary HP placed above the ground plane and horizon plane that forms the top most HP or roof plane in the system.

   The object and the observer are projected on to this plane, while getting the perspective view. Since the location of the object is below this AGP and behind the PP, which is a VP, this forms the reference plane system as per third-angle projection method. This is the fundamental difference in the perspective projection method.

Rays of vision These are imaginary rays emanating from the eye of the observer (station point, S), and joining with the various corners of the object. When these rays pierce or meet the PP, the image formed on it is known as perspective view.

Axis of the vision It is an imaginary perpendicular line from the S to the PP. It indicates the direction of sight.

Centre of vision This is the point at which the axis of the vision pierces the PP. It lies on the HL.

Central plane This is an imaginary VP passing through S and centre of vision.

15.4 METHODOLOGY OF OBTAINING PERSPECTIVE PROJECTION

The basic methodology of drawing the perspective projections can be made from the principles of orthographic projections studied so far. In the orthographic projections, a view will be formed when the projectors emanating from the observer positioned at infinite distance, join the various corners of the object and meet the reference plane. In the perspective projection, the projectors emanating from the observer positioned at a finite distance, join the various corners of the object and meet the vertical PP to form an image known as perspective view. Since the observer is at infinite distance from reference plane, the observer's position is not shown in the orthographic projections while in the perspective projections, the position of the observer will also be included while drawing the projections. The object and the observer are placed above the ground plane but commonly below auxiliary ground plane (AGP that resembles the conventional HP) as shown in Fig. 15.5. The object is behind the PP, and observer is in front of the PP, which resembles the conventional VP.

The object and the observer together are projected to AGP to obtain their top views and will be projected to PP to obtain the front view along with respective projectors. Since these views are in two different perpendicular planes, the PP will be retained and HP (here AGP) will be rotated and brought in alignment with it, to express the views with clarity as explained in Chapter 4. The AGP will be rotated about the intersection line at top, with the PP, such that the box containing the object is opened. This results in drawing a reference line (PP) and having top view of the object above it and top of the observer below it, respectively. Figure 15.6 shows the arrangement of the top and the front views of a rectangular pyramid (of dimensions l, b, and h) placed behind the picture plane and the observer (at a height h_0), in front of the PP and to the left of the object. Since the observer is known as the station point, the corresponding top view is shown as a point 's'. The projectors in the top view are also marked by joining 's' with the various corners of the object in the top view. The front view of the object and that of the observer are marked at their respective height from the ground plane. This results in drawing the GL below the PP line at any convenient distance and obtaining the front view of the object above it, by representing its various corners at their due height locations and directly below the top view. The height of the observer's eye level is marked by drawing the HL above GL, and the front view of the observer (s′) is marked on it by projecting 's' perpendicularly. The projectors in the front view are indicated by joining s' with the various corners of the object in the front view.

Since the perspective projection is the intersection of the rays of vision with the PP, it can be obtained by identifying the intersection point of the rays of the vision in the top view with the PP line and obtaining their corresponding locations with the rays of vision in the front view as shown in Fig. 15.6.

15.5 PARAMETERS AFFECTING PERSPECTIVE PROJECTION

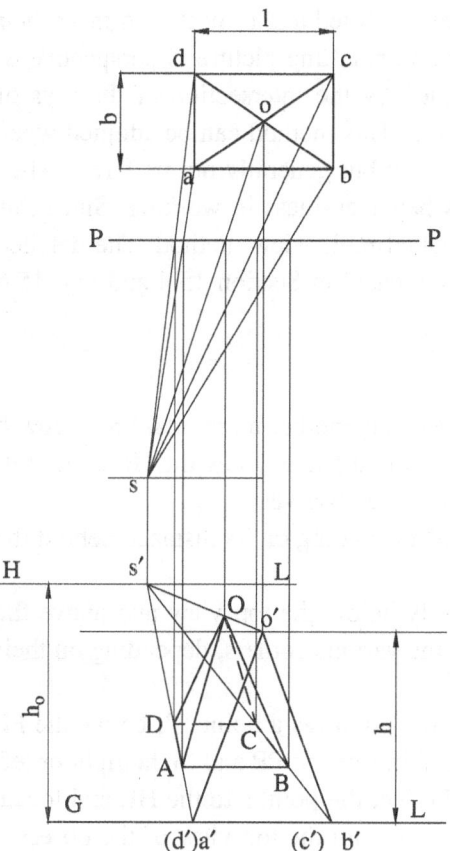

Fig. 15.6 Perspective projections of a rectangular pyramid

Since the perspective projection is object- and observer-dependent, the following facts are to be remembered while deciding the perspective of an object.

Position of object   Depending on the position of the object and PP, the size of the perspective will be decided. When the edges or faces of an object are on the PP, the perspective will show its true length or true size. In other words, when the picture plane coincides with the object, the perspective projection of the object will be of its exact size. When the object is behind PP, the picture will be showed in a reduced size, and if it is in front of the PP, the perspective will appear larger.

Position of station point   The position of the station point 'S' decides the general appearance of the perspective view. While an optimal location of an observer will give better appearance, a poor location will distort the shape itself. To have clear sharp images, a station point should be located at least twice the dimensions of the object facing the observer. The height of the station point is usually considered as the height of the observer or the eye level, which is around 1.5 to 1.8 meters for tall objects. For small objects, this is slightly above the top surface of the object, so that the top and the side surfaces become visual.

The distance of the station point from the object also sets up the limiting the angle of vision or the field of vision. For clear visibility of the object in space, the cone angle of the visual rays, therefore, generated is around 30° for sharper images. The width of the object and its distance in front of the observer should not exceed this angle to avoid distortions.

15.6 METHODS OF PERSPECTIVE PROJECTIONS

The following are the two methods by which the perspective projections of an object can be obtained.

1. Visual ray method
2. Vanishing point method

15.6.1 Visual Ray Method

Basic Principle

This method uses the conventional orthographic projections of an object (top view and front view, or top view and side view) and the associated rays of vision to obtain the perspective of a solid. The points at which the rays of vision joining the station point and the object pierces the PP in the top view are identified and projected to meet their corresponding rays of vision in the other view and resulting picture is perspective of the object. Since the perspective view is obtained by the intersection of the rays of vision, this method is known as visual ray method. This method can be adopted when the object is described in any orientation or position but generally not preferred when certain specific orientation of the object brings better easiness in working. Since this method uses two views at a time, this is also called multi-view method. The detailed procedure follows the general methodology as indicated in Section 15.4 and Fig. 15.6 and is presented in the following section.

General Procedure

Step 1: Draw a horizontal line to represent the PP and another horizontal line below it at an arbitrary distance, to represent the ground plane. Draw the HL above GL depending on the eye level or the height of the observer.

Step 2: Draw the top view of the solid above PP, depending on its distance behind the PP and mark the various corners.

Step 3: Draw the front view of the solid, directly below the top view and above the GL, using vertical projectors and mark the various corners, depending on their height.

Step 4: Mark the top view of the observer (station point) as a point 's', below the PP line, depending on the observer's position in front of PP and to the right or left of the object, as per the given details. Project the point s to the HL and locate s', to denote the front view of the observer. Join the top view of the observer with the top view of the object by means of straight lines, which are known as rays of vision in the top view. This is done by joining the point s, with the various corners of the object in the top view. Similarly, join the front view of the observer s' with the front view of the object by means of straight lines, which are known as rays of vision in the front view.

Step 5: Identify the meeting points of the rays of vision in the top view with the PP line and draw vertical projectors from those points, to meet the corresponding rays of vision in the front view. These points will yield the perspective view of the various corners of the object and by joining them sequentially, the perspective view of the solid can be obtained.

Step 6: The points in the top view that are closer to 's' and falling with in the extreme rays of vision in the top view are visible and the other points and their associated edges are marked in dots.

> **NOTE** *It can be noted that the perspective view of the various points of the object are represented by their capital letters, whereas their front, top, and views are denoted by the usual notations followed in the orthotropic projections.*

15.6.2 Vanishing Point Method

This method uses the principle of imaginary points at which the parallel edges or faces of an object converge, when viewed by the observer. These imaginary points of convergence are far away from the station point and are known as vanishing points. For example, in a railway track, the parallel rails (the width between two tracks) give an illusion of convergence at a single vanishing point as shown in Fig. 15.2. In fact, while drawing the perspective projections, the point at which the visual ray connecting the eye and the far of vanishing point pierces the picture plane is often known as the vanishing point, and it is always located on the HL. Since the 3D object has edges/faces inclined in all the three directions, it can result in three vanishing points, depending on its position with the PP and the observer. Based upon the number of vanishing points becoming active during any particular positioning arrangement of the object, the perspectives obtained can be classified as follows:

Single-point Perspective

When an object has one of its principal faces parallel to the PP, the vertical and horizontal edges (which are perpendicular to the PP) converge to a single point, and the perspective view obtained is known as single point (one-point) or parallel perspective and is represented as shown in Fig. 15.2. This method is used to draw the perspective view of the buildings or objects that have their front faces parallel to the observer and PP. The visual ray method explained in the previous section can also be recognized as a single-point perspective, as the perspective view converges to a single point as shown by the front view of the station point.

Two-point Perspective or Angular Perspective

When the object is placed such that two of its mutual vertical faces are inclined to the PP, the mutual edges converge to two vanishing points on either side of the observer, and such a perspective view is known as two-point or angular perspective. The perspective projection of objects that have distinct inclined faces on either side of the observer can be drawn using this principle.

Three-point Perspective or Oblique Perspective

When an object has all its three mutually perpendicular faces inclined to the PP (as in the case of any solid resting on one of its corners with all the faces inclined to

HP), then the edges mutually converge to three vanishing points, and hence, this perspective view is called three-point or oblique perspective. The horizontal edges of the object converge to two vanishing points and the vertical edges converge to a third vanishing point. The objects that are naturally in the inclined postures can be represented by this method.

The detailed explanation of the procedures involved in these methods is discussed in the individual examples.

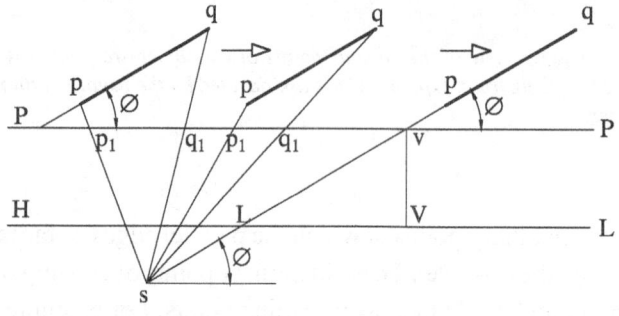

(a) With horizontal edge

Location of Vanishing Point for an Object With a Horizontal Edge

If an object has a horizontal edge inclined to PP, its vanishing point (V) can be located as per the following procedures:

Step 1: Draw a line parallel to the inclined edge appearing in the top view (line pq at an angle ϕ) from the top view of the station point and allow it to intersect the PP line at v.

Step 2: Project this intersecting point on to the HL to obtain the front view position of the vanishing point (V) as shown in the Fig. 15.7(a).

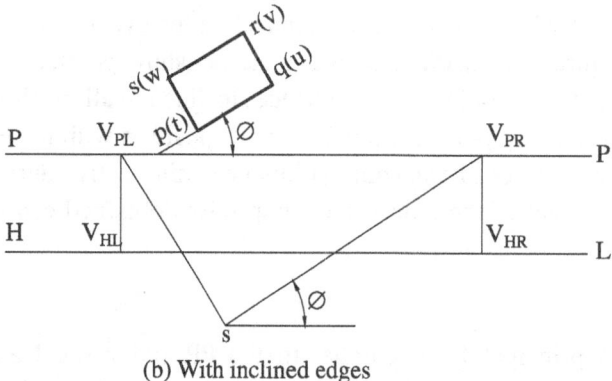

(b) With inclined edges

Fig. 15.7 Location of vanishing point for an object with

Location of Vanishing Point for an Object With Inclined Edges

If an object is bounded by a set of inclined edges in the top view, its vanishing points can be located as per the following procedure.

Figure 15.7(b) shows the top view of rectangular block with top edges p–q–r–s and bottom edges t–u–v–w. The block has its base on the ground plane and the vertical face p–q–u–t inclined at an angle ϕ to the picture plane.

Step 1: The vanishing points for the lines pq and for lines rs, tu, and vw (which are all parallel to pq) is obtained by drawing a line through s, parallel to pq and intersecting PP line at V_{PR}, as discussed in the previous section.

Step 2: Project this intersecting point on to the HL to obtain the front view position of the vanishing point V_{HR} as shown in the Fig. 15.7(b). The perspective views of the edges PQ, RS, TU, and VW will all converge to this vanishing point V_{HR}.

Step 3: Similarly, locate the vanishing point V_{HL} on the left-hand side, to which the perspective views of the edges PS, WT, UV, and QR will converge.

General Procedure for Canishing Point Method for Two-point Perspective

Step 1: Draw a horizontal line to represent the PP and another horizontal line below it at an arbitrary distance, to represent the ground plane (GL). Draw the HL above GL depending on the eye level or the height of the observer.

Step 2: Draw the top view of the solid above PP, depending on its distance behind the PP and mark the various corners.

Step 3: Mark the top view of the observer (station point) as a point s, below the PP line, depending on the observer's position in front of PP and to the right or left of the object, as per the given details. Join the top view of the observer with the top view of the object by means of straight lines, which are known as rays of vision in the top view. This is done by joining the point s, with the various corners of the object in the top view.

Step 4: Identify the set of inclined outer edges if any of the object in the top view. If the inclined edges do not exist in the top view, certain corner points in the top view can be suitably joined to obtain on an inclined line.

Step 5: From the point s, draw a line parallel to the right set of inclined edges to meet the PP line at a point called V_{PR} and project it on HL to get V_{HR}, which is known as right vanishing point. Repeat the same with the left set of inclined edge and get V_{PL} and V_{HL}.

Step 6: The right and left vanishing points (V_{HR} and V_{HL}) on HL are used to obtain the perspective view of the solid in the region between GL and HL.

Step 7: The perspective views of the corners (if any) located on the PP line can be obtained by projecting them vertically down to meet GL and by marking their respective heights. These perspective views can be joined with the right and left vanishing points to obtain the respective vanishing rays.

Step 8: The perspective views of the corners positioned on the right set of inclined edges can be located by joining these points with s and drawing perpendiculars from their meeting points on the PP line, to intersect the corresponding vanishing rays in the right. Similarly, the perspective views of the corners on the left set of inclined edges can be located on the vanishing rays in the left. This will result the perspective view of the right and the left inclined faces.

    The perspective views of the rear corner points can be located using (1) or (2) or (3).

1. The perspective view of the right inclined face can be joined with the left vanishing point V_{HL}. When the lines joining the top view of the station point 's' with the top view of the rear corner points, meet the PP line, the intersection points can be identified and perpendiculars can be drawn from them to meet their corresponding vanishing rays, resulting in the perspective view of the rear corner points.

2. The perspective view of the left inclined face can be joined with the right vanishing point V_{HR} and the perspective views of the rear corner points can be located, as discussed earlier.

3. Alternatively, the intersection of the vanishing rays of the right inclined face with V_{HL} and the left inclined face V_{HR} will also yield the perspective views of the rear corner points.

The visible and the invisible edges are marked in the perspective view in similar lines as discussed in section 15.6.1.

15.7 PROBLEMS ON PERSPECTIVE PROJECTIONS OF SOLID OBJECTS

Example 15.1

 A rectangular pyramid of sides of base 30 mm and 20 mm and height 40 mm rests with its base on the ground such that one of the longer base edges is parallel to the picture plane and 15 mm behind it. The observer is 50 mm in front of the picture plane, 25 mm to the left of the axis of the pyramid, and 50 mm above the ground. Draw the perspective view of the pyramid.

PROCEDURE (Refer Fig. 15.8)

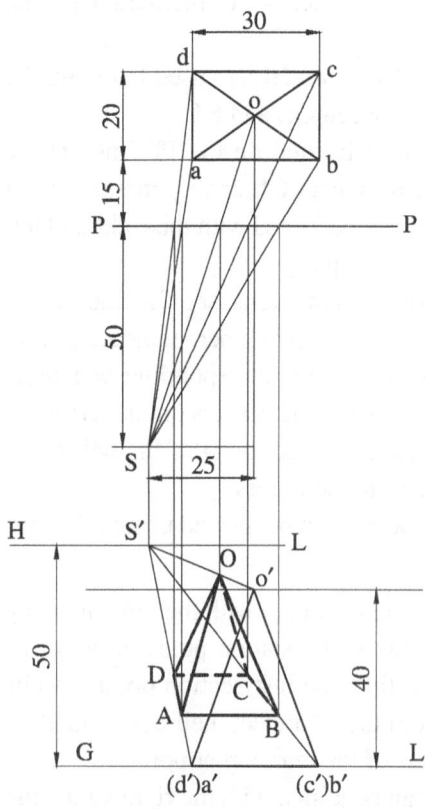

Fig. 15.8

Step 1: Draw a horizontal reference line (PP) and the ground line (GL) below it at an arbitrary distance. Mark HL above GL, at a distance of 50 mm, as given by observer's height.

Step 2: Construct a rectangle of 30 mm × 20 mm by placing the longer side parallel to PP line and 15 mm behind it to represent the top view of the solid. Mark the corner points 'a, ..., d' and the centre 'o' representing the apex.

Step 3: Project 'o' vertically down to meet the GL, mark the length of the axis (40 mm), and locate o'. Project all the base corner points from the top view to meet GL and name them as a', ..., d'. Join the base corner points with o' to obtain the front view.

Step 4: Locate the top view of the observer, as a point s, 50 mm below PP line and 25 mm to the left of the axis point 'o'. Project 's' to meet HL and obtain s', the front view of the observer. Draw lines joining 's' with all the points in the top view and s' with all the points in the front view.

Step 5: Find the meeting points of the lines sa, ..., so with the PP line and drop vertical lines from those points to meet s'a', ..., s'o' at A, B, ..., O. Join A, B, ..., O in the sequential order and obtain the perspective view of the solid.

Step 6: The points in the top view, which are closer to s, are observed, and points that are far away from 's' are hidden and marked as a dotted line. Since point 'c' is far away, all the edges (OC, OD, and BC) connected with that are marked as dotted lines as shown in Fig. 15.8.

Example 15.2

A rectangular pyramid of sides of base 30 mm and 20 mm and height 40 mm rests with its base on the ground such that one of the longer base edges is parallel to the picture plane and 5 mm in front of it. The observer is 50 mm in front of the PP, 25 mm to the left of the axis of the pyramid, and 50 mm above the ground. Draw the perspective view of the pyramid.

PROCEDURE (Refer Fig. 15.9)

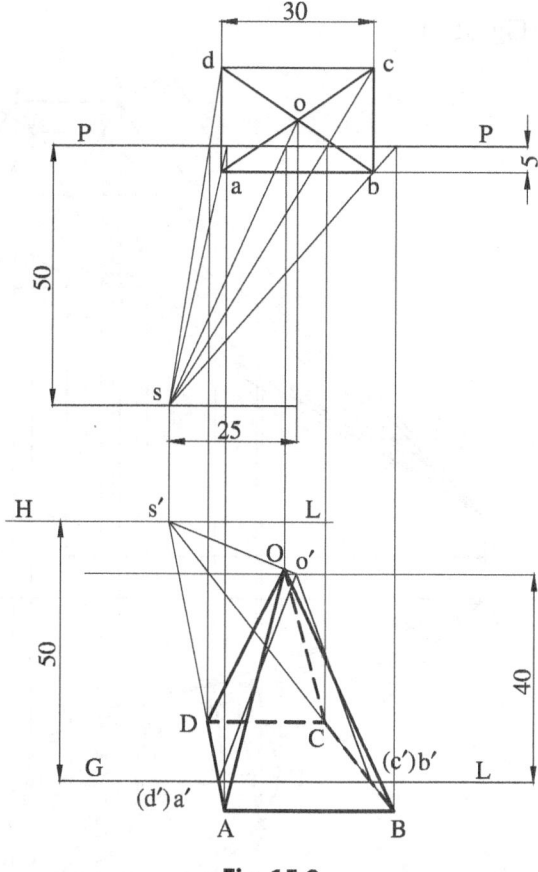

Fig. 15.9

The reader is advised to refer the previous Example 15.1. This problem is similar to the previous Example 15.1, except that the longitudinal edge in the top view is marked

5 mm below the PP line, as a part of the object is in front of PP. Due to this, the perspective views of the corners A and B are obtained, in the extension of the corresponding rays, beyond the GL and the edge AB appears larger than its length of 30 mm. This is because when the object is in front of PP, its perspective view appears larger than its original size. This is, in fact, the anomaly in human vision.   ▲

Example 15.3(a) ▲

A frustum of a hexagonal pyramid has a base of 30 mm sides, top of 20 mm sides, and height 50 mm long. The station point is 65 mm in front of PP and 120 mm from the axis and to its left. The horizon level is at 90 mm height. One of the base edges of the pyramid touches the picture plane. Draw the perspective view of the solid by the following methods.

Visual Ray Method Using the Front and the Top Views

PROCEDURE (Refer Fig. 15.10)

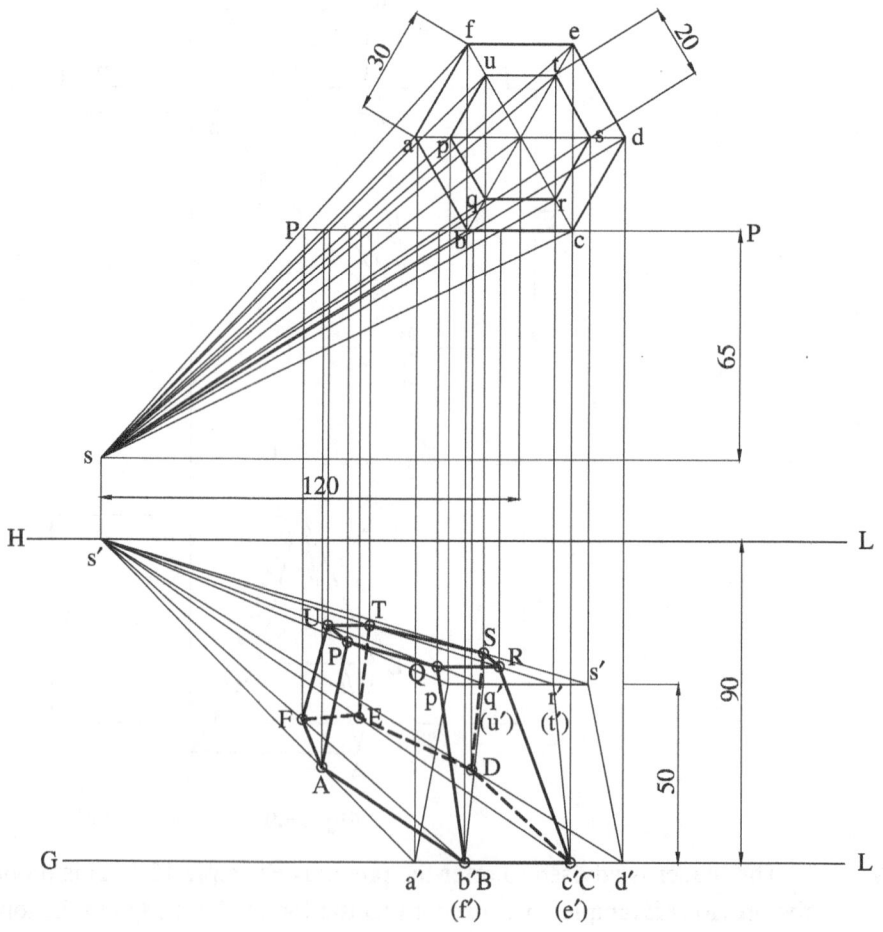

Fig. 15.10

Step 1: Draw PP line, ground line (GL), and horizon line (HL) as indicated.

Step 2: Draw the top view as a regular hexagon of 30 mm sides a, ..., f with one side bc touching the PP line and draw a concentric hexagon of 20 mm (p, ..., u) to represent the top surface. Locate axis 'o'.

Step 3: Project o and locate o' at a height 50 mm and mark the bottom and the top ends as a', ..., f' and p', ..., u' above GL.

Step 4: Locate s at 65 mm below PP line and 120 mm to the left of the axis and obtain s' on HL and join s and s' with the top view and front view corners, respectively.

Step 5: Identify the meeting points of sa, ..., su with the PP line, draw verticals from these points to meet s'a', ..., s'u', and obtain the perspective view as A, ..., F and P, ..., U.

Step 6: As the points d and e are far away from the sight of vision as indicated by the outer rays sc and sf, these points and their connecting edges are shown in dotted lines.   ▲

Example 15.3(b) ◣

Visual Ray Method Using the Front and the Side Views

PROCEDURE (Refer Fig. 15.11)

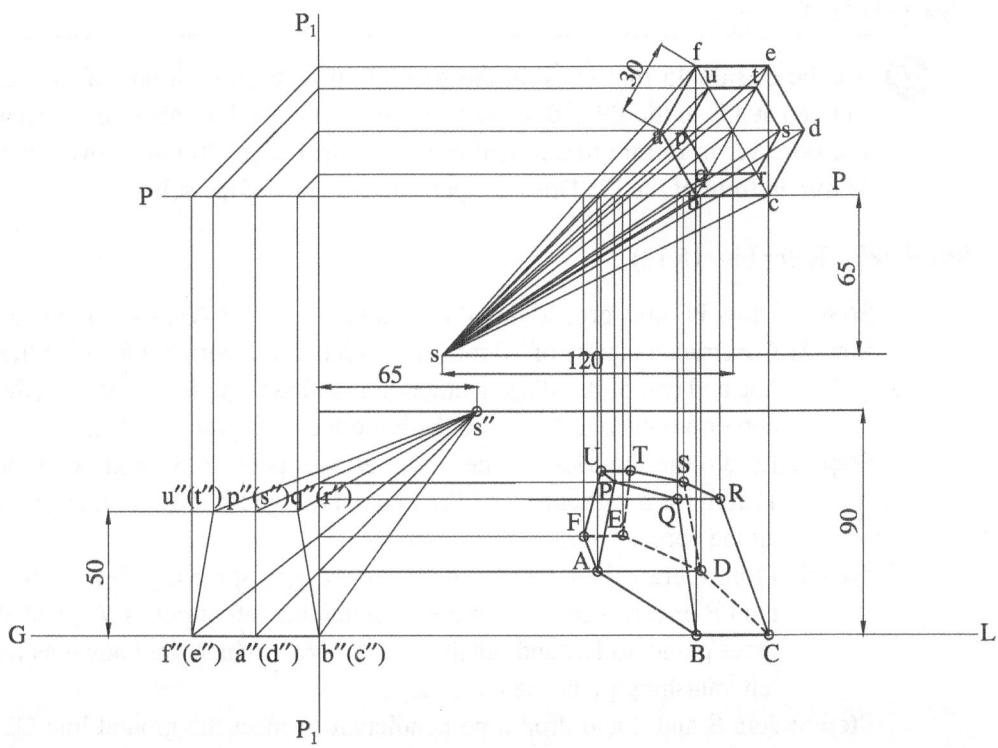

Fig. 15.11

Step 1: Draw PP line, ground line (GL), and horizon line (HL) as indicated.

Step 2: Draw the top view as a regular hexagon of 30 mm sides a, ..., f with one side bc touching the PP line and draw a concentric hexagon of 20 mm (p, ..., u) to represent the top surface. Locate the axis o.

Step 3: Draw the side view by drawing a vertical reference line P_1P_1. The side view of the object is drawn to the left at the required height and the corners are marked with usual notations.

Step 4: Locate s at 65 mm below the PP line and 120 mm to the left of the axis. Obtain the side view of the station point, s" on the right side of the vertical reference line (since the object and the observer are on either side of the picture plane), and at a distance of 65 mm, being its distance from PP. Join s and s" with the top view and the side view corners, respectively.

Step 5: Identify the meeting points of sa, ..., su with the PP line and draw verticals from these points, to meet the horizontal lines drawn from the intersection of s"a", ..., s"u" with the vertical reference line, and obtain the perspective view as A, ..., F and P, ..., U.

Step 6: As the points d and e are far away from the sight of vision as indicated by the outer rays sc and sf these points and their connecting edges are shown in dotted lines.   ▲

Example 15.4 ▲

 A cube of side 45 mm rests on the ground on its base with one of the vertical faces inclined at 45° to the PP and the vertical edge contained in this face is touching the PP. The observer is 15 mm to the right of this vertical edge, 70 mm above the ground, and 55 mm in front of the PP. Draw the perspective view of the solid.

PROCEDURE (Refer Fig. 15.12)

Step 1: Draw PP line, ground line (GL), and horizon line (HL) as indicated.

Step 2: Construct a square of 40 mm sides such that a corner edge b(2) lies on the PP line and two of its adjacent edges are inclined equally at 45° each. Mark the top corner points as a, ..., d and the bottom corner points as 1, ..., 4.

Step 3: Locate the top view of the observer as a point s, 55 mm below the PP line, 15 mm from vertical edge b(2), and to the right of it. Draw lines joining 's' with all the corner points in the top view.

Step 4: From s, draw lines parallel to the outer edges of the top view bc and ba to meet the PP line, respectively, at the right and the left at points V_{PR} and V_{PL}. Project these points to HL and obtain V_{HR} and V_{HL}, which are known as the right and left vanishing points, respectively.

Step 5: Join S and 2 and drop a perpendicular to meet the ground line GL at 2. This represents the perspective view of the base corner point 2, as it lies on the ground. Similarly, the perspective view of the top corner B is located at a height 45 mm above 2.

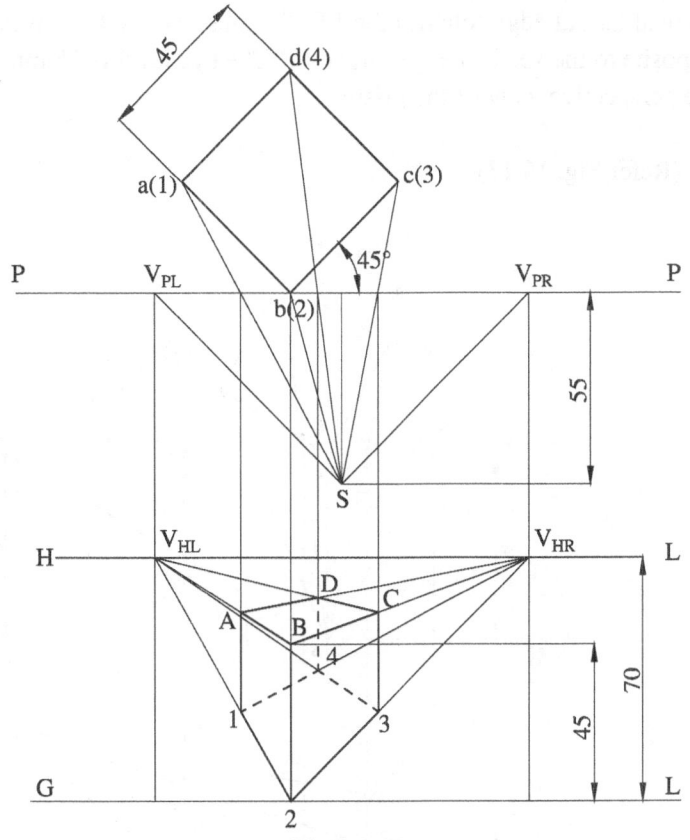

Fig. 15.12

Step 6: Join 2 with V_{HR} and B with V_{HR}. Since V_{HR} is the vanishing point at which the point 2 or any point along the line 2–3 or its extension will vanish, it can be used to locate the perspective view of such points. For example, when s–3 is joined and a perpendicular dropped from its meeting point on the PP line and when that meets the corresponding vanishing line to 2–V_{HR}, the perspective view of the corner 3 can be located. Similarly, the perspective view of the top corner C is located on the vanishing ray B–V_{HR}. Similarly, the perspective view of the corners 1 and A are located with the help of the left vanishing point V_{HL}.

Step 7: The perspective views of the rear corner points D and 4 are located when the vanishing rays connecting the right inclined face 2–B–C–3 with V_{HL} and the left inclined face 1–2–B–A with V_{HR} intersect each other.

Step 8: As the corner 4 is far away from s and is falling out of the extreme rays of vision in the top view, the corner 4 and its associated edges are marked dotted in the perspective view.   ▲

Example 15.5

A square prism of 30 mm side of base and axis 40 mm long rests with its base on the ground such that one of its rectangular faces is inclined at 30° to the PP. The nearest vertical lateral edge touches the PP. The observer is 45 mm in front of the PP and lies opposite to the vertical edge on the PP. The eye level is 65 mm above the ground. Draw the perspective view of the prism.

PROCEDURE   (Refer Fig. 15.13)

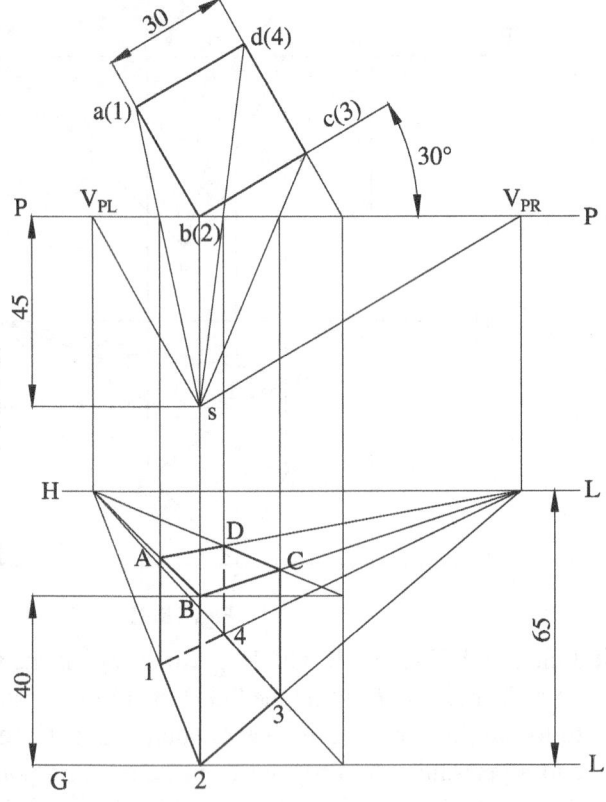

Fig. 15.13

This example is similar to the previous example 15.4, except the base dimensions, the height, and the angle of inclination of one face. The reader is advised to refer the step-by-step procedure indicated in the previous example.   ▲

Example 15.6   ▲

A hexagonal prism of 30 mm side edges and axis 60 mm long lies on the ground on its rectangular face with the axis inclined at 30° to the PP. Draw its perspective view when one of the hexagonal end corners touches the PP. The station point lies in the central plane which is bisecting the axis and is 160 mm in front of the PP. The horizon level is at 70 mm height. Use vanishing point method.

PROCEDURE (Refer Fig. 15.14)

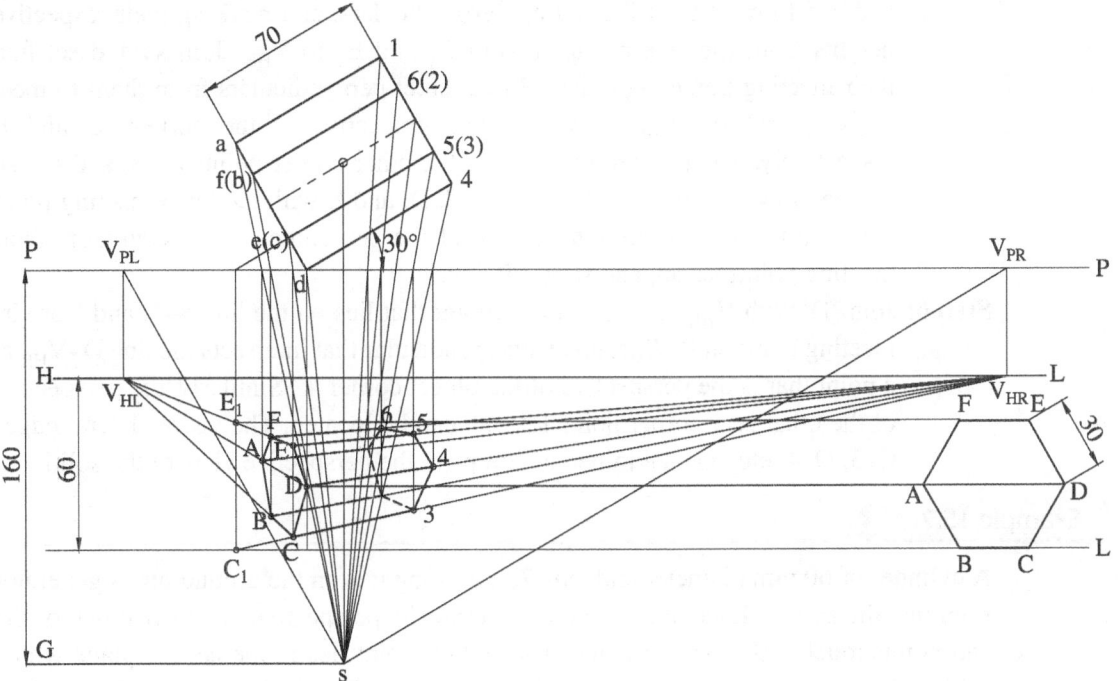

Fig. 15.14

Step 1: Draw PP line, ground line (GL), and horizon line (HL) as indicated.

Step 2: Construct a hexagon of 30 mm sides with one side flat on the GL at any convenient location to represent the cross section of the inclined solid. Measure the width across the corners A and D and draw a rectangle with this width and 60 mm length, such that its axis is inclined at 30° to PP and one corner d on it. Complete the top view by naming the front end as a, …, e and rear end as 1, …, 6.

Step 3: Locate the top view of the observer as a point 's', 160 mm below the PP line at the central axis. Draw lines joining 's' with all the corner points in the top view. It can be noted that since 's' is at a distance of 160 mm, the GL need not be located further below but can be located above the point 's' at any convenient distance. This will not affect the progress of the problem except the lines joining with 's' will also be available in the perspective drawing domain and the reader should not get confused with this.

Step 4: From 's', draw lines parallel to the outer edges of the top view d4 and da to meet the PP line, respectively, at the right and left at points V_{PR} and V_{PL}. Project these points to HL and obtain V_{HR} and V_{HL}, which are known as the right and left vanishing points, respectively.

Step 5: Obtain the perspective view of the corner 'D' below 'd' by marking its height above GL from the side hexagon. Obtain C_1 and E_1 by extending the line c–3 and e–5 to meet the PP, erecting perpendiculars, and marking their respective heights from the side hexagon. Join C_1 and E_1 to V_{HR}. Join sc and se, find their meeting points with the PP line, erect perpendiculars from them to meet C_1–V_{HR} and E_1–V_{HR} to obtain the perspective of the corners C and E respectively. The perspective view of the other corner points A, B, and F also can be drawn similarly or by joining C, D, and E with the left vanishing point V_{HL} and identifying the intersection of the perpendiculars erected from their meeting points sa, sb, and sf on PP line.

Step 6: Join 'D' with V_{HR}. Join 's' with 4 (point that lies on the line d–4) and from its meeting point on PP line, erect a perpendicular that intersects the line D–V_{HR} at a point that is the perspective of the object corner 4. Similarly, the perspective of the other rear corner point 1, ..., 6 can be located. The visible lateral edges C–3, D–4, etc., can be joined to complete the perspective view of the solid. ▲

Example 15.7 ◢

A cylinder of 60 mm diameter and axis 70 mm long lies on the ground on its generator such that the axis inclined at 30° to the PP. Draw its perspective view when one of the end points touches the picture plane. The station point lies in the central plane which is bisecting the axis and is 160 mm in front of the PP. The HL is at 70 mm height. Use vanishing point method.

PROCEDURE (Refer Fig. 15.15)

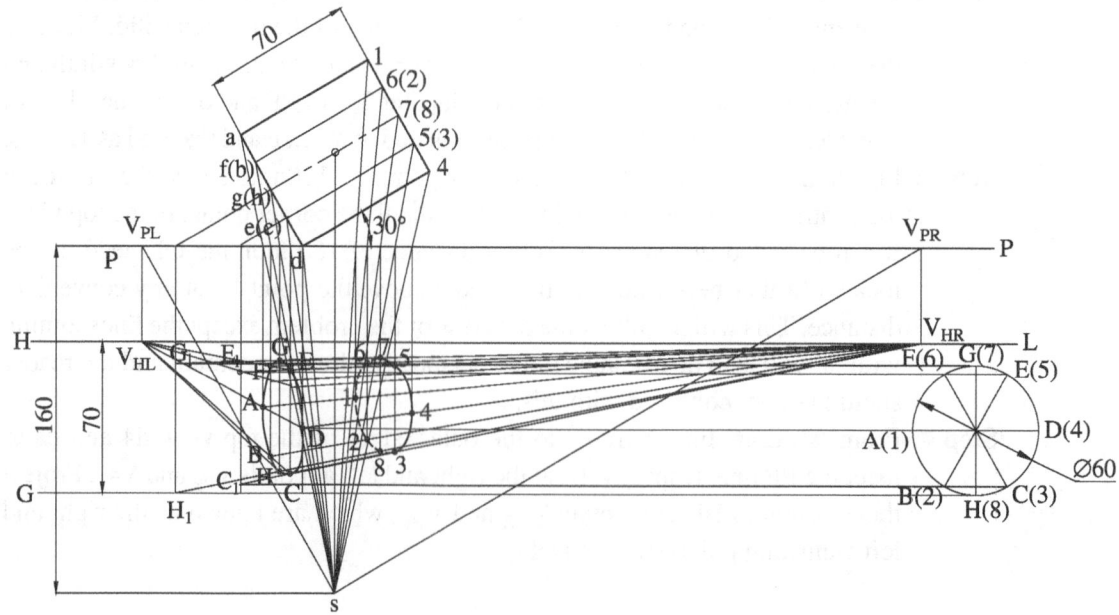

Fig. 15.15

This example is similar to the previous example 15.6, except the cross section of the solid is a circle. As mentioned earlier, the circle is drawn at any convenient distance on GL and the front end points A, ..., H and rear end points 1, ..., 8 are located. The perspective view is obtained by joining these points by means of smooth curve similarly as discussed in example 15.6.

Example 15.8

A brass model is in the shape of a square prism of base edge 30 mm and height of 40 mm. It tapers uniformly to a square of 20 mm at the top. The overall height of the brass model is 60 mm. The solid is kept on the ground with one of its base edges parallel and 15 mm behind the PP. The axis of the solid is 35 mm to the right of the station point, which is 60 mm in front of the PP and 80 mm above the ground. Draw the perspective view of the brass model.

PROCEDURE (Refer Fig. 15.16)

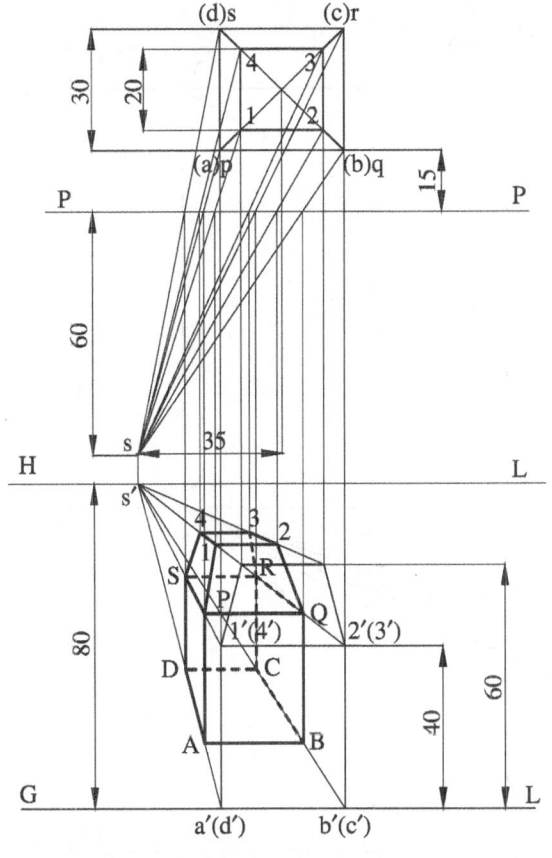

Fig. 15.16

Step 1: Draw PP line, ground line (GL), and horizon line (HL) as indicated.

Step 2: Draw the top view as an outer square of 30 mm sides to represent the base square prism dimensions (a, ..., d and p, ..., s) with one side 15 mm above PP line. Draw the concentric square (1, ..., 4) of 20 mm side to represent the top truncated square pyramid.

Step 3: Draw the front view consisting of two portions of the above arrangement for a height of 60 mm.

Step 4: Locate 's' 60 mm below PP line, 35 mm to the left of the axis, and obtain s′ on the HL and join 's' and 's''' with the top view and front view corners, respectively.

Step 5: Identify the meeting points of sa, ..., s4 with the PP line, draw verticals from these points to meet s′a′, ..., s′4′, and obtain the perspective view as ABCDPQRS1234.

Step 6: Since the points 'c' and 'r' are far off from 's', the edges connected with these points are not visible and hence shown in dotted lines.

Example 15.9

A combination solid consists of a cone of base diameter 25 mm and 40 mm height placed centrally on a square prism of 50 mm side and 35 mm height. The axis of the two solids is collinear. The combination is placed such that one of the vertical faces of the square prism is inclined at 30° to the PP and a vertical edge touches the PP. The station point 70 mm in front of PP and 100 mm above the ground and is exactly in front of the vertical edge touching the PP. Draw the perspective view of the combination solid.

PROCEDURE (Refer Fig. 15.18)

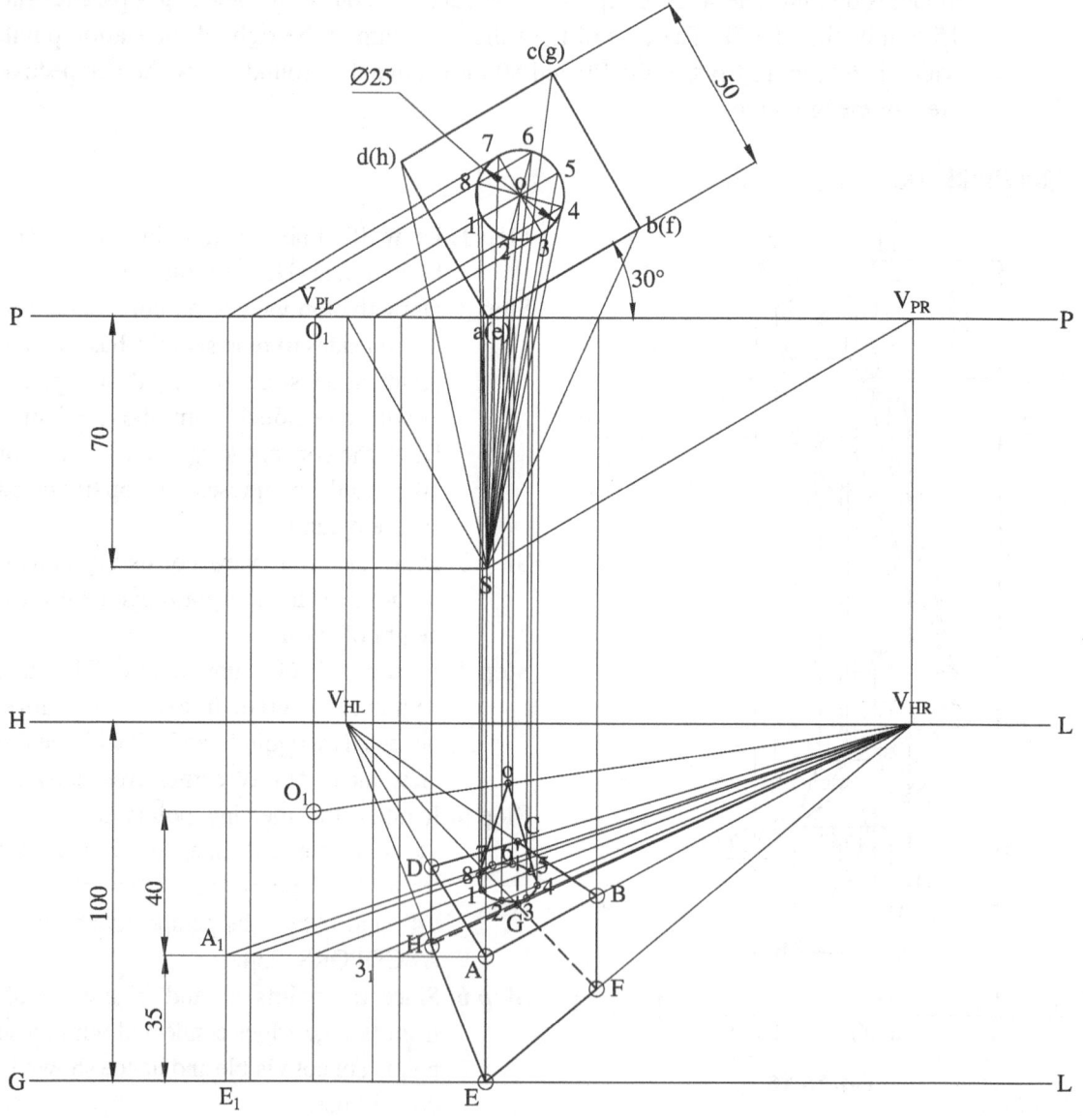

Fig. 15.17

Step 1: The base solid is a square prism. Its perspective view is drawn by using two vanishing points on the right and left in similar lines as discussed in Example 15.5.

Step 2: The top solid cone has its base situated on the top surface of the square prism. Its perspective view can be drawn by drawing a circle in the top view and identifying it with points 1, ..., 8. These points in the top view are extended to touch the PP line by drawing lines parallel to s–V_{PR}, so that the vanishing point s–V_{HR} can be used for getting the perspective projections of these points.

PROCEDURE FOR OBTAINING THE PERSPECTIVE VIEW OF THE CONE

Step 3: Choose any point on the base circle of the cone (point 3). Since the point 3 is behind the PP line, extend it to touch PP line as indicated earlier and draw a perpendicular from this point and mark its height from GL as shown by 3_1 at a height of 35 mm.

Step 4: Join 3_1 with the vanishing point V_{HR}, since the extension of point 3 is done in the top view to suit this.

Step 5: Join 's' and '3' in the top view and drop a perpendicular from its intersection with the PP line. When this perpendicular meets the vanishing ray 3_1–V_{HR}, the perspective view of point 3 is obtained.

Step 6: Similarly, the perspective view of the other points 1, ..., 8 of the base are located.

Step 7: The perspective view of the apex O can be obtained by extending 'o' to touch the PP line and marking its height as shown by O_1 at a distance of 40 mm above the base of the cone and joining it with V_{HR}. When 's' and 'o' are joined to meet PP line and a perpendicular is erected, the perspective O can be located on the vanishing ray O_1–V_{HR}.

15.8 ENGINEERING APPLICATIONS

Though the perspective projection is a general representation of a 3D object, as observed by a human eye, its engineering applications are mainly found in civil engineering where the presentation of architectural drawings play a vital role in advertising building or any facility before its formal existence. Usually, these dimensions are very much larger and since the orthographic drawings do not give the necessary feel or appeal in a place of non-technical interest, perspective drawings are used in these fields. Following these lines, even in applications of non-civil engineering, products that draw commercial attention are also advertised with perspective views as in automobile parts, interiors, consumer-oriented products, etc. Because of the mass appeal generated due to the optical illusion effects in the stage and theatre arts and settings, this principle is applied to a large extent. Due to the advent of computerization, when animation effects are incorporated into these drawings, a feeling of visual movement and dynamic feel are also set in these drawings.

Perspective drawings of small houses, buildings, roofs and sheds, pillars, lamp posts and sign posts, pedestals of statues/memorials, steps, cupboards, and castings of machine parts are some of the engineering applications that are often presented for student practice. The following are a few examples.

Example 15.10

A model of steps consists of three steps of 20 mm tread and rise of 15 mm. The width of the steps is 70 mm. Draw the perspective view of the model when placed with its first step 30 mm behind the PP and longer edge being parallel to it. The station point is situated at 90 mm from the PP, 70 mm above the ground and lies on the centre plane.

Visual Ray Method Using the Front and the Top Views

PROCEDURE (Refer Fig. 15.18)

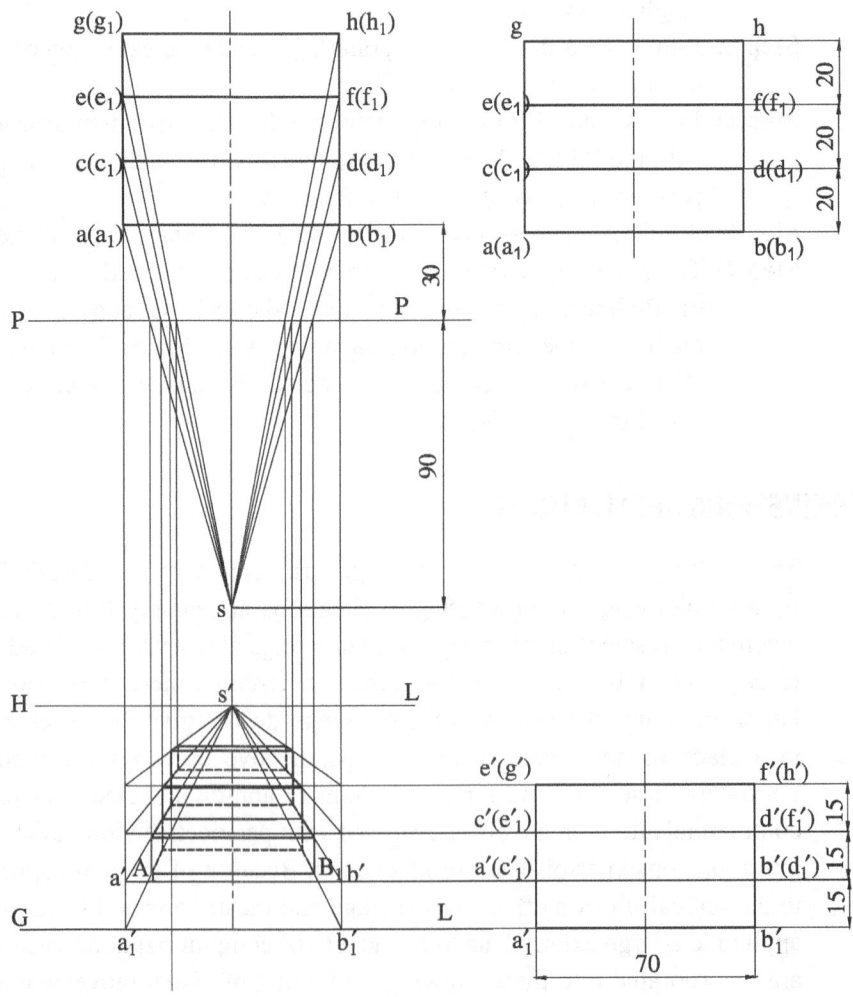

Fig. 15.18

Step 1: Draw PP line, GL, and HL as indicated.

Step 2: Draw the top view of the three steps with a length of 70 mm and treads of each 20 mm as shown by a, ..., g and a_1, ..., g_1, representing the top and bottom surfaces. The edge of the first step ab is located at 30 mm above the PP line.

Step 3: Project the front view of the three steps below the top view and mark points a', ..., g' and a_1', ..., g_1'. For clarity, a picture in the side can help.

Step 4: Draw the central axis line, locate s at 90 mm below PP line, and join s and s' with the top view and front view corners, respectively.

Step 5: Identify the meeting points of sa, ..., sg_1 with the PP line and draw verticals from these points to meet $s'a'$, ..., $s'g_1'$, obtain the perspective view as A, ..., G and A_1, ..., G_1, and join them in a sequential manner as shown in figure. ▲

Example 15.11 ▲

The isometric view and orthographic projections of an L-shaped metallic block with a cut-out and holes are given and shown in Fig. 15.19(a) and (b), respectively. Draw the perspective view of the block when the front vertical face is inclined at 45° to the PP and the corresponding front vertical edge AM is touching it. The station point is situated 120 mm from the PP and 135 mm above the ground. The station point lies on an imaginary VP perpendicular to the PP and passing through the edge AM that is lying on the PP.

PROCEDURE (Refer Fig. 15.19)

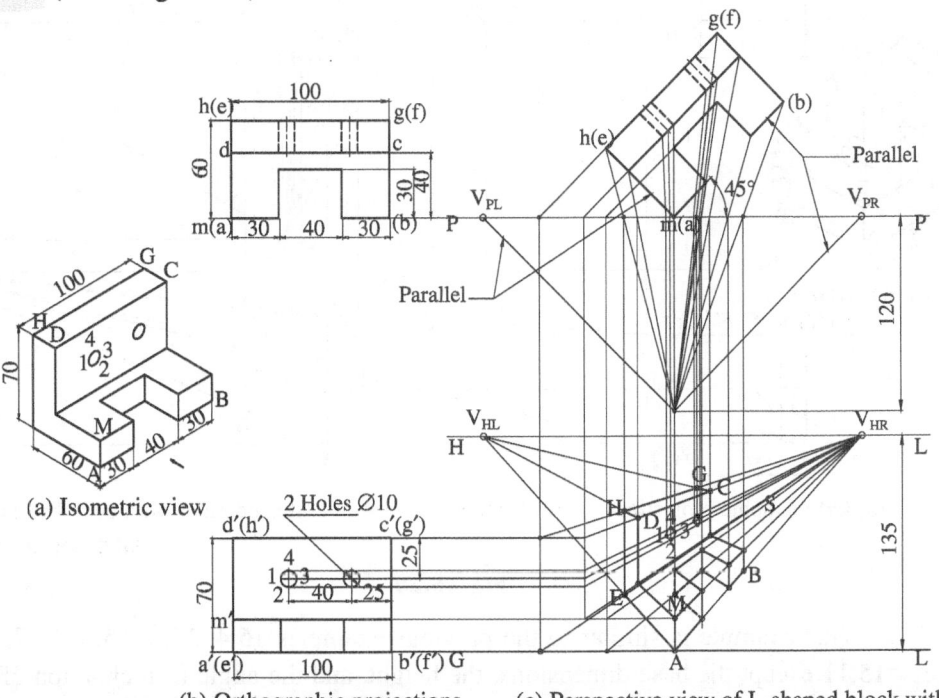

(a) Isometric view

(b) Orthographic projections

(c) Perspective view of L-shaped block with cutouts
All dimensions are in mm

Fig. 15.19

This example is similar to the previous examples 15.4, 15.5, 15.6, 15.7, and 15.9 except the base dimensions and the height and the angle of inclination of one of the faces. The reader is advised to refer the step-by-step procedure indicated in these previous Examples. ▲

Example 15.12 ▲

The isometric view and orthographic projections of a casting with cylindrical hole at the centre are shown in Fig. 15.20(a) and (b), respectively. Draw the perspective view of the block when the front vertical face is inclined at 30° to the PP and the corresponding front vertical edge AE is touching it. The station point is situated 100 mm from the PP, 20 mm right to the edge AE on an imaginary plane perpendicular to the PP, and 80 mm above the ground.

PROCEDURE (Refer Fig. 15.20)

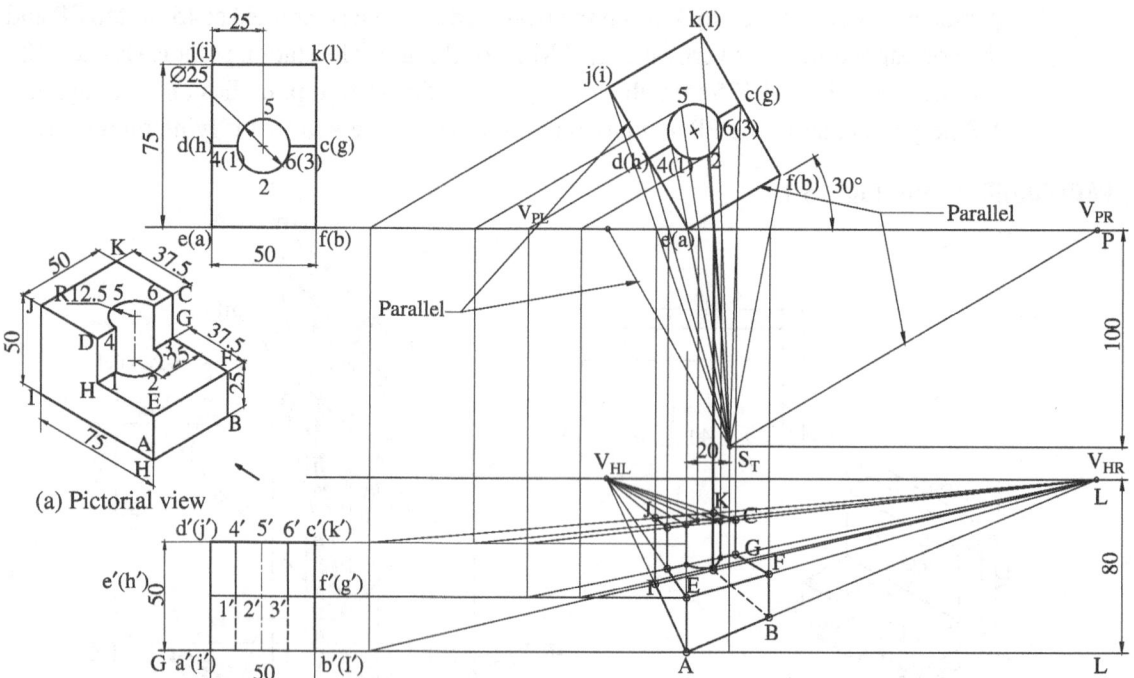

(a) Pictorial view

(b) Orthographic projections   (c) Perspective view of L-shaped block with cylindrical central hole

All dimensions are in mm

Fig. 15.20

This example is similar to the previous examples 15.4, 15.5, 15.6, 15.7, 15.9, and 15.11 except the base dimensions, the height, and the angle of inclination of one face. The reader is advised to refer the step-by-step procedure indicated in these previous Examples. ▲

- Perspective projection is the pictorial representation of a 3D object in a single picture obtained on a vertical imaginary plane known as the 'picture plane' (PP). This projection depends on the position of the observer as against the orthographic projection, where the observer is placed at infinity.
- The perspective view formed depends on the relative position and size of both the object and the observer, the observer size being the location of the eye level. Hence, for the same object, the perspective view will change depending on the location of the observer, and the picture is purely relative.
- In practice, a proper location of the observer is chosen such that it yields the better appearance of the object. The recommended distance of the observer is twice the largest dimension of the object, and the cone angle of the visual rays is around 30° for sharper images.
- There are two methods by which the perspective projections of an object can be obtained—visual ray method and vanishing point method.
- The visual ray method uses the conventional orthographic projections principle. The top views of the object and that of the observer (known as station point) are projected and placed on either side of the reference line (known as PP line). Similarly, the front views of the object and that of the observer are placed above the ground line, drawn parallel to the PP line and below it, depending on the height of the object and the observer. In some problems, the side views of the object and the observer are used. All the views are joined by lines that are known as visual rays.
- The points at which the rays of vision joining the station point and the object pierce the PP in the top view are identified and projected to meet their corresponding rays of vision in the other

view and the resulting picture is the perspective view of the object.
- The vanishing point method uses the principle of the imaginary points at which the parallel edges or faces of an object converge, when viewed by the observer. Since the 3D object has edges/faces inclined in all the three directions, it can result in three vanish points, depending on its position with the PP and the observer. Based upon the number of vanishing points becoming active during any particular positioning arrangement of the object, the perspectives obtained can be classified into the following:
  - Single-point perspective
  - Two-point or angular perspective
  - Three-point or oblique perspective
- The single-point perspective is used to draw the perspective view of the buildings or objects that have their front faces parallel to the observer and the picture plane.
- When the object is placed such that two of its mutual vertical faces are inclined to the picture plane (PP), the mutual edges converge to two vanishing points on either side of the observer and the two-point perspective method can be used.
- When an object has all its three mutually perpendicular faces inclined to the picture plane (as in the case of any solid resting on one of its corners with all the faces inclined to HP), then the edges mutually converge to three vanishing points and the three-point perspective method can be used.
- The perspective view in the respective cases mentioned earlier can be obtained by dropping perpendiculars from the PP line intersection points that are obtained by joining the top view of the object and that of the observer. When these perpendiculars meet the lines joining the vanishing rays for the above cases, the perspective projections are realized.

15.1 The perspective view of the object is dependent on
   (a) the position of the object alone
   (b) the position of the observer alone
   (c) the position of the object and the observer
   (d) the position of the observer at infinity

15.2 The perspective views exhibit the phenomenon of
   (a) reflection   (c) mirror image
   (b) optical illusion   (d) refraction

15.3 The perspective views of the objects are prepared for
   (a) manufacture   (c) appeal
   (b) design   (d) detailing

15.4 Horizon plane is an imaginary
   (a) vertical plane
   (b) ground plane
   (c) horizontal plane through the eye level of the observer
   (d) horizontal plane well above the eye level of the observer

15.5 Central plane is an imaginary that is
   (a) horizontal   (c) inclined to PP
   (b) parallel to PP   (d) perpendicular to PP

15.6 Axis of vision is an imaginary line from the observer and is
   (a) parallel to the visual rays
   (b) perpendicular to PP
   (c) inclined to PP
   (d) on the ground plane

15.7 An edge of an object is on the picture plane. The perspective view of this object will reveal
   (a) the true length   (c) an enlarged length
   (b) a reduced length   (d) one-fourth length

15.8 Perspective drawings of buildings that have their front faces parallel to the observer can be prepared by
   (a) two-point perspective
   (b) single-point perspective
   (c) three-point perspective
   (d) angular perspective

15.9 Centre of vision is a point on the
   (a) axis of vision
   (b) picture plane
   (c) horizon line
   (d) axis of vision, picture plane, and horizon line

15.10 When the object is placed in front of the PP, the size of perspective view will be
   (a) smaller
   (b) equal
   (c) non-existing
   (d) larger than the actual size of the object

15.1 A rectangular pyramid, with the base measuring 50 mm × 60 mm and axis 60 mm, rests with its base on the ground plane such that the longer base edge is parallel to the PP and 30 mm behind it. The station point is 70 mm in front of the PP, 40 mm to the left of the axis of the pyramid, and 70 mm above the ground plane. Draw the perspective view of the pyramid.

15.2 A pentagonal pyramid of 35 mm sides of base and axis 70 mm long is resting on GP on its base with a side of base parallel and 20 mm behind PP. The station point is 60 mm above GP and 80 mm in front of PP and lies in a central plane, which is 35 mm to the right of the axis of the pyramid. Draw perspective view of the pyramid.

15.3 A square pyramid 50 mm base edges and axis 70 mm long rests on its base on the ground such that two parallel base edges are receding at 40° to the right of the PP with the nearest corner of the base 15 mm behind PP. The station point is 50 mm in front of PP and 80 mm above the ground and 15 mm to the right of the nearest corner. Draw the perspective projection of the solid.

15.4 A frustum of a hexagonal pyramid of base 30 mm side, top 20 mm, and height 50 mm long rests with its base on ground with one of its base edges inclined at 45° to PP and one of the base corners touching it. The station point is 70 mm in front of PP, 65 mm above the ground and lies in front of the left extreme corner of the base. Draw the perspective projection of the frustum.

15.5 A cylinder of base diameter of 600 mm and axis 1100 mm long rests on the ground with its circular base. The axis of the solid is 400 mm behind the PP and 200 mm to the right of the observer. The observer is 1300 mm in front of the PP and 400 mm above the ground. Draw the perspective projection of the cylinder. Take suitable scale.

15.6 A frustum of a rectangular pyramid of base is 60 mm × 40 mm and the top is 40 mm × 30 mm. The height of the frustum is 70 mm. It rests on its base on the ground with the base edges equally inclined to PP. The axis of the frustum is 60 mm to the right of the eye. The eye is 90 mm in front of PP and 80 mm above the ground. The nearest base corner is 15 mm behind the PP. Draw the perspective projection of the frustum.

WORK PRACTICE LEVEL – 2

15.1 A rectangular block 35 mm × 25 mm × 20 mm high is lying on the ground on one of its largest faces. A vertical edge is in the PP, and the largest vertical rectangular faces make 35° with the PP. The station point is 60 mm in front of the PP, 35 mm above the ground, and lies in a central plane, which passes through the centre of the block. Draw the perspective projection of the block.

15.2 A hexagonal pyramid of sides of base 25 mm and height 40 mm rests on its base on ground with one of its base edges touching the picture plane. The station point 60 mm in front of the PP, 50 mm to the right of the axis of the pyramid, and 60 mm above the ground. Draw the perspective projection of the pyramid.

15.3 A cylinder 60 mm in diameter and 80 mm long is resting on the ground with its axis parallel to it and inclined to the PP at an angle of 40°. The centre of the axis is 60 mm from the PP. Draw the perspective projection of the cylinder if the station point is 65 mm from the PP, exactly in front of the centre of the axis of the cylinder and 45 mm above the ground.

15.4 A hexagonal prism of 35 mm base edges and the axis 60 mm long is resting on the ground with one of its rectangular faces such that its longer edges are positioned towards the right at an angle of 35°. The nearest corner is 30 mm behind the PP and directly in front of the observer. The observer is 110 mm in front of the PP and 30 mm above the ground. Draw the perspective projection of the prism.

15.5 A rectangular block 30 mm × 20 mm × 15 mm high is lying on the ground on one of its largest faces. A vertical edge is in the PP, and the vertical rectangular faces make 30° with the PP. The station point is 50 mm in front of the PP, 30 mm above the ground, and lies in a central plane, which passes through the centre of the block. Draw the perspective projection of the block.

15.6 A pillar is in the form of a frustum of a pyramid of 3 m high, 0.6 m square bottom, and 0.4 m square top. It is mounted centrally on a base which is a square prism of height 0.8 m and side of base 0.9 m. Draw the pillar in perspective, given that one side of the base is perpendicular to the PP and lower corner of base is nearest to the eye is 0.4 m behind the PP and 0.7 m to the right of the observer. The eye is 4 m from the PP and 3 m above the ground. Use the suitable scale.

15.7 A square pyramid of side of base 30 mm and height 50 mm rests centrally on top of a square prism, side of base 50 mm and height 20 mm with an edge of base of both the solids inclined at 30° to the picture plane and the nearest corner 20 mm from it. The station point is on the central plane passing through the apex of the pyramid and is located 80 mm in front of the PP and 60 mm above the ground. Draw the perspective projection of the solids.

15.8 A combination solid consists of a cylinder of 30 mm diameter and 50 mm height placed centrally on a square prism of 55 mm side and 40 mm height. The axes of the two solids are collinear.

The combination is placed such that one of the vertical faces of the square prism is inclined at 35° to the PP and a vertical edge touches the PP. The station point is 80 mm in front of the PP, 110 mm above the ground, and is exactly in front of the vertical edge touching the PP. Draw the perspective projection of the combination solid.

15.9 A statue model is in the form of a frustum of a square pyramid, measuring 60 mm at the bottom edges, 30 mm at the top edges, and 130 mm in height, placed centrally above the frustum of a cone 110 mm in diameter at the bottom, 90 mm in diameter at the top, and 50 mm height. A square pyramid with 30 mm base edge and 20 mm height is fixed at the top with the edges of its base coinciding with the top edges of the frustum of the pyramid. Draw the perspective projection

of the model when the base circle of the cone is touching the PP, the base edge of the pyramid near the PP is inclined at 35° to the PP, and is to the left of the observer. The station point is 170 mm in front of PP, 135 mm above the ground plane, and 20 mm to the right of the axis of the model.

15.10 The isometric view and orthographic projections of a block with a circular shaft are shown in Figs 15.21(a) and (b), respectively. Draw the perspective view of the block when the front vertical face is inclined at 45° to the picture plane. The station point is situated at 120 mm from the picture plane and 140 mm above the ground. The station point lies on an imaginary vertical plane perpendicular to the picture plane and passing through the front left vertical edge.

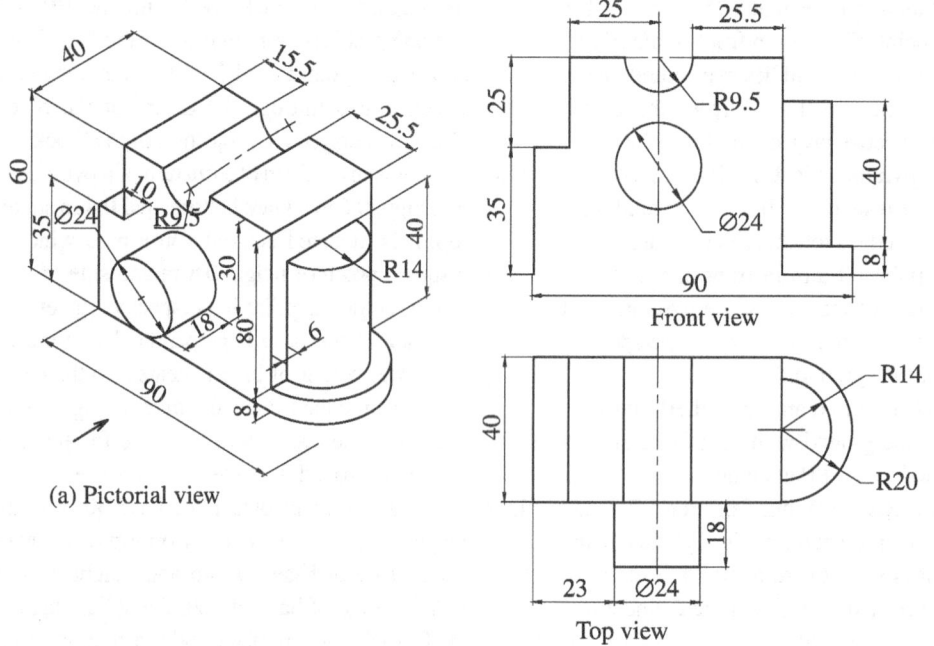

(a) Pictorial view

(b) Orthographic projections   All dimensions are in mm

Fig. 15.21 Orthographic projections of rectangular block with circular shaft

Computer-aided Design

16

OBJECTIVES

This chapter will help the reader to understand the following:

- Need for computer-aided drafting techniques and the practicing tools
- Familiarize one of the most powerful PC-based software known as 'AutoCAD' and develop skills to draw the orthographic views of an object using two-dimensional (2D) drawings in the computer and also to create isometric views of the three-dimensional (3D) object from its orthographic views
- Salient features of AutoCAD software and the hardware requirements for its installation in the computer
- Types of AutoCAD files, window size, the user coordinates, and the various command enablers
- Various entities of a drawing and the corresponding drawing commands, their dimensioning, and text-editing procedures
- Means of editing and display of the drawings using the relevant commands
- Drawing procedures of orthographic views of 3D objects
- Develop the isometric/pictorial drawings of 3D objects from their orthographic views
- Building exercise to think, understand and develop 2D and 3D drawings for products, processes, and in general, for any type of drawings using AutoCAD

16.1 INTRODUCTION

After studying all topics of engineering graphics, it is understood that the first course on 'Engineering Drawing' plays a dominant role in all the phases and steps in the process of engineering design. When computers came into existence, they changed the engineering practice by adopting the design methodologies too. In order to produce repeatable drawings with accuracy in a simple and faster manner, the need for computer-based graphic system was realized, since the manual drawings require large workspace and hardware, consume more time and labour, make editing more strenuous, and pose many storage/transfer problems. The computer-aided drafting (CAD) procedures provide all these benefits. Due to the rapid growth in mathematical procedures and computing algorithms, 3D shape formation has also become easier. Nowadays, the CAD becomes popular due to rapid technological developments in hardware due to miniaturization, higher hard disk and RAM capacity, and high-speed capacities of processors leading to the preparation of drawings without the drawing instruments such as drawing board, mini-drafter, set squares, dividers, and compasses in short period.

Any CAD software is based on interactive computer graphics (ICG) in which the user enters the data into the computer using input devices like keyboard, and the data are used by the software to create new graphics or modify the existing graphics. There are many CAD packages available in the market such as AutoCAD, CorelDraw, Micro Station, Versa-CAD, Unigraphics, Dogs, and so on. AutoCAD is the most powerful PC-based software and is relatively inexpensive, and hence, has been chosen for practising a drawing in this chapter.

16.2 COMPUTER SYSTEM AND AUTOCAD SOFTWARE

AutoCAD software installed in any workstation consists of a CPU (central processing unit), a keyboard, a mouse, and a monitor with latest graphics capabilities.

This chapter deals with the latest version of AutoCAD (AutoCAD 2015), it is powerful and capable of providing separate languages by which the computations and drafting activities can be carried out simultaneously. AutoCAD software supports the graphics with high resolution. AutoCAD uses support files to store menu definition, load AutoLISP programs, definite line types, define hatch patterns, and so on. A few advanced features of AutoCAD are listed as follows:

1. It is fully functional and compatible with the latest and most popular operating system from Microsoft to MAC.
2. Storage of drawings (softcopy) becomes very simple with AutoCAD. It is easy and convenient to locate and modify older drawings in its environment.
3. AutoCAD has a highly interactive and 'user-friendly' graphics user interface (GUI). It is basically command-driven, showing the result immediately on the graphic screen, and hence, making drafting a very smooth and easy-to-handle exercise.
4. It is one of the most powerful PC-based software available at very nominal price and is extensively used by all the branches of engineering.
5. Visual modelling in any 3D object or engineering component is possible in AutoCAD. In its latest edition, the facility of 3D viewing and editing has been enhanced effectively.
6. A built-in programming language AutoLISP/Visual LISP provides a programming environment facilitating many repetitive tasks.

The hardware components are as follows, which are needed to be used along with AutoCAD.

16.2.1 Mouse

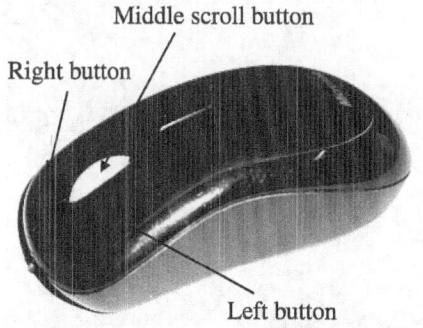

Middle scroll button

Right button

Left button

Fig. 16.1   Mouse

It is used as a pointing device and consists of three buttons as shown in Fig. 16.1. In this device, the left button is the pick button used to specify points on the screen or to select options and the right button is equivalent to the ENTER key on the keyboard. The middle scroll button is used for zoom in and zoom out by turning on either direction and used for panning by holding and moving too.

16.2.2 Keyboard

The keyboard is used for typing purposes such as to type commands, dimensional values, and any text. Any command to be executed is typed at the command prompt followed by pressing the ENTER key to instruct the computer. In AutoCAD, the additional tasks (Table 16.1) can be carried out by using F1 to F12 keys located on top of the keyboard.

Table 16.1   Keys and their purpose

Key	Purpose
F1	AutoCAD help
F2	AutoCAD text window (command history)
F3	Object snap ON/OFF
F4	3D Osnap ON/OFF
F5	Isoplane left/right/top
F6	Coordinate display ON/OFF
F7	Grid ON/OFF
F8	Ortho mode ON/OFF
F9	Snap ON/OFF
F10	Polar ON/OFF
F11	Object snap tracking (Otrack) ON/OFF
F12	Dynamic input ON/OFF
ENTER	To execute the last entered command
ESC	To cancel the current command

16.3 AUTOCAD WINDOWS AND COMMAND ENABLERS

When the user starts to work on the AutoCAD, the graphics window as shown in Fig. 16.2(a) is displayed where one can work on the drawings. By default, the background of the drawing area is displayed in black colour and the objects are drawn in white colour. The user can change the colour of the background or the drawing as desired. For

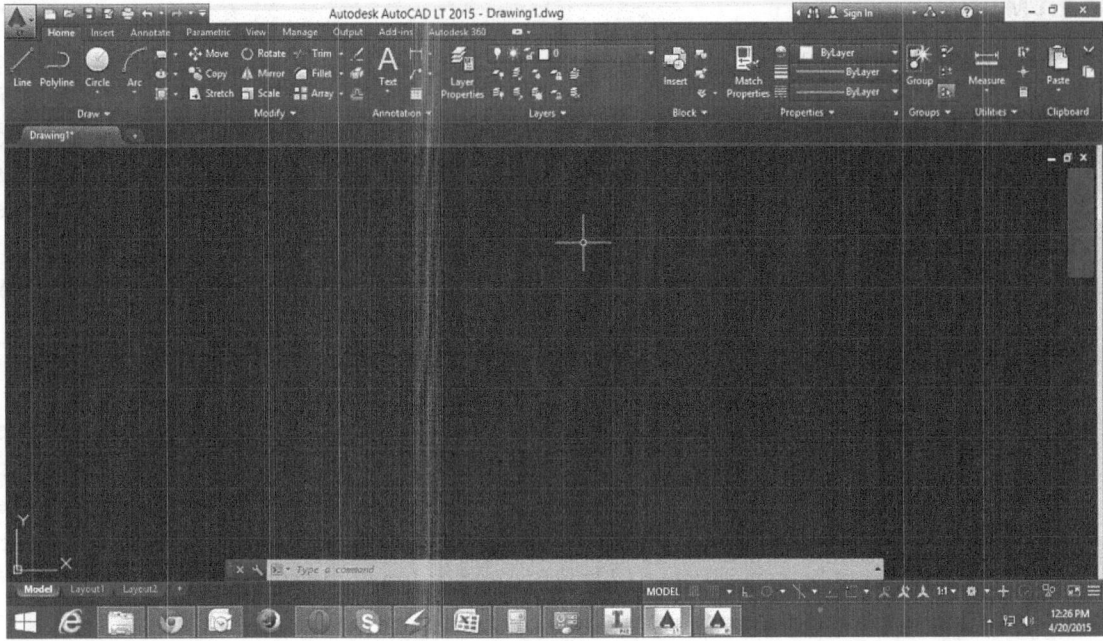

(a) Black background

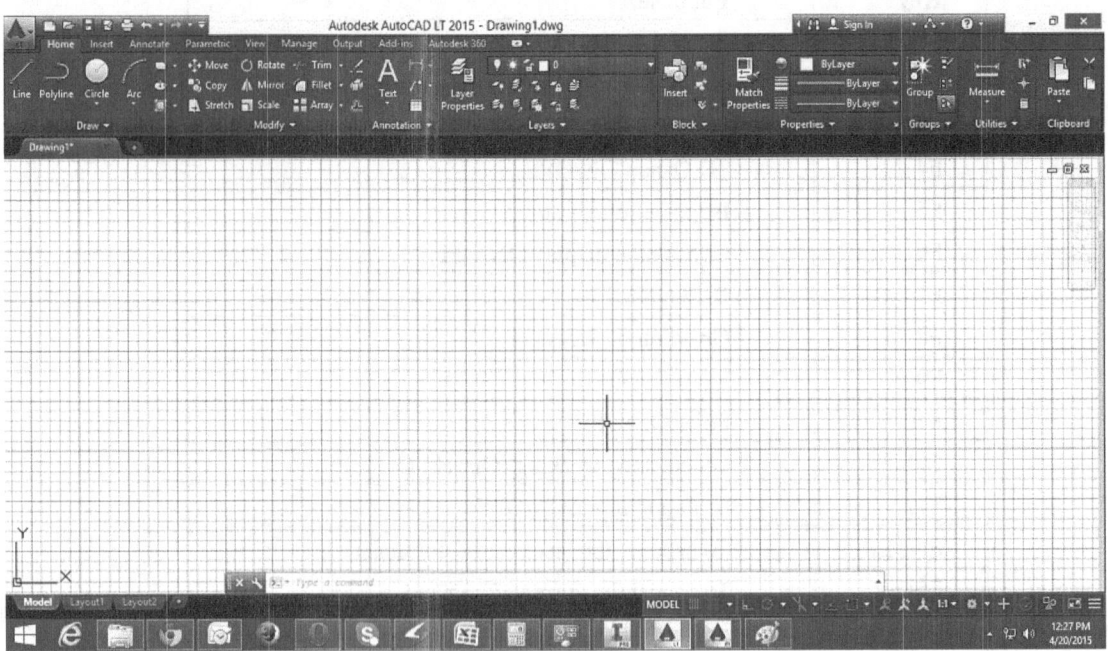

(b) White background and gridlines

Fig. 16.2 Graphics window

example, Fig. 16.2(b) shows AutoCAD graphics window with white colour background and grid lines. It is generally suggested to change the background of the window from black colour to white colour using the setting options.

'Units' module can de defined as the measuring units that can be used to draw the objects. The 'scale' determines the size of the unit when plotted on to the paper. In general, AutoCAD permits to draw the objects to full scale in the units that were initially set. The operation of the scale becomes active only at the time of plotting the drawings.

While preparing the drawings, various commands can be fed in one of the following ways:

1. Clicking the suitable command icon on the ribbon panel (or)
2. Typing the command in the command line (or)
3. Typing the command in the crosshair icon that appears in the graphic window.

If a particular command is to be activated, then it is to be typed in the command line followed by pressing the ENTER key and the command can be terminated by pressing again the ENTER key. If the ENTER key is pressed after the termination of a command, the command gets reactivated.

16.3.1 Ribbon Tabs and Panels

Under the ribbon, the tools are organized in logical groupings. The ribbon provides a compact set of all of the tools that are necessary to create or modify the drawing. They are placed in the following positions:

1. Docked horizontally at the top of the drawing area (default).
2. Docked vertically along the right or left edge of the drawing area.
3. Undocked, or floating, within the drawing area or on a second monitor.

The orientation of tabs changes when the ribbon is vertically docked or floating. The ribbon appears only when a new or existing document is opened in the drawing area.

The ribbon consists of a series of tabs, which are organized into sets or panels that contain many of the tools and controls available in the toolbars. Figure 16.3(a)–(f) shows various ribbon tabs and its panels. In Fig. 16.3(a), the tab and panel have been shown.

Ribbon panels can provide the access to a dialog box relevant to that panel. The related dialog box can be displayed by clicking the dialog box launcher, which is denoted by an arrow icon shown in the lower-right corner of the panel (Fig. 16.4).

To change the ribbon tabs and panels that are displayed, right click the ribbon and, on the shortcut menu, click or clear the names of tabs or panels.

16.3.2 Floating Panels

If the panel is turned on from the ribbon tab into the drawing area or onto another monitor, the panel remains floating in its place. The floating panel remains open until it is returned to the ribbon, even after the ribbon tabs are switched (Fig. 16.5).

16.3.3 Contextual Ribbon Tabs

When a particular type of object is selected or some of the commands are executed, a special contextual ribbon tab is displayed instead of a toolbar or dialog box. The contextual tab is automatically closed when the command is ended as shown in Fig. 16.6, where array creation tab closes when array command is terminated.

Tab

Panel

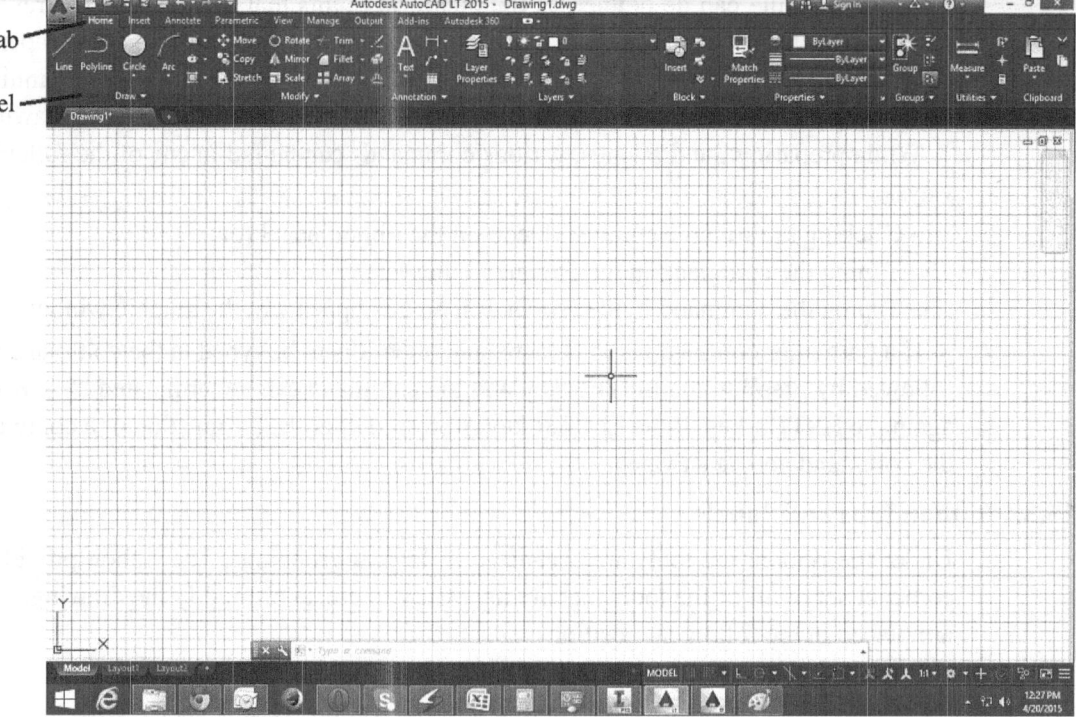

(a) Home Ribbon Tab and its panels

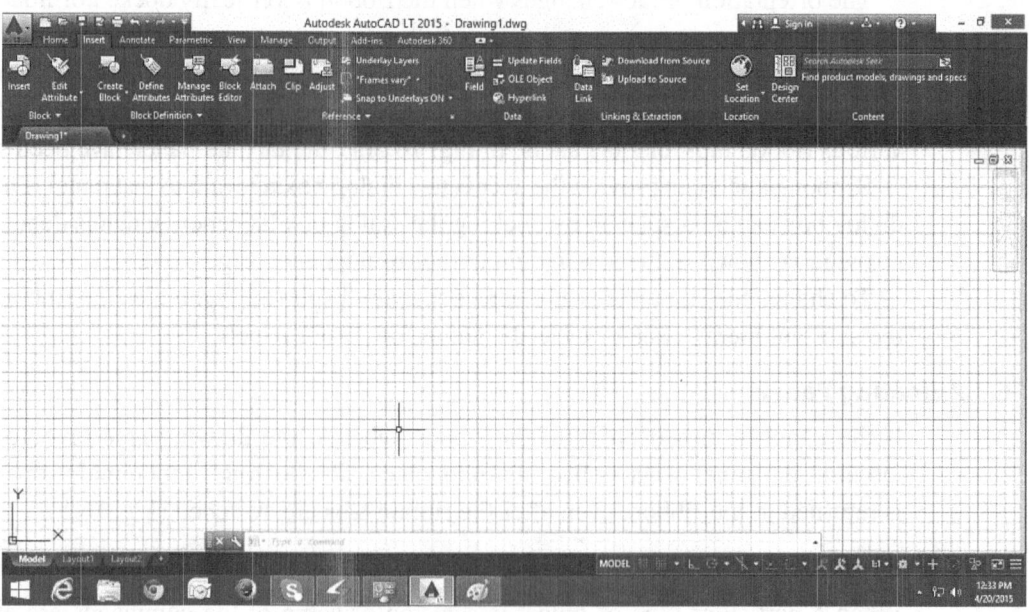

(b) Insert Ribbon Tab and its panels

Fig. 16.3 Ribbon tabs and its panels (*Contd*)

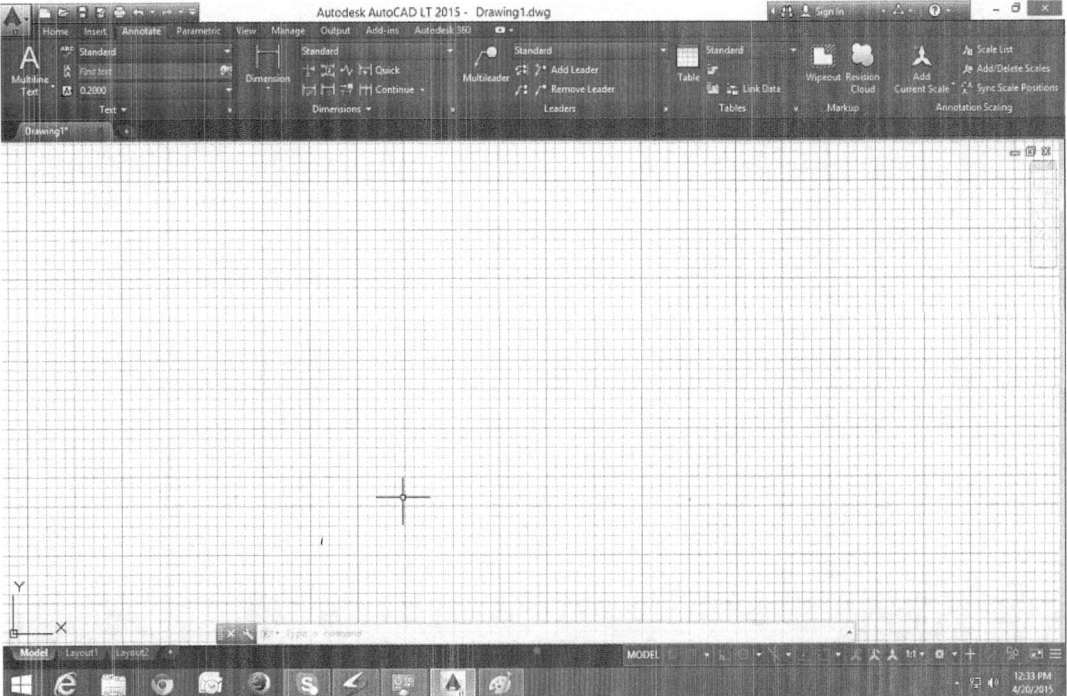

(c) Annotate Ribbon Tab and its panels

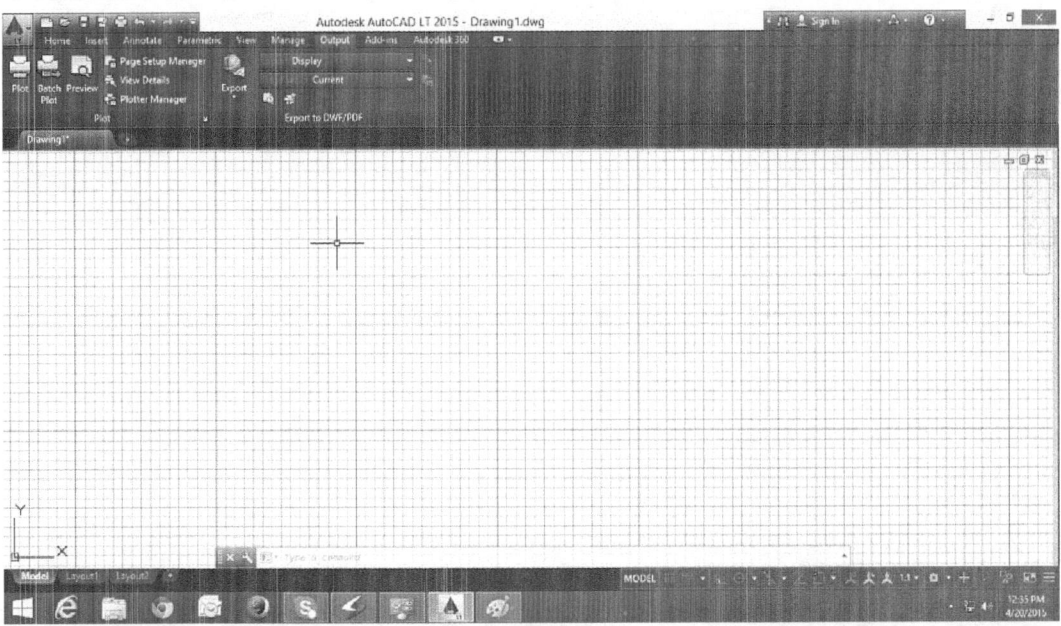

(d) Output Ribbon Tab and its panels

Fig. 16.3   (*Contd*)

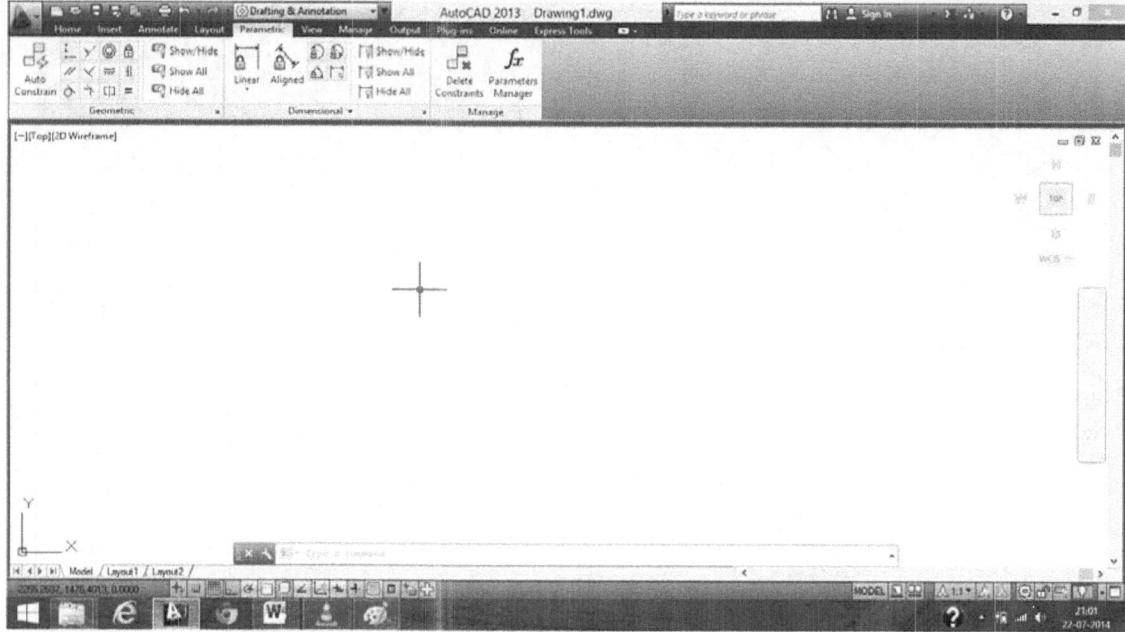

(e) Parametric Ribbon Tab and its panels

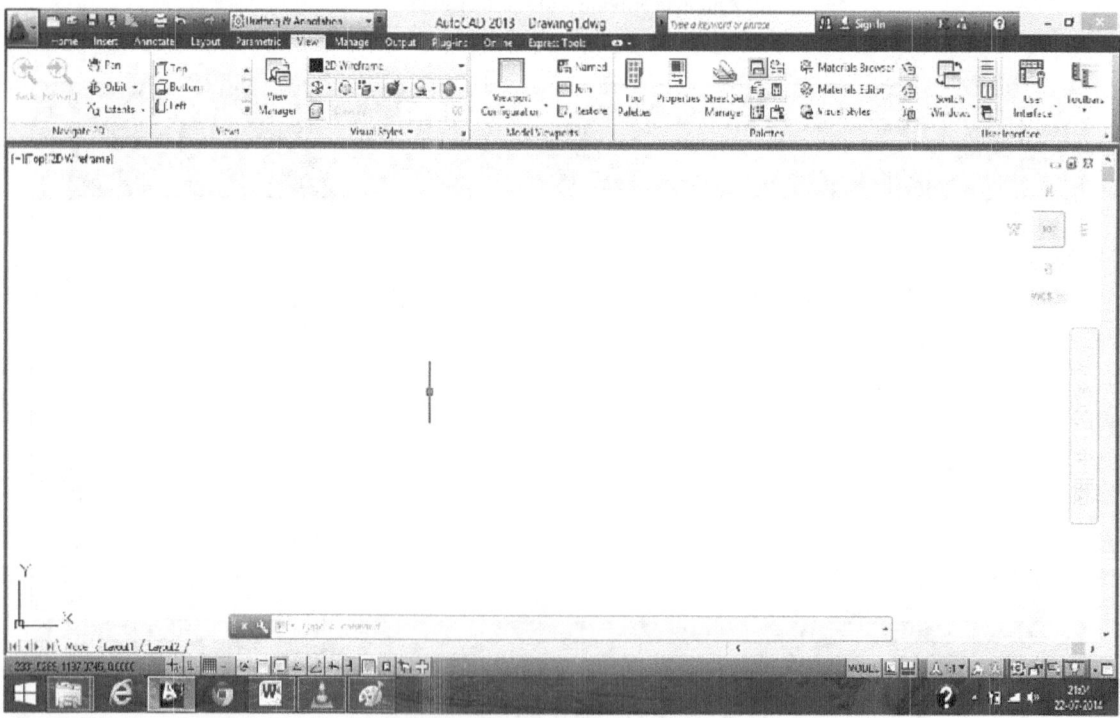

(f) View Ribbon Tab and its panels

Fig. 16.3 *(Contd)*

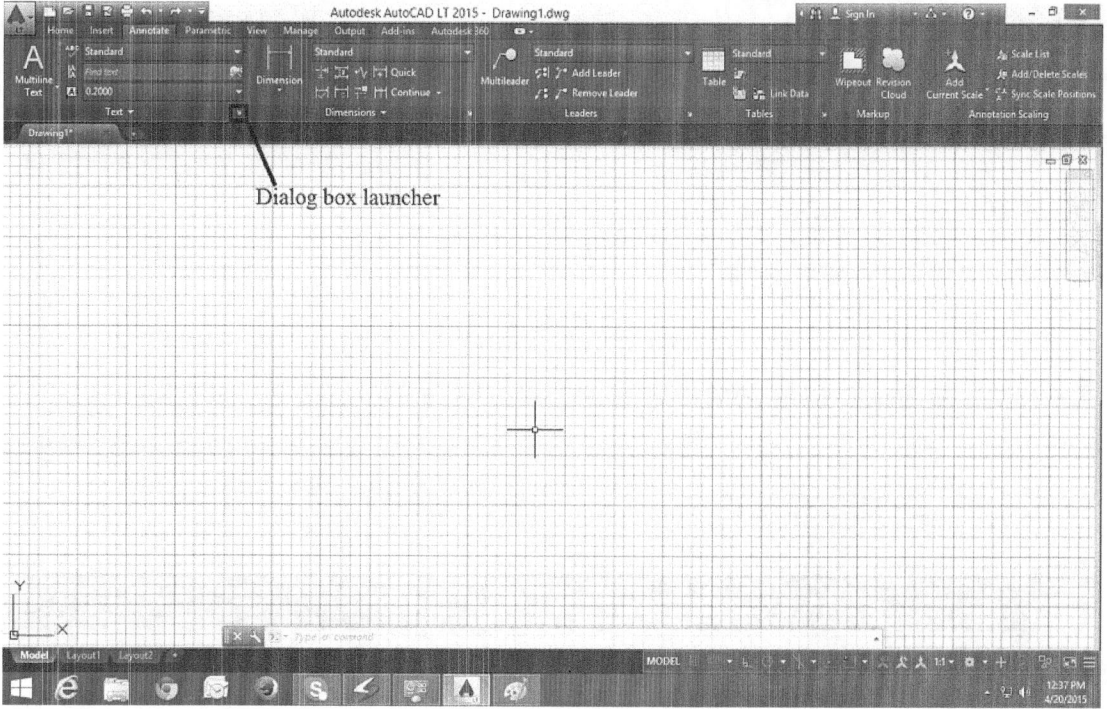

Fig. 16.4 Dialog box launcher

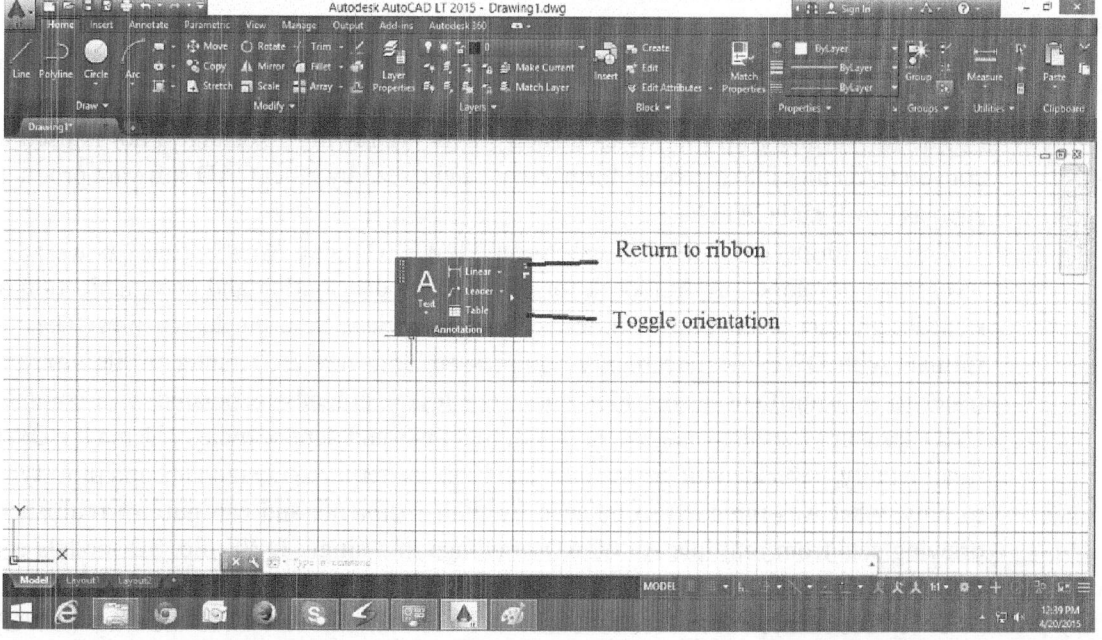

Fig. 16.5 Floating panel

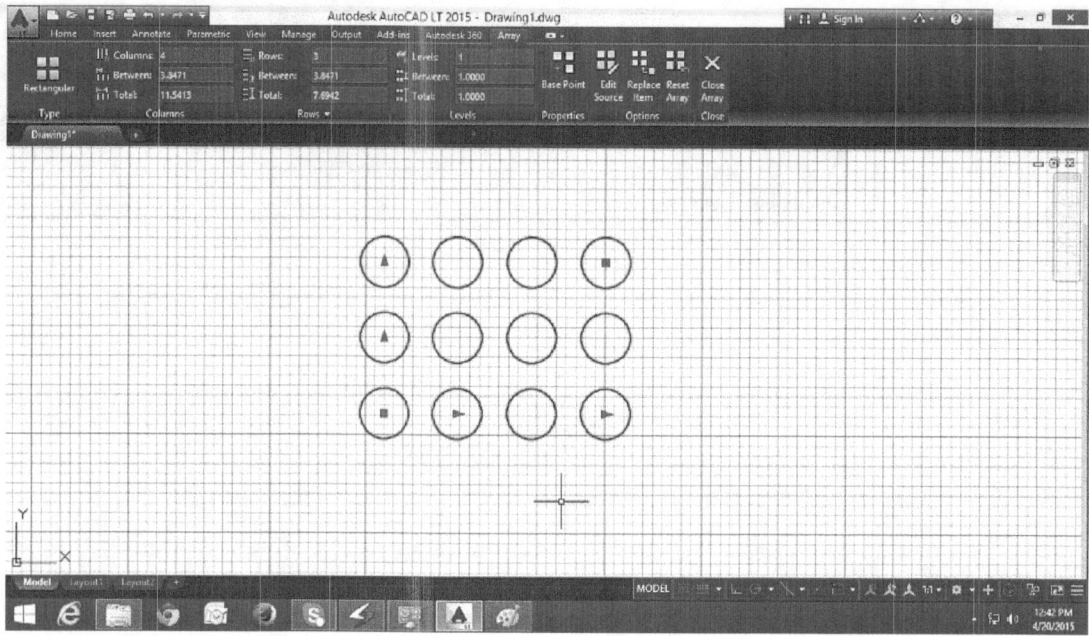

Fig. 16.6 Contextual ribbon

16.4 AUTOCAD FILES

An AutoCAD file is a drawing that is created with the file extension as '.dwg', which contains the binary data for the respective CAD design and is the drawing format for many CAD programs. DWG is the commonly used abbreviation for 'Drawing'. There are also other file formats such as DXF and DWF with which AutoCAD files can be saved.

The file format with the file extension '.DXF' (Drawing Exchange File) is used to convert CAD files into a generic format that can be read by other CAD software products such as Pro-E and CATIA.

The file format with the file extension '.DWF' (Design Web Format) is a secure file format developed by Autodesk for the efficient distribution and communication of rich design data to anyone who needs to view, review, or print design files. The file formats with DWF files are highly compressed, and they are smaller in file size and faster to transmit than design files with very less overheads associated with complex CAD drawings (or the management of external links and dependencies). Using the DWF functionality, publishers of design data can limit the required design data and plot styles to only, what was intended to be seen by their recipients and can publish multisheet drawing sets from multiple AutoCAD drawings in a single DWF file. 3D models can also be published from most of the Autodesk design applications.

DWF files cannot be replaced for native CAD formats like AutoCAD drawings (DWG). The main purpose of DWF of any component is to allow designers, engineers, project managers, and their colleagues to communicate the information related to the respective component design and its content to anyone who wishes to view, review, or print the same, without the knowledge of the team members or get trained on AutoCAD or other design software that are not in editable formats. Figure 16.7 shows various AutoCAD files of DWG, DXF, and DWF that can be converted into different non-editable formats such as JPEG, TIFF, and PDF using Acme CAD converter. Acme CAD converter can be used to support to open various file formats and convert file formats of editable DWG into non-editable file formats such as PDF, JPEG, TIFF, EPS, PLT, and SVG. It helps to convert DWF into DWG; it supports to save DWG and DXF into DWF, supports printing, supports the paper setting of each layout, and adjusts the size of output pages with its automatic layouts.

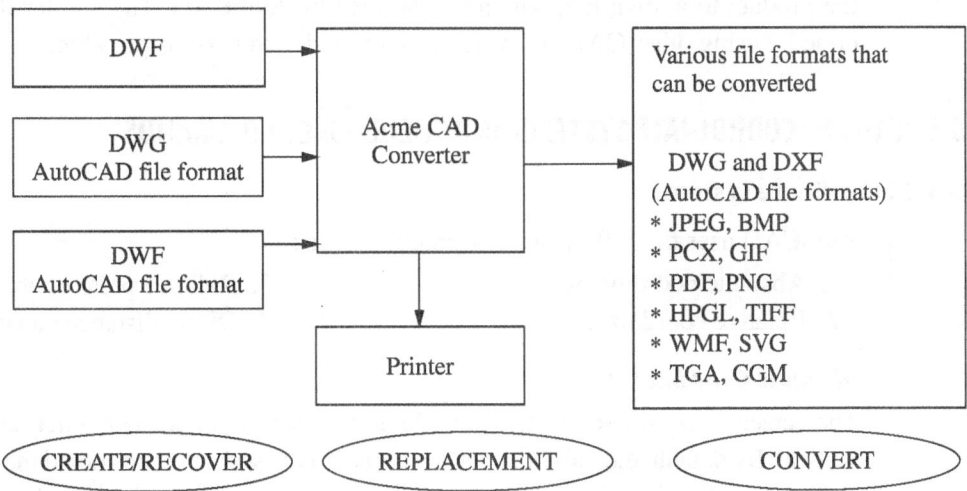

Fig. 16.7   Conversion of AutoCAD file formats (Editable native files) into other non-editable formats

16.4.1 Conversion of Other Format CAD Software Files into Native AutoCAD Files

There are many 3D modelling software products developed by different agencies and are commercially available such as Pro-E, CATIA, and Unigraphics. Each CAD modelling software has its own drawing file format that can be converted into AutoCAD file format with 'dwg' extension.

16.4.2 Importing CATIA File Using AutoCAD Plug-in

Through this plug-in, AutoCAD is capable to import the 3D data related to the drawings of the parts and assembly files that are developed using CATIA. The true solids in AutoCAD can also be created from the CATIA Import files that will be then edited using AutoCAD's solid modelling tools similar to any other 3D solid modelling software. For importing CATIA file formats, the plug-in does not need CATIA to be installed on the computer system. 3D InterOp technology from Spatial is being used in the CATIA Import for AutoCAD. 3D InterOp is the industry standard and proprietary CAD file format data exchange, which is used in almost all the major CAD systems.

The file format with file extension '.prt' is a CAD drawing file created by the Pro/ENGINEER program. The file contains the drawing of a part that is to be manufactured. A '.prt' file can be converted into a '.dwg' file using the appropriate software application. Using PTC Pro/ENGINEER (now renamed as PTC CREO) for 2D .prt files can be exported to a .dwg file.

In addition, if the .prt file that was created using the program other than Pro/ENGINEER, it still may have the capability to export as .dwg. The Sycode DWG Export is a plug-in for the Pro/ENGINEER program to allow it to export 3D .prt files as .dwg for 3D .prt files. The plug-in converts the .prt model into triangular meshes and exports the product to a .dwg file, which can be read by AutoCAD. The similar things can be done by using other CAD software products and vice versa is possible.

16.5 AUTOCAD COORDINATE SYSTEMS AND THEIR USAGE COMMANDS

16.5.1 Coordinate systems

AutoCAD uses the following coordinate systems:

1. Absolute coordinates
2. Relative coordinates
3. Relative polar coordinates
4. Direct distance entry

Absolute Coordinates

The screen is considered as the 'xy' plane with the 'x' values horizontal and 'y' values vertical. By default, the left-hand bottom corner of the screen is considered the origin $(0, 0)$.

To mark a point, the values for x, y coordinates are required, and Example 16.1 illustrates the procedure to mark various points on a plane.

Example 16.1

Using the coordinate system command, generate the line diagram given in Fig. 16.8.

PROCEDURE (Refer Fig. 16.8 and Table 16.2)

Step 1: In this example, a number of lines are required to be drawn using the absolute coordinate system with help of line command.

Step 2: As the coordinate distances are given, the steps given in Table 16.2 can be adopted.

Step 3: The drawing shown in Fig. 16.8 can be completed by specifying all the points. By pressing ESC, the line command is undone.

Relative Coordinates

The drawings can also be generated using relative coordinates. When relative coordinates are used, the line is drawn with reference to the previous point. Note that for using relative coordinates, the symbol '@' is required to be typed before typing the coordinate values. Example 16.2 explains the method of using relative coordinates.

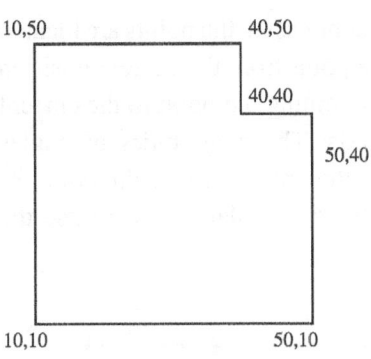

10,50   40,50

                    40,40

                        50,40

10,10   50,10

Fig. 16.8 Creating the line diagram using absolute coordinates

Table 16.2 List of AutoCAD commands to generate the drawing shown in Fig. 16.8 using the absolute coordinate system

Display in the command line window	Actions to be taken
Command:	Type "line" or l, Press Enter
Specify first pont:	10,10 Press Enter
Specify next point or [Undo]:	50,10 Press Enter
Specify next poit or [Undo]:	50,40 Press Enter
Specify next point or [Close/Undo]:	40,40 Press Enter
Specify next point or [Close/Undo]:	40,50 Press Enter
Specify next point or [Close/Undo]:	10,50 Press Enter
Specify next point or [Close/Undo]:	10,10 Press Enter
	Press ESC

Example 16.2

Draw Fig. 16.8 by using relative coordinates.

PROCEDURE (Fig. 16.9)

The drawing given in Fig. 16.8 can be drawn by using the commands given in the Table 16.3 and the resulting drawing is shown in Fig. 16.9.

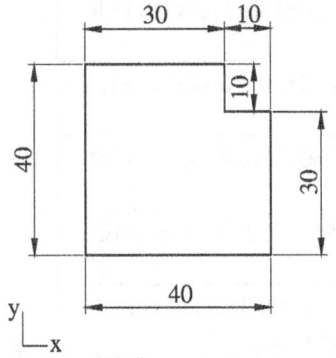

30   10

10

40   30

40

y
└x

Fig. 16.9 Creating the line diagram using relative coordinates

Table 16.3 List of AutoCAD commands to generate the drawing shown in Fig. 16.9 using the relative coordinate system

Display in the command line window	Actions to be taken
Command:	Type "line" press Enter
Specify first point:	10,10 Press Enter
Specify next point or [Undo]:	@40,10 Press Enter (Absolute X Coordinate being 50, relative coordinate from first point is 50-10=40)
Specify next point or [Undo]:	@0,30 Press Enter
Specify next point or [Close/Undo]:	@-10,0 Press Enter
Specify next point or [Close/Undo]:	@0,10 Press Enter
Specify next point or [Close/Undo]:	@-30,0 Press Enter
Specify next point or [Close/Undo]:	@0,-40 Press Enter
	Press ESC

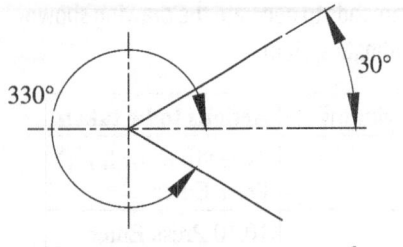

Fig. 16.10 Measurement of angles in AutoCAD

Relative Polar Coordinates

When relative polar coordinates are used, the points are located by defining the distance of the point from the current position and the angle made by the line joining the point to the current position and the positive x-axis. The magnitudes of angles in AutoCAD are measured in the anticlockwise direction, as shown in Fig. 16.10. Example 16.3 explains how to use the relative polar coordinates.

Example 16.3

Using relative polar coordinates, draw Fig. 16.9.

PROCEDURE The drawing can be drawn by following the commands given in Table 16.4. The resulting drawing is shown in Fig. 16.11.

Table 16.4 List of AutoCAD commands to generate the drawing shown in Fig. 16.11 using the relative polar coordinate system

Display in the command line window	Actions to be taken
Command	: Type "LINE", press ENTER
Specify first point	: 10,10, Press ENTER
Specify next point or [Undo]	: @40<0, Press ENTER
Specify next point or [Undo]	: @30<90, Press ENTER
Specify next point or [Close/Undo]	: @10<180, Press ENTER
Specify next point or [Close/Undo]	: @10<90, Press ENTER
Specify next point or [Close/Undo]	: @30<180, Press ENTER
Specify next point or [Close/Undo]	: @40<270, Press ENTER
Specify next point or [Close/Undo]	: Press ESC to undo the line command

Fig. 16.11 Creating the line diagram using relative polar coordinates

Direct Distance Entry Method

In this method, the point to which a line is to be drawn is located by entering the distance from the current point, and the direction is shown by the movement of the cursor. To get the exact horizontal and vertical line, 'Ortho' command is turned on by pressing F8 key. Example 16.4 explains the use of this method.

Example 16.4

Draw Fig. 16.9 using direct distance entry method.

PROCEDURE (Fig. 16.12)

The drawing can be drawn by executing the commands in Table 16.5. The resulting drawing is shown in Fig. 16.12.

Table 16.5 List of AutoCAD commands to generate the drawing shown in Fig. 16.12 using the direct distance entry method

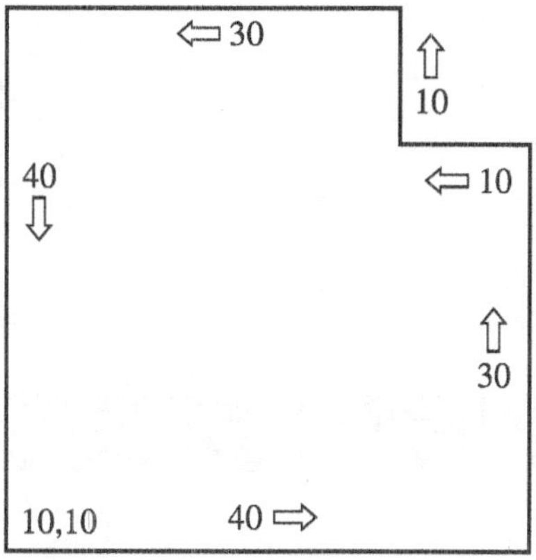

Fig. 16.12 Creating the line diagram using direct distance entry method

Display in the command line window	Actions to be taken
Command Line	: "Line", Press enter
Specify first point	: Press F8 key for <ortho on> 10,10, Press
Specify next point or [Undo]	: 40 (move mouse horizontally), Press enter
Specify next point or [Undo]	: 30 (move mouse vertically), Press enter
Specify next point or [Close/ Undo]	: 10 (move mouse horizontally), Press enter
Specify next point or [Close/ Undo]	: 10 (move mouse vertically), Press enter
Specify next point or [Close/ Undo]	: 30 (move mouse horizontally), Press enter
Specify next point or [Close/ Undo]	: 40 (move mouse vertically), Press enter
Specify next point or [Close/ Undo]	: Press ESC to undo the command.

16.5.2 User Coordinate System (UCS)

The drawing area displays an icon representing the xy axis of a rectangular coordinate system, which is known as the User Coordinate System or UCS and is shown in Fig. 16.13.

The user coordinate system (UCS) is a movable coordinate system and is a basic tool for 2D drawing and is used for the following:

1. Define the 'xy' plane known as the 'work plane', on which the objects are created, modified, and edited.
2. Invoke the 'ortho' mode for the horizontal and vertical directions, to use the 'polar tracking' for inclined direction features, and 'object snap tracking' to snap the object boundary points while generating the drawing.

Fig. 16.13 UCS icon

3. Define the alignment and angle of the grid, hatch patterns, text, and dimensioning the objects.
4. Define the origin and orientation for coordinate entry and absolute reference angles.

The location and orientation of the current UCS can be identified with the help of the UCS icon. The UCS can be manipulated by clicking the UCS icon and using its grips, or with the UCS command. The display options for the UCS icon are available with the UCSICON command.

16.6 AUTOCAD BASIC DRAWING COMMANDS

Any drawing in AutoCAD is created using various elements of the drawing, which are known as entities, such as a point, line, polyline, arc, circle, an ellipse, rectangle, and so on. Each entity has an individual existence and is also known as the object in AutoCAD. There are various drawing commands that are used in AutoCAD to draw the different features of a drawing. Some of the most commonly used features are explained as follows:

Point command This command creates a point object. Points can act as nodes to which one can snap the objects. To create a point, follow the command sequences are given in the following Table 16.6.

Table 16.6 List of AutoCAD commands to generate a point

Display in the command line window	Actions to be taken
Command	: point
Current point modes	: PDMODE=0 PDSIZE=0.0000
Specify a point	: 10,10

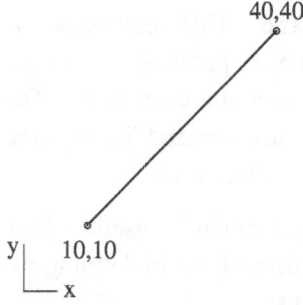

Line command This command creates straight line segments. With LINE command, a series of continuous line segments can be created. Each segment is a line object that can be edited separately. To create a line segment, follow the command sequences given in Table 16.7. After creating a line, the command asks for either continuing the line segment or ending it. Figure 16.14 shows line segment created using line command for the length using the position coordinates.

Fig. 16.14 Creating a line segment

Table 16.7 List of AutoCAD commands to generate a line

Display in the command line window	Actions to be taken
Command	: LINE, Press "Enter"
Specify first point	: 10,10, Press "Enter"
Specify next point or [Undo]	: @40,40, Press "Enter"
Specify next point or [Undo]	: Press "Enter"

Table 16.8 List of AutoCAD commands to generate an object using PLINE command

Display in the command line window	Actions to be taken
Command	: pline
Specify start point	: 10,10
Current line-width is 0.0000	
Specify next point or [Arc/Halfwidth/Length/Undo/Width]	: 15,10
Specify next point or [Arc/Close/Halfwidth/Length/Undo/Width]	: arc
Specify endpoint of arc or [Angle/CEnter/CLose/Direct/Halfwidth/Line/Radius/Second pt/Undo/Width]	: 15,5
Specify endpoint of arc or [Angle/CEnter/CLose/Direction/Halfwidth/Line/Radius/Second pt/Undo/Width]	: line

(Contd)

Table 16.7 *(Contd)*

Display in the command line window	Actions to be taken
Specify next point or [Arc/Close/Halfwidth/Length/Undo/Width]	: 10,5
Specify next point or [Arc/Close/Halfwidth/Length/Undo/Width]	: arc
Specify endpoint of arc or [Angle/CEnter/CLose/Direction/ Halfwidth/Line/Radius/Second pt/Undo/Width]	: 10,10
Specify endpoint of arc or [Angle/CEnter/CLose/Direction/ Halfwidth/Line/Radius/Second pt/Undo/Width]	: close

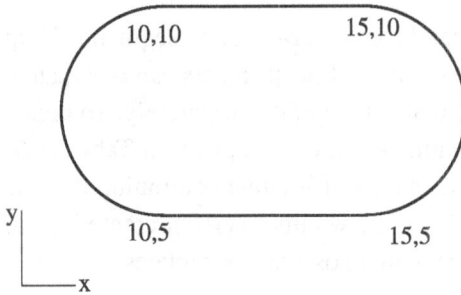

Fig. 16.15 Creating object using PLINE command

Polyline (PLINE) command This command is used for creating 2D polyline. A polyline is a single object that is composed of line and arc segments. The following Table 16.8 depicts the command list required to create the drawing shown in Fig. 16.15.

Rectangle command This command is used to draw a rectangular polyline. The various ways of defining the size of rectangle are as follows:

1. by entering the dimensions
2. by specifying the two corners
3. by specifying the area of rectangle.

Table 16.9 depicts the commands required to draw a rectangle using rectangle command and dimensions as shown in Fig. 16.16.

Table 16.9 List of AutoCAD commands to generate an object using RECTANGLE command

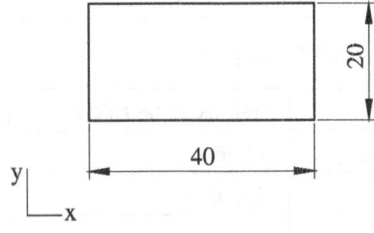

Fig. 16.16 Creating rectangle using RECTANGLE command

Display in the command line window	Actions to be taken
Command	: rectangle
Specify first corner point or [Chamfer/ Elevation/Fillet/Thickness/Width]	: 10,10
Specify other corner point or [Area/ Dimensions/Rotation]	: dimensions
Specify length for rectangles <40.0000>	: 40
Specify width for rectangles <60.0000>	: 20
Specify other corner point or [Area/ Dimensions/Rotation]	: 20,20

Polygon command This command is used to draw regular polygon with three to 1024 sides. The size of the polygon is defined by giving any one of the following:

1. Radius of the circle circumscribing the polygon
2. Radius of the circle inscribed in the polygon
3. Length of one edge of the polygon

   The polygon can be generated by using the commands as listed in Table 16.10 and dimensions as shown in Fig. 16.17.

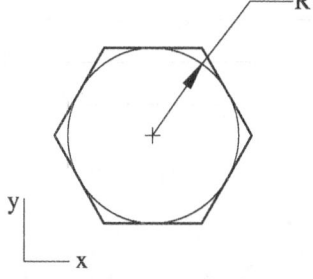

Fig. 16.17 Creating hexagon using POLYGON command

Table 16.10 List of AutoCAD commands to generate a hexagon using POLYGON command

Display in the command line window	Actions to be taken
Command	: polygon
Enter number of sides <4>	: 6
Specify center of polygon or [Edge]	: 20,20
Enter an option [Inscribed in circle/ Circumscribed about circle] <I>	: Circumscribed
Specify radius of circle	: 10

Circle command A circle can be drawn provided one of the following conditions is satisfied:

1. The centre point and the radius or diameter are given.
2. Any three points on the circle are given.
3. The diameter endpoints are given.
4. Two tangent lines to the circle and radius are given.
5. Three tangent lines to the circle are given.

   The following Table 16.11 uses the condition (b) to draw the circle and Fig. 16.18 depicts the same.

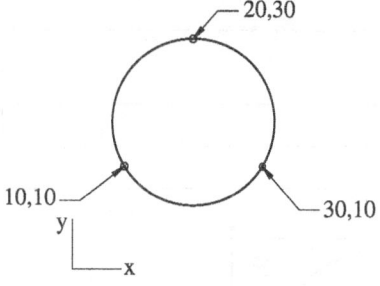

Fig. 16.18 Creating circle with three given points

Table 16.11 List of AutoCAD commands to generate a circle with three points given

Display in the command line window	Actions to be taken
Command	: circle
Specify center point for circle or [3P/2P/Ttr (tan tan radius)]	: 3p
Specify first point on circle	: 10,10
Specify second point on circle	: 30,10
Specify third point on circle	: 20,30

Arc command This command is used to draw circular arcs when the arc is to be drawn as follows:

1. Passing through three given points
2. Passing through two given points and having a given point as centre or given length as radius
3. Passing through a given point and subtending a fixed angle at the given centre

The commands for Case (a) are given in Table 16.12 and Fig. 16.19 depicts the same.

Table 16.12 List of AutoCAD commands to generate an arc with three given points

Display in the command line window	Actions to be taken
Command	: ARC
Specify start point of arc or [Center]	: 55,20
Specify second point of arc or [Center/End]	: 35,33
Specify end point or arc	: 15,20

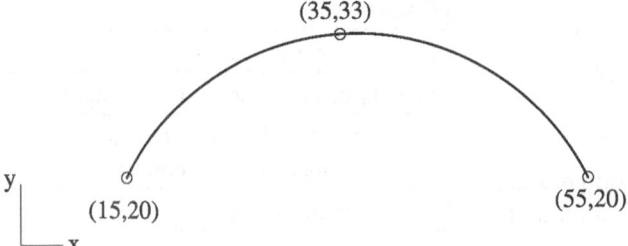

Fig. 16.19 Creating an arc with the given three points

Ellipse command This command is used to draw an ellipse. The ellipse can be drawn when both the endpoints of one axis, along with the distance of the other axis endpoint from the first, are known. The commands are given in Table 16.13 and Fig. 16.20 depicts the same.

Table 16.13 List of AutoCAD commands to generate an ellipse using ELLIPSE command

Display in the command line window	Actions to be taken
Command	: ELLIpse
Specify axis endpoint of ellipse or [Arc/Center]	: 10,10
Specify other endpoint of axis	: 60,10
Specify distance to other axis or [Rotation]	: 10

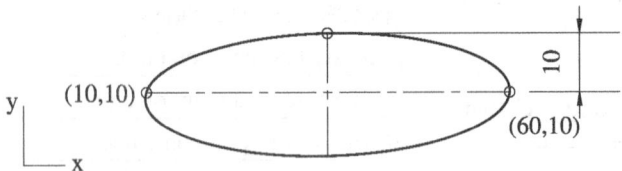

Fig. 16.20 Creating ellipse using ELLIPSE command

16.7 AUTOCAD BASIC EDITING COMMANDS

There are various AutoCAD editing commands used for editing a drawing. Some of the commonly used and important commands are described in the following sections.

16.7.1 Object Selection Methods (SELECT Command)

Objects that are drawn before can be selected by any one of the following methods after using SELECT command.

Select objects   Those objects that are clicked will be selected. Multiple objects can also be selected by simply clicking on it.

Window   The objects that are completely inside a rectangle defined by two points are selected. Specifying the corners from left to right creates a window selection.

Crossing   The objects within and crossing an area defined by two points are selected. Specifying the corners from right to left creates a crossing selection.

Last   Using this, the most recently created visible object gets selected. The object must be in the current space, that is, model space or paper space, and its layer must not be set to be frozen or off.

Box   This is equivalent to Window/Crossing.

All   This selects all objects in either model space or the current layout, except those objects on frozen or on locked layers.

Fence   This selects all objects crossing a selection fence. In Fence command, the fence is not closed, and a fence can cross itself.

Group   This selects all objects within one or more named or unnamed groups.

ERASE command   It is used to delete one or more unwanted objects from the drawing. It can be achieved by clicking the Erase icon or by typing 'E' (for Erase) in the command line. Then, user is prompted to select the object and then press DELETE key.

MOVE command   This command is used to move an object or a group of objects to a new location without any change in orientation or size. It moves the object from the original to a new location permanently.

COPY command   This command is used to copy an object or a group of objects. It allows creating several copies of the selected objects. The Copy command prompts the user to select the object to be copied and enter the base point, which can be any point serving as a reference. The next point is the point indicating the new location of the base point.

OFFSET command   This command constructs an object parallel to the selected object at a specified distance or through a specified point. If the selected circle was drawn using circle command, the whole circle will be drawn with sides parallel to the selected object. However, if the circle was constructed by drawing arcs using arc command, only arc parallel to the selected side will be redrawn. The commands are given in Table 16.14 and Fig. 16.21 shows two parallel circles obtained by the offset command from the selected middle circle.

Table 16.14 List of AutoCAD commands to generate a parallel object using OFFSET command

Display in the command line window	Actions to be taken
Command	: offset
Specify offset distance or [Through/Erase/Layer] <1.0000>	: 3
Select object to offset or [Exit/Undo] <Exit>	: 15,10 (Selecting the object)
Specify point on side to offset or [Exit/Multiple/Undo] <Exit>	: multiple
Specify point on side to offset or [Exit/Undo] <next object>	: 20,10 (Outer side point)
Specify point on side to offset or [Exit/Undo] <next object>	: 10,10 (Inner side point)

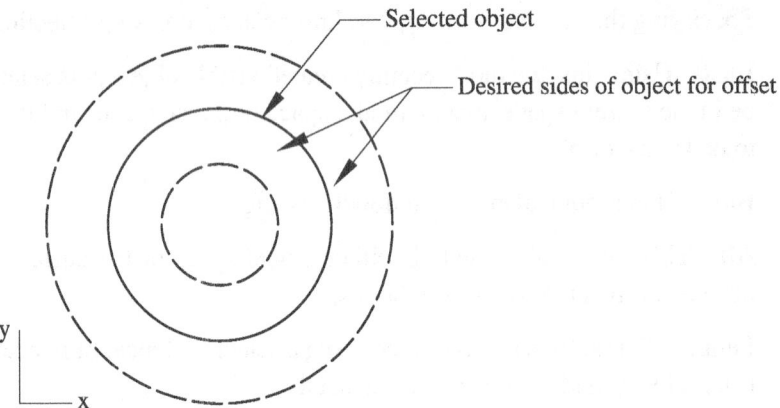

Fig. 16.21 Parallel circles obtained by offset command

FILLET command This command is used to join two non-parallel lines by an arc of a specified radius. The arc is joining two lines in such way that they become tangent to the arc. After clicking the fillet icon, the radius of the fillet and then the two objects, that is, the two lines to be connected by the tangent arc are required to be specified.

CHAMFER command This command is used to join two non-parallel lines with an intermediate inclined line, which is usually called a chamfer. The chamfer lengths on the two lines need to be specified. The first chamfer is taken on the first selected object and the followed on the second selected object.

HATCH command It is used to hatch a closed area. To use this, the user needs to follow the given steps:

1. Either click the hatch icon in the draw panel bar or type 'H' (for hatch) in the command line and then press the ENTER button. The hatch creation panel will open in panel bar.
2. In the Hatch tab, select the desired pattern and then click at select or pick points. Then, select the object by clicking on the boundary of the area to be hatched.
3. If the selected area is to be filled with colour, click on hatch colour in the hatch palette bar and select the desired colour.
4. After hatching, click on the close hatch creation to finish.

TRIM Command This command is used to cut drawn objects. The trimming is done up to the point where the objects intersect with the cutting edges. The commands are given in Table 16.15. Figure 16.22(a) and (b) shows the drawings before and after trimming, respectively.

Table 16.15 List of AutoCAD commands to generate the final object using TRIM command

Display in the command line window	Actions to be taken
Command	: TRIM
Select cutting edge ... Select objects or <select all>	: 1 found
Select objects	: 1 found, 2 total
Select objects	: Press Enter
Select object to trim or shift-select to extend or [Fence/Crossing/Project/Edge/eRase/Undo]	: Click on element ab
Select object to trim or shift-select to extend or [Fence/Crossing/Project/Edge/eRase/Undo]	: Click on element cd
Select object to trim or shift-select to extend or [Fence/Crossing/Project/Edge/eRase/Undo]	: Click on element ef
Select object to trim or shift-select to extend or [Fence/Crossing/Project/Edge/eRase/Undo] 3 constraint(s) removed	: Press Enter

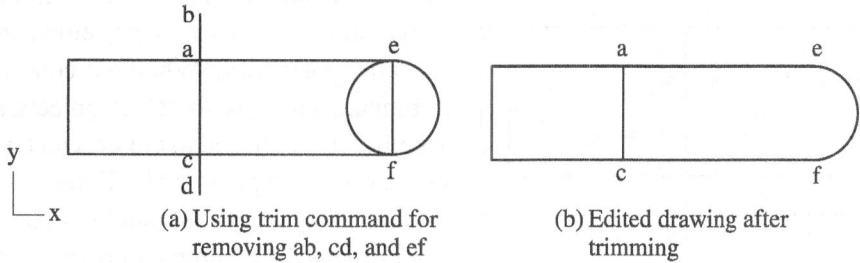

(a) Using trim command for removing ab, cd, and ef

(b) Edited drawing after trimming

Fig. 16.22 Trimming of the object by trim command

EXTEND command It is activated by clicking its icon, and on being prompted, the user has to select the boundary edge. After selecting it, and again on being prompted, the user has to click the object end nearer to the boundary edge and then press the ENTER key.

MIRROR command This command is used to create the mirror image copy of an object or a group of objects in a drawing. The mirror image is created about a specified axis. This axis is known as the mirror line. Any actual line or an imaginary line specified by two points can be selected as the mirror line. The original source object may be retained or deleted as desired. The sequence of commands is given in Table 16.16. Figure 16.23 shows the drawing created using mirror command.

Table 16.16   List of AutoCAD commands to generate an object using MIRROR command

Display in the command line window	Actions to be taken
Command	: MIRROR
Select objects 11 found	: all
Select objects	: Press enter
Specify first point of mirror line	: Select first point of mirror axis
Specify second point of mirror line	: select second point of mirror axis
Erase source objects? [Yes/No] <N>	: No

Table 16.17   List of AutoCAD commands to generate multiple copies of an object using rectangular ARRAY command

Display in the command line window	Actions to be taken
Command	: ARRAY
Select objects	: all
Select objects	: Press Enter
Enter array type [Rectangular/Path/Polar] <Rectangular>	: r for rectangular
Select grip to edit array or [Associative/Base point/Count/Spacing/COlumns/Rows/Levels/Exit] <exit>	: select number of columns and rows

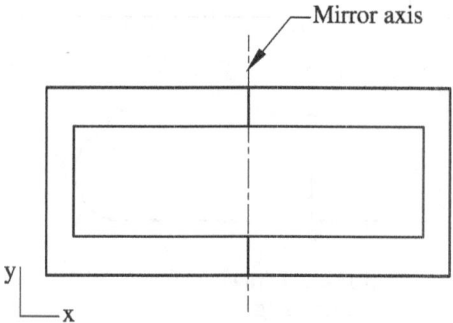

Fig. 16.23   Generating an object using mirror command

ARRAY command   This command is used to create multiple copies of an object (which may be a line, a polygon, or any other shape) either in the rectangular or in the polar form. When AR command is typed, the command prompts to select objects and then choose whether it is required to create a rectangular array, path curve arrayp or polar array. Upon selecting the option, a new array creation palette opens where various options such as column and rows for rectangular array and angle for polar array can be chosen. Tables 16.17 and 16.18 show the commands for rectangular and polar array, respectively, and the same are shown in Fig. 16.24(a) and (b), respectively.

Table 16.18   List of AutoCAD commands to generate multiple copies of an object using polar ARRAY command

Display in the command line window	Actions to be taken
Command	: ARRAY
Select objects	: 1 for last
Select objects	: press enter

(Contd)

Table 16.18   *(Contd)*

Display in the command line window	Actions to be taken
Enter array type [Rectangular/PAth/POlar] <Rectangular>	: Polar
Type = Polar Associative = Yes Specify center point of array or [Base point/ Axis of rotation]	: 50,25
Select grip to edit array or [Associative/Base point/Items/Angle between/Fill angle/Rows/ Levels/Rotate items/exit] <exit>	: select number of elements, number of rows

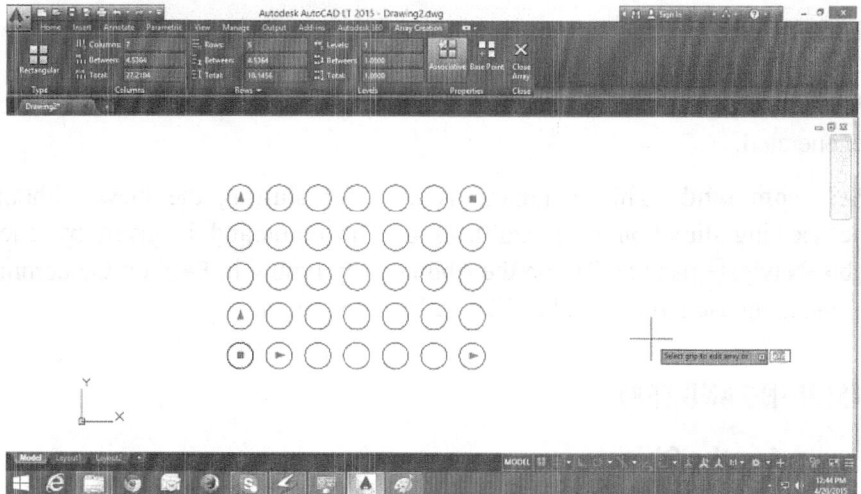

(a) Rectangular array

(b) Polar array

Fig. 16.24   Generating multiple copies of an object using

16.8 AUTOCAD DISPLAY COMMANDS

The following commands are used for viewing the drawing in required size and viewport. Some of the most commonly used commands are as follows.

ZOOM command This command is used to closely view a drawing or a particular part of it by zooming in. The Zoom command is given by clicking Zoom icon on the view tab in navigate panel or by typing in ZOOM or Z command in command box. Then select options like 'a' for the entire drawing or grid limits or 'w' for the window to enlarge a particular area and so on. The Zoom command has various options, namely All, Centre, Dynamic, Extents, Previous, Scale, Window, Object, Real time.

REGEN command It is used to regenerate the complete drawing. The command is given by clicking View tab>>Views panel>>Regen on the ribbon or by typing 're' (for REGEN) on the command line and then giving information about the object to be regenerated.

PAN command This command is used for shifting the view without changing the viewing direction or magnification. The command is given by clicking View tab>>Navigate panel>>Pan on the ribbon or by typing in PAN on the command line or by using mouse left or middle click button and drag it.

16.9 DIMENSIONING AND TEXT

16.9.1 Dimensioning the Object

Dimensioning the object is very easy in AutoCAD. Length of a line, radius/diameter of an arc/circle, and angle between two lines are automatically calculated. DIMALIGNED command is used to show the length of a line parallel to the object line. The dimension is placed when we select two points or the object line itself. DIMLINEAR command is used to show the perpendicular distance between two points. DIMANGULAR command dimensions the angle between two lines. DIMRADIUS and DIMDIAMETER commands are used to dimension radius and diameter, respectively. DIMARC command is used for measuring arc length. Figure 16.25 referred for more options.

The user may set the size and select the type of arrowheads, height of the dimension text, and other parameters of dimensioning before or after giving dimensions. DIMSTYLE command is used for this purpose. It opens 'Dimension Style Manager' Dialog box of Fig. 16.26. Click at 'New' button to open 'Create New Dimension Style' dialog box, which enables to set for the user's own dimension style. Clicking at 'Modify' button will open 'Modify Dimension Style: Standard' dialog box. Using the various option tabs, viz., lines, symbols and arrows, text, fit, primary units, and alternative units and tolerances, the user may set the required values/styles of various parameters.

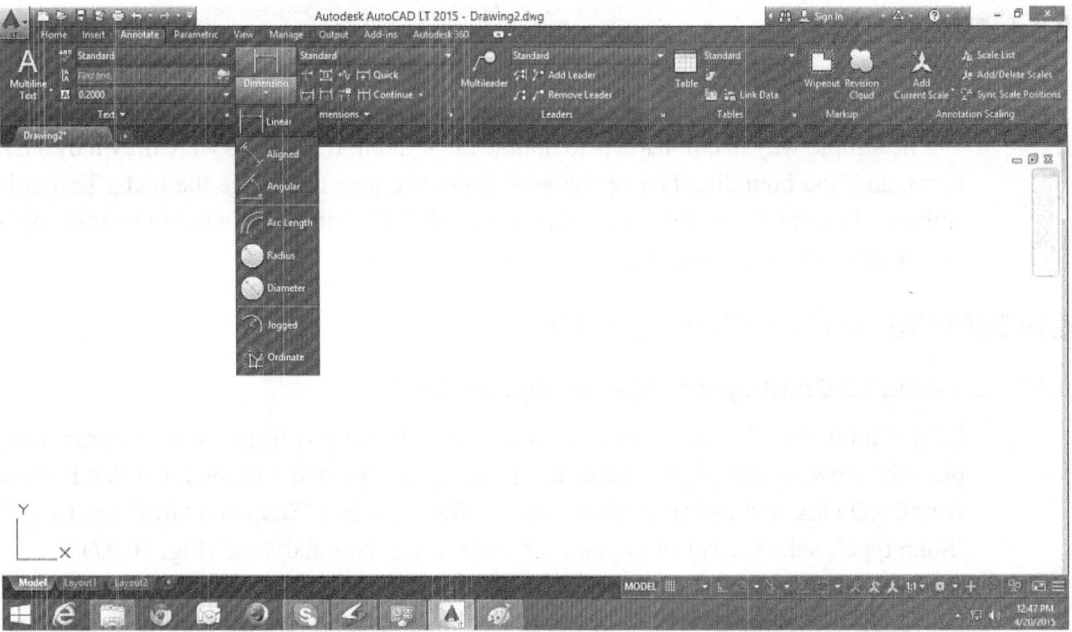

Fig. 16.25 Various dimensioning options

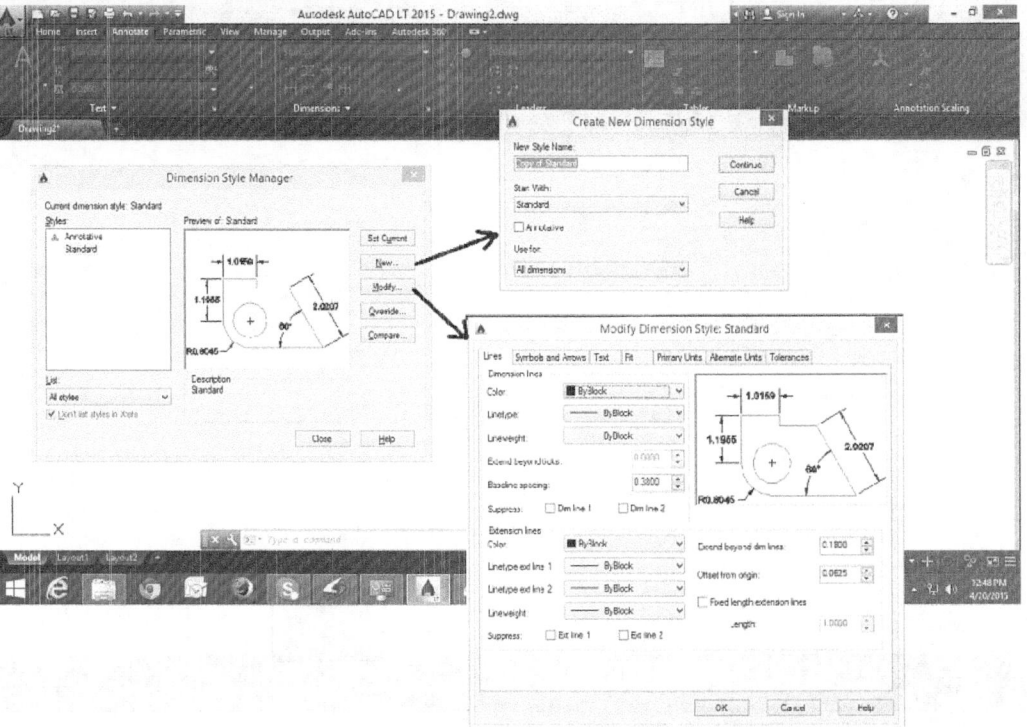

Fig. 16.26 Dimension style manager box with create new dimension style dialog box and modify dimension style dialog box

16.9.2 Text Editing

Existing texts can be edited using DDEDIT, FIND, STYLE, SCALETEXT, and JUSTIFYTEXT commands. All these commands are available in Annotate ribbon.

The simple way to edit texts is to double click them. If the texts were drawn by TEXT command, the bounding box opens and allows the user to change the texts. Text editor ribbon will open if the text were drawn by MTEXT command where changes to font size, font style, font face, justification, etc., can be done.

16.10 DRAWING PRACTICE USING AUTOCAD

16.10.1 Creating 2D Drawings of Object using AutoCAD

Using AutoCAD, 2D drawings or orthographic projections from the given isometric or pictorial views of any object can be generated by setting in the following order. From the AutoCAD classic drawing window, select 'Tools', select 'Snap and Grid' and from the 'Snap type', select 'Grid snap', and then select 'Rectangular snap'(Fig. 16.27).

If isometric or pictorial drawings of 3D objects are given, the reader should prepare the rough orthographic views of the object for his/her reference and then make them through AutoCAD by selecting the tools and invoking the necessary commands.

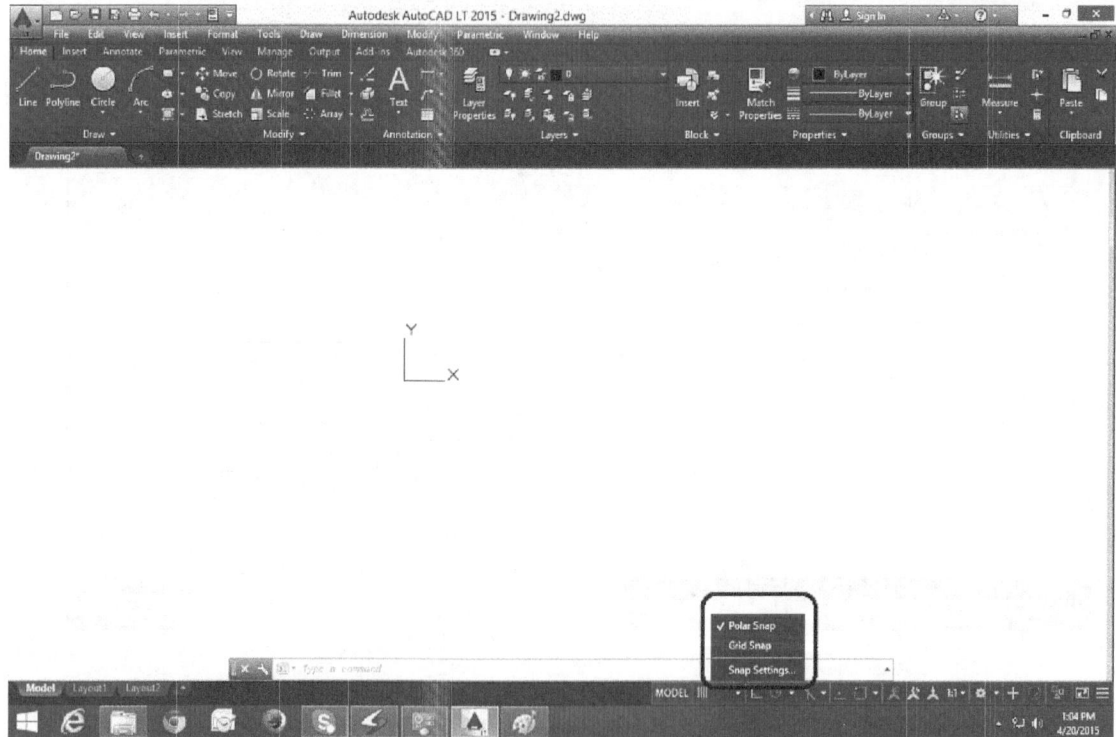

Fig. 16.27 Settings for generating 2D drawings using AutoCAD

The following examples illustrate the stage-by-stage procedure for making 2D drawings from the given 2D pictures or 3D pictorial projections.

Example 16.5

Using AutoCAD prepare the drawing shown in Fig. 16.28.

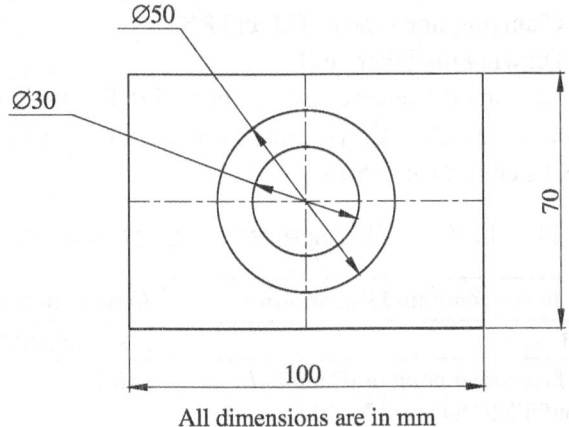

All dimensions are in mm

Fig. 16.28   Rectangle with concentric circles

PROCEDURE   (Refer Figs 16.29a–e)

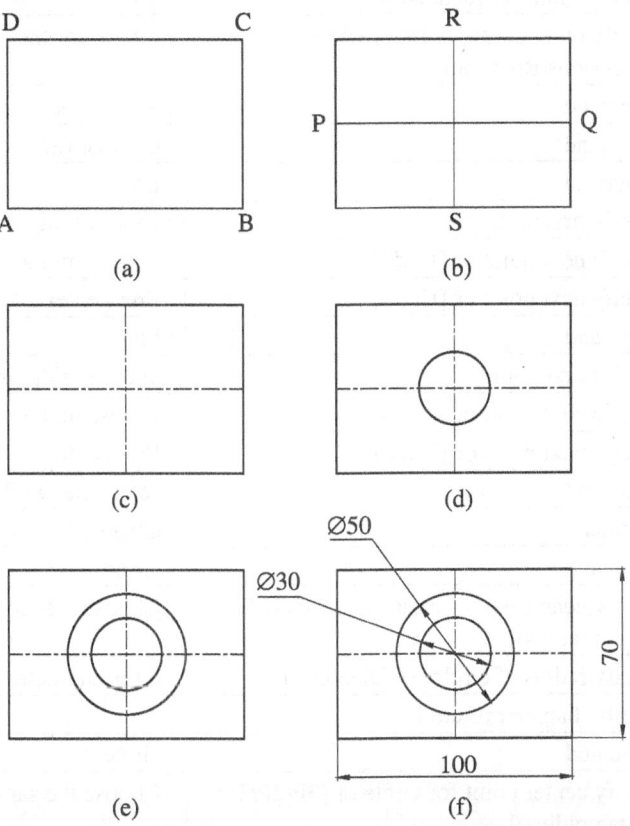

Fig. 16.29   Solution for Example 16.5 (a) Step 1 (b) Step 2 (c) Step 3 (d) Step 4 (e) Step 5

The step-by-step procedure is as follows:

Step 1: Drawing a rectangle and changing its line weight of ABCD.

Step 2: Locating centre point by drawing vertical and horizontal lines PQ and RS.

Step 3: Changing line type of PQ and RS.

Step 4: Drawing the inner circle.

Step 5: Drawing the outer circle and changing line weight of both circles.

The sequence of AutoCAD commands for generating the rectangle with two concentric circles is listed in Table 16.19.

Table 16.19 List of AutoCAD commands to generate the drawing shown in Fig. 16.28

Display in the command line window	Actions to be taken
Comand	: rectangle (ABCD)
Specify first corner point or [Chamfer/Elevation/Fillet/Thickness/Width]	: 10,10
Specify other corner point or [Area/Dimensions/Rotation]	: d
Specify length for rectangles	: 100
Specify width for rectangles	: 70
Specify other corner point or [Area/Dimensions/Rotation]	: 20,20 (for the side of plotting)
Command	: Select Rectangle ABCD
Command	: Change Line weigth in Properties ribbon bar
Command	: line
Specify first point	: choose midpoint of AB
Specify next point or [Undo]	: choose midpoint of CD
Specify next point or [Undo]	: Press enter
Command	: line
Specify first point	: choose midpoint of AD
Specify next point or [Undo]	: choose midpoint of BC
Specify next point or [Undo]	: Press enter
Command	: select the two lines
Command	: Change line type in properties ribbon bar
Command	: circle
Specify center point for circle or [3P/2P/Ttr (tan tan radius)]	: Choose intersection of lines PQ and RS
Specify radius of circle or [Diameter]	: d for Diameter
Specify diameter of circle	: 30
Command	: circle
Specify center point for circle or [3P/2P/Ttr (tan tan radius)]	: Choose the same center as previous

(Contd)

Table 16.19   *(Contd)*

Display in the command line window	Actions to be taken
Specify radius of circle or [Diameter] <15.0000>	: d
Specify diameter of circle <30.0000>	: 50
Command	: select the two circles
Command	: Change Line weigth in Properties ribbon bar

Example 16.6

Using AutoCAD, prepare the front view, top view, and side view of the rectangular block shown in Fig. 16.30.

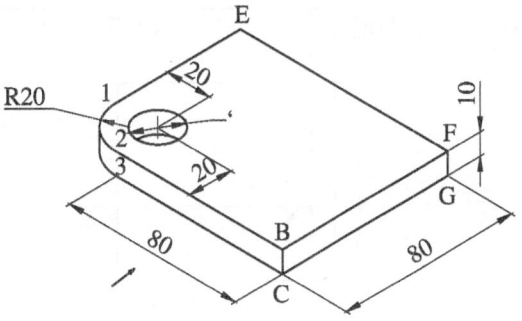

All dimensions are in mm

Fig. 16.30   Rectangular block with a corner fillet and a circular hole Example 1.6.

PROCEDURE   (Refer Figs 16.31a–h)

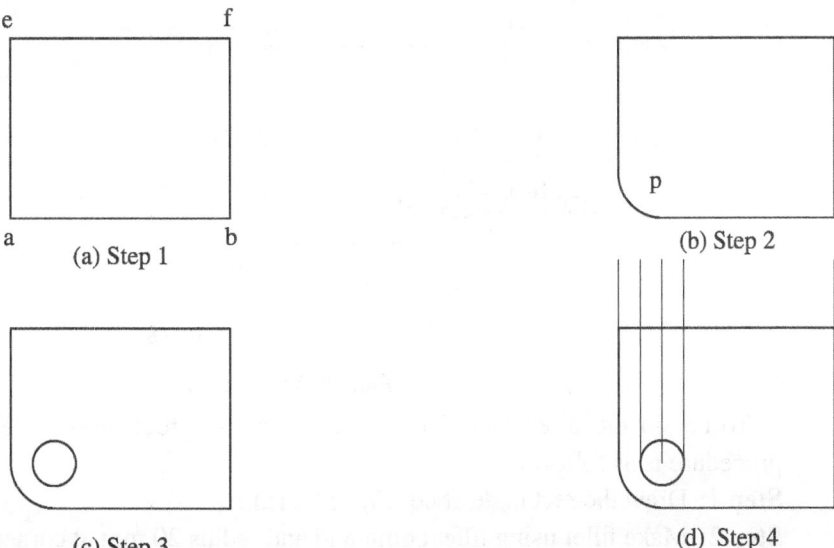

(a) Step 1

(b) Step 2

(c) Step 3

(d) Step 4

Fig. 16.31   Solutions for Example 16.6 *(Contd)*

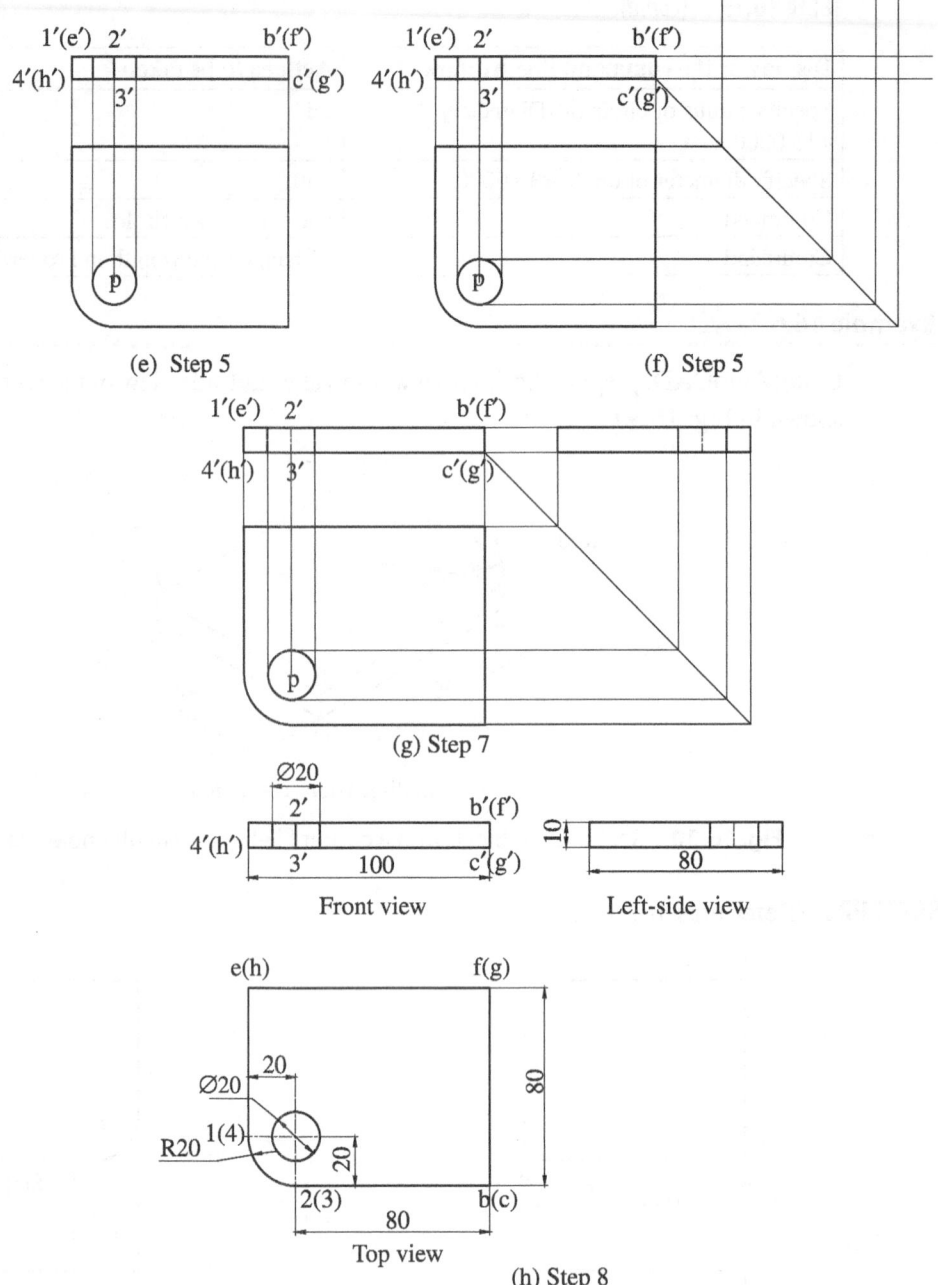

(e) Step 5   (f) Step 5

(g) Step 7

Front view   Left-side view

Top view

(h) Step 8

Fig. 16.31 (*Contd*)

To obtain the three views, there can be many procedures. One such step-by-step procedure is as follows.

Step 1: Draw the rectangle abcd (Fig. 16.31a).

Step 2: Make fillet using fillet command and radius 20 mm at corner 'a' (Fig 16.31b).

Step 3: With the fillet centre p as centre and diameter 20 mm draw a circle (Fig. 16.31c).

Step 4: From top view, draw ray lines from various points as shown in Fig. 16.31(d).

Step 5: Draw the front view as rectangle as shown in Fig. 16.31(e) intersecting rays from various top view points at respective points in the front view.

Step 6: Draw ray lines from various points of the top view and front view as shown in Fig. 16.31(f) and join the respective intersecting points to get the side view.

Step 7: Mark the hidden edges and axis with the appropriate line type as and trim all the excess ray lines to get the three views of the given block as in Fig. 16.31(g).

Step 8: Mark the dimensioning using standard procedure appropriately and complete the diagram as shown in Fig. 16.31(h). ▲

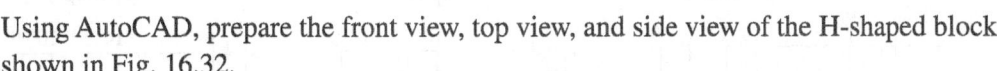

Using AutoCAD, prepare the front view, top view, and side view of the H-shaped block shown in Fig. 16.32.

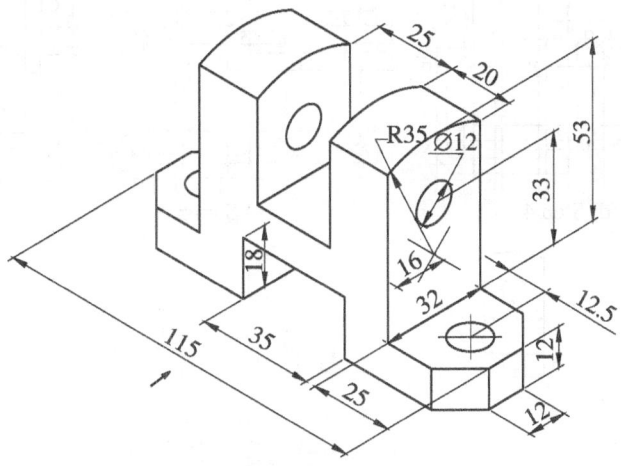

All dimensions are in mm

Fig. 16.32   H-shaped block Example 16.7

PROCEDURE Drawing the top, front, and side views of this block requires interference with all the three views. The step-by-step procedure is as follows:

Step 1: Draw the front view outline (1′–16′) with line command as shown in Fig. 16.33(a).

Step 2: Draw rays from 1′, 15′, 2′, 12′, 11′, 5′, 8′, and 6′ as shown in Fig. 16.33(b) and draw the outline of top view of the block.

Step 3: Draw circles with centres p and q (which can be located from the given diagram) and other intersecting lines. Also, change the properties of line if it is hidden (Fig. 16.33c).

Step 4: Draw rays to left side from various points as shown in Fig. 16.33(d) to obtain the right-side view.

Step 5: Draw the right-side view of the block with the intersecting rays as reference, with due interpretation from the given figure, and change line properties for the hidden lines (Fig. 16.33e).

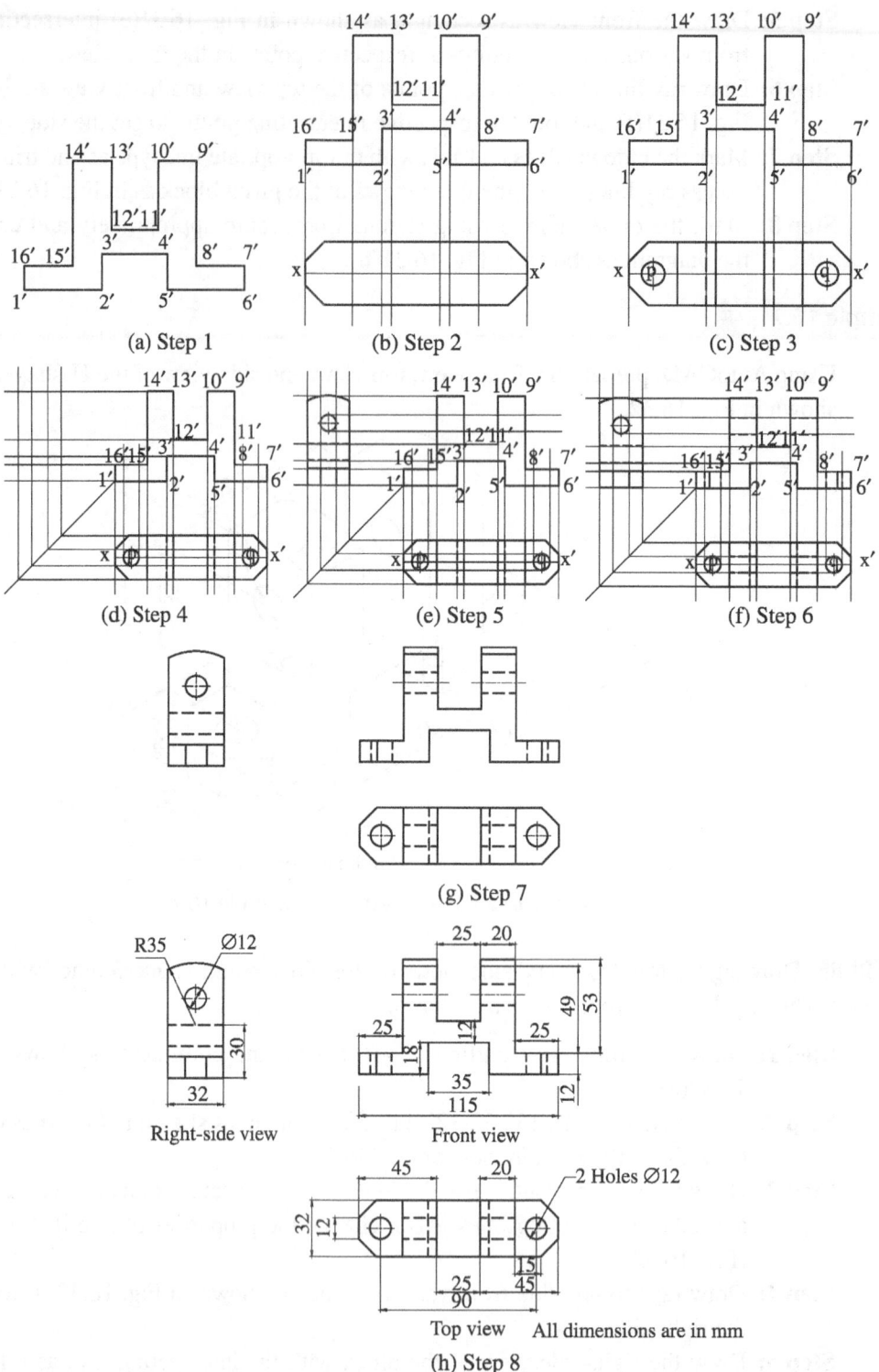

(a) Step 1 (b) Step 2 (c) Step 3

(d) Step 4 (e) Step 5 (f) Step 6

(g) Step 7

R35   Ø12

Right-side view

Front view

2 Holes Ø12

Top view   All dimensions are in mm

(h) Step 8

Fig. 16.33   Solution for Example 16.7

Step 6: Draw the rays from the ends of the arc located in the side view and obtain their positions in the front view. Draw the holes corresponding to the vertical legs in the side view as shown in Fig. 16.33(f).

Step 7: Identify the hidden and visible lines corresponding to the points in the side view and locate them in the front and top views as shown in Fig. 16.33(g). Erase all rays from the drawing to make the three views clear.

Step 8: Mark the dimensions appropriately and prepare the final solution as shown in Fig 16.33(h).

16.10.2 Creating Isometric Views of Objects using AutoCAD

Till now the isometric projection or the pictorial drawing of an object was given and its corresponding 2D orthographic views were obtained. Now, by using the first option of the AutoCAD, the isometric views of any object can be generated from their orthographic views by setting in the following order. From the AutoCAD classic drawing window, select 'Tools', then select 'Snap and Grid' and from the 'Snap type', select 'Grid snap', and then select 'Isometric snap' (Fig. 16.34).

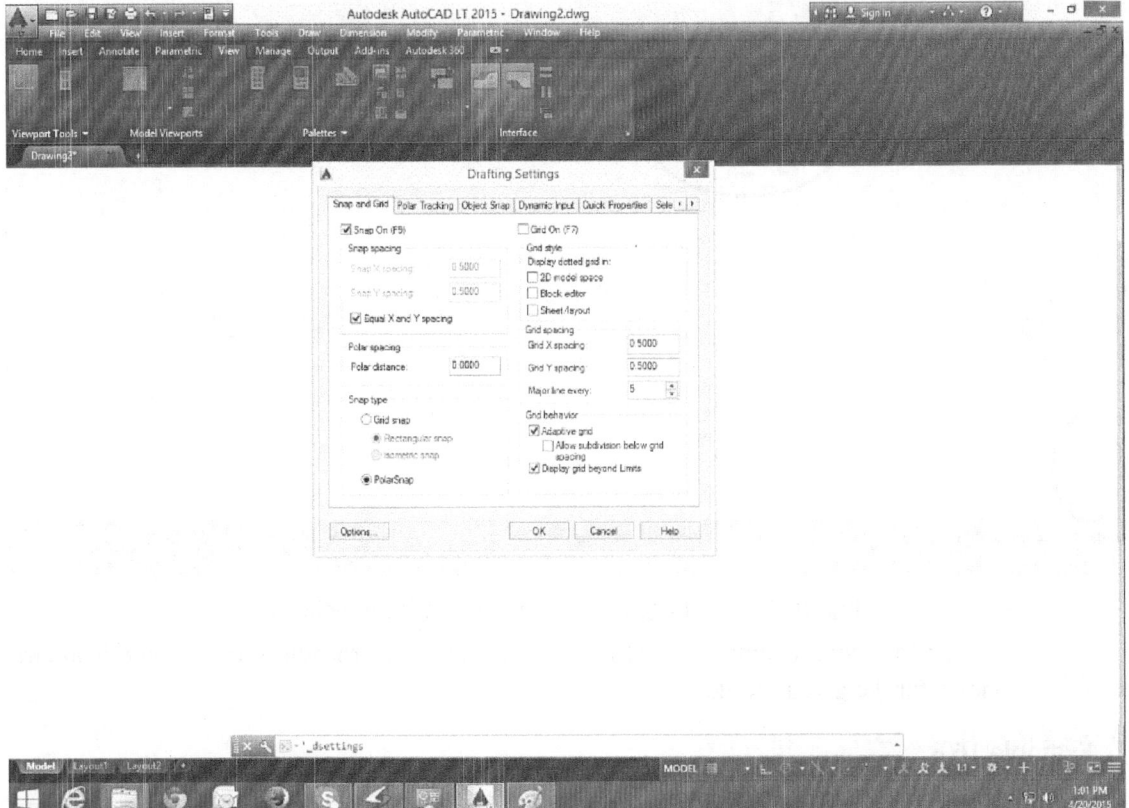

Fig. 16.34 Settings for generating isometric views of an object using AutoCAD

By pressing the function key F5, the current drawing plane say xy is changed to yz, similarly once again pressing the same function key F5 changes the drawing plane from yz to zx and so on.

By using the second option of the AutoCAD, the given orthographic projections of an object isometric views can be drawn by changing the initial settings as follows.

In the view ribbon and view panel, change the view to SE isometric as shown in Fig. 16.35. Pressing the Function key F5 once can help in changing the 'xy' isometric plane to 'yz' and again pressing F5 can lead to change the 'yz' isometric plane to 'zx' isometric plane.

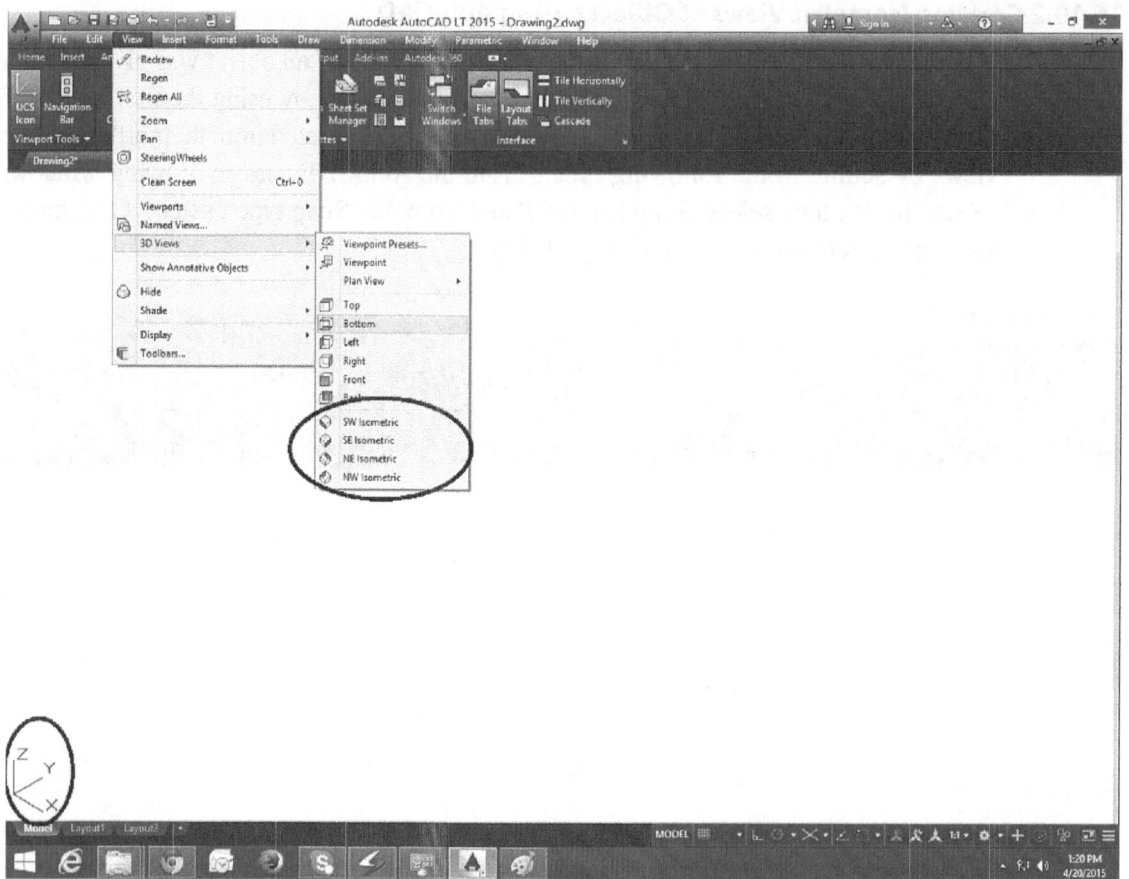

Fig. 16.35   Settings for drawing isometric projections

The following examples illustrate the step–by-step procedure to draw the isometric views for the given problem.

Example 16.8

Draw the isometric view from the front view, top view, and side views of the pillar given in Fig. 16.36.

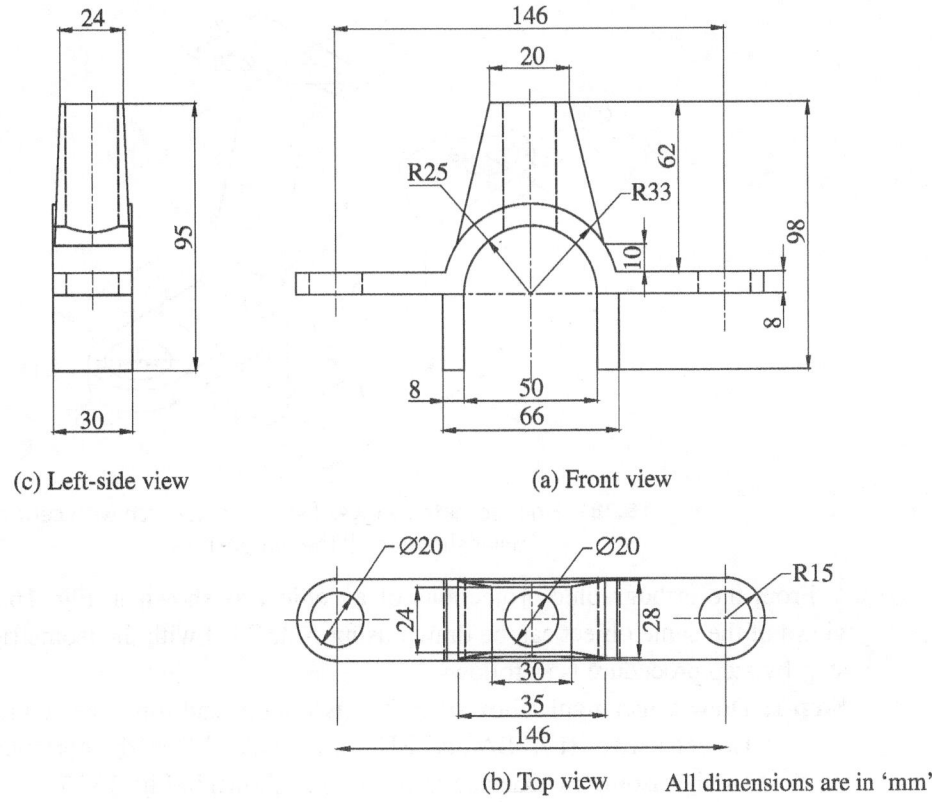

(c) Left-side view   (a) Front view

(b) Top view   All dimensions are in 'mm'

Fig. 16.36 Orthographic projections of semicircular arch with centre pillar

PROCEDURE (Refer Figs 16.37 and 16.38)

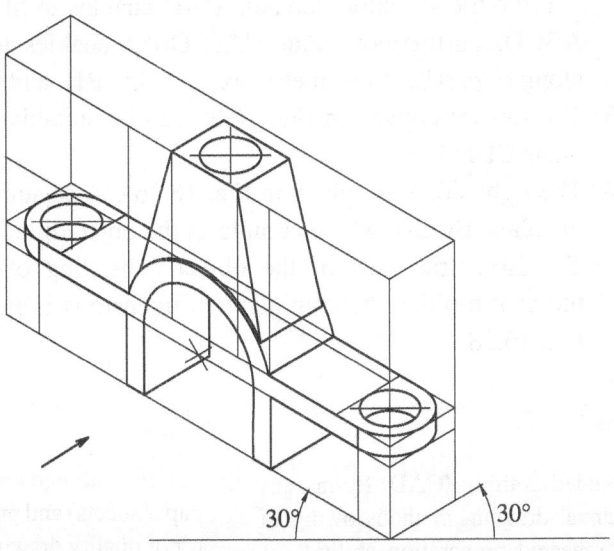

Fig. 16.37 Step 1 building of isometric views using rectangular box fitted
along parallel isometric lines

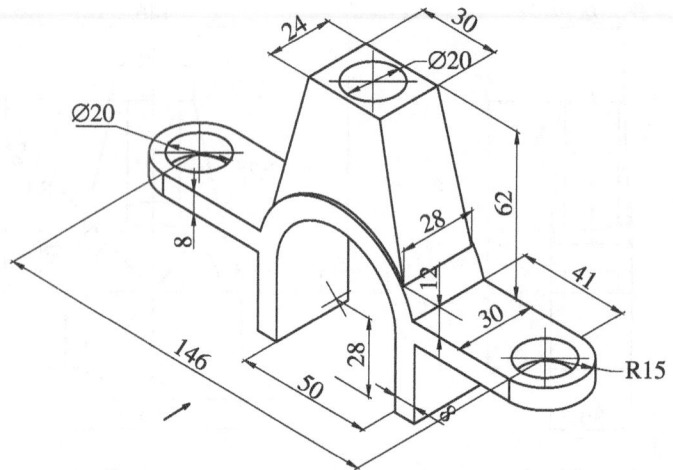

Fig. 16.38 Final isometric views of semicircular arch with centre pillar showing all the dimensions

From the orthographic projections of an object as shown in Fig. 16.36, isometric views of the same object can be drawn using AutoCAD with the isometric option. The step-by-step procedure is as follows.

Step 1: Draw a rectangular box such that its bottom and top edges coincide with the isometric axes (i.e., BA and BH which make 30° to the horizontal and BC is another isometric axis and is vertical) as shown in Fig. 16.37.

Step 2: The given front view of the object can be fitted in the plane ABCD by using AutoCAD with the following selection procedure:

From the Main Tool Bar → Select View option → Select SE Isometric

From the selection, the AutoCAD enables to fit the front view on the plane ABCD. Furthermore, AutoCAD Ortho enables to draw perpendicular lines along or parallel to isometric axes of BA, BH, and BC.

Step 3: The top view given in Fig. 16.36 can be suitably joined on or parallel to the plane CDFG.

Step 4: The right side view given in Fig. 16.36 can be suitably joined on or parallel to the plane BCGH, which completes the object fitted in the box.

Step 5: The isometric view of the object consisting of the semicircular arch and the centre pillar showing all the dimensions is presented in its final form in Fig. 16.38. ▲

RECAPITULATION

- The computer-aided drafting (CAD) techniques replace the manual drawing methods owing to their fast and repeated preparation abilities of the drawings, editing and add on conveniences, quicker display methods, less work space and minimal storage space (practically no hardware paper sheets) and production of accurate and fool proof quality drawings, and their fast transfer for 3D modelling and assembly.

- AutoCAD software, one of the most powerful PC-based software enables these facilities and is more user-friendly.
- AutoCAD has a flexible window for the preparation and operation of the drawings and is enhanced by ribbon tabs and panels for incorporating the commands for execution.
- AutoCAD has convenient file formats such as '.DWG', '.DXF', and '.DWF', which can be read by many 3D modelling software and vice versa.
- AutoCAD drawings can be drawn by using absolute or relative or polar coordinates or by using direct distance entry mode or in their combinations.
- The various entities such as point, polyline, arc, circle, ellipse, rectangle, and so on are made with respective drawing commands, while creating the drawings.
- The usage of various AutoCAD basic editing commands such as 'select objects', 'window', 'crossing', 'last', 'box', 'all', 'fence', 'group', 'erase', 'move', 'copy', 'offset', 'fillet', 'chamfer', 'hatch', 'trim', 'extend', 'mirror', and so on have been explained such that the user will be able to practice easily.
- The usage of the various AutoCAD display commands of 'zoom', 'regen', 'pan', etc., have been explained to gain the expertise of the users.
- Dimensioning the object using various dimensioning options of AutoCAD is explained. Dimensioning style manage module of AutoCAD has been introduced to create new dimension style and modify the existing dimension style.
- Text writing and editing in AutoCAD has been discussed to help the reader/user to use the AutoCAD to print the details on the drawing and for better communication.
- Sample problems are explained with step-by-step AutoCAD procedures for making 2D drawings with different entities.
- Development of isometric drawings of objects from their orthographic views is explained from the fundamental principles.

WORK PRACTICE LEVEL – 1

16.1 Draw the top view, front view, and side views for the following objects using AutoCAD.

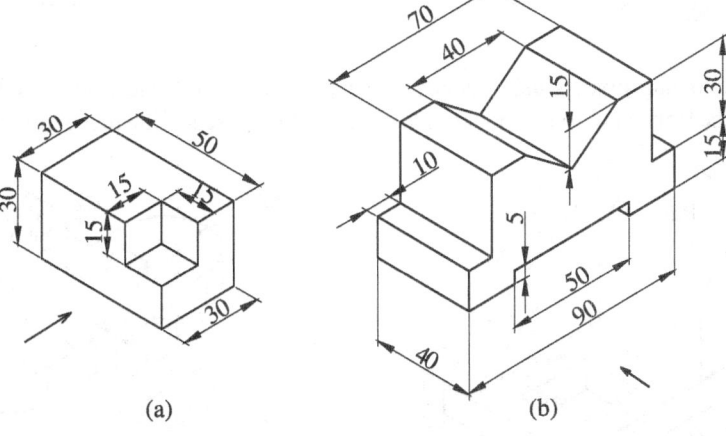

(a)   (b)

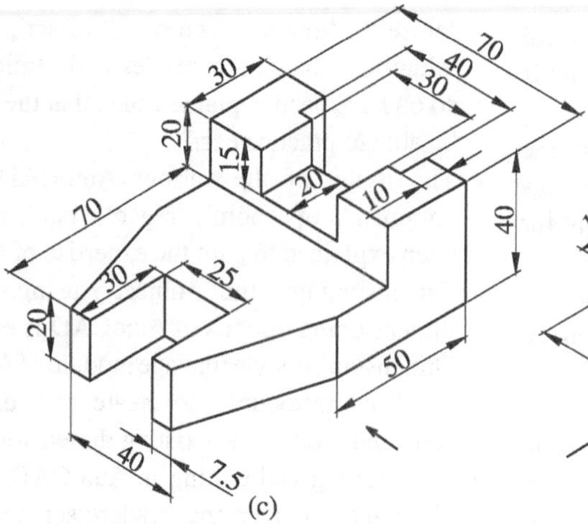

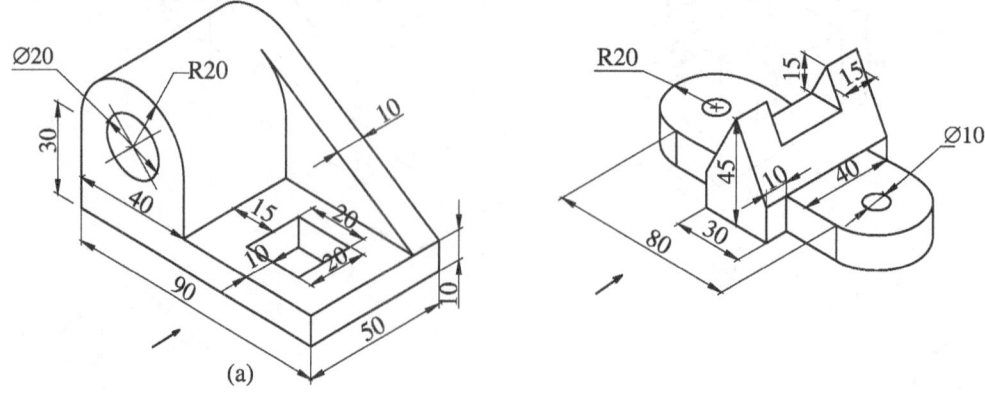

(c)

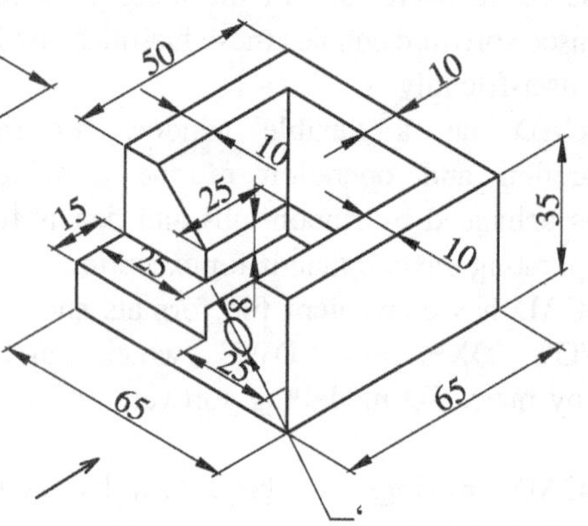

(d)

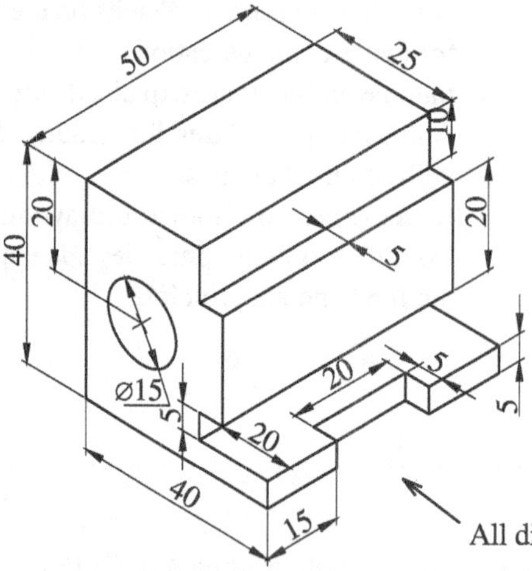

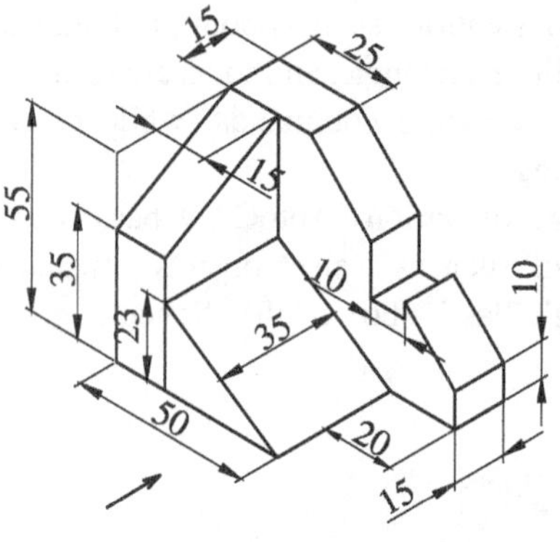

All dimensions in mm

WORK PRACTICE LEVEL – 2

16.1 Redraw the following objects using isometric snap of AutoCAD and also draw their orthographic projections—the front view, top view, and side views for the objects using rectangular snap of AutoCAD.

(a)

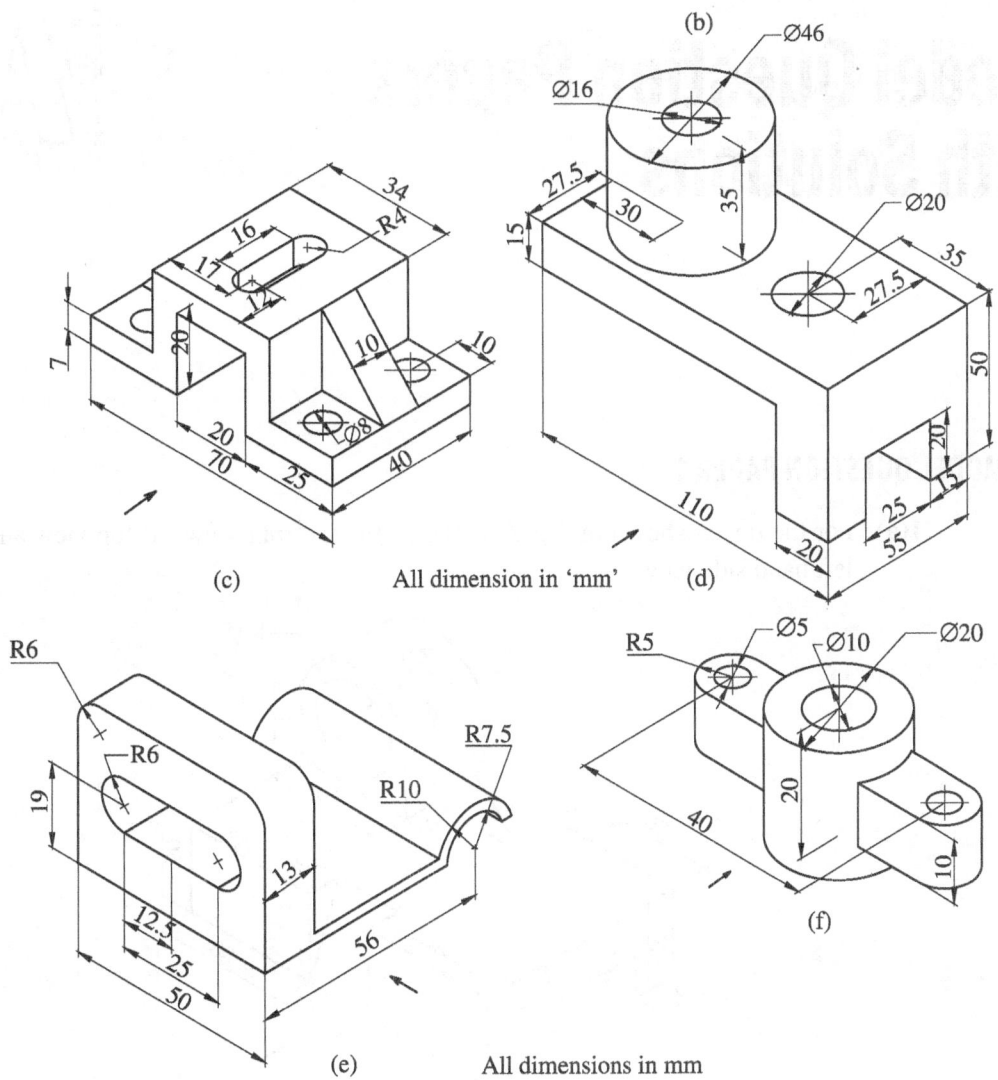

(b)

(c)   All dimension in 'mm'   (d)

(e)   All dimensions in mm   (f)

Model Question Papers with Solutions

APPENDIX **A**

A.1 MODEL QUESTION PAPER 1

1(a). For the object shown in Fig. A.1, sketch the (i) front view (ii) top view and (iii) left hand side view

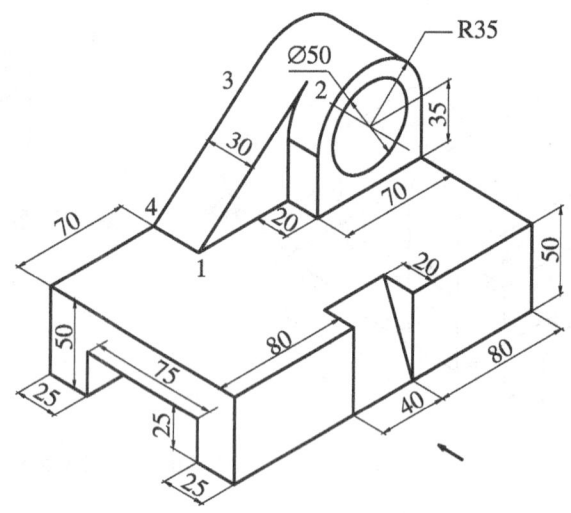

All dimensions are in mm

Fig. A.1

Solution   See worked-out Example 4.13 (Fig. 4.27).

[OR]

1(b). Construct a parabola, with the distance of the focus from the directrix as 50mm. Also, draw a normal and tangent to the curve at a point 40 mm from the directrix.
Solution   See worked-out Example 2.9 (Fig. 2.11).

2(a). The front view of the line AB of length 70 mm is inclined at 30o to xy line and measures 45 mm. The end A is 20 mm above HP and 25 mm in front of VP. Draw the projections of the line and find the inclinations with HP and VP.
Solution   See worked-out Example 7.8 (Fig. 7.23).

[OR]

2(b). A regular circular lamina of 60 mm diameter rests on HP such that the surface of the lamina is inclined at 30° HP. Obtain its projection when the top view of the diameter passing thro' the point on HP makes 45 ° to VP.

Solution   See worked-out Example 8.17 (Fig. 8.21).

3(a). A rectangular prism 50 × 25 mm base and length 70 mm, rests with one of its longer edges of the base on HP and the axis is inclined at 30° to HP and parallel to VP. Draw its projections.

Solution   See worked-out Example 9.12 (Fig. 9.32).

[OR]

3(b). A hexagonal prism of 30 mm base edges and axis 70 mm long, rests on one of its corners of base on HP. Draw its projections, when the lateral edge through that corner on HP, is inclined at 30° to HP and the vertical plane containing that lateral edge and the axis, is parallel to VP.

Solution   See worked-out Example 9.17 (Fig. 9.37).

4(a). A right regular hexagonal pyramid side of base 30 mm and height 80 mm is resting on its base on the HP with two of its adjacent lateral faces equally inclined to VP. It is cut by a horizontal section plane and an inclined section plane thereafter. The two section planes meet at the midpoint of the axis in the front view. The inclined section plane makes 70° with the HP and is perpendicular to the VP. Draw the projections indicating the cut surfaces. Also represent the true shape of the cut portion corresponding to the inclined section plane.

Solution   See worked-out Example 10.13 (Fig. 10.18).

[OR]

4(b). A lamp shade is formed by cutting a cone of base 144 mm diameter and 174 mm height by a horizontal plane at a distance of 72 mm from the apex and by an another plane inclined at 30° to HP and passing through one extremity of the base. Draw the development of the lamp shade.

Solution   See worked-out Example 12.9 (Fig. 12.13).

5(a). A frustum of the conical solid of base diameter 50 mm and top diameter 26 mm and 50 mm height is placed centrally over a cylindrical block of 76 mm base diameter and axis 25 mm long. The axes of the two solids are collinear. Draw the isometric view of the combined solid.

Solution   See worked-out Example 13.18 (Fig. 13.23).

[OR]

5(b). A cylinder of 60 mm diameter and axis 70 mm long lies on the ground on its generator such that the axis is inclined at 30° to the picture plane. Draw its perspective view when one of the end points touches the picture plane. The station point lies in the central plane which is bisecting the axis and is 160 mm in front of the picture plane. The horizon level is at 70 mm height.

Solution   See worked-out Example 15.7 (Fig. 15.15).

A.2 MODEL QUESTION PAPER 2

1(a). A circle of 60 mm diameter rolls on a horizontal line for half revolution and then on a vertical line for another half. Draw the curve traced out by a point lying on the circumference of the circle.

Solution This problem resembles Example 2.27 in Chapter 2, except that the second spell of the curve is on a vertical line, instead of a 60° line. The reader is advised to refer to the complete constructional procedure in Example 2.27.

[OR]

1(b). Draw the following views of the component shown in Fig. A.2.
(a) Front view   (b) Top view   (c) Right-side view

Solution Solution to this problem is given in Fig. A.3. The reader is advised to follow similar steps as given in Examples 4.13 and 4.14 in Chapter 4 for obtaining the various views as shown in Fig. A.3.

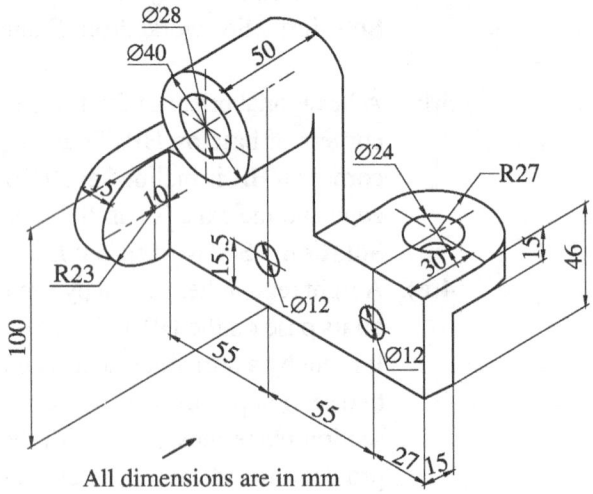

All dimensions are in mm

Fig. A.2

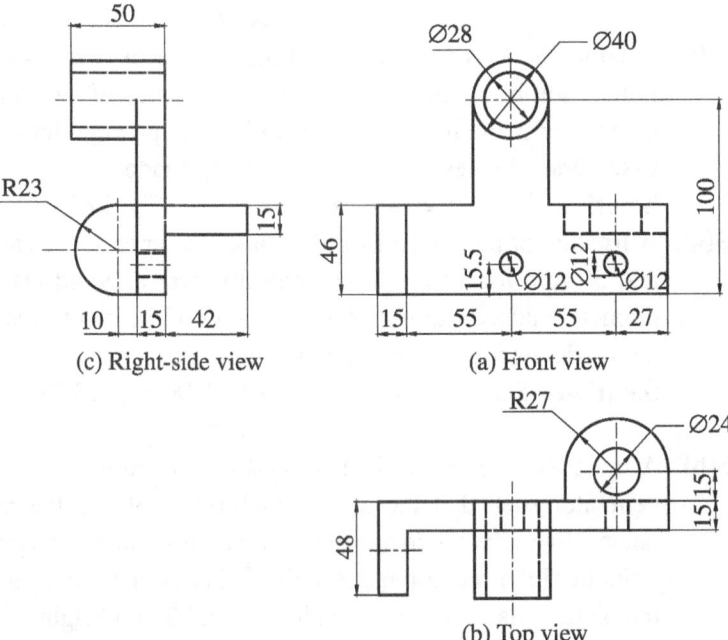

(c) Right-side view

(a) Front view

(b) Top view

Fig. A.3

2(a). A 75 mm long straight line PQ lying in the first quadrant has an end P in the HP and end Q in the VP. The line is inclined at 45° to HP and 30° to the VP. Draw the projections.

Solution This problem resembles Example 7.16 in Chapter 7, except that the length of the straight line and the angles of inclinations are different. The reader is advised to refer to the complete constructional procedure in Example 7.16.

[OR]

2(b). A thin circular plate of 60 mm diameter appears in the front view as an ellipse of major and minor axes, 60 mm and 40 mm in length, respectively. Draw its projections when one of the diameters is parallel to both the reference planes.

Solution *Hint*

NOTE *Decide the initial projections based on the surface inclination.*

- Since the plane surface appears as an ellipse in the front view (which is apparent), it is inclined to VP. Draw the given ellipse in the second stage front view keeping its major axis vertical. Therefore the circle appears in the first stage front view.

PROCEDURE (Refer Fig. A.4)

- Draw an ellipse a_2', ..., l_2' in the second stage front view keeping its major axis vertical using any procedure as discussed in Chapter 2.
- Project the point a_2' (which is in the VP) to meet the xy line at a_2 which is the corresponding top view of this point. Draw an arc taking 60 mm as radius (diameter of the circular plate), a_2 as centre to intersect the vertical projector drawn from the other end of the minor axis g_2' at point g_2.
- Join a_2–g_2 which is the diameter of the lamina and draw the vertical projectors from the other points of the second stage front view to locate the corresponding points in the second stage top view on this line.
- Measure the angle between xy and a_2–g_2 which is found to be 48° and is the inclination of the surface of the circular lamina with the VP.
- The first stage front view (which is the circle and true shape of the lamina) fitting in between the end horizontal projectors of the second stage front view and its corresponding first stage top view which are given to understand the problem clearly.
- Redraw the reduced shape a', ..., l' such that the front view of the diameter (d_2' j_2') is parallel to the reference line and draw vertical projectors.
- Draw locus line from the 2nd stage top view to meet the projectors and get final top view a...l.

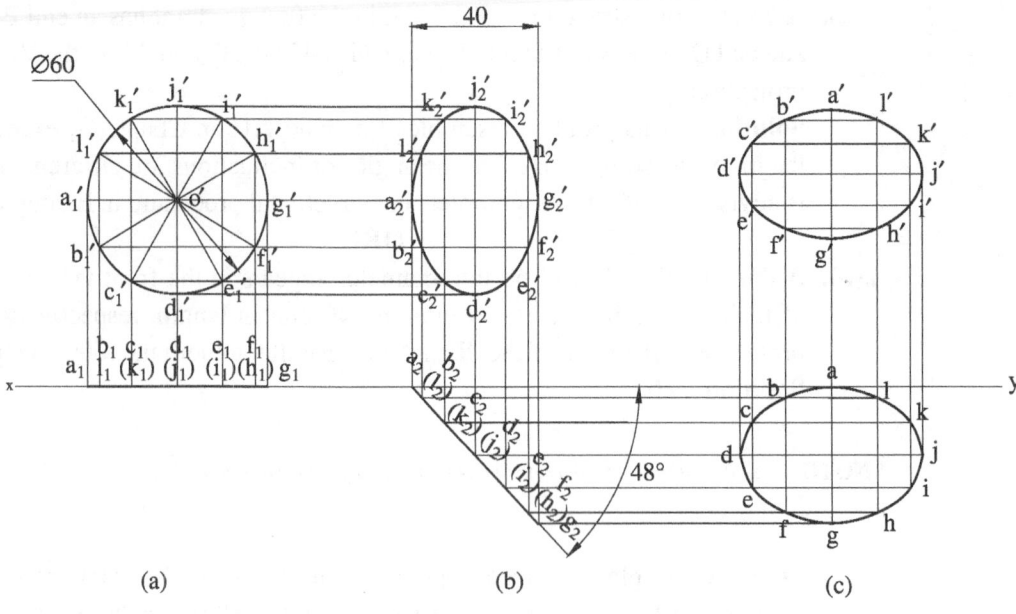

Fig. A.4

3(a). A pentagonal pyramid, having a base of 30 mm side and a 55 mm long axis, has a triangular face in the VP and the axis parallel to and 40 mm from the HP. Draw its projections.

Solution This problem resembles the worked Example 9.14 except that the triangular face is on VP instead of on HP. The initial projections and the second stage tilting diagrams are all to be reversed. The reader is advised to refer the complete constructional procedure in Example 9.14.

[OR]

3(b). A hexagonal prism of base edge 25 mm and axis height 65 mm is resting on HP on one of its base edges such that its axis is inclined at 30° to HP and parallel to VP. Draw the projections.

Solution This problem resembles Example 9.17 in Chapter 9, except that the hexagonal prism is tilted about an edge of the base instead of a corner of the base. In the initial projections, the hexagon will be laid in the top view with an edge perpendicular to the reference line and later tilted about this edge in the other view. The reader can also refer to a similar problem done with a pentagonal prism for this purpose (Example 9.13).

4(a). A hexagonal pyramid of base edge 25 mm and axis height 70 mm is resting on HP on its base such that two of its base edges are parallel to VP. It is cut by a section plane which is inclined at 45° to HP, perpendicular to VP and passing through a point 40 mm from the base along the axis. Draw its front view, sectional top view, and true shape of the section.

Solution *Hint*

- Since the section plane (SP) is perpendicular to VP, it is seen as a straight line in the front view and drawn as an inclined line to the reference line due to its inclination with HP.

PROCEDURE (Refer Fig. A.5)

- Complete the projections of the solid as per the details given and as discussed in Chapter 9.
- Mark the SP as a straight line, inclined at 45° to the reference line and passing through a point on the axis at distance of 40 mm above the base of the solid, in the front view and obtain the cutting points 1'...6', in the sequential order of the edges.
- Project the cutting points to meet the corresponding edges in the top view and join them and obtain a closed bounded area. Shade the bounded area by hatching lines and this is referred as the sectional top view.

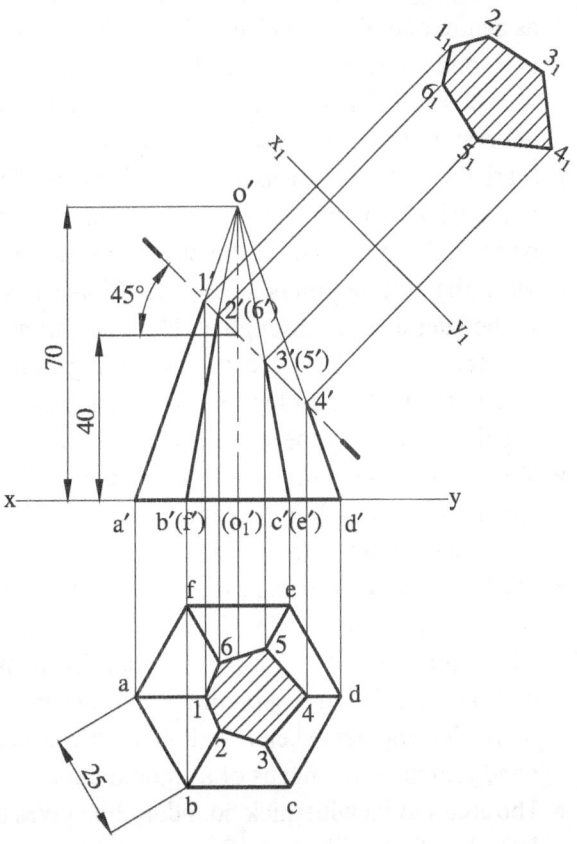

Fig. A.5

- Since the SP is inclined to HP, the sectional top view does not give the true shape of the cut portion. Therefore, draw a new reference line x_1y_1 parallel to the line representing the SP and draw it opposite to the part removed in the front view and draw projections to x_1y_1 as shown in Fig. A.5. As done in the auxiliary projections of the solids, measure the distances of the projections of these cutting points i.e. 1, 2, ..., 6 from xy line and mark along the new projectors from x_1y_1 line as 1_1... 6_1. The resulting shape obtained by joining these points gives the true shape of the section. Shade it by hatching lines.

[OR]

4(b). A cone of diameter 50 mm and axis height 70 mm is resting on HP on its base as shown in Fig. A.6. Draw the development of the lateral surfaces of the central portion of the cone.

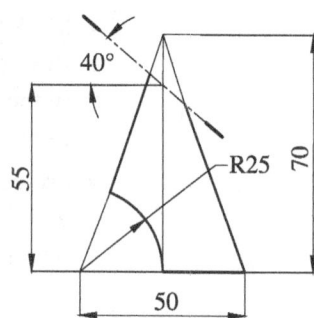

Fig. A.6

Solution   **PROCEDURE** (Refer Fig. A.7)

- Draw the top and front views of the cone and obtain the generators $o'a'$, $o'b'$, ..., $o'l'$. With O as centre and $o'a'$ or $o'g'$ as radius, draw an arc at a solid angle $\theta = (r/l) \times 360°$ or $\theta = (25/74) \times 360° = 121°$ and obtain the development sketch O–A–A–O with all the generators marked as shown in Fig. A.7(b)
- Mark the section plane as a straight line in the front view, as it is perpendicular to the VP and orient it at 40° to the reference line and passing through a point on the axis 55 mm above the base of the solid.
- Mark the cutting points $1'$, $2'$, ..., $12'$ as shown. As all of them are situated in the lateral edges through a', b', ..., l', transfer them to the true slant edge $o'a'$ (or $o'g'$) and set the corresponding radial distances on the respective generators in the development sketch and obtain the points 1, 2, ..., 12, and 1 and join them by means of a smooth curve.
- Mark another section plane as a curve of radius 25 mm taking a' as centre in the front view, as it is perpendicular to the VP and passing through a point $d'(j')$ on the base of the solid.
- Mark the cutting points i', ii', ..., vii' as shown. These cutting points (except iv' and v' which coincide with d' and j' respectively) are situated in the lateral edges through a', b', c', k', and l', transfer them to the true slant edge $o'a'$ (or $o'g'$) and set the corresponding radial distances on the respective generators in the development sketch and obtain the points i, ii, ..., v, and v, vi, vii, and i and join them by means of a smooth curve.
- The area shown with thick boundary line gives the development sketch for the lateral surface of the central portion of the cone.

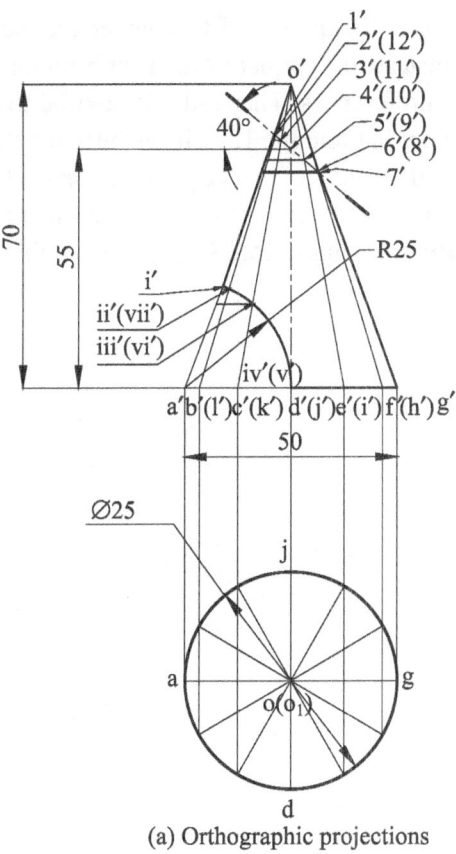

(a) Orthographic projections

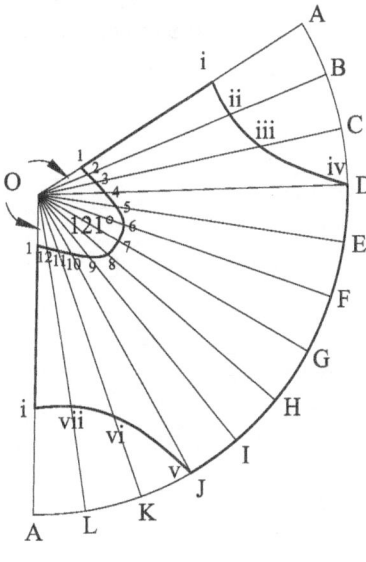

(b) Development of the lateral surface
of the central portion of the cone

Fig. A.7

5(a). A cylinder of diameter 50 mm and axis height 60 mm is resting on the ground on its base. It is cut by a section plane which is perpendicular to VP and inclined at 55° to HP. The section plane passes through a point at a distance 45 mm from the base along the axis. Draw the isometric view of the bottom portion of the cylinder.

Solution   This problem resembles Example 13.15 in Chapter 13, except that the dimensions of the solid and the angle of inclination of section plane are different. The reader is advised to refer to the complete constructional procedure in Example 13.15.

[OR]

5(b). A rectangular prism 80 mm × 60 mm × 30 mm is placed on the ground behind the PP with the longest edges vertical and the shortest edges receding to the left at an angle of 40° to the PP. The nearest vertical edge is 10 mm behind the PP and 15 mm to the left of the observer who is at a distance of 150 mm in front of the PP. The height of the observer above the ground is 120 mm. Draw the perspective view of the prism.

Solution This problem resembles Example 15.5 in Chapter 15, except that the solid is a rectangular prism, instead of a square prism with minor changes in the dimensions and the location references. The reader is advised to refer to the complete constructional procedure in Example 15.5. It can be noted that since the edge is behind the picture plane, the length of that edge in the perspective drawing will appear reduced in height. The reader can also note the facts mentioned about the change of edges from the PP in Example 15.2 and reason out this problem.

Bibliography

Booker, P.J. (1963), *A History of Engineering Drawing*, 1st edition, Chatto & Windus.

French, Thomas E., Vierck, Charles J., *et al.* (1953), *A Manual of Engineering Drawing for Students and Draftsmen*, 8th edition, McGraw Hill Book Company, New York, USA.

Gopalakrishna, K.R. (2014), *Engineering Drawing*, Vols 1 and 2, 23rd edition, Subhas Publications/Subhas Stores, Bangalore.

Loney, S.L. (Sidney Luxton) (1897), *The Elements Of Coordinate Geometry*, Macmillan and Co., London.

Luzadder, Warren J. and Duff, Jon M. (1992), *Fundamentals of Engineering Drawing: With an Introduction to Interactive Computer Graphics for Design and Production*, 11th edition, Peachpit Press.

Pickup, F. and Parker, M.A. (1970), *Engineering Drawing with Worked Examples* [in 2 Vols], 2nd edition, Revised and Metricated, Volume 1, Hutchinson Educational.

Simmons, Colin H., Dennis E. Maguire, and Neil Phelps (2012), *Manual of Engineering Drawing*, 4th edition, Elsevier.

Wellman, B. Leighton (1948), *Technical Descriptive Geometry*, McGraw Hill Book Company.

Index

About the Authors

Dr N.S. Parthasarathy, obtained his PhD degree in Mechanical Engineering from Bangalore University in 1999 for his work on 'Effective Stiffness and Slip—Damping of Stranded Cables', a topic of vital interest for the life estimation of overhead power transmission lines in the country. He obtained his Master of Engineering degree in Engineering Design and Bachelors in Mechanical Engineering, both from Government College of Technology, Coimbatore, under Madras University, in 1973 and 1971 respectively. He has 24 years of teaching experience in College of Engineering, Guindy, Anna University, Chennai from 1974, which includes 10 years as Professor of Mechanical Engineering. During his tenure as Professor, he was entrusted with additional responsibilities such as Head of Engineering Design Division, Director (Research), Director–AUFRG Institute for CAD/CAM, all in Anna University, Chennai.

His areas of expertise are Contact Mechanics, Vibrations, and Finite Element Analysis. He has attended many international conferences, visited various foreign universities, and published numerous papers in national and international journals of repute. He has guided four PhD students and is currently guiding two more.

His special subject of interest is Engineering Drawing/Graphics, which he has taught for more than 24 years by way of regular class lectures, special lectures in various colleges, web courses, and EDUSAT classes periodically arranged by Anna University, Chennai.

Dr Vela Murali obtained his PhD in Mechanical Engineering from National University of Singapore (NUS), Singapore in 2002. He has been awarded the Asian International Scholarship sponsored by NUS during his PhD course for his excellent academic background. He obtained his ME in Machine Design from B.M.S. College of Engineering, Bangalore in 1993 and BTech. in Mechanical Engineering from Jawaharlal Nehru Technological University College of Engineering, Anantapur in 1991. He has about 20 years of experience in teaching various undergraduate and postgraduate mechanical engineering courses. He has done extensive research in the areas of fracture, fatigue, composite materials, and applied mechanics and has published many articles in national and international journals of repute.

Dr Vela Murali is considered a resource person in Engineering Mechanics and has conducted several faculty development programmes on the subject. He has also authored a book on *Engineering Mechanics*, published in 2010 by Oxford University Press, India.

He served as the Head of the Engineering Design Division, Anna University, Chennai in 2008–2015 and he is currently Professor of Engineering Design Division, Department of Mechanical Engineering, College of Engineering, Guindy, Anna University, Chennai.

Related Titles

Engineering Mechanics
(9780198062240)

Vela Murali, Professor, Engineering Design Division, Anna University, Chennai.

Engineering Mechanics is a textbook specifically designed for a one-semester interdisciplinary course offered at the university level for undergraduate engineering programmes in India.

- Adopts simple algebraic methods of solving problems instead of the conventional vectorial method
- Uses a simple and direct style of writing with effective terminologies and nomenclatures to explain the concepts
- Incorporates 'quadrant approach' to resolve forces that act on planes
- Contains numerous solved problems and exercises with hints for practice
- Provides objective-type questions with answers

Engineering Physics
(9780199452811)

D.K. Bhattacharya, Head of Hydrophone, Ion-implantation, Microwave, and Instrumentation Groups, Solid State Physics Laboratory, New Delhi.
Poonam Tandon, Associate Professor, Department of Applied Sciences, Maharaja Agrasen Institute of Technology, IP University, New Delhi.

Engineering Physics is designed as a textbook for first-year undergraduate engineering students for a two-semester course in engineering physics.

- Provides more than 400 illustrations and solved examples, which aid in the easy understanding of concepts
- Contains a self-assessment section after each chapter, which includes numerous multiple-choice questions, review questions, and numerical problems
- Lists important applications in each chapter to help students explore the scope and applicability of the concepts discussed
- Provides appendices containing important physical and lattice constants, periodic table, and properties of semiconductors and relevant compounds for ready reference

Fundamentals of Computers
(9780199452729)

Reema Thareja, Assistant Professor, Department of Computer Science, Shyama Prasad Mukherjee College for Women, University of Delhi

Fundamentals of Computers has been specifically designed for those who want to learn the basic concepts of computers. It is an ideal text for self-learning basic computer concepts such as organization, architecture, input and output devices, primary and secondary memory, as well as advanced topics including operating systems, computer networks, and databases. The book also provides step-by-step procedures to learn different applications such as MS Word, MS PowerPoint, and MS Excel.

- Provides a comprehensive coverage of important topics ranging from the basics of computers to emerging technologies
- Supports numerous well-labelled diagrams and screenshots throughout the text
- Includes many solved examples and chapter-end exercises such as objective-type questions and review questions that enable students to check their understanding of concepts

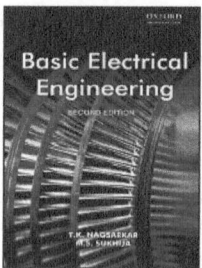

Basic Electrical Engineering (2e)
(9780198068907)

T.K. Nagsarkar, Retd Professor and Head, Dept of Electrical Engineering, Punjab Engineering College

M.S. Sukijha, Founder Principal, Guru Nanak Engineering College, Bidar, Karnataka

Basic Electrical Engineering (2e) provides a lucid exposition of the principles of electrical engineering for both electrical as well as non-electrical undergraduate students of engineering. Students pursuing diploma courses as well as those appearing for AMIE examinations would find this book extremely useful.

- New sections on electrostatics, Biot–Savart law, and synchronous generator connected to an infinite bus bar
- New chapter on single-phase induction motors and special machines
- Many new illustrations to supplement the text
- Several additional solved examples to enhance the understanding of new concepts
- Numerous new chapter-end exercises with answers and MCQs to stimulate student interest

Other related titles